Analysis of Crop Genetic and Germplasm Diversity

Analysis of Crop Genetic and Germplasm Diversity

Editors

Antonio M. De Ron
A. Paula Rodiño

MDPI

Basel • Beijing • Wuhan • Barcelona • Belgrade • Novi Sad • Cluj • Manchester

Editors
Antonio M. De Ron
Misión Biológica de Galicia (MBG)
Pontevedra, Spain

A. Paula Rodiño
Misión Biológica de Galicia (MBG)
Pontevedra, Spain

Editorial Office
MDPI
St. Alban-Anlage 66
4052 Basel, Switzerland

This is a reprint of articles from the Special Issue published online in the open access journal *Agronomy* (ISSN 2073-4395) (available at: https://www.mdpi.com/journal/agronomy/special_issues/germplasm_genetic_diversity).

For citation purposes, cite each article independently as indicated on the article page online and as indicated below:

Lastname, A.A.; Lastname, B.B. Article Title. *Journal Name* **Year**, *Volume Number*, Page Range.

ISBN 978-3-0365-8676-2 (Hbk)
ISBN 978-3-0365-8677-9 (PDF)
doi.org/10.3390/books978-3-0365-8677-9

Cover image courtesy of Antonio M. De Ron

Contents

About the Editors

Antonio M. De Ron

Antonio M. De Ron (Lugo, Spain, 1952) is Ad Honorem Professor at the Misión Biológica de Galicia (MBG), Spanish National Research Council (CSIC) in Pontevedra, Spain. He received both his graduate (1974) and doctorate (1987) in Biology from the University of Santiago de Compostela (USC, Spain). Initially he worked as a postgraduate student in plant protection at the National Institute for Agricultural Research (INIA-CSIC, Spain) and head of Biology at a Secondary School. In 1988, he gained a position as a tenured scientist at the MBG working on legume genetics, germplasm and breeding, and the symbiotic system plant–microorganisms for the mitigation of climatic change. He is currently the leader of the Biology of Agrosystems Research Group. He was also a part-time lecturer of genetics and breeding at the USC and is currently teaching a Master in Genomics and Genetics at this University and the University of Vigo (Spain). He served as the President of the Spanish Association for Legumes (AEL) and as member of the Coordinating Committee of the Bean Improvement Cooperative (BIC, USA). Currently, he is the leader of the Protein Crops Working Group of the European Association for Research on Plant Breeding (EUCARPIA, The Netherlands). He has published many scientific articles, books and monographs, as well as several educational publications. Prof De Ron has won awards for his work and is well recognized in the international scientific community for his achievements, especially in legume research.

A. Paula Rodiño

A. Paula Rodiño (Arcade, Pontevedra, 1972) is a researcher at the Misión Biológica de Galicia (MBG), Spanish National Research Council (CSIC) in Pontevedra, Spain. She received both her graduate (1995) and doctorate (2001) in Biology from the University of Santiago de Compostela (USC, Spain). Her scientific trajectory has focused on the genetic improvement of economically important traits in a species of agronomic value, such as the common bean. She has an Honorable mention by the Diputación Provincial de Pontevedra for the Doctoral Thesis "Morphoagronomic and biochemical characterization of common bean germplasm (*Phaseolus vulgaris*) from Spain". Her research line at that stage was detecting new sources of variation from the analysis of a large collection of common bean germplasm. In the period 2001–2012, she had Marie Curie-EU, I3P-CSIC and Parga Pondal postdoctoral fellowships at INRA (Montpellier, France) and MBG. Her research line at that stage was the study of the bean–rhizobia interactions from physiological and molecular point of views, identifying candidate genes involved in symbiotic nitrogen fixation. Currently, she is in the Biology of Agrosystems (BAS) research group, taking responsibility for the legume collection of the group, as well as field trials and the study of the symbiotic system legumes–rhizobia. The objectives of the BAS group are obtaining new varieties of crops, knowledge on the evolution of a protein crop (common bean), interactions of the crops with the agrosystems and the legume–rhizobia soil symbiotic system for the biological fixation of nitrogen.

Preface

The objective of this Special Issue was to publish research papers related to the genetics and diversity of cultivated plants, presented by researchers involved in the genetics and crop breeding of different plant species relevant to agriculture worldwide. The Special Issue includes 33 articles by 205 authors from 33 countries (Europe, Africa, Asia and America) involved in the research of crop genetic and germplasm diversity analysis.

Antonio M. De Ron and A. Paula Rodiño
Editors

agronomy

Editorial
Analysis of Crop Genetic and Germplasm Diversity

Antonio M. De Ron * and Ana Paula Rodiño *

Misión Biológica de Galicia (MBG), Spanish National Research Council (CSIC), 36143 Pontevedra, Spain
* Correspondence: amderon@mbg.csic.es (A.M.D.R.); aprodino@mbg.csic.es (A.P.R.);
Tel.: +34-986-854-800 (A.M.D.R. & A.P.R.)

Plant genetic resources are the basis for the genetic improvement of cultivated plants and future food and feed security. The interactions of plants with their agrosystem, the influence of different cropping systems, and the resistance of crops to biotic and abiotic stresses, as well as the adaptation of plants to different environmental conditions under the current changing conditions, are relevant aspects.

Agricultural biodiversity is the diversity of crops and their wild relatives, trees, animals, microbes, and other species relevant to agricultural production. Food and feed biodiversity is crucial to fight malnutrition and diet-related diseases. A diverse diet increases the likelihood of consuming adequate amounts of the full range of nutrients essential to human health.

The genetic diversity of several crop species is presented in this publication. This biodiversity was assessed by morphological and agronomical traits and nutritional characters, also including molecular diversity and tolerance/resistance to abiotic and biotic stresses.

The analysis of biodiversity in crops is focused on different goals: use of germplasm for breeding; adaptation of germplasm to different environments; evolution of adapted varieties and the nutritional and quality properties. The agromorphological variation, genetic variability, and quality characters were studied in species of the *Poaceae* family as fonio ((*Digitaria exilis* (Kippist) Stapf)) [1], maize (*Zea mays* L.) [2], rice (*Oryza sativa* L. spp. *Indica* and *O. sativa* spp. *Japonica*) [3,4]), and *Cynodon* [5]. Some horticultural crops such as eggplant (*Solanum melongena* L.) [6,7] and *Capsicum* (several species) [8,9] were studied, presented in this volume, with regard to morphological, sensorial, and chemical characterization and adaptation to different conditions. With regard to legume crops, the diversity of pigeonpea (*Cajanus cajan* (L.) Millspaugh) was highlighted [10].

Fonio is an ancient orphan cereal, cultivated by resource-poor farmers in arid and semi-arid regions of West Africa, who conserved and used the cereal for nutrition and income generation. Furthermore, this study highlighted agromorphological descriptors that discriminate fonio accessions and provide useful information for parental selection with economically important agronomic traits [1]. Seven maize races of the central high plateau of Mexico were characterized using a combined analysis of 13 morphological traits and 31 microsatellite loci. Principal component analysis separated the different accessions into well-defined groups using the first three principal components. The accessions of the Arrocillo Amarillo and Elotes Cónicos races did not exhibit a grouping pattern, indicating greater genetic complexity [2]. Rice (*Oryza sativa* L.) contains many high-value nutritional compounds, including nutraceutical lipid compounds that offer health benefits. Cluster analysis sorted the germplasm into nine clusters, based on their nutraceutical lipid content. TU-010, TU-027, TU-093, and TU-244 genotypes had the highest levels, making them a potentially useful genetic resource in breeding programs for nutraceutically improved rice. The findings of this study can support the introduction of novel rice varieties with high added-value bioactive properties [3]. The japonica rice variety in northeastern China is well known because of its high quality. Eating and cooking qualities (ECQs) are the most important factors that determine cooked rice quality. The japonica varieties in northeastern China exhibited a narrow genetic basis. These results further elucidate the genetic basis of ECQs

Citation: Ron, A.M.D.; Rodiño, A.P. Analysis of Crop Genetic and Germplasm Diversity. *Agronomy* **2022**, *12*, 91. https://doi.org/10.3390/agronomy12010091

Received: 15 December 2021
Accepted: 27 December 2021
Published: 30 December 2021

of japonica varieties in northeastern China and provide local breeders some assistance for improving ECQs of rice grain in rice breeding [4]. The Cynodon spp. collection maintained by the United States Department of Agriculture National Plant Germplasm System (USDA–NPGS) has limited information on nutritive value (NV) traits. Genetic variables revealed the presence of significant genetic variability, indicating potential improvements for NV through breeding. Specifically, P and IVDOM presented large variation, while NDF had lower diversity but some accessions exhibited lower NDF than Tifton 85. Low GHI, except for IVDOM, indicated genotypic stability and potential for selecting improved accessions under fewer harvests. Breeding line 240, PI-316510, and PI-3166536 presented superior NV than Tifton 85 [5]. The methodology and the approaches used may provide a model for the enhancement of other vegetable crop diversity toward adaptability to the cropping condition decision. This result revealed the importance of preserving eggplant germplasm for future varietal development and that open-field cropping condition is more suitable under Malaysia's agroecology [6]. Due to the particularity of environment and isolation from the mainland, Greek islands represent a natural laboratory for comparing the diversity of landraces from the islands with those of the Greek mainland. Based on the available data, the Greek eggplant landraces present considerable morphological and genetic diversity with some differentiation signatures between the island and the mainland accessions. These results have implications for the conservation of Greek landraces and suggest that Greece might be considered as part of a secondary center of diversity for eggplant in the Mediterranean basin [7]. Six distinct groups of these landraces and the joint analysis presented an entanglement rate of 0.58, offering evidence for the divergence of accessions between the groups of both dendrograms. The Bayesian analysis allowed the distinction of two clusters for SSR. Significant variability was observed among accession with the potential to integrate several breeding programs [8]. The highest capsaicin concentration was recorded in accessions 16,209 (55.90 mg/g) and 10,757 (48.80 mg/g). The highest antioxidant value was recorded in accessions 17,750 (90.85 mg/g) and 15,661 (87.03 mg/g). All these characteristics are important for increasing industrial use and genetic improvement processes. These results show the existence of significant genetic variability within the genus Capsicum [9]. Knowledge of the genetic diversity in pigeonpea is essential for an effective breeding program. In one of the presented studies, the objective was to assess the genetic diversity among diverse pigeonpea accessions to select complementary and unique genotypes for breeding. The selected high-yielding and early maturing genotypes may be recommended as parental lines for breeding and grain yield improvement in Malawi or similar agro-ecologies [10].

Some of the research included in this volume focused on molecular variation. Cereal as maize [11,12] and wheat (*Triticum aestivum* L.) [13–16] was studied in different ways, including nucleotide diversity and genetic association analysis, population structure, allelic variation, congruity of the polymorphisms in the expressed and noncoding parts of genes, and management of high-throughput genotyping data. Medicinal crops as ginseng (*Panax ginseng* C.A. Meyer) [17,18], laurel (*Laurus nobilis* L.) [19], and tea (*Camellia sinensis* (L.) Kuntze) [20] are also included in this volume. Genetic variation and population structure and phytochemical and molecular diversity were studied in these crops. The variation in a relevant tuber crop, sweet potato (*Ipomoea batatas* L. Lam) [21,22], was assessed using chloroplast SSR markers and sequence diversity of biosynthesis pathways. In legumes, the genetic diversity in cowpea (*Vigna unguiculata* (L.) Walp.) [23] was studied to identify an important hotspot of variation.

Root length is a determining factor of the root system architecture, which is essential for the uptake of water, nutrients, and plant anchorage. The results of the study by [11] reveal that ZmMADS60 may be involved in the elongation of primary and lateral roots in the seedling stage and that significant variants can be used to develop functional markers to improve root length in maize [11]. Adequate knowledge and understanding of the genetic diversity and inter-trait relationships among elite maize inbred lines are crucial for determining breeding strategies and predicting hybrid performance. A wide range of genetic variability was observed among the inbred lines, indicating that they are an invaluable resource for breeding for combined heat and drought (HD) tolerance in maize breeding programs, especially in West and Central Africa [12]. Wheat landraces are currently being explored mainly as a source of allelic variation related to crop resilience and low-input adaptation. Characterization of their high- and low-molecular-weight glutenin subunits can aid breeders to select as donor local materials those ensuring good end-use properties in the derived elite cultivars. Some of the landraces, showing outstanding values for the gluten quality parameters analyzed, might be directly used by farmers interested in the cultivation of traditional varieties for specialized food markets [13]. The objective of another presented study was to analyze the mutation frequency in *Pin* genes and their influence on grain hardness in 118 modern bread-wheat cultivars and breeding lines and 80 landraces from Poland. Based on grain hardness measured by a single-kernel characterization system (SKCS), the analyzed cultivars and lines were categorized into different classes according to a previously proposed classification system [14]. The previously defined pairs of primers GliB1.1 and GliB1.2 were found to produce three and four principal variants, respectively, of PCR sequence length for the γ-gliadin pseudogene in 46 *Triticum aestivum* L. cultivars from 15 countries carrying 19 known alleles at the *Gli-B1* locus. In addition to revealing its evolutionary significance, this information is of potential use in wheat breeding, and we believe that novel variants of the *Gli-B1* locus will likely be found in hitherto unstudied germplasm [15]. In this study, a core collection of Spanish landraces of bread wheat was created using high-throughput genotyping technologies (DArTseq), which yielded more than 50 K molecular markers. This core collection, which includes a broad range of adapted genotypes, can be efficiently utilized for mining new alleles for useful traits in wheat breeding [16]. Ginseng (*Panax ginseng* C.A. Meyer) has long been used as a medicinal herb in Korea and China. Since ginseng originated from wild ginseng with low genetic diversity, it is also expected to have low genetic diversity. In another study, the genetic diversity of 451 ginseng accessions conserved in the National Agrobiodiversity Center (NAC) in South Korea was analyzed. The results of this study suggest that NAC should increase the genetic diversity of ginseng accessions for breeding programs, and alternatives are needed for securing ginseng genetic resources [17]. There have been limited research efforts to analyze the genetic diversity and population structure of ginseng germplasm conserved in the National Agrobiodiversity Center (NAC). The results of this study provide molecular evidence for the narrow genetic base of ginseng germplasm in NAC. For the broad understanding and efficient use of ginseng germplasm, it is necessary to analyze functional factors and to evaluate morphological traits [18]. Laurel is a medicinally important plant and is known to the world for its essential oil. Turkey is the main market in the laurel leaf trade, sharing about 90% of the world trade. Here, the authors aimed to elucidate the genetic diversity and population structure of 94 Turkish laurel genotypes collected from 26 provinces and four geographical regions using inter-primer binding-site (iPBS) retrotransposon markers. The information provided in this work will be helpful for the scientific community to devote more attention to this forgotten but medicinally important plant [19]. For the direct use of tea germplasm in breeding programs, a target-oriented core collection (TOCC) that retains the genetic diversity and various phytochemicals in tea is needed. This is the first report describing the development of a TOCC retaining the diversity of phytochemicals in tea germplasm. This TOCC will facilitate the identification of the genetic determinants of trait variability and the effective utilization of phytochemical diversity in crop improvement programs [20]. Sweet potato is an important food crop, widely cultivated in the world. In one of the included studies, nine chloroplast simple sequence repeat (cpSSR) markers were used to analyze the genetic diversity and relationships of 558 sweet potato accessions in the germplasm collection of the National Agrobiodiversity Center (NAC). The results of this study confirm that the genetic diversity of the female parents of sweet potato accessions conserved at the NAC is low, and therefore, more sweet potato accessions need to be collected. These results will help to establish an efficient management plan for sweet potato genetic germplasms at the NAC [21]. Sweet potato is also an important starch-producing crop, but little is known about the genetic variations in starch biosynthesis and sucrose metabolism genes. In this research, sequence variations between germplasms were identified in 20 genes involved in starch biosynthesis and sucrose metabolism, demonstrating the diversity in intron-loss alleles among sweet potato germplasms. These findings provide critical genetic information and useful molecular markers for revealing the regulatory mechanism of starch properties [22]. Cowpea is a multiple-purpose, drought-tolerant leguminous pulse crop grown in several dry tropical areas. Its domestication center is believed to be East or West Africa, where a high level of genetic diversity is apparently still found. However, detailed genetic information is lacking in many African countries, limiting the success of breeding programs. The high genetic diversity found in Mozambique sustains the importance of local genetic resources and farm protection to enhance genetic diversity in modern varieties of cowpea worldwide [23].

The performance of crops under abiotic and biotic stresses was analyzed regarding pests, diseases, and abiotic factors. The cereals studied were triticale (×*Triticosecale* Wittmack) [24], wheat [25], rice [26,27], and barley (*Hordeum vulgare* L.) [28]. Resistance to diseases as rust and powdery mildew, together with blast resistance, phosphorous deficiency, and salt stress were evaluated in this group of crops. In legumes, studies on resistance to anthracnose in lentils (*Lens* spp.) [29], the inheritance and expressivity of neoplasm trait in crosses between the domestic pea (*Pisum sativum* subsp. *sativum*) and wild pea (*Pisum sativum* subsp. *elatius*) [30], and the implications of the symbiotic nitrogen fixation in common bean (*Phaseolus vulgaris* L.) [31] under water stress are presented. The isolate-dependent inheritance of resistance against *Pseudoperonospora cubensis* of a horticultural crop (*Cucumber*) [32] was also analyzed.

Hexaploid triticale is a cultivated hybrid that combines wheat and rye properties. It has a better ability to be grown on poor soils, compared to wheat. In a review study, the recent state of the art of pre-breeding studies is presented, which are focused on the transfer of leaf and stripe rust resistance genes from *Aegilops* species into cultivated triticale using the distant crossing and chromosome engineering approach [24]. Wheat-*Psathyrostachys huashanica*-derived lines were developed by crossing common wheat and *P. huashanica* Keng (2n = 2x = 14, NsNs) using embryo culture; line H5-5-4-2 was selected with immunity to powdery mildew at both growth stages. The results indicate that resistant line H5-5-4-2 was a wheat-*P. huashanica* 1Ns disomic addition line [25]. The objectives of another presented study were to evaluate the newly developed blast-resistant rice lines in varied environmental conditions, precisely measure the response of the advanced lines in multiple environments, and classify the genotypes into groups that could serve as varieties for commercial cultivation. A low genetic advance was observed in some agronomical traits. Cluster analysis classified the evaluated genotypes into six groups that were recommended as varieties for commercial cultivation in Malaysia and other rice-growing regions [26]. The yield of rice is severely limited by phosphorus deficiency. This study aimed to identify P-deficiency-responsive differentially expressed proteins in rice through analysis of leaf proteome of contrasting P-responsive rice cultivars under P-deficiency conditions because genetic variability has been found in rice cultivars for adaptive response to P deficiency, and a controlled regulatory system is involved in the P deficiency adaptation response. This study could help to unravel the complex regulatory process that is involved in adaptation to P deficiency in rice [27]. The interaction between salinity and nitrogen metabolism has been investigated in two barley landraces, one tolerant ("100/1B") and one susceptible to salinity ("Barley medenine") from the Middle East and North Africa region. Specific enzymatic activities and occurrence can be used to determine the stress responsiveness of different landraces. It is suggested that a rapid increase in the activities of G6PDH, APX, and nitrogen assimilation enzymes represents an index of tolerance in "100/1B" and a stress symptom in "Barley medenine" [28]. Anthracnose, caused by the fungal pathogen *Colletotrichum lentis*, is a severe disease in lentils, causing premature defoliation, necrotic stem lesions that lead to plant wilting, and death in susceptible varieties. Two races of

C. lentis (0 and 1) have been described so far. Most of the collection studied was highly susceptible, but some levels of resistance were identified in about 15% of the accessions. Selected accessions showed potential to integrate several breeding programs [29]. The Neoplasm trait in pea pods is reported to be due to the lack of ultraviolet (UV) light in glasshouse conditions or in response to pea weevil (*Bruchus pisorum* L.) damage. This pod deformation arises from the growth of non-meristematic tissue on pods of domesticated peas. There was no relationship between neoplasm and damage by pea weevil under heavy insect epidemics under field conditions. The neoplasm occurring under glasshouse conditions may be due to one or to a combination of environmental factors. Since wild peas are useful genetic resources for breeding programs aiming at fresh pea production that could be utilized under glasshouse conditions, negative selection could be considered in segregating populations [30]. The objective of this research was to study the performance of common bean genotypes under water-deficit stress, and how it affects their symbiotic relationship with different *Rhizobium* strains in both greenhouse and field conditions. The genotype–strain relationship was very specific, and the local strains achieved the greatest productivity with some genotypes in irrigated and drought conditions that enable their application as inoculating strains, particularly relevant for the environmental impact of agriculture [31]. Wild accessions of cucumber were evaluated for resistance against each of the 23 isolates of the downy mildew oomycete *Pseudoperonospora cubensis*. F2 progeny plants of the cross between two resistant accessions were resistant, except for a few plants that were partially susceptible, suggesting that some of the resistance genes in PI 197088 and PI 330328 are not allelic [32].

In conclusion, in agricultural production, agrobiodiversity supports long-term productivity, resilience, and multiple ecosystem services, boosting yields in quality and quantity, increasing soil and water quality, and reducing the need for synthetic fertilizers. It also makes farmer livelihoods more resilient, reducing yield losses due to climate change and pest damage. Broadening the types of cultivated plants also benefits the environment, increasing the abundance of ecological services, such as pollinators and beneficial soil organisms, and reducing the risk of pests and diseases. Agrobiodiversity, when adequately maintained, also could support options for unknown future needs; additionally, using biodiversity-based solutions on farms can also decrease emissions of greenhouse gases that contribute to climate change.

Author Contributions: A.M.D.R. and A.P.R. wrote the paper. All authors have read and agreed to the published version of the manuscript.

Funding: This research received no external funding.

Acknowledgments: The authors thank the support from the projects INIA-RFP2015-00008-C04-03 and CSIC-202040E190, and Diputación de Pontevedra (Spain) for their experimental work.

Conflicts of Interest: The authors declare no conflict of interest.

References

1. Ibrahim Bio Yerima, A.R.; Achigan Dako, E.G.; Aissata, M.; Sekloka, E.; Billot, C.; Adje, C.O.A.; Barnaud, A.; Bakasso, Y. Agromorphological Characterization Revealed Three Phenotypic Groups in a Region-Wide Germplasm of Fonio (*Digitaria exilis* (Kippist) Stapf) from West Africa. *Agronomy* **2020**, *10*, 1653. [CrossRef]
2. Rocandio Rodriguez, M.; Santacruz Varela, A.; Cordova Tellez, L.; Lopez Sanchez, H.; Hernandez Bautista, A.; Castillo Gonzalez, F.; Lobato Ortiz, R.; Garcia Zabala, J.J.; Lopez, P.A. Estimation of Genetic Diversity in Seven Races of Native Maize from the Highlands of Mexico. *Agronomy* **2020**, *10*, 309. [CrossRef]
3. Harakotr, B.; Prompoh, K.; Boonyuen, S.; Suriharn, B.; Letrrat, K. Variability in Nutraceutical Lipid Content of Selected Rice (*Oryza sativa* L. spp. indica) Germplasms. *Agronomy* **2019**, *9*, 823. [CrossRef]
4. Yang, Y.; Xu, X.; Zhang, M.; Xu, Q.; Feng, Y.; Yuan, X.; Yu, H.; Wang, Y.; Wei, X. Genetic Basis Dissection for Eating and Cooking Qualities of *Japonica* Rice in Northeast China. *Agronomy* **2020**, *10*, 423. [CrossRef]
5. de Souza, C.; Lopez, Y.; Munoz, P.; Anderson, W.; Dall'Agnol, M.; Wallau, M.; Rios, E. Natural Genetic Diversity of Nutritive Value Traits in the Genus *Cynodon*. *Agronomy* **2020**, *10*, 1729. [CrossRef]
6. Mat Sulaiman, N.N.; Rafil, M.Y.; Duanglit, J.; Ramiee, S.I.; Phumichal, S.; Oladosu, Y.; Datta, D.R.; Musa, I. Genetic Variability of Eggplant Germplasm Evaluated under Open Field and Glasshouse Cropping Conditions. *Agronomy* **2020**, *10*, 436. [CrossRef]

7. Gramazio, P.; Chatziefstratiou, E.; Petropoulos, C.; Chioti, V.; Mylona, P.; Kapotis, G.; Vilanova, S.; Prohens, J.; Papasotiropoulos, V. Multi-Level Characterization of Eggplant Accessions from Greek Islands and the Mainland Contributes to the Enhancement and Conservation of this Germplasm and Reveals a Large Diversity and Signatures of Differentiation between both Origins. *Agronomy* **2019**, *9*, 887. [CrossRef]
8. Alvares Bianchi, P.; Almeida da Silva, L.G.; Da Silva Alencar, A.A.; Araujo Diniz Santos, P.H.; Pimenta, S.; Pombo Sudré, C.; Dalia Corte, L.E.; Azeredo Gonçalves, L.S.; Rodrigues, R. Biomorphological Characterization of Brazilian Capsicum Chinense Jacq. Germplasm. *Agronomy* **2020**, *10*, 447. [CrossRef]
9. Paredes Andrade, N.J.; Monteros Altamirano, A.; Tapia Bastidas, C.G.; Serensen, M. Morphological, Sensorial and Chemical Characterization of Chilli Peppers (*Capsicum* spp.) from the CATIE Genebank. *Agronomy* **2020**, *10*, 1732. [CrossRef]
10. Yohane, E.N.; Shimelis, H.; Laing, M.; Mathew, I.; Shayanowako, A. Phenotypic Divergence Analysis in Pigeonpea [*Cajanus cajan* (L.) Millspaugh] Germplasm Accessions. *Agronomy* **2020**, *10*, 1682. [CrossRef]
11. Li, P.; Ge, Z.; Wang, H.; Wel, J.; Wang, Y.; Xu, Y.; Yang, Z.; Xu, C. Nucleotide Diversity and Association Analysis of *ZmMADS60* with Root Length in the Maize Seedling Stage. *Agronomy* **2020**, *10*, 342. [CrossRef]
12. Osuman, A.S.; Badu-Apraku, B.; Ifie, B.E.; Tongoona, P.; Obeng-Blo, E.; Garcia Oliveira, A.L. Genetic Diversity, Population Structure and Inter-Trait Relationships of Combined Heat and Drought Tolerant Early-Maturing Maize Inbred Lines from West and Central Africa. *Agronomy* **2020**, *10*, 1324. [CrossRef]
13. Chacon, E.A.; Vazquez, F.J.; Giraldo, P.; Carrillo, J.M.; Benavente, E.; Rodriguez Quijano, M. Allelic Variation for Prolamins in Spanish Durum Wheat Landraces and Its Relationship with Quality Traits. *Agronomy* **2020**, *10*, 136. [CrossRef]
14. Przyborowsky, M.; Gasparis, S.; Kala, M.; Orczyk, W.; Orczyk, A.N. The Variability of Puroindoline-Encoding Alleles and Their Influence on Grain Hardness in Modern Wheat Cultivars Cultivated in Poland, Breeding Lines and Polish Old Landraces (*Triticum aestivum* L.). *Agronomy* **2020**, *10*, 1075. [CrossRef]
15. Popovych, Y.; Chebotar, S.; Meinik, V.; Rodriguez Quijano, M.; Pascual, L.; Rogers, W.J.; Metakovsky, E. Congruity of the Polymorphisms in the Expressed and Noncoding Parts of the *Gli-B1* Locus in Common Wheat. *Agronomy* **2020**, *10*, 1510. [CrossRef]
16. Pascual, L.; Fernandez, M.; Aparicio, N.; Lopez Fernandez, M.; Fité, R.; Giraldo, P.; Ruiz, M. Development of a Multipurpose Core Collection of Bread Wheat Based on High-Throughput Genotyping Data. *Agronomy* **2020**, *10*, 534. [CrossRef]
17. Lee, K.J.; Sebastin, R.; Kim, S.H.; Yoo, E.; Lee, S.; Cho, G.T.; Kang, M.; Hyun, D.Y. Genetic Composition of Korean Ginseng Germplasm by Collection Area and Resource Type. *Agronomy* **2020**, *10*, 1643. [CrossRef]
18. Lee, K.J.; Lee, J.R.; Sebastin, R.; Cho, G.T.; Hyun, D.Y. Molecular Genetic Diversity and Population Structure of Ginseng Germplasm in RDA-Genebank: Implications for Breeding and Conservation. *Agronomy* **2020**, *10*, 68. [CrossRef]
19. Karik, U.; Nadeem, M.A.; Habyarimana, E.; Ercisli, S.; Yildiz, M.; Yilmaz, A.; Yang, S.H.; Chung, G.; Baloch, F.S. Exploring the Genetic Diversity and Population Structure of Turkish Laurel Germplasm by the iPBS-Retrotransposon Marker System. *Agronomy* **2019**, *9*, 647. [CrossRef]
20. Hyun, D.Y.; Gi, G.Y.; Sebastin, R.; Cho, G.T.; Kim, S.H.; Yoo, E.; Lee, S.; Son, D.M.; Lee, K.J. Utilization of Phytochemical and Molecular Diversity to Develop a Target-Oriented Core Collection in Tea Germplasm. *Agronomy* **2020**, *10*, 1667. [CrossRef]
21. Lee, K.J.; Lee, G.A.; Lee, J.R.; Sebastin, R.; Shin, M.J.; Cho, G.T.; Hyun, D.Y. Genetic Diversity of Sweet Potato (*Ipomoea batatas* L. Lam) Germplasms Collected Worldwide Using Chloroplast SSR Markers. *Agronomy* **2019**, *9*, 752. [CrossRef]
22. Zhang, K.; Luo, K.; Li, S.; Peng, D.; Tang, D.; Lu, H.; Zhao, Y.; Lv, C.; Wang, J. Genetic Variation and Sequence Diversity of Starch Biosynthesis and Sucrose Metabolism Genes in Sweet Potato. *Agronomy* **2020**, *10*, 627. [CrossRef]
23. Figueira Gomes, A.M.; Draper, D.; Talhinhas, P.; Batista Santos, P.; Simoes, F.; Nhantumbo, N.; Massinga, F.; Ramalho, J.C.; Marques, I.; Ribeiro-Barros, A.I. Genetic Diversity among Cowpea (*Vigna unguiculata* (L.) Walp.) Landraces Suggests Central Mozambique as an Important Hotspot of Variation. *Agronomy* **2020**, *10*, 1893. [CrossRef]
24. Ulaszewski, W.; Kwlatek, M.T. *Aegilops* Species for the Improvement of the Leaf and Stripe Rust Resistance in Cultivated Triticale (×*Triticosecale* Wittmack). *Agronomy* **2020**, *10*, 1991. [CrossRef]
25. Han, J.; Liu, Y.; Hou, C.; Li, J.; Wang, J.; Zhang, Q.; Yang, Q.; Chen, X.; Wu, J. A 1Ns Disomic Addition from *Psathyrostachys Huashanica* Keng Confers Resistance to Powdery Mildew in Wheat. *Agronomy* **2020**, *10*, 312. [CrossRef]
26. Sabri, R.S.; Rafii, M.Y.; Ismall, M.R.; Yusuff, O.; Chukwu, S.C.; Hasan, N. Assessment of Agro-Morphologic Performance, Genetic Parameters and Clustering Pattern of Newly Developed Blast Resistant Rice Lines Tested in Four Environments. *Agronomy* **2020**, *10*, 1098. [CrossRef]
27. Tantray, A.Y.; All, H.M.; Ahmad, A. Analysis of Proteomic Profile of Contrasting Phosphorus Responsive Rice Cultivars Grown under Phosphorus Deficiency. *Agronomy* **2020**, *10*, 1028. [CrossRef]
28. Azalez, F.E.B.; Ayadi, S.; Capasso, G.; Landi, S.; Paradisone, V.; Jallouli, S.; Hammami, Z.; Chamekh, Z.; Zouari, I.; Trifa, Y.; et al. Salt Stress Induces Differentiated Nitrogen Uptake and Antioxidant Responses in Two Contrasting Barley Landraces from MENA Region. *Agronomy* **2020**, *10*, 1426.
29. Barilli, E.; Moral, J.; Aznar Fernandez, T.; Rubiales, D. Resistance to Anthracnose (*Colletotrichum lentis*, Race 0) in *Lens* spp. Germplasm. *Agronomy* **2020**, *10*, 1799. [CrossRef]
30. Sari, H.; Sari, D.; Eker, T.; Aydinoglu, B.; Canci, H.; Ikten, C.; Gokturk, R.S.; Zeybek, A.; Bakir, M.; Smykal, P.; et al. Inheritance and Expressivity of Neoplasm Trait in Crosses between the Domestic Pea (*Pisum sativum* subsp. *sativum*) and Tall Wild Pea (*Pisum sativum* subsp. *elatius*). *Agronomy* **2020**, *10*, 1869.

31. Rodiño, A.P.; Riveiro, M.; De Ron, A.M. Implications of the Symbiotic Nitrogen Fixation in Common Bean under Seasonal Water Stress. *Agronomy* **2021**, *11*, 70. [CrossRef]
32. Chen, T.; Katz, D.; Naim, Y.B.; Hammer, R.; Daniel, B.H.B.; Rubin, A.E.; Cohen, Y. Isolate-Dependent Inheritance of Resistance Against *Pseudoperonospora cubensis* in Cucumber. *Agronomy* **2020**, *10*, 1086. [CrossRef]

agronomy

MDPI

Article

Implications of the Symbiotic Nitrogen Fixation in Common Bean under Seasonal Water Stress

Ana P. Rodiño [1], Manuel Riveiro [2] and Antonio M. De Ron [1,*]

[1] Biology of Agrosystems, Misión Biológica de Galicia (MBG)-National Spanish Research Council (CSIC), El Palacio-Salcedo, 36143 Pontevedra, Spain; aprodino@mbg.csic.es
[2] Experimental Agricultural Station Baixo Miño-INGACAL, 36470 Salceda de Caselas, Spain; manuel.riveiro.leira@xunta.es
* Correspondence: amderon@mbg.csic.es; Tel.: +34-986854800

Abstract: The objective of this research was to study the performance of 10 common bean genotypes under water deficit stress and how it affects to their symbiotic relationship with 10 *Rhizobium* strains in both greenhouse and field conditions. PHA-0471, a small seeded genotype had the best yield under irrigation and under water stress. Other genotypes with tolerance to drought were the large-seeded PHA-0432 and PHA-0683. In the *Rhizobium* inoculation tests it was observed that the increase of dry nodular weight produced less seed yield in beans. PHA-0683 genotype presented a great uniformity on nodule size and an association with yield when it displays the big nodule phenotype. Further research about this would be interesting because this fact could be due to the existence of a plant blocking mechanism for inefficient strain nodules. The inoculated plants were productive in irrigated fields and in drought ones and their productivity was the same or even better than the N supplemented plant control. The genotype-strain relationship was very specific and the local strains achieved the greatest productivity with some genotypes in irrigated and drought conditions that make possible their use as inoculating strains, with relevance for the environmental impact of agriculture.

Keywords: drought; genetic resources; *Phaseolus vulgaris* L.; plant breeding; rhizobia; stress

Citation: Rodiño, A.P.; Riveiro, M.; De Ron, A.M. Implications of the Symbiotic Nitrogen Fixation in Common Bean under Seasonal Water Stress. *Agronomy* **2021**, *11*, 70. https://doi.org/10.3390/agronomy11010070

Received: 24 November 2020
Accepted: 28 December 2020
Published: 31 December 2020

1. Introduction

The common bean (*Phaseolus vulgaris* L.) is produced worldwide as an important protein crop and an alternative to animal protein in the human diet [1]. This crop contains also vital nutrients including vitamins and minerals, and the stems are also used as fodder for livestock, especially in the dry spell following the main cropping season [2].

Drought stress is a major yield-limiting factor in dry bean production worldwide [3,4] especially in marginal, unfavorable environments [5]. Most common bean production in the developing world occurs under conditions where the risk of drought is high [3]. Over 60% of dry bean production worldwide is subject to water-deficit stress at some stage of growth [6], with yield losses up to 80% resulting from drought in some regions [7]. Therefore dry bean varieties that retain yield potential under drought stress are a focus of breeding programs [3,8,9]. Drought stress in beans can elicit numerous plant responses including reduced root, shoot, and leaf development, poor nutrient uptake, reduced photosynthesis, stomatal conductance, leaf area, and biomass, inefficient partitioning of carbohydrates, and reduced pod set and seed yield [3,8,10].

Dry bean yield is very sensitive to water deficit stress during the reproductive phase (flowering and pod elongation) [11,12]. Irrigation is often used in dry bean production to reduce drought effects or at least manage it to maximize bean yield or improve water use efficiency when rainfall is limited. The best option for reducing such yield gaps and realizing yield stability under unfavorable environments is the development of drought-tolerant genotypes. Drought tolerance, once genetically encoded in the seed of a genotype, can be used readily by many farmers for combating drought effects in common bean production [5]. Breeding for drought-tolerant crops is challenging and time-consuming, owing to the need for simultaneously considering multiple abiotic and biotic factors modulating the level of drought-tolerance.

As legumes, common bean plants, also contribute to soil fertility enhancement through symbiotic atmospheric nitrogen fixation [13]. The importance of the *Rhizobium*-bean symbiosis lies in the ability to transform atmospheric nitrogen into an assimilable one to the plant, through the nodules. Initially, all rhizobia found in beans were assigned to the species *R. leguminosarum* bv. *phaseoli*. Currently they are framed in the genus *Rhizobium* and *Sinorhizobium* but *R. etli* predominates in beans. Although, the symbiosis with other species and even other genera was described and it is possible that there exists a great diversity of rhizobium-beans interactions [14,15]. Five species have been identified that can nodule with the common bean in the soils of Spain [16]. Effective nodulation by the symbiotic nitrogen-fixing bacterium *Rhizobium* is difficult to achieve in certain soil types and environmentally stressed conditions [17]. Low humidity, soil salinity and high temperatures in the root zone adversely affect rhizobial survival and infection of bean roots, nodule formation, and development of the enzymatic system that participates in the symbiotic system for N2-fixation [18,19]. The high loss of plant water by evapotranspiration due to high temperatures is also stressful.

Previous efforts have been made to increase the drought tolerance and manage the growth and performance of legumes under conditions of drought stress [20]. These studies documented the important benefits of establishing an efficient *Rhizobium*-bean symbiosis, but lack the field assessments needed to evaluate, identify and recommend high-performing biofertilizer strains that can be used with confidence to alleviate drought when cultivated in various agroecosystems. Biofertilization of legumes improves soil fertility and decreases groundwater pollution by avoiding the excessive application of chemical fertilizers. Most importantly, description of comprehensive field inoculation experiments is needed to address many issues required to formulate and implement a biofertilization technology based in *Rhizobium* for field management of the bean crop under these stresses over broad agrosystems.

That lack of a proven rhizobia biofertilizer efficacy based on extensive agronomic field assessments justifies the present study. The general objective of this research was to study the performance of common bean genotypes under water deficit stress and how this stress affects the symbiotic relationship between the bean plants and different *Rhizobium* strains. For this goal, the experimental tasks were carried out in factorial trials in both controlled conditions in greenhouse and in an open field under water stress and without stress.

2. Materials and Methods

Experimental design. Two factorial trials including 10 bean genotypes inoculated with 10 strains of *Rhizobium* were carried out under conditions of water deficit stress and irrigation in a greenhouse and in an open field (Figure 1). The trials were conducted at the Baixo Miño Agricultural Experimental Station (Salceda de Caselas, Spain, 42°4′ N, 8°34′ W, 85 masl). The bean genotypes used belong to the germplasm collection at the MBG-CSIC (Pontevedra, Spain), while the *Rhizobium* strains were collected in different soils in the region (Galicia, NW of Spain) (Table 1).

The first treatment consisted of water supply both in a greenhouse and in an open field: non-stress (NS), in which the plants were irrigated to maintain between 80– and 100% of field capacity and drought stress (DS), in which it has been tried that the humidity did not fall below 40% of field capacity.

The second treatment was the inoculation with rhizobia. The *Rhizobium* strains were seeded in Petri dishes with YMA medium (Yeast Mannitol Agar) and incubated at 25 °C for three days. The rhizobia colonies were extracted and dissolved in distilled water at a concentration of 10^8 cells/mL. The concentration was determined by measuring the turbidity with a spectrophotometer (Nanodrop 2000, Thermo Fisher Scientific, Waltham,

Massachusetts, US). Before inoculation, the seeds were washed in a 10% bleach solution (5.25% sodium hypochlorite) to avoid contamination. The seeds, once inoculated, were stored in a polythene bag to retain moisture [21].

(**A**) (**B**)

Figure 1. (**A**) Greenhouse experiment and (**B**) field experiment.

Table 1. There were three genotypes of beans and two strains of rhizobia of reference (R).

A. Bean genotypes used in this study, name, commercial class and performance under drought.

Bean Genotypes	Name	Commercial Class	Drought Performance
PHA-0155	local variety	White Kidney	unknown
PHA-0432	local variety	Marrow	unknown
PHA-0471	local variety	Navy	unknown
PHA-0483	local variety	Guernikesa	unknown
PHA-0683	local variety	Cranberry	unknown
PMB-0220	Mattherhorn	Great Northern	tolerant
PMB-0222	Almonga	Great Northern	unknown
PMB-0244	Linex (R)	Cannellini	unknown
PMB-0285	L88-18 (R)	Black Turtle	sensitive
PMB-0286	L88-63 (R)	Black Turtle	tolerant

B. Rhizobia strains used.

Strains of Rhizobia	Type
SLL2	local strain
EXIC	local strain
EXIB	local strain
EG	local strain
EF	local strain
EPOB	local strain
APAFI	local strain
LTMF	local strain
CIAT899 (R)	*R. tropici*
CFN42 (R)	*R. etli*

According to this, the experiments both in the greenhouse and in the field included 100 combinations: 10 bean genotypes × 10 rhizobia strains each one.

Greenhouse trial. Coconut fiber and vermiculite were used as a substrate in a 2:1 ratio. The bean genotypes were seeded in 35 × 100 × 42 cm double layer polypropylene boxes. These boxes had four internal divisions of 200 μm polyethylene to isolate the experimental plots. The boxes had a drainage system with 20 holes of 20 mm in their lower part covered with acrylic textile to avoid the loss of substrate. In each plot of 0.35 × 0.20 m eight bean seeds were sown. This experiment included four replications.

Field trial. The soil texture was sandy loam (62.0% sand, 26.6% silt and 11.4% clay). The experimental plot had 0.75 m^2, the distance between rows was 0.50 m and between bean plants 0.10 m, 15 plants were grown in each plot, with two replications; the two replications that were inoculated with the same strain were sown side by side to avoid mixing of strains. Among the bean plots inoculated with different strains of rhizobia, three rows of maize were planted as a barrier to avoid the mixture of rhizobia strains, since the bacteria are not able to propagate across the maize roots. As controls, the 10 genotypes were grown without fertilization and with nitrogen fertilization (40 kg ha^{-1}).

The traits measured (Table 2) were: shoot dry weight (SDW); root dry weight (RDW); nodules number (NN); nodule dry weight (NDW); medium nodular dry weight (MNDW), calculated as MNDW = NDW/NN; 100 seed weight (100 SW) adjusted to 12% moisture content; seed dry matter (SDM); percentage of nitrogen in shoot dry matter (PNSDW), using the Kjeldahl method; quantity of nitrogen in shoot dry matter (NSDW), calculated as NSDW = (SDW × PNSDW)/100; percentage of nitrogen in seed dry matter (PNSEDM) and nitrogen content in seed dry matter (NSEDM), calculated as NSEDM = (SDM × PNSEDM)/100.

Table 2. Traits measured (F = field, G = greenhouse), acronym and units of measurement.

Traits Measured	Acronym	Units
Shoot dry weight, F, G	SDW	grams $plant^{-1}$
Root dry weight, G	RDW	grams $plant^{-1}$
Nodules number, F, G	NN	number $plant^{-1}$
Nodule dry weight, F, G	NDW	milligrams $plant^{-1}$
Medium nodular dry weight, F, G	MNDW	milligrams
100 seed weight, F	100SW	grams 100 $seeds^{-1}$
Seed dry matter, F	SDM	grams $plant^{-1}$
Percentage of nitrogen in shoot dry matter, F	PNSDW	%
Quantity of nitrogen in shoot dry matter, F	NSDW	grams $plant^{-1}$
Percentage of nitrogen in seed dry matter, F	PNSEDM	%
Nitrogen content in seed dry matter, F	NSEDM	grams $plant^{-1}$

Data analyses. Analysis of variance was performed for the quantitative characters measured in the trials using the SAS statistical package [22], by the General Linear Model (GLM). The model consists of 10 genotypes (G), two treatments (T) (without and with water deficit stress), 10 strains (I), genotype treatment interaction (G×T), genotype strain interaction (G×I), treatment strain interaction (T×I), genotype treatment strain interaction (G×T×I), replications (R) and the model error, where Y is any of the variables studied Y = (G + T + I + G×T + G×I + T×I + G×T×I + R + error). The "F" test of variance was carried out to show the existence of significant differences between the origins of variations studied [23]. For the comparison of means, the same GLM procedure was used and the least significant difference of Fisher (LSD) was calculated for the variables that displayed significant differences in the analysis of variance.

3. Results and Discussion

3.1. Greenhouse Experiment

The analysis of variance (ANOVA) included the independent variables, not MNDW since it is an index. Table 3 shows significant differences among all the genotypes for all the variables studied. The treatment (T) is high significant in all the variables except for dry nodular matter (NDW). There are no significant differences between strains in the variables studied. The interaction genotype and treatment (G×T) was significant in SDW, RDW and NN. Thus, the treatment influences the variables studied although it does not influence all the variables in the same way. It is remarkable that there is only G×I interaction in RDW. The rest of the interactions were not significant. In previous works [24,25], genotype-strain specificity was demonstrated. There are strains that increase the productivity of certain genotypes, and others do not have the same effect [14,24–27]. This does not agree with our results, since the G*I interaction only displayed significant differences for RDW.

Table 3. Analysis of variance in the greenhouse experiment, displaying genotypes (G), strains (I), treatments (T), genotype strain interaction (G×I), genotype treatment interaction (G×T), treatment strain interaction (T×I) and genotype treatment strain interaction (G×T×I).

Source of Variation	SDW	RDW	NN	NDW
G	***	***	**	**
T	***	***	**	ns
I	ns	ns	ns	ns
G×T	***	***	*	ns
G×I	ns	*	ns	ns
T×I	ns	ns	ns	ns
G×T×I	ns	ns	ns	ns

*** $p < 0.0001$, ** $p < 0.001$, * $p < 0.01$, ns = not significant.

The genotypes with the highest score for SDW without stress were PHA-0155, PMB-0222 and PHA-0483 (Table S1). Within each genotype the SDW vary significantly between the different strains used, a fact also observed by Rodiño et al. [25], who found large differences in the development of the plant depending on the inoculated rhizobia strain. PHA-0483 and PMB-0285 reach their highest SDW with strain CFN42 and PMB-0220 with strain EXIC. This great specificity in the strain-genotype makes it necessary to optimize the yield of the improved genotype with the specific strain [14,24–27]. The RDW is similar to previous works [25], with a range of 0.26 g plant^{-1} to 4.24 g plant^{-1} (Table S1). PMB-0285 (1.86 g plant^{-1}) and PMB-0286 (1.74 g plant^{-1}) have plants with superficial roots and with many adventitious roots being plants with a better efficiency in the adsorption of nutrients such as phosphorus [28].

The number of nodules per plant (NN) (Table S1) has great variability, with LSD (least significant difference) of 133.8 nodules, values similar to other studies [29]. Nodule formation is a plant-controlled process; therefore, the formation of new nodules in the younger roots is inhibited by existing nodules through systemic N signals [25,30,31]. The genotypes with higher nodulation were PMB-0285 and PMB-0286, with 241.6 and 140.0 nodules plant^{-1} respectively. The combination of EXIB—PMB-0285 displayed 1097.8 nodules plant^{-1}. This excessive nodulation could be attributed to a specific interaction. PMB-0285 and PMB-0286 were the genotypes with higher nodule dry weight (NDW) (Table S1), with 183.0 and 81.9 mg plant^{-1}, respectively. The EXIB strain has an NDW of 109.5 mg plant^{-1}, mainly due to the disproportionate growth of the PMB-0285 that reaches 937.8 mg plant^{-1} of NDW. There was a significant relationship ($p < 0.0001$) between RDW and NDW, which accounts for the 85% of the genotype-strain combinations of the trial. This could be due to the fact that a larger RDW supposes a larger surface that can be infected by rhizobia. In the case of PMB-0285 and PMB-0286, it may be due to an over nodulation. The mean nodular dry matter (MNDW) (Table S1) was between 0.000 and 3.895 mg nodules^{-1}, similar to that observed by Rodiño et al. [25]. Under irrigation, the genotypes with the highest MNDW mean values were PMB-0222 and PMB-0285 with 1.094 mg and 0.837 mg nodule^{-1}, and the EPOB and EF strains with 0.796 mg and 0.713 mg nodule^{-1}, respectively. In general, the plants showed very small nodules, with values < 1 mg nodule^{-1} in 90% of the plants.

The water stress in this experiment represented an average reduction of 32.6% in SDW. The genotype PHA-0155 had the highest SDW (Table S2) in conditions of water stress, but it does not indicate a tolerance to this stress, since it suffers a reduction of 48.3%, compared to the absence of stress. The genotypes that can be considered drought tolerant are PHA-0471 and PMB-0220 since they displayed around 15% more SDW in an environment with water stress than without stress. Possibly this increase is due to an adaptation to the drought of the *R. tropici* CIAT899 strain, which induced in these genotypes a considerable increase in SDW. The LTMF, EXIC and SLL2 strains can be defined as sensitive strains since the plants inoculated with these strains present a much lower SDW than the rest. In general terms, there was a significant 33% RDW reduction (Table S2) in plants subjected to water stress. This reduction occurs because the roots are an important carbon sink for the plant and if this is altered the RDW can decrease [32]. The strain-genotype combinations with the highest RDW were LTMF-PMB-0244 and CIAT899-PHA-0683, with values greater than 1.20 g plant^{-1}.

Water stress shows an average reduction of 57.2% in NN. The genotypes with the highest NN were PHA-0471 and PHA-0683, with 47.5 and 43.6 nodule plant^{-1}, respectively (Table S2). The genotype PHA-0471 nodules well in irrigation, and under water stress conditions it does not suffer a significant reduction. This stability in NN also occurs in some soybean genotypes, and it is a competitive advantage, since if the stress subsides, the plant can quickly regain nitrogen-fixing capacity [33]. The EPOB strain with PHA-0471 reaches maximum nodulation with 84.3 nodules plant^{-1}. The genotype with the highest NDW is PHA-0683 with 124.4 mg plant^{-1} (Table S2), due to the exceptional values of 428.0 and 327.6 mg plant^{-1} obtained with EG and EPOB strains, respectively. The plants inoculated with these strains have the highest NDW with 107.0 and 80.5 mg plant^{-1} respectively. This response of the bean to the water stress conditions, contrasts with that observed in other legumes such as pea, where the plants reduced their NDW by 62% under water stress [34,35]. This may be due to a worse adaptation of the strains, which would imply lower levels of infection or an early senescence of the nodules due to stress [36–39]. The Table S2 shows that the MNDW values under water stress conditions increase compared to the irrigated experiment. There were genotypes such as PHA-0483 that had a value five times higher for MNDW. The genotypes with the highest MNDW are PHA-0683 and PHA-0483, with 2.307 mg and 2.077 mg respectively. The strains that induce the highest MNDW are EF and EPOB, with 2.410 mg and 2.374 mg on average.

Voisin et al. [35] in their study of nodular self-regulation processes found that the NDW had a direct relationship with the SDW, independently of the number of nodules. Thus, it is logical to assume that if the number of nodules is less, they tend to be larger. PHA-0683 shows great size uniformity in its nodules (Figure 2). When it presents big nodules (BNO) (A), it maintains this size regardless of the NN, and the same happens when they have small nodules (SNO) (B). According to Mergaert et al. [40], BNO-inducing strains appear capable of overcoming plant defense mechanisms. The combination PMB-0222-EPOB presented BNO under irrigation and SNO in drought conditions. Instead, the combination PHA-0683-EG presented the opposite. Thus, it is shown that the infection mechanisms of the strain and the morphology of the root cells are affected by water stress. The greenhouse experiment was a preliminary study that showed the main trends in the performance of the symbiotic system that was studied in more detail in the field experiment.

3.2. *Field Experiment*

There are significant differences in all the variables studied between the bean genotypes and the treatments(Table 4). Between strains the difference was not significant in some variables related to nodulation (NN, NDW). It is surprising that the strains do not influence nodulation but influence nitrogen fixation, which may suppose that these characters are determined by the genetics of the plant. The interaction G×T was significant for all variables, except NDW, which showed that the genotypes had a differentiated response to water stress, both in dry matter and in the partition of N in the plant. The G×I and G×T×I interactions were only significant for SDW, NSDW, and 100SW. The interaction T×I was significant for all variables except for those related to nodulation.

Table S3 shows the average production of SDW that in irrigation was 34.2 g plant^{-1}. In the plants inoculated with the LTMF strain, SDW was significantly higher with 46.7 g plant^{-1}. The combinations LTMF-PHA-0471 and LTMF-PMB-0285 doubled the SDW, exceeding 70 g plant^{-1}. The mean value of 100SW under irrigation was 31.6 g (Table S3), but there are large differences among genotypes. Despite the fact that the size of the seed is genetically regulated [41,42], there is a certain range of variation within each genotype, since PHA-0155 had a coefficient of variation of 24%, PMB-0224 38% and PHA-0483 40%. The control plants cultivated with N supply had larger seeds than the rest, 47.2 g 100 seeds^{-1} highlighting PMB-0222, PMB-0285, PHA-0483 and PMB-0244, with increases of 65% with respect to the mean of the inoculated plants. However, the SDM performance was higher in the inoculated genotypes, except PMB-0244, which shows the efficacy of the symbiotic association.

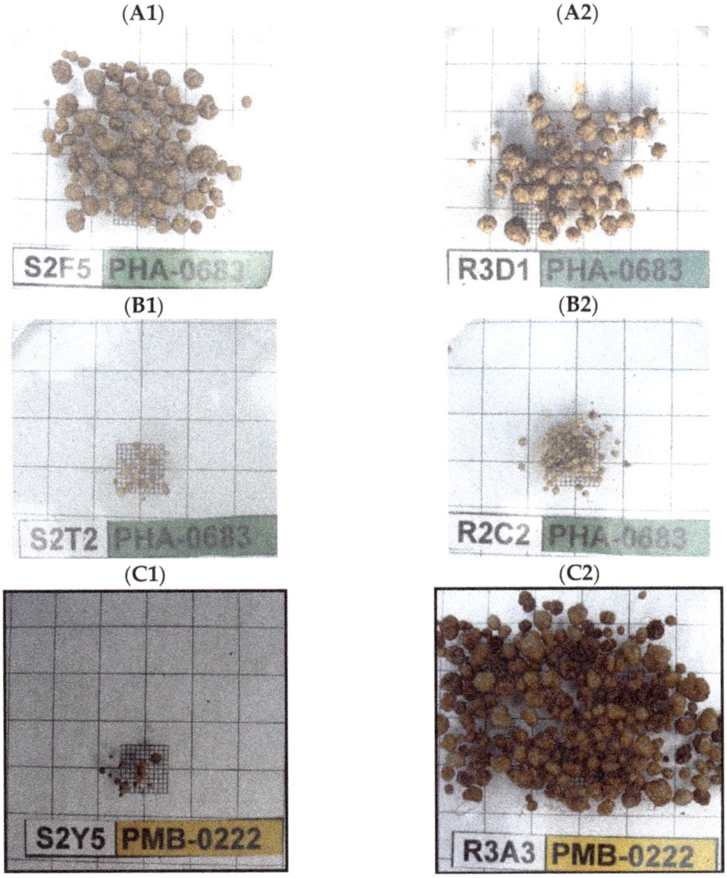

Figure 2. Variation in the size of nodules depending on the bean genotype and the strain. (**A1**,**A2**). BNO nodules in PHA-0683. (**B1**,**B2**). SNO nodules in PHA-0683. (**C1**,**C2**). PMB-0222-EPOB combination showing the difference in size in irrigation and drought.

The mean NN was 29.1 nodules per plant. PMB-0286 presents the highest nodulation, with 56.9 nodules plant^{-1}. The combination with the highest nodulation was PMB-0286-LTMF, with 122.7 nodules plant^{-1}. Given the low nodular specificity of the common bean [43], the zero control allows evaluating native populations of the rhizosphere with the capacity to generate nodules. The growth of the native populations of rhizobia is significant when the inoculated populations show a very low nodulation, and instead they can hardly be detected when they present high nodulation. Just as there are inoculated strains that compete well with native populations [44,45], there are also inoculated strains that do not adapt to the environment or do not compete with native ones. The number of nodules is determined by the rate of N supplied and N demanded [35]. The control plants with N presented nodulation, despite those at doses higher than 30 kg N ha^{-1}, early nodulation is reduced [46]. This shows that the degree of N tolerance is a variable factor linked to the genotype. Rennie and Kemp [47] applied a dose of 40 kg N ha^{-1} and detected genotypes that suffered only a 10% reduction in N fixation, while in others the difference was 60%. In the case of PMB-0286 and PMB-0244, this dose of N in the initial vegetative phase could have a "starter" effect, which allowed the development of the plant that favored better nodular growth [48]. The average NDW was 29.1 mg plant^{-1}. The genotypes with the highest NDW (Table S3), above 30 mg plant^{-1} were PMB-0286 and PMB-0285. According to Vessey [48], a higher value would be expected, since bean genotypes with indeterminate growth habit continue to accumulate NDW during the pod-filling phase. This low value could be because the plants were collected in the flowering stage and they did not reach their maximum potential. The LTMF-PMB-0286 combination showed a value of 94.1 mg plant^{-1}. There is a direct relationship between NDW and SDM, but the relationship varies based on SDM/NDW performance [49]. As NDW increases, SDM per mg of NDW decreases. Genotype-strain combinations with SDM/NDW between 2 and 7 g show a significant relationship ($p < 0.0001$). Some variation that appears in the nodulation could be due to the incomplete collection of the root system when extracting from the soil [46].

Table 4. Analysis of variance in the field experiment, with genotypes (G), treatments (T), strains (I), genotype treatment interaction (G×T), genotype strain interaction (G×I), treatment strain interaction (T×I), and genotype treatment strain interaction (G×T×I).

	SDW	NSDW	PNSDW	SDM	NSEDM	PNSEDM	NN	NDW	100SW
G	***	***	***	***	***	***	***	***	***
T	***	***	***	***	***	***	***	***	***
I	***	***	***	***	***	*	ns	ns	***
G×T	***	***	***	**	**	***	**	ns	*
G×I	***	***	ns	ns	ns	ns	ns	ns	*
T×I	***	**	***	***	***	*	ns	ns	***
G×T×I	***	***	ns	ns	ns	ns	ns	ns	***

*** $p < 0.0001$, ** $p < 0.001$, * $p < 0.01$, ns = not significant.

The first variable to be analyzed related to the partition of N in the plant is the percentage of N in the aerial dry matter (PNSDW), whose mean value in irrigation was 3.15% (Table S3). PHA-0483, PHA-0683 and PHA-0432 with 3.40%, were the genotypes with the highest PNSDW while PHA-0155 obtained the lowest percentage with 2.53%. Control plants with N and those inoculated with CIAT899 and CFN42 were significantly higher with 3.36%, while those inoculated with SLL2 and LTMF, with 2.90% had the lowest. PNSDW can be very stable with respect to SDW as it occurs in PMB-0222 or it can decrease with increasing SDW as observed in PHA-0683 and PMB-0220 (Figure 3). Thus, the stability of PNSDW is a function of how limiting N may be for each genotype. The mean NSDW in irrigation was 1.06 g plant^{-1}. PHA-0471 and PMB-0285 (Table S3) displayed values significantly higher, with 1.57 and 1.49 g plant^{-1} respectively. The plants inoculated with the LTMF strain obtained the highest NMSA (1.22 g plant^{-1}), while the zero and N control had a significantly lower value than the inoculated plants. PHA-0471 with LTMF was the combination with the highest accumulation of N (2.25 g plant^{-1}). N fixation can account for 35 to 70% of plant N [50].

PNSEDW is determined by genotype and agronomic conditions [51]. In irrigation it was 3.75% (Table S3). PHA-0471 and PMB-0285, were around 4% of PNSEDW, and were superior to the rest of the genotypes. The combination PHA-0471-EXIC, reached 4.28%, which was higher than the values by Escribano et al. [52]. The NSEDM ranges from 0.12 to 3.83 g plant^{-1} (Table S3), with an average value of 1.53 g plant^{-1}. PHA-0471 and PMB-0286 had means of 2.22 and 1.96 g plant^{-1}, respectively. The plants inoculated with CIAT899, EXIC, EPOB and N control had averages greater than 1.70 g plant^{-1}. The maximum value was 3.83 g plant^{-1} in the combination PHA-0471-CIAT899. N mobility is genetically regulated [53], but the greater or lesser mobilization of N is not correlated with seed production [54]. The maximum symbiotic nitrogen fixation (SNF) rate occurs before flowering, and it was in flowering when NSDW was determined, observing that the higher the NDW, the lower the NSDW. Logic would suggest that NDW would act as a sink for N and only when the filling phase begins, therefore N would be mobilized towards the seeds.

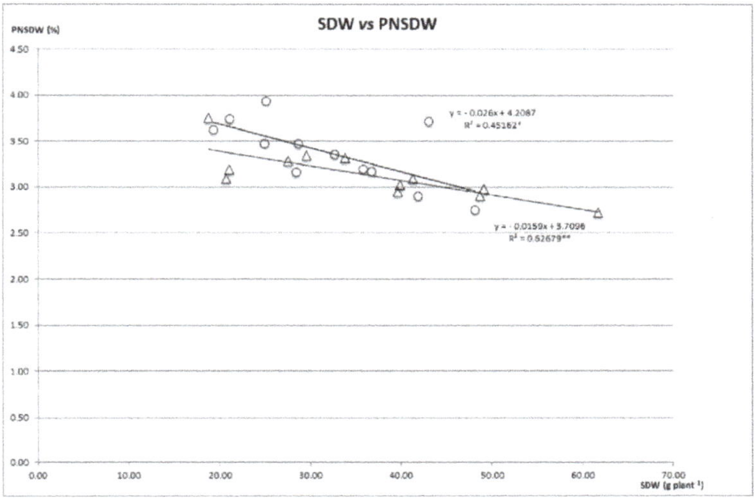

Figure 3. SDW vs. PNMSA ratio of genotypes PHA-0683 and PMB-0220 under irrigation in field. ** $p < 001$ and * $p < 0.01$.

Under water stress, the plant tends to close its stomata to reduce transpiration, which limits gas exchange, and therefore carbohydrate production decreases. This reduction of carbohydrate production directly affects the total biomass, and limits nitrogen fixation indirectly [10,33,55]. This would imply a reduction of N in the leaves, causing senescence [49] and a reduction in the SDW. Water stress represents a 46% decrease in SDW, with an average of 18.7 g plant^{-1} (Table S4). PHA-0471, despite suffering a reduction of 22.5 g plant^{-1}, remained, along with the genotype being PMB-0285, PHA-0155 and PMB-0286 the genotypes with the highest production of SDW. The plants inoculated with LTMF and APAFI presented the highest SDW (25 g plant^{-1}). Under water stress conditions, the 100SW values showed a 17% reduction (Table S4), although this was not uniform for all genotypes. PHA-0471, PHA-0683, PMB-0222 and PMB-0285, had a sTable 100SW under water stress. Singh [56] interprets that it is because water stress produces a partial abortion of the seeds, which generates a greater distribution of photosynthesis in the few seeds that remain. Sadras [57] observed that the small variability in the size of the seeds is associated with some plasticity in the number of seeds per pod. The 100SW stability is not tied to seed size. The plants inoculated with LTMF, SLL2, EPOB, APAFI and EF obtained the highest values of 100SW. In contrast, the control N lost 53% with 21.1 g 100 seeds^{-1}. In previous works, N-fertilized plants subjected to water stress also suffered similar reductions that were attributed to a change in direction in the balance of plant-soil N [58]. Plants under water stress conditions had a drought intensity index (DII) of 0.51, or what is the same, a reduction of SDM of 51% (Table S4). The genotype with the highest SDM was PHA-0471 with 27.4 g plant^{-1}. PHA-0432 was the most tolerant to water stress with only a 33% reduction, but despite this it had the lowest SDM value (13.0 g plant^{-1}) along with PMB-0244. PMB-0286 (24.7 g plant^{-1}) was the most sensitive genotype to water stress. The LTMF-PMB-0220 combination obtained the maximum SDM value with 52.9 g plant^{-1}.

Water stress causes a reduction in the number of absorbent hairs in the plant roots; therefore, by reducing the place where the primary infection occurs, a decrease in NN is logical [59,60]. This stress affects all states of symbiosis, in addition to limiting growth and soil survival of *Rhizobium* [61–63]. In this trial, nodulation was not affected by water stress, since it did not occur until 35 days after planting (DAS). The nodules of the primary nodulation can already be counted in 14 DAS [64]. Once the nodules are formed, water stress does not affect either NN or NDW, although there is a decrease in the SNF rate [65]. This is a competitive advantage, since after stress the plant can regain its N-binding activity [66]. The water stress supposes for all the bean genotypes an increase in the number of nodules per plant, although this increase can only be considered significant in three genotypes: PHA-0432, PHA-0683 and PMB-0285 (Table S4). The variability of the data for each strain was so great that, despite having a range between 45.7 and 69.8 plant^{-1} nodules, there are no significant differences among the inoculated and not inoculated control plants. The maximum value under water stress conditions was 243.0 nodules plant^{-1}, which was reached by the combination PHA-0683 with the EXIB strain. The water stress conditions also suppose an increase in NDW of 87%, although the increase in PHA-0683 can only be considered significant (Table S4). The genotypes with the highest NDW were PMB-0286 and PMB-0285, with values higher than 45 mg plant^{-1}. In the control N, PMB-0286 and PMB-0285 registered the highest values, 76.6 and 71.7 mg plant^{-1}, respectively. This indicated that instead of inhibiting the growth of nodules, it acted as a "starter". The combination EXIB-PHA-0683 with 161.1 mg plant^{-1} reached the highest NDW value. Water stress produces a drastic reduction in the production of SDM/NDW (g mg^{-1}) of 64%. Even so, the same trend was maintained as in irrigation, the higher the NDW, the lower the yield in the production of SDM.

Between 55 and 75 DAP, 88% of the N demands of the pods come from the biological fixation of N [67]. According to some authors [34,68], the plants that have undergone water stress in this interval of the plant cycle also experience a decrease in the dry matter of the plant, and in NSEDM due to the decrease in nitrogenase activity. The mean PNSDW in drought was 2.65% (Table S4), which represents a reduction of 16% compared to the irrigated experiment. PMB-0286, PHA-0471 and PMB-0220 were the bean genotypes that experienced the least reduction and those that had the highest PNSDW value under water stress conditions. There are no significant differences between the inoculated plants and the not inoculated controls. LTMF-PMB-0286 was the combination with the highest PNSDW (3.38%). The PNSDW of PMB-0222, PMB-0244, PMB-0285 and PMB-0286, were independent of SDW in the two treatments, so that N in these genotypes was a limiting factor. Water stress shows a reduction of more than 50% of the NSDW and affects almost all genotypes equally, due to the strong C/N interaction. The plant subject to water stress limits the flow of C towards the nodules, which causes a decline in N fixation [69]. Despite being PHA-0471 and PHA-0285, the genotypes with the highest drought losses still maintain the highest NSDW. LTMF-PMB-0285 was the combination with the highest NSDW (1.17 g plant^{-1}).

The PNSEDW was quite stable under water stress [70], and only presented a decrease of 4% (Table S4). PHA-0220, PHA-0432 and PHA-0471 were genotypes with PNSEDW greater than 3.75% while the combinations with the highest PNSEDW were SLL2-PMB-0220 and N control-PHA-0432, with values greater than 4.40%. Water stress implied a 53% reduction in NSEDM, mainly due to a 51% reduction in SDM, since the percentage of N in seed (PNSEDW) is hardly affected by water stress. PMB-0220- LTMF achieves the maximum value with 1.97 g N plant^{-1}. Under water stress conditions, there is less dispersion of the NSDW and NSEDM data, which indicated the linear correlation between the two variables (Figure 4). This may be because the shortage of N produced by water stress forces the plant to use N more efficiently. There is a significant relationship between SDM and NSEDM; Araujo and Grandi-Teixeira [71] demonstrated this relationship and that it also occurs with total P and SDM, so it can be said that great or less SNF will determine a greater or lesser production of SDM. The results from the field experiment consolidate the preliminary greenhouse data on symbiotic efficiency.

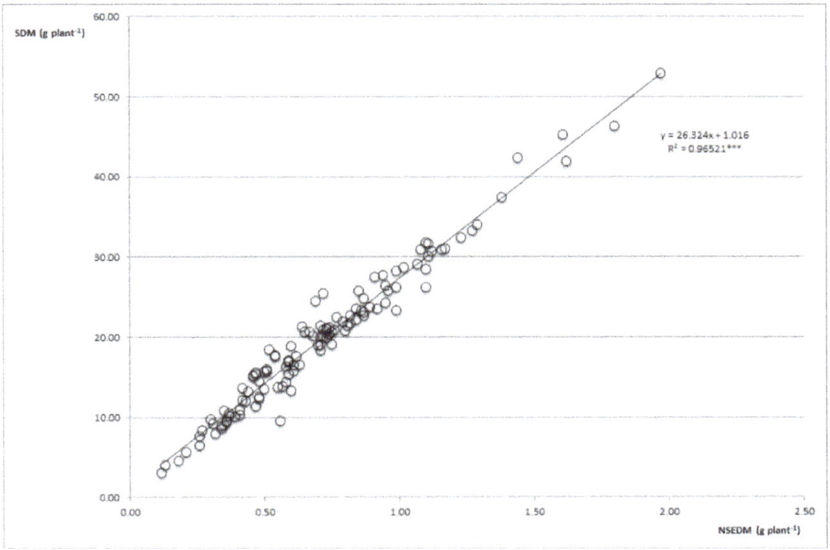

Figure 4. NSEDM vs. SDM ratio under water stress in the field. *** $p < 0.0001$.

4. Conclusions

None of the rhizobia strains provided the highest productivity in all genotypes. The main significant effect of the strains in the genotypes in the field implies important genotype-strain interspecificity.

The greenhouse experiment gives information on the symbiotic system under controlled conditions on some variables, which have been validated under field growing conditions. In the greenhouse, SNF was assessed by SDW. Neither the NN nor the NDW were correlated with the SDW due to the lack of homogeneity in the size of the nodules. The only exception was PHA-0683, which when the BNO phenotype presented its NDW if it was related to SDW. It is interesting to understand the mechanisms that produce BNO nodules and if the plant blocks non-functional nodules. The BNO phenotype may be more beneficial for plant growth and seed yield and it can be considered a trait for symbiosis efficiency.

These results were validated in the field. Thus, the best strains under irrigated conditions were LTMF, CIAT899 and EXIC and under water stress conditions they were APAFI, LTMF and EF. The best genotypes in irrigated conditions were PHA-0471, PMB-0285 and PMB-0286 and in drought conditions they were PHA-0471, PMB-0222 and PMB-0286. The growth of the inoculated strains is highly influenced by the pH and P of the soil [72–74]. Thus, some inoculum may have adaptation problems depending on the soil conditions. In low fertility soils, SNF is the main source of N in the plant. In the field, there was a relationship observed between NSEDM and SDM, so that N is a limiting factor for seed production. In general, the bean genotypes with the highest amount of nitrogen are the genotypes with the best seed production. The N content of the seed also indicates the amount of protein, important at a nutritional level. Under water stress, the inoculated genotypes with the strains LTMF, APAFI and EF presented high values of these variables. The genotypes with higher production and higher nitrogen fixation in drought conditions were PHA-0471, PHA-0683, PMB-0220, PMB-0222, PMB-0285 and PMB-0286. The best strain-genotype combination under drought conditions was APAFI-PMB-0286. Under water stress, the most affected strain and genotypes were CIAT 899 and PMB-0244, respectively.

Only PHA-0483 with control N presented a high value because there was no optimal strain for this genotype in this trial. PHA-0155, inoculated with rhizobia, have a nitrogen content higher than the control with N supplement, both under irrigated and water stress conditions. Local strains have achieved maximum yield with some varieties, thus highlighting their possible use as inoculants. Thus, the results of this work could be the basis for obtaining an effective rhizobia biofertilizer for beans, with benefits for bean production and the environment. The biofertilization implies a lower cost in nitrogen fertilizers and reduces water pollution and the emission of greenhouse nitrogen gases.

Supplementary Materials: The following are available online at https://www.mdpi.com/2073-4395/11/1/70/s1, Table S1: Mean values and mean comparison of the variables studied in greenhouse without water stress, Table S2: Mean values and mean comparison of the variables studied in greenhouse under water stress, Table S3: Mean values and mean comparison of the variables studied in field without water stress, Table S4: Mean values and mean comparison of the variables studied in field under water stress.

Author Contributions: A.P.R.: manuscript redaction and interpretation of results, M.R.: management of the experimental work and interpretation of results, A.M.D.R.: manuscript revision and interpretation of results. All authors have read and agreed to the published version of the manuscript.

Funding: Projects PGIDIT06RAG40301PR and INCITE07PXI403088ES, Xunta de Galicia, Spain.

Data Availability Statement: Data is contained within the article or supplementary material. The data presented in this study are available in: Implications of the Symbiotic Nitrogen Fixation in Common Bean under Seasonal Water Stress.

Acknowledgments: Xunta de Galicia (Spain) for the use of the experimental fields, greenhouses and laboratories at the Experimental Agricultural Station Baixo Miño-INGACAL, Salceda de Caselas, Spain.

Conflicts of Interest: The authors declare no conflict of interest.

References

1. Duke, J.A. *Handbook of Legumes of World Economic Importance*; Plenum Press: New York, NY, USA, 1981.
2. Wondatir, Z.; Mekasha, Y. Feed resources availability and livestock production in the central rift valley of Ethiopia. *Int. J. Livest. Prod.* **2014**, *5*, 30–35. [CrossRef]
3. Beebe, S.E.; Rao, I.M.; Blair, M.W.; Acosta-Gallegos, J.A. Phenotyping common beans for adaptation to drought. *Front. Physiol.* **2013**, *4*, 1–20. [CrossRef] [PubMed]
4. Sánchez-Reinoso, A.D.; Ligarreto-Moreno, G.A.; Restrepo-Díaz, H. Evaluation of drought índices to identify tolerant geno-types in common bean bush (*Phaseolus vulgaris*, L.). *J. Integr. Agric.* **2020**, *19*, 99–107. [CrossRef]
5. Asfaw, A.; Blair, M.W. Quantification of drought tolerance in Ethiopian common bean varieties. *Agric. Sci.* **2014**, *5*, 124–139. [CrossRef]
6. White, J.W.; Singh, S.P. Breeding for adaptation to drought. In *Common Beans: Research for Crop Improvement*; Van Shoonhoven, A., Voysest, O., Eds.; CABI: Wallingford, UK; CIAT: Cali, Colombia, 1991; pp. 501–560.
7. Cuellar-Ortiz, S.M.; Arrieta-Montiel, M.D.L.P.; Acosta-Gallegos, J.; Covarrubias, A.A. Relationship between carbohydrate partitioning and drought resistance in common bean. *Plant Cell Environ.* **2008**, *31*, 1399–1409. [CrossRef] [PubMed]
8. Asfaw, A.; Demissie, D.A.; Shah, T.; Blair, M. Trait associations in diversity panels of the two common bean (*Phaseolus vulgaris* L.) gene pools grown under well-watered and water-stress conditions. *Front. Plant Sci.* **2017**, *8*, 733. [CrossRef] [PubMed]
9. Miklas, P.N.; Kelly, J.D.; Beebe, S.E.; Blair, M.W. Common bean breeding for resistance against biotic and abiotic stresses: From classical to MAS breeding. *Euphytica* **2006**, *147*, 105–131. [CrossRef]
10. Muñoz-Perea, C.G.; Terán, H.; Allen, R.G.; Wright, J.L.; Westermann, D.T.; Singh, S.P. Selection for Drought Resistance in Dry Bean Landraces and Cultivars. *Crop. Sci.* **2006**, *46*, 2111–2120. [CrossRef]
11. Nielsen, D.C.; Nelson, N.O. Black Bean Sensitivity to Water Stress at Various Growth Stages. *Crop Sci.* **1998**, *38*, 422–427. [CrossRef]
12. Yonts, C.D. Deficit Irrigation of Dry Edible Beans during Early, Mid and Late Season. In Proceedings of the 5th National Decennial Irrigation Conference Proceedings, Phoenix, AZ, USA, 5–8 December 2010; American Society of Agricultural and Biological Engineers (ASABE): St. Joseph, MI, USA, 2010; p. 1.
13. Broughton, W.J.; Zhang, F.; Perret, X.; Staehelin, C. Signals exchanged between legumes and Rhizobium: Agricultural uses and perspectives. *Plant Soil* **2003**, *252*, 129–137. [CrossRef]
14. Martínez-Romero, E. Diversity of Rhizobium-Phaseolus vulgaris symbiosis: Overview and perspectives. *Plant Soil* **2003**, *252*, 11–23. [CrossRef]
15. Kuykendall, L.D. Family I Rhizobiaceae. In *Bergey's Manual of Systematic Bacteriology*, 2rd ed.; Garrity, G., Krieg, N.R., Holt, J.G., Eds.; Springer: New York, NY, USA, 2005; pp. 324–340, ISBN 9780387241456.

16. Herrera-Cervera, J.A.; Caballero-Mellado, J.; Laguerre, G.; Tichy, H.-V.; Requena, N.; Amarger, N.; Martínez-Romero, E.; Olivares, J.; Sanjuán, J. At least five rhizobial species nodulate Phaseolus vulgaris in a Spanish soil. *FEMS Microbiol. Ecol.* **1999**, *30*, 87–97. [CrossRef]
17. Drevon, J.J.; Alkama, N.; Bargaz, A.; Rodiño, A.P.; Sungthongwises, K.; Zaman-Allah, M. The Legume–Rhizobia Symbiosis. In *Grain Legumes*; De Ron, A.M., Ed.; Springer Science + Business Media: New York, NY, USA, 2015.
18. Kaymakanova, M.; Mincheva, T. Salinity and its effects on the physiological response of bean (*Phaseolus vulgaris* L.). *J. Cent. Eur. Agric.* **2008**, *9*, 749–755.
19. Wilmowicz, E.; Kucko, A.; Golinska, P.; Burchardt, S.; Przywieczerski, T.; Świdziński, M.; Brzozowska, P.; Kapuścińska, D. Abscisic acid and ethylene in the control of nodule-specific response on drought in yellow lupine. *Environ. Exp. Bot.* **2020**, *169*, 103900. [CrossRef]
20. Bouhmouch, I.; Souad-Mouhsine, B.; Brhada, F.; Aurag, J. Influence of host cultivars and Rhizobium species on the growth and symbiotic performance of Phaseolus vulgaris under water stress. *J. Plant Physiol.* **2005**, *162*, 1103–1113. [CrossRef]
21. Chemining, G.N.; Vessey, J.K. The abundance and efficacy of *Rhizobium leguminosarum* bv. *viciae* in cultivated soils of the eastern Canadian prairie. *Soil Biol. Biochem.* **2006**, *38*, 294–302.
22. SAS Institute Inc. *SAS/STAT 15.1*; SAS Institute Inc.: Cary, NC, USA, 2018.
23. Steel, R.G.D.; Torrie, J.H.; Dickey, D.A. *Principle and Procedures of Statistics: A Biometrical Approach*, 3rd ed.; Mc Graw Hill: New York, NY, USA, 1997; p. 666, ISBN 0070610282.
24. Hungria, M.; Franchini, J.C.; Campo, R.J.; Crispino, C.C.; Moraes, J.Z.; Sibaldelli, R.N.; Mendes, I.C.; Arihara, J. Nitrogen nutrition of soybean in Brazil: Contributions of biological N2 fixation and N fertilizer to grain yield. *Can. J. Plant Sci.* **2006**, *86*, 927–939. [CrossRef]
25. Rodiño, A.P.; De La Fuente, M.; De Ron, A.M.; Lema, M.J.; Drevon, J.-J.; Santalla, M. Variation for nodulation and plant yield of common bean genotypes and environmental effects on the genotype expression. *Plant Soil* **2011**, *346*, 349–361. [CrossRef]
26. Van Kessel, C.; Hartley, C. Agricultural management of grain legumes: Has it led to an increase in nitrogen fixation? *Field Crops Res.* **2000**, *65*, 165–181. [CrossRef]
27. Rengel, Z. Breeding for better symbiosis. *Plant Soil* **2002**, *245*, 147–162. [CrossRef]
28. Dorcinvil, R.; Sotomayor-Ramírez, D.; Beaver, J. Agronomic performance of common bean (*Phaseolus vulgaris* L.) lines in an Oxisol. *Field Crops Res.* **2010**, *118*, 264–272. [CrossRef]
29. Mostasso, L.; Mostasso, F.L.; Dias, B.G.; Vargas, M.A.; Hungria, M. Selection of bean (*Phaseolus vulgaris* L.) rhizobial strains for the Brazilian Cerrados. *Field Crops Res.* **2002**, *73*, 121–132. [CrossRef]
30. Oka-Kira, E.; Kawaguchi, M. Long-distance signaling to control root nodule number. *Curr. Opin. Plant Biol.* **2006**, *9*, 496–502. [CrossRef] [PubMed]
31. Bourion, V.; Laguerre, G.; Depret, G.; Voisin, A.S.; Salon, C.; Duc, G. Genetic variability in nodulation and root growth affects nitrogen fixation and accumulation in pea. *Ann. Bot.* **2007**, *100*, 589–598. [CrossRef] [PubMed]
32. Schultze, M.; Kondorosi, A. Regulation of symbiotic root nodule development. *Annu. Rev. Genet.* **1998**, *32*, 33–57. [CrossRef] [PubMed]
33. Sinclair, T.R.; Purcell, L.C.; King, C.A.; Sneller, C.H.; Chen, P.; Vadez, V. Drought tolerance and yield increase of soybean resulting from improved symbiotic N2 fixation. *Field Crops Res.* **2007**, *101*, 68–71. [CrossRef]
34. Mahieu, S.; Germon, F.; Aveline, A.; Hauggaard-Nielsen, H.; Ambus, P.; Jense, E.S. The influence of water stress on biomass and N accumulation, N partitioning between above and below ground parts and on N rhizodeposition during reproductive growth of pea (*Pisum sativum* L.). *Soil Biol. Biochem.* **2008**, *41*, 380–387. [CrossRef]
35. Voisin, A.S.; Munier-Jolain, G.; Salon, C. The nodulation process is tightly adjusted to plant growth an analysis using environmentally and genetically induced variation of nodule. *Plant Soil* **2010**, *337*, 399–412. [CrossRef]
36. Voisin, A.-S.; Bourion, V.; Duc, G.; Salon, C. Using an Ecophysiological Analysis to Dissect Genetic Variability and to Propose an Ideotype for Nitrogen Nutrition in Pea. *Ann. Bot.* **2007**, *100*, 1525–1536. [CrossRef]
37. Graham, P.H. Ecology of the Root-Nodule Bacteria of Legumes. In *Nitrogen-Fixing Leguminous Symbioses*; Springer: Berlin/Heidelberg, Germany, 2008; pp. 23–58. [CrossRef]
38. Muchero, W.; Ehlers, J.; Close, T.J.; Roberts, P.A. Mapping QTL for drought stress-induced premature senescence and ma-turity in cowpea. *Theor. Appl. Genet.* **2009**, *118*, 849–863. [CrossRef]
39. Muchero, W.; Ehlers, J.D.; Roberts, P.A. Restriction site polymorphism-based candidate gene mapping for seedling drought tolerance in cowpea [*Vigna unguiculata* (L.) Walp.]. *Theor. Appl. Genet.* **2010**, *120*, 509–518. [CrossRef] [PubMed]
40. Mergaert, P.; Uchiumi, T.; Alunni, B.; Evanno, G.; Cheron, A.; Catrice, O.; Mausset, A.-E.; Barloy-Hubler, F.; Galibert, F.; Kondorosi, A.; et al. Eukaryotic control on bacterial cell cycle and differentiation in the Rhizobium-legume symbiosis. *Proc. Natl. Acad. Sci. USA* **2006**, *103*, 5230–5235. [CrossRef] [PubMed]
41. Park, S.O.; Coyne, D.P.; Jung, G.; Skroch, P.W.; Arnaud-Santana, E.; Steadman, J.R.; Ariyarathne, H.; Nienhuis, J. Mapping of QTL for Seed Size and Shape Traits in Common Bean. *J. Am. Soc. Hortic. Sci.* **2000**, *125*, 466–475. [CrossRef]
42. González, A.; Rodiño, A.; Santalla, M.; De Ron, A.M. Genetics of intra-gene pool and inter-gene pool hybridization for seed traits in common bean (*Phaseolus vulgaris* L.) germplasm from Europe. *Field Crops Res.* **2009**, *112*, 66–76. [CrossRef]
43. Michiels, J.; Dombrecht, B.; Vermeiren, N.; Xi, C.; Luyten, E.; Vanderleyden, J. Phaseolus vulgaris is a non-selective host for nodulation. *FEMS Microbiol. Ecol.* **1998**, *26*, 193–205. [CrossRef]

44. Graham, P. Some problems of nodulation and symbiotic nitrogen fixation in *Phaseolus vulgaris* L.: A review. *Field Crops Res.* **1981**, *4*, 93–112. [CrossRef]
45. Romdhane, S.B.; Aouani, M.E.; Trabelsi, M.; de Lajudie, P.; Mhamdi, R. Selection of high nitrogen-fixing Rhizobia modulating chickpea (*Cicer arietinum*) for Semi-Arid Tunisia. *J. Agron. Crop Sci.* **2008**, *194*, 413–420.
46. Hungria, M.; Campo, R.J.; Mendes, I.C. Benefits of inoculation of the common bean (*Phaseolus vulgaris*) crop with efficient and competitive *Rhizobium tropici* strains. *Biol. Fertil. Soils* **2003**, *39*, 88–93. [CrossRef]
47. Rennie, R.J.; Kemp, G.A. N_2-Fixation in Field Beans Quantified by $_{15}N$ Isotope Dilution. II. Effect of Cultivars of Beans 1. *Agron. J.* **1983**, *75*, 645–649. [CrossRef]
48. Vessey, J.K. Cultivar differences in assimilate partitioning and capacity to maintain N2 fixation rate in pea during pod-filling. *Plant Soil* **1992**, *139*, 185–194. [CrossRef]
49. Araya, T.; Noguchi, K.; Terashima, I. Effect of nitrogen nutrition on the carbohydrate repression of photosynthesis in leaves of *Phaseolus vulgaris* L. *J. Plant Res.* **2009**, *123*, 371–379. [CrossRef] [PubMed]
50. Hardarson, G.; Atkins, G. Optimising biological N2 fixation by legumes in farming systems. *Plant Soil* **2003**, *252*, 41–54. [CrossRef]
51. Mutschler, M.A.; Bliss, F.A. Inheritance of bean seed globulin content and its relationship to protein content on quality. *Crop Sci.* **1981**, *21*, 289–294. [CrossRef]
52. Escribano, M.; Santalla, M.; De Ron, A.M. Genetic diversity in pod and seed quality traits of common bean populations from northwestern Spain. *Euphytica* **1997**, *93*, 71–81. [CrossRef]
53. Vasilas, B.L.; Nelson, R.L.; Fuhrmann, J.J.; Evans, T.A. Relationship of Nitrogen Utilization Patterns with Soybean Yield and Seed-Fill Period. *Crop Sci.* **1995**, *35*, 809–813. [CrossRef]
54. Kurdali, F.; Kalifa, K.; Al-Shamma, M. Cultivar differences in nitrogen assimilation, partitioning and mobilization in rain-fed grown lentil. *Field Crops Res.* **1997**, *54*, 235–243. [CrossRef]
55. Naya, L.; Ladrea, R.; Ramos, J.; González, E.M.; Arrese-Igor, C.; Minchin, F.R.; Becana, M. The response of carbon metabolism and antioxidant defenses of alfalfa nodules to drought stress and to the subsequent recovery of plants. *Plant Physiol.* **2007**, *144*, 1104–1114. [CrossRef]
56. Singh, S.P. Drought Resistance in the Race Durango Dry Bean Landraces and Cultivars. *Agron. J.* **2007**, *99*, 1219–1225. [CrossRef]
57. Sadras, V.O. Evolutionary aspects of the trade-off between seed size and number in crops. *Field Crops Res.* **2007**, *100*, 125–138. [CrossRef]
58. Harper, L.A.; Giddens, J.E.; Angdale, G.W.; Sharpe, R.R. Environmental effects on nitrogen dynamics in soybean under conservation and clean tillage systems. *Agron. J.* **1989**, *81*, 623–631. [CrossRef]
59. Boutraa, T.; Sanders, F.E. Effects of interactions of moisture regime and nutrient addition on nodulation and carbon partitioning in two cultivars of bean (*Phaseolus vulgaris* L.). *J. Agron. Crop Sci.* **2001**, *186*, 229–237. [CrossRef]
60. Silva, A.d.N.; Gerosa, M.L.; Ribeiro, W.Q.; Rodrigues, E.; Carvalhoda, P.; Andreade, C.; Cleo, C.; Vanderlei, M.A. Water stress alters physical and chemical quality in grains of common bean, triticale and wheat. *Agric. Water Manag.* **2020**, *231*, 106023. [CrossRef]
61. Hungria, M.; Vargas, M.A. Environmental factors affecting N2 fixation in grain legumes in the tropics, with an emphasis on Brazil. *Field Crops Res.* **2000**, *65*, 151–164. [CrossRef]
62. Mnasri, B.; Aouani, M.E.; Mhamdi, R. Nodulation and growth of common bean (*Phaseolus vulgaris*) under water deficiency. *Soil Biol. Biochem.* **2007**, *39*, 1744–1750. [CrossRef]
63. Güereña, D.T.; Lehmann, J.; Thies, J.E.; Vanek, S.; Karanja, N.; Neufeldt, H. Nodulation of beans with inoculant carriers from pyrolyzed and non-pyrolyzed sugarcane bagasse in response to different pre-planting water availability. *Appl. Soil Ecol.* **2019**, *143*, 126–133. [CrossRef]
64. Kipe-Nolt, J.A.; Vargas, H.; Giller, K.E. Nitrogen fixation in breeding lines of *Phaseolus vulgaris* L. *Plant Soil* **1993**, *152*, 103–106. [CrossRef]
65. Sinclair, T.R.; Muchow, R.C.; Ludlow, M.M.; Leach, G.J.; Lawn, R.J.; Foale, M.A. Field and model analysis of the effect of water deficits on carbon and nitrogen accumulation by soybean, cowpea and black gram. *Field Crops Res.* **1987**, *17*, 121–140. [CrossRef]
66. Pfeiffer, N.E.; Malik, N.S.A.; Wagner, F.W.; Wang, C.Y.; Adams, D.O. Reversible Dark-Induced Senescence of Soybean Root Nodules. *Plant Physiol.* **1983**, *71*, 393–399. [CrossRef]
67. Kumarasinghe, K.S.; Danso, S.K.A.; Zapata, F. Field evaluation of fixation and N partitioning in climbing bean (*Phaseolus vulgaris* L.) using 15N. *Biol. Fertil. Soils* **1992**, *13*, 142–146.
68. Ramos, M.L.G.; Gordon, A.J.; Minchin, F.R.; Sprent, J.I.; Parsons, R. Effect of water stress on nodule physiology and bio-chemistry of a drought tolerant cultivar of common bean (*Phaseolus vulgaris* L.). *Ann. Bot.* **1999**, *83*, 57–63. [CrossRef]
69. Gálvez, L.; González, E.M.; Arrese-Igor, C. Evidence for carbon flux shortage and strong carbon/nitrogen interactions in pea nodules at early stages of water stress. *J. Exp. Bot.* **2005**, *56*, 2551–2561. [CrossRef] [PubMed]
70. Peña-Cabriales, J.J.; Castellanos, J.Z. Effects of water stress on N2 fixation and grain yield of *Phaseolus vulgaris* L. In *Enhancement of Biological Nitrogen Fixation of Common Bean in Latin America*; Springer: Dordrecht, The Netherlands, 1993; Volume 52, pp. 151–155. [CrossRef]
71. Araújo, A.P.; Teixeira, M.G. Nitrogen and phosphorus harvest indices of common bean cultivars: Implications for yield quantity and quality. *Plant Soil* **2003**, *257*, 425–433. [CrossRef]

72. Vadez, V.; Drevon, J.J. Genotypic variability in phosphorus use efficiency for symbiotic N2 fixation in common bean (*Phaseolus vulgaris*). *Agronomie* **2001**, *21*, 691–699. [CrossRef]
73. Remans, R.; Croonenborghs, A.; Gutierrez, R.T.; Michiels, J.; Vanderleyden, J. Effects of plant growth-promoting rhizobacteria on nodulation of *Phaseolus vulgaris* L. are dependent on plant P nutrition. *Eur. J. Plant Pathol.* **2007**, *119*, 341–351. [CrossRef]
74. Wakelin, S.A.; Macdonald, L.; Rogers, S.; Gregg, A.; Bolger, T.; Baldock, J. Habitat selective factors influencing the structural composition and functional capacity of microbial communities in agricultural soils. *Soil Biol. Biochem.* **2008**, *40*, 803–813. [CrossRef]

agronomy

MDPI

Article

Genetic Diversity among Cowpea (*Vigna unguiculata* (L.) Walp.) Landraces Suggests Central Mozambique as an Important Hotspot of Variation

Ana Maria Figueira Gomes [1,2,3], David Draper [4,5], Pedro Talhinhas [6], Paula Batista Santos [6], Fernanda Simões [7], Nascimento Nhantumbo [2], Rafael Massinga [2], José C. Ramalho [1,8,*], Isabel Marques [1,*] and Ana I. Ribeiro-Barros [1,8,*]

1 Plant Stress and Biodiversity Lab, Forest Research Center, Instituto Superior de Agronomia, Universidade de Lisboa, 1349-017 Lisbon, Portugal; anagomes13@gmail.com
2 Divisão de Agricultura, Instituto Superior Politécnico de Manica (DivAG-ISPM), Chimoio 2202, Mozambique; tonhantumbo@gmail.com (N.N.); rafael.massinga@gmail.com (R.M.)
3 TropiKMan Doctorate Programme, Nova School of Business and Economics, Universidade NOVA de Lisboa, 2775-405 Carcavelos, Portugal
4 National Museum of Natural History and Science and Centre for Ecology, Evolution and Environmental Change, Universidade de Lisboa, 1250-102 Lisboa, Portugal; ddmunt@gmail.com
5 UBC Botanical Garden & Centre for Plant Research, Department of Botany, University of British Columbia, Vancouver, BC V6T 1Z4, Canada
6 Linking Landscape Environment, Agriculture and Food, Instituto Superior de Agronomia, Universidade de Lisboa, 1349-017 Lisbon, Portugal; ptalhinhas@isa.ulisboa.pt (P.T.); pbbsantos@isa.ulisboa.pt (P.B.S.)
7 Unidade de Investigação em Biotecnologia e Recursos Genéticos (UIBRG), Instituto Nacional de Investigação Agrária e Veterinária, I.P. (INIAV), 2784-505 Oeiras, Portugal; fernanda.simoes@iniav.pt
8 GeoBioSciences, GeoTechnologies and GeoEngineering (GeoBioTec), Faculdade de Ciências e Tecnologia, Universidade NOVA de Lisboa, 2829-516 Caparica, Portugal
* Correspondence: cochichor@isa.ulisboa.pt or cochichor@mail.telepac.pt (J.C.R.); isabelmarques@isa.ulisboa.pt (I.M.); aribeiro@isa.ulisboa.pt (A.I.R.-B.)

Received: 8 November 2020; Accepted: 25 November 2020; Published: 30 November 2020

Abstract: Cowpea is a multiple-purpose drought-tolerant leguminous pulse crop grown in several dry tropical areas. Its domestication center is thought to be East or West Africa, where a high level of genetic diversity is apparently still found. However, detailed genetic information is lacking in many African countries, limiting the success of breeding programs. In this work, we assessed the genetic variation and gene flow in 59 *Vigna unguiculata* (cowpea) accessions from 10 landraces spanning across six agro-ecological zones of Mozambique, based on nuclear microsatellite markers. The results revealed the existence of high genetic diversity between the landraces, even in comparison to other world regions. Four genetic groups were found, with no specific geographic pattern, suggesting the presence of gene flow between landraces. In comparison, the two commercial varieties had lower values of genetic diversity, although still close to the ones found in local landraces. The high genetic diversity found in Mozambique sustains the importance of local genetic resources and farm protection to enhance genetic diversity in modern varieties of cowpea worldwide.

Keywords: Africa; cowpea; genetic diversity; microsatellites

---

1. Introduction

Cowpea (*Vigna unguiculata* (L.) Walp.), also known as black eye pea, is a major annual pulse crop mostly grown in dry tropical areas of Latin America, South Asia, and Africa [1]. It is cultivated mainly for its seeds, which have a high dry matter content of proteins (20–32%) and carbohydrates (50–60%).

Both the grains and leaves are also rich in the amino acid lysine and tryptophan, vitamin C, iron, and zinc [2]. Cowpea therefore has an essential role in the human diet in many developing countries, being referred to as the "poor man's meat" [3]. As a legume, it is also an important component of traditional cropping systems since rhizobium facilitation fixes atmospheric nitrogen and contributes to soil fertility improvement, particularly in smallholder farming systems where little or no fertilizer is used [4]. The bulk of cowpea production and consumption occurs in sub-Saharan Africa, namely West and Central Africa [1], where its nutritional value and tolerance to drought place this crop in a unique position in the continent's efforts to establish nutrition-sensitive food systems. That is especially important to help curb malnutrition, particularly among the most vulnerable—pregnant or lactating women and children under five [5]. Although cowpea is known to be drought tolerant when compared to other crops, the productivity of cowpea varieties is hampered by erratic rainfall, and many are sensible to high temperatures [1,5]. Thus, appropriate agronomic practices could improve the performance of new varieties within different agro-ecological zones described by the Food and Agriculture Organizatio as a land resource mapping unit, defined in terms of climate, landform and soils, and/or land cover, and having a specific range of potentials and constraints for land use [6]. Indeed, physiological, and metabolic studies show a progressive acclimation of cowpea plants to stress [7] and differential drought responses of landraces with contrasting abiotic tolerance levels [8].

Despite being native to Africa [9,10], the domestication center of cowpea is unclear but thought to be in either East or West Africa, where high morphological and genetic diversity is found, followed by a sub-domestication region in India resulting in the selection/domestication of two additional subspecies, the yard-long bean (*Vigna unguiculata* subsp. *sesquipedalis* (L.) Verdc.) and the catiang (*V. unguiculata* subsp. *cylindrica* (L.) Verdc.) [9–11]. European accessions usually cluster together with those from West Africa and were likely imported from this region [11]. Breeding lines in America also show a high genetic similarity with African accessions, although local American landraces show a high genetic divergence with those of other regions such as East Asia and Oceania [11,12].

Because of this domestication history linked to a center of origin in Africa, cowpea research has been underway in several African countries for many years. Breeding activities in sub-Saharan Africa involving germplasm collection, evaluation, and screening for the identification of lines with high yield potential resulted in a diverse cowpea germplasm collection constituting more than 15,000 cultivated cowpeas from 89 different countries [1,11,12]. Additionally, a core collection of more than 2000 accessions based on geographical, agronomical, and botanical descriptors has been established in The International Institute for Tropical Agriculture (IITA) genebank with the aim of discovering new traits related with stress tolerance for the development of new breeding lines [13,14]. The United States Department of Agriculture—Genetic Resources Information Network (USDA-GRIN) at Griffin USA [10] and University of California, Riverside, USA [12] also hold cowpea collections with 7737 and 6000 germplasm accessions, respectively. Cowpea has several features of a classical model plant for genomic studies, such as a relatively small diploid ($2n = 2x = 22$ chromosomes) genome of ~613 Mbp, a short annual life cycle, and a highly selfing nature [15].

The limited number of cowpea breeding programs in Mozambique has contributed to the country's ineffectiveness in taking advantage of the continent's high genetic potential. A significant pool of cowpea landraces is thought to be available, but the limited detailed information available about their diversity and agronomic potential makes it difficult for breeding programs to thrive. Thus, the characterization of cowpea genetic resources available in Mozambique is of extreme importance for conservation and breeding, since it is the second most cultivated legume crop in the country, occupying an extension of ca. 380,000 ha, with an average yield of 0.275 t ha^{-1} [1]. Unlike commercial varieties, landraces maintained by farmers usually have high levels of genetic variability as they have evolved from years of uncontrolled cross-regional and infield genetic exchange, even between previously released and discontinued open pollinated varieties [16], not being subjected to selection over a long period of time [17–19]. However, knowledge about their variability is usually limited [20]. Therefore, the aim of this study was to assess the genetic diversity of 59 cowpea accessions from 10 landraces across six agro-ecological zones of Mozambique, using single sequence repeat (SSR) markers.

2. Materials and Methods

2.1. Plant Material

Fifty-nine cowpea accessions corresponding to 10 landraces were sampled in six agro-ecological zones (AEZs) in the provinces of Manica, Sofala, and Zambezia, where cowpea is grown as an integral component of local cereal–legume cropping systems (Figure 1): R3 (North and Central Gaza and Western Inhambane), R4 (medium-altitude areas of Central Mozambique), R5 (low-altitude areas of Sofala and Zambezia), R6 (dry areas of Zambezia and Southern Tete), R7 (mid-altitude areas of Zambezia, Nampula, Tete, Niassa, and Cabo Delgado), and R10 (high-altitude areas of Zambezia, Niassa, Angonia-Maravia, and Manica [21]). Additionally, two widely used commercial cultivars (IT16 and IT18) released by the Mozambican Institute of Agricultural Research (IIAM) and bred through a partnership with the International Institute of Tropical Agriculture (IITA) in Nigeria were also used in this study.

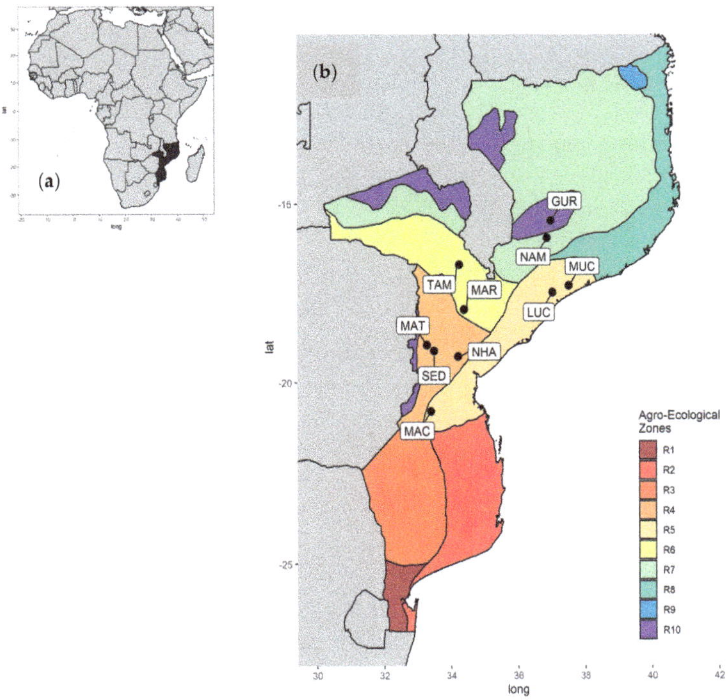

Figure 1. (**a**): Location of Mozambique in East Africa. (**b**): Dots indicate the studied landraces of *Vigna unguiculata* among the different eco-geographical zones (AEZs) existing in the country, based on [21]. Landraces code: Gurué (GUR), Namarroi (NAM), Muchela (MUC), Lucas Branco (LUC), Nhamatanda (NHA), Maringué (MAR), Tambara (TAM), Sede nova (SED), Matsinho (MAT), and Machaze (MAC).

2.2. DNA Extraction and nSSR Amplification

The 61 samples used in this study were genotyped based on nine polymorphic nuclear single sequence repeats (nSSRs) previously developed by [22]: VuUGM05, VuUGM22, VuUGM31, VuUGM33, VuUGM39, VuUGM40, VuUGM68, VuUGM71, and VuUGM74. Based on an initial survey, we selected these nSSR markers since they produced robust, highly polymorphic amplified bands among the entire collection of cowpea samples. Total genomic DNA was extracted from 50 mg of ground leaves using the InnuSPEED Plant DNA Kit (Analytik Jena Innuscreen GmbH, Germany) according to the manufacturer's protocol. The average yield and purity were assessed spectrophotometrically by OD230, OD260, and OD280 readings (Nanodrop 2000, Thermo Fisher Scientific, Waltham, MA, USA) and visualized by electrophoresis in 1% agarose gels under UV light. Amplifications were performed in 15 µL reactions containing 1.25U TaKaRa Hot startTaq polymerase, 1X Buffer I, 0.15 mM dNTPs, 0.2 µM Primer F-ROX and R, and 100 ng DNA under the following PCR conditions: an initial denaturation at 95 °C for 5 min, followed by 35 cycles of denaturation at 65 °C (20 s),annealing at 56 °C (30 s), and a final extension at 60 °C for 30 min. Allele sizes were determined using GeneMapper 3.2 (Applied Biosystems; UK).

2.3. Genetic Diversity and Population Structure

For each nSSR locus and landrace, genetic diversity was assessed by calculating the total number of alleles (N_a), mean expected heterozygosity (H_e), mean observed heterozygosity (H_o), allelic richness (A_R), and inbreeding coefficient (F_{IS}) using FSTAT 2.9.3.2 [23]). GenAlEx 6 software was used to estimate the mean expected heterozygosity (H_e) and mean observed heterozygosity (H_o) for each population, as well as the number of private alleles [24]. The selfing rate (s) was estimated as $s = 2FIS/(1 + FIS)$ [25]. Hardy Weinberg Exact (HWE) Tests and linkage disequilibrium among SSR markers were determined using the software program GENEPOP 4.7 [26,27] with dememorization numbers at 10,000 and performing 100,000 iterations for all permutation tests. Significant values were corrected for multiple comparisons by Bonferroni correction [28]. An analysis of variance was used to detect significant differences between sites for the measured genetic values. Grids for all significant genetic parameters were generated in R and were based on a grid with a cell size of 30 s (which corresponds to approximately 1 km in the study area), applying a 1.5-degree circular neighborhood diameter. The circular neighborhood was used to re-sample the genetic composition of a single sample to all surrounding grid cells, with a size of 30 s, within a diameter of 1.5 degrees around its location. In this way, the genetic composition of each sample is representative for the area within the defined buffer zone.

2.4. Population Structure and Differentiation

The Bayesian program STRUCTURE v.2.3.4 [29] was used to test whether any discrete genetic structure existed among the landraces and AEZs sampled. The analysis was performed assuming clusters from $K = 1$ to $K = 8$, with 10 repetitions per K. Models were run assuming ancestral admixture and correlated allele frequencies with 50,000 burn-in steps, followed by run lengths of 300,000 interactions for each K. The optimum K was determined using STRUCTURE HARVESTER [30], which identifies the optimal K based on both the posterior probability of the data for a given K and the ΔK [31]. To correctly assess the membership proportions (q values) for clusters identified in STRUCTURE, the results of the replicates at the best-fit K were post-processed using CLUMPP 1.1.2 [32]. POPULATION 1.2 [33] was used to calculate the Nei's genetic distance [34] among individuals and to construct an unrooted neighbor-joining (NJ) tree with 1000 bootstrap replicates. A principal component analysis (PCoA) was also constructed in GenAlEx6 [24] to detect the genetic relatedness among individuals based on Nei's genetic distance. We estimated genetic differentiation among locations via an analysis of molecular variance (AMOVA) using ARLEQUIN 3.11 [35]. Molecular variance was quantified among and within landraces, considering AEZs and wild cowpea versus cultivars, via an AMOVA using 10,000 permutations at the 0.95 significance level in ARLEQUIN 3.11 [35].

2.5. Spatial Analysis and Genetic Diversity Rarefaction

Grids for genetic parameters were generated in DIVA-GIS version 7.5 (www.diva-gis.org) based on a grid with a cell size of 2.5 min (which corresponds to approximately 4.5 km in the study area) and applying a circular neighborhood with a diameter buffer of one degree (corresponding to approximate

111 km). The circular neighborhood was used to illustrate the allelic composition of each sampled site representative for the area within the defined buffer zone. Genetic diversity rarefaction considered the spatial average of several population parameters such as the number of alleles (N_A), observed heterozygosity (H_o), inbreeding coefficient (F_{IS}), and selfing rate (s).

3. Results

3.1. Genetic Diversity

The total number of alleles varied between 49 in VuUGM74 and 145 in VuUGM40 (Table 1). For each locus, the observed heterozygosity values (H_o) ranged from 0.014 in VuUGM74 to 1 in VuUGM40, and expected heterozygosity (H_e) varied from 0.016 in VuUGM74 to 0.806 in VuUGM33. F_{IS} values varied between −0.008 and 0.857 (for loci VuUGM68 and VuUGM31, respectively; Table 1) across the loci studied.

Table 1. Characteristics and genetic diversity statistics of the nuclear single sequence repeat (nSSR) primers used in the genetic study of Vigna unguiculata. For each locus, the total number of alleles (Na), mean expected heterozygosity (He), mean observed heterozygosity (Ho), and fixation index (FIS) obtained from the 61 studied samples are shown.

Primer Name	Primer Sequence 5′-3′	N_a	H_o	H_e	F_{IS}
VuUGM33	F: AAAGGTGGGGGATTATGAGG R: TGTCCAATCCTGATGGATGA	83	0.907	0.806	−0.091
VuUGM71	F: TTCACAACCTGTCCACCTCA R: GGCGTCCCAACAGATAAGAA	125	0.143	0.548	0.783
VuUGM05	F: GCGGGATTCTATTCCAGTGA R: TCCATTGGGTTTCTCAACCT	82	0.174	0.617	0.767
VuUGM39	F: CGAAAAAGCATGATCAACCA R: CCCCTTTCGCTAAAATTTCC	97	0.149	0.749	0.851
VuUGM22	F: CAATCACCATTCACCAAACA R: TATTGGGACTCAGGTCTTGG	112	0.181	0.629	0.749
VuUGM31	F: TGGTTCACTTCCCATATTGTC R: AGGCAGAGACGAAGGAGTGA	122	0.136	0.711	0.857
VuUGM40	F: TTCTACATGGTTTTGGGGTCA R: GAGCTTGCCCTCAAGAATTG	145	1.003	0.671	−0.426
VuUGM68	F: TGATTGATGGTGGTGTAGCC R: GCACTTCACTCATCGTTGCT	59	0.415	0.397	−0.008
VuUGM74	F: GCCTCCTCTCACAAACTTGC	49	0.014	0.016	0.018

A total of 327 alleles were found among the set of *V. unguiculata* accessions, varying significantly between sites ($p < 0.001$; Table 2). The number of alleles varied geographically from 14 in the coastal area of Muchela to 71 in the dry western area of Tambara (Figure 2). Allelic richness varied between 1.250 in Muchela and 1.751 in Gurué, with no statistical differences being found between areas ($p = 0.452$; Table 1). However, the number of private alleles varied significantly across areas ($p < 0.001$; Table 2), with the highest numbers being found in Gurué, Tambara, and Machaze (Figure 3).

Table 2. Genetic diversity within the cowpea genotypes studied. The agro-ecological zone (AEZ), number of samples analyzed (N), total number of alleles (N_A), mean allelic richness (A_R), mean observed heterozygosity (H_o) and expected heterozygosity (H_e), inbreeding coefficient (F_{IS}), and selfing rate (s) as a percentage are shown for each landrace.

Landrace Identity	Province	AEZ	N	N_A	A_R	H_o	H_e	F_{IS}	s
Gurué (GUR)	North Zambezia	R10	6	46	1.751	0.389	0.688	0.506	60%
Namarroi (NAM)	North Zambezia	R7	4	23	1.534	0.379	0.454	0.250	25%
Muchela (MUC)	Central Zambezia	R7	4	14	1.250	0.222	0.535	−0.412	74%
Lucas Branco (LUC)	South Zambezia	R7	4	22	1.432	0.426	0.577	−0.032	41%
Nhamatanda (NHA)	Central Sofala	R4	4	31	1.682	0.278	0.479	0.707	59%
Maringué (MAR)	Central Sofala	R5	3	22	1.503	0.407	0.494	0.310	30%
Tambara (TAM)	North Manica	R6	23	71	1.612	0.320	0.654	0.592	68%
Sede nova (SED)	North Manica	R6	3	23	1.562	0.222	0.451	0.323	69%
Matsinho (MAT)	Central Manica	R4	3	23	1.577	0.222	0.451	0.156	67%
Machaze (MAC)	South Manica	R3	5	29	1.555	0.267	0.500	0.549	64%
IT-16	Commercial cultivar	R4	1	12	1.333	0.333	0.167	-	
IT-18	Commercial cultivar	R6	1	11	1.222	0.222	0.111	-	

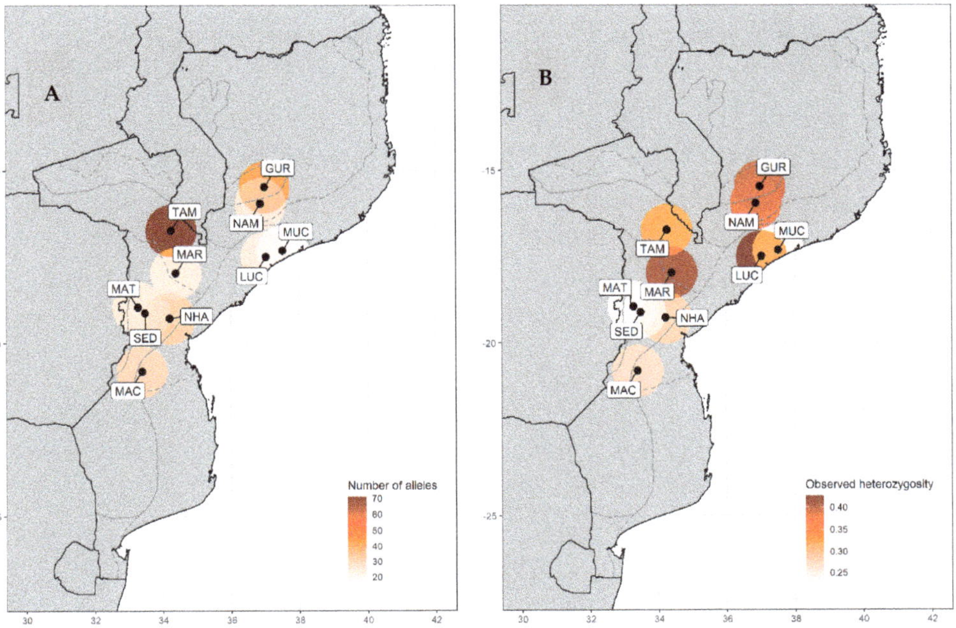

Figure 2. Map of the number of alleles (**A**) and observed heterozygosity (**B**) in 30 s (1 km) grid cells applying a 1-degree circular neighborhood. Dashed lines indicate the borders of agro-ecological zones [16].

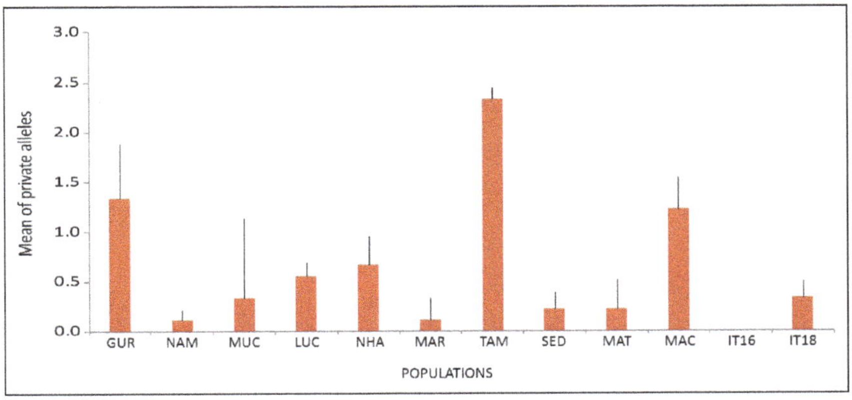

Figure 3. Population structure of *Vigna unguiculata* based on nine SSRs and using the best assignment result retrieved by STRUCTURE ($K = 4$). Each individual sample is represented by a thin vertical line divided into K colored segments that represent the individual's estimated membership fractions in the K cluster. Distribution of private alleles among the different landraces of *Vigna unguiculata*. Landraces code follows Table 2. The two cultivars are also indicated.

The mean observed heterozygosity varied significantly between 0.222 (Muchela, Sede Nova, and Matsinho) and 0.426 (Lucas Blanco) ($p < 0.001$; Figure 2), and the mean expected heterozygosity varied between 0.451 (Sede Nova, Matsinho) and 0.688 (Gurué) without statistically significant differences ($p = 0.481$; Table 2). F_{IS} values varied significantly between sites ($p < 0.001$; Table 2), ranging from negative values of −0.412 in the central area of Zambezia to positive values of 0.707 in the central area of Nhamatanda (Figure 4). The rate of self-fertilization in *V. unguiculata* also varied significantly between sites ($p < 0.001$; Table 2), with the lowest values found in the northern region of Namarroi (25%) and the highest in the coastal area of Muchela (74%) (Figure 4).

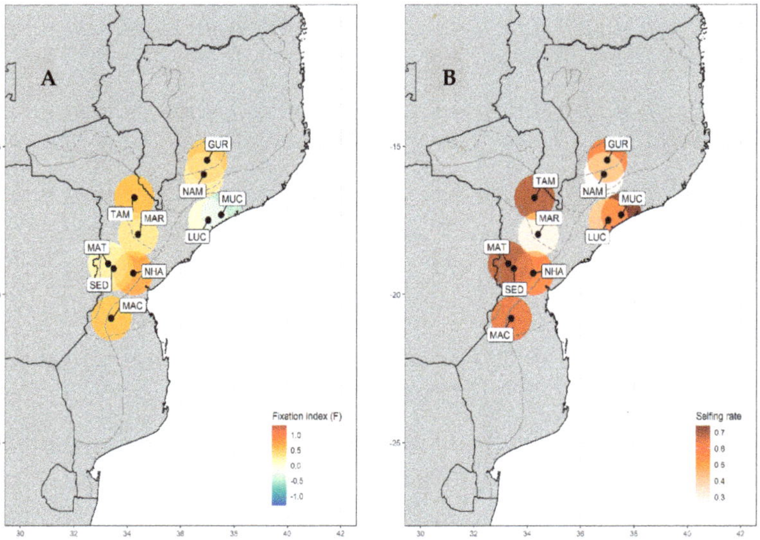

Figure 4. Map of the fixation index (**A**) and selfing rate (**B**) in 30 s (1 km) grid cells applying a 1-degree circular neighborhood. Dashed lines indicate the borders of agro-ecological zones [16].

The two cultivars had a low number of alleles (IT-16: 11 and IT-18: 2) and low allelic richness (IT-16: 1.333 and IT-18: 1.222) constrained by the small sampling size. However, although the observed heterozygosity (IT-16: 0.333 and IT-18: 0.222) was higher than expected in both cultivars (IT-16: 0.167 and IT-18: 0.111; $p < 0.001$ in both cases), it was also lower than that found in most local accessions (Table 2; $p < 0.001$).

From all landraces sampled, only one was not at HWE (GUR) at the 5% level after the sequential Bonferroni correction (Table S1). Pairwise comparisons between loci revealed no significant linkage disequilibrium at $p = 5\%$, suggesting that alleles are assorted independently at different loci.

3.2. Genetic Structure of *V. unguiculata*

The Bayesian clustering program STRUCTURE found the highest LnP(D) and ΔK values for $K = 4$ (Figure S1). The results showed a high degree of admixture between landraces without any specific geographic pattern or clustering when considering the different AEZs (Figure 5). One cluster was predominant and grouped all accessions from North Zambezia and most accessions from Sofala and Central Manica; the second cluster characterized Central and South Zambezia accessions; the third clustered accessions from North Manica as well as Central Sofala; and the fourth cluster was exclusively composed of accessions from South Manica (Figure 5). The two cultivars were assigned to the first cluster, although with signs of admixture with other clusters.

In accordance with these results, the NJ tree separated all the groups assigned by STRUCTURE, revealing again no general correlation with the geographical distribution of accessions (Figure 6). All individuals from R3 and R7 were clustered into two different clades, one with 65% and the other with 34% bootstrap support (BS) value (Figure 6). Most individuals from R6 were clustered in the same group (57% BS), while R4, R5, and R10 were clustered into two different groups. The two cultivars were nested within the landraces, although in two different separated groups.

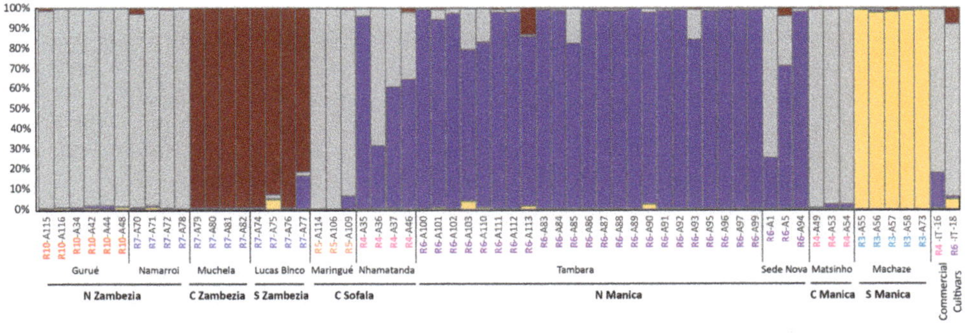

Figure 5. Population structure of *Vigna unguiculata* based on nine SSRs and using the best assignment result retrieved by STRUCTURE ($K = 4$). Each individual sample is represented by a thin vertical line divided into K colored segments that represent the individual's estimated membership fractions in the K clusters. Landraces and provinces are indicated below. AEZs are indicated in individual labels with different colors for better visualization. The two cultivars are also indicated.

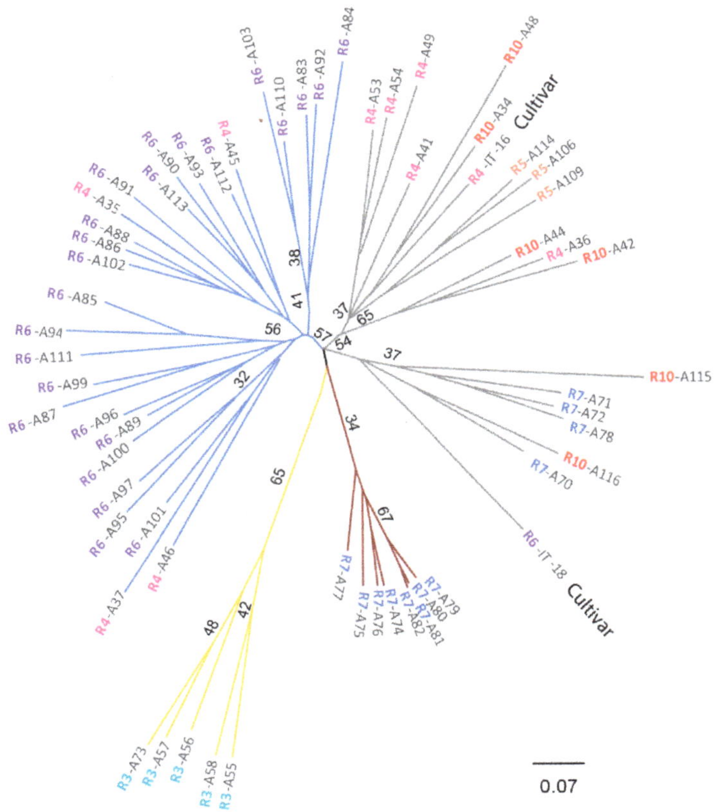

Figure 6. Unrooted neighbor-joining (NJ) tree of the studied *Vigna unguiculata* accessions, including the two cultivars, based on Nei's Da genetic distance. Numbers associated with branches indicate bootstrap values (BS) based on 1000 replications. Only BS values above 30 are shown. The colors of branches indicate the four genetic groups found in STRUCTURE. AEZs are indicated in branch labels with different colors, following Figure 3.

The PCoA spatially separated the accessions analyzed into three main groups (Figure 7). In accordance with the NJ tree, the accessions from R3 and R7 were separated from the main group: the first being on the upper left of axis 2, accumulating 21.14% of variance, with the second on the lower left of axis 2. All remaining accessions were clustered in a heterogeneous group also containing the two cultivars.

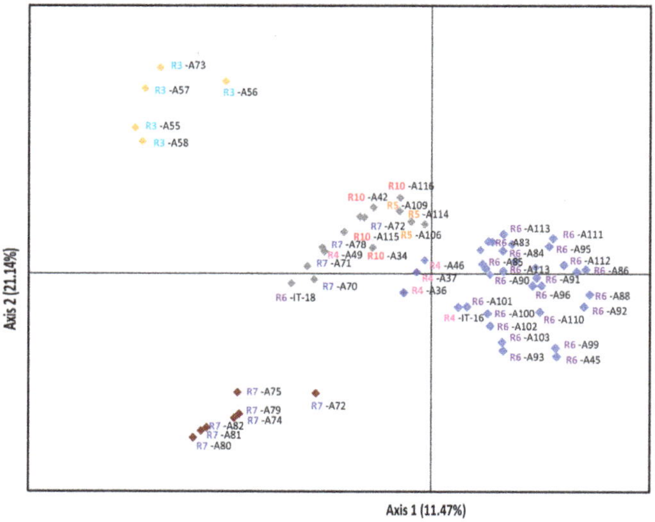

Figure 7. Principal Coordinate analysis (PCoA) of the studied *Vigna unguiculata* landraces using the scored nSSRs markers. Percentage of explained variance of each axis is given in parentheses. Population symbols and colors are shown in the chart. Colors of symbols indicate the four genetic groups found in STRUCTURE.

3.3. Genetic Differentiation between Landraces

Overall, genetic differentiation was significantly low (AMOVA FST = 0.199, $p < 0.001$). The analysis performed over the landraces sampled indicated that only 19.92% of the genetic variation could be attributed to among the AEZs (Table 3). The highest molecular variance was found among genotypes within accessions (47.39%), followed by that found within genotypes (32.69%; $p < 0.001$; Table 3). Remarkably, a very low molecular variance was found between wild cowpea versus the cultivars (0.12%), most of the variance being found among individuals within samples (65.58%; Table 3).

Table 3. Analysis of molecular variance (AMOVA) for the sampled landraces of *Vigna unguiculata*.

Source of Variance		Degrees of Freedom	Sum of Squares	% of Variance
Among landraces				
	Among AEZs	6	77.612	19.92
	Among genotypes within landraces	54	207.109	47.39
	Within genotypes	61	60.001	32.69
Among cowpea accessions vs. cultivars				
	Among accessions	1	4.772	0.12
	Among accessions	58	279.949	65.58
	Within accessions	61	60.000	34.30

4. Discussion

Landraces harbor a gene pool of unexplored alleles that constitute a unique set of genetic resources for breeding to improve productivity, nutritional value, adaptation, and resilience to climate change [36–39]. Given their evolutionary history and adaptation to local conditions, they usually harbor higher genetic diversity and environmental resilience than modern varieties [40–43]. However, such richness tends to be lost because most of the current intensive agricultural systems are based on few high-input and high-yielding cultivars [44]. Thus, a comprehensive characterization of landraces towards the development of conservation and breeding strategies is among the main clues that will allow us to face the major agricultural challenges related to population growth and environmental risks.

Despite the ongoing agricultural changes in Africa, according to our data, the nine microsatellites employed in this study were highly polymorphic and revealed the existence of high genetic diversity between landraces of *V. unguiculata* from Mozambique (Tables 1 and 2). A total of 327 alleles were found among the 59 cowpea accessions, which can be attributed to high genetic heterogeneity (Table 2). Indeed, the genetic diversity values found within the studied Mozambique landraces (Ho: 0.222–0.426; He: 0.451–0.654) were much higher than those reported in other cowpea studies. For instance, high-density single-nucleotide polymorphism (SNP) genotyping using the Cowpea iSelect Consortium Array was used to study the population structure and genetic diversity in a set of 96 worldwide cowpea accessions, and average PIC and He values of 0.25 and 0.31, respectively, were found [45]. Similar results were obtained by [10,12] using 422 cowpea landraces and 768 cowpea genotypes, respectively, collected in 56 countries. A genotyping by sequencing (GBS)study of 298 lines from the mini core collection of 360 landraces maintained at The International Institute of Tropical Agriculture revealed a low number of robust SNP markers with an observed genetic distance of between 0.0096 and 0.462 [11]. The studied accessions of the world cowpea collection maintained at IITA were grouped into three main clusters, agreeing with the groups also reported by [10]. Genetic diversity using SSR markers was also assessed in 105 selected cowpea genotypes from the National Genebank of China at the Institute of Crop Science (ICS), with a low level of genetic diversity found among accessions [46]. Genetic diversity was also reported to be low among cowpea populations collected from Benin Republic [47], Ghana [48], and Sudan [49]. In comparison, in our study, the two commercial cultivars (IT-16 and IT-18) had a very low number of alleles and low heterozygosity values. Cluster analyses (PcoA or NJ tree) showed no clear differentiation between these modern varieties and wild accessions. The pairwise genetic distances reported in other studies have also shown that African landraces are close to wild cowpea samples [11,12]. This suggests that the genetic diversity of these two commercial varieties is still close to that of the wild accessions, although more individuals are needed to accurately determine if genetic erosion is occurring.

Population structure analysis using worldwide cowpea samples usually delineates African landraces into two major gene pools separated by the Congo River basin, the East/South, and West Africa [8–10], although nothing has been reported in terms of the cowpea genetic structure within these regions. Our study, focused on Mozambican (East Africa) landraces, found four genetic groups with a high degree of admixture (Figure 5). No specific geographic pattern or clustering was found considering the different AEZs in either the NJ tree or the PcoA (Figures 6 and 7), which supports the presence of gene flow between these regions. The rate of self-fertilization in *V. unguiculata* varied across landraces (25–74%; Table 2; Figure 4), supporting the possibility of gene flow between individuals. In fact, two landraces (Lucas and Muchela) exhibited negative F_{IS} values, indicating that these landraces are less related than expected under random mating, which could imply fewer homozygotes and consequent crossbreeding. Nonetheless, most of the remaining landraces had low F_{IS} values (0.1–0.3), which indicates that inbreeding might not be prevalent.

The analysis of genetic differentiation indicated that most of the genetic variation was explained by differences among genotypes within landraces (Table 3), which also supports the hypothesis of gene flow. This low genetic differentiation and the absence of a geographical pattern associated with AEZs might be due to crossbreeding between individuals but also to seed exchange by farmers.

Seed exchange is a common practice between African farmers of neighboring areas [50] and could explain the specific genetic cluster found in the isolated accessions of South Manica that show no admixture with the remaining ones. It is economical unfeasible for seed companies to distribute small amounts of seeds over long rural distances in Africa, and therefore, certified, commercial seeds do not reach the farmers [51]. In addition, certified seeds are generally expensive, and farmers are unwilling to buy them at a cost twice (or more) that of grain [51]. Nonetheless, continuous recycling of seeds results in poor grain yields [50], highlighting the importance of conserving wild accessions and their seed stock.

The high genetic diversity found in Mozambique in comparison to other world regions reinforces the importance of local landraces to widen the genetic base of modern varieties of cowpea. The results of this study underline the hidden genetic diversity in local landraces, which should be conserved as sublines in genebanks to avoid the expected reduction of genetic diversity with successive regeneration of bulk samples. The high levels of genetic differentiation found within landraces (but not among AEZs) could imply the presence of different phenotypes, which might exhibit desired traits for exploitation in future breeding programs. In fact, according to [5], accessions A55 from R3, A80 from R7, and A116 from R10, which were clustered into different genetic groups (Figure 6), revealed contrasting responses, respectively leading to high sensitivity, mild sensitivity, and high tolerance to drought stress related to the regulation of photosynthesis, C/N metabolism, and antioxidative status [5]. Compared to close relatives and other crops, cowpea is well adapted to semi-arid and arid regions because of its ability to fix nitrogen, but local genotypes still display significantly different levels of tolerance [52,53]. The same plasticity has been observed in response to virus and fungal diseases, with impact in terms of yield losses ranging from 20 to 100% [54]. Such a diverse genetic pool could be used to investigate the genetic basis of tolerance to abiotic and biotic stress.

A priority for in situ, on-farm conservation should be given to the landraces of Gurué, Tambara, and Machaze, which showed a high number of private alleles (Figure 3) and belong to different genetic groups according to STRUCTURE analysis (Figure 5). On-farm conservation allows the concomitant evolution and retention of potentially useful genetic variation needed to maintain crop ability to adapt to changes [55]. However, genetic diversity conserved *on the farm* is complementary to that found in the genebank, and both systems are required for efficient conservation of cowpea. Thus, further to molecular tools, farmers' knowledge should be employed to optimize the sampling of sublines within landraces for ex situ conservation. A core germplasm collection should include most of the cowpea's genetic diversity, for which the results outlined in this study can be used. The results of this work encourage a broad network of on-farm activities that should be enrolled in a socio-economic framework to complement genebank collections. This is also the best way to prevent genetic erosion in the genebank while maintaining and expanding the cultivation of cowpea in a wide range of environmental conditions.

Supplementary Materials: The following are available online at http://www.mdpi.com/2073-4395/10/12/1893/s1, Table S1: Results of the Hardy Weinberg exact tests retrieved by GENEPOP for the sample landraces of Vigna unguiculata. *p*-value (0.05) associated with the null hypothesis of random union of gametes estimated with a Markov chain algorithm. Figure S1: STRUCTURE analysis of *Vigna unguiculata*. Above: mean log probability of data LnP(D) over 10 runs for each *K* value as a function of *K* (error bars represent standard deviation). Below: Evanno's ad hoc statistic; DK as a function of K.

Author Contributions: Conceptualization, A.M.F.G., N.N., R.M., J.C.R., I.M., A.I.R.-B.; methodology, A.M.F.G., D.D., P.B.S., P.T., F.S.; software, D.D.; I.M.; validation, A.M.F.G., D.D., P.T., F.S. and I.M.; formal analysis, A.M.F.G., D.D., P.T., F.S. and I.M.; investigation, A.M.F.G., D.D., P.B.S., I.M. and A.I.R.-B.; resources, N.N., P.T., F.S., R.M., J.C.R. and A.I.R.-B.; data curation, D.D., P.T., J.C.R., I.M. and A.I.R.-B.; writing—original draft preparation, A.M.F.G., I.M. and A.I.R.-B.; writing—review and editing, all co-authors; visualization, D.D. and I.M.; supervision, R.M., J.C.R., I.M. and A.I.R.-B.; project administration, R.M., J.C.R. and A.I.R.-B.; funding acquisition, N.N., R.M. and A.I.R.-B. All authors have read and agreed to the published version of the manuscript.

Funding: This research received funds from the Mozambican Fundo Nacional de Investigação (Project 201-Inv-FNI), NUFFIC, the Netherlands (Project NICHE-Moz-151), and Fundação para a Ciência e a Tecnologia, I.P., through the Ph.D. fellowship SFRH/BD/113952/2015 (A.M.F.G.) and the post-doctoral fellowship SFRH/BPD/100384/2014

(D.D), the research units UID/04129/2020 (LEAF), UIDP/04035/2020 (GeoBioTec), and UIDB/00239/2020 (CEF), and the APC.

Conflicts of Interest: The authors declare no conflict of interest.

References

1. Boukar, O.; Belko, N.; Chamarthi, S.; Togola, A.; Batieno, J.; Owusu, E.; Haruna, M.; Diallo, S.; Umar, M.L.; Olufajo, O.; et al. Cowpea (*Vigna unguiculata*): Genetics, genomics and breeding. *Plant Breed.* **2019**, *138*, 415–424. [CrossRef]
2. Goncalves, A.; Goufo, P.; Barros, A.; Dominguez-Perles, R.; Trindade, H.; Rosa, E.A.; Ferreira, L.; Rodrigues, M. Cowpea (*Vigna unguiculata* L. Walp), a renewed multipurpose crop for a more sustainable agri-food system: Nutritional advantages and constraints. *J. Sci. Food Agric.* **2016**, *96*, 2941–2951. [CrossRef] [PubMed]
3. Jayathilake, C.; Visvanathan, R.; Deen, A.; Bangamuwage, R.; Jayawardana, B.C.; Nammi, S.; Liyanage, R. Cowpea: An overview on its nutritional facts and health benefits. *J. Sci. Food Agric.* **2018**, *98*, 4793–4806. [CrossRef] [PubMed]
4. Bado, B.; Bationo, A.; Cescas, M. Assessment of cowpea and groundnut contributions to soil fertility and succeeding sorghum yields in the Guinean savannah zone of Burkina Faso (West Africa). *Biol. Fertil. Soils* **2006**, *43*, 171–176. [CrossRef]
5. Gomes, A.M.F.; Nhantumbo, N.; Ferreira-Pinto, M.; Massinga, R.; Ramalho, J.C.; Ribeiro-Barros, A. Breeding Elite Cowpea [*Vigna unguiculata* (L.) Walp] Varieties for Improved Food Security and Income in Africa: Opportunities and Challenges. In *Legume Crops*; El-Esawi, M.A., Ed.; IntechOpen: London, UK, 2019; p. 636.
6. FAO. *Agro-Ecological Zoning. Guidelines. FAO Soils Bulletin Soil Resources, Management and Conservation Service, FAO Land and Water Development Division*; Food and Agriculture Organization of the United Nations: Rome, Italy, 1996; pp. 73–78. ISBN 92-5-103890-2.
7. Goufo, P.; Moutinho-Pereira, J.M.; Jorge, T.F.; Correia, C.M.; Oliveira, M.R.; Rosa, E.A.S.; António, C.; Trindade, H. Cowpea (*Vigna unguiculata* L. Walp.) metabolomics: Osmoprotection as a physiological strategy for drought stress resistance and improved yield. *Front. Plant Sci.* **2017**, *8*, 586. [CrossRef] [PubMed]
8. Gomes, A.M.F.; Rodrigues, A.P.; António, C.; Rodrigues, A.M.; Leitão, A.E.; Batista-Santos, P.; Nhantumbo, N.; Massinga, R.; Ribeiro-Barros, A.I.; Ramalho, J.C. Drought response of cowpea (*Vigna unguiculata* (L.) Walp.) landraces at leaf physiological and metabolite profile levels. *Environ. Exp. Bot.* **2020**, *175*, 104060. [CrossRef]
9. Carvalho, M.; Lino-Neto, T.; Rosa, E.; Carnide, V. Cowpea: A legume crop for a challenging environment. *J. Sci. Food Agric.* **2017**, *97*, 4273–4284. [CrossRef]
10. Xiong, H.; Shi, A.; Mou, B.; Qin, J.; Motes, D.; Lu, W.; Ma, J.; Weng, Y.; Yang, W.; Wu, D. Genetic diversity and population structure of cowpea (*Vigna unguiculata* L. Walp). *PLoS ONE* **2016**, *11*, e0160941. [CrossRef]
11. Fatokun, C.; Girma, G.; Abberton, M.; Gedil, M.; Unachukwu, N.; Oyatomi, O.; Yusuf, M.; Rabbi, I.; Boukar, O. Genetic diversity and population structure of a mini-core subset from the world cowpea (*Vigna unguiculata* (L.) Walp.) germplasm collection. *Sci. Rep.* **2018**, *8*, 16035. [CrossRef]
12. Huynh, B.; Close, T.J.; Roberts, P.A.; Hu, Z.; Wanamaker, S.; Lucas, M.R.; Chiulele, R.; Cissé, N.; David, A.; Hearne, S.; et al. Gene pools and the genetic architecture of domesticated cowpea. *Plant Genome* **2013**, *6*, 1–8. [CrossRef]
13. Fang, J.; Chao, C.T.; Roberts, P.A.; Ehlers, J.D. Genetic diversity of cowpea [*Vigna unguiculata* (L.) Walp.] in four West African and USA breeding programs as determined by AFLP analysis. *Genet. Resour. Crop Evol.* **2007**, *54*, 1197–1209. [CrossRef]
14. Mahalakshmi, V.; Ng, Q.; Lawson, M.; Ortiz, R. Cowpea [*Vigna unguiculata* (L.) Walp.] core collection defined by geographical, agronomical and botanical descriptors. *Plant Genet. Resour.* **2007**, *5*, 113–119. [CrossRef]
15. Ehlers, J.D.; Hall, A.E. Cowpea (*Vigna unguiculata* (L.) Walp.). *Field Crops Res.* **1997**, *53*, 187–204. [CrossRef]
16. Ficiciyan, A.; Loos, J.; Sievers-Glotzbach, S.; Tscharntke, T. More than yield: Ecosystem services of traditional versus modern crop varieties revisited. *Sustainability* **2018**, *10*, 2834. [CrossRef]
17. De Luca, D.; Cennamo, P.; Del Guacchio, E.; Di Novella, R.; Caputo, P. Conservation and genetic characterisation of common bean landraces from Cilento region (southern Italy): High differentiation in spite of low genetic diversity. *Genetica* **2018**, *146*, 29–44. [CrossRef]

18. Alipour, H.; Bihamta, M.R.; Mohammadi, V.; Peyghambari, S.A.; Bai, G.; Zhang, G. Genotyping-by-sequencing (GBS) revealed molecular genetic diversity of Iranian wheat landraces and cultivars. *Front. Plant Sci.* **2017**, *8*, 1293. [CrossRef]
19. Ram, S.G.; Thiruvengadam, V.; Vinod, K.K. Genetic diversity among cultivars, landraces and wild relatives of rice as revealed by microsatellite markers. *J. Appl. Genet.* **2007**, *48*, 337–345. [CrossRef]
20. Hagenblad, J.; Zie, J.; Leino, M.W. Exploring the population genetics of genebank and historical landrace varieties. *Genet. Resour. Crop Evol.* **2012**, *59*, 1185–1199. [CrossRef]
21. Ministry of Agriculture and Fisheries (MAF). Agroecological zones and production systems. In *Working Document 2/B, Program of Investment in Agricultural Extension, Process of the Formulation of Proagri*; Ministry of Agriculture and Fisheries: Maputo, Mozambique, 1996; p. 72.
22. Gupta, S.K.; Gopalakrishna, T. Development of unigene-derived SSR markers in cowpea (*Vigna unguiculata*) and their transferability to other *Vigna* species. *Genome* **2010**, *53*, 508–523. [CrossRef]
23. Goudet, J. FSTAT, a Program to Estimate and Test Gene Diversities and Fixation Indices (Version 2.9.3.2). 2002. Available online: http://www.unil.ch/izea/softwares/fstat (accessed on 3 November 2019).
24. Peakall, R.; Smouse, P.E. GenAlEx 6 genetic analysis in Excel. Population genetic software for teaching and research. *Mol. Ecol. Notes* **2006**, *6*, 288–295. [CrossRef]
25. Ritland, K. Inferences about inbreeding depression based on changes of the inbreeding coefficient. *Evolution* **1990**, *44*, 1230–1241. [CrossRef] [PubMed]
26. Raymond, M.; Rousset, F. GENEPOP (version 1.2): Population genetics software for exact tests and ecumenicism. *J. Hered.* **1995**, *86*, 248–249. [CrossRef]
27. Rousset, F. Genepop'007: A complete reimplementation of the Genepop software for Windows and Linux. *Mol. Ecol. Resour.* **2008**, *8*, 103–106. [CrossRef] [PubMed]
28. Rice, W.R. Analyzing tables of statistical tests. *Evolution* **1989**, *43*, 223–225. [CrossRef]
29. Pritchard, J.K.; Stephens, M.; Rosenberg, N.A.; Donnelly, P. Association mapping in structured populations. *Am. J. Hum. Genet.* **2000**, *67*, 170–181. [CrossRef] [PubMed]
30. Earl, D.A.; von Holdt, B.M. STRUCTURE HARVESTER: A website and program for visualizing STRUCTURE output and implementing the Evanno method. *Conserv. Genet. Resour.* **2012**, *4*, 359–361. [CrossRef]
31. Evanno, G.; Regnaut, S.; Goudet, J. Detecting the number of clusters of individuals using the software structure: A simulation study. *Mol. Ecol.* **2005**, *8*, 2611–2620. [CrossRef]
32. Jakobsson, M.; Rosenberg, N.A. CLUMPP: A cluster matching and permutation program for dealing with label switching and multimodality in analysis of population structure. *Bioinformatics* **2007**, *23*, 1801–1806. [CrossRef]
33. Langella, O. *Populations 1.2.30. Laboratoire Populations. Genetique et Evolution. Centre National de la Recherche Scientifique*; CNRS UPR9034: Gif Sur Yvette, France, 2007; Available online: http://bioinformatics.org/populations/ (accessed on 3 November 2019).
34. Nei, M.; Tajima, F.; Tateno, Y. Accuracy of estimated phylogenetic trees from molecular data. II. Gene frequency data. *J. Mol. Evol.* **1983**, *19*, 153–170. [CrossRef]
35. Excoffier, L.; Laval, G.; Schneider, S. ARLEQUIN, Ver 3.0. An integrated software package for population genetic data analysis. *Evol. Bionf. Online* **1983**, *1*, 47–50.
36. Lopes, M.S.; El-Basyoni, I.; Baenziger, P.S.; Singh, S.; Royo, C.; Ozbek, K.; Aktas, H.; Ozer, E.; Ozdemir, F.; Manickavelu, A.; et al. Exploiting genetic diversity from landraces in wheat breeding for adaptation to climate change. *J. Exp. Bot.* **2015**, *66*, 3477–3486. [CrossRef] [PubMed]
37. Dwivedi, S.L.; Ceccarelli, S.; Blair, M.W.; Upadhyaya, H.D.; Are, A.K.; Ortiz, R. Landrace germplasm for improving yield and abiotic stress adaptation. *Trends Plant Sci.* **2016**, *21*, 31–42. [CrossRef] [PubMed]
38. Monteagudo, A.; Casas, A.M.; Cantalapiedra, C.P.; Contreras-Moreira, B.; Gracia, M.P.; Igartua, E. Harnessing novel diversity from landraces to improve an elite barley variety. *Front. Plant Sci.* **2019**, *10*, 434. [CrossRef] [PubMed]
39. Hour, A.; Hsieh, W.; Chang, S.; Wu, Y.; Chin, H.; Lin, Y. Genetic diversity of landraces and improved varieties of rice (*Oryza sativa* L.) in Taiwan. *Rice* **2020**. [CrossRef]
40. Ganeva, G.M.; Korzun, V.; Landjeva, S.; Popova, Z.; Christov, N.K. Genetic diversity assessment of Bulgarian durum wheat (*Triticum durum* Desf.) landraces and modern cultivars using microsatellite markers. *Genet. Resour. Crop Evol.* **2010**, *57*, 273–285. [CrossRef]

41. Kumbhar, S.D.; Kulwal, P.L.; Patil, J.V.; Sarawate, C.D.; Gaikwad, A.P.; Jadhav, A.S. Genetic diversity and population structure in landraces and improved rice varieties from India. *Rice Sci.* **2015**, *22*, 99–107. [CrossRef]
42. Soriano, J.M.; Villegas, D.; Aranzana, M.J.; García del Moral, L.F.; Royo, C. Genetic structure of modern durum wheat cultivars and Mediterranean landraces matches with their agronomic performance. *PLoS ONE* **2016**, *11*, e0160983. [CrossRef] [PubMed]
43. Pascual, L.; Ruiz, M.; López-Fernández, M.; Pérez-Peña, H.; Benavente, H.; Vázquez, J.F.; Sansaloni, C.; Giraldo, P. Genomic analysis of Spanish wheat landraces reveals their variability and potential for breeding. *BMC Genom.* **2020**, *21*, 122. [CrossRef]
44. FAO/IPGRI. *Review and Development of Indicators for Genetic Diversity, Genetic Erosion and Genetic Vulnerability (GDEV): Summary Report of a Joint FAO/IPGRI Workshop*; FAO/IPGRI: Rome, Italy, 2002.
45. Carvalho, M.; Muñoz-Amatriaín, M.; Castro, I.; Lino-Neto, T.; Matos, M.; Egea-Cortines, M.; Rosa, E.; Close, T.; Carnide, V. Genetic diversity and structure of Iberian Peninsula cowpeas compared to world-wide cowpea accessions using high density SNP markers. *BMC Genom.* **2017**, *18*, 891. [CrossRef]
46. Chen, H.; Chen, H.; Hu, L.; Wang, L.; Wang, S.; Wang, M.L.; Cheng, X. Genetic diversity and a population structure analysis of accessions in the Chinese cowpea [*Vigna unguiculata* (L.) Walp.] germplasm collection. *Crop J.* **2017**, *5*, 363–372. [CrossRef]
47. Zannouou, A.; Kossou, D.K.; Ahanchede, A.; Zoundjihékpon, J.; Agbicodo, E.; Struik, P.C.; Sanni, A. Genetic variability of cultivated cowpea in Benin assessed by random amplified polymorphic DNA. *Afr. J. Biotechnol.* **2008**, *7*, 24.
48. Egbadzor, K.F.; Danquah, E.Y.; Ofori, K.; Yeboah, M.; Offei, S.K. Diversity in 118 cowpea (*Vigna unguiculata* (L.) Walp.) accessions assessed with 16 morphological traits. *Int. J. Plant Breed. Genet.* **2014**, *8*, 35–43. [CrossRef]
49. Ali, Z.B.; Yao, K.N.; Odeny, D.A.; Kyalo, M.; Skilton, R.; Eltahir, I.M. Assessing the genetic diversity of cowpea (*Vigna unguiculata* (L.) Walp.) accessions from Sudan using simple sequence repeat (SSR)markers. *Afr. J. Plant Sci.* **2015**, *9*, 293–304. [CrossRef]
50. Njonjo, M.W.; Muthomi, J.W.; Mwang'ombe, A.W. Production practices, postharvest handling, and quality of cowpea seed used by farmers in Makueni and Taita Taveta counties in Kenya. *Int. J. Agron.* **2019**. [CrossRef]
51. Kimani, P.; Kamundia, D.K.; Narla, R.D.; Mwang'ombe, A.W. An integrated seed delivery system and seed research in Kenya: Case studies of pigeon pea, onion and dry bean. In *Seed Systems, Science, and Policy in East and Central Africa*; Francis, J.A., Ed.; CTA: The Hague, The Netherlands, 2014; pp. 73–88.
52. Hall, A. Breeding for adaptation to drought and heat in cowpea. *Eur. J. Agron.* **2004**, *21*, 447–454. [CrossRef]
53. Muchero, W.; Ehlers, J.D.; Roberts, P.A. Seedling stage drought-induced phenotypes and drought-responsive genes in diverse cowpea genotypes. *Crop Sci.* **2008**, *48*, 541–552. [CrossRef]
54. Mbeyagala, E.K.; Mukasa, B.S.; Tukamuhabwa, P.; Bisikwa, J. Evaluation of cowpea genotypes for virus resistance under natural conditions in Uganda. *J. Agric. Sci.* **2014**, *6*, 176–187. [CrossRef]
55. Bellon, M.R.; van Etten, J. *Climate change and on-farm conservation of crop landraces in centres of diversity In Plant Genetic Resources and Climate Change*; CABI Publishing: Egham, UK, 2014; pp. 137–150.

agronomy

MDPI

Article

Inheritance and Expressivity of Neoplasm Trait in Crosses between the Domestic Pea (*Pisum sativum* subsp. *sativum*) and Tall Wild Pea (*Pisum sativum* subsp. *elatius*)

Hatice Sari [1,*], Duygu Sari [1], Tuba Eker [1], Bilal Aydinoglu [1], Huseyin Canci [1], Cengiz Ikten [2], Ramazan S. Gokturk [3], Ahmet Zeybek [4], Melike Bakir [5], Petr Smykal [6] and Cengiz Toker [1]

1 Department of Field Crops, Faculty of Agriculture, Akdeniz University, 07070 Antalya, Turkey; duygusari@akdeniz.edu.tr (D.S.); ekertuba07@gmail.com (T.E.); aydinoglu@akdeniz.edu.tr (B.A.); huseyincanci@akdeniz.edu.tr (H.C.); toker@akdeniz.edu.tr (C.T.)
2 Department of Plant Protection, Faculty of Agriculture, Akdeniz University, 07070 Antalya, Turkey; cikten@akdeniz.edu.tr
3 Department of Biology, Faculty of Science, Akdeniz University, 07070 Antalya, Turkey; gokturk@akdeniz.edu.tr
4 Department of Biology, Faculty of Science, Mugla Sitki Kocman University, 48000 Mugla, Turkey; azeybek@mu.edu.tr
5 Department of Plant Biotechnology, Faculty of Agriculture, Erciyes University, 38280 Kayseri, Turkey; melikecu@gmail.com
6 Department of Botany, Faculty of Science, Palacky University, 78371 Olomouc, Czech Republic; petr.smykal@upol.cz
* Correspondence: haticesari@akdeniz.edu.tr; Tel.: +90-242-310-37-42

Received: 24 September 2020; Accepted: 23 November 2020; Published: 27 November 2020

Abstract: The Neoplasm trait in pea pods is reported to be due to the lack of ultraviolet (UV) light in glasshouse conditions or in response to pea weevil (*Bruchus pisorum* L.) damage. This pod deformation arises from the growth of non-meristematic tissue on pods of domesticated peas (*Pisum sativum* L. subsp. *sativum*). Neither expressivity, nor the effect of pea weevil on neoplasm in the tall wild pea (*P. sativum* L. subsp. *elatius* (M. Bieb.) Asch. & Graebn.), have been adequately studied. We aimed to study the expression and inheritance of neoplasm in the tall wild pea and crosses between domesticated and tall wild peas grown in the glasshouse (without pea weevils) and in the field (with pea weevils) under natural infestation conditions. Neoplasm was found in all pods in tall wild peas when grown in the glasshouse, while it was not detected on pods of field-grown plants despite heavy pea weevil damage. In inter-subspecific crosses between *P. sativum* subsp. *sativum* and *P. sativum* subsp. *elatius*, all F_1 plants had neoplastic pods, and the F_2 populations segregated in a good fit ratio of 3 (neoplasm): 1 (free from neoplasm) under glasshouse conditions, which suggests that neoplasm on pods of the tall wild pea was controlled by a single dominant gene. Expressivity of neoplasm in the progeny differed from parent to parent used in inter-subspecific crosses. There was no relationship between neoplasm and damage by pea weevil under heavy insect epidemics under field conditions. The neoplasm occurring under glasshouse conditions may be due to one or to a combination of environmental factors. Since wild peas are useful genetic resources for breeding programs aiming at fresh pea production that could be utilized under glasshouse conditions, negative selection could be considered in segregating populations.

Keywords: tall wild pea; *Pisum sativum* subsp. *elatius*; neoplasm; pea weevil; *Bruchus pisorum*; expressivity; inheritance

1. Introduction

The genus *Pisum* L. consists of the following species, subspecies and varieties: *P. sativum* L. subsp. *sativum* var. *sativum*, *P. sativum* L. subsp. *sativum* var. *arvense* (L.) Poir., *P. sativum* L. subsp. *elatius* (M. Bieb.) Aschers. & Graebn. var. *elatius*, *P. sativum* L. subsp. *elatius* (M. Bieb.) Aschers. & Graebn. var. *brevipedunculatum* Davis & Meikle, *P. sativum* L. subsp. *elatius* (M. Bieb.) Aschers. & Graebn. var. *pumilio* Meikle, *P. fulvum* Sibth. & Sm. and *P. abyssinicum* A. Br. [1–5]. The former two varieties are under cultivation as the garden pea and field pea, respectively; while *P. sativum* subsp. *elatius* and *P. fulvum* are wild species [3–5]. *P. sativum* subsp. *abyssinicum* is referred to as dekoko or Abyssinian pea and grown in eastern Africa and the Arabian Peninsula [3,6]. *P. sativum* subsp. *elatius*, known as the tall wild pea, is native to the Europe–Mediterranean region, the Balkans, the Crimean and Caucasian region, the Middle East and northwest Asia [3,5,7], whereas *P. fulvum* is limited to the Middle East [3] and Turkey [1].

Domesticated peas are not only important protein sources in the world but also essential rotation crops, especially in cereal-based cropping systems for sustainable agriculture due to fixation of atmospheric nitrogen [8–11]. The field pea is grown as a fodder crop for animal feeding, while the Abyssinian and garden peas are grown for dry seeds as a food legume. Dried seeds of the domesticated pea had a production of 13.5 million t from an area of 7.9 million ha with an average seed yield of 1718 kg per ha, while 21.2 million t of green peas as a vegetable were produced from 2.7 million ha with an average yield of 7735 kg per ha in the world in 2018 [12]. The statistics given for the domesticated pea place it in the second rank among food legumes based on production quantity [12].

Garden peas are often used as green peas (frozen and canned) or as fresh vegetables for fresh pods [3,13], but disease and insect damage on pea pods are undesirable for fresh use. Some genotypes of the *Pisum* species, when deprived of ultraviolet (UV) light in glasshouse conditions, form neoplasm on the pods. Neoplasm emerges as a response to the lack of UV light [14,15] on the surface of young pods with the growth of non-meristematic tissue. It is also stated that neoplasm is triggered by the pea weevil (*Bruchus pisorum* L.) [16,17]. Neoplasm found in domesticated plants occurs as a result of mutation, and a smooth pod without neoplasm is governed by a recessive "*np*" gene, while neoplasm is controlled by a single dominant gene "*Np*" [14,15]. Pods with neoplasm are not preferred by consumers due to the unpleasant image of the pods.

Expressivity, which is the degree of phenotypic visibility of neoplasm in the progeny in domesticated peas, is influenced not only by environmental factors such as lack of UV light and the pea weevil [16–18], but also by dominant genes responsible for homozygous (*Np/Np*) or heterozygous (*Np/np*) mutations [17,19]. Neoplasm has also been reported to occur in wild species such as *P. elatius* M. Bieb. and *P. humile* Boiss. & Noe. when grown under glasshouse conditions [15]. However, the expression and inheritance of neoplasm have not been adequately studied in progeny obtained from inter-subspecific crosses between *P. sativum* subsp. *sativum* and *P. sativum* subsp. *elatius* species. Since wild peas are sources for improvement of biotic and abiotic stresses in pea breeding programs, it is important to understand the inheritance and expressivity of an undesirable characteristic such as neoplasm in these germplasms. Therefore, the aims of this study were (i) to study the inheritance of neoplasm and (ii) to determine the expression of neoplasm in progeny derived from inter-subspecific crosses between *P. sativum* subsp. *sativum* and *P. sativum* subsp. *elatius* species.

2. Materials and Methods

2.1. Plant Materials

Three accessions (AWP 442, AWP 449 and AWP 451) of the tall wild pea (*P. sativum* subsp. *elatius* (M. Bieb.) Asch. & Graebn.) formed neoplasm (*Np*), while four accessions (ACP 13, ACP 14, ACP 20 and ACP 773) of domesticated species (*P. sativum* L. subsp. *sativum*) were free from the neoplasm (*np*) trait when grown in a glasshouse.

The original numbers of AWP 442, AWP 449 and AWP 451 were IG 52442, IG 52459 and P 51 (ICARDA, The International Center for Agricultural Research in the Dry Areas, germplasm collection), respectively. AWP 442 was reported to be resistant to the pulse beetle (*Callosobruchus chinensis* L.), while AWP 449 was susceptible [20]. The important traits of the *Pisum* subspecies are detailed in Table 1. Crosses' populations (ACP 13 × AWP 442, ACP 773 × AWP 451, ACP 14 × AWP 449 and ACP 20 × AWP 442) were used for neoplasm observations under glasshouse conditions, whereas 210 accessions including the parents of these crosses were screened for neoplasm and pea weevil under field conditions.

Table 1. Important traits of *Pisum* subspecies used in inter-subspecific crosses.

Subspecies	Accessions	Landrace/Wild	*Np* in Field	*Np* in Glasshouse (*Np*/*np*)	Flower Color	100-Seed Weight (g)	Tolerance to Cold	Resistance to Seed Beetle	Resistance to Pea Weevil
P. s. ssp. *sativum*	[1] ACP 13	Landrace	No	*np*	Pink	23.5	Medium	Susceptible	Susceptible
P. s. ssp. *sativum*	ACP 14	Landrace	No	*np*	Pink	29.7	Medium	Susceptible	Susceptible
P. s. ssp. *sativum*	ACP 20	Landrace	No	*np*	White	34.1	Medium	Susceptible	Susceptible
P. s. ssp. *sativum*	ACP 773	Landrace	No	*np*	White	28.9	*	*	*
P. s. ssp. *elatius*	[2] AWP 442	Wild	No	*Np*	Lilac purple	9.8	Tolerant	Resistant	Susceptible
P. s. ssp. *elatius*	AWP 449	Wild	No	*Np*	Lilac purple	10.6	Tolerant	Susceptible	Susceptible
P. s. ssp. *elatius*	AWP 451	Wild	No	*Np*	Lilac purple	11.2	*	*	*

[1] Akdeniz University cultivated *Pisum*, [2] Akdeniz University wild *Pisum*, * not evaluated.

2.2. Field Screening for Pea Weevil

A total of 210 accessions of *Pisum* ssp. was evaluated for resistance to pea weevil (*Bruchus pisorum* L.) under field conditions for seven years, from 2014 to 2020 at Antalya (30°38′ E, 36°53′ N and 51 m above sea level), Turkey. ACP 13, ACP 14, ACP 20, ACP 773, AWP 442, AWP 449 and AWP 451 were grown every year. Incidence of pea weevil for each accession was scored using a 1–9 visual scale, based on percent damage under natural insect infestation in the field, where 1 = free from any damage by pea weevil, 9 = damage in more than 91% of seeds (Table 2). According to the scale, the accessions having a rate between 1 and 4 were considered to be resistant and those having a rate between 5 and 9 were regarded as susceptible. Damage evaluation of the infested seeds for pea weevil in field trials was carried out after harvesting and threshing. One hundred seeds selected randomly from four plants were assessed for seed damage evaluation in each accession and replication. Mean of percent seed damage was used for data analyses. Percent seed damage was used for resistance to pea weevil by Teshome et al. [21] and Aznar-Fernandez et al. [22].

Table 2. A visual 1–9 scale for resistance to pea weevil in *Pisum* species under field conditions.

Score	Response to Pea Weevil	Damages on Seeds
1	Very Highly Resistant	Free from any seed damage after careful observation
2	Highly Resistant	Damage present in 1 to 10% of the seeds
3	Resistant	Damage present 11 to 20% of the seeds
4	Moderately Resistant	Damage present in 21 to 30% of the seeds
5	Less susceptible	Damage present in 31 to 40% of the seeds
6	Moderately susceptible	Damage present in 41 to 50% of the seeds
7	Susceptible	Damage present in 51 to 70% of the seeds
8	Highly susceptible	Damage present in 71 to 90% of the seeds
9	Very highly susceptible	Damage present in more than 91% of the seeds

2.3. Plant Sowing and Growing

In the glasshouse, which was at the same location as the field experiments, seeds of the parent plants were sown from row to row at 100 cm and from plant to plant at 20 cm in 2016, and the same sowing norm was used for progeny in the glasshouse. Plants were watered with a drip irrigation system and weeds were removed by hand.

In the field, the experiments were conducted as a randomized complete block design with three replications from October to June. Plots were arranged as a single row of 2 m length with inter-row spacing of 50 cm and intra-row spacing of 10 cm. Plants were grown under rainfed conditions, and weeds were cleaned by hand during the seedling stage. In addition, harvesting and threshing were done manually.

The same soil was used in glasshouse and field. According to the soil analysis results of the experimental field, the amount of organic matter and nitrogen in the soil was low and the soil texture was loam. Although the plant nutrient elements in the soil were generally balanced, it had been detected to be low in iron and zinc [23].

2.4. Climatic Conditions in Field and Glasshouse

The field experiments for pea weevil resistance screening were conducted from October to June in the years between 2014 and 2020. During this period, long-term average monthly maximum temperatures were recorded as 32.7 °C, 26.3 °C, 21.5 °C, 19.4 °C, 23.3 °C, 24.3 °C, 29.3 °C, 31.9 °C and 41.4 °C from October to June, respectively. Long-term average monthly minimum temperatures were recorded as 14.0 °C in October, 10.3 °C in November, 5.5 °C in December, 3.1 °C in January, 6.8 °C in February, 5.7 °C in March, 11.6 °C in April, 14.2 °C in May and 19.4 °C in June. The long-term average total rainfall was 26.5 mm in October, 83.3 mm in November, 145.7 mm in December, 207.1 mm in January, 66.7 mm in February, 119.2 mm in March, 25.0 mm in April, 31.6 mm in May and 18.6 mm in June (Figure 1A). Weather conditions were also given by Kivrak et al. [24].

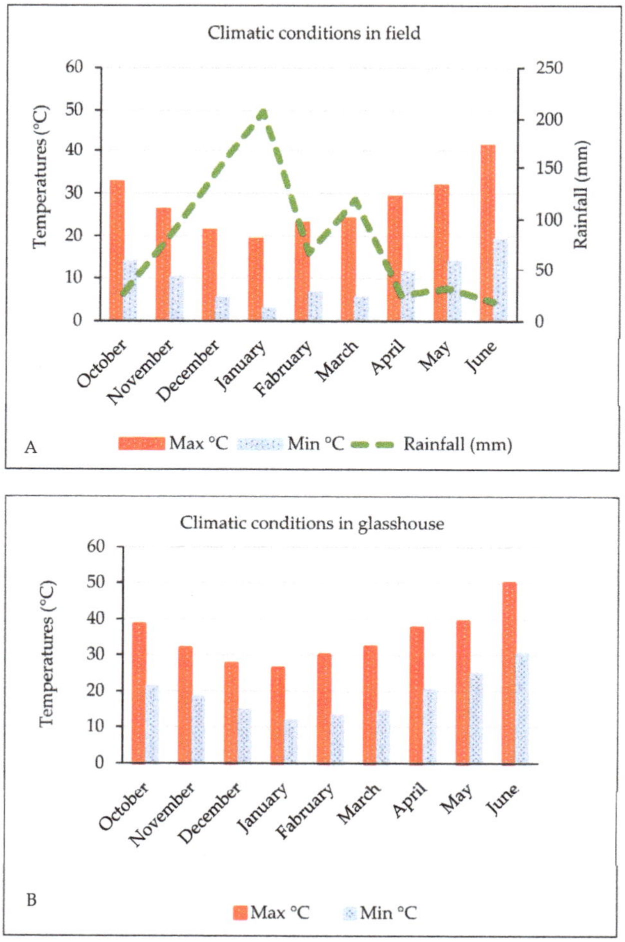

Figure 1. Long-term climatic conditions in field (**A**) and glasshouse (**B**) from 2016 to 2020.

Under glasshouse conditions, average monthly maximum temperatures from 2016 to 2020 were recorded as 38.4 °C in October, 31.7 °C in November, 27.5 °C in December, 26.2 °C in January, 29.8 °C in February, 32.1 °C in March, 37.6 °C in April, 39.3 °C in May and 49.8 °C in June. The monthly average minimum temperatures were recorded as 21.3 °C in October, 18.4 °C in November, 14.7 °C in December, 11.8 °C in January, 13.3 °C in February, 14.5 °C in March, 20.4 °C in April, 24.8 °C in May and 30.2 °C in June (Figure 1B).

2.5. *Evaluation of Neoplasm*

Presence (*Np*) and absence (*np*) of neoplasm on pods of accessions was checked weekly starting with pod formation until harvest under both glasshouse and field conditions. In glasshouse conditions, the density of neoplasm on pods of accessions and F_1 to F_3 progeny was visually assessed in accordance with previously reported dominant homozygous (*Np/Np*) or heterozygous (*Np/np*) alleles [14,17,19].

2.6. *Expressivity of Neoplasm Trait*

In the F_2 population, expressivity (E) of neoplasm was calculated according to the formula proposed by Yasar et al. [25]. According to this,

$$E\ (\%) = (\text{No of } Np \text{ plants/No of total plants}) \times 100 \quad (1)$$

2.7. *Plant Crosses and Progeny*

In the spring of 2016 on campus, the following inter-subspecific crosses were made between ACP 13 (*P. sativum* subsp. *sativum*) and AWP 442 (*P. sativum* subsp. *elatius*), ACP 14 (*P. sativum* subsp. *sativum*) and AWP 449 (*P. sativum* subsp. *elatius*), ACP 20 (*P. sativum* subsp. *sativum*) and AWP 442 (*P. sativum* subsp. *elatius*), and between ACP 773 (*P. sativum* subsp. *sativum*) and AWP 451 (*P. sativum* subsp. *elatius*). That is, domesticated peas that were free from neoplasm were used as female parent (♀), while wild tall peas having neoplasm were used as male parent or pollen donor (♂). Progeny derived from the inter-subspecific cross between ACP 13 and AWP 442 were advanced from F_1 to F_3 as one generation per year. Only one of the inter-subspecific crosses between ACP 13 (*np*) and AWP 442 (*Np*) was advanced to F_3 from 2016 to 2019, whereas the inter-subspecific crosses between ACP 773 (*np*) and AWP 451 (*Np*) were advanced to F_2 in 2020, while the inter-subspecific crosses between ACP 14 (*np*) and AWP 449 (*Np*), and between ACP 20 (*np*) and AWP 442 (*Np*), were used for confirmation and advanced to F_1. From F_1 to F_3, progeny were advanced as single plant progeny grown in the same glasshouse.

2.8. *Data Analysis*

Progeny having neoplasm were counted in segregating populations derived from inter-subspecific crosses between ACP 13 (*np*) and AWP 442 (*Np*), and also between ACP 773 (*np*) and AWP 451 (*Np*). The chi-square test (χ^2) [26] was used to test the expected 3:1 ratio of segregation in the F_2 population:

$$\chi^2 = \frac{(O - E)^2}{E} \quad (2)$$

where O and E are the observed and expected values, respectively. The data on percent seed damage were subjected to analysis of variance (ANOVA) using SPSS 22.0 software (SPSS: Chicago, IL, USA).

3. Results

3.1. *Relationships between Neoplasm and Pea Weevil*

The screening test for seed damage indicated that three of the accessions were found to be very highly resistant with a score of one (free from any seed damage), three were highly resistant with a score of two (seed damage of 1–10%), and forty-four accessions were resistant with a score of three (seed damage of 11–20%). On the other hand, 78 accessions had considerable seed damage with a score of four (seed damage of 21–30%). ACP 13, ACP 14, ACP 20 and AWP 442 accessions were found to be moderately susceptible (seed damage of 31–40%), while AWP 449 was susceptible with a score of six (seed damage of 41–50%) under natural pea weevil infestation conditions in the field (Figures 2 and 3). Significant differences among the accessions were determined for percent seed damage at a probability level of $p \leq 0.05$ but accession-by-year interaction was not significant ($p \leq 0.05$), which means that the accessions exhibited stable reaction over years regarding seed damage.

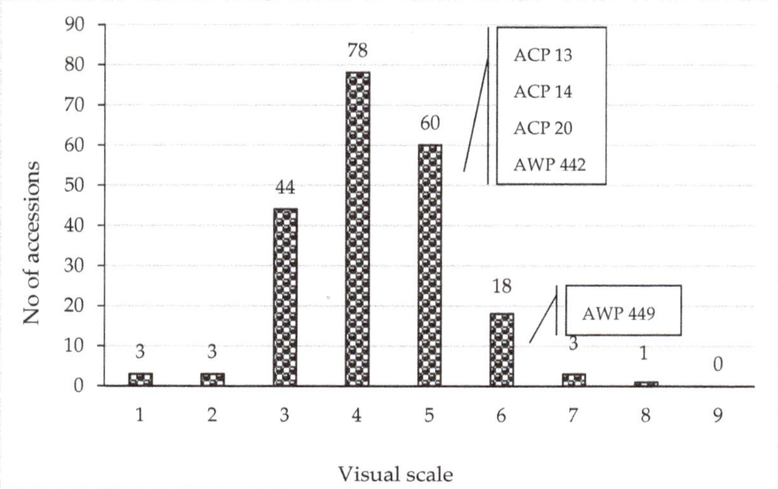

Figure 2. Resistance classes of pea accessions for the pea weevil based on a visual scale of 1–9 under field conditions.

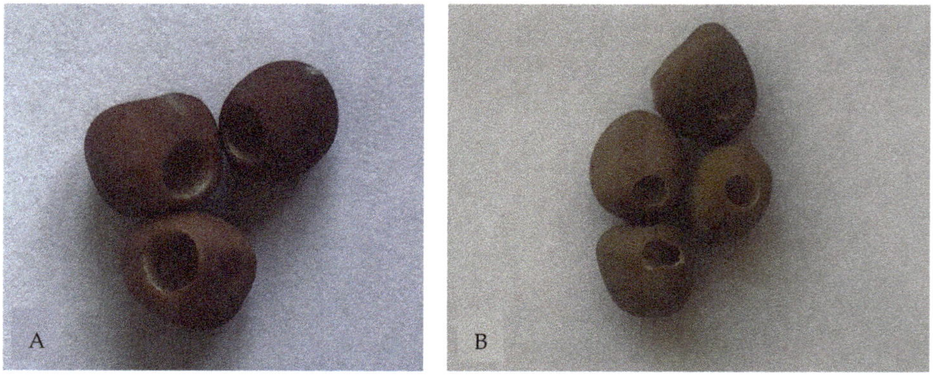

Figure 3. Seed damage holes of pea weevil in AWP 442 (**A**) and AWP 449 (**B**) under field conditions.

Accessions (ACP 13, ACP 14 and ACP 20) of *P. sativum* subsp. *sativum* were free from neoplasm both in field and glasshouse conditions (Figure 2 and Table 1). Despite the natural pea weevil infestation under field conditions, neoplasm did not occur on pods of accessions (AWP 442 or AWP 449) of *P. sativum* subsp. *elatius*, indicating no relationship between neoplasm and presence of pea weevil (Figures 2 and 3A,B). However, neoplasm on pods of the tall wild pea was observed under glasshouse conditions (Figure 4). In addition, neoplasm occurrence was frequent in young pods, while density of neoplasm decreased in mature pods (Figure 4A,B). Neoplasm was also observed on young and matured pods in the natural habitat of the tall wild pea in the Taurus Mountains, Antalya, Turkey (30°24′ E, 36°52′ N and 1071 m above sea level) during expeditions from May to July of 2020 (Figure 4C,D).

Figure 4. Neoplasm on a young pod (**A**) and two mature pods (**B**) of a progeny derived from inter-subspecific crosses between *P. sativum* subsp. *sativum* (ACP 13) and *P. sativum* subsp. *elatius* (AWP 442) in the F_2 population under glasshouse conditions in 2019. Neoplasm on a young (**C**) and mature pod (**D**) of a tall wild pea (*P. sativum* subsp. *elatius*) in its natural habitat in the Taurus Mountains, Antalya, Turkey in 2020.

3.2. Expressivity of Neoplasm

Phenotypic occurrence of neoplasm (expressivity) in the F_2 population obtained from inter-subspecific crosses between ACP 13 (*P. sativum* subsp. *sativum*) and AWP 442 (*P. sativum* subsp. *elatius*) was 77.2% (Table 3). In inter-subspecific crosses between ACP 773 (*P. sativum* subsp. *sativum*) and AWP 451 (*P. sativum* subsp. *elatius*), expressivity was found to be 67.9% (Table 3). However, no difference was seen between dominant homozygous (*NpNp*) or heterozygous (*Npnp*) progeny in F_2 and F_3 populations, while differences in density of neoplasm were observed among the pods of a progeny. Expressivity of neoplasm was found to be higher in inter-subspecific crosses between ACP 13 (*P. sativum* subsp. *sativum*) and AWP 442 (*P. sativum* subsp. *elatius*) than that in inter-subspecific crosses between ACP 773 (*P. sativum* subsp. *sativum*) and AWP 451 (*P. sativum* subsp. *elatius*), as shown in Table 3.

Table 3. Expressivity of neoplasm in F_2 populations derived from inter-subspecific crosses between *P. sativum* subsp. *sativum* and *P. sativum* subsp. *elatius*.

Inter-Subspecific Crosses	Expected Ratio of Neoplasm	Observed Neoplasm Progeny	Expressivity (%)
ACP 13 (*np*) × AWP 442 (*Np*)	3/4 (82.5)	85	77.2
ACP 773 (*np*) × AWP 451 (*Np*)	3/4 (98.3)	89	67.9

3.3. Inheritance of Neoplasm in *P. sativum subsp. elatius*

A total of 11, 12 and 15 progeny of three F_1s obtained from inter-subspecific crosses between ACP 13 (*P. sativum* subsp. *sativum*) and AWP 442 (*P. sativum* subsp. *elatius*), ACP 14 (*P. sativum* subsp. *sativum*) and AWP 449 (*P. sativum* subsp. *elatius*), ACP 20 (*P. sativum* subsp. *sativum*) and AWP 442 (*P. sativum* subsp. *elatius*), and between ACP 773 (*P. sativum* subsp. *sativum*) and AWP 451 (*P. sativum* subsp. *elatius*), was produced, respectively. All F_1 plants had neoplasm on their pods.

There were no differences of neoplasm on the pods of both inter-subspecific crosses in F_1 plants. In the F_2 population derived from the inter-subspecific cross between ACP 13 (*P. sativum* subsp. *sativum*) and AWP 442 (*P. sativum* subsp. *elatius*), neoplasm was observed on pods of 85 progeny, while 25 progeny were free from neoplasm. In inter-subspecific cross between ACP 773 (*P. sativum* subsp. *sativum*) and AWP 451 (P. *sativum* subsp. *elatius*), neoplasm was seen in 89 of 131 progeny. Based on the chi-square test, the segregations in both F_2 populations were found to have a good fit ratio of 3:1 (Table 4).

Table 4. Chi-square (χ^2) analyses of neoplasm (*Np*) vs. free-from-neoplasm (*np*) progeny derived from inter-subspecific crosses between *P. sativum* subsp. *sativum* and *P. sativum* subsp. *elatius*.

Inter-Subspecific Crosses	Phenotype of F_{1s}	F_2				
		No of Plants	Observed	Expected	χ^2	*p*
ACP 13 × AWP 442	*Np*	85 25	*Np* *np*	3:1	0.3	0.9-0.1
ACP 773 × AWP 451		89 42	*Np* *np*	3:1	3.5	0.9-0.1

F_{1s}: First filial generation, F_2: Second filial generation, *p*: Probability value.

3.4. Relationships between Neoplasm and Climatic Conditions

Maximum and minimum temperatures were higher in the glasshouse than those of the field conditions (Figure 1). Although light intensity was not recorded, neoplasm occurred on pods under the glasshouse conditions and pods in plants that are shaded under natural habitat conditions of the tall wild pea.

4. Discussion

ACP 13, ACP 14, ACP 20 and AWP 442 were moderately susceptible to pea weevil with seed damage of 31–40%, whereas AWP 449 had seed damage of 41–50% under natural pea weevil infestation conditions in the field (Figures 2 and 3). On the other hand, some accessions of *P. fulvum* were very highly resistant to pea weevil (Figures 2 and 3), as reported in previous studies [22,27–30].

Although neoplasm was observed on pods of AWP 442 and AWP 449 accessions of *P. sativum* subsp. *elatius* in the glasshouse, it did not arise under field conditions even when there was pea weevil damage on pea pods (Table 1, Figures 3 and 4). Neoplasm in peas with the *Np* gene has been reported to occur under glasshouse conditions with reduced UV light, as well as on pods that are shaded under natural habitat conditions [31]. Our findings on the shaded pods in the glasshouse and natural habitats are in agreement with those of Snoad and Matthews [31]. Although there is no clear evidence as to whether there is resistance to pea weevil in plants with the *Np* gene, it has been reported that pea weevils spend more time during oviposition in pods with neoplasm than in non-neoplastic pods [32], and that pea weevils prefer to oviposit in pods with neoplasm less than in non-neoplastic pods [21,30,33]. Neoplasm was not only reported by Nuttal and Lyall [14] in domesticated peas (PI 206988, PI 244219 and PI 261668), but the occurrence of neoplasm in wild *Pisum* species including *P. humile* Mill., *P. elatius* and *P. fulvum* was also stated by Dodds and Matthews [15]. Additional accessions of field pea (*P. sativum*) (32433A, 203084A, 235899A, 237065A, 226037A, 226037B, 226037C, 226037D and 226037E) having neoplasm were revealed by Teshome et al. [18]. In addition to these neoplastic peas, some accessions of *Pisum*, including *P. elatius* and *P. sativum* originating from different countries, were reported to express neoplasm on pods when plants were grown in the glasshouse [16]. Neoplasm was stated to be induced by light quality (lack of UV light) and started to appear on the surface of young pods of peas with the growth of non-meristematic tissue [14,15]. It was also reported to be related to the pea weevil or bruchins production [16–18,22,30–36]. Neoplasm formation occurs due to light quality such as insufficient UV light or as an antibiosis mechanism in pea weevil resistance [17,33]. Aznar-Fernandez et al. [22] revealed that neoplasm was formed in the pods of the P669 accession without pea weevil eggs in their field trials for resistance to pea weevil in peas. Neoplasm did not occur in field conditions despite pea weevil damage in our findings. On the other hand, it occurred without any pea weevil damage in the glasshouse in the present study. Neoplasm in these accessions of *P. sativum* subsp. *elatius* depended on reduced light in the glasshouse and shadow in natural habitat in the Taurus Mountains (Figure 4). In addition, there may be a relationship between occurrence of neoplasm and high temperature because temperatures during pod formation stages were higher (about 20 °C) in the glasshouse than those of the field conditions (Figure 1). Thus, temperature deserves attention in future studies. Teshome et al. [21] and Aznar-Fernandez et al. [22] emphasized that neoplasm formation was observed more frequently under glasshouse conditions. Aznar-Fernandez and Rubiales [37] did not distinguish whether the reduced pea weevil infestation was due to neoplasm that reduced ovulation or the inhibition of penetration on the pods. With the application of bruchin to pea pods, the expression of *CYP93C18*, a putative isoflavone synthase gene, increased, followed by an increase in the level of pisatin, an isoflavone phytoalexin [34]. Neoplasm formation on pods of the domesticated pea was negatively correlated with oviposition by pea weevil [30]. It was pointed out that wall thickness of the pod and trichomes on the pod of the domesticated pea might have affected the oviposition preference of the weevils [30].

Nevertheless, in the present study, neoplasm in accessions of *P. sativum* subsp. *elatius* was not triggered by pea weevil, indicating that the gene *Np* in these accessions could be different from those in previous studies [14,15,18]. Although neoplasm was related with pea weevil damage in previous studies [16–18,31,33], there was no relationship between neoplasm and pea weevil damage in accessions of *P. sativum* subsp. *elatius* (Table 1) in the present study. Doss et al. [17] outlined that peas having the *Np* gene were more resistant to pea weevil than others, and this was supported by Teshome et al. [18]. Pea weevil damage in *Np* accessions was found to be lower when compared to neoplasm-free ones, and so, a method of enhancing *Np* expression under field conditions via intercropping with sorghum and maize has been proposed, which can serve as part of an integrated pea weevil management strategy, especially for small-scale farming systems [18]. Based on the results found in the present study, neoplasm in accessions of the wild pea (*P. sativum* subsp. *elatius*) was symbolized as Np^+, as neoplasm in the tall wild pea could be a novel gene due to the fact that the gene was not affected by pea weevil (Table 1).

The expected number of neoplastic plants in F_2 progeny was 82.5 for ACP 13 (*P. sativum* subsp. *sativum*) × AWP 442 (*P. sativum* subsp. *elatius*), in agreement with the dominant single gene, while the observed number of neoplasms was detected as 85 progeny. Moreover, the expected number of neoplastic plants in ACP 773 (*P. sativum* subsp. *sativum*) × AWP 451 (*P. sativum* subsp. *elatius*) was 98.3, while the observed number of neoplasms was detected as 89 progeny (Table 3). Expressivity of neoplasm in F_2 progeny obtained from an inter-subspecific cross between ACP 13 (*P. sativum* subsp. *sativum*) and AWP 442 (*P. sativum* subsp. *elatius*) was found to be 77.2% (Table 3), while it was determined to be 67.9% in F_2 progeny obtained from a cross between ACP 773 (*P. sativum* subsp. *sativum*) and AWP 451 (*P. sativum* subsp. *elatius*) (Table 3). Expressivity of neoplasm in the tall wild pea was higher in the former inter-subspecific cross than that in the latter one (Table 3), explaining that it might be influenced by accessions used in inter-subspecific crosses or by year/environment. Aznar-Fernandez et al. [22] reported the effect of environment as the major factor in the development of neoplasm.

Neoplasm on pods of all progeny in F_1 populations obtained from inter-subspecific crosses between ACP 13 (*P. sativum* subsp. *sativum*) and AWP 442 (*P. sativum* subsp. *elatius*), ACP 14 (*P. sativum* subsp. *sativum*) and AWP 449 (*P. sativum* subsp. *elatius*), and between ACP 20 (*P. sativum* subsp. *sativum*) and AWP 442

(*P. sativum* subsp. *elatius*), was evident, indicating that neoplasm in these three *P. sativum* subsp. *elatius* accessions was dominant. In F_2 populations derived from crosses between ACP 13 (*P. sativum* subsp. *sativum*) and AWP 442 (*P. sativum* subsp. *elatius*), and between ACP 773 (*P. sativum* subsp. *sativum*) and AWP 451 (*P. sativum* subsp. *elatius*), progeny carrying neoplasm (*Np*) and free from neoplasm (*np*) were counted as 85 and 25 plants, respectively (Table 4), showing that neoplasm in accession (AWP 442) of *P. sativum* subsp. *elatius* was controlled by a single dominant gene (Table 4). Similar findings were also found in the inter-subspecific crosses among domesticated peas prior to the present study [14,15,18]. Expressivity and inheritance of neoplasm in the tall wild pea have not been adequately studied until now. This is the first study on expressivity of neoplasm in progeny derived from inter-subspecific crosses between *P. sativum* subsp. *sativum* and *P. sativum* subsp. *sativum* according to the available literature [38]. However, the linkage segment containing neoplasm (*Np*) in the domesticated pea was tagged as a part of LG (Linkage group) III by Weeden et al. [39] and Prioul et al. [40].

No differences in density of neoplasm on pods of homozygous (*NpNp*) or heterozygous (*Npnp*) F_2 progeny was found, while neoplasm was reported to be reduced in the heterozygous (*Npnp*) compared to the homozygous (*NpNp*) dominant progeny [15,18]. Since there was no difference in neoplasm density between the heterozygous (*Npnp*) and the homozygous (*NpNp*) dominant progeny, the neoplasm encoding gene in AWP 442, AWP 449 and AWP 451 accessions of *P. sativum* subsp. *elatius* could be novel (Figure 4). However, this needs to be confirmed by an allelism test, although findings in the natural habitat of the tall wild pea in the Taurus Mountains may support this (Figure 4C,D). Differences in density of neoplasm on pods of the same progeny were considered to be due to shade, since neoplasm in the natural habitat was found in shady places (Figure 4).

Wild species were pointed out to be important potential resources for breeding [41–43]. The pea's wild relatives, including *P. fulvum* and *P. sativum* subsp. *Elatius*, were reported to be resistant to many biotic and abiotic stressors such as pea weevil [28,29,44–46], powdery mildew [47], rust [48,49], fusarium wilt [50,51], ascochyta blight [52–54] and drought [55].

5. Conclusions

All F_1 plants had neoplastic pods, and the F_2 populations segregated in a good fit ratio of 3:1 under glasshouse conditions, which suggests that neoplasm on pods of the tall wild pea was controlled by a single dominant gene. The male parents carrying the neoplasm gene in AWP 442, AWP 449 and AWP 451 accessions of *P. sativum* subsp. *elatius* did not have different densities of neoplasm on pods, and the homozygous dominant (*NpNp*) progeny derived from inter-subspecific crosses did not have a distinct appearance from the heterozygous (*Npnp*) progeny, indicating that neoplasm in the tall wild pea could be under the control of a different gene or allele than the *Np* gene. No neoplasm occurred in AWP 442, AWP 449 and AWP 451 due to pea weevil damage in field conditions, but *Np* formation was observed in the same accessions and populations derived from them, although no pea weevil damage was present. It was understood that the neoplasm occurring in the accessions, and in the F_2 and F_3 populations, used in the study was not caused by pea weevil damage. The neoplasm occurring under glasshouse conditions could be due to one or to a combination of factors such as light intensity, humidity and temperature. Since wild tall pea accessions are potential genetic sources for breeding programs aimed at fresh pea production that could be utilized under glasshouse conditions, a negative selection scheme should be incorporated into breeding programs.

Author Contributions: Conceptualization, C.T.; methodology, H.S.; software, H.S.; validation, D.S. and T.E.; formal analysis, H.S.; investigation, H.S.; resources, C.T.; data curation, D.S. and T.E.; writing—original draft preparation, H.S., P.S. and C.T.; writing—review and editing, B.A., H.C., C.I., R.S.G., A.Z., M.B. and P.S.; visualization, P.S.; consultancy, P.S. and C.T.; project administration, C.T.; funding acquisition, C.T. All authors have read and agreed to the published version of the manuscript.

Funding: Inter-subspecific crosses of this research was funded by TUBITAK 119O060. The APC was funded by authors.

Conflicts of Interest: We declare that there are no conflicts of interest.

References

1. Davis, P.H. *Flora Turkey East Agean Islands 3*; University of Edinburgh: Edinburgh, UK, 1970; pp. 370–372.
2. Maxted, N.; Ambrose, N. Peas (*Pisum* L.) . In *Plant Genetic Resources of Legumes in the Mediterranean*; Maxted, N., Bennett, S.J., Eds.; Kluwer Academic Publishers: Dordrecht, The Netherlands, 2000; Chapter 10, pp. 181–190.
3. Ladizinsky, G.; Abbo, S. The *Pisum* Genus. In *the Search for Wild Relatives of Cool Season Legumes*; Springer: Cham, Switzerland, 2015; pp. 55–69.
4. Smykal, P.; Aubert, G.; Burstin, J.; Coyne, C.J.; Ellis, N.T.H.; Flavell, A.J.; Ford, R.; Hybl, M.I.; Macas, J.; Neumann, P.; et al. Pea (*Pisum sativum* L.) in the genomic era. *Agronomy* **2012**, *2*, 74–115. [CrossRef]
5. Smykal, P.; Coyne, C.J.; Ambrose, M.J.; Maxted, N.; Schaefer, H.; Blair, M.W.; Berger, J.; Greene, S.L.; Nelson, M.N.; Besharat, N.; et al. Legume crops phylogeny and genetic diversity for science and breeding. *Crit. Rev. Plant Sci.* **2015**, *34*, 43–104. [CrossRef]
6. Trneny, O.; Brus, J.; Hradilova, I.; Rathore, A.; Das, R.R.; Kopecky, P.; Coyne, C.J.; Reeves, P.; Richards, C.; Smykal, P. Molecular evidence for two domestication events in the pea crop. *Genes* **2018**, *9*, 535. [CrossRef]
7. Smykal, P.; Trneny, O.; Brus, J.; Hanacek, P.; Rathore, A.; Roma, R.D.; Pechanec, V.; Duchoslav, M.; Bhattacharyya, D.; Bariotakis, M.; et al. Genetic structure of wild pea (*Pisum sativum* subsp. *elatius*) populations in the northern part of the Fertile Crescent reflects moderate cross-pollination and strong effect of geographic but not environmental distance. *PLoS ONE* **2018**, *13*, e0194056.
8. Yang, C.; Fan, Z.; Chai, Q. Agronomic and economic benefits of pea/maize intercropping systems in relation to N fertilizer and maize density. *Agronomy* **2018**, *8*, 52. [CrossRef]
9. Shiwakoti, S.; Zheljazkov, V.D.; Gollany, H.T.; Xing, B.; Kleber, M. Micronutrient Concentrations in Soil and Wheat Decline by Long-Term Tillage and Winter Wheat–Pea Rotation. *Agronomy* **2019**, *9*, 359. [CrossRef]
10. Andersen, B.J.; Samarappuli, D.P.; Wick, A.; Berti, M.T. Faba bean and pea can provide late-fall forage grazing without affecting maize yield the following season. *Agronomy* **2020**, *10*, 80. [CrossRef]
11. Zhao, C.; Fan, Z.; Coulter, J.A.; Yin, W.; Hu, F.; Yu, A.; Fan, H.; Chai, Q. High Maize Density Alleviates the Inhibitory Effect of Soil Nitrogen on Intercropped Pea. *Agronomy* **2020**, *10*, 248. [CrossRef]
12. FAOSTAT. Food and Agriculture Organization Statistical Databases. Available online: http://www.fao.org/faostat/en/ (accessed on 17 July 2020).
13. Duke, J.A. *Hand Book of Legumes of World Economic Importance*; Plenum Press: New York, NY, USA, 1981; pp. 99–265.
14. Nuttall, V.W.; Lyall, L.H. Inheritance of neoplastic pod in the pea. *J. Hered.* **1964**, *55*, 184–186. [CrossRef]
15. Dodds, K.S.; Matthews, P. Neoplastic pod in the pea. *J. Hered.* **1966**, *57*, 83–85. [CrossRef]
16. Berdnikov, V.A.; Trusov, Y.A.; Bogdanova, V.S.; Kosterin, O.E.; Rozov, S.M.; Nedelkina, S.V.; Nikulina, Y.N. The neoplastic pod gene (*Np*) may be a factor for resistance to the pest *Bruchus pisorum* L. *Pisum Genet.* **1992**, *24*, 37–39.
17. Doss, R.P.; Oliver, J.E.; Proebsting, W.M.; Potter, S.W.; Kuy, S.; Clement, S.L.; Willianmson, R.T.; Carney, J.R.; DeVilbiss, E.D. Bruchins: Insect-derived plant regulators that stimulate neoplasm formation. *Proc. Natl. Acad. Sci. USA* **2000**, *97*, 6218–6223. [CrossRef] [PubMed]
18. Teshome, A.; Bryngelsson, T.; Mendesil, E.; Marttila, S.; Geleta, M. Enhancing neoplasm expression in field pea (*Pisum sativum*) via intercropping and its significance to pea weevil (*Bruchus pisorum*) management. *Front. Plant Sci.* **2016**, *7*, 654. [CrossRef] [PubMed]
19. Burgess, J.; Fleming, E.N. The structure and development of a genetic tumour of the pea. *Protoplasma* **1973**, *76*, 315–325. [CrossRef]
20. Esen, A.; Sari, H.; Erler, F.; Adak, A.; Sari, D.; Eker, T.; Canci, H.; Ikten, C.; Kahraman, A.; Toker, C. Screening and selection of accessions in the genus *Pisum* L. for resistance to pulse beetle (*Callosobruchus chinensis* L.). *Euphytica* **2019**, *4*, 82. [CrossRef]
21. Teshome, A.; Mendesil, E.; Geleta, M.; Andargie, D.; Anderson, P.; Rämert, B.; Seyoum, E.; Hillbur, Y.; Dagne, K.; Bryngelsson, T. Screening the primary gene pool of field pea (*Pisum sativum* L. subsp. *sativum*) in Ethiopia for resistance against pea weevil (*Bruchus pisorum* L.). *Genet. Resour. Crop Evol.* **2015**, *62*, 525–538. [CrossRef]
22. Aznar-Fernández, T.; Carrillo-Perdomo, E.; Flores, F.; Rubiales, D. Identification and multi-environment validation of resistance to pea weevil (*Bruchus pisorum*) in *Pisum* germplasm. *J. Pest Sci.* **2018**, *91*, 505–514. [CrossRef]

23. Koseoglu, K.; Adak, A.; Sari, D.; Sari, H.; Ceylan, F.O.; Toker, C. Transgressive segregations for yield criteria in reciprocal interspecific crosses between *Cicer arietinum* L. and *C. reticulatum* Ladiz. *Euphytica* **2017**, *213*, 116. [CrossRef]
24. Kivrak, K.G.; Eker, T.; Sari, H.; Sari, D.; Akan, K.; Aydinoglu, B.; Catal, M.; Toker, C. Integration of Extra-Large-Seeded and Double-Podded Traits in Chickpea (*Cicer arietinum* L.). *Agronomy* **2020**, *10*, 901. [CrossRef]
25. Yasar, M.; Ceylan, F.O.; Ikten, C.; Toker, C. Comparison of expressivity and penetrance of the double podding trait and yield components based on reciprocal crosses of kabuli and desi chickpeas (*Cicer arietinum* L.). *Euphytica* **2014**, *3*, 331–339. [CrossRef]
26. Steel, R.G.; Torrie, J.H. *Principles and Procedures of Statistics, a Biometrical Approach*, 2nd ed.; McGraw-Hill: New York, NY, USA, 1980.
27. Hardie, D.C.; Baker, G.J.; Marshall, D.R. Field screening of *Pisum* accessions to evaluate their susceptibility to the pea weevil (Coleoptera: Bruchidae). *Euphytica* **1995**, *84*, 155–161. [CrossRef]
28. Clement, S.L.; Hardie, D.C.; Elberson, L.R. Variation among accessions of *Pisum fulvum* for resistance to pea weevil. *Crop Sci.* **2002**, *42*, 2167–2173. [CrossRef]
29. Aryamanesh, N.; Byrne, O.; Hardie, D.C.; Khan, T.; Siddique, K.H.M.; Yan, G. Large-scale density-based screening for pea weevil resistance in advanced backcross lines derived from cultivated field pea (*Pisum sativum*) and *Pisum fulvum*. *Crop Pasture Sci.* **2012**, *63*, 612–618. [CrossRef]
30. Mendesil, E.; Ramert, B.; Marttila, S.; Hillbur, Y.; Anderson, P. Oviposition preference of pea weevil, *Bruchus pisorum* L. among host and non-host plants and its implication for pest management. *Front. Plant Sci.* **2016**, *6*, 1186. [CrossRef] [PubMed]
31. Snoad, B.; Matthews, P. Neoplasms of the pea pod. *Chromosom. Today* **1969**, *2*, 126–131.
32. Doss, R.P. Treatment of pea pods with Bruchin B results in up-regulation of a gene similar to *MtN19*. *Plant Physiol. Biochem.* **2005**, *43*, 225–231. [CrossRef]
33. Doss, R.P.; Proebsting, W.M.; Potter, S.W.; Clement, S.L. Response of *Np* mutant of pea (*Pisum sativum* L.) to pea weevil (*Bruchus pisorum* L.) oviposition and extracts. *J. Chem. Ecol.* **1995**, *21*, 1–10. [CrossRef]
34. Cooper, L.D.; Doss, R.P.; Price, R.; Peterson, K.; Oliver, J.E. Application of Bruchin B to pea pods results in the up-regulation of *CYP93C18*, a putative isoflavone synthase gene, and an increase in the level of pisatin, an isoflavone phytoalexin. *J. Exp. Bot.* **2005**, *56*, 1229–1237. [CrossRef]
35. Ketter, A.P. Cellular Events Conditioned by the *Np* Gene of *Pisum sativum* L. in Response to Reduced UV Light, Weevil Oviposition, and Bruchins. Master's Thesis, Oregon State University, Corvallis, OR, USA, 2007.
36. Reddy, G.V.; Sharma, A.; Gadi, R.L. Biology, ecology, and management of the pea weevil (Coleoptera: Chrysomelidae). *Ann. Entomol. Soc. Am.* **2018**, *111*, 161–171. [CrossRef]
37. Aznar-Fernandez, T.; Cimmino, A.; Masi, M.; Rubiales, D.; Evidente, A. Antifeedant activity of long-chain alcohols, and fungal and plant metabolites against pea aphid (*Acyrthosiphon pisum*) as potential biocontrol strategy. *Nat. Prod. Res.* **2019**, *33*, 2471–2479. [CrossRef]
38. WOS. Web of Science. Available online: http://apps.webofknowledge.com/ (accessed on 18 August 2020).
39. Weeden, N.F.; Ellis, T.H.N.; Timmerman-Vaughan, G.M.; Swiecicki, W.K.; Rozov, S.M.; Berdnikov, V.A. A consensus linkage map for *Pisum sativum*. *Pisum Genet.* **1998**, *30*, 1–4.
40. Prioul, S.; Frankewitz, A.; Deniot, G.; Morin, G.; Baranger, A. Mapping of quantitative trait loci for partial resistance to *Mycosphaerella pinodes* in pea (*Pisum sativum* L.), at the seedling and adult plant stages. *Theor. Appl. Genet.* **2004**, *108*, 1322–1334. [CrossRef] [PubMed]
41. Singh, K.B.; Saxena, M.C. *Breeding for Stress Tolerance in Cool-Season Food Legumes*; Wiley: Chichester, UK, 1993; p. 52.
42. Ali, S.M.; Sharma, B.; Ambrose, M.J. Current status and future strategy in breeding pea to improve resistance to biotic and abiotic stresses. In *Expanding the Production and Use of Cool Season Food Legumes*; Springer: Dordrecht, The Netherlands, 1994; pp. 540–558.
43. Toker, C.; Uzun, B.; Ceylan, F.O.; Ikten, C. Chickpea. In *Alien Gene Transfer in Crop Plants*; Springer: New York, NY, USA, 2014; Volume 2, pp. 121–151.
44. Clement, S.L.; McPhee, K.E.; Elberson, L.R.; Evans, M.A. Pea weevil, *Bruchus pisorum* L. (Coleoptera: Bruchidae), resistance in *Pisum sativum* × *Pisum fulvum* interspecific crosses. *Plant Breed.* **2009**, *128*, 478–485. [CrossRef]

45. Byrne, O.M.; Hardie, D.C.; Khan, T.; Yan, G. Genetic analysis of pod and seed resistance to pea weevil in a *Pisum sativum* x *P. fulvum* interspecific cross. *Aust. J. Agric. Res.* **2008**, *59*, 854–862. [CrossRef]
46. Aryamanesh, N.; Zeng, Y.; Byrne, O.; Hardie, D.C.; Al-Subhi, A.M.; Khan, T.; Siddique, K.H.M.; Yan, G. Identification of genome regions controlling cotyle don, pod wall/seed coat and pod wall resistance to pea weevil through QTL mapping. *Theor. Appl. Genet.* **2014**, *127*, 489–497. [CrossRef] [PubMed]
47. Fondevilla, S.; Torres, A.M.; Moreno, M.T.; Rubiales, D. Identification of a new gene for resistance to powdery mildew in *Pisum fulvum*, a wild relative of pea. *Breed. Sci.* **2007**, *57*, 181–184. [CrossRef]
48. Barilli, E.; Sillero, J.C.; Moral, A.; Rubiales, D. Characterization of resistance response of pea (*Pisum* spp.) against rust (*Uromyces pisi*). *Plant Breed.* **2009**, *128*, 665–670. [CrossRef]
49. Barilli, E.; Satovic, Z.; Rubiales, D.; Torres, A.M. Mapping of quantitative trait loci controlling partial resistance against rust incited by *Uromyces pisi* (Pers.) Wint. in a *Pisum fulvum* L. intraspecific cross. *Euphytica* **2010**, *175*, 151–159. [CrossRef]
50. McPhee, K.E.; Tullu, A.; Kraft, J.M.; Muehlbauer, F.J. Resistance to *Fusarium* wilt race 2 in the *Pisum* core collection. *J. Am. Soc. Hortic. Sci.* **1999**, *124*, 28–31. [CrossRef]
51. Hance, S.T.; Grey, W.; Weeden, N.F. Identification of tolerance to *Fusarium solani* in *Pisum sativum* ssp. *elatius*. *Pisum Genet.* **2004**, *36*, 9–13.
52. Wroth, J.M. Possible role of wild genotypes of *Pisum* spp. to enchance ascochyta blight resistance in pea. *Aust. J. Exp. Agric.* **1998**, *38*, 469–479. [CrossRef]
53. Fondevilla, S.; Ávila, C.M.; Cubero, J.I.; Rubiales, D. Response to *Mycosphaerella pinodes* in a germplasm collection of *Pisum* spp. *Plant Breed.* **2005**, *124*, 313–315. [CrossRef]
54. Carrillo, E.; Rubiales, D.; Pérez-de-Luque, A.; Fondevilla, S. Characterization of mechanisms of resistance against *Didymella pinodes* in *Pisum* spp. *Eur. J. Plant Pathol.* **2013**, *135*, 761–769. [CrossRef]
55. Naim-Feil, E.; Toren, M.; Aubert, G.; Rubinstein, M.; Rosen, A.; Eshed, R.; Sherman, A.; Ophir, R.; Saranga, Y.; Abbo, S. Drought response and genetic diversity in *Pisum fulvum*, a wild relative of domesticated pea. *Crop Sci.* **2017**, *57*, 1145–1159. [CrossRef]

agronomy

MDPI

Article

Resistance to Anthracnose (*Colletotrichum lentis*, Race 0) in *Lens* spp. Germplasm

Eleonora Barilli [1,*,†], Juan Moral [2,†], Thaïs Aznar-Fernández [1] and Diego Rubiales [1]

[1] Institute for Sustainable Agriculture, CSIC, Av. Menéndez Pidal s/n, 14004 Córdoba, Spain; tisseta@gmail.com (T.A.-F.); diego.rubiales@ias.csic.es (D.R.)

[2] Department of Agronomy, University of Córdoba, Campus Rabanales, 14071 Córdoba, Spain; juan.moral@uco.es

* Correspondence: ebarilli@ias.csic.es; Tel.: +34-957-499-211

† Both authors equally contributed to this work.

Received: 9 October 2020; Accepted: 13 November 2020; Published: 16 November 2020

Abstract: Anthracnose, caused by the fungal pathogen *Colletotrichum lentis*, is a severe disease of lentil (*Lens culinaris*) causing premature defoliation, necrotic stem lesions that lead to plant wilting and death in susceptible varieties. Two races of *C. lentis* (0 and 1) have been described so far. Race 0 is the most virulent one and limited genetic resistance is available to date. To address this scarcity of resistance, we screened a germplasm collection covering different *Lens* spp. originating from different countries for *C. lentis* race 0 resistance. Leaf and stem damage and plant mortality were assessed on seedlings inoculated under controlled conditions. A significant variability was observed among accession. Most of the collection studied was highly susceptible, but some levels of resistance were identified in about 15% of the accessions. The highest levels of resistance were identified in *L. ervoides* accessions PI572330, PI572334 and PI572338. Moderate resistance was also identified in 10 *L. culinaris* ssp. *culinaris* accessions but not in the remaining species studied. Selected accessions showed potential to integrate several breeding programs.

Keywords: anthracnose; *Colletotrichum lentis*; disease screening; lentil; plant resistance

1. Introduction

Lentil (*Lens culinaris* Medik.) is an annual, herbaceous, self-pollinating cool-season pulse crop with appreciable importance as food. In 2018, lentils were cultivated worldwide in more than 6 million hectares [1]. The largest producers are Canada, India, Turkey and the USA [1]. Lentil production can be constrained by a number of pests and diseases [2]. Among these, anthracnose is a globally important fungal disease. The causing agent was named until 2014 as *Colletotrichum truncatum* (Schwein.) Andrus & W.D. Moore. Then, it was renamed as *C. lentis* Damm based on (i) its host specialization on *Lens* and *Vicia* species [3]; (ii) the morphology of the conidium and (iii) the sequence of ITS (internal-transcribed-spacer) region of ribosomal DNA [4–6]. During the last decade, anthracnose has turned into one of the most damaging lentil diseases in countries such as Canada and USA [6–8]. Yield losses up to 70% on susceptible lines can easily occur under high disease pressure, favorable climatic conditions and absence of chemical control [9].

The species *C. lentis* can survive for several years on plant debris that remain in the field after harvest, such as microsclerotia and conidia [10]. The pathogen requires more than 16 h of leaf wetness in combination with temperatures between 20 and 30 °C to infect the host plant [10]. Initial symptoms on leaves are small yellow spots that enlarge into brown-colored lesions with a distinct dark margin. This might result in premature leaf drop. In the stem, the first lesions appear in its base from where they progress upwards [9]. Large stem lesions can surround the whole stems and penetrate the vascular tissue, causing wilting with subsequent plant death. As soon as lesions become necrotic, acervuli develop, producing abundant conidia [10] and then microsclerotia [6]. In susceptible lentil lines, more than 10% of the harvested seeds could show necrotic lesions affecting their quality and market sale. During the growing season, the inoculum is primarily spread by rain splash and secondarily by windblown infected debris or during the harvesting process [7,11]. Under field conditions, *C. lentis* can severely affect other legume species such as faba bean (*Vicia faba* L.) and vetch (*Vicia sativa* L.) and, with minor severity, field pea (*Pisum sativum* L.) and chickpea (*Cicer arietinum* L.) [3].

The pathogen *C. lentis* is a hemibiotrophic fungus that causes intracellular colonization of plant tissues [7]. The fungus has to go through a short biotrophic phase with a length of several days depending on environmental conditions, before change to a destructive necrotrophic growth [12]. Conidia develop a germ tube that originates a melanized functional appressorial, essential to successfully penetrate the host [8]. Penetration pegs, which form on the appressoria, shortly penetrate the epidermis and subsequently, infection vesicles appear in the apoplastic space of epidermal cells below the penetration sites [6]. The primary hypha, essential for successful infection and symptom development, invaginates the plasma membrane of epidermal cells and plasmolysis starts [4]. Pathogens move from the symptomless biotrophic phase to a necrotrophic destructive phase, during which the fungus invades and kills host cells [4,6]. When environmental conditions are favorable to pathogen development, the latent period for *C. lentis* infection is about one week, allowing the fungus to accomplish several cycles of infection during the field growing season [10].

Based on severity of both leaf and stem lesions, two pathogenic races (named races 0 and 1) of *C. lentis* have been described. These races have different virulence on lentil differentials both under controlled and field conditions, race 0 being more aggressive than race 1 [7]. The search for sources of resistance in the cultivated lentils (*L. culinaris* subsp. *culinaris*) led to the identification of several accessions with high levels of partial resistance to race 1 [7,11,13]. This resistance to race 1 is conferred by a combination of recessive (*ctr1* and *ctr2*) and dominant (*CtR3*, *CtR4* and *CtR5*) genes [6,14]. On the contrary, little resistance has been identified so far in cultivated lentils to race 0 [7], and the genetics of resistance to race 0 in *L. culinaris* have not yet been examined [6]. However, some effective resistance to both races, and particularly the virulent race 0, has already been identified in *L. ervoides* (Brign.) Grande and *L. lamottei* Czefr. [7,13]. Thereafter, lentil wild relatives could be valuable sources for disease resistance to anthracnose [15]. The objective of our studies was to identify additional sources of resistance to *C. lentis* race 0 in order to increase the genetic background for resistance for future breeding programs.

2. Materials and Methods

2.1. Plant Material and Growing Conditions

A collection of 250 accessions of species and subspecies of the genus *Lens* was screened for resistance to *C. lentis* race 0 under controlled conditions. The collection consisted of *L. culinaris* subsp. *culinaris* (204 accessions), *L. culinaris* subsp. *orientalis* Boiss. (Ponert) (10 accessions), *L. culinaris* subsp. *odemensis* (Ladiz.) M.E. (5 accessions), *L. ervoides* (8 accessions), *L. nigricans* (M. Bieb.) Godr. (21 accessions) and *L. lamottei* (2 accessions). Accessions originated from 37 countries were provided by CRF-INIA (BGE-code; 84 accessions) and USDA (PI- and W6- codes; 166 accessions).

Seeds were stored at 4 °C until usage. Seeds were surface sterilized by immersion in a 10% solution of commercial bleach (NaClO) in sterile water for 2 min and allowed to air dry on a laboratory bench before sowing. Disinfested seeds were scarified by nicking with a razor blade and pregerminated on wet filter paper in Petri dishes for 72 h at 4 °C to ensure experiments with a uniform plant development stage. Petri dishes were then transferred to 20–22 °C for 5–7 days. Thereafter, the seeds were planted in plastic pots (15 × 15 × 15 cm with 250 cm^3 each of 1:1 sand and peat mixture) under controlled conditions, 20 ± 2 °C, 65% relative humidity and a photoperiod of 14/10 h day/night regime with 150 $\mu mol\ m^{-2}\ s^{-1}$ irradiance at plant canopy. The experimental design consisted of three randomized blocks with six seedlings per accession and block sown in two pots (three seedlings per pot). Each block consisted of a separate 80 cm-high bench located in the controlled chamber.

Pots were placed in plastic trays with a capacity of 48 pots each. In each tray, two pots with the local susceptible "S12" accession (*L. culinaris* subsp. *culinaris*) were used as a susceptible control to verify adequate disease density during inoculation and uniformity of infection among trays. Plants were irrigated by manually filling the trays with ±1 cm water every 48 h.

2.2. Fungal Isolates

A previously well-characterized isolate of *C. lentis* race 0 (isolate A8, race 0) [16] was used for the study. Single-spore isolates were cultured in Petri dishes containing potato dextrose agar (PDA) (Sigma Aldrich, Saint-Quentin Fallavier, France) with chloramphenicol (5% (*v/v*) at 12 mL L^{-1} of medium). Petri dishes were incubated at 20 ± 2 °C under a 12 h photoperiod at 150 µmol m^{-2} s^{-1} photon flux density for 48 h.

2.3. Fungal Inoculation and Incubation

Lentil seedlings were inoculated 3 weeks after sowing by spraying with a conidial suspension of 5×10^5 conidia mL^{-1} in sterile water with Tween-20 (0.03%, *v/v*) at a rate of 1 mL per plant [10]. Thereafter, pots were placed into a dark humidity chamber with 100% relative humidity for 48 h incubation, at a temperature of 20 ± 2 °C. After this time, the plants were returned to growth chamber conditions as described in Section 2.1.

To test the plant infection by *C. lentis* and exclude other pathogens, random samples of 5–10 symptomatic leaves and stems were collected at 120 hpi, washed for 1 min in 10% bleach and dried in a flow chamber. Then, plant tissues were transferred to PDA with chloramphenicol (Figure 1) until *C. lentis* was reisolated. The species *C. lentis* was identified according to the type of conidia and its formation using a bright field microscope at 40× magnification (Leica Microsystems, Wetzlar, Germany).

Figure 1. Damage produced by *C. lentis* race 0 on the susceptible check (S12) by 120 hpi: (**a**) whole plant's appearance; (**b**) damages on leaves; (**c**) damages on stems; (**d**) infested leaf dried in flow chamber; (**e**) PDA + CLF Petri dish with fungus developed from infected leaf segment (upper part) and stem (lower part).

2.4. Disease Evaluation and Statistical Analysis

Anthracnose incidence and severity of damage produced on plant leaves and stems were assessed at 120 and 192 h post-inoculation (hpi). Plant mortality (PM) was assessed as percentage (%) of dead seedlings. Damages were separately assessed in leaves (leaf damage, LD) and stems (stem damage, SD) with the help of a 0 (no symptoms) to 5 (deep and coalescent lesions) rating scale developed by Buchwaldt et al. [11].

Area under the disease progress curves were calculated using both LD and SD dependent variables (AULDPC and AUSDPC, respectively) by trapezoidal integration [17]. Data of AULDPC and AUSDPC were standardized considering the reaction of the susceptible control S12 and the number of days alive (sAULDPC and sAUSDPC, respectively) [18].

The mean and standard error of the mean of the dependent variables (sAULDPC, sAUSDPC and PM) were calculated by accession, subspecies, and species. The relationships between the sAULDPC, sAUSDPC and PM were calculated using Spearman's correlation. All the data analyses were conducted using Statistix 10 Software (Analytical Software, Tallahassee, FL, USA).

The phenotypic reaction of the *Lens* accessions against *C. lentis* infection was classified using the aforementioned 0–5 rating scale [11]. Thus, lentil accessions were classified as highly resistant (HR), resistant (R), moderately resistant (MR), moderately susceptible (MS), susceptible (S) and highly susceptible (HS), when their inoculated replicated plants showed an average SD of 0 (no lesions on the stem), 1 (just superficial lesions on the stem), 2 (a few deep lesions on the stem base), 3 (a mixture of superficial and deep lesions on the whole length of the stem, no die-back), 4 (deep lesions accompanied by shoot die-back or partial wilting), and 5 (deep and coalescent lesions with partial or total wilting), respectively [11].

3. Results and Discussion

In the present study, 250 lentil accessions were screened according to their resistance to *C. lentis* race 0 (COL-428). In all the performed inoculations, we used the susceptible control S12, which showed PM over 91%, AULDC and AUSDC values of 110.6 and 129.9, respectively (Table S1). Previously, Buchwaldt et al. [11] classified lentil accessions according to the severity of symptoms on the seedling stems using the 0–5 rating scale, since this dependent variable was well correlated with the plant mortality. Because there was a significant a positive correlation between both variables PM and sAUSDC (Spearman's correlation, $r = 0.401$; $p < 0.001$), we used the classification method proposed by the above-mentioned authors [11]. As an exception, accessions 1318 (*L. ervoides*) only showed superficial stem lesions but all its replicated plants died at 192 hpi. For this reason it was classified as HS. Moreover, clustering analyses were used for classification of the lentil accessions considering sAULDPC, sAUSDPC and PM as dependent variables, individually or as a whole, following Moral et al. [19], butresults were inconclusive, since there was dissimilitude among the formed groups.

Although a wide range of disease response was found, the distribution was markedly skewed toward high PM and disease severities on stem and leaf lesions (HS and S responses). Most of the accessions (about 75% of the total) displayed PM values higher than the susceptible control S12, indicating their elevated susceptibility and high virulence of the used *C. lentis* isolate.

Although susceptible (HS, S or MS) lentil accessions were common, some accessions showed some levels of resistance (Table S1). In fact, 11 accessions (4.4% of the collection) showed just superficial lesions on the stems and were classified as MR, as well as three accessions (1.2% of the collection) showed PM lower than 10% and scarce symptoms on the stems (classified as R). Accessions free (immune) of disease symptoms were not found.

Availability of resistance to *C. lentis* race 0 varied with *Lens* species and subspecies (Figure 2). Even when scarce, levels of resistance were identified in *L. culinaris* subsp. *culinaris* with 10 (5%) accessions out of 204 that classified as MR (Figure 2). These included the accessions BGE001814, BGE001882, W627758, PI283604 and PI432005, which showed reduced lesions at the stem and foliar level at least half that of the susceptible control, and a PM < 33% (Table S1, Figure 3). Remaining *L. culinaris* subsp. *culinaris* accessions were HS, S or MS.

Resistance was more frequent in *L. ervoides*, with three accessions out of eight that rated as R (PI572330, PI572334 and PI572338) and one as MR (PI572327) (Figure 2). Resistant accessions of *L. ervoides* showed very limited and superficial damages at both the stem and foliar level, leading to values of plant mortality ranging between 0–9%. Resistance response of accession PI572330 was in line with previous reports by Tullu et al. [15]. On the contrary, resistance was not identified in *L. culinaris* subsp. *orientalis*, *L. culinaris* subsp. *odemensis*, *L. lamottei* nor in *L. nigricans*, in which all studied accessions were rated as HS, S or MS (Table S1; Figure 2), in agreement with previous findings [13,15].

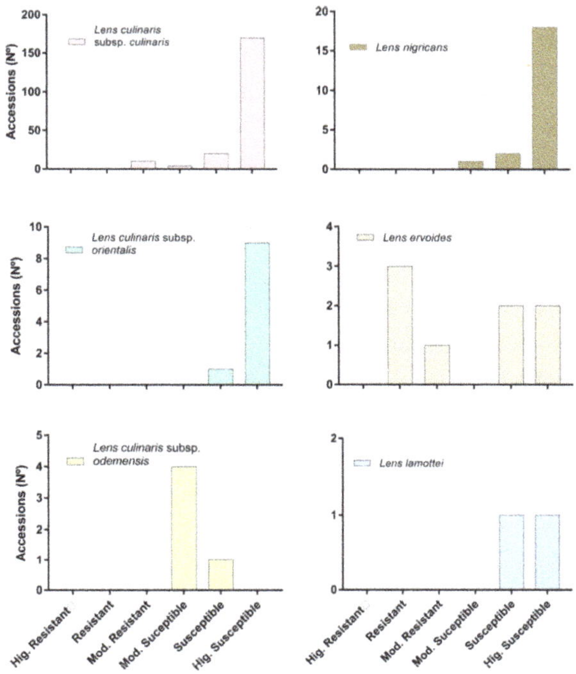

Figure 2. Plant response to *C. lentis* race 0 in each group of *Lens* spp. Macroscopical symptom appearance lead to plant classification into four groups as resistant, moderately resistant, moderately susceptible and susceptible.

(**a**) (**b**)

Figure 3. Symptoms in the leaves and stem of *Lens culinaris* spp. *culinaris* inoculated with *C. lentis* race 0 at 8 days after inoculation under controlled conditions: susceptible control S12 (**a**) and moderately resistant W627758 (**b**).

Regarding geographic origin, 12 of the moderately resistant or resistant accessions originated in the Mediterranean Basin and the Fertile Crescent (4 from Iran, 2 from Spain, 2 from Turkey, 1 from Algeria, 1 from Israel, 1 from Jordan, and 1 from Portugal), 1 was from Serbia and 1 had unknown origin (Table 1).

Table 1. Lentil accessions screened for *C. lentis* race 0 resistance, following their geographical origin.

Country of Origin	N° of Accessions	Accessions Resistant (R)	Accessions Moderately Resistant (MR)
Afghanistan	1	-	-
Algeria	2	-	PI297287
Bosnia	1	-	-
Brazil	1	-	-
Bulgaria	1	-	-
Canada	1	-	-
Chile	7	-	-
Czech Republic	2	-	-
Egypt	3	-	-
Ethiopia	2	-	-
France	4	-	-
Germany	2	-	-
Guatemala	1	-	-
India	10	-	-
Iran	43	-	PI431809, PI432005, PI432033, PI432071
Israel	4	PI572330	-
Italy	3	-	-
Jordan	3	-	PI477921
Lebanon	1	-	-
Mexico	4	-	-
Montenegro	1	-	-
Morocco	3	-	-
North Macedonia	2	-	-
Pakistan	4	-	-
Peru	1	-	-
Portugal	1	-	PI283604
Republic of Cyprus	3	-	-
Russia	2	-	-
Serbia	5	-	PI572327
Syria	5	-	-
Spain	88	-	BGE001814, PI298644
State of Palestine	1	-	-
Tajikistan	1	-	-
Turkey	21	PI572334, PI572338	-
Ukraine	1	-	-
United States	1	-	-
Unknown	13	-	W627756

The Mediterranean Basin is considered the primary centre of diversity and domestication of *Lens* species [20,21], and may also be the centre of diversity for resistance to diseases, as has been demonstrated for other crops.

For breeders, the identification of new sources of resistance from germplasm of different origins and their subsequent inclusion into an elite cultivated background is an essential step that needs to be efficient to develop varieties with the desired characteristics of interest. Accordingly, identification of novel sources of resistance to *C. lentis* race 0 is of great importance since resistance found until now within the *Lens culinaris* subsp. *culinaris* gene pool is very limited [22,23], with fewer than 10 accessions with slight resistance [9,24] which mainly originate from Eastern Europe. Here, we identified for the first time 10 accessions of *L. culinaris* subsp. *culinaris* with MR to *C. lentis* race 0. These accessions, which had both reduced plant mortality and low symptom development on leaves and stems by the pathogen have diverse geographical origin, enlarging the set of possible donors of *Colletotrichum* resistance available for lentil breeders.

Three accessions from wild *L. ervoides* from Israel and Turkey with a highly resistant response to *C. lentis* race 0 were also described here. According to previous studies, this wild species has the highest frequency of resistant accessions for both races of *C. lentis* [11,15,24]. Although with some difficulties, interspecific hybridization with *L. ervoides* is possible. In fact, resistant progenies have already been generated by crossing a resistant wild accession of *L. ervoides* with a susceptible *L. culinaris* one, followed by embryo rescue [24–26].

Resistant accessions found here from both *L. culinaris* subsp. *culinaris* and *L. ervoides* are new sources of resistance to *C. lentis* race 0, which are currently included in local breeding programs with the aim to originate material well adapted to Mediterranean rain-fed conditions. Research is now in progress to determine the number, allelism and dominance of these resistance genes.

Supplementary Materials: The following are available online at http://www.mdpi.com/2073-4395/10/11/1799/s1, Table S1: Lentil accessions screened for *C. lentis* race 0 resistance, following their species and subspecies classification, as well as their geographical origin. Plant mortality (%), area under the disease progress curves with leaf disease and stem disease values (AUSLPC and AUSDPC), standard errors (SE) and classification for *C. lentis* response.

Author Contributions: Conceptualization, D.R.; methodology, D.R., J.M., T.A.-F., formal analysis, E.B., J.M.; genetic resources, D.R.; data curation, E.B., J.M., T.A.-F.; supervision, D.R.; writing—original draft preparation, E.B., J.M., D.R.; writing—review and editing, E.B., J.M., D.R.; funding acquisition, D.R. All authors have read and agreed to the published version of the manuscript.

Funding: The authors are greatly indebted to the Spanish Project AGL2017-82907R, which has funded this research.

Conflicts of Interest: The authors declare no conflict of interests.

References

1. FAOSTAT-Food and Agriculture Organization. Available online: http://www.fao.org/faostat/en/#home (accessed on 20 September 2020).
2. Chen, W.; Sharma, H.C.; Muehlbauer, F.J. *Compendium of Chickpea and Lentil Diseases and Pests*; APS Press: St. Paul, MN, USA, 2011; pp. 6–98. [CrossRef]
3. Gossen, B.D.; Anderson, K.L.; Buchwaldt, L. Host specificity to *Colletotrichum truncatum* from lentil. *J. Plant Pathol.* **2009**, *31*, 65–73. [CrossRef]
4. Bhadauria, V.; Banniza, S.; Vandenberg, A.; Selvaraj, G.; Wei, Y. Cataloging proteins putatively secreted during the biotrophy-necrotrophy transition of the anthracnose pathogen *Colletotrichum truncatum*. *Plant Signal. Behav.* **2011**, *6*, 1457–1459. [CrossRef]
5. Damm, U.; O'Connell, R.J.; Groenewald, J.Z.; Crous, P.W. The *Colletotrichum destructivum* species complex—Hemibiotrophic pathogens of forage and field crops. *Stud. Mycol.* **2014**, *79*, 49–84. [CrossRef] [PubMed]
6. Buchwaldt, L.; Dzanovic, E.; Durkin, J. Lentil anthracnose: Epidemiology, fungicide decision support system, resistance and pathogen races. *Can. J. Plant Pathol.* **2018**, *40*, 189–198. [CrossRef]
7. Banniza, S.; Warale, R.; Menat, J.; Cohen-Skali, A.; Armstrong-Cho, C.; Bhadauria, V. The long path to understanding the host-pathogen interactions of *Colletotrichum lentis* on lentil. *Can. J. Plant Pathol.* **2018**, *40*, 199–209. [CrossRef]
8. Chongo, G.; Gossen, B.D.; Bernier, C.C. Infection by *Colletotrichum truncatum* in resistant and susceptible lentil genotypes. *Can. J. Plant Pathol.* **2002**, *24*, 81–85. [CrossRef]
9. Shaikh, R.; Diederichsen, A.; Harrington, M.; Adam, J.; Conner, R.L.; Buchwaldt, L. New sources of resistance to *Colletotrichum truncatum* race Ct0 in *Lens culinaris* Medikus subsp. *culinaris* obtained by single plant selection in germplasm accessions. *Genet. Resour. Crop Evol.* **2013**, *60*, 193–201. [CrossRef]
10. Chongo, G.; Bernier, C.C. Effects of host, inoculum concentration, wetness duration, growth stage and temperature on anthracnose of lentil. *Plant Dis.* **2000**, *84*, 544–548. [CrossRef]
11. Buchwaldt, L.; Anderson, K.L.; Morrall, R.A.A.; Gossen, B.D.; Bernier, C.C. Identification of lentil germ plasm resistant to *Colletotrichum truncatum* and characterization of two pathogen races. *Phytopathology* **2004**, *94*, 236–243. [CrossRef]
12. Armstrong-Cho, C.; Wang, J.; Wei, Y.; Banniza, S. The infection process of two pathogenic races of *Colletotrichum truncatum* on lentil. *Can. J. Plant Pathol.* **2012**, *34*, 58–67. [CrossRef]

13. Vail, S.; Strelioff, J.V.; Tullu, A.; Vandenberg, A. Field evaluation of resistance to *Colletotrichum truncatum* in *Lens culinaris, Lens ervoides* and *Lens ervoides* × *Lens culinaris* derivatives. *Field Crops Res.* **2012**, *126*, 145–151. [CrossRef]
14. Buchwaldt, L.; Shaikh, R.; Adam, J.; Tullu, A.; Slinkard, A.E. Recessive and dominant genes confer resistance to *Colletotrichum truncatum* in cultivated lentil. *Can. J. Plant Pathol.* **2013**, *35*, 222–231. [CrossRef]
15. Tullu, A.; Buchwaldt, L.; Lulsdorf, M.; Banniza, S.; Barlow, B.; Slinkard, A.E.; Sarker, A.; Tar'an, B.; Warkentin, T.; Vandenberg, A. Sources of resistance to anthracnose (*Colletotrichum truncatum*) in wild lens species. *Genet. Resour. Crop Evol.* **2006**, *53*, 111–119. [CrossRef]
16. Ford, R.; Banniza, S.; Photita, W.; Taylor, P.W.J. Morphological and molecular discrimination of *Colletotrichum truncatum* causing anthracnose on lentil in Canada. *Australas. Plant Pathol.* **2004**, *33*, 559–569. [CrossRef]
17. Jeger, M.J.; Viljanen-Rollinson, S.L. The use of the area under the disease-progress curve (AUDPC) to assess quantitative disease resistance in crop cultivars. *Theor. Appl. Genet.* **2001**, *102*, 32–40. [CrossRef]
18. Madden, L.V.; Hughes, G.; van den Bosch, F. *The Study of Plant Disease Epidemics*; APS Press: St. Paul, MN, USA, 2007; pp. 11–31. [CrossRef]
19. Moral, J.; Xaviér, C.J.; Viruega, J.R.; Roca, L.F.; Caballero, J.; Trapero, A. Variability in susceptibility to anthracnose in the world collection of olive cultivars of Cordoba (Spain). *Front. Plant Sci.* **2017**, *8*, 1892. [CrossRef] [PubMed]
20. Barulina, H. Lentils of the USSR and other countries. *Bull. Appl. Genet. Plant Breed.* **1930**, *40*, 1–319.
21. Coyne, C.; McGee, R. *Lentil. Genetic and Genomic Resources of Grain Legume Improvement*; Elsevier: Amsterdam, The Netherlands, 2013; pp. 157–180. [CrossRef]
22. Buchwaldt, L.; Diederichsen, A. New disease resistant lentil germplasm identified at Plant Gene Resources of Canada (PGRC). In Proceedings of the 5th Canadian Pulse Research Workshop, London, ON, Canada, 28–30 November 2004; p. 204.
23. Vail, S.; Vandenberg, A. Evaluation of a clonal propagation protocol to obtain replicated disease data on infection by *Colletotrichum truncatum* in *Lens culinaris. Crop Sci.* **2010**, *50*, 926–932. [CrossRef]
24. Gela, T.S.; Banniza, S.; Vandenberg, A. Lack of effective resistance to the virulent race of *Colletotrichum lentis* in *Lens culinaris* Medikus subsp. *culinaris. Plant Gen. Res.* **2020**, *18*, 81–87. [CrossRef]
25. Fiala, J.V.; Tullu, A.; Banniza, S.; Séguin-Swartz, G.; Vandenberg, A. Interspecies transfer of resistance to anthracnose in lentil (*Lens culinaris* Medic.). *Crop Sci.* **2009**, *49*, 825–830. [CrossRef]
26. Tullu, A.; Bett, K.E.; Banniza, S.; Vail, S.; Vandenberg, A. Widening the genetic base of cultivated lentil through hybridization of *Lens culinaris* 'Eston' and *L. ervoides* accession IG 72815. *Can. J. Plant Sci.* **2013**, *93*, 1037–1047. [CrossRef]

agronomy

MDPI

Article

Morphological, Sensorial and Chemical Characterization of Chilli Peppers (*Capsicum* spp.) from the CATIE Genebank

Nelly Judith Paredes Andrade [1], Alvaro Monteros-Altamirano [2], Cesar Guillermo Tapia Bastidas [2] and Marten Sørensen [3,*]

[1] INIAP, Estación Experimental Central de la Amazonía, Vía Sacha San Carlos km 3 de la Entrada a la Parker, Escuela Superior Politécnica del Chimborazo, Extensión Norte Amazónica, Orellana 170518, Ecuador; nelly.paredes@iniap.gob.ec

[2] INIAP, Estación Experimental Santa Catalina, Panamericana sur km 1, Quito 170401, Ecuador; alvaro.monteros@iniap.gob.ec (A.M.-A.); cesar.tapia@iniap.gob.ec (C.G.T.B.)

[3] Department of Plant and Environmental Sciences, Faculty of Science, University of Copenhagen, Sobi/Plen-KU, Thorvaldsensvej 40, 3, DK-1871 Frederiksberg C, Denmark

* Correspondence: ms@plen.ku.dk

Received: 17 September 2020; Accepted: 29 October 2020; Published: 6 November 2020

Abstract: In order to assess the potential of 192 accessions of *Capsicum* L., from 21 countries, a morphological and agronomic characterization was carried out by applying 57 qualitative and quantitative descriptors. Multivariate analyses identified two large groups: the first including *C. annuum* (G3, G5, G7 and G8) and the second *C. frutescens*, *C. baccatum*, *C. chinense* and *C. pubescens* (G1, G2, G4, G6 and G9). The discriminant qualitative descriptors were the colour of the corolla, the colour of the anthers and position of the flower. The quantitative discriminant characteristics were length, weight and width of the fruit. The participatory selection identified 15 materials by colour, aroma, texture, flavour, size and thickness of fruits. Chemical analyses determined the highest concentration of flavonoids in the accessions 10,757 (16.64 mg/g) and 15,661 (15.77 mg/g). Accessions 17,750 (11.68 mg/g) and 10,757 (11.41 mg/g) presented the highest polyphenol contents. The highest capsaicin concentration was recorded in accessions 16,209 (55.90 mg/g) and 10,757 (48.80 mg/g). The highest antioxidant value was recorded in accessions 17,750 (90.85 mg/g) and 15,661 (87.03 mg/g). All these characteristics are important with a view to increasing industrial use and genetic improvement processes. These results show the existence of significant genetic variability within the genus *Capsicum*.

Keywords: germplasm; genetic resources; accessions; descriptors

1. Introduction

Chilli pepper (*Capsicum* L.) is one of the most important spice and vegetable crops in the world in agricultural, cultural and economic terms [1–3]. The genus *Capsicum*, native to Tropical America, comprises 27 species [4,5]. The five most widely used species worldwide are: *C. annuum* L., *C. chinense* Jacq., *C. frutescens* L., *C. baccatum* L. and *C. pubescens* Ruíz & Pav. [6]. *Capsicum annuum* and *C. frutescens*—which according to The Plant List [7] are considered conspecific, but which are here maintained as separate taxa as they are in The Tropical Agricultural Research and Higher Education Center (CATIE) genebank—are known to have been domesticated in Mesoamerica [5], however the original description of *C. frutescens* by Linneaus was based on a cultivated specimen from India. In the Andean region, chilli peppers have been consumed for more than 5000 years; *C. baccatum*, *C. pubescens* and *C. chinense* are believed to be native to South America. *Capsicum chinense* was domesticated in the tropical lowlands; while *C. baccatum* possibly was domesticated in the valleys of Bolivia, while the domestication site for *C. pubescens*, which is a tall chilli, remains unknown [8,9]. Although Ecuador has not been considered a centre of origin of the species, archaeological *Capsicum* starches have been found that date back to 6250 years before our era [10].

The five cultivated species are diploid and self-compatible and the number of chromosomes of the species is 2n = 24 [11]. Cross-pollination rate among *Capsicum* spp. is highly variable, ranging from 2 to 90% [12]. The species *C. annuum*, *C. chinense* and *C. frutescens*, form a morphological complex of overlapping characters from a common base of genes; qualitative characters, such as floral characteristics, differentiate the three species morphologically. These are solitary flowers, creamy white in *C. annuum*; the *C. chinense* species presents two or three flowers per leaf node of a whitish-green colour and with a constriction at the base of the calyx at the junction with the pedicel; while the *C. frutescens* species presents an erect white-green flower without calyx constriction. The other species are relatively easy to distinguish: *Capsicum baccatum* has yellow or white-yellow flowers with greenish spots towards the basal part of the petals and *C. pubescens* has either uniformly purple or purple flowers with a white base and black-rough seeds [13,14].

To promote the use of germplasm collections, morphological characterizations are performed to describe the existing genetic diversity, as an example, intraspecific diversity studies for *C. annuum* in Mexico [15] and in India [16,17] are cited. Morphological characterization has allowed identifying intra and inter-specific variability in the case of *Capsicum* [18–22], identifying resistance to pathogens such as *Phytophthora capsici* [23] or insects such as the *Bemisia tabaci* Begomo-virus complex [24]. A combination of morphological and molecular descriptors was used in India by Yumnam et al. [25].

The objectives of this study were: (a) to determine the level of morphological diversity present in 192 accessions of *Capsicum*; (b) to identify the qualitative and quantitative characters with high discriminating capacity; (c) to classify *Capsicum* species into groups based on their quantitative and qualitative characteristics; (d) to identify germplasm through participatory sensory evaluation; and (e) to determine the chemical characteristics of promising materials.

Note: For abstract in Spanish, see Appendix A.

2. Methodology

2.1. Morphological Characterization

The morphological characterization was set up in a field experiment at CATIE location, in Turrialba, Costa Rica (9° 54′ North Latitude, 83° 40′ West Longitude, altitude 625 m a.s.l.). The average annual precipitation is 2700 mm, and the annual average temperature is 22 °C according to CATIE [Centro Agronómico de Investigación y Enseñanza Tropical, CR] [26]. For morphological characterization, we worked with 192 accessions of chilli (*Capsicum* spp.). These accessions come from various collections carried out in Central America and Mexico between 1976 and 2006 and germplasm exchanges carried out between institutions from five continents corresponding to 21 countries (Table 1).

Table 1. Origin and number of accessions of each of the species of *Capsicum* spp.

	Capsicum Species					
Origin	***C. annuum***	***C. frutescens***	***C. chinense***	***C. baccatum***	***C. pubescens***	**Not Identified**
Peru	1	--	--	5	--	--
Panama	3	5	2	--	--	2
Costa Rica	8	30	2	1	--	4
Mexico	21	4	--	1	--	--
Honduras	--	7	2	--	--	1
Guatemala	30	6	--	--	1	6
El Salvador	4	8	--	--	--	--
Ethiopia	5	2	--	--	--	1
Zimbabwe	1	--	--	--	--	--
Spain	2	--	--	--		

Table 1. *Cont.*

Origin	*Capsicum* Species					
	C. annuum	*C. frutescens*	*C. chinense*	*C. baccatum*	*C. pubescens*	Not Identified
Malaysia	10	--	--	--	--	2
U.S.A.	1	--	--	--	--	--
Syria	1	--	--	--	--	--
Russia	2	--	--	--	--	--
Ecuador	--	--	--	2	--	--
Brazil	1	--	--	2	--	1
Philippines	--	--	--	--	--	1
Cuba	--	--	--	--	--	1
Colombia	--	1	--	--	--	--
Nicaragua	--	1	--	--	--	--
Maldives	1	--	--	--	--	--

Twenty plants per accession were installed in the field and the middle 10 evaluated (to eliminate border effect). The descriptors jointly defined by IPGRI [International for Plant Genetic Resources Institute, now BIOVERSITY, AVRDC and CATIE [27] were used, which comprise 57 qualitative and quantitative descriptors (Tables 2 and 3). The data obtained from the morphological characterization were analysed using the InfoStat/Professional program version 2011 [28].

Table 2. *Capsicum* morphological qualitative descriptors.

Descriptor	State	Descriptor	State
Hypocotyl colour	White, Green, Purple	Corolla spot colour	White, Yellow, Yellowish Green, Green, Purple, Other
Hypocotyl pubescence	Sparse, Medium, Dense	Anther colour	White, Yellow, Pale Blue, Blue, Purple, Other
Cotyledon leaf colour	Light Green, Green, Dark Green, Light Purple, Purple, Dark Purple, Variegated, Yellow, Other	Anthocyanin spots or stripes	Absent, present
Cotyledon leaf shape	Deltoid, Oval, Lanceolate, Elongated-deltoid	Intermediate state fruit colour	White, Yellow, Green, Orange, Purple, Dark Purple, Other
Stem colour	Green, Green with purple stripes, Purple, Other	Mature state fruit colour	White, Lemon-Yellow, Pale Yellow-Orange, Yellow-Orange, Pale Orange, Orange, Light Red, Red, Dark Red, Purple, Brown, Black, Other
Node anthocyanin	Green, Light purple, Purple, Dark purple	Fruit shape	Elongated, Almost round, Triangular, Campanulate, Blocky, Other
Stem shape	Cylindrical, Angled, Flattened	Fruit shape at junction with pedicel	Acute, obtuse, truncated, chordate, lobate
Stem pubescence	Sparse, Medium, Dense	Neck at fruit base	Abscent, present
Plant growth habit	Prostrate, Intermediate, Erect	Shape fruit apex	Pointed, Blunt, Sunken, Sunken and pointed, Other
Branching density	Sparse, Medium, Dense	Appendix in the fruit	Abscent, present
Tillering	Sparse, Medium, Dense	Fruit cross-sectional corrugation	Slightly corrugated, Inter-mediate, Very corrugated
Leaf density	Sparse, Medium, Dense	Fruit epidermis type (surface)	Smooth, Semi-wrinkled, wrinkled
Leaf colour	Yellow, Light Green, Green, Dark Green, Light Purple, Purple, Heather, Other	Persistence of pedicel with fruit	Slight, Intermediate, Persistent
Leaf shape	Deltoid, Oval, Lanceolate	Persistence of pedicel with stem	Slight, Intermediate, Persistent

Table 2. *Cont.*

Descriptor	State	Descriptor	State
Leaf blade margin	Whole, Wavy, Ciliated	Seed colour	Light Yellow, Dark Yellow, Black
Leaf pubescence	Escasa, Intermedia, Densa	Seed surface	Smooth, Rough, Wrinkled
Flower position	Pendant, Intermediate, Erect	Susceptibility to biological stress	Very low or no visible signs of incidence, Low, Inter-mediate, High, Very high
Corolla colour	White, Light yellow, Yellow, Greenish yellow, Purple with white base, White with purple base, White with purple margin, Purple, Other	Chilli flavour	Sweet, spicy

Table 3. *Capsicum* morphological quantitative descriptors.

Descriptor	Unit of Measure	Descriptor	Unit of Measure
Plant height	cm	Fruit length	cm
Plant width	cm	Fruit width	cm
Plant height/width ratio	cm	Fruit weight	g
Stem length	cm	Fruit pedicel length	cm
Stem width	cm	Fruit wall thickness	mm
Length mature leaf	cm	Days to fruiting	number
Width mature leaf	cm	Weight of 1000 seeds	g
Leaf l/w ratio	number	Placenta length	cm
Days to flowering	number	Seed size	mm
No. flowers leaf axil^{-1}	number	Classification seed no. fruit^{-1}	number
Fruit set	number		

Qualitative and quantitative variables were used in a hierarchical cluster by using Ward's method and Gower distance, also, contingency tables were used, using Chi-square, and multiple correspondence analyses were made.

2.2. Participatory Sensory Evaluation

For this stage the 192 morphologically characterized accessions were evaluated by 50 invited untrained participants representing producers, industrialists, chefs, scientists and ordinary consumers from Costa Rica, Mexico, Colombia, Ecuador and Peru. The evaluation was carried out in four phases. In the first phase, farmers/producers were invited to the field trial in CATIE to select accessions based on criteria such as colour, shape, size and production. In the second phase, with previously selected accessions, participants made a second evaluation related to shape, colour and size of the fruit. In the third phase, a sensory evaluation was carried out using the 9-point hedonic test proposed by Lawless [29]. However, this scale caused confusion for the attendees, so we decided to change to a friendlier scale, in which six criteria were recorded colour, smell, texture, flavour, size and thickness of the pulp. Each criterion was evaluated based on a 5-point scale: Excellent 5; Very Good 4; Good 3; Regular 2; Bad 1, proposed by Hernández [30]. In the fourth phase, the samples were tasted to determine the degree of fruit spiciness as reported by Hernández [30], the following scale was used: Very spicy 3, Medium spicy 2 and Regular 1. In each scale the assistants marked with an X the criterion they considered adequate based on their preferences. For the analysis of the results, frequency tables were prepared.

2.3. Chemical Characterization

The chemical analysis of the samples was carried out on the best 15 accessions selected by the morphological characterization and the participatory process. One kilogram of fruits, at the same maturity state, was harvested per accession from different parts of the plants to avoid differences in biochemical properties due to plant position according to Zewdie et al. [31]; Kirschbaum-Titze et al. [32]; Mueller-Seitz [33]. These samples were dried out in oven at 60 °C for 30 h, after which 100 g per accession were sent for analysis. The determination of flavonoid content was conducted using the method proposed by Miean and Mohamed [34]. The determination of antioxidant activity was completed using the method of measurement of the absorption capacity of oxygen radicals proposed by Álvarez-Parrilla et al. [35]. The determination of the content of total phenolic compounds was done according to the method proposed by Blainski et al. [36]. The determination of the concentration of capsaicin and dihydrocapsaicin was completed according to the method proposed by Juangsamoot et al. [37]. Finally, the analysis of the results was carried out through multivariate descriptive statistics.

3. Results

3.1. Descriptive Analysis of Morphological Characters

Evaluation of the characteristics of stem, leaves, flowers and fruits from 192 accessions are summarized follow.

3.2. Stem

The purple hypocotyl character was present in 70 accessions, purple stem in 5 accessions, purple knot anthocyanin in 78 accessions, light purple anthocyanin in 46 accessions and dark purple in one accession. Having accessions with purple characteristics determines the presence of anthocyanins, which are classified as nutraceuticals and appetizing agents. Bhattacharya et al. [38] indicate that anthocyanins minimize the proliferation of cancer cells, prevent lipid damage in food and protect against diseases of the heart. Likewise, Rodríguez and Kimura [39] mention that antioxidants can neutralize or reduce the activity of free radicals, associated with cardiovascular diseases.

3.3. Leaves

The variable leaf shape of the individual cotyledon was recorded according to the lanceolate, elongated-deltoid and oval categories, where the oval category was the most dominant since it was found in 134 accessions, corresponding to 69.79% of the materials evaluated. Likewise, the colour of the mature leaf was mostly green in 159 accessions corresponding to 82.81%. The majority of our accessions (170) had reduced leaf pubescence corresponding to 88.54%. This is in agreement with the results reported by Smith and Heiser [40] who mention that for *C. frutescens* leaf pubescence tends to be scarce.

3.4. Flowers

The colour of the corolla, the colour of the anther, position of the flower, the length of the placenta, and the pubescence of the stem had a marked influence on the discrimination of species. To the extent that the *C. annuum* were characterized by presenting white flowers and a single flower per leaf axil and *C. frutescens* presented a greenish-yellow flower without calyx constriction. It is worth noting that in our study the length of the ovary influenced the grouping of the samples, which differs from IPGRI [27], which do not consider it as highly discriminating for species differentiation. According to Sreelathakumary and Rajamony [41], the length of the ovary is correlated with the size and weight of the fruits and the most extended shelf-life at the market. This descriptor is more significant in *C. frutescens* than in *C. annuum*, and capsaicin is stored in it, also, both weight and size influence good filling of fruit cavities and seed production.

In the same way, this study agrees with the results reported by Hernández et al. [42] who mention that the characteristics of the flowers per leaf axil are an important variable to establish differences between the *C. annuum* and *C. frutescens* species. Note that the two taxa here are treated as separate taxa, though taxonomists in general consider them as being conspecific, as already mentioned; however, for practical reasons they are here considered as representing separate cultivar groups. *Capsicum annuum* accessions were characterized by having solitary flowers and *C. frutescens* for presenting more than one flower per leaf axil. The white colour of the corolla appeared in 47.40% of the accessions evaluated mainly in the G3 and G7, represented by the *C. annuum* species. The corolla's greenish-yellow colour appeared in 36.46% of the accessions, specifically in the G4, G9 and G6 constituted by the species *C. frutescens*. In comparison, the colour of the corolla was light yellow at 7.29%, and in G1 mainly made up of the *C. baccatum* species (Figure 1). The above agrees with Pickersgill [43], who mentions that in *Capsicum*, two groups of flowers are defined: white and purple. In the group of white flowers, there are two subgroups, the one made up of *C. baccatum* and a second that groups *C. annuum*, *C. chinense*, *C. frutescens*. The group of purple flowers are the species *C. eximium* Hunz., *C. cardenasii* Heiser & P.G.Sm., and *C. pubescens*. On the other hand, Smith and Heiser [35] reported that in *C. frutescens*, the flowers are greenish-yellow, and for *C. annuum*, they are white.

3.5. Fruits

According to Andrews [43], the accessions belonging to the *C. annuum* species are characterized by having small, ovoid fruits with two locules, the fruit—a bloated berry—with different colours, e.g., light green, green, purple, yellow, orange and deep red. The *C. frutescens* species is characterized by presenting elongated fruits ending in a blunt point, with two locules per fruit, which agrees with the results found in this work where most of the accessions belong to *C. frutescens* and *C. annuum*.

In the characterized accessions, the following high variation coefficients were registered for the fruit characteristics:

- anthocyanin spots or streaks on the fruit,
- colour of the fruit in the intermediate state,
- colour of the fruit in the mature state,
- the shape of the fruit,
- the shape of the fruit at the attachment of the pedicel,
- fruit apex shape,
- fruit appendix,
- traces of petals and anthers,
- fruit transverse ridges,
- type of fruit epidermis,
- pedicel persistence with fruit,
- pedicel persistence with stem,
- seed colour
- seed surface.

These characters indicate the importance of the descriptor to discriminate variability within a collection. The results corroborate those reported by Smith and Heiser [40,44] who maintain that in each species of chilli pepper there are various fruit shapes and colour of immature fruits. In the accessions evaluated, the shape of the fruit was mostly triangular and elongated; however, small-fruited species tended to be round and conical, especially sweet pepper species. The variability of the genus is mainly due to the characteristics of the fruit, followed by the architecture of the plant, flower structure and the number of flowers per leaf axil [16].

Pickersgill [45] mentions that in *Capsicum*, the annular constriction of the calyx is characteristic of *C. chinense* and is absent in the other four species. For the colour of the immature fruits, it is typical for the fruits among the *Capsicum* spp. to start with a green colour before reaching the final colour at full maturity; however, the fruits in a mature state have mostly red tones and an elongated shape, while *C. chinense* matures with fruit shades of yellow and orange. Yellow colours of fruits in the intermediate maturity stage, i.e., apparent from the results obtained, were recorded in some accessions of both

C. chinense (7300, 12,154, 5489), and *C. frutescens* (10,946, 10,793, 165,654). Furthermore, the shape of the fruit apex was mostly pointed in the accessions studied and the epidermis of the fruit was smooth, i.e., characteristics that correspond to the *C. frutescens* and *C. annuum* species.

3.6. Grouping of Accessions Based on Morphological Variables

The hierarchical cluster, using the qualitative and quantitative variables identified significant differences among nine accession groups of *Capsicum* spp. Significance ($p < 0.0001$) was obtained employing a multivariate analysis and the differences between mean vectors were obtained using the Hotelling comparison test corrected by Bonferroni [46,47].

3.7. Combined Analysis of Qualitative and Quantitative Variables

The result of the grouping of the accessions—obtained with the Ward method and the Gower distance—allowed us to identify the taxonomic structure of the collection, where the relationship between the groupings could be seen, i.e., *C. annuum* (groups 7, 8, 5 and 3), *C. frutescens* (groups 9, 6, 4 and 2). However, within these groups there are also accessions of *C. chinense* and *C. pubescens* and group 1 formed by the *C. baccatum* species. Group 3 contains the highest number of accessions (36); while Group 1 is made up of nine accessions. The groups with the highest similarity for the qualitative and quantitative variables are Group 1 and Group 2 (Table 4, Figure 1).

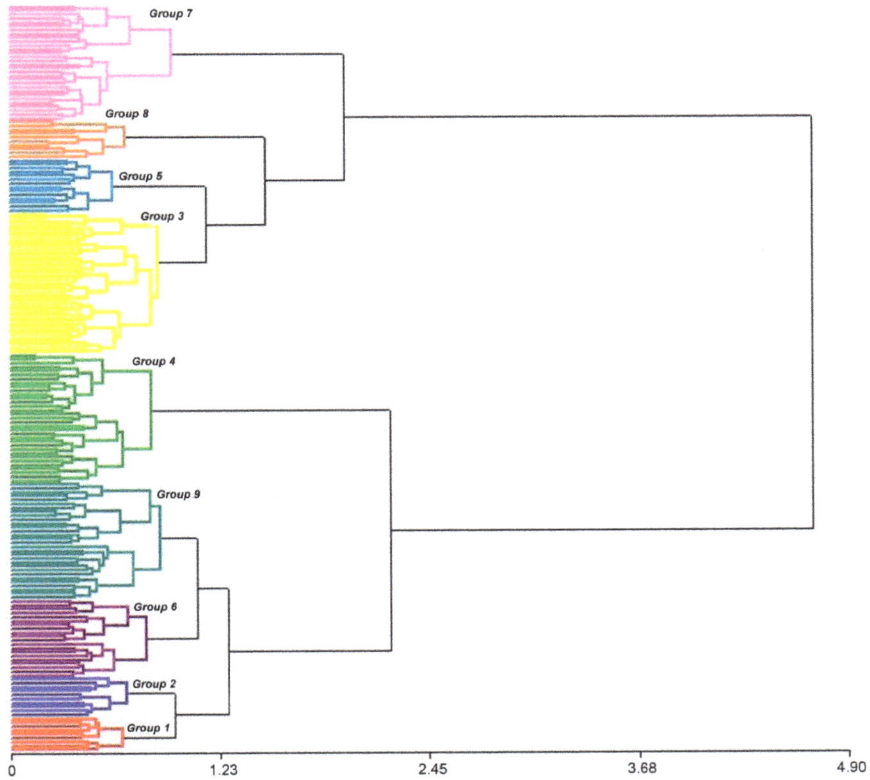

Figure 1. Dendrogram obtained by hierarchical cluster analysis (Ward's method, Gower's distance) with qualitative and quantitative variables from the morphological characterization of 192 accessions of chilli (*Capsicum* spp.).

Table 4. Description of the 9 groups of accessions from Figure 1, including species identification.

Group 1	5440 [b], 7203 [b], 7209 [b], 7417 [b], 16,463 [b], 16,209 [b], 17,268 [s], 18,060 [a], 18,645 [b]
Group 2	7300 [c], 8386 [f], 9892 [f], 10,757 [c], 10,792 [f], 10,793 [f], 14,757 [a], 16,308 [p], 22,115 [s], 22,119 [s]
Group 3	5445 [b], 6123 [a], 7816 [a], 8058 [s], 9135 [a], 9139 [a], 9186 [a], 10,886 [s], 11,757 [f], 12,911 [f], 14,376 [a], 15,239 [a], 15,407 [a], 15,412 [a], 15,422 [a], 15,449 [a], 15,587 [a], 15,632 [a], 15,646 [a], 15,653 [a], 15,661 [a], 15,983 [a], 16,270 [a], 16,297 [a], 16,304 [a], 16,458 [a], 16,462 [a], 16,467 [a], 17,151 [a], 17,294 [a], 18,314 [a], 18,631 [a], 18,651 [a], 18,660 [a], 18,757 [a], 19,259 [a]
Group 4	8248 [f], 8394 [a], 8998 [s], 8999 [f], 9015 [f], 9140 [a], 9159 [f], 9201 [a], 9777 [f], 9781 [f], 9801 [f], 9811 [f], 9832 [f], 9837 [f], 9902 [f], 9923 [f], 10,003 [c], 10,730 [f], 10,762 [f], 10,862 [f], 10,909 [f], 10,951 [f], 11,073 [f], 11,198 [f], 11,744 [f], 11,745 [f], 12,017 [f], 12,097 [f], 12,154 [c], 12,156 [c], 12,910 [b], 13,328 [s]
Group 5	9038 [a], 9096 [f], 9115 [a], 9183 [a], 9226 [a], 11,303 [a], 15,395 [f], 16,450 [a], 16,452 [a], 16,453 [s], 16,454 [a], 16,456 [a], 16,457 [a], 16,460 [a]
Group 6	6126 [f], 7216 [s], 8534 [s], 8567 [a], 9095 [f], 9200 [f], 9204 [f], 10,760 [f], 10,871 [f], 10,903 [f], 10,946 [f], 11,050 [f], 11,717 [s], 15,237 [a], 15,654 [f], 15,932 [a], 16,273 [a], 16,308 [a], 16,513 [f], 18,229 [f]
Group 7	6143 [a], 7818 [a], 7819 [a], 8047 [a], 8055 [s], 8064 [a], 9053 [f], 9110 [b], 9131 [a], 10,630 [a], 10,691 [a], 11,204 [s], 11,232 [a], 11,305 [a], 11,795 [a], 13,963 [a], 14,751 [a], 14,756 [a], 15,389 [s], 15,434 [f], 15,440 [a], 15,640 [a], 15,641 [a], 15,651 [a], 15,658 [a], 15,976 [s], 16,276 [a], 16,451 [s], 16,521 [a], 20,029 [a]
Group 8	9269 [a], 16,459 [a], 16,461 [a], 17,867 [a], 18,156 [a], 18,776 [a], 18,787 [a], 18,788 [a], 18,804 [a], 18,815 [a]
Group 9	6586 [f], 7218 [c], 7257 [a], 8395 [c], 8994 [s], 9016 [f], 9037 [f], 9040 [a], 9043 [f], 9079 [f], 9097 [f], 9103 [f], 9803 [f], 9835 [f], 9839 [f], 9841 [f], 9916 [f], 9917 [f], 9921 [f], 10,005 [f], 10,015 [f], 11,755 [f], 14,776 [f], 15,914 [f], 16,275 [a], 16,280 [s], 17,247 [a], 17,750 [c], 18,778 [a], 20,016 [s]

[a] *C. annuum*; [b] *C. baccatum*; [c] *C. chinense*; [f] *C. frutescens*; [p] *C. pubescens*; [s] *Capsicum* spp.

3.8. Discriminant Values

Regarding the qualitative descriptors, from the 39 characters analysed using the X^2 test, 30 of them were identified with high significance ($p < 0.0001$) (**), and nine were not significant (ns). These results indicated the presence of a large number of descriptors making an essential contribution to separate the nine genetic groups (Table 5), moreover they presented high association coefficients. Likewise, 15 characters with the highest discriminant value were recorded, which can be used to establish genetic differences between groups.

The colour of the corolla and the colour of the anthers were the characters with the highest discriminant value (325.34 and 323.7 respectively) and presented the highest association coefficients. The position of the flower presented an X^2 value of 166.04 and the highest value according to the Cramer test (0.54); therefore, it has a high contribution to discriminate between genetic groups, as has the colour of the corolla that also provides a discriminating value. These results indicate that the G4 and G9 groups are associated with the white character of the corolla.

Regarding the quantitative descriptors, six were identified with the highest discriminant value: leaf length/width ratio, width mature leaf, fruit length, fruit width, fruit wall thickness, plant height; these descriptors allowed to differentiate the nine groups (Table 6). In addition, we determined that accessions within the groups maintain a close relationship, once there is not much variation since they present small values of standard deviation.

Table 5. Qualitative descriptors with the highest discriminant value influencing the genetic groups' separation of *Capsicum* species.

Character	X^2	Coeficient (P)	Cramer (V)
Colour of the corolla	325.34 **	0.79	0.58
Colour of the anthers	323.70 **	0.79	0.65
Position of the flower	166.04 **	0.68	0.54
Leaf distance	144.82 **	0.66	0.50
Branching distance	141.30 **	0.65	0.50
Susceptibility to biological stress	129.04 **	0.63	0.37
Corollar spot colour	126.36 **	0.63	0.47
Tillering	123.08 **	0.63	0.46
Anthocyanin of the node	106.07 **	0.60	0.37
Cross wrinkling of the fruit	102.59 **	0.59	0.42
Fruit shape at the junction with the pedicel	97.38 **	0.58	0.32
Stem pubescence	95.95 **	0.58	0.41
Leaf pubescence	86.83 **	0.56	0.39
Persistence of the pedicel with the stem	82.99 **	0.55	0.38
Fruit colour in the intermediate state	75.09 **	0.53	0.28
Type of epidermis of the fruit	73.85 **	0.53	0.36
Persistence of the pedicel with the fruit	70.41 **	0.52	0.35
Fruit shape	69.66 **	0.52	0.27
Shape of cotyledonal leaf	66.37 **	0.51	0.34
Colour of the hypocotyl	64.24 **	0.50	0.33
Shape of fruit tip	63.75 **	0.50	0.29
Leaf margin	63.14 **	0.50	0.41
Growth habit	62.76 **	0.50	0.33
Pubescence of hypocotyl	62.48 **	0.50	0.33
Leaf colour	59.32 **	0.49	0.32
Ovary length	51.49 **	0.46	0.30
Seed size	38.59 **	0.41	0.32
Neck at the base of the fruit	33.99 **	0.39	0.30
Cotyledonal leaf colour	31.91 **	0.38	0.24
Anthocyanin stains or streaks of the fruits	27.14 **	0.35	0.27

** = high significance with ($p < 0.0001$).

Table 6. Eigenvalues determined by the canonical discriminant function discriminating grouping of *Capsicum* spp. accessions.

Variables	Axis 1	Axis 2
Leaf length/width ratio	2.35	−0.81
Width mature leaf	0.81	0.09
Fruit length	0.36	−0.08
Fruit width	0.23	0.82
Fruit wall thickness	0.09	−0.21
Plant height	0.03	−0.04
Days to flowering	0.01	−0.01
Fruit weight	1.80×10^{-3}	-2.20×10^{-3}
Plant width	−0.02	0.06
Stem length	−0.02	0.02
Days to fruiting	−0.02	−0.02
No. flowers leaf axil-1	−0.51	−0.38
Fruit pedicel length	−0.55	0.45
Weight of 1000 seeds	−0.60	0.19
Length mature leaf	−0.73	−0.10
Plant height/width radio	−1.07	0.47
Stem width	−1.22	0.60

The discriminant analysis found less distance between the species of *C. annuum*, *C. frutescens* and *C. chinense*. These three taxa are separated from *C. baccatum* and *C. pubescens*, because these two species grow in highlands (2800 m a.s.l.), where the climatic conditions differ from the low altitudes (200 m a.s.l.) where *C. pubescens* is distributed through the middle region of the Andes mountain range (1300 m a.s.l.). *Capsicum baccatum* is widely distributed throughout the lowlands of South America, as mentioned by Pickersgill [43]. The studies by García [48] corroborate the previous results; García [48] points out that the morphological characterization did not allow the species of *C. annuum*, *C. chinense* and *C. frutescens* to be differentiated. These observations are in agreement with Pardey et al. [18] who concluded that the species *C. annuum*, *C. chinense* and *C. frutescens* make up the same morphological group; like Vallejo et al. [49] who managed to discriminate the *C. pubescens* and *C. baccatum* species, but were unable to discriminate *C. annuum*, *C. frutescens*, *C. chinense*.

The *Capsicum* population presented morphological variation in the qualitative characteristics of the *C. annuum*, *C. chinense* and *C. frutescens* species because of the shared morphological features among the three species, making taxonomic classification difficult. This agrees with the results of García [48] and Palacios [50] who confirm this intraspecific variability; they also mention that as a result of the morphological description, it could be assumed that the three taxa constitute the same group. Similarly, the studies by Vallejo et al. [49], and Palacios Castro and García [51] managed to discriminate the species of *C. pubescens* and *C. baccatum*, but not between the species *C. annuum*, *C. frutescens* and *C. chinense*. The results of this study continue to corroborate the hypothesis that these three species are a large group in the process of differentiation, which is consistent with the studies conducted by Pickersgill [52].

The results of the present study are in agreement with those by Chávez-Servia [53] and Chávez-Servia and Castillo [54], who reported that variables such as length, width and shape of the *Capsicum* fruit showed considerable genetic variation. The purple colour of the anther was observed in the *C. chinense* species; while *C. baccatum* anthers presented a yellow colouration. On the other hand, *C. annuum* had pale blue anthers and *C. frutescens* blue. The species tended to have no stain on the corolla, except for *C. baccatum*, which is the typical characteristic of this species. The flowers in *C. frutescens* are erect; while in *C. annuum* and *C. chinense* the position of the flowers varied between intermediate and/or hanging.

According to Martín and González [55] and Fernández [56], chilli peppers with large-sized fruits and a thick epidermis tend to be less pungent. In contrast, in the accessions with smaller fruits where also the epidermis is thinner, the concentration of capsaicinoids increases, which is consistent with the results of the investigation, here accession 16,209 presented higher capsaicin content (5590 ug/g), while accession 16,450 registered low capsaicin content (200 ug/g).

3.9. Participatory Sensory Evaluation

The 50 people attending the workshop represented the following categories: producers (30), industry (5), chefs (4), scientists (5) and people who like to consume chilli (6). In the first phase, 134 accessions were selected as 'Very Good', presenting characteristics such as fruits with characteristics such as colour (pale orange, red, dark red and orange), shape (bell-shaped, triangular, elongated, bell-shaped and thick), size (medium and large) and fruit production. In the second phase, participants chose 64 accessions using the shape, colour and size of the fruit as selection criteria. In the third phase, 34 samples were selected with the criteria: taste, odour, texture, pulp thickness and size (Appendix B, Table A1); and, in the fourth phase, the most relevant fruits with orange, pale orange and red epidermis colours were selected. The shapes of the fruit are bell-shaped, and triangular, or bell-shaped and thick; the epidermis are of the fruit is semi-wrinkled and rough and fruits of medium to large size.

At the end of the process, the samples of the accessions were rated as follows:

- As excellent—accession 15,661 (dark red fruit colour, bell-shaped and thick shape, large size)
- As very good—accession 7818 (dark red fruit colour, triangular shape, medium size)

- As good—accessions 16,304 (red fruit colour, flared shape, large size), 10,757 and 22,119 (red fruit colour, flared shape and compact, small size), 9892 (red fruit colour, flared shape, medium size), 9916 and 17,750 (orange fruit colour, bell-shaped, large size), 8994 and 16,209 (red fruit colour, triangular shape, medium size), 17,268 and 9902 (red fruit colour, bell-shaped, medium size).

3.10. Chemical Characterization

The nutritional value was determined in 15 accessions, selected as promising in morphological characterization, corresponding to the taxa *C. annuum*, *C. chinense*, *C. frutescens*, *C. baccatum* and *Capsicum* spp. Furthermore, these accessions were rated as excellent in the participatory selection. The accessions that had the highest concentration of capsaicin (5.59 mg/g), polyphenol (18.68 mg/g) and flavonoid (16.64 mg/g) were 16,209, 17,750 and 10,757, respectively (Table 7). The content of total polyphenols, flavonoids and capsaicinoids varied in the accessions evaluated, with a tendency to present an association with the morphological classification described by Morán et al. [57]. Appedino [58] studied 13 cultivars of *C. annuum*, finding concentration levels of flavonoids (0.028 and 0.551 mg/g) lower than those found in the accessions in the present study.

Table 7. Flavonoid, polyphenol and capsaicin concentrations in seven accessions representing four species of *Capsicum* in the CATIE genebank.

Species	Accession No.	Flavonoid Concentration (mg/g)	Polyphenol Concentration (mg/g)	Capsaicin Concentration (mg/g)
C. chinense	10,757	16.64 [fh]	11.41	4.88
	15,661	15.77	10.69	0.32
C. annuum	7818	11.41	7.15	2.64
	14,757	11.90	6.91	1.37
	16,304	10.74	5.97	0.99
	16,450	8.94 [fl]	4.52 [pl]	0.20 [cl]
	16,454	11.68	7.83	1.34
	16,457	12.41	7.98	1.65
	16,462	11.48	6.79	1.10
	17,750	13.59	11.68 [ph]	4.60
C. frutescens	7816	12.05	7.46	4.61
	9892	13.30	8.56	3.14
	9902	11.09	5.12	2.80
C. baccatum	16,209	10.30	6.73	5.59 [ch]
Capsicum spp.	11,204	10.04	5.45	2.02

[fh]—highest flavonoid concentration; [fl]—lowest flavonoid concentration; [ph]—highest polyphenol concentration; [pl]—lowest polyphenol concentration; [ch]—highest capsaicin concentration; [cl]—lowest capsaicin concentration.

The results in Table 7 coincide with those found by Cázares et al. [59], who report that the populations of Ma'x ik and Sukurre belonging to the *C. chinense* species presented the highest capsaicin values (2.93 and 4.35 mg/g); while the lowest values reported for *C. annuum* sweet pepper populations (0.20 mg/g).

Additionally, Antonious and Jarret [60] studied different species of *Capsicum*, finding low concentration levels of capsaicin (0.0009 to 0.002 mg/g). Estrada et al. [61] reported increasing levels of capsaicinoids as maturation progresses, finding total capsaicinoid concentrations of between 0.15 to 0.70 mg/g (ps). However, these values are lower than those reported in the accessions studied here, where values between 5.59 to 0.20 mg/g were found (Table 8).

Table 8. Average, standard deviation, coefficient of variation, minimum and maximum values for the characteristics of the nutritional value of 15 accessions of chilli (*Capsicum* spp.).

Variable	Average	Standard Deviation	Coefficient of Variation	Minimum Values	Maximum Values
Humidity %	6.48	2.87	44.32	2.77	13.75
Polyphenols mg/g	7.62	2.18	28.69	4.52	11.68
Flavonoids mg/g	12.09	2.06	17.02	8.94	16.64
Capsaicinoids mg/g	2.48	1.74	70.12	0.20	5.59
Antioxidants mg/g	68.66	13.69	19.93	47.84	90.85

The Biplot from the principal component analysis of the 15 selected accessions shows the first two principal components explaining variance greater than 92% (Figure 2). The contents of polyphenols, antioxidants and flavonoids are highly correlated with each other and these in turn are correlated with humidity. In the case of capsaicin, its content is independent of the amounts of polyphenols, antioxidants and flavonoids.

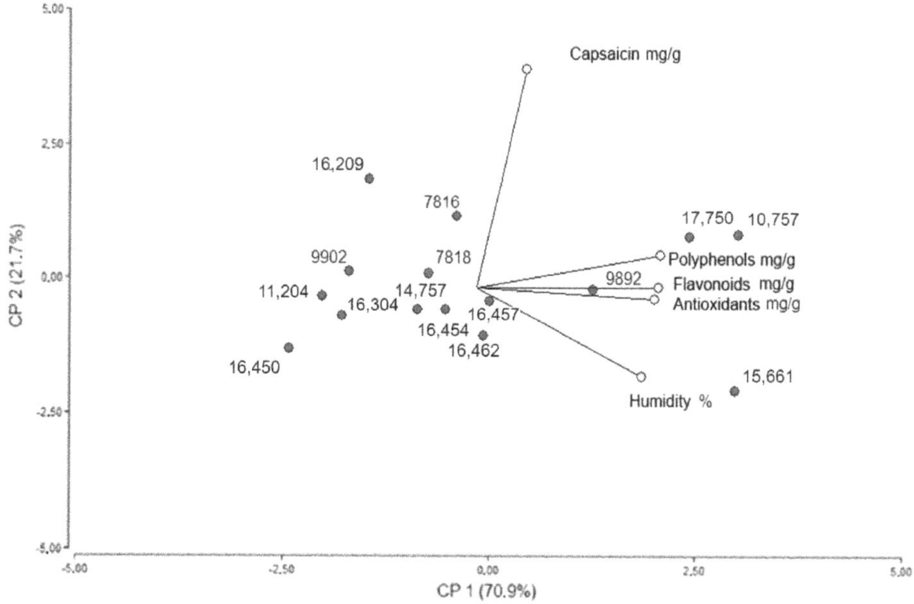

Figure 2. Biplot graph obtained from the principal component analysis of the 15 Capsicum accessions.

Accessions 17,750, 10,757, 9892 and 15,661 are those with the highest amount of polyphenols, antioxidants and flavonoids, in turn these four accessions together with accessions 16,209 and 17,262 were the ones that obtained the highest score in the selection made by the producers. Accessions 17,750, 10,757 contain the most capsaicin content, being only surpassed by accession 16,209. On the opposite side, accession 16,450 is the one with the least content of nutritional values and in turn has a low capsaicin content

Of the four selected accessions with the best nutritional content, two of them belong to the genus *C. chinense* (17,750 and 10,757) and the accession (15,661) belongs to the genus *C. annuum*, accession 9892 belongs to *C. frutescens*. Regarding the classification by conglomerates, these four accessions belong to groups G2, G3 and G9. The highest capsaicin content was recorded in accession 16,209 belonging to the species *C. baccatum*.

4. Discussion and Conclusions

The most discriminant qualitative characteristics were colour of the corolla, the colour of the anthers and position of the flower; while the most discriminant quantitative characteristics were leaf length/width ratio, width mature leaf, fruit length, fruit width, fruit wall thickness and plant height. This is similar to previous studies by Medina et al. [62], Pardey et al. [18] Ortiz et al. [20] and Castañón et al. [63] who have also identified colour of the corolla, corollar spot colour, fruit width and fruit length as discriminant variables within other *Capsicum* spp. collections. A practical morphological characteristic found in our *C. annuum* and *C. frutescens* accessions is the persistence of the pedicel with the fruit, According to Sreelathakumary et al. [41] this characteristic, along with the length of the placenta, is correlated with the mass of the fruits and, therefore, longer shelf life.

The agronomic characterization allowed classifying the genetic variability of the *Capsicum* germplasm collection into two large groups and nine subgroups. The two large groups are formed by the subgroups G7, G8, G5 and G3 represented by *C. annuum*, and the second group formed by subgroups G4, G9, G6, G2 having mostly *C. frutescens* accessions plus G1, represented mainly by *C. baccatum*. Subgroup 1 (G1) differentiates from the other subgroups because of the presence of spots on the corolla (Corollar spot colour) similarly observed by García [48], Palacios & García [51]. and Walsh & Hoot [64]. It is important to notice that within all subgroups (except G8) there are few intercalated accessions of *C. baccatum*, *C. pubescens*, *C. chinense* and *C.* spp. (Table 4). *Capsicum* phylogeny determined closer relation among *C. annuum*, *C. frutescens* and *C. chinense* which is known as *C. annuum* complex Pickersgill, [43], Vallejo et al. [49], Pardey et al. [18] and Palacios and García [51].

The description of CATIE's *Capsicum* spp. international collection permitted to identify promising materials e.g., after the morphological and participatory characterization, the following accessions were identified as promising: *C. chinense*—10,757 and 17,750; *C. frutescens*—9892; *C. annuum*—15,661 and 7816; and *C. baccatum*—16,209. Also, four accessions were selected with the highest chemical concentration (polyphenols, flavonoids, and capsaicinoids), two of them belong to the species *C. chinense* (17,750 and 10,757), the accession (15,661) belongs to *C. annuum*, and accession 9892 is referred to *C. frutescens*. These selected materials or the collection as a whole could be used by interested scientist as well as farmers.

Author Contributions: N.J.P.A. was responsible for the conceptualization, data curation, analyses and methodology, A.M.-A. and C.G.T.B. contributed further analyses and writing, M.S. contributed to the translation, writing and editing. All authors have read and agreed to the published version of the manuscript.

Funding: This research was funded exclusively by Instituto National de Investigaciones Agropecuarias INIAP [the Ecuadorian National Institute of Agricultural Research].

Acknowledgments: The authors express their gratitude to the Tropical Agricultural Research and Training Center CATIE, for allowing the research to be completed at the Genebank and, in turn, to the entire technical, field and teaching team that supported the execution. The National Institute of Agricultural Research (INIAP) is gratefully acknowledged for providing the funding to make the research possible.

Conflicts of Interest: The authors declare no conflict of interest.

Appendix A

Resumen

Con el objetivo de evaluar las potencialidades de 192 accesiones de *Capsicum* L., procedentes de 21 países, se realizó la caracterización morfológica y agronómica mediante la aplicación de 57 descriptores cualitativas y cuantitativas. Los análisis multivariados identificaron dos grandes grupos: *C. annuum* (G3, G5, G7 y G8) y *C. frutescens*, *C. baccatum*, *C. chinense* y *C. pubescens* (G1, G2, G4, G6 y G9). Los descriptores cualitativos discriminantes fueron color de la corola, color de las anteras y posición de la flor. Las características cuantitativas discriminante fueron longitud, peso y ancho del fruto. La selección participativa identificó 15 materiales por color, olor, textura, sabor, tamaño y grosor de frutos. Los análisis químicos determinaron la mayor concentración de flavonoides en las accesiones

10,757 (16.64 mg/g) y 15,661 (15.77 mg/g). Los contenidos de polifenoles más alto presentaron las accesiones 17,750 (11.68 mg/g) y 10,757 (11.41 mg/g). La mayor concentración de capsaicina se presentó en la accesión 16,209 (55.90 mg/g) y 10,757 (48.80 mg/g). El valor más alto de antioxidante se registró en las accesiones 17,750 (90.85 mg/g) y 15,661 (87.03 mg/g). Todas estas son características importantes con vistas a incrementar el uso industrial y procesos de mejora genética. Los resultados muestran la existencia de una variabilidad genética significativa en *Capsicum*.

Palabras claves: germoplasma; recursos genéticos; accesiones; descriptores.

Appendix B

Table A1. Results of the participatory sensory evaluation for 34 selected accessions of *Capsicum* spp. CATIE. Costa Rica.

Accession	Taste *	Odour *	Texture *	Pulp Thickness *	Colour *	Size *	Total
15,661	5	5	5	5	5	5	30
7818	4	4	4	4	4	4	24
9139	3	3	4	5	5	-- **	20
16,460	3	4	3	5	5	--	20
17,151	4	3	5	4	4	--	20
10,886	4	4	4	4	4	--	20
12,911	3	4	4	4	5	--	20
14,757	3	3	4	5	5	--	20
9916b	3	3	3	3	3	3	18
16,462	3	4	3	4	4	--	18
16,454	4	4	3	4	3	--	18
16,458	4	4	4	2	4	--	18
8994	3	3	3	3	3	3	18
10,757	3	3	3	3	3	3	18
16,209	3	3	3	3	3	3	18
22,119	3	3	3	3	3	3	18
17,268	3	3	3	3	3	3	18
9892	3	3	3	3	3	3	18
17,750	3	3	3	3	3	3	18
16,304	3	3	3	3	3	3	18
9902	3	3	3	3	3	3	18
11,204	4	3	3	4	3	--	17
16,450	4	3	3	4	3	--	17
16,457	4	3	3	3	3	--	16
7816	--	3	--	3	4	--	10
9186	--	--	--	5	3	--	8
11,745	--	1	--	2	3	--	6
16,521	--	--	3	--	3	--	6
18,757	--	--	--	2	3	--	5
20,016	1	--	--	2	2	--	5
16,275	--	1	1	--	1	1	4
18,778	1	--	1	1	1	--	4
9183	--	--	1	--	2	--	3
7819	--	--	--	1	--	--	1
Total							510

* Excellent 5; Very Good 4; Good 3; Regular 2; Poor 1; ** Variables that were not of interest.

References

1. Alberto, R.A. *Planeación de Cultivos Hortícolas. Basada en la Estacionalidad de Precios;* Ministerio de Agricultura y Ganadería: San Salvador, El Salvador, 2000; p. 55.
2. Bolaños, A. *Introducción a la Olericultura;* UNED: San José, Costa Rica, 2001; p. 380, ISBN 978-9977-64-967-2.
3. Molina Moreno, J.C.; Córdova Téllez, L. (Eds.) *Recursos Fitogenéticos de México para la Alimentación y la Agricultura: Informe Nacional 2006;* Secretaría de Agricultura, Ganadería, Desarrollo Rural, Pesca y

Alimentación; Sociedad Mexicana de Fitogenética: Chapingo, Mexico, 2006; p. 172. Available online: http://www.fao.org/3/i1500e/Mexico.pdf (accessed on 19 November 2019).
4. Nuez, F.; Gil, R.; Costa, J. *El Cultivo de Pimientos, Chiles y Ajíes*; Mundiprensa: Madrid, Spain, 1996; p. 607, ISBN 8471146096.
5. Ibiza, V.P.; Blanca, J.M.; Cañizares, J.; Nuez, F. Taxonomy and genetic diversity of domesticated *Capsicum* species in the Andean region. *Genet. Resour. Crop Ev.* **2012**, *59*, 1077–1088. [CrossRef]
6. Milla, A. *Capsicum* de capsa, cápsula el pimiento. In *Pimientos: Compendios de Horticultura. Capitulo 2*; Universidad de Rioja: La Rioja, Spain, 2006; pp. 21–31. Available online: http://www.horticom.com/tematicas/pimientos/pdf/ (accessed on 28 October 2018).
7. The Plant List Homepage. Available online: http://www.theplantlist.org/tpl1.1/record/kew-2698515 (accessed on 19 November 2019).
8. Pickersgill, B. The domestication of chile peppers. In *The Domestication and Exploitation of Plants and Animals*; Ucko, P.J., Dimbleby, G.W., Eds.; Duckworth and Co.: London, UK, 1969; pp. 443–450. [CrossRef]
9. Pickersgill, B. Domestication of plants in the Americas: Insights from Mendelian and molecular genetics. *Ann. Bot.* **2007**, *100*, 925–940. [CrossRef] [PubMed]
10. Perry, L.; Dickau, R.; Zarillo, S.; Holst, I.; Pearsall, D.; Piperno, D.; Berman, M.J.; Cooke, R.; Rademarker, K.; Ranere, A.; et al. Starch fossils and the domestication and dispersal of chilli peppers (*Capsicum* spp. L). in the Americas. *Science* **2007**, *315*, 986–988. [CrossRef] [PubMed]
11. Hernández-Verduga, S.; Dávila Aranda, P.; Oyama, K. Síntesis del conocimiento taxonómico, origen y domesticación del género *Capsicum*. *Boletín de la Soc. Botánica de México* **1999**, *64*, 65–84. [CrossRef]
12. Justino, E.V.; Fonseca, M.E.N.; Ferreira, M.E.; Boiteux, L.S.; Silva, P.P.; Nascimento, W.M. Estimate of natural cross-pollination rate of *Capsicum annuum* using a codominant molecular marker associated with fruit pungency. *Genet. Mol. Res.* **2018**, *17*. [CrossRef]
13. Melgarejo, L.; Hernández, M.; Barrera, J.; Bardales, X. *Caracterización y Usos Potenciales del Banco de Germoplasma de Ají Amazónico*; Universidad Nacional de Colombia: Bogotá, Colombia, 2004; p. 107, ISBN 958-97420-4-1.
14. Chávez-Servia, J.L.; Sevilla-Panizo, R. (Eds.) *Seminario: Fundamentos Genéticos y Socioeconómicos para Analizar la Agrobiodiversidad en la Región de Ucayali, 16 de enero de 2003 Pucallpa, Perú*; Bioversity International: Cali, Colombia, 2006; p. 93, ISBN 92-9043-738-3.
15. Hill, T.A.; Ashrafi, H.; Reyes-Chin-Wo, S.; Yao, J.Q.; Stoffel, K.; Truco, M.-J.; Kozik, A.; Michelmore, R.W.; Van Deynze, A. Characterization of *Capsicum annuum* genetic diversity and population structure based on parallel polymorphism discovery with a 30k unigene pepper genechip. *PLoS ONE* **2014**, *8*, e56200. [CrossRef] [PubMed]
16. Sood, S.; Kumar, N. Morphological studies of bell pepper germplasm. *Int. J. Veg. Sci.* **2011**, *17*, 144–156. [CrossRef]
17. Datta, S.; Das, L. Characterization and genetic variabilidty analysis in *Capsicum annuum* L. germplasm. *SAARC J. Agr.* **2013**, *11*, 91–103. [CrossRef]
18. Pardey, R.C.; García, D.M.A.; Vallejo, C.F.A. Caracterización morfológica de cien introducciones de *Capsicum* del Banco de Germoplasma de la Universidad Nacional de Colombia sede Palmira. *Acta Agron.* **2006**, *55*, 1–10. [CrossRef]
19. Pardey-Rodriguez, C. Caracterización y Evaluación de Accesiones de *Capsicum* del Banco de Germoplasma de la Universidad Nacional de Colombia Sede Palmira y Determinación del Modo de Herencia de la Resistencia a Potyvirus (PepDMV). Ph.D. Thesis, Universidad Nacional de Colombia Sede Palmira, Palmira, Colombia, 2008; p. 118.
20. Ortiz, R.; Delgado de la Flor, F.; Alvarado, G.; Crossa, J. Classifying vegetable genetic resources. A case study with domesticated *Capsicum* spp. *Sci. Hortic.* **2010**, *126*, 186–191. [CrossRef]
21. Olatunji, T.L.; Afolayan, A.J. Evaluation of genetic relationship among varieties of *Capsicum annuum* L. and *Capsicum frutescens* L. in West Africa using ISSR markers. *Heliyon* **2019**, *5*, e01700. [CrossRef] [PubMed]
22. Olatunji, T.L.; Afolayan, A.J. Comparison of nutritional, antioxidant vitamins and capsaicin contents in *Capsicum annuum* and *C. frutescens*. *Int. J. Veg. Sci.* **2020**, *26*, 190–207. [CrossRef]
23. Egea-Gilabert, C.; Bilotti, G.; Requena, M.E.; Ezziyyani, M.; Vivo-Molina, J.; Candela, M.E. Pepper morphological traits related with resistance to *Phytophthora capsici*. Brief communication. *Biol. Plant.* **2008**, *52*, 105–109. [CrossRef]

24. Ballina-Gomez, H.; Latournerie-Moreno, L.; Ruiz-Sánchez, E.; Pérez-Gutiérrez, A.; Rosado-Lugo, G. Morphological characterization of *Capsicum annuum* L. accessions from southern Mexico and their response to the *Bemisia tabaci-Begomovirus* complex. *Chil. J. Agr. Res.* **2013**, *73*, 329–338. [CrossRef]
25. Yumnam, J.; Tyagi, W.; Pandey, A.; Meetei, T.; Rai, M. Evaluation of genetic diversity of chilli landraces from North Eastern India based on morphology, ssr markers and the pun 1 locus. *Plant Mol. Biol. Rep.* **2012**, *30*, 1470–1479. [CrossRef]
26. CATIE. *Datos Meteorológicos (en Línea)*; CATIE: San José, Costa Rica, 2009; Available online: http://www.catie.ac.cr (accessed on 26 October 2011).
27. IPGRI; AVRDC; CATIE. *Descriptors for Capsicum (Capsicum Spp.)*; Bioversity: Rome, Italy, 1995; p. 114. Available online: https://www.bioversityinternational.org/fileadmin/user_upload/online_library/publications/pdfs/345.pdf (accessed on 26 October 2011).
28. Balzarini, M.; Di Rienzo, J.A.; Tablada, M.; González, L.; Bruno, C.; Córdoba, M.; Robledo, W.; Casanoves, F. *Introducción a la Bioestadística: Aplicaciones con InfoStat en Agronomía*; Brujas: Córdoba, Argentina, 2011; p. 383, ISBN 9875912514, 9789875912519.
29. Lawless, H.T.; Heymann, H. *Sensory Evaluation of Food: Principles and Practices*, 2nd ed.; Food Science Texts Series; Springer: New York, NY, USA, 2010; p. 596. [CrossRef]
30. Hernández, E. *Evaluación Sensorial*; Centro Nacional de Medios para el Aprendizaje: Bogotá, Colombia, 2005; p. 128. Available online: https://www.academia.edu/22625186/EVALUACION_SENSORIALfile:///C:/Users/zdt536/AppData/Local/Temp/EVALUACION_SENSORIAL.pdf (accessed on 28 October 2011).
31. Zewdie, Y.; Bosland, P.W. Pungency of chile (*Capsicum annuum* L.) fruit is affected by node position. *HortScience* **2000**, 3, 1174. [CrossRef]
32. Kirschbaum-Titze, P.; Mueller-Seitz, E.; Petz, M. Pungency in paprika (*Capsicum annuum*). 2. Heterogeneity of capsaicinoid content in individual fruits from one plant. *J. Agric. Food Chem.* **2002**, *50*, 1264–1266. [CrossRef]
33. Mueller-Seitz, E.; Hiepler, C.; Petz, M. Chili Pepper Fruits: Content and Pattern of Capsaicinoids in Single Fruits of Different Ages. *J. Agric. Food Chem.* **2008**, *56*, 12114–12121. [CrossRef]
34. Miean, K.H.; Mohamed, S. Flavonoid (myricetin, quercetin, kaempferol, luteolin, and apigenin) content of edible tropical plants. *J. Agric. Food Chem.* **2001**, *49*, 3106–3112. [CrossRef]
35. Álvarez-Parrilla, E.; de la Rosa, L.A.; Amarowicz, R.; Shahidi, F. Antioxidant activity of fresh and processed jalapeño and serrano peppers. *J. Agric. Food Chem.* **2011**, *59*, 163–173. [CrossRef]
36. Blainski, A.; Lopes, G.C.; Palazzo de Melo, J.C. Application and analysis of the Folin Ciocalteu Method for the determination of the total phenolic content from *Limonium brasiliense* L. *Molecules* **2013**, *18*, 6852–6865. [CrossRef]
37. Juangsamoot, J.; Ruangviriyachai, C.; Techawongstien, S.; Chanthai, S. Determination of capsaicin and dihydrocapsaicin in some hot chilli varieties by RPHPLC-PDA after magnetic stirring extraction and clean up with C18 cartridge. *Int. Food Res. J.* **2012**, *19*, 1217–1226. [CrossRef]
38. Bhattacharya, A.; Chattopadhyay, A.; Mazumdar, D.; Chakravarty, A.; Pal, S. Antioxidant constituents and enzyme activities in chilli peppers. *Int. J. Veg. Sci.* **2010**, *16*, 201–211. [CrossRef]
39. Rodríguez, D.; Kimura, M. HarvestPlus for carotenoid análisis. In *Series: HarvestPlus Technical Monographs No. 2*; International Food Policy Research Institute (IFPRI), International Center for Tropical Agriculture (CIAT): Washington, DC, USA, 2004; p. 57.
40. Smith, P.G.; Heiser, C.B. Taxonomic and genetic studies of the cultivated peppers, *Capsicum annuum* L. and *Capsicum frutescens* L. *Am. J. Bot.* **1951**, *38*, 362–368. [CrossRef]
41. Sreelathakumary, I.; Rajamony, L. Correlation and path coefficient analysis for yield in hot chilli (*Capsicum chinense* Jacq.). *Capsicum Eggplant Newslett.* **2004**, *23*, 53–56.
42. Hernández, V.S.; Luna, R.R.; Sánchez, C.; González, R.A.; Rivera, B.R.R.; Guevara, G.R.; Sánchez, G.P.; Casas, A.; Oyama, K. Variación genética en la resistencia a virus en poblaciones silvestres de chile (*Capsicum annuum* L.). In Proceedings of the Memoria de la 1a Convención Mundial del Chile, León, Guanajuato, Mexico, 10–14 July 2004; pp. 447–453. Available online: http://ww2.oikos.unam.mx/CIEco/gencon/images/stories/pdf_7.pdf (accessed on 17 November 2011).
43. Andrews, J.P. *Peppers: The Domesticated Capsicums*, New edition; University of Texas Press: Austin, TX, USA, 1995; p. 186, ISBN 0292704674, 9780292704671.
44. Smith, P.G.; Heiser, C.B. Taxonomy of *Capsicum sinense* Jacq. and the geographic distribution of the cultivated *Capsicum* species. *Bull. Torrey Bot. Club* **1957**, *84*, 413–420. [CrossRef]

45. Pickersgill, B. Some aspects of interspecific hybridization in *Capsicum*. In Proceedings of the Eucarpia IVth Meeting on Genetics and Breeding of *Capsicum* and Eggplant, Wageningen, The Netherlands, 14–16 October 1980; p. 15-15C.
46. Di Rienzo, J.A.; Guzman, A.W.; Casanoves, F. A multiple comparisons method based on the distribution of the root node distance of a binary tree obtained by average linkage of the matrix of Euclidean distances between treatment means. *JABES* **2002**, *7*, 129–142. [CrossRef]
47. Valdano, S.; Di Rienzo, J. Discovering meaningful groups in hierarchical cluster analysis. An extension to the multivariate case of a multiple comparison method based on cluster analysis. *InterStat* **2007**. Available online: http://interstat.statjournals.net/YEAR/2007/abstracts/0704002.php (accessed on 17 November 2011).
48. García, M.A. Estudio de la Diversidad Genética de las Accesiones de *Capsicum* spp. del Banco de Germoplasma de la Universidad Nacional de Colombia. Ph.D. Thesis, Universidad Nacional de Colombia Sede Palmira, Palmira, Colombia, 2006; p. 102.
49. Vallejo, F.A.; García, M.; Duran, T.; Pardey, C. *Caracterización Morfológica de 195 Accesiones de Capsicum del Banco de Germoplasma de Universidad Nacional de Colombia Sede Palmira*; Universidad de Colombia: Palmira, Colombia, 2006; p. 260.
50. Palacios, S. Caracterización Morfológica de Accesiones de *Capsicum* spp. Master's Thesis, Universidad Nacional de Colombia Sede, Palmira, Colombia, 2007; p. 89.
51. Palacios Castro, S.; García Dávila, M.A. Caracterización morfológica de 93 accesiones de *Capsicum* spp. del banco de germoplasma de la Universidad Nacional de Colombia Sede Palmira. *Acta Agron.* **2008**, *57*, 247–252. [CrossRef]
52. Pickersgill, B. Genetic resources and breeding of *Capsicum* spp. *Euphytica* **1997**, *96*, 129–133. [CrossRef]
53. Chávez-Servia, J.L. Diversidad Morfológica e Isoenzimática del Chile Manzano (*Capsicum pubescens* R. y P.) en México. Ph.D. Thesis, Colegio de Postgraduados, Texcoco, Mexico, 1999; p. 138.
54. Chávez-Servia, J.L.; Castillo, G.F. Variabilidad en características morfológicas de colectas de chile manzano (*Caspicum pubescens* R. y P.). *Rev. Fitotec. Mex.* **1999**, *22*, 27–41.
55. Martín, N.C.; González, W.G. Caracterización de Accesiones de Chile (*Capsicum* spp.). *Agron. Mesoam.* **1991**, *2*, 31–39. [CrossRef]
56. Fernández, S.S. Caracterización Química y Agronómica Preliminar de 73 "Tipos" de Chile Picante (*Capsicum* spp.) de la Colección del CATIE. Licentiate Thesis, Tec. Alimentos, San José, Costa Rica, 1984; p. 70.
57. Morán Bañuelos, S.H.; Ramírez Vallejo, P.; Chávez Servia, J.L.; Rivero Borja, M. Diversidad isoenzimática en chiles de Yaxcabá, Yucatán. In Proceedings of the Contribución al XIX Congreso Nacional de Fitogenética, Saltillo Coahuila, Mexico, 22 November 2002. [CrossRef]
58. Appendino, G. Capsaicin and Capsaicinoids. In *Modern Alkaloids: Structure, Isolation, Synthesis and Biology*; Fattorusso, E., Taglialatela-Scafati, O., Eds.; Wiley-VCH Verlag GmbH & Co.: Weinheim, Germany, 2008; pp. 73–110, ISBN 978-3-527-31521-5.
59. Cázares, E.; Ramírez, P.; Castillo, F.; Soto, R.; Rodríguez, M.; Chávez, L. Capsaicinoids and preference of use in different morphotypes of chilli peppers (*Capsicum annuum* L.) of East-Central Yucatán. *Agrociencia* **2005**, *39*, 627–638.
60. Antonious, G.F.; Jarret, R.L. Screening *Capsicum* accessions for capsaicinoids content. *J. Environ. Sci. Health B* **2006**, *41*, 717–729. [CrossRef] [PubMed]
61. Estrada, B.; Bernal, M.A.; Federico, J.D.; Pomar, F.; Merino, F. Fruit development in *Capsicum annuum*: Changes in capsaicin, lignin, free phenolics, and peroxidase patterns. *J. Agric. Food Chem.* **2000**, *48*, 6234–6239. [CrossRef]
62. Medina, C.I.; Lobo, M.; Gómez, A.F. Variabilidad fenotípica en poblaciones de ají y pimentón de la colección colombiana del género *Capiscum*. *Cienc. Tecnol. Agropecu.* **2006**, *7*, 25–39. [CrossRef]

63. Castañón, N.G.; Latournerie, M.L.; Mendoza, E.M.; Vargas, L.A.; Cárdenas, M.H. Colección y caracterización de Chile (*Capsicum* spp.) en Tabasco, Mexico. *Phyton-Int. Exp. Bot.* **2008**, *77*, 189–202.
64. Walsh, B.M.; Hoot, S.B. Phylogenetic relationships of *Caspicum* (Solanaceae), using DNA sequences from two noncoding regions: The chloroplast atpB-rbcL spacer region and nuclear waxy introns. *Int. J. Plant. Sci.* **2001**, *162*, 1409–1418. [CrossRef]

agronomy

MDPI

Article

Natural Genetic Diversity of Nutritive Value Traits in the Genus *Cynodon*

Cleber de Souza [1], Yolanda Lopez [1], Patricio Munoz [2], William Anderson [3], Miguel Dall'Agnol [4], Marcelo Wallau [1] and Esteban Rios [1,*]

1 Agronomy Department, University of Florida, Gainesville, FL 32603, USA; c.lopesdesouza@ufl.edu (C.d.S.); ylopezb@ufl.edu (Y.L.); mwallau@ufl.edu (M.W.)
2 Horticultural Sciences Department, University of Florida, Gainesville, FL 32603, USA; p.munoz@ufl.edu
3 USDA-ARS, Crop Genetics and Breeding Research Unit, 115 Coastal Way, Tifton, GA 31793, USA; bill.anderson@usda.gov
4 Forage Department, Federal University of Rio Grande do Sul, Porto Alegre 90040-060, Brazil; migueld@ufrgs.br
* Correspondence: estebanrios@ufl.edu

Received: 9 October 2020; Accepted: 3 November 2020; Published: 6 November 2020

Abstract: The *Cynodon* spp. collection maintained by United States Department of Agriculture National Plant Germplasm System (USDA-NPGS) has limited information on nutritive value (NV) traits. In this study, crude protein (CP), phosphorous concentration (P), in vitro digestible organic matter (IVDOM), and neutral detergent fiber (NDF) were determined to (i) estimate genetic parameters for NV, (ii) obtain genetic values for the whole population across two harvests, (iii) estimate genotype by harvest interaction (GHI) for NV traits, and (iv) select accessions exhibiting improved NV traits compared to 'Tifton 85'. The experiment was setup as a row-column design with two replicates and augmented representation of controls: Tifton 85, 'Jiggs', and 'Coastal'. The whole-population was harvested twice, and data were analyzed using linear mixed models with repeated measures. In addition, a selected population of 15 genotypes were evaluated across 11 harvests to determine the extent of GHI. Genetic parameters revealed the presence of significant genetic variability, indicating potential improvements for NV through breeding. Specifically, P and IVDOM presented large variation, while NDF had lower diversity but some accessions exhibited lower NDF than Tifton 85. Low GHI, except for IVDOM, indicated genotypic stability and potential for selecting improved accessions under fewer harvests. Breeding line 240, PI-316510, and PI-3166536 presented superior NV than Tifton 85.

Keywords: bermudagrass; forage breeding; genetic parameters; genotype by harvest interaction; Tifton 85

1. Introduction

The investigation of natural genetic diversity present in germplasm banks is a key step to improve traits with narrow genetic variability. The use of germplasm banks is even more important for perennial forages, as several important releases worldwide originated from selections of plant introductions made in large collections [1]. Bermudagrass (*Cynodon* spp.) is the most widely used warm-season perennial forage for hay and pasture in the southeastern United States, covering more than 12 million hectares [2]. Its popularity among livestock and hay producers lays on high biomass production, nutritive value, animal performance, fast-curing for hay production, and drought tolerance [3–7]. The genus *Cynodon* is composed of genetically diverse species of variable ploidy level [8,9]. The most agronomically valuable species are *Cynodon dactylon* Pers. and *Cynodon nlemfuensis* Vanderyst. Both of these are cross-compatible, and several improved interspecific hybrids have been developed, selected, and released commercially: Coastcross I, Tifton 85, Florakirk [10–14].

The collection of United States Department of Agriculture National Plant Germplasm Systems (USDA-NPGS) for *Cynodon* is maintained by the Agricultural Research Service Coastal Plain Experimental Station in Tifton, GA, USA. A core collection of 160 accessions was developed based on 22 phenotypic traits collected among 600 accessions [15], and its genetic diversity was assessed through amplified fragment length polymorphism markers [16]. Besides, the core collection was studied to estimate biochemical conversion to ethanol [17]. Recently, part of the bermudagrass core collection [15] was included in an experiment studying nitrogen use efficiency (NUE) in bermudagrass [18]. Authors found that several traits related to NUE had large genetic variability. Nevertheless, the genetic diversity for other nutritive value (NV) traits, as well as the determination of genetic parameters, remain unknown for bermudagrass [15–18].

Nutritive value traits, especially digestibility and crude protein (CP), are main targets in forage breeding [11]. Improving NV can increase animal performance, reduce need for supplemental feed and thus cost of production, and can help mitigate some of the current environmental challenges, such as greenhouse gas emissions and eutrophication of surface waters [19–21]. Greater digestibility, for example, can lead to increase in dry matter intake and animal performance, therefore reducing methane emission per kg of animal output [22,23]. Because of high nutrient absorption capacity, bermudagrass pastures are widely used for nitrogen and phosphorus loss mitigation in waste management lands, e.g., for application of liquid and solid cattle manure, broiler litter, and other industrial water [24–27].

Phenotypic improvement requires genetic variability and stability of traits. Estimated genetic variance for dry matter digestibility in bermudagrass has been reported high (coefficient of genetic variation: 4.1–8.5%; broad sense heritability: 0.27 to 0.78), showing the potential for selection and improvement [10]. Although gains are slow, the final impact of improving NV traits can be significant. "Grazer" bermudagrass showed between 3.7 and 4.8% improvement in digestibility, which represented between 6 to 11% increase in live weight gain [28]. Increasing NV without reducing forage yield is a constant challenge for forage breeders, especially when targeting improvement for multiple traits. In 60 bermudagrass accessions from different geographic regions in China, [29] reported a phenotypic correlation of −0.37 between CP and forage yield. Tifton 85, for example, has higher digestibility and higher dry matter yield compared to Coastal, Tifton 44, 'Tifton 78', 'Jiggs', and 'Vaquero' [4,5,30–33]. However, it has lower CP compared to Jiggs [33] and greater neutral detergent fiber (NDF) concentration, which can result in low voluntary dry matter intake [34,35]. In complement, the target to increase CP, results in a decrease in NDF, whereas CP and dry matter digestibility shows a positive correlation [23].

The USDA-NPGS *Cynodon* germplasm collection has phenotypic data that have never been used for estimating variance components and genetic parameters. The estimation of genetic parameters, such as broad sense heritability (H^2), genotype by harvest interaction (GHI), and *type-A* genetic correlations, are fundamental to define breeding strategies [36]. The potential genetic diversity present in bermudagrass germplasm for traits with moderate-high H^2 and low GHI might expand the use of this germplasm by forage breeders. Hence, the objectives of this study were: (i) estimate genetic parameters for NV in the USDA-NPGS *Cynodon* collection, (ii) predict genetic values for four NV traits for the whole population across two harvests, (iii) estimate GHI for four NV traits across 11 harvests in a selected population, and (iv) select accessions exhibiting improved NV traits compared to Tifton 85.

2. Materials and Methods

2.1. Germplasm

A set of 286 bermudagrass accessions were screened, including 146 *Cynodon* clonal accessions from the USDA-NPGS forage core collection maintained in Tifton, GA; and 137 from the USDA-NPGS *Cynodon* collection maintained in Griffin, GA, and 3 commercial cultivars: Tifton 85, Coastal, and Jiggs.

Planting material obtained from Griffin and Tifton were grown in a greenhouse in Gainesville, FL, USA, and the plugs were allowed to established in 5 × 5 cm containers prior to planting.

2.2. Location Description

The experiment was conducted at the Plant Science Research and Education Unit located in Citra, Florida (29°24′16″ N and 82°10′17″ W), at 60 m elevation, from 2015 to 2016. Historical weather data was extracted online from the Florida Automated Weather Network (https://fawn.ifas.ufl.edu) and is summarized in Figure 1.

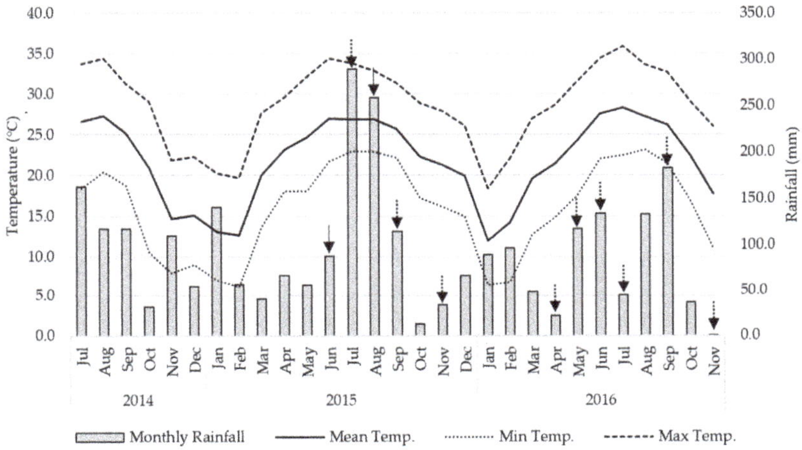

Figure 1. Maximum (Max Temp.), Minimum (Min Temp.), and Mean (Mean Temp.) monthly temperature (°C) and rainfall (mm) during the experimental period (April 2015 to November 2016). The arrows show the nutritive values sampling dates. The integer line refers to the whole population assessment, whereas the dashed line shows sampling date for the selected population.

The soil was a Chipley sand (thermic, coated Aquic Quartzipsamments) with a pH of 6.9 and characterized by high P_2O_5 content (164 kg ha^{-1}), and low K_2O (38.2 kg ha^{-1}), S (5.6 kg ha^{-1}), and Mg (44.8 kg ha^{-1}) content. The experimental plot size was 1.8 m × 3.0 m. Planting was done on 2 July 2014 using a single plug 5 × 5 cm planted in the center of the plot, and it was allowed to grow up to cover the plot. The long period between planting and the beginning of evaluations was necessary to guarantee a suitable plot establishment. The plants were fertilized with 40 g plant^{-1} with a mix of nitrogen (N), phosphorous (P_2O_5) and potassium (K_2O) 15-0-15 and micronutrients. An additional 75 kg ha^{-1} of nitrogen (N) and 45 kg ha^{-1} of K_2O were applied two months after planting. In early spring 2015, 90 kg ha^{-1} of N and K_2O were applied to promote spring regrowth. The field was re-fertilized with 90 kg ha^{-1} of N and 45 kg ha^{-1} of K_2O after each harvest, except for the last harvests of the growing seasons.

2.3. Experimental Design and Data Collection

The trial was established as a row-column design with two replicates and augmented representation of three controls, and the cultivars Tifton 85 and Jiggs were replicated thirteen times, and Coastal twelve times, in total. The plots were mowed to a stubble height of 10 cm in the beginning of each year (24 March 2015, and 3 March 2016). Plots were harvested every five weeks from April to November in both years. Biomass was collected to a 5-cm stubble height from a 1.2 × 3.0 m area in each plot. The remaining areas non-harvested were mowed to the same stubble after data collection. The fresh samples were weighed, and sub-samples (approximately 450 g) were taken, dried in a forced-air oven at 55 °C for 72 h, and reweighed to estimate forage harvested (FH). The samples were cleaned using sieves to avoid sand contamination and ground to 1 mm using a Wiley Mill (Model 4, Thomas Scientific, Swedesboro, NJ, USA) for NV analysis.

The analyses were performed using two subsets: (i) Whole-population: all accessions were harvested twice, first on June 1st and then in 11 August 2015; and (ii) Selected-population: a selected group of 15 accessions was sampled nine extra times between 1 June 2015 to 1 November 2016. The selected group resulted from a multi-location evaluation for biomass yield and other agronomic traits. The FH predictor values only were used to elucidate genetic correlations with NV traits. The NV traits included concentration of crude protein (CP), phosphorous (P), in vitro digestible organic matter (IVDOM) and neutral detergent fiber (NDF). These traits were determined by wet chemistry at the Forage Evaluation Support Laboratory, University of Florida (FESL). Crude protein was calculated as nitrogen multiplied by 6.25. Nitrogen and P samples were digested using a modification of the aluminum block digestion procedure [37]. Sample weight was 0.25 g, catalyst used was 1.5 g of 9:1 K_2SO_4:$CuSO_4$, and digestion was conducted for at least 4 h at 375 °C using 6 mL of H_2SO_4 and 2 mL H_2O_2. Nitrogen in the digestate was determined by semiautomated colorimetry [38]. In vitro digestible organic matter was performed by a modification of the two-stage technique [39]. Neutral detergent fiber was determined using the filter bag technique according to ANKOM^{200} (ANKOM Technology, Macedon, NY, USA) procedure [40].

2.4. Statistical Analyses

A dataset for the whole-population recorded traits of the two harvests were analyzed using linear mixed models with repeated measures implemented in ASReml-R (VSNI, Hemel Hempstead, Hertfordshire, UK) [41] using the R software [42]. The following model was fitted for single-trait and multi-harvest data:

$$y = 1\mu + X_1\alpha + X_2\beta_\alpha + Z_1 g + Z_2 \alpha g + Z_3 r_\beta + Z_4 c_\beta + e, \tag{1}$$

where μ is the overall population mean; 1 is a vector of ones; X_1, X_2, Z_1, Z_2, Z_3, and Z_4 are design matrices; α is the fixed harvest effect; β is the fixed effect of blocks; g is the random effects of entries, with g ~ multivariate normal distribution (MVN) $(0,\sigma^2_g I)$; $g\alpha$ is the random interaction effect between entry and harvest, with $g\alpha$ ~ MVN $(0,\sigma^2_{g\alpha} I)$; r_β and c_β are the random effects of row and column nested into block, with r_β ~ MVN $(0,\sigma^2_{r\beta} I)$ and c_β ~ MVN $(0,\sigma^2_{c\beta} I)$; and e is the random errors, with e ~ MVN $(0,\sigma^2_e I)$.

For each trait, the genotypic values for the entries were predicted and the variance components estimated: genotypic variance (s^2_g); variance of the genotype-harvest interaction (s^2_{gh}). The statistical significance of variance components were tested using the likelihood ratio test (LRT) with a Chi-square test with 1 degree of freedom [43]. Then, variance components estimates were used to calculate broad sense heritability (H^2) for each trait as

$$H^2 = \frac{s^2_g}{s^2_p} = \frac{s^2_g}{s^2_g + s^2_{gh} + s^2_e}, \tag{2}$$

and genotype by harvest correlation (r_{gh}) was calculated for each trait as

$$r_{gh} = \frac{s^2_g}{s^2_g + s^2_{gh}}, \tag{3}$$

where s^2_g is the estimated genotypic variance, s^2_p is the total phenotypic variance, and s^2_{gh} is the variance for the genotype-harvest interaction. The random error variance s^2_e for the multiharvest model was an average for both harvests (i.e., $s^2_e = (s^2_{e2} + s^2_{e5})/2$).

The genetic coefficient of variation (CV_g) was assessed for each trait using the following equation:

$$CVg = \left(\sqrt[2]{s_g^2 \frac{1}{\overline{X}}} \right) \times 100, \quad (4)$$

where s_g^2 is the estimated genotypic variance, and $\overline{X}$ is the mean value of the trait [10].

Accuracy refers to the correlation between the parametric genetic value and predicted genetic value, and it considers the residual variation, the settled experimental design and the proportion between the genetic and residual variations associated with the trait under evaluation [44]. Accuracy (*Acc*) was estimated as an average based on the standard error of predicted genotypic value (*PVse*) of each genotype [44]. *PVse* is related to accuracy through the equation:

$$Acc = \sqrt[2]{\left(1 - \left(\left(PVse^2\right)/s_g^2\right)\right)}, \quad (5)$$

where s_g^2 is the genotypic variance. Reliability (Rel) was obtained as average of genotypes *Acc* elevated to square. Overall mean, maximum, and minimum of predicted genotypic value were also computed. Additionally, predicted genetic values obtained with single-trait models were used to rank populations for each trait, which are given in Supplementary Table S1. A Principal Component Analyses (PCA) was performed with the *prcomp* function in R using a correlation matrix of the genotypic values obtained with the multiharvest model and predicted values of FH. These values were also used to obtain genetic correlations by Pearson method using *cor* function in R.

For the subset of selected population composed by 15 genotypes evaluated across 11 harvests, a linear model was fit for analysis of variance (ANOVA) using the package *agricolae* [45], with genotype, harvest, genotype-harvest interaction, and replicate factors considered fixed effects. A Tukey-Honest Significant Test was used to separate treatment means at $p = 0.05$ using the package *agricolae* [45], and graphs were created with the package *ggplot2* [46] in R. The coefficient of variation was calculated as standard deviation divided by mean multiplied by 100 and expressed as percentage. Additionally, a performance index was calculated by counting the number of times each genotype placed in the top of the ranking for each trait, and their mean response was statistically superior compared to any other genotype.

3. Results

3.1. Whole-Population

The whole *Cynodon* population showed significant genetic variability for all NV traits. Genetic variances for each trait were higher than zero ($p < 0.001$) based on LRT (Table 1), while GHI was only significant for IVDOM ($p < 0.001$). Genotype by harvest correlations were high for all traits (Table 1), and the H^2 ranged from low (CP and IVDOM) to moderate (P and NDF). The genetic coefficient of variation was low and exhibited a range from 2.6 to 9.6% (Table 1). Accuracy were moderate for CP, and high for other traits, whereas reliability was low for CP and moderate for the other traits.

Some traits exhibited large genotypic variability (Figure 2). For instance, CP ranged from 116 g kg^{-1} to 157 g kg^{-1} (Figure 2A), where PI 2922601, PI 292508, PI 547109, and breeding line 319 exhibited CP higher than 150 g kg^{-1}. The highest predicted value for p was almost double its lowest concentration. However, only four accessions, including PI 316507, Breeding line 240, 'Florakirk', and PI 364485 showed p higher than 3.5 g kg^{-1}, while Tifton 85 and Jiggs presented 2.9 and 3.4 g kg^{-1}, respectively, and all above the population mean (Figure 2B). Similarly, accessions exhibited wide IVDOM values from 363 to 563 g kg^{-1} (Figure 2C). 'Tifton 84' presented the highest IVDOM followed by Florakirk (555 g kg^{-1}), PI 316507 (549 g kg^{-1}), PI 255450 (548 g kg^{-1}), PI 204438 (546 g kg^{-1}), and Tifton 85 (546 g kg^{-1}). The NDF ranged from 651 g kg^{-1} to 767 g kg^{-1} (Figure 2D), where only PI 297827, PI 287156, 319, and Florakirk showed values below 660 g kg^{-1}. Despite the fact that NDF presented the lowest genotypic range and lowest genetic coefficient of variation (low genetic variability), NDF had medium broad sense heritability and high genotype by harvest correlation. Comparing the population mean with Tifton 85, CP, P, and NDF values were similar, but Tifton 85 exhibited a much higher IVDOM compared to the population mean and most other accessions (Figure 2).

Table 1. Genetic parameters for the United States Department of Agriculture National Plant Germplasm System (USDA-NPGS) bermudagrass germplasm collection evaluated in Citra, FL. Genetic coefficient of variation (CV_g), broad sense heritability ($H^2 \pm$ SE), and genotype by harvest correlation (r_{gh}) for crude protein (CP), phosphorous (P), In Vitro Organic Matter Digestibility (IVDOM), and Nutrient Detergent Fiber (NDF) expressed at grams per kilogram. Likelihood Ratio Test of Genotype and Genotype by Harvest, Reliability Mean, and Accuracy Mean for nutritive value (NV) traits.

	Nutritive Value Traits			
	CP g.kg^{-1}	P g kg^{-1}	IVDOM g kg^{-1}	NDF g kg^{-1}
CV_g	5.9	9.6	8.1	2.6
$H^2 \pm$ SE	0.20 ± 0.04	0.41 ± 0.03	0.36 ± 0.04	0.40 ± 0.03
r_{gh}	0.81 ± 0.15	0.97 ± 0.07	0.81 ± 0.07	0.99 ± 0.00
Genotype (LRT)	25.9 ***	105.6 ***	72.4 ***	113.5 ***
Genotype × Harvest (LRT)	0.91 ^{ns}	0.13 ^{ns}	9.93 ***	-7.23×10^{-6} ^{ns}
Reliability Mean	0.35	0.61	0.57	0.60
Accuracy Mean	0.57	0.79	0.75	0.77

*** Significative at 0.1% by Likelihood ratio test. ^{ns}, non-significative.

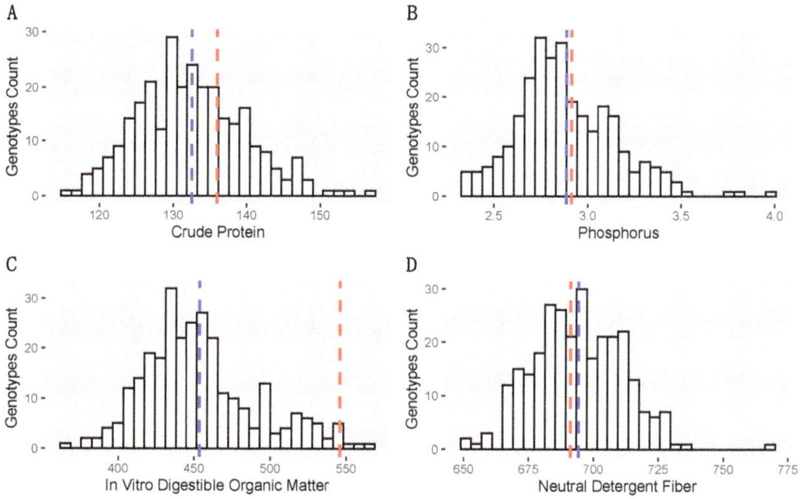

Figure 2. Histogram for four nutritive values traits using predictive values estimated for the USDA-NPGS bermudagrass germplasm collection and three checks: Tifton 85, Jiggs and Florida 44, evaluated in Citra, FL. (**A**) Crude Protein concentration, (**B**) Phosphorus concentration, (**C**) In Vitro Organic Matter Digestibility, and (**D**) Neutral Detergent Fiber, all expressed in g kg^{-1}. The dashed blue line shows the means, while the dashed red line shows the values for Tifton 85 of each NV trait.

The Pearson genetic correlations among the NV traits were all significant ($p < 0.05$), although FH was only correlated with IVDOM (Table 2). The NDF presented moderate negative correlations with CP, P, and IVDOM (Table 2). In contrast, positive correlations were observed among CP, P, and IVDOM, as well as between IVDOM and FH, in a moderate to low magnitudes.

Table 2. Pearson genetic correlations among forage harvested (FH) and nutritive value traits: CP—Crude Protein concentration, P—Phosphorus concentration, IVDOM—In Vitro Digestible Organic Matter, and NDF—Neutral Detergent Fiber; estimated for the USDA-NPGS bermudagrass germplasm collection and checks: Tifton 85, Jiggs and Florida 44, evaluated in Citra, FL. Color code: red and blue indicate negative/positive significant ($p < 0.05$) correlations, while white shows non-significative correlations.

	CP	P	IVDOM	FH
NDF	−0.68	−0.62	−0.41	−0.01
CP		0.60	0.29	−0.12
P			0.46	0.14
IVDOM				0.46

The PCA using predictive values for all accessions explained a large part of the variation found for NV traits and FH in this collection (Figure 3A). The first two PCAs accounted for 77.8% of the existing variation in the bermudagrass germplasm collection. The first principal component (PC) (eigenvalue of 2.57) explained 51.4% of variance, where the NV traits contributed more than the FH, and exhibited similar magnitudes (Figure 3A). The second PC exhibited an eigenvalue of 1.31, explained 26.41% of the variation, and the main contribution was FH (0.77), whereas other traits were less determinant, such as IVDOM (0.46), CP (−0.37), NDF (0.21), and P (−0.04). Other remaining PCs accounted for 22.2% of variation and had eigenvalues lower than 0.40.

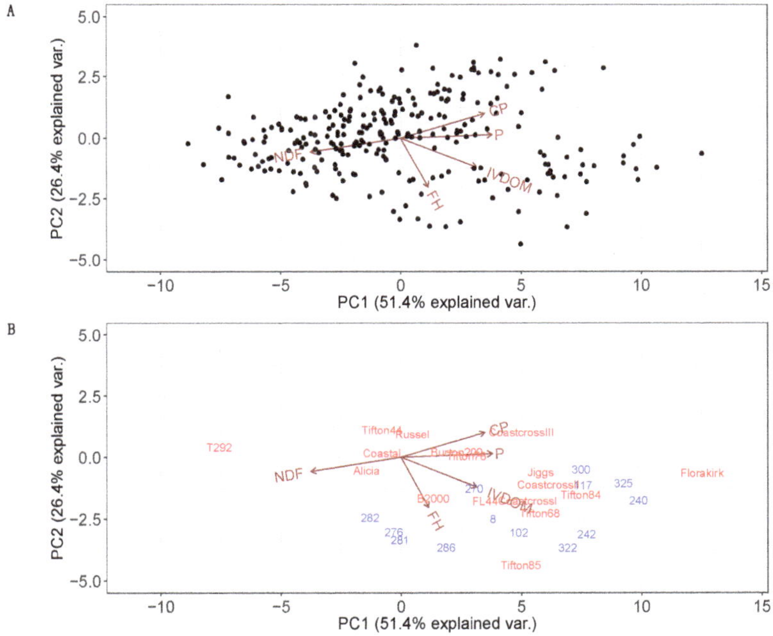

Figure 3. Principal Components Analysis (PCA) estimated for the USDA-NPGS bermudagrass germplasm collection and commercial cultivars. Black loadings represent predicted values for five traits estimated for the whole collection (**A**), and loadings colored only for 17 cultivars (red) and 15 selected accessions (blue) (**B**). The plants were grown at Citra, FL, and five traits were determined using wet chemistry: crude protein concentration—CP, phosphorus concentration—P, in vitro digestible organic matter—IVDOM, neutral detergent fiber—NDF, and forage harvested—FH. The selected accessions were identified by field ID instead of plant introduction numbers to improve visibility in the graph.

Commercial cultivars (red) and the selected accessions (blue) also exhibited broad variation (Figure 3B). Two cultivars, T292 and Florakirk, exhibited contrasting performance for NV and FH

(Figure 3B). Besides, most commercial cultivars showed lower NDF and higher values for CP, P and IVDOM in general. Tifton 85 exhibited the highest yield and IVDOM across cultivars in those two harvests. Comparing accessions and cultivars, some selected accessions exhibited higher combining of FH, IVDOM, CP, and P than commercial varieties, except by Tifton 85 and Florakirk. Tifton 85 presented high FH, while Florakirk presented high P and IVDOM. The accessions PI 316510 (322) and PI 255450 (242) grouped near Tifton 85, exhibited similar FH but higher IVDOM. In addition, some accessions exhibited high CP and P, such as PI 316507 (323), Breeding line 240 (240), PI 316536 (325), and PI 364484 (117).

3.2. Selected-Population

Significant genotype and harvest main effects ($p < 0.001$) were observed for all traits, but two-way interactions were non-significant for all traits (Table 3). For CP and IVDOM, the mean square values for harvest were considerably larger than for genotype, suggesting a higher environmental influence for these traits without influencing the genotypic response. However, the mean square value of harvest for P was lower than for genotype, which means a greater influence of genotype on P. The CV was higher for CP and P than for IVDOM and NDF but lower than 30%, which is the maximum threshold for field experiments [47]. Additionally, NDF exhibited CV lower than 5%, which was similar to the CV_g observed in the whole population.

Table 3. Analysis of variance showing sources of variation, degrees of freedom (df), and mean square values for each nutritive value trait.

Source of Variation	df	Mean Square Values			
		CP g kg^{-1}	P g kg^{-1}	IVDOM g kg^{-1}	NDF g kg^{-1}
Replicate	1	0.15 ^{ns}	0.00019 ^{ns}	1.2 ^{ns}	44.47 ***
Genotype	14	16.66 ***	0.05361 ***	284.2 ***	98.38 ***
Harvest	10	91.29 ***	0.02211 ***	511.4 ***	85.93 ***
Harvest by Genotype	140	2.46 ^{ns}	0.00178 ^{ns}	14.0 ^{ns}	2.51 ^{ns}
Error	170	3.64	0.0017	14.3	3.78

*** $p < 0.001$; ^{ns}, non-significant differences.

Tifton 85 and Florida 44 were not different for CP between themselves or from other accessions (Table 4). However, PI 308193 had higher CP compared to PI 255456, PI 255450, PI 290813, PI 290664, and PI 295114 (Table 4). PI 316510 had similar CP than PI 308193, and it was higher than PI 290813, PI 290664, and PI 295114. For phosphorus content, Tifton 85 and Florida 44 performed similarly, and Tifton 85 was not statistically different from the PIs with lowest P. Additionally, Breeding line 240 and PI 255450 presented the highest P.

The accessions PI 316510, Breeding line 240, Breeding line 8, Tifton 85, and PI 316536 had similar IVDOM. Finally, breeding line 240 and PI 364484 exhibited the lowest NDF but were not significantly different from PI 316510, PI 316536, PI 255456, and Florida 44. All these PIs presented lower NDF than Tifton 85. Based on a performance index (Table 4), breeding line 240 was placed at the top of the ranking for the four NV traits, while PI 316510, PI 316536, and Breeding line 8 performed well for CP, IVDOM, and NDF. Breeding line 240 presented higher P than PI 316510, PI 316536, and Breeding line 8, as well as lower NDF than Breeding line 8. These accessions had lower NDF and higher P than Tifton 85. The PI 255450 showed high P concentration and exhibited high IVDOM similar to PI 316510. Finally, NDF presented the largest difference between the selected accessions and Tifton 85 (Table 4).

Table 4. Average nutritive value for selected bermudagrass plant introductions (PI), breeding lines, and cultivars across eleven harvests evaluated in Citra, FL.

Entries	CP g kg^{-1}		P g kg^{-1}		IVDOM g kg^{-1}		NDF g kg^{-1}		Performance Index *
PI 308193 (300)	145	a *	3.2	bc	467	fg	678	cd	1
PI 316510 (322)	139	ab	3.2	bc	557	a	662	de	3
PI 316536 (325)	139	abc	3.1	bcd	529	abcd	670	de	3
PI 364484 (117)	134	abcd	3.4	b	503	cdef	656	e	1
Breeding line 240 (240)	133	abcd	4.0	a	551	ab	654	e	3
Breeding line 8 (8)	133	abcd	2.5	ef	543	abc	691	bc	1
PI 292143 (102)	127	abcd	3.2	bc	516	bcde	679	cd	0
PI 294467 (276)	125	abcd	2.6	ef	488	efg	708	ab	0
PI 255456 (270)	124	bcd	3.4	b	499	defg	669	de	1
PI 255450 (242)	122	bcd	3.9	a	564	a	680	cd	2
PI 290813 (286)	119	cd	2.4	f	483	efg	708	ab	0
PI 290664 (281)	115	d	2.5	ef	479	efg	716	a	0
PI 295114 (282)	115	d	2.4	f	463	fg	719	a	0
Florida 44	127	abcd	2.9	cde	461	g	667	de	1
Tifton 85	133	abcd	2.8	def	541	abc	693	bc	1
C.V. (%)	19		22		12		5		

Mean values with same letter do not differ statistically ($p \leq 0.05$) by Tukey Test. * Performance index: number of times a genotype appear in the top statistical group for each trait.

The NV traits varied greatly across harvests (Table 5). The NV fluctuated within the year, with higher CP and lower NDF in the April and September harvests. Higher IVDOM values were measured in the spring and beginning of summer for both years, coinciding with high volumes of rainfall and mild temperature (Figure 1). Lower values for CP, IVDOM, NDF, and P were observed in November, with scarce rainfall and low temperature inducing dormancy. The NV measured on the same month on both years were similar, and for some traits, there were no significant differences (Table 5). Although the fluctuations in temperature observed during both years were similar, the rainfall regime was different (Figure 1). In 2015, July and August had 250 mm per month, which was above average, whereas the rainfall the same period in 2016 was less than 133 mm per month. The rainfall in June was 290 mm in 2015, whereas it was 43 mm in 2016.

Table 5. Average nutritive value by harvest (11) for the selected population (15 entries).

Harvest	CP g kg^{-1}		P g kg^{-1}		IVDOM g kg^{-1}		NDF g kg^{-1}	
2015 June 1	124	d *	2.9	cd	556	a	694	bc
2015 July 7	117	d	3.0	abc	536	ab	692	bcd
2015 August 11	85	e	2.4	e	535	ab	699	b
2015 September 22	151	a	3.3	ab	518	bc	675	ef
2015 November 3	128	bcd	3.0	abc	448	e	686	bcde
2016 April 14	150	a	3.0	bcd	558	a	652	g
2016 May 19	120	d	3.1	abc	549	a	719	a
2016 June 23	129	bcd	3.3	a	509	bcd	678	def
2016 July 28	140	ab	3.2	abc	488	cd	681	cdef
2016 September 13	138	abc	3.2	abc	483	d	670	f
2016 November 1	125	cd	2.6	de	436	e	677	def
C.V. (%)	19		22.3		12.4		4.6	

* Mean values with same letter do not differ statistically ($p \leq 0.05$) by Tukey Test.

4. Discussion

The genetic parameters indicated that genetic gains can be expected in all NV traits in this bermudagrass germplasm, as most traits presented moderate H^2. Broad sense heritability explains the magnitude of phenotypic variance due to the genetic variance [36]. Indeed, a higher genetic variance than GHI was observed for all traits, and the GHI was only significant for IVDOM. GHI effect was considerably lower than the main genotypic effect, similar to previous reports in maize (*Zea mays* L.) and timothy grass (*Phleum pratense* L.) [48,49], if we treat harvest as an environment. Thus, high stability of genotypes for these traits across harvests would be expected as supported by the high estimates of r_{GH}. The use of r_{GH}, although it does not elucidate the significance of variance sources, indicates the stability of genotypes among harvests in a given location, meaning that traits with high stability can be subjected to selection with fewer harvests. Given the high r_{GH} for all traits, using estimates from fewer harvests can be a reliable NV measure to screen large nurseries in forage breeding programs. However, considering a selection for multiple locations, other environmental variables, such as soil type, crop management, and climate variability, might diverge and cause genotype by environment interaction (GEI). Therefore, we encourage other studies to elucidate GEI for NV traits. Despite that, using only two harvests for the whole population, it was possible to achieve prediction accuracies higher than 0.70 for IVDOM, NDF, and P, considered appropriate for forage breeding experiments [50]. In general, an increase in the number of harvests leads to an increase in precision; however, it adds costs and labor when dealing with large populations [51].

Genetic parameters, such as H^2, r_{GH}, and CV_g, are variable among species, germplasm, traits, and experiments [52]. Medium to high broad-sense heritability for IVDOM have been reported for other forage grasses, 0.56 to 0.93 [53–55]. Previous efforts with bermudagrass showed broad-sense heritability ranging between 0.27 and 0.78 across years and experiments [11], denoting that H^2 is inherent to the experimental conditions, as well as in where moderate to high estimates can be achieved for those species. In our study, H^2 estimates were within the range of previous reports. The magnitude of H^2 for P in the bermudagrass collection was similar to the broad-sense heritability of 0.46 reported for reed canarygrass (*Phalaris arundinacea* L.) [56] but lower than the 0.72 estimate reported in tall fescue (*Schedonorus arundinacus* (Schreb.) Dumort.) [57].

In turn, NDF had lower H^2 compared to previous studies, and lower variation (CV_g) in the population. Broad-sense heritability in NDF for populations of tall fescue [54] and koronivia grass (*Urhochloa humidicola* (Rendle), Morrone and Zuloaga [58]) were much greater (0.85 and 0.58, respectively) compared to the bermudagrass germplasm. Other reports also showed high narrow-sense heritability for NDF in tall fescue (0.6; Reference [59]) and signal grass (*Urochloa decumbens* (Stapf) R. Webster) (0.74; Reference [51]). Moreover, r_{GH} across multiple harvests was also lower for the referred studies (0.36 in koronivia grass [58]; 0.82 in signal grass [51]). Thus, the efforts to genetically improve NDF from this germplasm would be similar to the above-mentioned species but fewer evaluations are needed since it has higher stability.

Crude protein exhibited lower H^2 than the estimates for the other NV traits, but similar H^2 estimates were found in other species. Low broad-sense heritability was reported for CP in meadow fescue (0.21) [53], as well as low narrow-sense heritability (0.07) in koronivia grass [58], 0.14 in congo grass [60], and 0.18 in tall fescue [59]. The estimates of H^2, also, can be influenced by environment. In tall fescue, narrow heritability ranged from 0.18 to 0.54, with the winter harvest exhibiting a higher estimate than that for the fall harvest [59]. Thus, the choice of the harvest to characterize CP is important because managing data collection can minimize undesirable effects attributed to the large environmental variance compared to measured genetic variance in this study. Although CP had low H^2 that will potentially result in lower genetic gains for this trait, any gain in protein content in grasses can result in significant reductions of the use of external protein sources. The high stability of this trait over the two annual harvests (high r_{GH}), and non-significant GHI indicated that most of the variation in CP in this population was indeed due to genetics.

The genetic coefficient of variation complements the broad sense heritability estimates, and it allows the comparison of genetic variability among populations and traits [10]. The CV_g for IVDOM was 8.1%, which is higher than values reported for meadow fescue (4%) [53], congo grass (2.2%) [61] and *Urochloa* spp. (5.8%) [55]. Digestibility has been a major target in previous bermudagrass breeding efforts (e.g., Coastcross I; Reference [11]. In an effort to improve digestibility in bermudagrass, approximately 500 accessions showed great variability for in vitro dry matter digestibility (IVDMD) and

CV_g ranged from 6.1 to 28.9%, and the two-year average for IVDMD ranged from 400 to 690 g kg^{-1} [10]. This large genetic variation can be exploited in breeding programs. In our study, the PI 255450 (242), which is a parental line for Kenya 61 [11], exhibited the highest IVDOM. Although it could be used in crosses aiming to combine higher digestibility with other traits, the ploidy level of this accession can be a limitation because the progeny from Coastal and Kenya 61 were all male sterile and not able to produce seeds [11]. Other Pis, such as Tifton 84, Florakirk, and PI 316507, also showed high IVDOM.

Phosphorous had the largest CV_g among the NV traits in this population. Breeding line 240 and PI 316510 (322) had an average 30% more P concentration than Tifton 85, as well as desired levels of CP, IVDOM, and NDF. This genetic variability can be explored to generate information about P uptake and its relationship with forage yield, as previous research reported for N [18]. Accessions with high uptake P and N that combine reasonable agronomic performance can be used as phytoremediation agents for dairy farms [27,62]. On the other hand, genetic variability for NDF was the lowest among evaluated traits, concurrent with other reports from the literature, like 1.4 and 2.9% in tall fescue [59], 3.8 to 4.2% in leaves of *Brachiaria* spp. [55] 4.32, and 4.43% in leaves and stems of *Arachis* spp. [59]. These aspects confirm that difficulty in improving NDF by conventional breeding in forage species.

Entries with high NV traits were already included in the bermudagrass core collection except for four accessions (PI 292508, PI 547109, PI 297827, and PI 287156-01), as selection was based on FH across several environments [15]. Breeding line 240, PI 316510 (322) and PI 3166536 (325) presented lower NDF and greater P concentration than Tifton 85, and similar IVDOM and CP. Breeding for improved CP has received little attention in the past as nitrogen fertilizer has been an affordable and effective way to increase increasing CP concentration, IVDOM, and productivity in grasses [17,18,23]. Few studies looked at nitrogen use efficiency in bermudagrass [18], which found high variability for the trait, and a negative correlation to CP concentration especially at low N fertilization rates.

In general for bermudagrass, CP has negative correlation with biomass and NDF (−0.32 and −0.36, Reference [29]), but positive correlation with dry matter digestibility (0.34; [11]). Similar results were observed on a review of over one hundred forage species, where digestibility had a positive correlation with CP (0.62) and negative with NDF (−0.68) [63]. Our results showed similar trends, although FH was only significantly correlated to IVDOM. The moderate to low genetic correlations found in the current germplasm suggest genes related to the expression of these traits have higher independence or belong in different metabolic routes. This can be due to the genetic diversity present in this germplasm, particularly related to having different species and ploidy levels in the core collection [15,16]. Thus, breeding multiple nutritive value traits in *Cynodon* ssp. should be achievable, and there is still potential for NV to be improved with forage yield.

Selection on a single trait provides higher genetic progress but can be detrimental if some unwanted correlation exists with the trait under improvement. On the other hand, genetic correlations can be used to identify potential indirect selections for simultaneous breeding [36]. Although there was no clear cluster of genotypes in the PCA, it was possible to identify genotypes that presented superior NV traits and FH. For instance, except for Tifton 85 and Florakirk, other accessions in the selected population presented higher P, IVDOM, and FH than most commercial cultivars. In general, commercial varieties and the selected population showed variability and different patterns of NV performance. Accessions such the breeding line 240, PI 255450, and PI 316510 exhibited high FH and NV. Thus, they can be selected as parental lines in future crosses and public cultivar releases.

5. Conclusions

The results presented in this study complement previous findings and provide useful information for the entire forage bermudagrass collection, aiming at developing cultivars with improved NV. The high significance of the genotypic factor evaluated during eleven harvests showed differences among the accessions in the selected-population for all NV traits. These accessions have a good combination of NV and biomass production, and some of them had improved NV compared to Tifton 85. The lack of significant two-way interactions between genotype and harvest for NV traits confirmed the high stability of genotypes across harvests. Thus, selection for NV can be assessed with fewer harvests, resulting in savings for time, labor, and resources. Genetic parameters revealed that P has higher potential to be explored as breeding target, along with CP, while narrow variance for NDF, and the availability of varieties with high IVDOM, would require more effort than conventional breeding for improving those traits in bermudagrass. Breeding line 240, PI 316510 (322), and PI 3166536 (325) presented superior nutritive value than Tifton 85 and will be considered for public cultivar releases in the United States.

Supplementary Materials: The following are available online at http://www.mdpi.com/2073-4395/10/11/1729/s1, Table S1: Predicted genetic values estimated of nutritive value traits: crude protein (CP), phosphorous (P), in vitro digestible organic matter (IVDOM) and nutrient detergent fiber (NDF) in g.kg^{-1}, for the USDA-NPGS bermudagrass germplasm collection evaluated in Citra, FL.

Author Contributions: Conceptualization, P.M., W.A., and E.R.; Data curation, P.M., Y.L., and E.R.; Formal analysis, C.d.S. and E.R.; Funding acquisition, P.M.; Investigation, P.M., Y.L., and E.R.; Methodology P.M., Y.L., and E.R.; Project administration, P.M., Y.L., W.A., and E.R.; Resources, P.M., Y.L., W.A., and E.R.; Software, C.d.S. and E.R.; Supervision, P.M., Y.L., and E.R.; Visualization, C.d.S., M.W. and E.R.; Writing—original draft, C.d.S., M.W., M.D., and E.R.; Writing—review & editing, C.d.S., Y.L., P.M., W.A., M.D., M.W., and E.R. All authors have read and agreed to the published version of the manuscript.

Funding: This work was supported by the USDA National Institute of Food and Agriculture, Hatch project 1018058.

Acknowledgments: The authors greatly thank all the Forage Breeding and Genetics Lab members and staff at the University of Florida Plant Science Research and Education Unit, Citra, FL, USA, for providing help during the field trial and data collection.

Conflicts of Interest: The authors declare no conflict of interest.

References

1. Jank, L.; Valle, C.; Resende, R. Breeding Tropical Forages. *Crop Breed. Appl. Biotechnol.* **2011**, *11*, 27–34. [CrossRef]
2. Taliaferro, C.M.; Rouquette, F.M.; Mislevy, P. Bermudagrass and Stargrass. In *Agronomy Monographs*; Wiley: Hoboken, NJ, USA, 2016; pp. 417–475.
3. Ball, D.M.; Hoveland, C.S.; Lacefield, G.D. *Southern Forages: Modern Concepts for Forage Crop Management*; IPNI: Saskatoon, SK, Canada, 2007.
4. Hill, G.M.; Gates, R.N.; Burton, G.W. Forage Quality and Grazing Steer Performance from Tifton 85 and Tifton 78 Bermudagrass Pastures2. *J. Anim. Sci.* **1993**, *71*, 3219–3225. [CrossRef] [PubMed]
5. Mandebvu, P.; West, J.; Hill, G.M.; Gates, R.N.; Hatfield, R.D.; Mullinix, B.G.; Parks, A.H.; Caudle, A.B. Comparison of Tifton 85 and Coastal Bermudagrasses for Yield, Nutrient Traits, Intake, and Digestion by Growing Beef Steers. *J. Anim. Sci.* **1999**, *77*, 1572–1586. [CrossRef] [PubMed]
6. Pequeno, D.N.L.; Pedreira, C.G.S.; Sollenberger, L.E.; De Faria, A.F.G.; Silva, L.S. Forage Accumulation and Nutritive Value of Brachiariagrasses and Tifton 85 Bermudagrass as Affected by Harvest Frequency and Irrigation. *Agron. J.* **2015**, *107*, 1741–1749. [CrossRef]
7. Marcum, K.; Pessarakli, M. Salinity Tolerance and Salt Gland Excretion Efficiency of Bermudagrass Turf Cultivars. *Crop. Sci.* **2006**, *46*, 2571–2574. [CrossRef]
8. Harlan, J.M.J.; De Wet, J.R.; Richardson, W.L. Hybridization Studies with Species of *Cynodon* from East Africa and Malagasy. *J. Bot.* **1969**, *56*, 944–950. [CrossRef]
9. Ohmoto, T.; Ikuse, M.; Natori, S. Triterpenoids of the Gramineae. *Phytochemistry* **1970**, *9*, 2137–2148. [CrossRef]
10. Burton, G.W.; Monson, W.G. Inheritance of Dry Matter Digestibility in Bermudagrass, *Cynodon dactylon* (L.) Pers. 1. *Crop Sci.* **1972**, *12*, 375–378. [CrossRef]
11. Burton, G.W.; Hart, R.H.; Lowrey, R.S. Improving Forage Quality in Bermudagrass by Breeding 1. *Crop Sci.* **1967**, *7*, 329–332. [CrossRef]
12. Burton, G.W.; Monson, W.G. Registration of Tifton 68 Bermudagrass. *Crop Sci.* **1984**, *24*, 1211. [CrossRef]
13. Burton, G.W.; Gates, R.N.; Hill, G.M. Registration of 'Tifton 85' Bermudagrass. *Crop Sci.* **1993**, *33*, 644–645. [CrossRef]
14. Mislevy, P.; Brown, W.F.; Dunavin, L.S. Florakirk. Bermudagrass. Circ S-395. 1995. Available online: https://edis.ifas.ufl.edu/aa221 (accessed on 30 May 2020).

15. Anderson, W.F. Development of a Forage Bermudagrass (*Cynodon* sp.) Core Collection. *Grassl. Sci.* **2005**, *51*, 305–308. [CrossRef]
16. Anderson, W.F.; Maaß, A.; Ozias-Akins, P. Genetic Variability of a Forage Bermudagrass Core Collection. *Crop Sci.* **2009**, *49*, 1347–1358. [CrossRef]
17. Anderson, W.F.; Dien, B.S.; Jung, H.-J.G.; Vogel, K.P.; Weimer, P.J. Effects of Forage Quality and Cell Wall Constituents of Bermuda Grass on Biochemical Conversion to Ethanol. *Bioenergy Res.* **2009**, *3*, 225–237. [CrossRef]
18. Schneider-Canny, R.; Chekhovskiy, K.; Muñoz, P.; Kwon, S.; Saha, M.C. Characterization of Bermudagrass (*Cynodon dactylon* L.) Germplasm for Nitrogen Use Efficiency. *Euphytica* **2019**, *215*, 40. [CrossRef]
19. Gerber, P.J.; Hristov, A.N.; Henderson, B.; Makkar, H.; Oh, J.; Lee, C.; Meinen, R.; Montes, F.; Ott, T.; Firkins, J.; et al. Technical Options for the Mitigation of Direct Methane and Nitrous Oxide Emissions from Livestock: A Review. *Animal* **2013**, *7*, 220–234. [CrossRef]
20. Teague, W.; Delaune, P.B.; Dowhower, S.L. Impacts of over-Seeding Bermudagrass Pasture with Multispecies Cover Crops on Soil Water Availability, Microbiology, and Nutrient Status in North Texas. *Agric. Ecosyst. Environ.* **2019**, *273*, 117–129. [CrossRef]
21. Glendining, M.J.; Dailey, A.; Williams, A.G.; Van Evert, F.K.; Goulding, K.; Whitmore, A. Is It Possible to Increase the Sustainability of Arable and Ruminant Agriculture by Reducing Inputs? *Agric. Syst.* **2009**, *99*, 117–125. [CrossRef]
22. Knapp, J.; Laur, G.; Vadas, P.; Weiss, W.; Tricarico, J. Invited Review: Enteric Methane in Dairy Cattle Production: Quantifying the Opportunities and Impact of Reducing Emissions. *J. Dairy Sci.* **2014**, *97*, 3231–3261. [CrossRef]
23. Casler, M.D.; Vogel, K.P. Accomplishments and Impact from Breeding for Increased Forage Nutritional Value. *Crop Sci.* **1999**, *39*, 12–20. [CrossRef]
24. Sanderson, M.A.; Jones, R.M. Forage Yields, Nutrient Uptake, Soil Chemical Changes, and Nitrogen Volatilization from Bermudagrass Treated with Dairy Manure. *J. Prod. Agric.* **1997**, *10*, 266–271. [CrossRef]
25. Brink, G.E.; Rowe, D.E.; Sistani, K.R. Broiler Litter Application Effects on Yield and Nutrient Uptake of 'Alicia' Bermudagrass. *Agron. J.* **2002**, *94*, 911. [CrossRef]
26. Macoon, B.; Woodard, K.R.; Sollenberger, L.E.; French, E.C.; Portier, K.M.; Graetz, D.A.; Prine, G.M.; Van Horn, H.H. Dairy Effluent Effects on Herbage Yield and Nutritive Value of Forage Cropping Systems. *Agron. J.* **2002**, *94*, 1043. [CrossRef]
27. Read, J.J.; Lang, D.J.; Adeli, A.; Jenkins, J.N. Harvest Management Effects on 'Tifton 44' Bermudagrass Phosphorus Removal and Nutritive Value. *Agron. J.* **2018**, *110*, 879–889. [CrossRef]
28. Eichhorn, M.M.; Oliver, W.M.; Hallmark, W.B.; Young, W.A.; Davis, A.V.; Nelson, B.D. Registration of 'Grazer' Bermudagrass. *Crop Sci.* **1986**, *26*, 835. [CrossRef]
29. Gitau, M.M.; Fan, J.; Xie, Y.; Fu, J. Genetic Diversity and Association Mapping of Forage Quality in Diverse Bermudagrass Accessions. *Euphytica* **2017**, *213*, 234. [CrossRef]
30. Vendramini, J.M.B. Forage Evaluation and Quality in Florida. In *Proceedings of the 21st Annual Florida Ruminant Nutrition Symposium, Gainesville, FL, USA, 2–3 February 2010*; Department of Animal Sciences University of Florida, Institute of Food and Agricultural Sciences: Gainesville, FL, USA, 2010; pp. 92–106.
31. Burns, J.C.; Fisher, D.S. Dry Matter Intake and Digestibility of "Coastal", "Tifton 44", and "Tifton 85" Bermudagrass Hays Grown in the U.S. Upper South. *Crop Sci.* **2007**, *47*, 795–810. [CrossRef]
32. Pedreira, C.G.S.; Silva, V.J.; Newman, Y.C.; Sollenberger, L.E. Yearling Cattle Performance on Continuously Stocked 'Tifton 85' and 'Florakirk' Bermudagrass Pastures. *Crop Sci.* **2016**, *56*, 3354–3360. [CrossRef]
33. Kohmann, M.M.; Sollenberger, L.E.; Vendramini, J.M.B.; Silveira, M.L.; Moreno, L.S.B. Harvest Atubble Height and K Fertilization Affect Performance of "Jiggs" and "Tifton 85" bermudagrasses. *Crop Sci.* **2017**, *57*, 3352–3359. [CrossRef]
34. van Soest, P.J. Symposium on Factors Influencing the Voluntary Intake of Herbage by Ruminants: Voluntary Intake in Relation to Chemical Composition and Digestibility. *J. Anim. Sci.* **1965**, *24*, 834–843. [CrossRef]
35. Mertens, D.R. Predicting Intake and Digestibility Using Mathematical Models of Ruminal Function. *J. Anim. Sci.* **1987**, *64*, 1548–1558. [CrossRef] [PubMed]
36. Frankham, R. Introduction to Quantitative Genetics (4th edn). *Trends Genet.* **1996**, *12*, 280. [CrossRef]
37. Gallaher, R.N.; Weldon, C.O.; Futral, J.G. An Aluminum Block Digester for Plant and Soil Analysis. *Soil Sci. Soc. Am. J.* **1975**, *39*, 803–806. [CrossRef]

38. Hambleton, L.G. Semiautomated Method for Simultaneous Determination of Phosphorus, Calcium, and Crude Protein in Animal Feeds. *J. Assoc. Off. Anal. Chem.* **1977**, *60*, 845–852. [CrossRef]
39. Moore, J.; Mott, G. Recovery of Residual Organic Matter from in Vitro Digestion of Forages. *J. Dairy Sci.* **1974**, *57*, 1258–1259. [CrossRef]
40. Ankom Technology. Neutral Detergent Fiber in Feeds—Filter Bag Technique (for A2000 and A2000I). Method 6 2011; 10–11. Available online: http://www.ssco.com.tw/Ankom/PDF_file/NDF%20Method%20A200.pdf (accessed on 30 May 2020).
41. Butler, D.G.; Cullis, B.R.; Gilmour, A.R. ASReml-R Reference Manual Version 4. 2017. 176. Available online: https://asreml.kb.vsni.co.uk/wp-content/uploads/sites/3/2018/02/ASReml-R-Reference-Manual-4.pdf (accessed on 30 May 2020).
42. R Development Core Team. R: A Language and Environment for Statistical Computing. R Foundation for Statistical Computing. 2018; 1: 409. Available online: https://www.r-project.org/ (accessed on 30 May 2020).
43. Satorra, A.; Saris, W.E. Power of the Likelihood Ratio Test in Covariance Structure Analysis. *Psychometrika* **1985**, *50*, 83–90. [CrossRef]
44. de Resende, M.D.V. *Genética Biométrica e Estatística no Melhoramento de Plantas Perenes*; Embrapa Informação Tecnológica: Brasília, Brazil, 2002. (In Portuguese)
45. de Mendiburu, F. Agricolae: Statistical Procedures for Agricultural Research. R Packag Version. Available online: https://cran.r-project.org/web/packages/agricolae/index.html (accessed on 30 May 2020).
46. Wickham, H. *ggplot2: Elegant Graphics for Data Analysis*; Springer: New York, NY, USA, 2016.
47. Brown, C.E. Applied Multivariate Statistics in Geohydrology and Related Sciences. In *Applied Multivariate Statistics in Geohydrology and Related Sciences*; Springer Science and Business Media LLC: Berlin/Heidelberg, Germany, 1998; pp. 155–157. [CrossRef]
48. Argillier, O.; Barrière, Y.; Traineau, R.; Émile, J.C.; Hebert, Y. Genotype X Environment Interactions for Digestibility Traits in Silage Maize Estimated from in Vivo Measurements with Standard Sheep. *Plant Breed.* **1997**, *116*, 423–427. [CrossRef]
49. McElroy, A.R.; Christie, B.R. Genotype × Environment Interactions for in Vitro Digestibility of Timothy (*Phleum pratense* L.) Genotypes. *Can. J. Plant Sci.* **1986**, *66*, 315–321. [CrossRef]
50. Resende, M.D.V.; De Duarte, J.B. Precisão E Controle de Qualidade em Experimentos de avaliação de Cultivares. *Agric. Res. Trop.* **2007**, *37*, 182–194.
51. Matias, F.I.; Barrios, S.C.L.; Valle, C.B.D.; Mateus, R.G.; Martins, L.B.; Moro, G.V. Estimate of Genetic Parameters in *Brachiaria decumbens* Hybrids. *Crop Breed. Appl. Biotechnol.* **2016**, *16*, 115–122. [CrossRef]
52. Cruz, C.D.; Regazzi, A.J.; Carneiro, P.C.S. *Modelos Biométricos Aplicados ao Melhoramento Genético*, 3rd ed.; UFV: Viçosa, Brazil, 2004.
53. Kanapeckas, J.; Tarakanovas, P.; Lemeþienë, N. Variability, Heritability and Correlations of Genetic Resources in Meadow Fescue. *Biologija* **2005**, *3*, 10–14.
54. De Santis, G.; Chiaravalle, A.E. Heritabilities of Nutritive Quality Factors and Interrelationships with Yield in Selected Progenies of Tall Fescue. *Plant Breed.* **2001**, *120*, 337–343. [CrossRef]
55. Thaikua, S.; Ebina, M.; Kouki, K.; Inafuku, M.; Akamine, H.; Shimoda, K.; Suenaga, K.; Kawamoto, Y. Preliminary Evaluation on Digestibility and the Relation to Morphology and Water Content of *Brachiaria* spp. And Their Heritability. *Grassl. Sci.* **2015**, *61*, 92–100. [CrossRef]
56. Casler, M.D.; Undersander, D.J.; Jokela, W.E. Divergent Selection for Phosphorus Concentration in Reed Canarygrass. *Crop Sci.* **2008**, *48*, 119–126. [CrossRef]
57. Nguyen, H.T.; Sleper, D.A. Genetic Variability of Mineral Concentrations in *Festuca arundinacea* Schreb. *Theor. Appl. Genet.* **1981**, *59*, 57–63. [CrossRef] [PubMed]
58. De Assis, G.M.L.; Valle, C.B.D.; De Andrade, C.M.S.; Valentim, J.F. Selecting New Brachiaria Humidicola Hybrids for Western Brazilian Amazon. *Trop. Grassl. Forrajes Trop.* **2013**, *1*, 42. [CrossRef]
59. Annicchiarico, P.; Romani, M. Genetic Variation, Heritability and Genetic Correlations for Forage Quality and Yield Traits of Mediterranean Tall Fescue Germplasm. *Plant Breed.* **2005**, *124*, 99–101. [CrossRef]
60. Simeão, R.M.; Silva, A.; Valle, C.; Resende, M.D.; Medeiros, S. Genetic Evaluation and Selection Index in Tetraploid *Brachiaria ruziziensis*. *Plant Breed.* **2016**, *135*, 246–253. [CrossRef]
61. Simeão, R.M.; Assis, G.M.L.; Montagner, D.B.; Ferreira, R.C.U. Forage Peanut (*Arachis* spp.) Genetic Evaluation and Selection. *Grass Forage Sci.* **2016**, *72*, 322–332. [CrossRef]

62. Corriher, V.; Redmon, L.; Baxter, L.L. Genetic Resources and Genetic Transformation in bermudagrass—A Review. *Crop Sci.* **2009**, *47*, 1–9.
63. Lee, M.A. A Global Comparison of the Nutritive Values of Forage Plants Grown in Contrasting Environments. *J. Plant. Res.* **2018**, *131*, 641–654. [CrossRef]

Publisher's Note: MDPI stays neutral with regard to jurisdictional claims in published maps and institutional affiliations.

agronomy

MDPI

Article

Phenotypic Divergence Analysis in Pigeonpea [*Cajanus cajan* (L.) Millspaugh] Germplasm Accessions

Esnart Nyirenda Yohane *, Hussein Shimelis, Mark Laing, Isack Mathew and Admire Shayanowako

African Centre for Crop Improvement, School of Agricultural, Earth and Environmental Sciences, University of KwaZulu-Natal, Private Bag X01, Scottsville 3201, Pietermaritzburg 3209, South Africa; Shimelish@ukzn.ac.za (H.S.); laing@ukzn.ac.za (M.L.); isackmathew@gmail.com (I.M.); Shayanowako@googlemail.com (A.S.)

* Correspondence: nyirendaesnart@yahoo.com; Tel.: +26-5999104624

Received: 11 September 2020; Accepted: 9 October 2020; Published: 30 October 2020

Abstract: Pigeonpea (*Cajanus cajan* (L.) Millspaugh) is an important source of grain protein for low-income countries such as Malawi. Knowledge of the genetic diversity in pigeonpea is essential for an effective breeding program. The study objective was to assess the genetic diversity among diverse pigeonpea accessions to select complementary and unique genotypes for breeding. Eighty-one pigeonpea accessions were evaluated in six environments in Malawi using a 9 × 9 alpha-lattice design with two replications. The cross-tabulation analysis revealed a significant genotype variation on plant growth, flower, and seed traits. The combined analysis of variance identified genotypes MWPLR 14, ICEAP 01170, ICEAP 871091, and ICEAP 01285 as early maturing varieties, while Kachangu, MWPLR 16, TZA 5582, No. 40, and MWPLR 14 were identified as high-yielding genotypes. The correlation analysis revealed a significant positive correlation between grain yield and a hundred seed weight (HSWT) (r = 0.50, $p < 0.01$), suggesting the usefulness of this trait for selection. The nonlinear principal component analysis identified grain yield (GDY), days to 50% flowering (DTF), days to 75% maturity (DTM), number of pods per plant (NPP), number of racemes per plant (NRP), 100 seed weight (HSWT), leaf hairiness (LH), and number of seeds per pod (NSP) as the most discriminated traits among the test genotypes. The cluster analysis using morphological traits delineated the accessions into three clusters. The selected high-yielding and early-maturing genotypes may be recommended as parental lines for breeding and grain yield improvement in Malawi or similar agro-ecologies.

Keywords: agronomic performance; correlation analysis; malawi; pigeonpea; yield stability

1. Introduction

Pigeonpea (*Cajanus cajan* (L.) Millspaugh, 2n = 2x = 22) is an essential cash and food crop in the tropical and subtropical regions of the world. It is a multi-purpose crop that is cultivated mainly for its edible grains that are high in dietary protein and essential amino acids such as leucine (16.48 g/kg), tyrosine (14.77 g/kg), and arginine (13.51 g/kg) [1].

Pigeonpea is an essential component of the agriculture systems in semi-arid ecologies due to its adaptation to growing with relatively low rainfall and with poor soil fertility. It has a deep root system and a unique ability to maintain optimal osmotic adjustment under limited water conditions [2]. Pigeonpea can fix atmospheric nitrogen in the soils through symbiosis with species of *Rhizobium* bacteria depositing up to 200 kg of nitrogen per hectare in agricultural lands [3,4]. Thus, pigeonpea has important roles in enhancing food security and livelihoods, especially during drought years, and providing ecosystem services through nitrogen fixation and soil health improvement.

Pigeonpea accounts for 5% of the world's pulse production [5]. India is the largest producer of pigeonpea, accounting for 25% of the world's production, followed by Myanmar and Malawi [6]. In Malawi, pigeonpea accounts for more than 22% of total legume production and ranks as the 3rd most important legume crop after groundnut and common beans. The grain productivity of pigeonpea in Malawi is low (≈700 kg ha^{-1}) compared to its potential yield of 2500 kg ha^{-1} [7]. The yield gap is due to various constraints, including insect pests and diseases, drought stress, and a lack of improved cultivars. The breeding and deployment of improved cultivars can enhance pigeonpea production and productivity. The successful development of improved cultivars with the client and market-preferred traits depends on the availability of adequate genetic variation.

Reportedly, modern pigeonpea cultivars and varieties exhibit relatively low levels of genetic diversity [8]. The loss of genetic diversity is due to continuous artificial selection and breeding for a few targeted economic traits to meet the market requirements [9]. Hence, there is a need to initiate pre-breeding programs in the target production environments through divergence breeding involving modern and obsolete cultivars, landraces, and wild relatives that possess desirable traits. This will broaden the genetic diversity of pigeonpea through gene recombination and effective selection [9]. The International Crops Research Institute for the Semi-Arid Tropics (ICRISAT) and various national and regional improvement programs are actively involved in genetic improvement and conservation of the pigeonpea. Diverse pigeonpea collections are preserved globally, including by ICRISAT, the International Institute of Tropical Agriculture (IITA), and the Svalbard Global Seed Vault in Norway. These genetic resources can be used for pigeonpea improvement and breeding programs globally [10].

To date, only seven pigeonpea cultivars have been released in Malawi. These cultivars were introductions from ICRISAT [7] developed in Kenya with germplasm from eastern and southern Africa (ESA). The ESA region is recognized as a secondary center of genetic diversity for pigeonpea. The introduced cultivars are poorly adapted to local farming conditions in Malawi and lack farmer-preferred traits such as good cooking quality, resistance to pod borers, and high yield potential. Therefore, the development of high performance, locally adapted pigeonpea cultivars is an important target in Malawi. This requires a range of genetic resources and crosses to integrate adaptive and functional traits, according to the needs and preferences of farmers and the value chain. Introduced germplasm can provide useful genetic resources that can be introgressed into locally adapted germplasm to improve economic traits such as high yield, early maturity, and pest and disease resistance, among others [9]. Evaluating accessions maintained by the public and private breeding sectors within the ESA region provides an opportunity to identify stable and high-yielding genotypes for selection.

Many pigeonpea genotypes have been collected and maintained at the Department of Agricultural Services in Malawi for breeding purposes. The genotypes are adapted to the ESA region and possess valuable attributes including good cooking quality, insect pests, and disease resistance, but they are limited by their poor yield performance. The key traits present in the local and introduced germplasm should be assessed for pre-breeding and breeding purposes. Hence, the objectives of the study were to determine the genetic diversity among pigeonpea accessions in selected target production environments in Malawi to select complementary and unique genotypes for breeding.

2. Materials and Methods

2.1. Plant Materials

The study evaluated 81 pigeonpea genotypes comprising 28 landraces, 6 released cultivars, and 47 advanced elite lines (Table 1), which were sourced from the Department of Agricultural Research Services (DARS)/Malawi and the Tanzania Agriculture Research Institute (TARI) and the International Crop Research Institute for the Semi-Arid Tropics (ICRISAT)/Kenya. The landraces were included as checks for adaptation to local conditions and possessing farmers' traits, while the elite lines provide important genetic resources, since Tanzania and Kenya have more advanced pigeonpea-breeding programs. The released cultivars provided a benchmark against commercial standards that are currently in production.

Table 1. Description of the pigeonpea genotypes used in the study.

Code	Genotype Designation	Description	Source	Origin	Code	Genotype Designation	Description	Source	Origin
G1	ICEAP 0673/1	Breeding line	ICRISAT	Kenya	G42	ICEAP 87105	Cultivar	ICRISAT	Kenya
G2	ICEAP 00554	Breeding line	ICRISAT	Kenya	G43	MWPLR 16	Landrace	GENEBANK	Malawi
G3	ICEAP 01164/1	Breeding line	ICRISAT	Kenya	G44	TZA 2496	Landrace	TARI	Tanzania
G4	MWPLR 19	Landrace	GENEBANK	Malawi	G45	TZA 5582	Landrace	TARI	Tanzania
G5	MWPLR 22	Landrace	GENEBANK	Malawi	G46	TZA 5596	Landrace	TARI	Tanzania
G6	ICEAP 01170	Breeding line	ICRISAT	Kenya	G47	Chitedze Pigeonpea 2	Cultivar	DARS	Malawi
G7	ICEAP 01169	Breeding line	ICRISAT	Tanzania	G48	MWPLR 7	Landrace	GENEBANK	Malawi
G8	TZA 2439	Landrace	TARI	Tanzania	G49	Babati	Landrace	TARI	Tanzania
G9	MWPLR 9	Landrace	GENEBANK	Malawi	G50	TZA 5557	Landrace	TARI	Tanzania
G10	MWPLR 6	Landrace	GENEBANK	Malawi	G51	MWPLR 14	Landrace	ICRISAT	Kenya
G11	MWPLR 17	Landrace	GENEBANK	Malawi	G52	ICEAP 01101/1	Breeding line	ICRISAT	Kenya
G12	TZA 253	Landrace	TARI	Tanzania	G53	TZA 2456	Landrace	TARI	Tanzania
G13	MWPLR 1	Landrace	GENEBANK	Malawi	G54	TZA 5464	Landrace	TARI	Tanzania
G14	MWPLR 18	Landrace	GENEBANK	Malawi	G55	ICEAP 01101/2	Breeding line	ICRISAT	Kenya
G15	TZA 2464	Landrace	TARI	Tanzania	G56	ICEAP 01285	Breeding line	ICRISAT	Kenya
G16	ICEAP 00604	Breeding line	ICRISAT	Kenya	G57	MWPLR 25	Landrace	GENEBANK	Malawi
G17	TZA 2509	Landrace	GENEBANK	Malawi	G58	ICEAP 87091	Breeding line	ICRISAT	Kenya
G18	ICEAP 01146/1	Breeding line	ICRISAT	Kenya	G59	TZA 2692	Landrace	TARI	Tanzania
G19	MWPLR 11	Landrace	GENEBANK	Malawi	G60	TZA 2807	Landrace	TARI	Tanzania
G20	TZA 5555	Landrace	TARI	Tanzania	G61	ICEAP 00068	Breeding line	ICRISAT	Kenya
G21	No. 40	Landrace	TARI	Tanzania	G62	TZA 2785	Landrace	TARI	Tanzania
G22	ICEAP 01150	Breeding line	ICRISAT	Kenya	G63	MWPLR 10	Landrace	GENEBANK	Malawi
G23	MZ2/9	Breeding line	TARI	Tanzania	G64	ICEAP 00612	Breeding line	ICRISAT	Kenya
G24	ICEAP 01172/1	Breeding line	ICRISAT	Kenya	G65	MWPLR 21	Landrace	GENEBANK	Malawi
G25	ICEAP 01103/1	Breeding line	ICRISAT	Kenya	G66	TZA 2514	Landrace	TARI	Tanzania
G26	MWPLR 24	Landrace	GENEBANK	Malawi	G67	TZA 2466	Landrace	TARI	Tanzania
G27	ICEAP 01155	Breeding line	ICRISAT	Kenya	G68	ICEAP 01179	Breeding line	ICRISAT	Kenya
G28	ICEAP 01180/2	Breeding line	ICRISAT	Malawi	G69	MWPLR 13	Landrace	GENEBANK	Malawi
G29	MWPLR 4	Landrace	GENEBANK	Malawi	G70	MWPLR 2	Landrace	GENEBANK	Malawi
G30	Kachangu	Cultivar	DARS	Malawi	G71	TZA 250	Landrace	DARS	Malawi
G31	Mwayiwathualimi	Cultivar	DARS	Kenya	G72	MWPLR 3	Landrace	GENEBANK	Malawi
G32	MWPLR 8	Landrace	ICRISAT	Malawi	G73	TZA 5541	Landrace	TARI	Tanzania
G33	ICEAP 01154/2	Breeding line	ICRISAT	Kenya	G74	MWPLR 23	Landrace	GENEBANK	Malawi
G34	Chitedze Pigeonpea 1	Cultivar	DARS	Malawi	G75	ICEAP 00979/1	Breeding line	ICRISAT	Kenya
G35	ICEAP 01164	Breeding line	ICRISAT	Kenya	G76	TZA 197	Landrace	TARI	Tanzania
G36	Bangili	Landrace	TARI	Tanzania	G77	MWPLR 20	Landrace	GENEBANK	Malawi
G37	ICEAP 00053	Breeding line	ICRISAT	Kenya	G78	HOMBOLO	Landrace	TARI	Tanzania
G38	MWPLR 12	Landrace	GENEBANK	Malawi	G79	ICEAP 86012	Breeding line	ICRISAT	Kenya
G39	TZA5463	Landrace	TARI	Tanzania	G80	ICEAP 01106/1	Breeding line	ICRISAT	Kenya
G40	MWPLR 5	Landrace	GENEBANK	Malawi	G81	Sauma	Cultivar	DARS	Malawi
G41	MWPLR 15	Landrace	GENEBANK	Malawi					

ICRISAT = International Crops Research Institute for the Semi-Arid Tropics, DARS = Department of Agricultural Research Services, TARI = Tanzania Agricultural Research Institute.

2.2. Study Sites

Field experiments were conducted in Malawi at three sites, Bvumbwe, Chitedze, and Makoka Research Stations, during the 2017/18 and 2018/19 cropping seasons. The geographic location, altitude, weather, and soil characteristics of the study locations are presented in Table 2. Each season and site combination presented unique environmental conditions due to variations in temperature, rainfall, and agronomic practices. Therefore, due to site × season combinations, six environments were identified for evaluating the genotypes: Bvumbwe during 2017/18 (Environment 1), Bvumbwe in 2018/19 (Environment 2), Chitedze in 2017/18 (Environment 3), Chitedze in 2018/19 (Environment 4), Makoka in 2017/18 (Environment 5), and Makoka in 2018/19 (Environment 6).

Table 2. Physical and weather characteristics of the study locations.

Site	Latitude	Longitude	Altitude (Masl)	Soil Texture	Rainfall (mm)		Min Temp (°C)		Max Temp (°C)	
					2017/18	2018/19	2017/18	2018/19	2017/18	2018/19
Bvumbwe	15°55′ S	35°04′ E	1228	Sandy clay loam	975.2	1442	16.2	17.9	22.6	24.9
Chitedze	13°59′ S	33°38′ E	1146	Sandy clay	929.8	693.4	18.5	20.2	24.7	29.4
Makoka	15°32′ S	35°11′ E	1029	Sandy clay loam	566.6	1184.8	16.3	15.6	23.2	28.2

Masl = meters above sea level, mm = millimeters, min = minimum, max = maximum, temp = temperature, °C = degrees Celsius.

2.3. Experimental Design and Data Collection

The experiment at each site was laid out in an alpha-lattice design with two replications. Each genotype was planted on a plot consisting of two rows. The rows were 5 m in length and 0.90 m apart, giving a plot size of 4.5 m^2. Seeds were planted at 0.75 m apart within a row. Three seeds were planted per planting station and thinned to one plant two weeks after emergence. All agronomic practices were applied following standard practices for pigeonpea production in Malawi [7]. Both qualitative and quantitative phenotypic traits' data were collected as presented in Table 3 according to pigeonpea descriptors of the International Board for Plant Genetic Resource (IBPGR) and International Centre for Research Institute for Semi-Arid Tropics (ICRISAT) [11].

2.4. Statistical Analysis

Data collected on qualitative traits (Table 3) were subjected to frequency distribution and cross-tabulation analyses using SPSS for Windows 25.0 [12].

The quantitative data from each variable were tested for homogeneity of variances using Bartlett's test and data normality using the Shapiro–Wilkes test before the analysis of variance (ANOVA). Subsequently, the data were pooled across sites and subjected to a combined analysis of variance following the alpha lattice procedure in Genstat 18th edition [13]. The total variance was partitioned into genotype (σ^2g), environment (σ^2e), and genotype by environment (σ^2ge) components based on the mean squares derived from the partial analysis of variance adapted from [14]. Correlation and principal component analyses were performed using Genstat 18th edition [13] to determine influential components and trait relationships. A nonlinear principal component analysis was conducted in SPSS (SPSS 2016). The nominal variables (qualitative traits) were transformed using the categorical principal component analysis (CATPCA) procedure described by [15]. The nonlinear PCA can standardize both quantitative and qualitative data to deduce their associations and identified the most important components.

Table 3. Descriptors for the pigeonpea qualitative and quantitative traits.

Traits	Code	Description
		Qualitative Traits
Plant habit	PH	1 = Compact (erect), 2 = semi-spreading (semi-erect) or 3 = spreading
Flower streak pattern	FSP	0 = no streaks, 1 = sparse, 2 = medium and 3 = dense streaks, 4 = uniform coverage of second color
Flower base/main color	FBC	1 = ivory (green white), 2 = light yellow, 3 = yellow, 4 = orange, 5 = red, 6 = purple
Leaf shape	LS	1 = ovate, 2 = triangular, 3 = trullate
Leaf hairiness	LH	1 = hairy, 2 = non-hairy
Pod form	PF	1 = flat, 2 = cylindrical
Pod color	PC	1 = green, 2 = purple, 3 = mixed (green +purple) and 4 = dark purple
Seed color pattern	SCP	1 = plain, 2 = mottled, 3 = speckled, 4 = mottled and speckled, 5 = ringed
Seed main color	SMC	1 = white (yellow white), 2 = cream (gray white), 3 = orange, 4 = brown, 5 = grey, 6 = purple, 7 = black
Seed eye color	SEC	1 = purple, 2 = light brown, 3 = reddish brown, 4 = gray/dark, 5 = cream/white
Seed shape	SSH	1 = Oval, 2 = pea-shape, 3 = square/angular, 4 = elongate
		Quantitative Traits
Plant height	PH	Measured in cm from plant base to the tip of the main stem
Days to 50% flowering	DTF	Number of days from sowing until when 50% of the plants have at least one open flower
Primary branches	PBR	The average number of primary branches of 10 randomly selected and tagged plants
Secondary branches	NSB	The average number of secondary branches of 10 randomly selected and tagged plants
Days to 75% maturity	DTM	Number of days from sowing until when 75% of the pods in a plot turn brown
Number of seeds per pod	NSP	The average number of pods per plant from 10 randomly selected and tagged pods
Number of pods per plant	NPP	The average number of pods from 10 randomly selected and tagged plants
Number of racemes per plant	NRP	The average number of racemes from 10 randomly selected and tagged plants
Grain yield (t/ha)	GYD	Weight of the grain harvested in a plot extrapolated to t/ha
100 seed weight (g)	HSWT	Weight of a random sample of 100 grain

3. Results

3.1. Genotype Variation Based on Qualitative Traits

Significant variations were exhibited among genotypes for all assessed qualitative traits ($p < 0.001$) such as growth habit, flower main color, flower streak pattern, pod color, and seed traits (Table 4, Figure 1A–D). A large proportion of test genotypes (61.9%) were semi-spreading, followed by spreading (26.6%) and compact (11.5%) in growth habits. A majority of the test genotypes (64.9%) had yellow flower color (Table 4, Figure 1A), while 16.8% had purple flowers, 13.6% had ivory flowers, and 7.4%

had light yellow flowers (Table 4, Figure 1A). A large population of the genotypes (60.5%) had no flower streaks, and the rest of the genotypes had sparse, medium, dense, and uniform coverage streaks at 8.1%, 1.9%, 14.5%, and 15%, respectively (Table 4, Figure 1B). About 48.7% of the genotypes had a green pod color, while 33.9% had a mixed pod color and 7.1% had purple pods (Table 4, Figure 1C). A majority of the genotypes (76.8%) had a cream seed coat color, while 11% had a brown seed coat color and the rest had gray, orange, and purple seed coat colors (Table 4, Figure 1D). About 70.2% of the test genotypes had a brown seed eye, and 20.7% had a purple seed eye, while the remainder had gray or cream seed eyes. The most common seed shape was square or angular shapes, which were exhibited by 69.3% of the test genotypes.

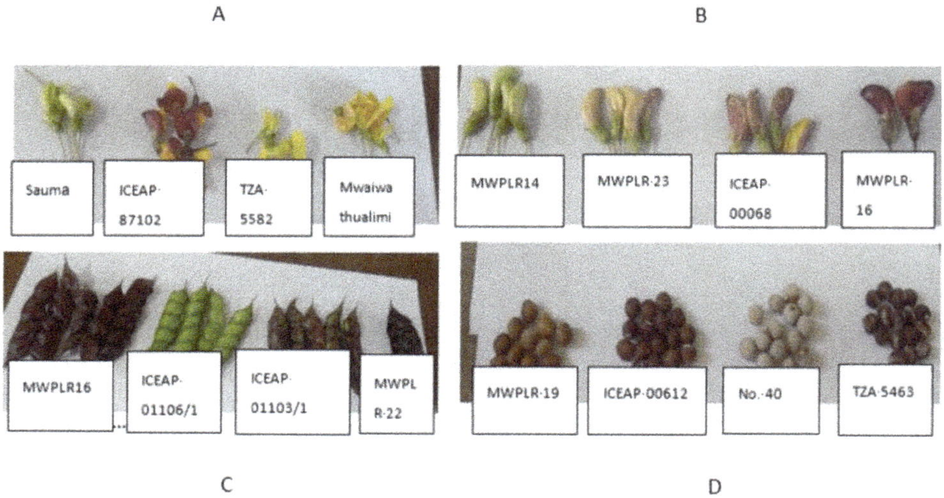

Figure 1. Genetic variability for some qualitative traits in pigeonpea genotypes: (**A**) flower color: genotype Sauma (ivory), ICEAP 87105 (purple), TZA 5582 (yellow), Mwaiwathualimi (light yellow); (**B**) flower streak pattern: genotype MWPLR 14 (no streak), MWPLR 23 (medium streaks), ICEAP 00068 (dense streaks), MWPLR 16 (uniform coverage); (**C**) pod color: genotype MWPLR 16 (purple), ICEAP 01106/1 (green), ICEAP 01103/1 (mixed), MWPRL 22 (dark purple) and (**D**) seed coat color: genotype MWPLR 19 (orange), ICEAP 00612 (brown), No. 40 (cream), and TZA 5463 (purple).

3.2. Genotype and Environment Variances for Quantitative Traits

The quantitative agronomic data were pooled across sites after applying tests for homogeneity of variance and normality. The genotype × environment interaction effects were significant ($p < 0.001$) for grain yield (GYD), days to 50% flowering (DTF), days to 75% maturity (DTM), plant height (PH), number of primary branches (NPB), number of pods per plant (NPP), number of racemes per plant (NRP), 100 seed weight (HSWT), and number of secondary branches per plant (NSB) (Table 5). The genotype and environment had significant ($p < 0.001$) effects on all assessed traits except the NSP. The site × season × type interaction effects were significant ($p \leq 0.001$ and $p = 0.05$, respectively) for DTM, PH, and NSP (Table 6). The site × type interaction effects were only significant ($p = 0.01$ and $p = 0.05$) for DTM, NPP, NRP, and HSWT, respectively. However, season × type interaction effects were not significant for all the variables except for GYD, which was highly significant.

Table 4. Frequency distribution and significance tests among 81 pigeonpea genotypes assessed based on qualitative traits.

Trait	Description	Frequency (%)	DF	Chi-Square	Genotype Code [a]
Growth habit	Compact	11.5	160	304.52 **	G53, G2, G1, G27, G26
	Semi-spreading	61.9			G63, G50, G28, G70, G76, G80, G51, G78, G49, G32, G62, G39, G67, G5, G8, G13, G72, G24, G74, G3,32, G22, G4, G40, G30, G52, G56, G48, G79, G36, G23,G16, G77, G7, G71, G44, G67, G46, G69, G33, G54, G20, G43, G42, G71, G62, G65,G39, G69, G17, G18, G59
	Spreading	26.6			G45, G41, G29, G49, G56, G64, G37, G60, G15, G11, G65, G75, G81, G44, G67, G11, G46
Flower color	Ivory	13.6	240	910.08 ***	G78, G40, G36, G27, G33, G80, G51
	Light yellow	7.4			G13, G5, G31
	Yellow	64.9			G50, G45, G70, G53, G76, G72, G24, G74, G3, G22, G4, G58, G68, G18, G19, G17, G9, G62, G29, G32, G65, G21, G52, G1, G56, G37, G48, G79. G23, G16, G61, G77, G7, G71, G44, G15, G67, G11, G69, G65, G75, G20, G43, G26, G71, G44, G15, G67, G62, G11, G46, G65
	Purple	16.8			G63, G28, G41, G56, G60, G25, G46, G54, G26, G42
Flower streak pattern	No streaks	60.5	320	589.69 ***	G17, G53, G36, G12, G15, G37, G20, G60, G9, G54, G11, G66, G55, G80, G81, G71, G73, G23, G1, G65, G21, G18, G7, G13, G51, G62, G48, G49, G58, G14, G32, G16, G2, G27, G22, G6, G57, G10, G31, G8, G39, G30
	Sparse streaks	8.1			G49, G69, G42, G33, G28, G5, G70
	Medium sparse	1.9			G72, G74
	Dense streaks	14.5			G47, G61, G29, G60, G34, G40, G45, G67, G45, G68, G63, G77, G19
	Uniform coverage	15			G79, G50, G76, G59, G25, G46, G78, G38, G51, G75, G26, G35, G52, G56, G41, G43
Pod color	Green	48.7	240	647.43 ***	G73, G42, G1, G24, G74, G75, G52, G16, G65, G21, G18, G7, G13, G62, G17, G47, G61, G15, G20, G29, G44, G72, G60, G64, G9, G11, G66, G55, G80, G71, G58, G14, G27, G6, G57, G10, G8, G19
	Purple	7.1			G76, G45, G67, G38
	Mixed (green + purple)	33.9			G81, G70, G53, G36, G61, G43, G37, G34, G54, G79, G50, G40, G25, G33, G46, G42, G51, G4, G68, G26, G49, G3, G35, G32, G69, G2, G63, G22, G56, G77, G41, G30
	Dark purple	10.3			G31, G28, G39, G48, G59, G43

Table 4. *Cont.*

Trait	Description	Frequency (%)	DF	Chi-Square	Genotype Code [a]
Seed color pattern	Plain	56.6	240	841.57 ***	G59, G80, G5, G18, G6, G53, G65, G62, G35, G34, G67, GG4, G60, G66, G21, G70, G36, G42, G40, G14, G50, G66, G20, G79, G49, G2, G3, G69, G56, G81, G47, G72, G15, G44
	Mottled	15.3			G41, G25, G34, G48, G28, G78, G23, G31, G9, G37, G57
	Speckled	22.2			G75, G68, G43, G38, G10, G19, G52, G58, G51, G73, G59, G76, G16, G29, G13, GG3, G17, G8, G54, G1, G24, G7, G71, G27, G12, G22, G55, G77
	Mottled + speckled	5.9			G46, G33, G30, 632, G39, G45, G26
Seed main color	Cream	76.8	320	1049.31 ***	G75, G68, G59, G43, G5, G18, G6, G38, G10, G53, G65, G63, G35, G19, G34, 52, G72, G15, G44, G22, G55, G57, G77, G60, G58, G78, G32, G73, G51, G70, G36, G16, G29, G42, G40, G23, G14, G17, G8, G50, G66, G20, G49, G54, G2, G3, G69, G1, G24, G45, G7, G9, G71, G81, G12, G47
	Orange	3			G4, G46, G25
	Brown	11			G64, G76, G63, G30, G34, G48, G28, G31, G37, G26
	Gray Purple	6.2 3			G80, G66, G67, G56 G39, G33, G41
Seed shape	Oval	30.7	80	480.21 ***	G75, G22, G5, G25, G38, G53, G35, G34, G28, G73, G51, G70, G36, G29, G42, G40, G31, G8, G18, G49, G3, G45, G37, G28, G27, G12, G55, G57
	Square/angular	69.3			G15, G44, G22, G77, G68, G59, G43, G46, G80, G18, G33, G30, G41, G6, G10, G65, G62, G19, G34, G67, G4, G52, G48, G60, G58, G66, G32, G64, G76, G21, G16, G13, G23, G14, G63, G17, G39, G52, G66, G79, G54, G2, G69, G1, G24, G56, G7, G9, G71, G81
Seed eye color	Purple	20.7	240	848.32 ***	G68, G5, G34, G25, G60, G78, G51, G64, G76, G21, G16, G29, G42, G40, G31, G50, G49, G2, G69, G24, G81, G55, G57
	Light brown	70.2			G75, G59, G43, G46, G18, G33, G30, G41, G6, G10, G53, G65, G62, G35, G19, G34, G67, G52, G48, G58, G28, G66, G32, G73, G36, G23, G14, G17, G39, G74, G20, G79, G54, G1, G46, G45, G9, G71, G37, G27, G12, G47, G15, G44, G22
	Gray/dark	1.2			G25
	Cream	7.5			G80, G38, G63, G8, G7, G26

DF = degrees of freedom, ** and *** = significance at 0.01 and 0.001 levels, respectively; [a] see genotype codes (G1–G81) in Table 1.

Table 5. Mean squares and significant tests for grain yield and yield components measured in 81 pigeonpea genotypes across six environments in Malawi.

Source of Variation	DF	DTF	DTM	PH	NPB	NSB	NRP	NPP	NSP	GYD	HSWT
Location	2	9024.2 ***	8735.4 ***	54,965 ***	114.4 ***	93.7 *	226.9 ***	3236 **	22.5 ***	5,968,860 ***	1008.1 ***
Replication (Rep)	1	701.9 ns	289 ns	118 ns	1.2 ns	105.4 *	14,646 ns	9810 *	0.45 ns	1,663,232 *	9.5 ns
Block (Rep)	8	3168.5 ***	5703.4 ***	7710.9 ns	52.9 *	93.7 *	9099 *	6433.6 **	2.4 *	16,534,356.5 ***	72.2 **
Genotype (G)	80	879.2 ***	1234.9 ***	2137 ***	12.5 *	30.9 *	5004.9 *	1990.3 *	0.8 ns	351,745.3 *	16.8 *
Season (S)	1	3370.5 **	2945.3 *	447 ns	409.6 ***	650.1 ***	2,023,492 ***	437.5 ***	31.5 ***	30,308,789 ***	50.2 *
G × L	160	243 *	361.9 *	1106 *	18 *	35.6 *	6150.9 *	1916.1 *	0.9 *	360,816.9 *	20.7 **
G × S	80	3610.3 ns	606.9 ns	1198 ns	17.9 *	34.7 *	4642.7 ns	1060.3 *	0.9 ns	400,468.2 *	14.9 ns
G × L × S	160	330.6 ns	484.9 ns	744 ns	15.2*	34.5 *	6110.9 ns	1502.8 *	0.7 ns	919,105.3 ns	16.2 ns
Residual	469	345.4	585.8	1243.1	14.5	11.8	5822.9	5667.2	0.8	313,554	15.4

DF = degrees of freedom, Rep = replication, DTF = days to 50% flowering, DTM = days to 75% maturity, PH = plant height, NPB = number of primary branches, NSB = number of secondary branches per plant, NRP = number of racemes per plant, NPP = number of pods per plant, NSP = number of seeds per pod, GYD = grain yield, HSWT = 100 seed weight, *, ** and *** = significance at 0.05, 0.01 and 0.001 probability levels, respectively.

Table 6. Mean squares and significant tests for grain yield and yield components among the three categories of pigeonpea genotypes.

Source of Variation	DF	DTF	DTM	PH	NPB	NSB	NRP	NPP	NSP	GYD	HSWT
Site	2	9167 ***	8020 ***	55,114 ***	108.6 ***	93 *	1,309,332 ***	118,174 ***	22.54 ***	1.658 ***	80.16 ***
Rep	1	717	328	111	2.7	109	14184	9836	0.45	1.651	0.26
Rep (Block)	16	123	82	9769	33.3	71	14908	21,509 ***	1.35	0.206	11.78
Season	1	3797 **	3625 *	407	433.3 ***	4672 ***	2,043,617 ***	440,237 ***	31.51 ***	3.092 ***	0.04
Type	2	1629 *	4725 **	44,433 **	18.5	14	4686	5891 *	40.2 ***	2.087 ***	20.09
Site*Season	2	2523 **	700	55,081 ***	910.5 ***	253 ***	1,018,464 ***	149,039 ***	36.09 ***	6.38 ***	39.42 *
Site*Type	4	114	4385 **	1257	17	29	988,914 *	8167 ***	0.41	0.078	45.83 *
Season*Type	2	161	388	3023	9.9	2	642	380	1.94	2.006 ***	6.44
Site*Season*Type	4	676	7883 ***	65,810 ***	3.1	44	1771	2032	30.17 *	0.2	1.82
Residual	937	386	619	1177	15.5	31	5823	1682	0.79	0.149	12.45

DF = degrees of freedom, Rep = replication, DTF = days to 50% flowering, DTM = days to 75% maturity, PH = plant height, NPB = number of primary branches, NSB = number of secondary branches per plant, NRP = number of racemes per plant, NPP = number of pods per plant, GYD = grain yield, HSWT = 100 seed weight, *, ** and *** = significance at 0.05, 0.01 and 0.001 probability level.

3.3. Mean Performance of Pigeonpea Genotypes across the Test Environments

Tables 7–10 summarize the mean values and statistics for eight quantitative traits recorded from three locations in two seasons. The tables presents the best ten, and the bottom five genotypes on DTF, DTM, PH, NPB, NRP, NPP and HSWT ranked on grain yield response. The mean DTF and DTM were 112 and 157 days, respectively (Table 7). Genotype MWPLR 14 was the earliest to attain 50% flowering and maturity at 74 and 113 days, which was followed by ICEAP 01170 at 85 and 125 days, ICEAP 87091 at 85 and 132 days, ICEAP 01285 at 87 and 133 days, and ICEAP 01169 at 91 and 137 days, respectively. Sauma was among the latest genotypes to flower and mature at 145 and 205 days, respectively. There were marked genotype differences in plant height that varied from 125.3 to 202.4 cm (Table 8). The mean plant height of the test genotypes was 167.5 cm. The shortest genotype across the testing environments was ICEAP 87105. The tallest genotypes with plant heights exceeding 180 cm were Kachangu, No. 40, ICEAP 01106/3, ICEAP 00068, TZA 5596, MWPLR 6, Sauma, and ICEAP 00053. The mean number of the primary branch of the test genotypes was 15 (Table 8). The most productive genotypes with many primary branches per plant were MWPLR 12, MWPLR 20, ICEAP 01170, and MWPLR 23, with 19, 18, 17, and 17 primary branches per plant, in that order. The mean number of pods per plant varied from 67 to 144, with a grand mean of 94 pods per plant (Table 9). The highest number of pods per plant was 144, 134, 126, 124, and 123 observed on the genotypes Kachangu, MWPLR 16, TZA 5582, No. 40, and MWPLR 14, in that order. The number of seeds per pod exhibited non-significant differences among the assessed genotypes. The mean number of grains per pod was five. There was a wide genetic variation for grain yield ranging from 0.5 to 1.8 t ha^{-1} with a mean of 1.1 t ha^{-1} (Table 10). Accessions No. 40, MWPLR 14, and MWPLR 16 were the three best performing genotypes with mean yields of 1.8, 1.7, and 1.7 t ha^{-1}, respectively. The lowest grain yield response was 0.5 t/ha recorded for the genotypes ICEAP 00604 and ICEAP 01285. The 100 seed weight ranged from 11.0 to 17.3 g/100 seed (Table 10). Accessions MWPLR 22, TZA 5582, and MWPLR 14 expressed the highest HSWT ≥ 17 g/100 seed.

Table 7. Mean values for 10 quantitative traits among the ten top best and five bottom performing genotypes after evaluating 81 genotypes in six environments in Malawi.

Genotype	DTF							DTM						
	Y1			YII			Mean	YI			YII			Mean
	S1	S2	S3	S1	S2	S3		S1	S2	S3	S1	S2	S3	
Top Ten Genotypes														
21	129	131	141	124	131	132	131	173	191	211	158	176	176	181
43	125	105	119	117	105	105	113	177	166	172	156	161	154	164
51	63	65	64	87	67	98	74	95	105	102	127	116	132	113
30	100	97	118	128	116	118	113	133	150	164	159	159	164	155
45	107	96	91	128	101	124	108	143	158	146	170	153	165	156
81	163	127	155	132	165	130	145	215	201	254	171	211	178	205
17	147	120	125	109	120	106	121	182	167	174	156	160	147	164
66	120	95	115	116	108	116	111	155	151	170	154	158	161	158
74	118	78	123	113	115	118	110	163	145	166	153	165	163	159
20	116	120	129	122	120	127	122	143	163	175	156	160	172	161
Bottom Five Genotypes														
39	113	90	131	85	90	88	99	149	144	195	127	150	122	147
13	126	117	109	116	107	115	115	167	166	153	145	154	155	156
50	117	77	107	116	77	115	101	141	136	156	155	137	149	145
42	114	102	127	120	102	120	114	145	154	172	164	166	162	160
79	124	101	122	117	127	119	118	168	153	165	152	179	161	163
Mean	117.8	102.8	115.5	110.6	106.1	113.1	110.6	154.7	156.5	163.2	148.7	155.7	154.3	155.3
STD	17.9	18.2	15.1	13.0	16.9	12.3	10.5	22.0	22.0	21.1	13.7	18.4	14.9	11.9
SED±	2.0	2.0	1.7	1.4	1.9	1.4	1.2	2.4	2.4	2.3	1.5	2.0	1.7	1.3
CV (%)	15.2	17.7	13.1	11.8	15.9	10.8	9.5	14.2	14.0	12.9	9.2	11.8	9.6	7.7

STD = standard deviation, SED = standard error of difference, CV = coefficient of variation, S1 = site 1 (Bvumbwe), S2 = site 2 (Chitedze), S3 = site 3 (Makoka), Y1 = year 1 (2017/18), Y11 = year 2 (2018/19), DTF = days to flowering, DTM = days to 75% maturity, See genotype codes (G1–G81) in Table 1.

Table 8. Mean values for plant height and number of primary branches among the ten top best and five bottom performing genotypes after evaluating 81 genotypes in six environments in Malawi.

Genotype	PH							NPB						
	Y1			Y11			Mean	Y1			Y11			Mean
	S1	S2	S3	S1	S2	S3		S1	S2	S3	S1	S2	S3	
Top Ten Genotypes														
21	166.5	220.0	193.0	160.0	212.8	193.0	190.9	19	19	17	14	18	12	16
43	113.5	147.5	127.5	96.5	146.7	148.0	163.7	14	15	17	14	17	11	15
51	151.5	109.0	158.0	234.5	209.4	149.0	168.6	13	12	14	18	13	11	13
30	229.5	188.5	204.0	170.0	218.5	204.0	202.4	15	13	18	15	16	15	15
45	139.5	144.5	173.0	161.5	169.4	197.5	164.2	15	13	22	15	17	14	16
81	163.0	222.0	191.0	160.5	168.1	194.5	183.2	13	17	19	18	12	14	15
17	163.5	164.0	163.5	100.0	152.1	156.0	149.9	15	14	21	17	16	13	16
66	181.5	177.5	164.0	161.5	156.8	149.5	165.1	12	13	13	14	16	12	13
74	156.0	195.0	185.5	124.5	178.7	164.0	167.3	15	18	17	20	18	12	17
20	152.5	163.0	168.5	138.5	247.5	166.5	172.8	10	12	20	12	18	11	14
Bottom Five Genotypes														
39	203	154.5	174	157.5	200	151.5	173.4	16	18	17	15	12	12	15
13	169	171.5	134	134.5	203.3	156.5	161.5	18	12	18	15	10	15	14
50	119	101.5	149.5	130.5	218.5	166.5	147.6	18	13	14	15	17	13	15
42	140	153	175.5	104.5	207.7	120	125.3	14	9	16	14	13	13	13
79	174	165.5	167.5	120.5	201.4	148	162.8	11	18	23	13	14	13	15
Mean	168.0	166.7	166.2	143.4	195.5	166.1	167.3	14.6	13.6	18.0	14.9	14.6	12.8	14.5
STD	23.9	34.5	22.1	23.0	27.0	23.1	12.6	2.7	4.4	2.7	2.4	3.2	2.0	1.3
SED±	2.7	3.8	2.5	2.6	3.0	2.6	1.4	0.3	0.5	0.3	0.3	0.4	0.2	0.1
CV (%)	14.2	20.7	13.3	16.0	13.8	13.9	7.5	18.7	32.1	15.0	16.3	22.0	15.6	9.1

STD = standard deviation, SED = standard error of difference, CV = coefficient of variation, S1 = site 1 (Bvumbwe), S2 = site 2 (Chitedze), S3 = site 3 (Makoka), Y1 = year 1 (2017/18), Y11 = year 2 (2018/19), PH = plant height (cm), NPB = number of primary branches, see genotype codes (G1–G81) in Table 1.

Table 9. Mean values for number of racemes and number of pods per plant among the ten top best and five bottom performing genotypes after evaluating 81 genotypes in six environments in Malawi.

Genotype	NRP							NPP						
	Y1			Y11			Mean	Y1			Y11			Mean
	S1	S2	S3	S1	S2	S3		S1	S2	S3	S1	S2	S3	
Top Ten Genotypes														
21	214	402	71	130	61	47	154	157	270	66	61	92	98	124
43	138	173	97	117	95	58	113	119	315	98	72	110	90	134
51	260	155	146	113	80	51	134	167	231	109	65	76	90	123
30	178	430	134	132	73	52	166	127	362	95	97	83	101	144
45	191	647	160	151	88	83	220	96	261	106	81	92	122	126
81	200	536	85	89	69	40	170	140	240	69	61	70	82	110
17	184	258	96	139	94	61	139	102	158	65	35	112	89	93
66	148	168	108	119	76	49	111	69	186	82	26	78	64	84
74	196	414	98	125	84	81	166	128	112	64	46	40	94	81
20	126	259	106	148	130	73	140	115	177	78	38	157	45	101
Bottom Five Genotypes														
39	161	465	103	145	55	60	165	128	125	93	38	61	82	88
13	155	228	80	119	99	52	122	98	195	55	37	60	95	90
50	116	321	199	195	81	46	159	79	78	60	59	96	84	76
42	122	150	87	151	80	62	109	99	78	90	62	67	70	78
79	98	552	70	131	163	54	178	53	226	51	26	165	90	102
Mean	174.1	312.3	99.0	161.6	91.8	58.9	149.4	114.6	148.2	80.0	51.0	80.9	86.7	93.4
STD	43.9	146.5	27.7	39.8	30.0	12.1	26.2	30.5	56.7	22.1	16.1	33.4	19.7	14.1
SED±	4.9	16.3	3.1	4.4	3.3	1.3	2.9	3.4	6.3	2.5	1.8	3.7	2.2	1.6
CV (%)	25.2	46.9	28.0	24.7	32.7	20.6	17.5	26.6	38.2	27.7	31.5	41.3	22.8	15.1

STD = standard deviation, SED = standard error of difference, CV = coefficient of variation, S1 = site 1 (Bvumbwe), S2 = site 2 (Chitedze), S3 = site 3 (Makoka), Y1 = year 1 (2017/18), Y11 = year 2 (2018/19), NRP = number of racemes per plant, NPP = number of pods per plant, see genotype codes (G1–G81) in Table 1.

Table 10. Mean values for grain yield and hundred seed weight among the ten top best and five bottom performing genotypes after evaluating 81 genotypes in six environments in Malawi.

Genotype	GYD							HSWT						
	Y1			Y11			Mean	Y1			Y11			Mean
	S1	S2	S3	S1	S2	S3		S1	S2	S3	S1	S2	S3	
Top Ten Genotypes														
21	2.1	0.9	2.3	2.4	1.3	1.7	1.8	16.0	16.5	10.0	10.5	12.5	15.5	13.5
43	1.7	1.7	1.6	1.8	1.6	1.9	1.7	17.0	14.5	14.0	17.0	22.5	13.0	16.3
51	1.8	1.0	2.1	2.1	1.7	1.7	1.7	16.5	17.5	14.5	18.5	21.5	13.5	17.0
30	2.3	1.6	1.2	1.2	1.4	1.8	1.6	17.5	17.0	15.0	16.0	16.0	12.0	15.6
45	1.5	0.9	1.4	1.5	2.3	1.9	1.6	18.4	19.0	15.5	16.0	16.5	18.0	17.2
81	1.3	0.5	1.5	1.6	2.3	2.3	1.6	19.5	16.0	15.5	19.0	15.0	11.0	16.0
17	1.1	0.5	0.7	1.4	2.5	3.0	1.5	18.5	14.0	11.0	17.5	20.0	15.5	16.1
66	2.4	1.5	1.2	1.2	1.4	1.5	1.5	15.5	15.5	15.0	17.5	17.5	13.5	15.8
74	2.2	1.6	1.1	1.0	1.1	1.8	1.5	14.5	14.5	15.5	16.9	20.0	13.5	15.8
20	1.2	0.9	1.7	1.7	1.7	1.2	1.4	16.0	12.5	15.0	18.5	15.0	14.0	15.2
Bottom Five Genotypes														
39	0.4	0.4	1.1	1.1	1.2	0.9	0.8	15.5	14.5	14.5	16	15	16	15.3
13	0.8	0.2	0.5	1.4	0.5	0.3	0.6	12.5	15	14	15	16.5	16	14.8
50	0.9	0.5	0.4	0.5	0.4	0.7	0.6	13	10.5	17.5	21	19	14.5	15.9
42	0.6	0.4	0.9	0.5	0.4	0.4	0.5	12	12.5	14	19	20	14.5	15.3
79	0.8	0.3	0.4	0.5	0.3	0.5	0.5	13	16.5	14.5	17.5	17.5	14	15.5
Mean	1.1	0.6	1.3	1.3	1.5	1.3	1.2	15.9	13.9	13.5	17.6	12.9	14.2	14.7
STD	0.4	0.3	0.4	0.4	0.5	0.4	0.2	2.4	3.2	2.4	2.3	4.5	2.5	1.3
SED±	0.0	0.0	0.0	0.0	0.1	0.0	0.0	0.3	0.4	0.3	0.3	0.5	0.3	0.1
CV (%)	37.3	43.3	32.8	32.1	31.3	33.5	20.5	15.1	22.9	18.0	13.2	35.1	17.5	8.9

STD = standard deviation, SED = standard error of difference, CV = coefficient of variation, S1 = site 1 (Bvumbwe), S2 = site 2 (Chitedze), S3 = site 3 (Makoka), Y1 = year 1 (2017/18), Y11 = year 2 (2018/19), GYD = grain yield (t ha^{-1}), HSWT = 100 seed weight (g), see genotype codes (G1–G81) in Table 1.

3.4. Correlation Analysis among Phenotypic Traits

Assessed traits exhibited variable degrees of associations with grain yield (Table 11). Grain yield was moderately correlated with HSWT (r = 0.50, $p < 0.01$). A number of secondary traits exhibited variable pairwise correlations. DTF and DTM exhibited the strongest correlation (r = 0.79, $p < 0.01$). There were moderate correlations between DTF and PH (r = 0.44, $p < 0.01$), NPB and NSP (r = 0.41, $p < 0.01$), and DTM and PH (r = 0.41, $p < 0.01$). Relatively, HSWT exhibited weak correlations (r < 0.30) with NPB and NPP.

Table 11. Phenotypic correlation coefficients among the ten quantitative traits of 81 pigeonpea genotypes evaluated in six environments.

Trait	DTF	DTM	PH	NPB	NSB	NRP	NPP	NSP	GYD	HSWT
DTF	1	0.787 **	0.442 **	0.069	0.006	0.063	0.121	−0.134	0.232 *	−0.021
DTM		1	0.409 **	0.066	0.037	0.034	0.121	−0.020	0.131	0.023
PH			1	0.057	0.149	0.249 *	0.190	−0.123	0.123	0.021
NPB				1	0.044	0.261 *	0.145	0.406 **	0.174	0.350 **
NSB					1	0.024	0.152	−0.101	0.214	0.090
NRP						1	0.191	0.262 *	0.177	0.124
NPP							1	0.099	0.354 **	0.307 **
NSP								1	0.051	0.173
GYD									1	0.498 **
HSWT										1

**. Correlation is significant at the 0.01 level, *. Correlation is significant at the 0.05 level (2-tailed), DTF = days to 50% flowering, DTM = days to 75% maturity, PH = plant height, NPB = number of pods per plant, NSB = number of secondary branches per plant, NRP = number of racemes per plant, NPP = number of primary branches per plant, GYD = grain yield, HSWT = 100 seed weight.

3.5. Nonlinear Principal Component (PC) and Cluster Analysis

The nonlinear principal component analysis was performed to identify the most discriminative variables among the pigeonpea genotypes. A total of 98% of the variation explained by the qualitative and quantitative traits were explained by the first three principal components (Table 12). In general, traits such as GYD, DTF, DTM, leaf hairiness (LH), leaf shape (LS), and NRP contributed much to the phenotypic variation in the PCs. However, GYD, LH, NPP, HSWT, and NSP were the highest contributors (with contributions of 0.86, 0.63, 0.63, 0.51, and 0.45, respectively) on PC1. The second principal component accounted for 73% of the total variation, with NRP and pod form (PF) being the highest (0.74 and 0.62) positive contributors. Conversely, traits including LS, flowering pattern (FP), and flower main color (FMC) negatively correlated with PC2 exhibiting negative (−0.63, −0.57, and −0.44, respectively) PC scores. DTF and DTM were the positive contributors to the observed phenotypic variation on PC3 with PC loadings of 0.83 and 0.79, respectively.

Table 12. Principal components showing variation and contribution by 24 phenotypic traits among 81 pigeonpea genotypes assessed in six environments in Malawi.

Trait	Dimension		
	1	2	3
FMC	−0.026	−0.435	−0.269
FP	0.016	−0.568	0.050
FSC	−0.101	−0.172	0.291
FSP	0.357	−0.095	−0.080
GH	−0.077	0.227	0.151
LH	0.629	−0.435	−0.112
LS	−0.386	−0.626	0.143
PC	0.010	0.175	−0.310
PF	0.203	0.616	0.043
SCP	−0.050	0.327	−0.235
SEC	−0.038	0.236	−0.345
SMC	0.020	−0.144	−0.060
SSH	0.082	−0.134	0.186
STC	−0.042	0.000	0.023
DTF	0.186	−0.069	0.827
DTM	0.236	−0.190	0.793
PH	0.294	−0.357	0.118
NPB	0.037	0.208	0.239
NSB	0.160	0.398	0.204
NPP	0.626	0.086	−0.073
PL	0.252	0.353	0.104
NRP	0.001	0.735	−0.117
NSP	0.476	0.076	−0.075
HSWT	0.508	0.219	−0.084
GYD	0.863	0.146	−0.109
Eigen value	3.404	2.967	2.163
Variance %	39	34	25
Cumulative	39	73	98

DTF = days to 50% flowering, DTM = days to 75% maturity, PH = plant height, NPB = number of primary branches, NSB = number of secondary branches per plant, NRP = number of racemes per plant, NPP = number of pods per plant, GYD = grain yield, HSWT = 100 seed.

Figure 2 shows the variable correlation plot showing positive association between PC1 and DTF traits such as DTM, GYD, HSWT, NSP, plant height (PH), and pod length (PL). Conversely, seed eye color (SEC), pod color (PC), number of secondary branches per plant (NSB), and FMC exhibited negative associations with PC1. Quantitative traits such as plant habit (PH), NPP, HSWT, DTF, DTM, NRP, and GYD were positively correlated to each other as exhibited by their vectors, which were in the same direction and separated by acute angles between them. Similarly, the qualitative traits such as SEC, PC, and FMC were positively correlated to each other. However, the quantitative traits were positively correlated with GYD, while the qualitative traits were negatively correlated with GYD.

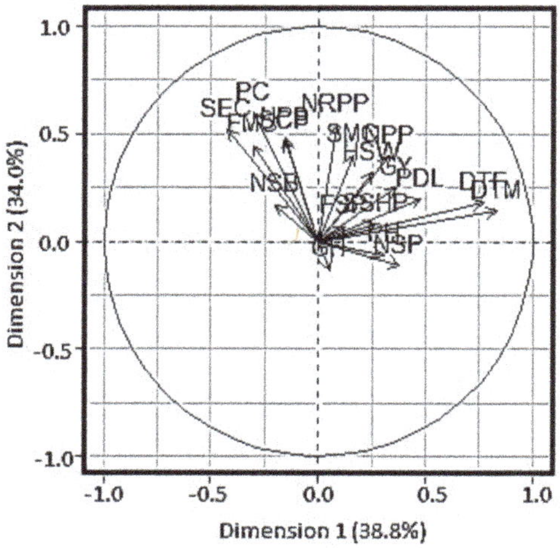

Figure 2. Trait biplot showing the relationship among quantitative and qualitative traits in 81 pigeonpea genotypes evaluated in six environments in Malawi. Dim 1 = dimension 1, Dim2 = dimension 2, PCA = principal component analysis. For trait code description, refer to Table 3.

Assessment of the phenotypic diversity using morphological attributes delineated the genotypes into three distinct clusters (Figure 3). The first cluster had the highest number (51) of genotypes. The second cluster had 27 genotypes, and the third cluster had three genotypes. However, the composition of genotypes in all the three clusters consisted of mixtures of landraces, breeding lines, and cultivars.

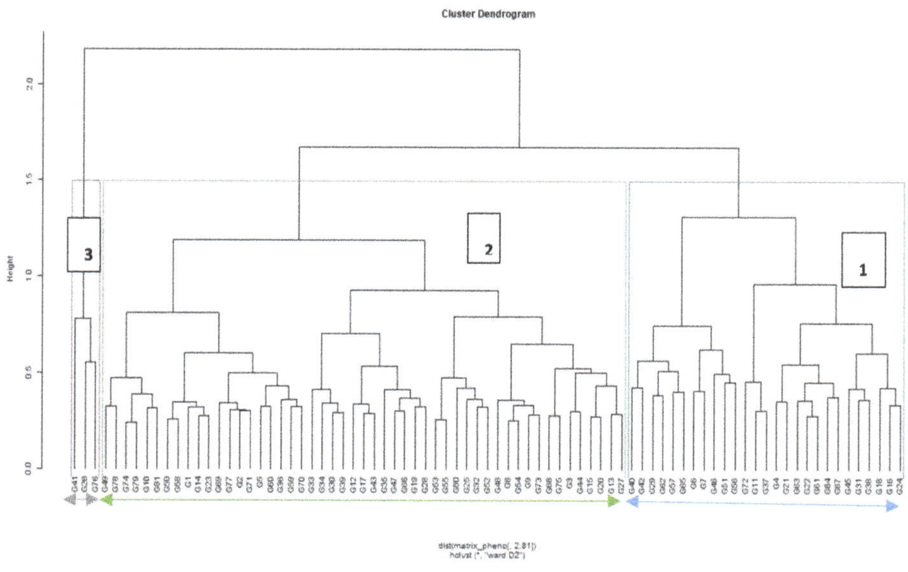

Figure 3. Hierarchical cluster dendogram showing genetic similarity matrix of 81 genotypes evaluated in six environments in Malawi based on phenotypic traits. Cluster 1, in pink color, cluster 2 in green color and cluster 3 in blue color. See Table 1 for the genotypes codes.

4. Discussion

The current study evaluated 81 pigeonpea genotypes across six environments to assess the genetic diversity and yield stability, and to select complementary and unique genotypes for breeding. The genotypes exhibited wide and significant variation in qualitative traits (Table 4), which indicated that the tested germplasm could harbor important genetic variation that underpins the morphological variation. Similarly, [16] reported significant variation in qualitative traits among pigeonpea accessions sourced from ICRISAT's international genebank. The variation in qualitative traits such as growth habit and seed color is important for breeding cultivars that meet farmer expectations. For instance, the variation present in growth habit is important to identify genotypes with compact growth habit for intercropping to maximize space utilization and productivity in moisture-limited environments. Farmers often intercrop pigeonpea with cereal crops such as maize and sorghum, and legumes such as groundnuts. Hence, pigeonpea genotypes with a spreading growth habit may not be suitable for mixed cropping systems [17]. The diversity in pigeonpea seed color helps to identify genotypes that are preferred by local farmers. For instance, farmers in Malawi prefer pigeonpea varieties with a cream seed color, which they associate with good cooking quality. Similar findings were reported by [18], who reported a predominance of cream and light gray pigeonpea varieties in Benin, reflecting the farmers' color preferences. Knowledge of variability in qualitative traits among the accessions and understanding farmer preferences are important as a basis for the development of direct breeding objectives and appropriate breeding strategies.

The significant genetic variation exhibited in the quantitative traits (Table 5) highlights the genetic diversity available for exploitation during cultivar development. The genotype performances were also affected by significant genotype × environment interactions, suggesting that genotype performances were not consistent in all the environments. Genotypic variation is underpinned by differences in genetic constitution among the genotypes, which is important for crop improvement [19]. The environment influences phenotypic expression through variation in factors such as temperature, humidity, and soil fertility. The significant impact of the environment on phenotypic expression is known to reduce genotype–phenotype correlation [20], which complicates the identification of stable and superior genotypes. However, significant genotype × environment interaction on yield and yield components of legumes such as common bean, cowpea, and pigeonpea has been previously reported [21–23]. In the present study, the genotypes that matured early were shorter with low numbers of branches and pods per plant and low grain yields compared to the medium to late maturing genotypes that grew taller, produced more branches and pods per plant, and had higher grain yields. Similarly, [24] reported that cultivars with higher numbers of primary branches, secondary branches, number of pods per plant, and taller plant height had higher grain yields.

There was limited genetic variation among pigeonpea landraces, cultivars, and breeding lines in this study (Table 6). This could be attributed to gene flow arising from the exchange of germplasm between Malawi and Tanzania. In addition, there could be high level of genotype relatedness since the breeding lines and cultivars were developed from the landraces collected from Malawi and Tanzania by ICRISAT. However, the genotype performance in terms of days to maturity and plant height were affected by significant genotype × environment interactions (Tables 7 and 8), suggesting that genotype performances were not consistent in all the environments. This could be because the landraces, cultivars, and breeding lines belong to three maturity groups: early, medium, and late duration. The early maturity exhibited by the ICRISAT genotypes could be a result of selection for earliness at ICRISAT in Kenya, which has advanced pigeonpea breeding programs and has developed a number of elite breeding lines that have been distributed in several East and Southern African countries for evaluation. The TARI and DARS genotypes are comprised of landraces and cultivars that are medium to late maturing. Similarly, [21] also reported that traditionally grown pigeonpea cultivars and landraces are represented by varieties from medium to long maturity groups (150 to 280 days), which are high yielding but very sensitive to photoperiod.

The positive and moderate correlation between GYD and HSWT (r = 0.50, $p < 0.01$) (Table 11) indicated that HSWT could be used for the direct selection for GYD. The moderate positive correlation between DTF, DTM, and PH revealed that selection for earliness can be based on the plant height. Although pigeonpea is relatively drought-tolerant, there is a need to develop early flowering and maturing cultivars to fit in the cropping cycles of sub-Saharan Africa, which are becoming progressively shorter due to climate change. The positive correlations exhibited by most secondary traits show that multiple trait selection would be possible. However, the weak correlations among the traits would result in an inefficient selection or low genetic gains. A strong correlation (r = 0.858) between grain yield and the number of pods per plant was reported by [25]. In addition, [26] reported moderate to weak correlations between grain yield and days to 50% flowering (r = 0.58), days to maturity (r = 0.59), and plant height (r = 0.42). Conversely, [27] and [28] reported a negative association between 100 seed weight and grain yield. The significant relationship between DTF, DTM, HSWT, PH, NPP, and GYD is useful when selecting for high grain yield [16]. Direct selection for these traits would result in yield improvement in pigeonpea.

The nonlinear principal component analysis enabled the identification of important traits with high variability among the genotypes. In this study, GDY, DTF, DTM, NPP, NRP LH, HSWT, and NSP were identified as the most important traits due to their high contribution on PC1 and PC2 (Table 12). This suggests that these traits are useful for selection. Accessions that exhibit high and desirable mean performances based on the target traits would be selected for improvement. Other reports indicated that trait contribution to different PCs varies with genetic diversity within the tested germplasm and the number of traits evaluated [16,25]. The results further revealed that DTF, NPP, NPB, NSB, PH, PL, and HSWT are important secondary traits for the indirect selection for GYD due to their positive association with GYD and their high contribution on the PCs.

The cluster analysis delineated the accessions into three groups (Figure 3), suggesting the presence of considerable genetic variation among the genotypes. However, a mixture of breeding lines, landraces, and cultivars in each group could be attributed to the geographical proximity between Malawi and Tanzania (where some of the landraces were collected). The level of natural outcrossing in pigeonpea is very high and varies from 5 to 70% depending on the prevailing weather conditions and insect activities for pollination [29]. In addition, the breeding lines from ICRISAT were developed using some parents selected from the landraces from Tanzania and Malawi. East Africa is known as a center of diversity for pigeonpea—hence the close genetic relatedness. The present finding is in agreement with [30], who reported little variation among the cultivated pigeonpea collected in Africa based on diversity array technology (DArT) markers.

5. Conclusions

The study examined 81 pigeonpea genotypes for their diversity and yield stability. The genotypes exhibited a wide genetic variation in qualitative traits such as growth habit, flower main color, flower streak pattern, pod color, and seed traits. The combined analysis revealed significant genotype × environment interaction effects for most traits, suggesting the need for selection for specific adaptation. A lack of significant variation in quantitative traits among landraces, cultivars, and breeding lines indicate that there is potentially high gene flow among the different categories of germplasm, which could present genetic bottlenecks during breeding. Traits such as GDY, DTF, DTM, NPP, NRP, HSWT, LH, and NSP with high scores on PC1 and PC2 are useful selection indices for pigeonpea improvement. Accessions that exhibited high and desirable mean performances in the target traits such as early maturing (MWPLR 14, ICEAP 01170, ICEAP 871091, ICEAP 01285) and high yielding (Kachangu, MWPLR 16, TZA 5582, No. 40, and MWPLR 14) would be recommended as parental lines for the breeding program. The genetic diversity analysis using morphological traits has enabled the identification of promising parents and heterotic clusters for breeding.

Author Contributions: Each author has participated sufficiently in the completion of this work. H.S., M.L., I.M., and A.S. contributed to the experimental design, data analysis, and review of this manuscript. E.N.Y. implemented the experiments and wrote up the manuscript. All authors have read and agreed to the published version of the manuscript.

Funding: This research was supported by the Alliance for Green Revolution in Africa (AGRA) through the African Centre for Crop Improvement (PASS030), University of KwaZulu-Natal.

Acknowledgments: The authors acknowledge the station managers and technical staff of the Chitedze, Bvumbwe, and Makoka Research Stations in Malawi for technical assistance and for the overall research support. The Department of Agricultural Research Services (DARS)/Malawi, Tanzania Agricultural Research Institute, International Crops Research Institute for Semi-Arid Tropics, and Malawi Genebank are sincerely thanked for providing the germplasm used in the study.

Conflicts of Interest: The authors declare no conflict of interest.

References

1. Ade-Omowaye, B.; Tucker, G.; Smetanska, I. Nutritional potential of nine underexploited legumes in Southwest Nigeria. *Int. Food. Res. J.* **2015**, *22*, 798.
2. Subbarao, G.; Chauhan, Y.; Johansen, C. Patterns of osmotic adjustment in pigeonpea—Its importance as a mechanism of drought resistance. *Eur. J. Agron.* **2000**, *12*, 239–249. [CrossRef]
3. Giller, K.E. *Nitrogen Fixation in Tropical Cropping Systems*; Cabi: Wallingford, UK, 2001.
4. Kwena, K.; Karuku, G.; Ayuke, F.; Esilaba, A. Nitrogen Deficiency in Semi-Arid Kenya: Can Pigeonpea fix it? *East Afr. Agric For. J* **2019**, *83*, 322–340. [CrossRef]
5. Mula, M.; Saxena, K. *Lifting the Level of Awareness on Pigeonpea—A Global Perspective*; International Crops Research Institute for the Semi-Arid Tropics: Patancheru, India, 2010.
6. Food and Agriculture Organisation of the United Nations. FAOSTAT. Crop Statistics. 2017. Available online: www.fao.org/faostat/en/#data/QC (accessed on 20 January 2019).
7. Kananji, G.; Yohane, E.; Mviha, P.J.Z.; Siambi, M.; Silim, S. *A Guide to Pigeonpea Production in Malawi*; Department of Agricultural Research Services: Lilongwe, Malawi, 2016.
8. Bohra, A.; Dubey, A.; Saxena, R.K.; Penmetsa, R.V.; Poornima, K.; Kumar, N.; Farmer, A.D.; Srivani, G.; Upadhyaya, H.D.; Gothalwal, R. Analysis of BAC-end sequences (BESs) and development of BES-SSR markers for genetic mapping and hybrid purity assessment in pigeonpea (Cajanus spp.). *Biomed. Cent. BMC Plant Bio.* **2011**, *11*, 56. [CrossRef] [PubMed]
9. Saxena, R.K.; Von Wettberg, E.; Upadhyaya, H.D.; Sanchez, V.; Songok, S.; Saxena, K.; Kimurto, P.; Varshney, R.K. Genetic diversity and demographic history of Cajanus spp. illustrated from genome-wide SNPs. *PLoS ONE* **2014**, *9*, e88568. [CrossRef]
10. Upadhyaya, H.D.; Reddy, K.N.; Sharma, S.; Dwivedi, S.L.; Ramachandran, S. Enhancing the value of genetic resources for use in pigeonpea improvement. *Legume Persp.* **2016**, *11*, 13–16.
11. IBPGR; ICRISAT. *Descriptors for Pigeonpea [Cajanus cajan (L.) Millsp.]*; International Board of Plant Genetic Resources: Rome, Italy; International Crops Research Institute for Semi-Arid Tropics: Patancheru, India, 1993.
12. George, D.; Mallery, P. *IBM SPSS Statistics 26 Step by Step: A Simple Guide and Reference*; Routledge: Abbingdon, UK, 2019.
13. Payne, R.; Murray, D.; Harding, S.; Baired, D.; Soultar, D. *Genstat Windows*, 17th ed.; VSN International: Hemel Hempstead, UK, 2017.
14. Shimelis, H.; Shiringani, R. Variance components and heritabilities of yield and agronomic traits among cowpea genotypes. *Euphytica* **2010**, *176*, 383–389. [CrossRef]
15. Meulman, J.J.; Van der Kooij, A.J.; Heiser, W.J. Principal components analysis with nonlinear optimal scaling transformations for ordinal and nominal data. *Sage Handb. Quant. Methodol. Soc. Sci.* **2004**, 3, 49–72.
16. Upadhyaya, H.; Reddy, K.; Gowda, C.; Singh, S. Phenotypic diversity in the pigeonpea (Cajanus cajan) core collection. *Genet. Res. Crop Evol* **2007**, *54*, 1167–1184. [CrossRef]
17. Manyasa, E.; Silim, S.; Christiansen, J. Variability patterns in Ugandan pigeonpea landraces. *J. Sat. Agric. Res.* **2009**, *7*, 1–9.
18. Ayenan, M.A.T.; Ofori, K.; Ahoton, L.E.; Danquah, A. Pigeonpea [(Cajanus cajan (L.) Millsp.)] production system, farmers' preferred traits and implications for variety development and introduction in Benin. *Agric. Food Sec.* **2017**, *6*, 48. [CrossRef]
19. Grausgruber, H.; Sailer, C.; Ghambashidze, G.; Bolyos, L.; Ruckenbauer, P. Genetic variation for plant breeding. In Proceedings of the 17th EUCARPIA General Congress, Tulln, Austria, 8–11 September 2004.

20. Bustos-Korts, D.; Romagosa, I.; Borràs-Gelonch, G.; Casas, A.; Slafer, G.; Van Eeuwijk, F. *Genotype by Environment Interaction and Adaptation; In Encyclopedia of Sustainability Science and Technology*; Springer: New York, NY, 2018; pp. 29–71.
21. Vales, M.; Srivastava, R.; Sultana, R.; Singh, S.; Singh, I.; Singh, G.; Patil, S.; Saxena, K. Breeding for earliness in pigeonpea: Development of new determinate and nondeterminate lines. *Crop Sci.* **2012**, *52*, 2507–2516. [CrossRef]
22. Kimaro, D. Genetic Improvement of Pigeonpea (Cajanus cajan (L.) Millsp.) for Fusarium wilt resistance in Tanzania. Ph.D. Thesis, University of Kwazulu-Natal, Pitermaritzburg, South Africa, December 2016.
23. Gerrano, A.S.; Van Rensburg, W.S.J.; Mathew, I.; Shayanowako, A.I.T.; Bairu, M.W.; Venter, S.L.; Swart, W.; Mofokeng, A.; Mellem, J.J.; Labuschagne, M. Genotype and genotype 3 environment interaction effects on the grain yield performance of cowpea genotypes in dryland farming system in South Africa. *Euphytica* **2020**, *216*, 80. [CrossRef]
24. Rekha, R.; Prasanthi, L.; Sekhar, M.; Priya, M. Studies on selection indices in pigeonpea [Cajanus cajan (L.) Millsp]. *Int. J. Appl. Biol. Pharm.* **2013**, *4*, 291–294.
25. Saroj, S.; Singh, M.; Kumar, R.; Singh, T.; Singh, M. Genetic variability, correlation and path analysis for yield attributes in pigeonpea. *Bioscan* **2013**, *8*, 941–944.
26. Sreelakshmi, C.; Shivani, D.; Kumar, C. Genetic divergence and stability analysis in Pigeonpea (Cajanus cajan L.). *Electr. J. Plant Breed* **2010**, *1*, 530–535.
27. Hemavathy, A.T.; Bapu, J.; Priyadharshini, C. Principal component analysis in pigeonpea (Cajanus cajan (L.) millsp.). *Electr. J. Plant Breed* **2017**, *8*, 1133–1139. [CrossRef]
28. Narayanan, S.L.; Manivannan, N.; Mahalingam, A. Correlation and Path Analyses of Yield and its Component Traits in Pigeonpea [Cajanus cajan(L.) Millsp.]. *Int. J. Curr. Microbiol. App. Sci.* **2018**, *7*, 614–618. [CrossRef]
29. Choudhary, A.K.; Sultana, R.; Ontagodi, T.; Singh, I.; Bhatt, B. Recent advances in breeding pigeonpea [Cajanus cajan (L.) Millsp.]. In Proceedings of the National Conference on Global Research Initiatives for Sustainable Agriculture & Allied Sciences (GRISASS-2015), Gwalior, MP, India, 12–13 December 2015.
30. Yang, S.; Pang, W.; Ash, G.; Harper, J.; Carling, J.; Wenzl, P.; Huttner, E.; Zong, X.; Kilian, A. Low level of genetic diversity in cultivated pigeonpea compared to its wild relatives is revealed by diversity arrays technology. *Appl. Gen.* **2006**, *113*, 585–595. [CrossRef] [PubMed]

agronomy

MDPI

Article

Utilization of Phytochemical and Molecular Diversity to Develop a Target-Oriented Core Collection in Tea Germplasm

Do Yoon Hyun [1], Gwang-Yeon Gi [2], Raveendar Sebastin [1], Gyu-Taek Cho [1], Seong-Hoon Kim [1], Eunae Yoo [1], Sookyeong Lee [1], Dong-Mo Son [2] and Kyung Jun Lee [1,*]

1 National Agrobiodiversity Center, National Institute of Agricultural Sciences (NAS), RDA, Jeonju, Jeollabuk-do 54874, Korea; dyhyun@korea.kr (D.Y.H.); raveendars@gmail.com (R.S.); gtcho@korea.kr (G.-T.C.); shkim0819@korea.kr (S.-H.K.); eung77@korea.kr (E.Y.); xsanta7@korea.kr (S.L.)

2 Tea Industry Institute, Jeollanamdo Agricultural Research and Extension Service, Boseong-gun, Jeollanam-do 59455, Korea; gygi4975@korea.kr (G.-Y.G.); dmson0810@korea.kr (D.-M.S.)

* Correspondence: lkj5214@korea.kr; Tel.: +82-63-238-4861

Received: 10 October 2020; Accepted: 26 October 2020; Published: 29 October 2020

Abstract: Tea has received attention due to its phytochemicals. For the direct use of tea germplasm in breeding programs, a core collection that retains the genetic diversity and various phytochemicals in tea is needed. In this study, we evaluated the content of eight phytochemicals over two years and the genetic diversity through 33 SSR (simple sequence repeats) markers for 462 tea accessions (entire collection, ENC) and developed a target-oriented core collection (TOCC). Significant phytochemical variation was observed in the ENC between genotypes and years. The genetic diversity of ENC showed high levels of molecular variability. These results were incorporated into developing TOCCs. The TOCC showed a representation of the ENC, where the mean difference percentage, the variance difference percentage, the variable rate of coefficient of variance percentage, and the coincidence rate of range percentage were 7.88, 39.33, 120.79, and 97.43, respectively. The Shannon's diversity index (I) and Nei's gene diversity (H) of TOCC were higher than those of ENC. Furthermore, the accessions in TOCC were shown to be selected proportionally, thus accurately reflecting the distribution of the overall accessions for each phytochemical. This is the first report describing the development of a TOCC retaining the diversity of phytochemicals in tea germplasm. This TOCC will facilitate the identification of the genetic determinants of trait variability and the effective utilization of phytochemical diversity in crop improvement programs.

Keywords: catechin; genetic diversity; phytochemicals; SSR; targeted-oriented core collection; tea germplasm

1. Introduction

Since the International Board for Plant Genetic Resources (IBPGR) was established in 1974 to coordinate the global efforts to systematically collect and conserve the world's threatened genetic plant diversity, many countries and organizations have founded gene banks, and millions of crop resources have been preserved [1,2]. As a result of the global efforts to conserve plant genetic resources for food and agriculture, the number and scale of ex situ germplasm collection has increased tremendously in the last 40 years [3]. However, the large sizes of redundant collections, either individually or collectively, for particular species have become an obstacle to the characterization, evaluation, utilization, and maintenance of those species [2,3]. As part of this solution, the authors in [4] proposed that collections could be pruned to core collections, which could "represent with a minimum of repetitiveness, the genetic diversity of a crop species and its relatives." This core collection serves as a working collection that can be extensively examined, while the accessions excluded from such collections can be preserved as preliminary collections [5]. Therefore, core collections can facilitate the use of crop germplasm and manage the entire collections [2].

Tea (*Camellia sinensis* (L.) Kuntze) is a woody evergreen plant in the family Theaceae and is native to the region covering the northern part of Myanmar, as well as the provinces of Yunnan and Sichuan in China. It is one of the most popular beverages and has become a daily drink for many people around the world [6]. To combat climate change, biological threats, and market fluctuations, the main tea-producing countries of China, Sri Lanka, and India have managed and preserved their tea and genetic resources both in situ and ex situ [7]. In addition, they have developed core collections or subsets of tea germplasm that maintain the original diversity of the collections but at a size that facilitates the evaluation, use, and conservation of the entire collections using geographical origin, phenotypic traits, and molecular markers [8–12]. Knowledge and understanding of the genetic background, genetic diversity, relationships, and identification are important for the collection, preservation, characterization, and utilization of tea resources [13]. The proper characterization and evaluation of genetic resources via systematic preservation and maintenance is the most important factor in utilizing such resources for improving crops [14]. The characterization of germplasm can be carried out using morphological, biochemical, and molecular descriptors according to the standard criteria contained in the tea descriptors [15]. Among the characteristics in tea descriptors, morphological traits and phytochemical content tend to be most affected by environmental factors. In addition, since phytochemical content can have very large variations depending on the environment, these characteristics need to be evaluated for multiple years, yielding more precise data. On the other hand, molecular markers are rarely influenced by the environment and thus directly offer an observation of genomic diversity.

The phytochemical characterization of plant germplasm is an acceptable method to define biochemical diversity [16]. The composition of phytochemicals in tea is important, as these chemicals contribute to tea's quality and pharmacological properties [15]. Tea consists of compounds rich in polyphenols, theanine, and caffeine, which not only determine the quality of tea but also provide tremendous health benefits [17]. Among tea polyphenols, catechins account for 8% to 26% of the tea leaves' dry weight [18]. Previous studies reported that because each catechin monomer has a different chemical structure they each have unique bioactivity, bioavailability, and physiological pharmacokinetic properties [19,20]. In addition, the origin and growing conditions of the tea plant affect the contents of the tea's phytochemicals, which changes bioactivity [21,22]. The leaves of the tea tree have been primarily cultivated as a source of tea beverages, in which phytochemicals such as catechin and caffeine are the main functional compounds. The development of a new variety that contains enhanced phytochemical contents (qualitatively or quantitatively) is the ultimate objective of tea breeding programs. Therefore, along with the evaluation of phytochemical diversity, the development of a core collection that can represent the diversity of the entire germplasm is very important not only for the conservation and management of germplasm but also for tea breeding programs.

To assess the genetic diversity and/or develop new cultivars in many countries, molecular markers such as restriction fragment of length polymorphism RFLP) [23,24], random amplified polymorphic DNA (RAPD) [25–27], amplified fragment length polymorphism (AFLP) [24], and simple sequence repeats (SSR) [11,28–30] were used. [31] reported that morphological traits have drawbacks such as the influences of environment on trait expressions, epistatic interactions and pleitrophic effects among others despite the value of their advantages. On the other hand, molecular markers are used because they are least affected by environmental factors and are almost unlimited in number. In addition, they offer a possibility to observe the genome directly, and thus eliminate the shortcomings inherent in a phenotype observation [32].

In our previous study, we analyzed the genetic diversity of tea accessions collected in Korea using 21 SSRs [28]. In this study, we evaluated the content of eight phytochemicals over two years (2018 and 2019) and analyzed the genetic diversity through 33 SSR markers for 462 tea accessions collected from

Korea, China, Japan, and Indonesia. In addition, a target-oriented core collection was developed using both the phytochemical content and genetic diversity. This core collection will be used to efficiently preserve, manage, and evaluate tea germplasm in the genebank of Korea and to be provided to the tea breeding program as breeding materials.

2. Materials and Methods

2.1. Plant Material

A total of 462 tea accessions were obtained from the National Agrobiodiversity Center (NAC) at the Rural Development Administration in South Korea (Table S1). These accessions are currently preserved as genetic resources in the Tea Industry Institute (34″46′ N, 127″5′ E) and are maintained through similar horticultural practices. Fresh tea buds and young leaves of the first flush were harvested between 09:00 a.m. and 12:00 a.m. on 24 May in 2018 and 2019. All samples were stored in a freezer at −80 °C until analysis.

2.2. Phytochemical Analysis

The powdered tea samples (0.1 g) were extracted by intermittent shaking with 1 mL of 70% (*v*/*v*) methanol at 70 °C, and the mixture was centrifuged at 12,000 rpm for 10 min. The supernatant was then diluted 1:10 (sample: 70% methanol) and filtered through a 0.45-µm syringe filter. The diluted samples were analyzed with an Agilent 1260 Infinity HPLC system (Agilent Technology, Santa Clara, CA, USA). The analysis was performed using a COSMOSIL 2.5 Cholester (2.5 µm, 2.0 × 50 mm, NACALAI TESQUE, INC., Kyoto, Japan). The HPLC conditions were as follows: solvent A, Acetonitrile/20 mmol/L Phosphate buffer (pH2.5) = 10/90; solvent B, Acetonitrile/ 20 mmol/L Phosphate buffer (pH2.5) = 30/70; B concentration, 0% to 100% 5 min linear gradient; column temperature, 40 °C; and flow rate, 0.6 mL/min. The filter detector was set to 280 nm.

All the data collected from three replicate experiments. The summarized phytochemicals in the tea accessions were calculated, and evaluation of the annual variation in the phytochemicals under consideration was conducted through a multivariate analysis of variance (MANOVA) using PAST 3 [33]. Hierarchical clustering was performed using the R statistical software (http://www.r-project.org).

2.3. DNA Extraction

Genomic DNA was extracted from the leaves of the tea accessions using a Qiagen DNA extraction kit (Qiagen, Hilden, Germany). The DNA quality and quantity were measured using 1% (*w*/*v*) agarose gel and spectrophotometry (Epoch, BioTek, Winooski, VT, USA). The extracted DNA was diluted to 30 ng/uL and stored at −20 °C until further PCR amplification.

2.4. SSR Genotyping

For the SSR analysis, a total of 33 SSRs were selected from previous studies [11,34] based on linkage groups and PIC value (Supplementary Table S2). They were fluorescently labelled (6-FAM, HEX, and NED) and used to facilitate the detection of the amplification products. The PCR reactions were carried out in a 25 uL reaction mixture containing 30 ng template DNA, 1.5 mM $MgCl_2$, 0.2 mM of each dNTP, 0.5 um of each primer, and 1 U Taq polymerase (Inclone, Korea). Amplification was performed with the following cycling conditions: initial denaturation at 94 °C for 5 min, followed by 35 cycles of denaturation at 95 °C for 30 s, annealing at 55–62 °C (depending on the primers, Table S2) for 30 s, extension at 72 °C for 1 min, and a final extension step at 72 °C for 10 min. Each amplicon was resolved on an ABI prism 3500 DNA sequencer (ABI3500, Thermo Fisher Scientific Inc., Wilmington, DE, USA) and scored using the Gene Mapper Software (Version 4.0, Thermo Fisher Scientific Inc.).

2.5. Genetic Diversity and Population Structure

The number of alleles (Na), number of genotypes (Ng), Shannon–Wiener index (S), Expected heterozygosity (He), and Evenness were calculated using the poppr package for the R software [35]. An analysis of molecular variance (AMOVA) within and between the gene pools was performed using the GenAlEx software v. 6.5 [36].

The population structure was analyzed using STRUCTURE v.2.3.4 [37] and DAPC. In the STRUCTURE analysis, Bayesian-based clustering was performed, testing three independent runs, with K ranging from 1 to 10. Each run had a burn-in period of 50,000 iterations and 500,000 Monte Carlo Markov iterations, assuming an admixture model. The output was subsequently visualized with the STRUCTURE HARVESTER v.0.9.94 [38]. The most likely number of clusters was inferred according to Evanno [39]. The DAPC analysis was performed using the adegenet package for the R software [40,41] according to Lee et al. [28]. A Mantel test was performed using R software [40] in order to investigate the relationship between the genetic and phytochemical distances of tea accessions.

2.6. Development and Evaluation of the Core Collection

The POWERCORE program [42] was used to develop the independent core collection using the phytochemical data for two years (2018 and 2019) and the genotypic data of 33 SSR markers. The mean difference percentage (MD%), variance difference percentage (VD%), variable rate of coefficient of variance (VR%), and coincidence rate of range (CR%) were calculated to assess the level of diversity captured in the core collection compared to the entire collection [43]. In addition, the representation of the core collection was evaluated by estimating Shannon's diversity index (I) and Nei's diversity index (H). The distance matrix was used to construct a dendrogram via the neighbor-joining (NJ) method with 1000 bootstrap replicates. The principal coordinate analysis (PCoA) was performed using DARwin v. 6.0 [44].

3. Results

3.1. Phytochemical Diversity of 462 Tea Accessions

The variation and distribution of the catechin and caffeine (Caf) content in the 462 tea accessions are summarized in Table 1. Wide variation and diversity of the phytochemicals in the tea germplasms were observed. The average total catechin (TC) content in 2018 and 2019 was 70.04 ± 15.37 mg/g and 89.77 ± 19.04 mg/g, varying from 34.14 to 125.25 and 42.03 to 152.30 mg/g, respectively. Epigallocatechin 3-gallate (EGCG) was the most abundant catechin (63.9% in 2018 and 55.9% in 2019), averaging 44.74 ± 12.31 mg/g in 2018 and 50.20 ± 15.01 mg/g in 2019; epigallocatechin 3-gallate (ECG), epicatechin (EC), gallocatechin (GC), catechin (C), catechin 3-gallate (CG), and gallocatechin 3-gallate (GCG) were the next most common in abundance.

The levels of Caf, EC, ECG, EGCG, and TC in the 462 tea accessions between 2018 and 2019 demonstrated a normal distribution, and their H′ levels were high (≥2.00). In contrast, minor components such as GC, C, CG, and GCG did not show a normal distribution, and their H′ values were lower, although their coefficients of variation were high.

All nine phytochemicals showed highly significant differences between tea accessions ($p < 0.001$) and experimental years ($p < 0.001$) (Table S3). In the year × accessions interactions, C and CG did not show significant differences, while the other phytochemicals showed highly significant differences ($p < 0.001$).

Table 1. Variation and distribution of phytochemical contents (mg/g) in the 462 tea accessions.

Phytochenical	Year	Min	Max	Mean	SD	Median	Skewness	Kurtosis	CV (%)	H′ [1]
C	2018	0.32	42.53	3.14	4.38	2.03	5.12	31.94	139.37	1.14
	2019	1.05	19.13	4.76	2.87	4.04	1.75	4.34	60.26	1.86
Caf	2018	0.44	36.64	17.42	5.20	16.96	−0.02	1.41	29.83	2.00
	2019	0.39	28.79	15.95	5.24	16.29	−0.46	0.15	32.86	2.07
CG	2018	0.15	5.61	1.30	1.03	0.97	1.27	1.49	79.22	1.88
	2019	0.34	21.76	2.47	1.72	2.06	4.60	39.43	69.53	1.65
EC	2018	3.42	26.58	10.40	3.21	10.07	1.05	2.50	30.83	2.02
	2019	1.98	22.88	8.12	2.97	7.73	0.98	1.88	36.59	2.00
ECG	2018	1.96	17.98	7.88	2.84	7.51	0.71	0.53	36.01	2.03
	2019	3.15	34.42	14.68	4.89	14.64	0.38	0.51	33.30	2.07
EGCG	2018	13.15	95.90	44.74	12.31	43.74	0.39	0.40	27.51	2.06
	2019	11.49	91.34	50.20	15.01	50.38	0.02	−0.05	29.90	2.08
GC	2018	0.67	13.71	3.05	1.86	2.61	2.22	6.74	60.88	1.74
	2019	3.21	15.57	7.72	1.84	7.53	0.53	0.73	23.87	2.03
GCG	2018	0.01	3.49	0.50	0.43	0.38	2.84	12.26	85.69	1.67
	2019	0.33	9.75	1.89	1.07	1.62	1.81	6.92	56.70	1.87
TC	2018	34.14	125.25	70.04	15.37	68.50	0.53	0.23	21.94	2.05
	2019	42.03	152.30	89.77	19.04	89.00	0.08	0.07	21.21	2.06

[1] *H′*, Shannon-Weave index; C, (+)-Catechin; Caf, Caffeine; CG, (−)-Catechin 3-gallate; EC, (−)-Epicatechin; ECG, (−)-Epigallocatechin 3-gallate; EGCG, (−)-Epigallocatechin 3-gallate; GC, (−)-Gallocatechin; GCG, (−)-Gallocatechin 3-gallate; TC, total catechin.

3.2. Clustering Analysis

In total, 462 tea accessions were classified into four clusters according to their phytochemicals (Table 2 and Figure 1.). Cluster I contained 108 tea accessions and had higher content of Caf, ECG, EGCG, GCG, and TC between 2018 and 2019. Cluster II had 111 accessions and showed lower contents of C, EC, GC, and GCG in 2018 and higher contents of Caf, ECG, EGCG, GC, and TC in 2019. Cluster III consisted of 59 tea accessions and showed higher C and EC and lower Caf, ECG, and EGCG content between the two years. Cluster IV had 184 tea accessions with lower C and GCG contents between 2018 and 2019.

Table 2. Average cluster values of the phytochemicals in the 462 tea accessions for the two years (mg/g).

Year	Group	N [1]	C	Caf	CG	EC	ECG	EGCG	GC	GCG	TC
2018	I	108	3.66b [2]	19.83a	1.95a	11.16ab	10.94a	55.78a	3.55a	7.99a	92.18a
	II	111	2.06c	17.84b	1.22b	8.75c	7.10b	45.62b	2.40c	4.63bc	65.56b
	III	59	10.71a	12.88c	1.50b	11.62a	5.70c	33.59d	3.21ab	5.66b	67.46b
	IV	184	2.63bc	17.21b	1.27b	10.55b	7.24b	41.25c	3.09b	3.95c	65.36b
2019	I	108	5.01b	18.66a	2.35b	7.67b	18.34a	59.04a	7.83b	25.25a	124.85a
	II	111	4.70b	19.80a	2.25b	7.43b	17.54a	61.10a	8.89a	20.74b	121.84a
	III	59	8.60a	9.37c	4.32a	11.11a	9.64c	32.21c	6.68c	19.52b	91.59b
	IV	184	3.50c	13.76b	2.21b	7.91b	12.15b	43.23b	7.24bc	13.89c	89.60b

[1] N, number of accessions; C, (+)-Catechin; Caf, Caffeine; CG, (−)-Catechin 3-gallate; EC, (−)-Epicatechin; ECG, (−)-Epigallocatechin 3-gallate; EGCG, (−)-Gallocatechin 3-gallate; GC, (−)-Gallocatechin; GCG, (−)-Gallocatechin 3-gallate; TC, total catechin. [2] The same letter in each column indicates no significant difference according to a least significant difference test, $p < 0.05$.

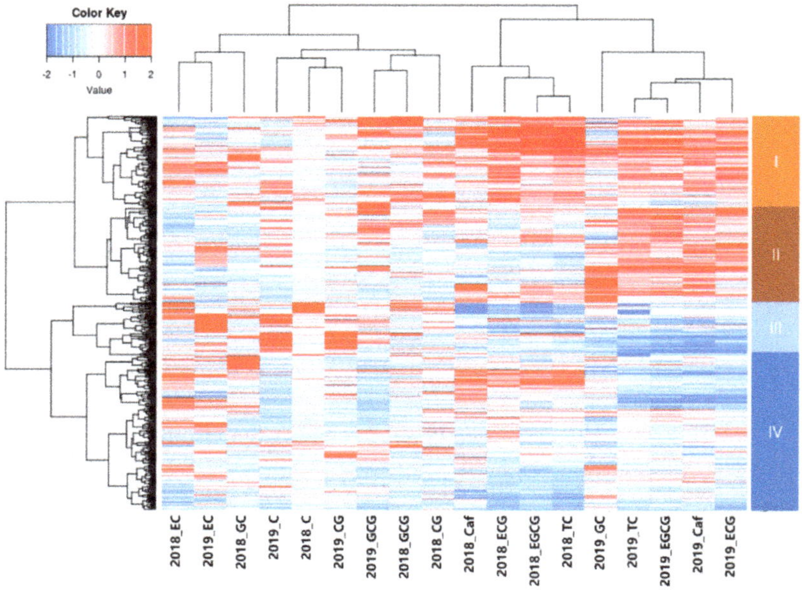

Figure 1. Hierarchical clustering analysis of the phytochemicals in the 462 tea accessions. The colours in the heatmap indicate the z-score which was calculated by subtracting the mean of phytochemicals across different samples and dividing it by the standard deviation of the phytochemicals across all the samples. The red color indicates positive z-score, the white color indicates zero z-score, whereas the blue colour indicates negative z-score. Higher intensity of the color in the scale indicates a higher magnitude of the z-score. The dendrogram on the x-axis indicates the degree of similarity between the phytochemicals, the closer the phytochemicals the higher the level of similarity in them and the phytochemicals have been clustered using hierarchical clustering. Similarly, the dendrogram on the y-axis indicates the degree of similarity between the different samples, the closer the samples the higher the level of similarity in them and they have been clustered using hierarchical clustering (Ward, Euclidean distance).

3.3. SSR Fingerprinting

A total of 428 alleles were detected in 33 SSR loci among the 462 tea accessions (Table 3). The number of observed alleles (Na) and the number of genotypes (Ng) ranged from 5 (TM324 and TM480) to 23 (MSE0083), with an average of 13.0, and 10 (TM324 and TM480) to 103 (MSE0083), with an average of 50.2. The Shannon–Wiener index (S) and expected heterozygosity (He) ranged from 0.92 (TM461) to 2.42 (TM422), with an average of 1.78, and 0.54 (MSE0237) to 0.88 (MSE0083 and TM422), with an average of 0.77. The evenness was calculated from 0.57 (TM576) to 0.90 (TM351), with an average of 0.75.

The diversity indices among the four origins are calculated in Table 4. The Na and Ng contents ranged from 3.2 (IDN) to 11.8 (KOR) and 2.4 (IDN) to 43.6 (KOR), respectively. The S and He contents were calculated to be 1.05 (IDN) to 1.91 (CHN) and 0.73 (JPN) to 0.81 (CHN), respectively. The Evenness ranged from 0.76 (KOR) to 0.90 (IDN), with an average of 0.79.

Table 3. Genetic diversity parameters of the 33 SSR markers in the 462 tea accessions.

Locus	Na [1]	Ng	S	He	Evenness	Locus	Na	Ng	S	He	Evenness
MSE0029	14	63	2.25	0.86	0.79	MSG0699	20	101	1.02	0.86	0.70
MSE0083	23	103	1.92	0.88	0.74	TM241	9	30	2.03	0.75	0.81
MSE0107	19	66	2.03	0.82	0.66	TM324	5	10	2.05	0.56	0.84
MSE0113	18	63	1.91	0.84	0.76	TM337	14	68	1.83	0.87	0.83
MSE0173	12	47	2.40	0.82	0.80	TM341	9	22	1.48	0.71	0.76
MSE0237	9	19	1.15	0.54	0.65	TM351	6	13	1.09	0.7	0.90
MSE0291	13	59	1.35	0.78	0.69	TM382	11	49	2.17	0.82	0.85
MSE0313	17	59	2.30	0.83	0.73	TM422	22	94	2.42	0.88	0.73
MSE0403	16	58	2.15	0.79	0.64	TM428	13	59	1.17	0.8	0.67
MSG0258	18	101	1.79	0.87	0.79	TM447	7	22	1.44	0.75	0.88
MSG0361	18	68	1.82	0.87	0.74	TM461	8	28	0.92	0.77	0.78
MSG0380	15	79	2.18	0.86	0.77	TM480	5	10	1.93	0.63	0.85
MSG0423	20	76	1.88	0.84	0.70	TM530	7	25	1.20	0.68	0.73
MSG0429	13	44	2.26	0.76	0.61	TM576	7	14	1.56	0.56	0.57
MSG0470	17	58	1.82	0.81	0.74	TM581	7	16	1.65	0.62	0.71
MSG0610	10	40	2.16	0.78	0.73	TM604	7	16	1.26	0.65	0.85
MSG0681	19	78	2.14	0.87	0.82	Mean	13.0	50.2	1.78	0.77	0.75

[1] Na, Number of observed alleles; Ng, Number of genotypes; S, Shannon–Wiener index; He, Expected heterozygosity.

Table 4. Genetic diversity parameters of the four origins in the 462 tea accessions using 33 SSRs.

Origin	N [1]	Na	Ng	S	He	Evenness
KOR	408	11.8	43.6	1.73	0.76	0.76
JPN	13	5.9	6.8	1.42	0.73	0.78
IDN	3	3.2	2.4	1.05	0.76	0.90
CHN	38	10.5	18.4	1.91	0.81	0.79

[1] N, Number of accessions; Na, Average number of alleles per accessions; Ng, Number of genotypes; S, Shannon2013Wiener index; He, Expected heterozygosity; KOR, Korea; JPN, Japan; IDN, Indonesia; CHN, China.

Genetic and phytochemical distance differences among tea accessions were concordant based on the Mantel test ($r = 0.0899$, $p = 0.017$) indicating that these two analyses (genetic and phytochemicals) grouped the genotypes in a different manner.

3.4. Population Structure

The relatedness among genotypes and their rooting with geographical designation were studied using a population structure analysis. Determination of the log mean probability and change in the log probability (ΔK) (following [29]) provided two subpopulations (K = 2) (Figure 2A). STR_C1 was dominated by genotypes belonging to 191 accessions from KOR, 12 accessions from JPN, 3 accessions from CHN, and 2 accessions from IDN (Figure 2B). STR_C2 contained 217 accessions from KOR, 35 accessions from CHN, and one accession from JPN and IDN. The mean alpha value (an estimate of the degree of admixture) for the analyzed samples was 0.2759.

To understand the genetic relationship among the 462 tea accessions, a DAPC analysis was performed (Figure 3). Four clusters were detected in coincidence with the lowest BIC values using the *find.clusters* function. The DAPC analysis was carried out using the detected number of clusters. Typically, the 50 first PCs (60.3% of variance conserved) of PCA and three discriminant eigenvalues were retained. These values were confirmed via a cross-validation analysis. The four clusters were titled D1-4. A major shift in accessions from STR_C1 to D1 and D3 was observed, and the main tea accessions of D2 and D4 were located in STR_C2. D1 contained 132 (126 accessions from STR_C1 and sic accessions from STR_C2) accessions, with 123 (117 from STR_C1 and 6 from STR_C2) from KOR, 6 (STR_C1) from JPN, two (STR_C1) from IDN, and one (STR_C1) from CHN. The major accessions of D2 included 43 accessions (42 from STR_C2 and one from STR_C1) from KOR and 32 accessions (STR_C2) from CHN, with one accession from JPN. D3 consisted of 83 accessions, with 75 accessions (73 in STR_C1 and two in STR_C2) from KOR, six (STR_C1) from JPN, and two (STR_C1) from CHN.

D4 comprised 171 accessions, with 167 accessions (STR_C2) from KOR, three (STR_C2) from CHN, and one (STR_C2) from IDN.

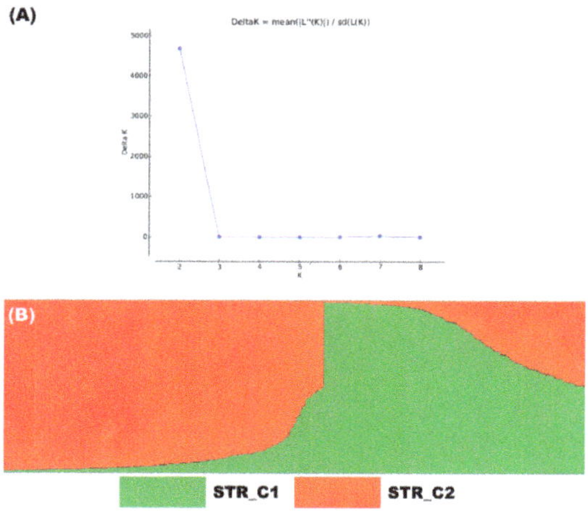

Figure 2. Model-based populations in association panels consisted of 462 tea accessions: (**A**) Delta K values for different numbers of populations (K) assumed in analysis completed with the STRUCTURE software (**B**) Classification of 462 tea accessions into two populations using STRUCTURE Version 2.3.4, where the numbers on the y-axis show the subgroup membership, and the x-axis shows the different accession. The distribution of accessions into different populations is indicated by the color coding (Cluster 1, STR_C1, is red round; and Cluster 2, STR_C2, is green).

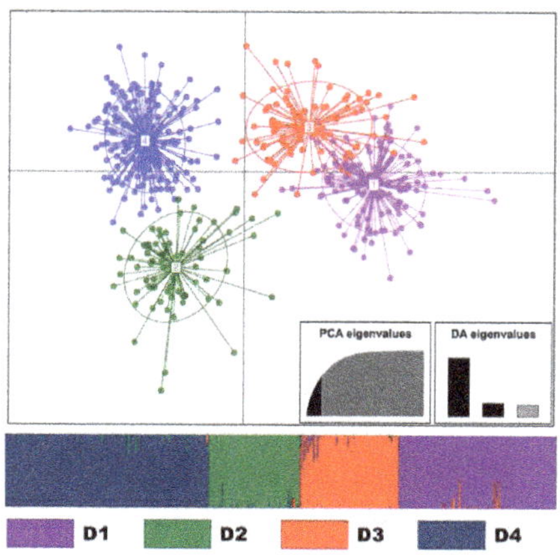

Figure 3. Discriminant analysis of the principal components (DAPC) for 462 tea accessions. The axes represent the first two Linear Discriminants (LD). Each circle represents a cluster, and each dot represents an individual. The numbers represent different subpopulations identified by the DAPC analysis.

The molecular variance within and between the regional pools, as well as the sub-populations derived from the clustering analysis, STRUCTURE, and DAPC analysis, was evaluated (Table 5). For the regional gene pools, percentage of variance within and among populations was found to be 94% and 6% of the total variation, respectively. The clustering analysis and DAPC provided a variance of 99% and 1% for within and among sub-populations, respectively, while the two sub-population derived from STRUCTURE showed only 100% of total variance within the groups. Among the four AMOVA results, regional pools showed minimum within-population variance (94%) and maximum among-population variance (6%), indicating that the regional pools are fairly structured groups for the panel under consideration. The genetic differentiation (PhiPT) of the four subpopulations showed a range from 0.005 (STRUCTURE) to 0.056 (regional pools).

Table 5. Analysis of the molecular variance (AMOVA) among and within populations in the regional pools and sub-populations derived from the clustering analysis, STRUCTURE, and DAPC.

	df [1]	SS	MS	Est. Var.	%	PhiPT	*p* Value
			Regional pools				
Among Populations	3	326.490	108.830	2.199	6%	0.056	0.001
Within Populations	458	16,882.545	36.861	36.861	94%	-	-
Total	461	17,209.035	-	39.061	100%	-	-
		Sub-population derived from clustering analysis					
Among Populations	3	211.739	70.580	0.305	1%	0.008	0.001
Within Populations	458	16,997.296	37.112	37.112	99%	-	-
Total	461	17,209.035	-	37.417	100%	-	-
		Sub-population derived from STRUCTURE					
Among Populations	1	77.344	77.344	0.175	0%	0.005	0.003
Within Populations	460	17,131.690	37.243	37.243	100%	-	-
Total	461	17,209.035	-	37.418	100%	-	-
		Sub-population derived from DAPC					
Among Populations	3	263.207	87.736	0.456	1%	0.012	0.001
Within Populations	458	16,945.828	37.000	37.000	99%	-	-
Total	461	17,209.035	-	37.456	100%	-	-

[1] df, degree of freedom; SS, sum of squares; MS, mean squares; Est. Var., estimates of variance; %, percent of variance.

3.5. Development and Evaluation of a Core Collection

The MANOVA analysis indicated significant year effects, as well as significant interaction effects between the year and accession effects by considering all quantitative traits together (Table S4). Therefore, the phytochemical data for both years (2018 and 2019) and molecular marker data were treated independently for the development of the core collection. The target-oriented core collection (TOCC) was developed with phytochemicals and molecular data using POWERCORE. TOCC included 100 accessions (21.6% of the entire collection) belonging to four origins, with 73 accessions from KOR, 22 from CHN, 4 from JPN, and 1 from IDN.

Differences between the means of the entire collection (ENC) and TOCC were found to be not significant for all traits (Table 6). The mean difference percentage (MD%), coincidence rate of range (CR%), variance difference percentage (VD%), and variable rate of the coefficient of variance (VR%) were used to comparably evaluate the properties of TOCC with ENC (Table 7). Overall, the nine phytochemicals, MD%, VD%, VR%, and CR% were 7.88%, 39.33%, 120.79%, and 97.43%, respectively.

Table 6. Phytochemical diversity comparison between the entire collection and core collection of tea accessions (mg/g).

		C [1]	Caf	CG	EC	ECG	EGCG	GC	GCG	TC
2018	Entire Collection	3.14	17.42	1.30	10.40	7.88	44.74	3.05	0.50	70.04
	Target-oriented core collection	3.93	16.71	1.26	10.45	7.43	43.42	2.91	0.47	68.58
		ns	ns	ns	ns	ns	ns	ns	ns	ns
2019	Entire Collection	4.76	15.95	2.45	8.12	14.68	50.20	7.72	1.89	89.78
	Target-oriented core collection	4.61	15.91	2.45	8.04	14.53	50.73	7.56	1.91	89.85
		ns	ns	ns	ns	ns	ns	ns	ns	ns

[1] C, (+)-Catechin; Caf, Caffeine; CG, (−)-Catechin 3-gallate; EC, (−)-Epicatechin; ECG, (−)-Epigallocatechin 3-gallate; EGCG, (−)-Gallocatechin 3-gallate; GC, (−)-Gallocatechin; GCG, (−)-Gallocatechin 3-gallate; TC, total catechin; ns, not significant at $p = 0.05$.

Table 7. Evaluation indices for the developed core collection.

	N [1]	MD% [1]	VD%	VR%	CR%	Na	*I*	*H*
Entire Collection (ENC)	462	-	-	-	-	2.11	0.308	0.195
Target-oriented core collections (TOCC)	100	7.88	39.33	120.79	97.43	2.11	0.335	0.209

[1] N, number of accessions; MD%, mean difference percentage; VD%, variance difference percentage; VR, variable rate of coefficient of variance; CR%, coincidence rate of range; *I*, Shannon's diversity index; *H*, Nei's diversity index; Na, Average number of alleles per locus.

To evaluate the quality of TOCC, Shannon's diversity index (I), Nei's diversity index (H), and the number of alleles (Na) were calculated using the molecular data (Table 7). The number of alleles (Na) in TOCC was the same as that in ENC. The genetic diversity of TOCC revealed by these markers was compared with that of ENC. The I and H of TOCC were higher (0.335, 0.209) than those of ENC (0.308, 0.195).

The distribution of tea accessions in TOCC was determined via a line graph obtained through the phytochemicals of ENC, along with a Principal coordinate analysis (PCoA) and Neighbor Joining (NJ) obtained through the genetic analysis of ENC (Figure 4 and Supplementary Figure S1). The accessions in TOCC were shown to be selected proportionally, accurately reflecting the distribution of the overall accessions for each phytochemical. For the distribution of molecular data, TOCC showed a balanced distribution in PCoA and NJ.

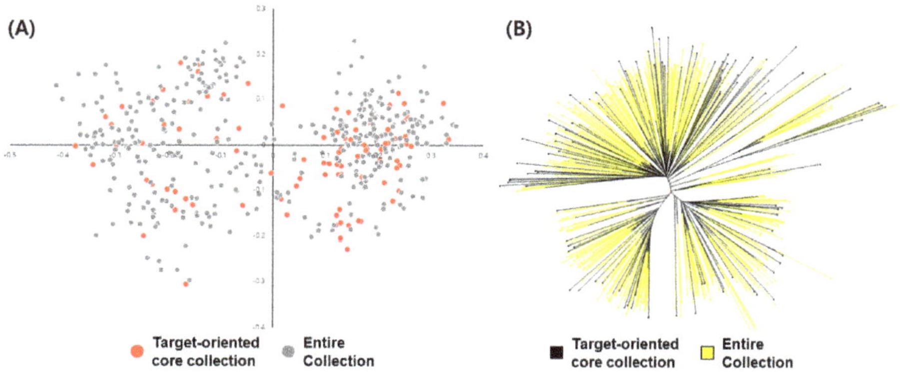

Figure 4. Distribution of tea accessions in the core collections using molecular data. (**A**) Principal coordinate analysis; (**B**) Neighbor-Joining Tree.

4. Discussion

A vast collection consisting of 15,234 accessions of tea is available in 23 gene banks around the world [7]. The biochemical characterization of tea germplasm in earlier studies demonstrated significant variability [18,45–48]. Despite the substantial diversity of compounds in tea germplasm, the development of tea cultivars was limited due to bottlenecks in tea breeding, such as long gestation periods, high inbreeding depression, and self-incompatibility [49]. In addition, the tea quality and yield in the main tea producing countries, such as China, India, Sri Lanka, Kenya, Japan, etc., were significantly improved with an increase in the ratio of clonal tea acreage [50]. Breeding strategies often focus on a limited set of target traits, resulting in cultivars with a narrow genetic base. Yao et al. [51] reported that the developed tea cultivars from China, Japan, and Kenya have a narrow genetic basis due to the popularity of only a few cultivars for breeding and planting. This has produced several problems, such as the spread of specific diseases and insects, the concentration of plucking time in the tea season, the non-uniformity of taste and flavor, and susceptibility to environmental changes [40,51]. Meegahakumbura et al. [29] noted that a molecular analysis that can discern not only patterns of lineage, but the origin of tea germplasm is also required because the morphological characteristics that are traditionally used to define cultivars are highly plastic and easily influenced by environmental conditions. The present study attempted to address the above issue by generating a core collection of tea germplasm that includes data on the molecular variability of the crop, in addition to biochemical characterization.

4.1. Phytochemical Diversity of Tea Germplasm

Significant variation was observed among the 462 tea accessions for catechin and caffeine content in this study (Table 1). In addition, significant differences between the two years were observed (Table S3). Catechins and caffeine serve as secondary metabolite defense compounds in tea plants. They provide sessile plants with protection against pathogens and predators, oxidative stress, and other environmental variables. Thus, the content of catechins and caffeine varied in the tea samples based on environmental variability [45]. Many previous studies reported a large variation in catechin and caffeine contents in tea accessions [15,18,52,53]. The authors in [54] noted that a biochemical characterization with different proportions of total catechins and their components would be a useful tool for the development of quality-tea clones. The authors in [55] reported that differences between locations were far larger than the variations among cultivars, implying that environmental effects should be taken into consideration when total catechin and its component contents are utilized as biochemical markers in tea breeding programs.

There are six major catechins in tea leaf: (+)-catechin (C), (−)-epicatechin (EC), (−)-epicatechin gallate (ECG), (+)-gallocatechin (GC), (−)-epigallocatechin (EGC), and (−)-epigallocatechin gallate (EGCG) [56]. The concentration of catechins in tea was determined as follows: EGCG>ECG>EGC>EC>GC>C [52,53,57,58]. In addition, the authors determined antioxidant activity in the following order: ECG>EGCG>EC>EGC [59]. The variation of catechin contents in tea accessions depends on the condition of the tea germplasm, such as the number of samples and the origin of the tea accessions, in each study. The range of each catechin's content in the previous studies was as follows: EGCG, 13.0 to 139.0 mg/g; ECG, 3.2 to 89.1 mg/g; EGC, 2.1 to 249 mg/g; EC, 2.0 to 54.5 mg/g; GC, 1.4 to 22.7 mg/g; and C, 0.3 to 30.9 mg/g [52,54,57,59,60]. In this study, the 462 tea accessions also showed a similar level of catechin content to that in previous studies (Table 1). The concentration of catechins in tea germplasm is important for tea quality. For instance, the ratio of (EGCG + ECG) × 100/EGC has been suggested as a quality index for measuring the difference in the catechin levels of fresh tea shoots across growing seasons [60]. In addition, the catechin index (CI)) (EC + ECG)/(EGC + EGCG)) has been used as a biochemical marker for studying the genetic diversity of tea germplasms [54]. The tea accessions with desirable compositions of catechins in this study could be incorporated into breeding programs for crop improvement.

Caffeine is the most abundant alkaloid in tea, with content usually between 15 and 50 mg/g [15]. In this study, the caffeine content of 462 tea accessions ranged from 0.4 to 36.6 mg/g (2018) and 0.4 to 28.8 mg/g (2019) (Table 1). Kottawa-Arachchi et al. [15] noted that various amounts of caffeine have been observed in different tea growing countries. Due to the pharmacological properties of caffeine on the central nervous system, the demand for low-caffeine tea is increasing greatly, from 2% of total tea consumption in 1980 to 15% in the early twenty-first century [61]. Although many countries have invested in methods and techniques to make decaffeinated tea, such techniques can remove the tea's unique aroma and taste, which will worsen the quality. It is thus important to develop low caffeine clones through breeding and selection, as such clones could be a solution to the problem of high caffeine levels and contribute tremendously to the provision of natural low-caffeine tea [18]. The tea accessions with a lower caffeine content in this study could be used as naturally low-caffeine genetic resources for crossbreeding parents.

4.2. Genetic Diversity of Tea Germplasm

In our previous study, we analyzed the genetic diversity and population structures of 410 tea accessions collected from South Korea using 21 SSR markers and revealed the narrow genetic base of South Korean tea accessions [28]. In the present study, the genetic diversity and population structure of 462 tea accessions from China, Japan, Indonesia, and Korea (conserved in NAC) were analyzed using 33 SSR markers. As shown in Table 4, higher diversity was detected among the tea accessions in China (H = 1.91, I = 0.81) than among those in Korea (H = 1.73, I = 0.76), Japan (H = 1.42, I = 0.73), and Indonesia (H = 1.05, I = 0.76). Other studies also similarly reported that the Chinese tea population exhibited a higher level of genetic diversity than that of other tea populations from other countries [24,51]. In general, China is thought to be the origin of tea, so Chinese tea populations are the most likely to account for the largest proportion of diversity [51]. Our previous study noted that Korean tea germplasm showed low genetic diversity because of limitations in the gene stock from China, political and religious reasons, and extreme environmental conditions [45]. Tanaka et al. [62] reported that the tea plant in Japan was first introduced from China about 1200 years ago and that the country's original tea populations were established based on only a few of seeds from a restricted source. In addition, the authors in [23,25] suggested that the low genetic diversity of tea accessions in Japan could be attributed to long and intensive selection and breeding from the genetically limited tea stock in Japan.

It is important to identify the correspondence between the genetic diversity of tea accessions and their origins. In this study, the different approaches (STRUCTURE and DAPC) used to analyze the population structures of the 462 tea accessions were able to provide complementary information. However, the structuring of tea accessions at K = 2 (based on the estimated ΔK value in STRUCTURE) and K = 4 (based on the BIC and DAPC) clearly did not segregate the accessions based on geographical distinctions. The Evanno method is artificially maximal at K = 2, in some cases, because it finds the highest level of structure in the data by focusing only on the changes in slope [39,63]. Similar results were obtained in previous studies on tea germplasm structures based on SSR (K = 2) [11,30,64,65]. The DAPC method does not require that populations be in HW equilibrium and can handle large sets of data without using parallel processing software, so it provides an interesting alternative to the STRUCTURE software [66]. In addition, the DAPC analysis provided more detailed clusters compared to the STRUCTURE analysis in previous analyses using SSR [28,66,67]. Our results also agree with those of previous studies where the DAPC analysis (K = 4) provided more detail than STRUCUTRE (K = 2). However, these results indicated lower genetic differentiation (PhiPT, DAPC = 1.2%; Clustering analysis of phytochemicals = 0.8%; STRUCTURE = 0.5%) than that in the collection area (5.6%). This might be due to an imbalance in the distribution of tea accessions used in this study, as 88.3% of tea accessions in this study were collected from South Korea. In our previous study, the genetic differentiation in the DAPC analysis of Korean tea germplasm was 1.4% [23]. This affected the low genetic differentiation between groups resulting from an analysis of the population structure, although the genetic differentiation of tea origins was also shown to be low (5.6%).

4.3. Development of a Target-Oriented Core Collection

To develop core collections, various methods, such as phenotypes, proteins, and molecular markers, have been used. However, there is no universally accepted method to construct a core collection because every method has advantages and disadvantages [68]. Previous studies have proven that phenotypes are useful parameters for developing core collections [2,12,69]. Kumar et al. [70] reported that the use of molecular markers in the development of a core collection is more effective than the use of other data, such as morphological traits sensitive to environmental effects. In addition, molecular markers are more effective in identifying and minimizing redundancy. Le et al. [71] suggested that the use of both phenotypic and molecular data together is more effective than their use individually when constructing a core collection. In this study, molecular markers and biochemical contents were utilized for the construction of a core collection in tea germplasm using the POWERCORE program, which was successfully used to build a core collection for various plant species, including olive [69], safflower [71], and tea [9].

In this study, seasonal data sets were handled independently to develop the core collections because the MANOVA analysis presented noticeable Genotype X environmental interactions. In addition, the evaluation indices (MD%, VD%, VR%, CR%) were comparable and reflected their effectiveness in capturing diversity to validate the core collection. MD%, VD%, and VR% were used to evaluate the statistical consistency between the core and entire collections [42], while MD% was used to represent the difference in the accession averages between the core and entire collections, which should be <20% for a representative core collection. VD% indicates the variance captured by the core collection, and VR% indicates a comparison between the coefficient of variation values present in the core and entire collections. CR% indicates whether the distribution ranges of each variable in the core set are well represented when compared to the entire collection, which should be greater than 80% [12,42,43,70]. In this study, the core collections yielded a CR% of more than 80% (97.43%) and an MD% of less than 20% (7.88%) (Table 7). Similar results for other species were reported in core collections developed with a lower MD% or higher CR%, which were more representative of the entire collections [72,73]. In addition, the distributions of each phytochemical in the tea accessions showed similarities to those of the entire collection (Figure S1). In general, the core collections can be classified into three types or categories: core collections representing (1) individual accessions, (2) extremes, and (3) the distribution of accessions in the entire collection [3]. Odong et al. [3] suggested that a core collection of type 3 (distribution of accessions) is only of interest if the aim is to provide an overview of the composition of the whole collection using only a part of the collection. The authors in [23,74] suggested that this type of core collection can be obtained by maximizing the representativeness of the pattern of trait variations in the whole collection. Considering these reports, the core collection developed in this study showed a similar pattern of type 3, which could represent the entire collection.

By integrating genetic diversity and phytochemical content, we developed a target-oriented core collection that we have not tried before in tea germplasm. The main targets for tea breeding and use are mostly related to catechin content; therefore, the phytochemical analysis and development of TOCC allow us to extend the use of tea germplasm broadly. Furthermore, the TOCC retained the phytochemical and genetic diversity of ENC, as we extracted the accessions after analyzing the variation of the content over two years using molecular marker data. The genetic diversity indices (I and H) and the distribution of accessions (NJ and PCoA) also indicate that the TOCC is well developed and reflects the whole diversity of ENC. Throughout this process, we developed a greater value-added core collection, which will not only provide useful materials to breeders but also aid in the efficient management of genebank. This target-oriented core collection is distinguished from the previous core collection in which accessions were selected based on their agronomic traits and molecular markers. Our upgraded core collection focused on the phytochemical content in tea germplasm suggests new directions for the use and conservation of tea germplasm.

5. Conclusions

Evaluating a plant germplasm and establishing a core collection will enhance the proper utilization of plant genetic resources [73]. Especially, core collections have been developed in various crop collections because they have a size that facilitates evaluation, use, and conservation while maintaining existing genetic diversity [7]. In this study, phytochemicals content and genetic diversity on 462 tea accessions were evaluated and the target-oriented core collection was constructed based on these results. The phytochemical contents of 462 tea accessions showed varying distributions, although the genetic diversity was low. In addition, this is the first attempt to combine molecular diversity data with phytochemicals to develop a core collection of tea germplasm conserved in NAC. This target-oriented core collection will provide access to genetic diversity and phytochemical traits, which will be useful for characterizing the genetic determinants of the traits of interest. Furthermore, it could be used to design more effective breeding programs to increase the global utility of tea as a functional crop.

Supplementary Materials: Supplementary materials can be found at http://www.mdpi.com/2073-4395/10/11/1667/s1, Figure S1. Distribution of phytochemicals between the entire collection and the core collection in 2018 (A) and 2019 (B). Blue line, entire collection; Orange line, core collection. C, (+)-Catechin; Caf, Caffeine; CG, (−)-Catechin 3-gallate; EC, (−)-Epicatechin; ECG, (−)-Epigallocatechin 3-gallate; EGCG, (−)-Gallocatechin 3-gallate; GC, (−)-Gallocatechin; GCG, (−)-Gallocatechin 3-gallate; TC, total catechin. Table S1. List and the content of phytocemicals of 462 tea accessions in this study. Table S2. List of 33 SSR primers in this study. Table S3. Mean squares for phytochemicals according to year, accessions, and year x accession interactions. Table S4. Multivariate analysis of variance (MANOVA) to study yearly differences in all quantitative traits together.

Author Contributions: Conceptualization, K.J.L. and D.Y.H.; data curation, K.J.L. and D.Y.H.; formal analysis, K.J.L. and R.S.; resources, G.-Y.G. and D.-M.S.; investigation, S.-H.K., E.Y., and S.L.; writing—original draft, K.J.L.; writing—review and editing, D.Y.H.; funding acquisition, G.-T.C. All authors have read and agreed to the published version of the manuscript.

Funding: This research was funded by the Research Program for Agricultural Science and Technology Development, National Institute of Agricultural Sciences, Rural Development Administration, Republic of Korea, grant number [PJ01355702].

Conflicts of Interest: The authors declare no conflict of interest.

Abbreviations

TOCC	Target-oriented core collection
ENC	Entire collection
C	(+)-Catechin
CG	(−)-Catechin 3-gallate
EC	(−)-Epicatechin
ECG	(−)-Epigallocatechin 3-gallate
EGCG	(−)-Epigallocatechin 3-gallate
GC	(−)-Gallocatechin
GCG	(−)-Gallocatechin 3-gallate
TC	Total catechin

References

1. Engels, J.; Fassil, H. Plant and Animal Genebanks. In *Encyclopdia of Life Support Systems*; UNESCO, Ed.; Eolss Publishers: Paris, France, 1999; p. 34.
2. Qi-Lun, Y.; Fan, P.; Zou, S.-X. Constructing a Core Collection for Maize (*Zea mays* L.) Landrace from Wuling Mountain Region in China. *Agric. Sci. China* **2008**, *7*, 1423–1432. [CrossRef]
3. Odong, T.L.; Jansen, J.; Van Eeuwijk, F.A.; Van Hintum, T.J.L. Quality of core collections for effective utilisation of genetic resources review, discussion and interpretation. *Theor. Appl. Genet.* **2012**, *126*, 289–305. [CrossRef]
4. Frankel, O.H. Genetic perspectives of germplasm conservation. In *Genetic Manipulation: Impact on Man and Society*; Arber, W., Illemensee, K., Peacock, W.J., Starlinger, P., Eds.; Cambridge University Press: Cambridge, UK, 1984; pp. 161–170.
5. Brown, A.H.D. Core collections: A practical approach to genetic resources management. *Genome* **1989**, *31*, 818–824. [CrossRef]
6. Wambulwa, M.C.; Meegahakumbura, M.K.; Kamunya, S.; Muchugi, A.; Möller, M.; Liu, J.; Xu, J.-C.; Ranjitkar, S.; Olinirina, N.; Gao, L.-M. Insights into the Genetic Relationships and Breeding Patterns of the African Tea Germplasm Based on nSSR Markers and cpDNA Sequences. *Front. Plant Sci.* **2016**, *7*, 1244. [CrossRef]
7. Bramel, P.; Chen, L. *A Global Strategy for the Conservation and Use of Tea Genetic Resources*; The Crop Trust: Bonn, Germany, 2019.
8. Chen, L. Tea genetic resources in China. *Int. J. Tea Sci.* **2012**, *8*, 1–10.
9. Raina, S.N.; Ahuja, P.S.; Sharma, R.K.; Das, S.C.; Bhardwaj, P.; Negi, R.; Sharma, V.; Singh, S.S.; Sud, R.K.; Kalia, R.K.; et al. Genetic structure and diversity of India hybrid tea. *Genet. Resour. Crop. Evol.* **2011**, *59*, 1527–1541. [CrossRef]
10. Ranatunga, M.; Gunasekare, M. Stratification of Camellia Germplasm to Facilitate Construction of Core Collection; A Prerequisite for Tea Crop Improvement. *Sri Lanka J. Tea Sci.* **2009**, *74*, 62–73.

11. Taniguchi, F.; Kimura, K.; Saba, T.; Ogino, A.; Yamaguchi, S.; Tanaka, J. Worldwide core collections of tea (*Camellia sinensis*) based on SSR markers. *Tree Genet. Genomes* **2014**, *10*, 1555–1565. [CrossRef]
12. Wang, X.; Chen, L.; Yang, Y. Establishment of core collection for Chinese tea germplasm based on cultivated region grouping and phenotypic data. *Front. Agric. China* **2011**, *5*, 344–350. [CrossRef]
13. Ni, S.; Yao, M.-Z.; Chen, L.; Zhao, L.-P.; Wang, X.-C.; Apostolides, Z.; Chen, Z.-M. Germplasm and Breeding Research of Tea Plant Based on DNA Marker Approaches. In *Uncertainty Modeling for Data Mining*; Springer Science and Business Media LLC: Berlin, Germany, 2012; pp. 361–376.
14. Gunasekare, M.; Ranatunga, M.; Piyasundara, J.H.N.; Kottawa-Arachchi, J. Tea genetic resources in Sri Lanka: Collection, Conservation and Appraisal. *Int. J. Tea Sci.* **2012**, *8*, 51–60.
15. Kottawa-Arachchi, J.D.; Gunasekare, M.T.K.; Ranatunga, M.A.B. Biochemical diversity of global tea [*Camellia sinensis* (L.) O. Kuntze] germplasm and its exploitation: A review. *Genet. Resour. Crop. Evol.* **2018**, *66*, 259–273. [CrossRef]
16. Ullah, J.; Shah, A.; Nisar, M.; Khan, U.; Khan, J.; Ahmad, H.; Rokhan, G.; Jabalok, B. Biochemical characterization of lentil germplasm for genetic diversity. *Plant Cell Biotech. Mol. Biol.* **2016**, *17*, 7–13.
17. Xia, E.-H.; Tong, W.; Wu, Q.; Wei, S.; Zhao, J.; Zhang, Z.-Z.; Wei, C.; Wan, X. Tea plant genomics: Achievements, challenges and perspectives. *Hortic. Res.* **2020**, *7*, 1–19. [CrossRef] [PubMed]
18. Chen, L.; Zhou, Z.X.; Chen, L.; Zhou, Z.X. Variations of main quality components of tea genetic resources [*Camellia sinensis* (L.) O. Kuntze] preserved in the China National Germplasm Tea Repository. *Plant Foods Hum. Nutr.* **2005**, *60*, 31–35. [CrossRef]
19. Yü, Y.; Deng, Y.; Lu, B.-M.; Liu, Y.-X.; Li, J.; Bao, J.-K. Green tea catechins: A fresh flavor to anticancer therapy. *Apoptosis* **2013**, *19*, 1–18. [CrossRef]
20. Higdon, J.V.; Frei, B. Tea Catechins and Polyphenols: Health Effects, Metabolism, and Antioxidant Functions. *Crit. Rev. Food Sci. Nutr.* **2003**, *43*, 89–143. [CrossRef] [PubMed]
21. Musial, C.; Kuban-Jankowska, A.; Gorska-Ponikowska, M. Beneficial Properties of Green Tea Catechins. *Int. J. Mol. Sci.* **2020**, *21*, 1744. [CrossRef] [PubMed]
22. Masek, A. Antioxidant and Antiradical Properties of Green Tea Extract Compounds. *Int. J. Electrochem. Sci.* **2017**, *12*, 6600–6610. [CrossRef]
23. Matsumoto, S.; Kiriiwa, Y.; Yamaguchi, S. The Korean Tea Plant (*Camellia sinensis*): RFLP Analysis of Genetic Diversity and Relationship to Japanese Tea. *Breed. Sci.* **2004**, *54*, 231–237. [CrossRef]
24. Wachira, F.; Tanaka, J.; Takeda, Y. Genetic variation and differentiation in tea (*Camellia sinensis*) germplasm revealed by RAPD and AFLP variation. *J. Hortic. Sci. Biotechnol.* **2001**, *76*, 557–563. [CrossRef]
25. Kaundun, S.S.; Zhyvoloup, A.; Park, Y.-G. Evaluation of the genetic diversity among elite tea (*Camellia sinensis* var. sinensis) accessions using RAPD markers. *Euphytica* **2000**, *115*, 7–16. [CrossRef]
26. Chen, L.; Yamaguchi, S. RAPD markers for discriminating tea germplasms at the inter-specific level in China. *Plant Breed.* **2005**, *124*, 404–409. [CrossRef]
27. Young-Goo, P.; Kaundun, S.S.; Zhyvoloup, A. Use of the bulked genomic DNA-based RAPD methodology to assess the genetic diversity among abandoned Korean tea plantations. *Genet. Resour. Crop. Evol.* **2002**, *49*, 159–165. [CrossRef]
28. Lee, K.J.; Lee, J.-R.; Raveendar, S.; Shin, M.-J.; Kim, S.-H.; Cho, G.-T.; Hyun, D.Y. Assessment of Genetic Diversity of Tea Germplasm for Its Management and Sustainable Use in Korea Genebank. *Forest* **2019**, *10*, 780. [CrossRef]
29. Meegahakumbura, M.K.; Wambulwa, M.C.; Li, M.-M.; Thapa, K.K.; Sun, Y.; Moller, M.; Xu, J.-C.; Yang, J.-B.; Liu, J.; Liu, B.-Y.; et al. Domestication Origin and Breeding History of the Tea Plant (*Camellia sinensis*) in China and India Based on Nuclear Microsatellites and cpDNA Sequence Data. *Front. Plant Sci.* **2018**, *8*. [CrossRef]
30. Fang, W.; Cheng, H.; Duan, Y.; Jiang, X.; Li, X. Genetic diversity and relationship of clonal tea (*Camellia sinensis*) cultivars in China as revealed by SSR markers. *Plant Syst. Evol.* **2011**, *298*, 469–483. [CrossRef]
31. Korir, N.K.; Han, J.; Shangguan, L.; Wang, C.; Kayesh, E.; Zhang, Y.; Fang, J. Plant variety and cultivar identification: Advances and prospects. *Crit. Rev. Biotechnol.* **2012**, *33*, 111–125. [CrossRef] [PubMed]
32. Lai, J.A.; Yang, W.C.; Hsiao, J.Y. An assessment of genetic relationships in cultivated tea clones and native wild tea in Taiwan using RAPD and ISSR markers. *Bot. Bull. Acad. Sin.* **2001**, *42*, 93–100.
33. Hammer, O.; Harper, D.A.T.; Ryan, P.D. PAST: Paleontological statistics software package for education and data analysis. *Palaeontol. Electron.* **2001**, *4*, 1–9.

34. Ori, F.; Ma, J.-Q.; Gori, M.; Lenzi, A.; Chen, L.; Giordani, E. DNA-based diversity of tea plants grown in Italy. *Genet. Resour. Crop. Evol.* **2017**, *64*, 1905–1915. [CrossRef]
35. Kamvar, Z.N.; Tabima, J.F.; GrünwaldN, J. Poppr: An R package for genetic analysis of populations with clonal, partially clonal, and/or sexual reproduction. *PeerJ* **2014**, *2*, e281. [CrossRef]
36. Peakall, R.; Smouse, P.E. GenAlEx 6.5: Genetic analysis in Excel. Population genetic software for teaching and research—An update. *Bioinformatics* **2012**, *28*, 2537–2539. [CrossRef]
37. Pritchard, J.K.; Stephens, M.; Donnelly, P. Inference of population structure using multilocus genotype data. *Genetics* **2000**, *155*, 945–959. [PubMed]
38. Earl, D.A.; Vonholdt, B.M. Structure Harvester: A website and program for visualizing structure output and implementing the Evanno method. *Conserv. Genet. Resour.* **2011**, *4*, 359–361. [CrossRef]
39. Evanno, G.; Regnaut, S.; Goudet, J. Detecting the number of clusters of individuals using the software structure: A simulation study. *Mol. Ecol.* **2005**, *14*, 2611–2620. [CrossRef] [PubMed]
40. Ivandic, V.; Hackett, C.A.; Nevo, E.; Keith, R.; Thomas, W.T.; Forster, B.P. Analysis of simple sequence repeats (SSRs) in wild barley from the Fertile Crescent: Associations with ecology, geography and flowering time. *Plant Mol. Biol.* **2002**, *48*, 511–527. [CrossRef]
41. Jombart, T. Adegenet: A R package for the multivariate analysis of genetic markers. *Bioinformatics* **2008**, *24*, 1403–1405. [CrossRef] [PubMed]
42. Kim, K.-W.; Chung, H.-K.; Cho, G.-T.; Ma, K.-H.; Chandrabalan, D.; Gwag, J.-G.; Kim, T.-S.; Cho, E.-G.; Park, Y.-J. PowerCore: A program applying the advanced M strategy with a heuristic search for establishing core sets. *Bioinformatics* **2007**, *23*, 2155–2162. [CrossRef] [PubMed]
43. Hu, J.; Xu, H.M.; Zhu, J. Methods of constructing core collections by stepwise clustering with three sampling strategies based on the genotypic values of crops. *Theor. Appl. Genet.* **2000**, *101*, 264–268. [CrossRef]
44. Perrier, X.; Jacquemoud-Collet, J.P. DARwin Software. 2006. Available online: http://darwin.cirad.fr (accessed on 9 September 2020).
45. Ahmed, S.; Stepp, J.R. Chapter 5—Pu-erh Tea: Botany, Production, and Chemistry. In *Tea in Health and Disease Prevention*; Preedy, V.R., Ed.; Academic Press: Cambridge, MA, USA, 2013; pp. 59–71.
46. Anand, J.; Upadhyaya, B.; Rawat, P.; Rai, N. Biochemical characterization and pharmacognostic evaluation of purified catechins in green tea (*Camellia sinensis*) cultivars of India. *3 Biotech* **2014**, *5*, 285–294. [CrossRef]
47. Feng, L.; Gao, M.-J.; Hou, R.-Y.; Hu, X.-Y.; Zhang, L.; Wan, X.-C.; Wei, S. Determination of quality constituents in the young leaves of albino tea cultivars. *Food Chem.* **2014**, *155*, 98–104. [CrossRef] [PubMed]
48. Gai, Z.; Wang, Y.; Jiang, J.; Xie, H.; Ding, Z.; Ding, S.; Wang, H. The Quality Evaluation of Tea (*Camellia sinensis*) Varieties Based on the Metabolomics. *HortScience* **2019**, *54*, 409–415. [CrossRef]
49. Mondal, T.K. *Breeding and Biotechnology of Tea and Its Wild Species*; Springer Science and Business Media LLC: Berlin, Germany, 2014; pp. 1–167.
50. Chen, Z.-M.; Chen, L.; Apostolides, Z. Delicious and Healthy Tea: An Overview. In *Uncertainty Modeling for Data Mining*; Springer Science and Business Media LLC: Berlin, Germany, 2012; pp. 1–11.
51. Yao, M.Z.; Chen, L.; Liang, Y.R. Genetic diversity among tea cultivars from China, Japan and Kenya revealed by ISSR markers and its implication for parental selection in tea breeding programmes. *Plant Breed.* **2008**, *127*, 166–172. [CrossRef]
52. Jin, J.-Q.; Ma, J.-Q.; Ma, C.-L.; Yao, M.-Z.; Chen, L. Determination of Catechin Content in Representative Chinese Tea Germplasms. *J. Agric. Food Chem.* **2014**, *62*, 9436–9441. [CrossRef] [PubMed]
53. Punyasiri, P.N.; Jeganathan, B.; Kottawa-Arachchi, J.D.; Ranatunga, M.A.B.; Abeysinghe, I.S.B.; Gunasekare, M.T.K.; Bandara, B.M.R. New Sample Preparation Method for Quantification of Phenolic Compounds of Tea (*Camellia sinensis* L. Kuntze): A Polyphenol Rich Plant. *J. Anal. Methods Chem.* **2015**, *2015*, 964341. [CrossRef] [PubMed]
54. Gulati, A.; Rajkumar, S.; Karthigeyan, S.; Sud, R.K.; Vijayan, D.; Thomas, J.; Rajkumar, R.; Das, S.C.; Tamuly, P.; Hazarika, M.; et al. Catechin and Catechin Fractions as Biochemical Markers to Study the Diversity of Indian Tea (*Camellia sinensis* (L.) O. Kuntze) Germplasm. *Chem. Biodivers.* **2009**, *6*, 1042–1052. [CrossRef] [PubMed]
55. Wei, K.; Wang, L.; Zhou, J.; He, W.; Zeng, J.; Jiang, Y.; Cheng, H. Catechin contents in tea (*Camellia sinensis*) as affected by cultivar and environment and their relation to chlorophyll contents. *Food Chem.* **2011**, *125*, 44–48. [CrossRef]
56. Robertson, A. The chemistry and biochemistry of black tea production—The non-volatiles. In *Tea: Cultivation to Consumption*; Willson, K.C., Clifford, M.N., Eds.; Springer: Dordrecht, The Netherlands, 1992; pp. 555–601.

57. Koch, W.; Kukula-Koch, W.; Komsta, Ł.; Marzec, Z.; Szwerc, W.; Głowniak, K. Green Tea Quality Evaluation Based on Its Catechins and Metals Composition in Combination with Chemometric Analysis. *Molecules* **2018**, *23*, 1689. [CrossRef]
58. Leung, L.K.; Su, Y.; Chen, R.; Zhang, Z.; Huang, Y.; Chen, Z.-Y. Theaflavins in Black Tea and Catechins in Green Tea Are Equally Effective Antioxidants. *J. Nutr.* **2001**, *131*, 2248–2251. [CrossRef]
59. Chen, Z.-Y.; Zhu, Q.Y.; Tsang, D.; Huang, Y. Degradation of Green Tea Catechins in Tea Drinks. *J. Agric. Food Chem.* **2001**, *49*, 477–482. [CrossRef]
60. Yao, L.; Caffin, N.; D'Arcy, B.; Jiang, Y.; Shi, J.; Singanusong, R.; Liu, X.; Datta, N.; Kakuda, Y.; Xu, Y. Seasonal Variations of Phenolic Compounds in Australia-Grown Tea (*Camellia sinensis*). *J. Agric. Food Chem.* **2005**, *53*, 6477–6483. [CrossRef] [PubMed]
61. Gill, M. Speciality and herbal teas. In *Tea*; Springer Science and Business Media LLC: Berlin, Germany, 1992; pp. 513–534.
62. Tanaka, J.; Chen, L.; Apostolides, Z.; Chen, Z.-M. *Japanese Tea Breeding History and the Future Perspective*; Springer Science and Business Media LLC: Berlin, Germany, 2012; pp. 227–239.
63. Vigouroux, Y.; Glaubitz, J.C.; Matsuoka, Y.; Goodman, M.M.; Sánchez, G.J.; Doebley, J. Population structure and genetic diversity of New World maize races assessed by DNA microsatellites. *Am. J. Bot.* **2008**, *95*, 1240–1253. [CrossRef]
64. Yao, M.-Z.; Ma, C.-L.; Qiao, T.-T.; Jin, J.-Q.; Chen, L. Diversity distribution and population structure of tea germplasms in China revealed by EST-SSR markers. *Tree Genet. Genomes* **2011**, *8*, 205–220. [CrossRef]
65. Zhang, Y.; Zhang, X.; Chen, X.; Sun, W.; Li, J. Genetic diversity and structure of tea plant in Qinba area in China by three types of molecular markers. *Hereditas* **2018**, *155*, 22. [CrossRef]
66. Campoy, J.A.; Lerigoleur, E.; Christmann, H.; Beauvieux, R.; Girollet, N.; Quero-García, J.; Dirlewanger, E.; Barreneche, T. Genetic diversity, linkage disequilibrium, population structure and construction of a core collection of Prunus avium L. landraces and bred cultivars. *BMC Plant Biol.* **2016**, *16*, 49. [CrossRef] [PubMed]
67. Lee, K.J.; Lee, J.-R.; Sebastin, R.; Cho, G.-T.; Hyun, D.Y. Molecular Genetic Diversity and Population Structure of Ginseng Germplasm in RDA-Genebank: Implications for Breeding and Conservation. *Agronomy* **2020**, *10*, 68. [CrossRef]
68. Liu, W.; Shahid, M.Q.; Bai, L.; Lu, Z.; Chen, Y.; Jiang, L.; Diao, M.; Liu, X.; Lu, Y. Evaluation of Genetic Diversity and Development of a Core Collection of Wild Rice (Oryza rufipogon Griff.) Populations in China. *PLoS ONE* **2015**, *10*, e0145990. [CrossRef]
69. Belaj, A.; Dominguez-García, M.D.C.; Atienza, S.G.; Urdíroz, N.M.; De La Rosa, R.; Satovic, Z.; Martín, A.; Kilian, A.; Trujillo, I.; Valpuesta, V.; et al. Developing a core collection of olive (*Olea europaea* L.) based on molecular markers (DArTs, SSRs, SNPs) and agronomic traits. *Tree Genet. Genomes* **2011**, *8*, 365–378. [CrossRef]
70. Kumar, S.; Ambreen, H.; Variath, M.T.; Rao, A.R.; Agarwal, M.; Kumar, A.; Goel, S.; Jagannath, A. Utilization of Molecular, Phenotypic, and Geographical Diversity to Develop Compact Composite Core Collection in the Oilseed Crop, Safflower (*Carthamus tinctorius* L.) through Maximization Strategy. *Front. Plant Sci.* **2016**, *7*. [CrossRef]
71. Le, D.S.; Pagès, J. Analyse factorielle multiple hiérarchique. *Rev. Stat. Appl.* **2003**, *51*, 47–73.
72. Yanfang, Z.; DeChang, H.; Jincheng, Z.; Ping, Z.; Zhaohong, W.; Chuanjie, C. Development of a mulberry core collection originated in China to enhance germplasm conservation. *Crop. Breed. Appl. Biotechnol.* **2019**, *19*, 55–61. [CrossRef]
73. Lee, H.-Y.; Ro, N.-Y.; Jeong, H.-J.; Kwon, J.-K.; Jo, J.; Ha, Y.; Jung, A.; Han, J.-W.; Venkatesh, J.; Kang, B.-C. Genetic diversity and population structure analysis to construct a core collection from a large Capsicum germplasm. *BMC Genet.* **2016**, *17*, 1–13. [CrossRef]

74. Galwey, N.W. Verifying and validating the representativeness of a core collection. In *Core Collections of Plant Genetic Resources*; Hodgkin, T.B.A., van Hintum, T.J.L., Morales, E.A.V., Eds.; John Wiley and Sons: Chichester, UK, 2005; pp. 187–198.

agronomy

MDPI

Article

Agromorphological Characterization Revealed Three Phenotypic Groups in a Region-Wide Germplasm of Fonio (*Digitaria exilis* (Kippist) Stapf) from West Africa

Abdou R. Ibrahim Bio Yerima [1,2], Enoch G. Achigan-Dako [1,*], Mamadou Aissata [2], Emmanuel Sekloka [3], Claire Billot [4,5], Charlotte O. A. Adje [1], Adeline Barnaud [6] and Yacoubou Bakasso [7]

1. Laboratory of Genetics, Biotechnology and Seed Sciences (GBioS), Faculty of Agronomic Sciences (FSA), University of Abomey-Calavi (UAC), Cotonou 01 BP 526, Benin; abdourazakoui01@gmail.com (A.R.I.B.Y.); abikecharlotte@gmail.com (C.O.A.A.)
2. Department of Rainfed Crop Production (DCP), National Institute of Agronomic Research of Niger (INRAN), Niamey BP 429, Niger; intaaoudi@gmail.com
3. Laboratory of Phytotechny, Plant Breeding and Plant Protection (LaPAPP), Department of Sciences and Techniques of Vegetal Production (STPV), Faculty of Agronomy, University of Parakou, Parakou BP 123, Benin; emmanuelsekloka@hotmail.com
4. Unité Mixte de Recherche Amélioration Génétique et Adaptation des Plantes (AGAP), Agricultural Research Centre for International Development (CIRAD), F-34398 Montpellier, France; claire.billot@cirad.fr
5. Amélioration Génétique et Adaptation des Plantes (AGAP), University of Montpellier, CIRAD, Institut National de la Recherche Agronomique (INRA), l'Institut Agro/Montpellier SupAgro, 2 Place Pierre Viala, 34060 Montpellier, France
6. Diversité, Adaptation et Développement des Plantes (DIADE), Institut de Recherche pour le Développement (IRD), University of Montpellier, 34060 Montpellier, France; adeline.barnaud@ird.fr
7. Faculty of Science and Techniques, University of Abdou Moumouni of Niamey, Niamey BP 10662, Niger; bakasso@yahoo.com

* Correspondence: e.adako@gmail.com; Tel.: +229-95-393283 or +227-96-401486

Received: 5 August 2020; Accepted: 17 September 2020; Published: 27 October 2020

Abstract: Fonio is an ancient orphan cereal, cultivated by resource-poor farmers in arid and semi-arid regions of West Africa, who conserved and used the cereal for nutrition and income generation. However, the extent and patterns of phenotypic variation within the fonio germplasm remain scarcely evaluated to inform breeding decisions. In this study, we used alpha lattice design to assess the phenotypic variability of 180 fonio accessions using 20 agro morphological descriptors, including both qualitative and quantitative traits at Bengou research station in 2018 and 2019. Descriptive statistics, combined analysis and multivariate analyses, including principal components analysis (PCA), hierarchical clustering, and multiple factor analysis (MFA) were performed. As results, we found significant differences among fonio accessions and among agro ecological zones of provenance for most of the quantitative traits. Furthermore, highly significant positive correlations were found between grain yield and other yield-related attributes such as harvest index and thousand seeds weight, whereas significant negative correlations were found between grain yield and flowering and maturity times. Clusters analysis revealed three phenotypic groups. Group 1 was characterized by early maturing fonio accessions with higher grain yield. Groups 2 and 3 were characterized by late maturing accessions with intermediate to moderate grain yield. The accessions from Group 1 are candidate for yield improvement and development of fonio lines with enhanced lodging resistance. Accessions from Group 2 and 3 can be improved for yield through marker-assisted selection of best thousand seed weight. This study highlights how traits are correlated within the major phenotypic groups of fonio in West Africa, and we discussed how these groups could be further exploited for improving traits of economic importance. Furthermore, this study exhibited agro morphological descriptors that discriminate fonio accessions and provide useful information for parental selection with economically important agronomic traits.

Keywords: fonio; fonio millet; white fonio; *Digitaria exilis*; agro morphological descriptors; phenotypic diversity; neglected and underutilized species (NUS); genetic improvement

---

1. Introduction

Fonio (*Digitaria exilis* (Kippist) Stapf), also referred to as white fonio or fonio millet, is one of the oldest cereal crops domesticated by farmers in West Africa. Its cultivation seems to have started about 7000 years ago [1]. Fonio plays a critical role for food and nutrition security for several millions of people in this region, especially during shortage period due to short time life cycle [2]. Besides, its grain is nutritiously valuable due to the presence of two human-vital amino acids methionine and cysteine, which are absent in other staple food cereals such as wheat (*Triticum aestivum* L.), rice (*Oryza sativa* L.), maize (*Zea mays* L.), or sorghum (*Sorghum bicolor* (L.) Moench) [3]. Additionally, some antithyroid properties have been reported due to high flavonoid content in the crude fonio grain [4]. The grains are poor in gluten and beneficial for people suffering from coeliac disease and overweight people, pregnant and breast feeding women, as well as diabetic people [5]. Furthermore, an efficient formulation of fonio flour enriched with local food resources for the complementary feeding of children under age two has been reported [6]. Due to its C4 metabolism, the crop resists to drought and is well adapted to local pedo-climatic conditions [7].

In traditional fonio farming systems of West Africa, all cultivated varieties are landraces [8]. Their productivity is limited due to the lack of improved varieties and good agronomic practices, lack of institutional support, lack of knowledge about some biotic and abiotic stresses, and tedious post-harvest activities [9]. The development of improved varieties well adapted to different agro ecological zones and preferred by farmers and consumers is of paramount importance in fonio cultivation systems [10]. This can only be achieved by implementing efficient genetic resources management, conservation and valorization, a prerequisite for any crop improvement. This requires a comprehensive knowledge of uses and of diversity patterns. Plant genetic diversity helps breeders in developing improved varieties with high quality, tolerant or resistant to biotic and abiotic stresses, and well adapted to different climate conditions [11].

Morphological markers based on phenotypic characterization of traits are among methods used for assessing diversity within and between plant populations because they are simple, direct, easy and cheap to characterize germplasm accessions. Besides, they are directly related to farmers' perception. Even though these type of markers may present a low level of polymorphism, heritability, and expression and are often susceptible to phenotypic plasticity [12], they were used by many authors to study agro morphological diversity in other small grain cereals [13–16]. Various studies reported the existence of a wide range of diversity in fonio germplasm accessions [17–21] for quantitative traits (plant height, number of leaves per plant, internode length, days to 50% flowering, days to 50% maturity, leaf width, panicle leaf length, stem girth, raceme length, dry biomass yield and grain yield, seeds size, thousand seed weight) and for qualitative traits (collar color, green color of foliar limb, anthocyanin coloration and distribution in different aerial organs, type of panicle and panicle exertion, and grain color). However, previous works on fonio agromorphological characterization limited their collection area to only one or two countries, or used few accessions, overlooking the wider distribution range of the crop. Hence, this limits breeders' chance to get genes of interest, particularly in an orphan crop with a narrow knowledge of the genetic base.

So far, and for the last two decades, the following questions remain unresolved: (a) how is the phenotypic variation distributed among West African fonio accessions? (b) Which agro morphological traits contribute to distinctiveness in fonio germplasm? (c) How are they related? (d) How many phenotypic groups can be identified? The objectives of this study are to assess phenotypic variability among fonio accessions using agro morphological descriptors, to investigate association patterns among quantitative traits and to identify useful phenotypic groups for future improvement. We hypothesize that exploring a large set of fonio accessions collected from various origins and adaptation areas will enrich the existing germplasm. This will equally help breeders in identifying useful landraces with key agronomic traits for future crop improvement.

2. Materials and Methods

2.1. Plant Material

A total of 180 fonio accessions were used in this study. They were collected from five Western African countries including Benin (50 accessions), Burkina Faso (five accessions), Guinea (33 accessions), Mali (49 accessions), and Niger (43 accessions), maintained and available in the genebank of the Laboratory of Genetics, Biotechnology and Seed Science (GBioS) from University of Abomey-Calavi (Benin) in 2018. A total of 57 accessions were collected from the arid region, 93 from the semi-arid region, and 30 from the sub-humid region (Figure 1). The arid zone includes the Sahel with up to 750 mm of rainfall in a single short season followed by a long dry season. The vegetation is made up of grass, bush and thicket. In the semi-arid region, the annual rainfall varies from 750 mm to 1250 mm on average in one season followed by a dry season; the vegetation is a savanna grassland. The sub humid region is a deciduous forest and woodland savanna with an average annual rainfall varying between 1250 mm to 1500 mm in one season.

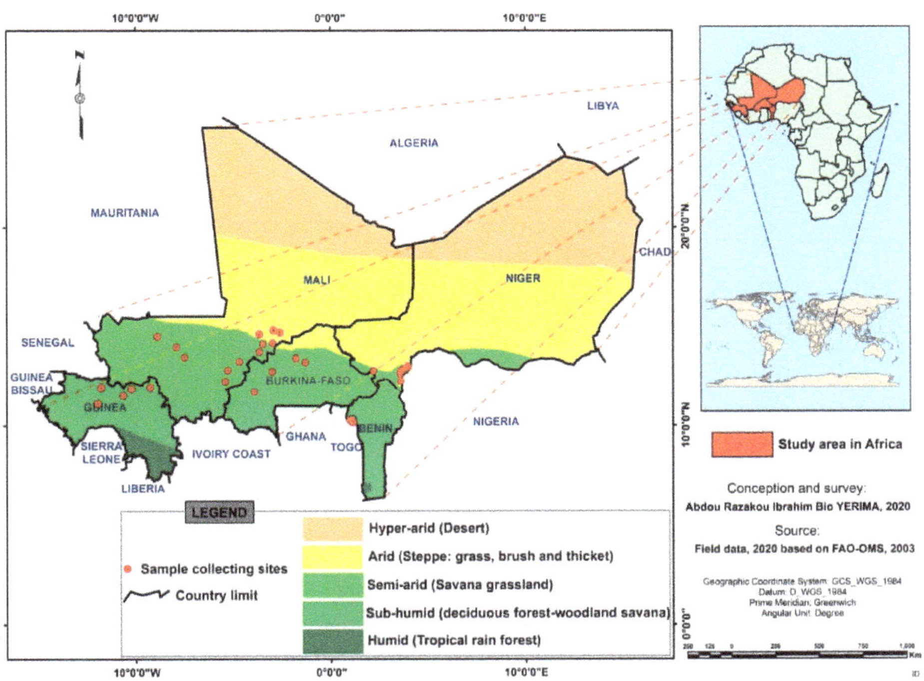

Figure 1. Distribution of collected fonio accessions according to countries and agro ecological zones.

2.2. Description of Site

The field trial was carried out at Bengou research station (National Institute of Agronomic Research of Niger) during the rainy seasons 2018 and 2019. This site is in Northern Sudan Savanna agroecological zone located at 11°58′44.4″ N, 3°32′51.6″ E, at 172 m above sea level and is at 260 km from Niamey and at 17 km form Republic of Benin border. The soil was sandy loam with basic pH medium. Annual rainfall was 1047.62 mm and 1211.2 mm in 2018 and 2019 respectively (Figure 2). The temperature varied between 22.90 °C and 35.20 °C with an average of 29.05 °C. The District of Bengou offers rich and varied vegetation where we can meet almost all the plant species of Niger. The plant population can be subdivided according to the topography and the nature of the soils into three groups: (i) the group of lateritic plateaus is made up of shrub from savannah dominated by species such as *Combretum nigricans* [22], *Combretum micranthum* G.Don [23], *Acacia macrostachya* Rchb. ex DC [24], etc.; (ii) the group of valleys mainly consisted of wooded savanna and classified forests dominated by *Borassus aethiopium*, *Hyphaene thebaica* (L.) Mart [25], *Diospyros mespiliformis* Hochst. ex A.DC, *Parinari macrophylla* (Sabine) Prance ex F.White [26], etc.; and (iii) the group of the old fixed dunes.

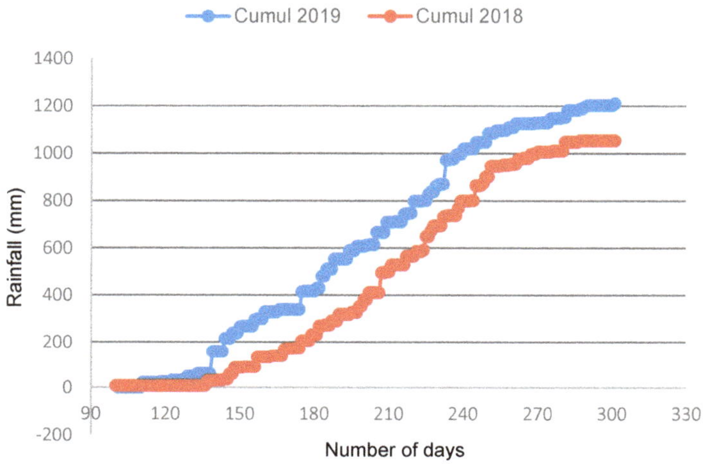

Figure 2. Cumulative rainfall at Bengou research station in 2018–2019.

2.3. *Experimental Design and Management*

The experiment design used was alpha lattice (18 × 10) with two replicates and each accession had four rows per plot of 1 m length and 0.20 m apart. The plot area was 1 m^2 and distance between blocks and a replicate was 0.5 m and 1 m respectively. The observations were taken from randomly selected ten sampled plants from the four rows of a plot. Prior to planting, land was ploughed, harrowed and fonio seeds were sown (line sowing) at the rate of 2 g per plot on 19 July 2018 and 18 July 2019. Weeds were rouged out manually three weeks after planting and no fertilizer was applied. Weeding was subsequently done in order to keep the trials clean till harvest. No irrigation was applied during the experiment.

2.4. *Data Collection*

Data were collected on twenty traits including five qualitative (vigour at seedling. panicle exertion, panicle type, panicle leaf attitude and grain color) and 15 quantitative (plant height, number of internodes per plant, number of leaves per plant, leaf length, leaf width, culm diameter, days to 50% maturity, raceme length, number of racemes per panicle, number of grain par raceme, panicle weight, shoot dry matter yield, grain yield. harvest index and thousand seeds weight) (Table 1). These data were recorded following the Bioversity International rice descriptors list [27].

Table 1. Qualitative and quantitative descriptors for agro morphological characterization of fonio accessions.

	Qualitative	
Descriptors	**Code**	**Description**
Vigour at seedling	1 = good establishment 2 = moderate establishment 3 = poor establishment	Recorded three weeks after planting
Panicle exertion	1 = well exerted 3 = moderately exerted 5 = lightly exerted 7 = partially exerted	Recorded near maturity
Panicle type	1 = compact 3 = intermediate 5 = open	Recorded near maturity
Panicle leaf attitude	1 = erect 3 = intermediate 5 = horizontal 7 = descending	Recorded after heading
Phenotypic grain colour	1 = brown 2 = greyed orange	Recorded after harvesting
	Quantitative	
Descriptors	Code	Description
Plant height (cm)	PHT	Measured from the soil level to the tip of longest panicle
Number internode	NIN	Mean number of internodes on the main stem from 10 plants
Internode length (cm)	INL	Measured from the 2nd internode to last internode below the panicle
Number of leaves/plant	NFP	Mean number of leaves from 10 plants
Culm diameter (mm)	CD	Measured at the mid portion of the culm.
Days to 50% flowering (days)	FLO	Date on which 50% of the plants are flowering
Days to 50% maturity (days)	MAT	Date on which 50% of the plants are maturing
Number of racemes/panicle	NRP	Mean number of raceme from 10 panicles
Raceme length (cm)	RLT	Length of main axis measured from the raceme base to the tip
Number of grains/raceme	NGR	Mean number of grain from 10 racemes
Panicle length (cm)	PLT	Measured from panicle leaf insertion to the tip of longest raceme
Dry biomass yield (kg.ha^{-1})	DMY	Ratio of shoot dry weight (kg) to the plot area (ha)
Grain yield (kg.ha^{-1})	GRY	Ratio of grain weight (kg) to the plot area (ha)
Harvest index (%)	HI	Ration of grain yield to total dry biomass × 100
Thousand seeds weight (g)	TSW	Counting and weighting of 1000 seeds

2.5. Data Analysis

Descriptive statistics were used to depict variations that emerge from qualitative and quantitative traits. When conditions are met, analysis of variance was performed for quantitative traits in order to firstly verify differences among accessions from various provenances and thereafter from agro ecological zones. Pearson's correlation analysis was carried out to investigate association patterns among quantitative traits. In order to describe the variation of the set of variables together principal components analysis (PCA) [28] was carried out using packages "FactoMineR" [29] and Factoextra [30] in R software version 3.5.1 [31]. Moreover, hierarchical clustering on principal component (HCPC) [32] analysis and linear discriminant analysis (LDA) [33] were used to define hierarchical typologies of fonio accessions and to describe phenotypic groups obtained. Multiple factor analysis (MFA) [34] was used for synthetic comparison of the clusters or phenotypic groups obtained.

3. Results

3.1. Variation of Qualitative Traits among Fonio Accessions

Proportion distributions of the qualitative traits (Table 2) revealed that three weeks after planting 30% of accession showed good seedling vigor, while 63% and 7% showed moderate and poor vigor, respectively. Three types of panicle were observed among accessions: compact type (50%), intermediate type (48%), and open type (2%). Panicle leaf attitude was erect (48%), intermediate (45%), or horizontal (7%). Moreover, three types of panicle exertion were noticed with 65% of accessions showing well exerted panicle, 27% moderately exerted, and 8% lightly exerted. Majority of fonio accessions' seed had 84% of brown color, while 16% had greyed orange color (Figure 3a,b).

Table 2. Qualitative parameters variations among fonio accessions.

Descriptors	Modalities	Number of Accessions	Percentage (%)
Vigour at seedling	Good	54	30
	Moderate	114	63
	Poor	12	7
Panicle leaf attitude	Erect	86	48
	Intermediate	82	45
	Horizontal	12	7
Panicle type	Compact	90	50
	Intermediate	86	48
	Open	4	2
Panicle exertion	Well exerted	117	65
	Moderately exerted	48	27
	Lightly exerted	15	8
Phenotypic grain color	Brown	152	84
	Greyed orange	28	16

Figure 3. Grain color in fonio accessions, (**a**): Greyed orange, (**b**): Brown.

3.2. Fonio Performance Analysis Based on Country of Provenance

Table 3 presents the mean values of quantitative traits of fonio from the different countries. The accessions from Guinea produced significantly taller plant (81 ± 4.92 cm), higher number of internode (11 ± 2.04), and higher internode length (24.24 ± 11.6 cm). This was followed by accessions from Benin with 72 ± 4.62 cm plant height and 10 ± 1.76 internodes. Similar trend was observed for raceme length trait distribution. Accessions with longer racemes were frequent in Guinea (5.52 ± 0.59 cm) and Benin (5.10 ± 1.70 cm) whereas accessions with lower raceme were from Niger and Mali with 4.50 ± 0.60 cm and 4.41 ± 0.93 cm length respectively. However, shorter plants with reduced number of internodes were more predominant among accessions from Niger and Mali with respectively 70 ± 4.98 cm plant height and 7 ± 0.62 internodes, and 68 ± 6.27 cm plant height and 6 ± 0.78 internodes. Accessions from Burkina exhibited medium raceme length (4.80 ± 1.16 cm).

Flowering times (50% of plants flowering) in Guinean fonio accessions was 85 ± 7.75 days, and maturity happens after 105 ± 4.57 days. Flowering time in Malian accession is shorter with 69 ± 1.06 days only. The accessions from Benin exhibited higher number of grain per raceme (105 ± 12.03 grains) and higher dry biomass yield (5095 ± 145.3 $kg.ha^{-1}$) followed by Guinea with 94 ± 11.9 grains per raceme and 5075 ± 96.47 $kg.ha^{-1}$ of dry biomass yield, respectively. Accessions of Burkina Faso had 86 ± 22.3 grains per raceme and 4063± $kg.ha^{-1}$ of dry biomass. Accessions from Niger and Mali exhibited lower values with respectively 79 ± 9.18 and 76 ± 11.18 grains per raceme, and 4536 ± 125.8 $kg.ha^{-1}$ and 3398 ± 98.27 $kg.ha^{-1}$ for dry biomass yield. In contrast, accessions from Mali and Niger had higher number of leaves per plant with respectively 212 ± 70.02 leaves and 186 ± 64.9 leaves. In addition, accessions from Niger produced higher grain yield (1936 ± 62.25 $kg.ha^{-1}$) compared to those from Mali (1910 ± 68.75 $kg.ha^{-1}$), Burkina (1063 ± 70.53 $kg.ha^{-1}$), Benin (1057 ± 45.99 $kg.ha^{-1}$), and Guinea (924 ± 45.15 $kg.ha^{-1}$). Higher harvest index and larger seed size were observed in accessions from Mali and Niger (Table 3).

Table 3. Mean performances of fonio accessions in different countries.

Quantitative Traits	Countries							
	Benin	Burkina Faso	Guinea	Mali	Niger	Mean	SD	Pr (>F)
Plant height (cm)	72 ± 4.62b	71 ± 12.29bc	81 ± 4.92a	68 ± 6.27c	70 ± 4.98bc	72	5.03	***
Number of internodes	10 ± 0.85ab	10 ± 1.76b	11 ± 2.04a	6 ± 0.78c	7 ± 0.62c	9	2.17	***
Internode length (cm)	21.42 ± 6.1ab	22.62 ± 9.6ab	24.24 ± 11.6a	19.44 ± 9.3b	18.44 ± 3.9b	20.73	2.34	**
Number of leaves/plant	92 ± 10.8c	81 ± 10.7c	79 ± 8.1c	212 ± 70.2a	186 ± 64.9b	144	63.85	***
Culm diameter (mm)	1.48 ± 0.55a	1.40 ± 0.13ab	1.36 ± 0.20b	1.21 ± 0.11c	1.29 ± 0.09b	1.33	0.10	**
Days to flowering	81 ± 3.24b	80 ± 2.30b	85 ± 7.95a	69 ± 1.06d	74 ± 2.20c	77	6.30	***
Days to 50% Maturity	105 ± 3.75a	98 ± 3.87b	105 ± 4.57a	85 ± 5.07c	84 ± 1.94c	95	10.36	***
Number of racemes/panicle	4 ± 0.21a	4 ± 0.11a	3 ± 0.14b	3 ± 0.32b	3 ± 0.25b	3	0.55	***
Raceme length (cm)	5.10 ± 1.70b	4.80 ± 1.16bc	5.52 ± 0.59a	4.41 ± 0.93c	4.50 ± 0.60c	4.82	0.46	***
Number of grains/raceme	105 ± 12.03a	86 ± 22.3bc	94 ± 11.9b	76 ± 11.18c	79 ± 9.18c	88	11.77	***
Panicle length (cm)	15.1 ± 2.56c	15.9 ± 1.73bc	15.8 ± 1.07c	17.2 ± 3.2ab	17.4 ± 1.40a	16.4	0.98	**
Dry biomass yield ($kg.ha^{-1}$)	5095 ± 145.3a	4603 ± 159.3ab	5075 ± 96.47a	3398 ± 98.27c	4536 ± 125.8b	4490	689.64	***
Grain yield ($kg.ha^{-1}$)	1057 ± 45.99b	1063 ± 70.53b	924 ± 45.15b	1910 ± 68.75a	1936 ± 62.25a	1474	500.69	***
Harvest index (%)	26 ± 4.90c	27 ± 8.98c	25 ± 6.23c	59 ± 9.48a	54 ± 7.18b	41	16.81	***
Thousand seed weight (g)	0.59 ± 0.05d	0.63 ± 0.10c	0.68 ± 0.04b	0.82 ± 0.04a	0.69 ± 0.03b	0.70	0.09	***

SD = standard deviation. Means followed by the same letter are not significantly different at $p = 0.05$, *** = significant at 0.001, ** = significant at 0.01.

3.3. Fonio Performance Analysis Based on Ecological Zones

Fonio accessions from arid zone produced significantly higher grain yield (2937 ± 327.81 kg.ha^{-1}), higher harvest index (60%), and higher thousand seed weight (0.53 ± 0.07 g) (Figure 4m–o). Accessions from sub humid zones revealed lower value with 1364 ± 246.58 kg.ha^{-1} grain yield, 30% harvest index, and 0.42 ± 0.05 g thousand seed weight. Furthermore, 50% flowering and 50% maturity times were shorter in accessions from arid zone (72 ± 2.95 days and 93 ± 4.03 days, respectively), whereas accessions from sub humid areas exhibited longer 50% flowering (86 ± 4.55 days) and 50% maturity (108 ± 5.19 days) times (Figure 4f,g). Accessions from the semi-arid zone showed relatively an intermediate 50% flowering and 50% maturity times. Highest number of internode (8 ± 1.19) and the higher internode length (25.62 ± 5.56 cm) were observed in accessions from the sub humid zone. A similar trend was observed for raceme length (9.37 ± 3.97 cm), the number of grain per raceme (88 ± 13.60 grains), panicle length (18.3 ± 1.16 cm), and dry biomass yield (5818 ± 543.87 kg.ha^{-1}). However, accessions with wider culm diameter (1.32 ± 0.26 mm) were abundant in the semi-arid zone compared to those from sub humid and arid zones (Figure 4e).

3.4. Relationship among Quantitative Traits

We observed highly positive correlations between grain yield (GRY) and harvest index (HI) (r = 0.89), between maturity time (MAT) and flowering time (FLO) (r = 0.80), and between maturity (MAT) and raceme length (RLT) (r = 0.81). Likewise, strong significant positive correlations were observed between number of internode (NIN) and maturity (MAT) (r = 0.77) and between grain yield (GRY) and number of leaves per plant (NLP) (r = 0.68), and thousand seeds weight (TSW) (r = 0.60). However, negative correlations (highly significant and moderately significant) were observed between grain yield (GRY) and flowering time (FLO) (r = −0.79), maturity time (MAT) (r = −0.77), number of internode (NIN) (r = −0.76), and plant height (PHT) (r = −0.35). Moreover, strong significant negative correlations were found between harvest index (HI) and flowering (FLO) (r = −0.84), and maturity (MAT) (r = −0.79) (Table 4).

Table 4. Pearson's correlation matrix on fifteen quantitative traits used in fonio genotypes description.

	PHT	NIN	INL	NLP	CD	MAT	FLO	PLT	RLT	NRP	NGR	DBY	GRY	HI	TSW
PHT	1														
NIN	0.53	1													
INL	0.55	0.77	1												
NLP	−0.38	−0.75 *	−0.70	1											
CD	0.35	0.53	0.51	−0.51	1										
MAT	0.36	0.77 **	0.70	−0.71	0.51	1									
FLO	0.35	0.77	0.64	−0.71	0.56	0.80 **	1								
PLT	0.46	0.25	0.34	−0.19	0.02	0.24	0.05	1							
RLT	0.57	0.74	0.73	−0.64	0.42	0.81 **	0.66	0.49	1						
NRP	−0.02	0.21	0.16	−0.27	0.31	0.27	0.23	−0.21	0.13	1					
NGR	0.43	0.69	0.64	−0.60	0.50	0.64	0.67	0.15	0.61	0.30	1				
DBY	0.48	0.65	0.52	−0.50	0.43	0.51	0.66	0.17	0.53	0.11	0.59	1			
GRY	−0.35 **	−0.76 **	−0.70	0.68 *	−0.42	−0.77 *	−0.79 **	−0.18	−0.69	−0.17	−0.56	−0.43	1		
HI	−0.37	−0.76	−0.74	0.69	−0.48	−0.79 **	−0.84 **	−0.21	−0.69	−0.16	−0.61	−0.62	0.89 **	1	
TSW	−0.24	−0.63	−0.45	0.61	−0.53	−0.64	−0.79	0.08	−0.49	−0.34	−0.62	−0.60	0.60 **	0.66	1

PHT = plant height. NIN = number of internodes, INL = internode length, NFP = number of leaves per plant, CD = culm diameter, FLO = 50% flowering, MAT = 50% maturity, NRP = number of racemes per panicle, PLT = panicle length, RLT = raceme length, NGR = number of grain raceme, DBY = dry biomass yield, GRY = grain yield, HI = harvest index, TSW = thousand seeds weight, * = significance at 0.05, ** = significance at 0.01.

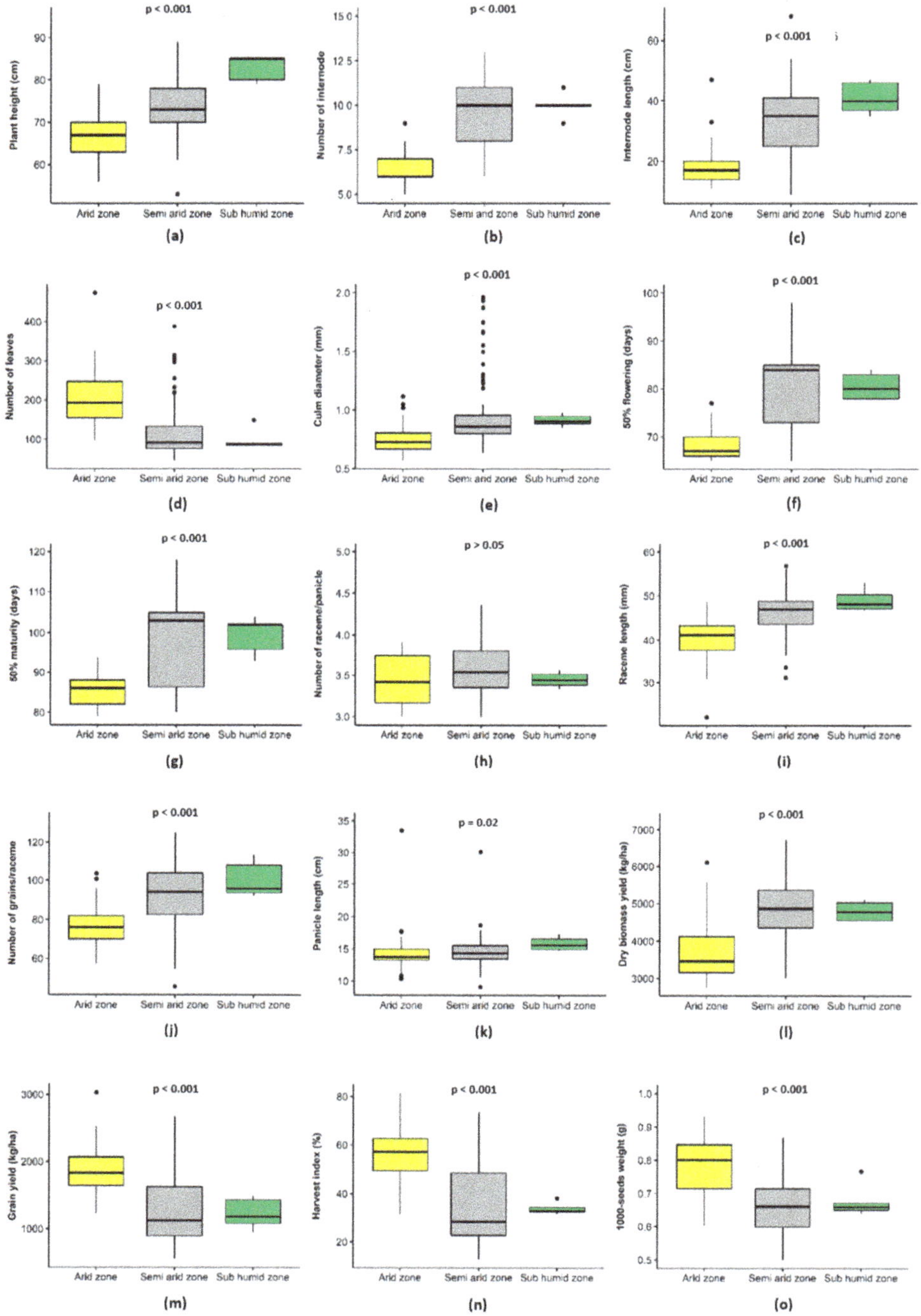

Figure 4. Boxplot showing fonio accessions performance based on ecological zones. Variation of: (**a**) plant height, (**b**) number of internodes, (**c**) internodes length, (**d**) number of leaves, (**e**) culm diameter, (**f**) 50% flowering, (**g**) 50% maturity, (**h**) number of racemes per panicle, (**i**) raceme length, (**j**) number of grains per raceme, (**k**) panicle length, (**l**) dry biomass yield, (**m**) grain yield, (**n**) harvest index, (**o**) 1000 seeds weight.

3.5. Clustering of Accessions Based on Country of Provenance

The results of PCA showed that the first two dimensions accounted for 50.90% of the total variation among fonio accessions. The first dimension (Dimension 1) was highly correlated with flowering time, number of internodes, maturity time, internode length, raceme length, number of grain per raceme, dry biomass yield, thousand seed weight, number of leaves per plant, grain yield and harvest index while the second dimension (Dimension 2) was moderately correlated to number of raceme per panicle, panicle length, phenotypic grain color, culm diameter, and plant height (Figure 5). Furthermore, the results of principal component analysis (PCA) showed that fonio accessions from Benin, Burkina Faso, and Guinea were characterized by late flowering and maturing accessions, with taller plants, long raceme length and higher number of grains per raceme compared the group of accessions from Mali and Niger that were characterized by early flowering and early maturing accessions, with higher grain yield, higher thousand seeds weight and higher number of leaves of per plant. Moreover, the PCA results showed that Dim1 differentiates two groups of fonio accessions: one group composed essentially of accessions from Mali and Niger, while the second group was composed of accessions from Benin, Burkina, and Guinea. Likewise, Dimension 2 differentiates two groups of accessions, the first group was composed of accessions from Guinea, Benin, and Burkina Faso, while the second group was again a mixed of accessions from these three countries (Figure 5).

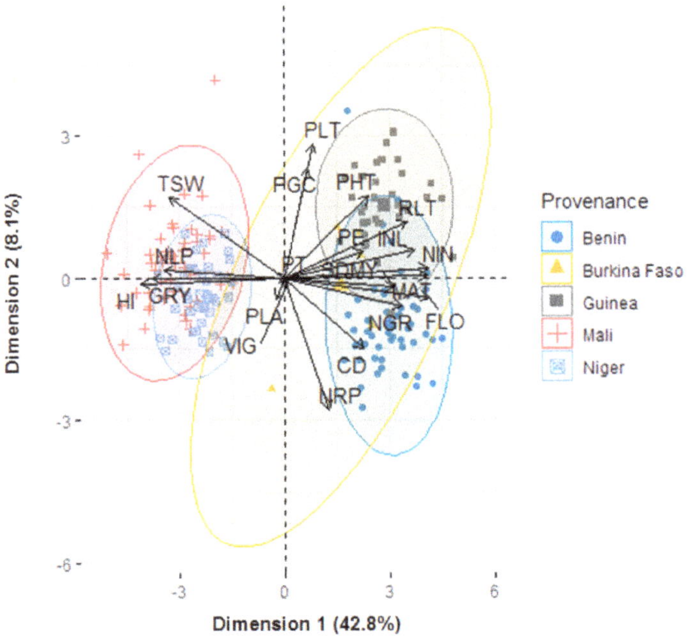

Figure 5. Principal components analysis biplot showing variables and individuals on Dimensions 1 and 2.

3.6. Partition of Accessions into Phenotypic Groups

The hierarchical clustering on principal components analysis grouped fonio accessions into three different clusters or phenotypic groups (Figure 6). Cluster 1 (50.56% of total accessions), essentially gathered accessions from Mali and Niger, characterized by early flowering and early maturing genotypes (respectively means of 69 ± 3.06 days and 86 ± 3.99 days), and short plants (PHT = 69 ± 5.76 cm), and high number of leaves per plant (NLP = 202 ± 69.74 leaves per plant). Accessions of Cluster 1 produced higher average grain yield (GRY = 1280 ± 660.10 kg.ha^{-1}), with higher harvest index (HI = 50%) and the higher thousand seeds weight (TSW = 0.76 ± 0.07 g). Cluster 1 was predominantly composed of brown color accessions with well exerted panicle. Even though, accessions from this phenotypic group showed high performance in terms of yield, they produced less vigorous plant (CD = 0.77 ± 0.11 mm), and recorded lower values of number of grains per raceme (NGR = 77 ± 10.39 grains), with intermediate raceme length (8 ± 0.79 cm) and panicle length (14 ± 1.63 cm) (Table 5). Cluster 2 (23.89% of total accessions) and Cluster 3 (25.55% of total accessions) were composed of late flowering (FLO = 84 ± 5.21 days for Cluster 2 and 86 ± 3.38 days for Cluster 3) and late maturing accessions (MAT = 104 ± 5.50 days for Cluster 2 and 105 ± 3.38 days for Cluster 3). These phenotypic groups (Cluster 2 and 3) were predominantly composed of accessions with long to intermediate plant height (PHT = 79 ± 5.42 cm for Cluster 2 and PHT = 72 ± 5.33 cm for Cluster 3), with more vigorous plants (CD = 1.14 ± 0.38 mm for Cluster 3 and CD = 0.87 ± 0.09 mm for Cluster 2), and with long racemes (RLT = 9.90 ± 0.61 cm for Cluster 2 and RLT = 9.40 ± 0.58 cm for Cluster 3) and long panicles (PLT = 17 ± 3.54 cm for Cluster 2, PLT = 14 ± 1.05 cm for Cluster 3). Accessions of these two Cluster recorded intermediate number of leaves per plant and intermediate grain yield (GRY = 2048 ± 450.40 kg.ha^{-1} for Cluster 3 and GRY = 2019 ± 550.80 kg.ha^{-1} for Cluster 2). However, accessions from Cluster 3 recorded higher number of grains per raceme (NGR = 104 ± 14.27 grains) than accessions from Cluster 2 (NGR = 96 ± 13.22 grains). The same trend was observed for number of racemes per panicle (NRP = 4 ± 0.21 racemes for Cluster 3 and NRP = 3 ± 0.23 racemes for Cluster 2).

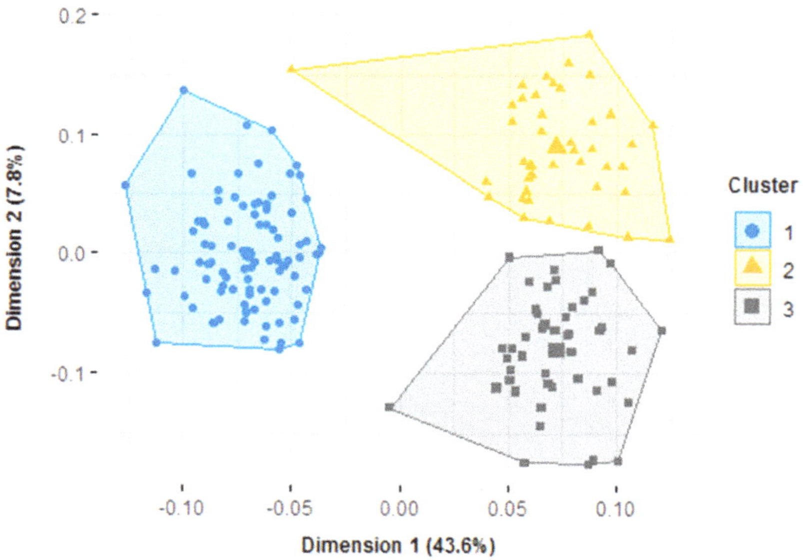

Figure 6. Hierarchical clustering on principal components analysis (HCPC) showing number of clusters and individuals within clusters in fonio accessions.

Table 5. Description of phenotypic groups in fonio accessions.

Variables	Cluster 1, N = 91 ML = 48 and NG = 43	Cluster 2, N = 43 BN = 6, BF = 2, GN = 33, ML = 2	Cluster 3, N = 46 BN = 43 and BF = 3	Pr (>F)
	Mean ± SD	Mean ± SD	Mean ± SD	
Plant height (cm)	69 ± 5.76	79 ± 5.42	72 ± 5.33	***
Number of internodes/plant	7 ± 0.77	11 ± 1.20	10 ± 1.02	***
Internode length (cm)	19 ± 7.69	39 ± 6.16	37 ± 6.68	**
Number of leaves/plant	202 ± 69.74	90 ± 23.84	83 ± 17.75	***
Culm diameter (mm)	0.77 ± 0.11	0.87 ± 0.09	1.14 ± 0.38	***
Flowering (days)	69 ± 3.06	84 ± 5.21	86 ± 3.38	***
Maturity (days)	86 ± 3.99	104 ± 5.50	105 ± 3.38	**
Panicle exertion	1 ± 0.68	3 ± 1.42	3 ± 1.48	***
Raceme length (cm)	8.1 ± 0.79	9.90 ± 0.61	9.4 ± 0.58	***
Panicle length (cm)	14 ± 1.63	17 ± 3.54	14 ± 1.05	***
Number of racemes/panicle	3 ± 0.29	3 ± 0.23	4 ± 0.21	***
Number of grains/raceme	77 ± 10.39	96 ± 13.22	104 ± 14.27	**
Grain yield (kg/ha)	3840 ± 660.10	2019 ± 550.80	2048 ± 450.40	**
Harvest index (%)	57 ± 8.78	26 ± 7.68	26 ± 5.43	**
Thousand seeds Weight (g)	0.76 ± 0.07	0.68 ± 0.06	0.59 ± 0.04	ns

SD = standard deviation, BN = Benin, BF = Burkina Faso, GN = Guinea, ML = Mali, NG = Niger. *** = significance at 0.001, ** = significance at 0.01, ns = not significant.

The HCPC analysis was supported by linear discriminant analysis (LDA) (Figure 7). The LDA results showed that linear discriminators LD1 and LD2 clearly separate the three phenotypic groups without admixture, while LD1 and LD3, LD1 and LD4 are fairly good at separating two to three Clusters. The result of multiple factor analysis (MFA) showed a synthetic comparison of the three phenotypic groups. This method groups variables in different classes of morphology, phenology and yield before making comparison between phenotypic groups. Dimension 1 clearly differentiates Cluster 1 from Clusters 2 and 3, while Dimension 2 differentiates Cluster 1 from Cluster 2.

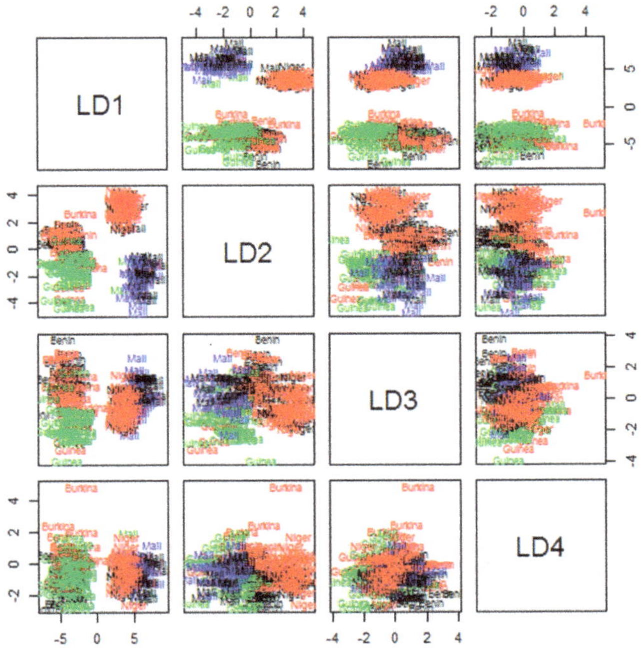

Figure 7. Linear discriminant analysis (LDA) in fonio.

4. Discussion

4.1. Qualitative Traits Variation in Fonio

Qualitative traits analysis revealed that most of the fonio accessions had well exerted panicle and brown grain color. Panicle exertion is important in cereal research though this was rarely documented in fonio for which a persistent interest for higher yielding cultivars is nowadays expected. Panicle exertion was reported as a crucial morphological marker that could be used for identification of hybrid plants derived from crosses between cultivated rice and African wild rice (*Oryza longistaminata* A. Chev. & Roehrich) [35]. The effect of genetic and abiotic factors on the panicle exertion in fonio is a topic of interest for future investigations. Seed color is one of the most important traits that guide breeders, farmers, processors, as well as consumers for cultivar preference [36,37]. In fonio, seed color was often associated with other traits including crop cycle, yield, tolerance to biotic and abiotic stresses, harvesting and processing abilities [3], though this needs to be fully documented. The brown seeded accessions, commonly known as white fonio (reference to *D. exilis*), is the most diverse and widely cultivated cultivar in the region [38]. Except in Niger where the majority of fonio accessions had greyed orange color [21], the predominance of brown seeded color in West Africa could be explained by its well adaptability to various pedo-climatic conditions and its association with other desirable agronomic traits. Lule et al. [39] reported that farmer's preferences of grain color (purple black) in finger millet (*Eleusine coracana* (L.) Gaertn.) was associated with adaptability to environment factors, yielding ability, major use of the grain, or cultural role and ease of production in northern Ethiopia. Therefore, future research should investigate the genetic basis of the seed color and its association of functional agronomic traits.

4.2. Performance of Fonio Accessions Based on Provenance

Our findings revealed that grain yield in fonio from various countries and ecological zones were higher than that reported previously [18]. Similar trend was observed for number of internode and internode length which were also higher compared to results obtained by Saidou et al. on Niger accessions only [21]. The mean thousand seeds weight recorded by Nyam [20] was also lower than that obtained in this study. However, higher values for plant height, number of internode, number of raceme par panicle, raceme length, and panicle length were reported by Sekloka et al. [19] in the characterization at the University of Parakou (Rep. of Benin) of twenty fonio accessions collected from Boukoumbe. This is an illustration of the morphological diversity among fonio accessions which could be due to the expression of genes controlling complex traits, or environmental factors including agricultural practices [10,18,40]. The significant differences observed among fonio accessions for most of the quantitative traits could also be attributed to the variability in rainfall patterns because the cumulative rainfall recorded in 2019 (1211.20 mm) was much higher than that of 2018 (1047.62 mm). This suggests that the amount and distribution of rainfall impact the performance of fonio accessions during the two consecutive years of experiment. Dehgahi et al. [41] reported also that grain yield was positively correlated with the average annual rainfall in wheat. That variability of rainfall can trigger the disruption in crop production and food insecurity [42].

Differences observed among accessions of various provenances indicated that the variation of most quantitative traits may be likely due to adaptation to specific environment and sowing date. Differences for flowering and maturity times were conspicuous among accessions. The early maturing accessions were more frequent in relatively hot and dry areas of arid zone from Mali and Niger whereas the late maturing accessions occurred abundantly in semi-arid and sub-humid areas from Benin, Burkina Faso and Guinea. This suggests that fonio accessions from the arid zone might have lost their photoperiod sensitivity in trying to adapt to prolonged dry spells and intensified heat waves compare to those from sub-humid zone which might be photoperiod sensitive with less drought spells and heat waves. Earlier works reported highly significant differences for number of days to flowering as affected by photoperiods within *D. exilis* [43]. In pearl millet (*Pennisetum glaucum* (L.) R. Br.), one of major crops cultivated in the dryland areas of West Africa, similar results were reported by Haussmann et al. [44] that long-cycle millets showed a stronger sensitivity to photoperiod than shorter-cycle cultivars in West and Central African pearl millet landraces.

In this study, fonio accessions were sown around mid-July with the consequence of exposing them to flower late in long days. Ordinary, farmers from semi-arid and sub humid zones sow fonio earlier in May. This shift in sowing date might have led to the lower grain yield observed in fonio accessions from sub-humid zone compared to those from arid zones. Gueye et al. [45] also reported that the delay in sowing dates reduced grain yield by 50% to 87% in fonio. Our result was in accordance with those of Wolabu and Tadege [46], who reported that regardless of the day length, non-photoperiod sorghum genotypes flower earlier than photoperiod sensitive genotypes. Future research will certainly decipher the effect of specific environment on the yield and yield components through a multiple environmental trial in order to reveal the ability of each ecological zone to contribute to yield accumulation in Fonio. This will clearly highlight the vulnerability and adaptability of fonio along the ecological gradient of west Africa as this was investigated for wheat in Parkistan [47].

4.3. Quantitative Traits Association and Partition of Phenotypic Groups in Fonio

Correlation analysis revealed highly significant positive relationships between grain yield and harvest index, thousand seed weight, number of leaves. Traits with high significant positive correlations would contribute to grain yield improvement in fonio. This implies that those traits can be used as proxy for yield in indirect selection and harvest index could be recommended as a selection criterion for increasing yield in fonio. In finger millet, similar results were reported that grain yield was positively associated with harvest index [48]. However, number of internode and plant height negatively impacted grain yield. Likewise, long flowering and maturing times negatively influenced grain yield. Similar results were found on wheat (*Triticum aestivum*) [49]. However, some findings showed that grain yield had highly significant positive correlation with maturity and flowering times in finger millet [50]. In addition, grain yield exhibited moderate negative correlation with plant height indicating that higher grain yield is associated with shorter fonio plants. This is an important feature in tackling lodging problem which is the main cause of yield losses and quality reduction in fonio production in West Africa. So, breeding for shorter plant could enhance lodging resistance in fonio varieties. Similar results were reported in tef (*Eragrotis tef* (Zucc.) Trotter) [51].

Yield, yield components, and phenological traits were the contributors to the variation observed in this study. Clusters analysis and linear discriminant analysis (LDA) showed three clusters of fonio accessions. A previous report indicated four clusters while characterizing 20 fonio accessions only from Benin [19]. Another report revealed two groups in the evaluation of 30 fonio accessions from Nigeria [20]. Our findings cover accessions from five countries (Benin, Burkina Faso, Guinea, Mali, and Togo) and three phytogeographical areas. However, accessions from Nigeria should be added for a broader perspective on the morphological variation in fonio in relation to different agro ecological zones in order to enrich the cultivated germplasm from West Africa. With the clear separation of phenotypic groups, it becomes critical to revisit the level of genetic diversity within fonio germplasm [18] and the association between genetic and phenotypic features for proper improvement of the crop. We found that fonio accessions clustered in phenotypic groups based on their provenance; though accessions from Benin and Mali that were distributed throughout Cluster 1 and Cluster 2. Amplified fragment length polymorphisms (AFLPs) analysis of 122 fonio accessions [18] revealed similar trends where fonio accessions from Benin, Burkina Faso, Guinea, and Togo clustered in identical group based on their origin except accessions from Guinea and Mali. In addition to site specificity, genes flow is a powerful mechanism that contribute to homogenization of variability distribution. For instance, grouping of accessions regardless of boundaries reflect the importance of farmers' seed exchange in fonio seed systems. As informal seed systems, seed exchange is commonly practiced by fonio farmers as reported by previous works [2,10,52].

Group 1 was composed essentially of early flowering and maturing accessions, and recorded higher values for grain yield, harvest index and thousand seeds weight. The accessions of this group are adequate candidates for developing early maturing fonio lines for arid and semi-arid regions where drought stress occurs frequently. Furthermore, this group is composed of fonio accessions with shorter plants that could be used as source for lodging resistance since short stature plants have been reported in cereals to be preferred in varietal development for that trait. Phenotypic Groups 2 and 3 were composed predominantly of late maturing accessions with long to intermediate plant height, with higher values of culm diameter. Moreover, these groups were composed of accessions with long racemes and long panicles. The difference observed in the groups could be likely due to the genetic make-up of each genotype and prevailing environmental factors. However, despite their phenotypical similitude, accessions from Group 3 recorded higher number of grains per raceme and number raceme per panicle than accessions from Group 2 revealing another agronomic trait of importance for the improvement of the crop. Marker assisted index selection of fonio lines with long raceme, and long panicle alongside with high number of grains per raceme and high number of racemes per panicle can be explored as crop a improvement avenue. Moreover, genome-wide association studies (GWAS) are needed to understand the genetic basis of key important agronomic traits in fonio.

5. Conclusions

This study assessed phenotypic variability among 180 fonio accessions from five West African countries. It revealed the existence of wide range of agro morphological variability. The results revealed interesting correlations between yield and yield components. Furthermore, clusters analysis revealed three phenotypic groups. Group 1 was composed of early flowering and maturing accessions with higher grain yield, higher harvest index, higher thousand seeds weight, and higher number of leaves per plant. Accessions from Group 2 and 3 were characterized by late flowering and maturing times, with intermediate grain yield, thousand seeds weight, and number of leaves per plant, with taller plants, higher values of number of grains per raceme, and number of racemes per panicle. Accessions from Group 1 are candidates for yield improvement and enhancing lodging resistance, whereas accessions from Group 2 and Group 3 are candidate also for yield improvement via indirect selection and good parent for hybrids development. The study exhibited agro morphological descriptors that discriminate fonio accessions and provide also useful information for parental selection with key economically important agronomic traits.

Author Contributions: A.R.I.B.Y. and E.G.A.-D. conceptualized the study. A.R.I.B.Y. and M.A. conducted the experiments. A.R.I.B.Y. and E.G.A.-D. wrote the manuscript. A.R.I.B.Y. and E.G.A.-D. edited the manuscript. E.G.A.-D., M.A., E.S., C.O.A.A., C.B., A.B., and Y.B. critically revised the draft and updated the manuscript for publication. All authors have read and agreed to the published version of the manuscript.

Funding: Ibrahim Bio Yerima, from the National Institute of Agronomic Research of Niger (INRAN), is a scholar of the "Intra-Africa Academic Mobility Scheme" under the project grant number 2016–2988 on "Enhancing training and research mobility for novel crops breeding in Africa (MoBreed)" funded by the Education, Audiovisual and Culture Executive Agency (EACEA) of the European Commission. The project provided a scholarship for academic training and research mobility and a research grant to the first author to complete a PhD degree at the University of Abomey-Calavi (Republic of Benin).

Acknowledgments: The authors wish to thank MoBreed program for the financial support. We are thankful to all farmers and agriculture extension services in Benin, Burkina Faso, Guinea, Mali, and Niger. We also thank Dèdéou Tchopkonhoue, Djamilou Ibrahim, Doumbia Zan Ibrahima, Diarra Cheik, and Habou Kano for assistance during the survey, field activities and data analysis.

Conflicts of Interest: Authors declare no conflict of interest.

References

1. Garí, J.A. Review of the African millet diversity. In *International Workshop on Fonio, Food Security and Livelihood among the Rural Poor in West Africa*; IPGRI/IFAD: Bamako, Mali, 2002.

2. Adoukonou-Sagbadja, H.; Dansi, A.; Vodouhè, R.; Akpagana, K. Indigenous knowledge and traditional conservation of fonio millet (*Digitaria exilis, Digitaria iburua*) in Togo. *Biodivers. Conserv.* **2006**, *15*, 2379–2395. [CrossRef]
3. Ballogou, V.Y.; Soumanou Mohamed, M.; Toukourou, F.; Hounhouigan Joseph, D. Indigenous knowledge on landraces and fonio-based food in Benin. *Ecol. Food Nutr.* **2014**, *53*, 390–409. [CrossRef]
4. Sartelet, H.; Serghat, S.; Lobstein, A.; Ingenbleek, Y.; Anton, R.; Petitfrere, E.; Aguie-Aguie, G.; Martiny, L.; Haye, B. Flavonoids extracted from fonio millet (*Digitaria exilis*) reveal potent antithyroid properties. *Nutrition* **1996**, *12*, 100–106. [CrossRef]
5. Vodouhè, S.; Achigan Dako, E.G. *Digitaria exilis (Kippist) Stapf*; Plant Resources of Tropical Africa 1. Cereals and Pulses; Brink, M., Belay, G., Eds.; PROTA Foundations: Wageningen, The Netherlands; Backhuys Publishers: Leiden, The Netherlands, 2006; Volume 1.
6. Fogny, N.F.; Madode, E.Y.; Laleye, F.F.; Amoussou-Lokossou, Y.; Kayode, A.P. Formulation de farine de fonio enrichie en ressources alimentaires locales pour l'alimentation complémentaire des jeunes enfants au Bénin. *Int. J. Biol. Chem. Sci.* **2017**, *11*, 2745–2755. [CrossRef]
7. Cruz, J.; Beavogui, F.; Dramé, D. *Fonio, an African Cereal [Le Fonio, une Céréale Africaine]*; CTA: Wageningen, The Netherlands, 2011.
8. Sekloka, E.; Adoukonou-Sagbadja, H.; Paraïso Armand, A.; Yoa Brigitte, K.; Bachabi, F.-X.; Zoumarou-Wallis, N. Evolution de la diversité des cultivars de fonio pratiqués dans la commune de Boukoumbé au Nord-Ouest du Bénin. *Int. J. Biol. Chem. Sci.* **2015**, *9*, 2446–2458. [CrossRef]
9. Ayenan, M.A.T.; Sodedji, K.A.F.; Nwankwo, C.I.; Olodo, K.F.; Alladassi, M.E.B. Harnessing genetic resources and progress in plant genomics for fonio (*Digitaria* spp.) improvement. *Genet. Resour. Crop Evol.* **2017**, *65*, 373–386. [CrossRef]
10. Dansi, A.; Adoukonou-Sagbadja, H.; Vodouhè, R. Diversity, conservation and related wild species of Fonio millet (*Digitaria* spp.) in the northwest of Benin. *Genet. Resour. Crop Evol.* **2010**, *57*, 827–839. [CrossRef]
11. Bhanu, A.N. Assessment of Genetic Diversity in Crop Plants—An Overview. *Adv. Plants Agric. Res.* **2017**, *7*, 279–286. [CrossRef]
12. Mondini, L.; Noorani, A.; Pagnotta, M. Assessing Plant Genetic Diversity by Molecular Tools. *Diversity* **2009**, *1*, 19–35. [CrossRef]
13. Odjo, T.C.; Dossou-Aminon, I.; Dansi, A.; Bonou-Gbo, Z.; Kombaté, K. Agro-Morphological Characterization and Assessment of Variability within a Germplasm of Benin Rice (*Oryza sativa* L.) Varieties. *Int. J. Curr. Res. Biosci. Plant Biol.* **2017**, *4*, 1–16.
14. Babu, B.K.; Sood, S.; Agrawal, P.; Chandrashekara, C.; Kumar, A.; Kumar, A. Molecular and Phenotypic Characterization of 149 Finger Millet Accessions Using Microsatellite and Agro-Morphological Markers. *Proc. Natl. Acad. Sci. India Sect. B Biol. Sci.* **2016**, *87*, 1217–1228. [CrossRef]
15. Sattler, F.; Sanogo, M.; Kassari, I.; Angarawai, I.; Gwadi, K.; Dodo, H.; Haussmann, B. Characterization of West and Central African accessions from a pearl millet reference collection for agro-morphological traits and Striga resistance. *Plant Genet. Resour. Charact. Util.* **2017**, *16*, 260–272. [CrossRef]
16. Naoura, G.; Sawadogo, N.; Atchozou, E.A.; Emendack, Y.; Hassan, M.A.; Reoungal, D.; Amos, D.N.; Djirabaye, N.; Tabo, R.; Laza, H. Assessment of agro-morphological variability of dry-season sorghum cultivars in Chad as novel sources of drought tolerance. *Sci. Rep.* **2019**, *9*, 19581. [CrossRef]
17. Clottey, V.; Agyare, W.; Bayorbor, T.; Abanga, J.; Kombiok, J. Genetic relatedness of fonio (*Digitaria* spp.) landraces assembled in Ghana. *Plant Genet. Resour. Newsl. (Bioversity Int. FAO)* **2006**, *147*, 6–11.
18. Adoukonou-Sagbadja, H.; Wagner, C.; Dansi, A.; Ahlemeyer, J.; Dainou, O.; Akpagana, K.; Ordon, F.; Friedt, W. Genetic diversity and population differentiation of traditional fonio millet (*Digitaria* spp.) landraces from different agro-ecological zones of West Africa. *Theor. Appl. Genet.* **2007**, *115*, 917–931. [CrossRef]
19. Sekloka, E.; Kanlindogbe, C.; Biaou Samadori, S.H.; Adoukonou-Sagbadja, H.; Kora, A.; Motouama, F.T.; Seidou, M.; Zinsou Valérien, A.; Afouda, L.; Baba-Moussa, L. Agro-morphological characterization of fonio millet accessions (*Digitaria exilis* Stapf.) collected from Boukoumb, Northwest of Benin. *J. Plant Breed. Crop Sci.* **2016**, *8*, 211–222.
20. Nyam, D.; Kwon-Ndung, E.; Wuyep, A. Genetic affinity and breeding potential of phenologic traits of acha (fonio) in Nigeria. *J. Sci. Eng. Res.* **2017**, *4*, 91–101.

21. Saidou, S.I.; Bakasso, Y.; Inoussa, M.M.; Zaman-Allah, M.; Atta, S.; Barnaud, A.; Billot, C.; Saadou, M. Diversité agro-morphologique des accessions de fonio [*Digitaria exilis* (Kippist.) Stapf.] au Niger. *Int. J. Biol. Chem. Sci.* **2014**, *8*, 1710. [CrossRef]
22. Peyre, D. *Catalogue des Plantes Vasculaires du Niger*; Institut d'Elevage et de Médecine Vétérinaire des Pays Tropicaux: Paris, France, 1976; Volume 1, pp. 405–419.
23. Peyre de Fabregues, B.; Lebrun, J. *Catalogue des Plantes Vasculaires du Niger (Etude Botanique n 3.) IEMVT*; Laboratoire National d'Elevage: Maisons-Alfort, France; Niamey, Niger, 1976; p. 433.
24. White, F. *The Vegetation of Africa*; Natural Resources Research; UNESCO: Paris, France, 1983; Volume 20, p. 356.
25. Dobignard, A.; Chatelain, C. *Index Synonymique de la Flore d'Afrique du Nord*; Éditions des Conservatoire et Jardin Botaniques de la Ville de Genève: Geneve, Switzerland, 2010; Volume 1, p. 455.
26. Prance, G.; Sothers, C. *Chrysobalanaceae 1 & 2, Species Plantarum: Flora of the World, Parts 9 &10*; Australian Biological Resources Study: Canberra, Australia, 2003.
27. Bioversity International, I.W. *Descriptors for Wild and Cultivated Rice (Oryza spp.)*; Bioversity International: Rome, Italy, 2007.
28. Kassambara, A. *Practical Guide to Principal Component Methods in R (Multivariate Analysis)*; CreateSpace Independent Publishing Platform: Scotts Valley, CA, USA, 2017; Volume 2.
29. Lê, S.; Josse, J.; Husson, F. FactoMineR: An R package for multivariate analysis. *J. Stat. Softw.* **2008**, *25*, 1–18. [CrossRef]
30. Kassambara, A.; Mundt, F. Package 'Factoextra'. Extract and Visualize the Results of Multivariate Data Analyses; R pachage. 2017. Available online: http://www.sthda.com/english/rpkgs/factoextra (accessed on 5 August 2020).
31. Team, R.C. Version 3.5. 1. R Foundation for Statistical Computing, Vienna, Austria. Available online: https://www.r-project.org (accessed on 3 November 2018).
32. Argüelles, M.; Benavides, C.; Fernández, I. A new approach to the identification of regional clusters: Hierarchical clustering on principal components. *Appl. Econ.* **2014**, *46*, 2511–2519. [CrossRef]
33. Rencher, A. Discriminant analysis: Description of group separation. In *Methods of Multivariate Analysis*, 2nd ed.; John Wiles and Sons: New York, NY, USA, 2002; pp. 270–298.
34. Pagès, J. *Multiple Factor Analysis by Example Using R*; CRC Press: Boca Raton, FL, USA, 2014.
35. Kanya, J.I.; Hauser, T.P.; Kinyamario, J.I.; Amugune, N.O. Hybridization potential between cultivated rice *Oryza sativa* and African wild rice *Oryza longistaminata*. *Int. J. Agric. Res.* **2012**, *7*, 291–302. [CrossRef]
36. Govindaraj, M.; Rao, A.S.; Shivade, H.; Rai, K. Effect of grain colour on iron and zinc density in pearl millet. *Indian J. Genet. Plant Breed.* **2018**, *78*, 247. [CrossRef]
37. Garg, M.; Chawla, M.; Chunduri, V.; Kumar, R.; Sharma, S.; Sharma, N.K.; Kaur, N.; Kumar, A.; Mundey, J.K.; Saini, M.K.; et al. Transfer of grain colors to elite wheat cultivars and their characterization. *J. Cereal Sci.* **2016**, *71*, 138–144. [CrossRef]
38. Blench, R.M. Vernacular names for African millets and other minor cereals and their significance for agricultural history. *Archaeol. Anthropol. Sci.* **2016**, *8*, 1–8. [CrossRef]
39. Lule, D.; Tesfaye, K.; Fetene, M. Qualitative traits diversity and eco-geographical distribution in finger millet (*Eleusine coracana*) landraces from eastern and south eastern Africa: An implication for germplasm collection and conservation. *Afr. J. Plant Sci.* **2012**, *6*, 346–354.
40. Vigouroux, Y.; Barnaud, A.; Scarcelli, N.; Thuillet, A.-C. Biodiversity, evolution and adaptation of cultivated crops. *Comptes Rendus Biol.* **2011**, *334*, 450–457. [CrossRef]
41. Dehgahi, R.; Joniyas, A.; Latip, S.N.H.B.M. Rainfall Distribution and Temperature Effects on Wheat Yield in Torbate Heydarie. *Int. J. Sci. Res. Knowl. ISSN* **2014**, *2*, 2322–4541.
42. Kyei-Mensah, C.; Kyerematen, R.; Adu-Acheampong, S. Impact of Rainfall Variability on Crop Production within the Worobong Ecological Area of Fanteakwa District, Ghana. *Adv. Agric.* **2019**, *2019*. [CrossRef]
43. Aliero, A.; Morakinyo, J. Photoperiodism in *Digitaria exilis* (Kipp) Stapf accessions. *Afr. J. Biotechnol.* **2005**, *4*, 241–243.
44. Haussmann, B.I.; Fred Rattunde, H.; Weltzien-Rattunde, E.; Traoré, P.; Vom Brocke, K.; Parzies, H.K. Breeding strategies for adaptation of pearl millet and sorghum to climate variability and change in West Africa. *J. Agron. Crop Sci.* **2012**, *198*, 327–339. [CrossRef]

45. Gueye, M.; Kanfany, G.; Fofana, A.; Noba, K.; Grove, J. Effect of planting date on growth and grain yield of fonio millet (*Digitaria exilis* Stapf) in the Southeast of Senegal. *Int. J. Biol. Chem. Sci.* **2015**, *9*, 581–592. [CrossRef]
46. Wolabu, T.W.; Tadege, M. Photoperiod response and floral transition in sorghum. *Plant Signal. Behave.* **2016**, *11*, e1261232. [CrossRef]
47. Sultana, H.; Ali, N.; Iqbal, M.M.; Khan, A.M. Vulnerability and adaptability of wheat production in different climatic zones of Pakistan under climate change scenarios. *Clim. Chang.* **2009**, *94*, 123–142. [CrossRef]
48. Jyothsna, S.; Patro, T.; Ashok, S.; Rani, Y.S.; Neeraja, B. Studies on genetic parameters, character association and path analysis of yield and its components in finger millet (*Eluesine coracana* L. Gaertn). *Int. J. Theor. Appl. Sci.* **2016**, *8*, 25.
49. Assefa, E. Correlation and Path Coefficient Studies of Yield and Yield Associated Traits in Bread Wheat (*Triticum aestivum* L.) Genotypes. *Adv. Plants Agric. Res.* **2017**, *6*, 1–10. [CrossRef]
50. Chavan, B.; Jawale, L.N.; Shinde, A.V. Correlation and path analysis studies in finger millet for yield and yield contributing traits (*Eleusine coracana* L. Gaertn). *Int. J. Chem. Stud.* **2020**, *8*, 2911–2914.
51. Jifar, H.; Dagne, K.; Tesfaye, K.; Assefa, K.; Tadele, Z. Agro-Morphological Traits Diversity in Tef [*Eragrostis Tef* (Zucc.) Trotter] Genotypes from Various Sources. *Ethiop. J. Agric. Sci.* **2018**, *28*, 131–148.
52. Sidibé, A.; Meldrum, G.; Coulibaly, H.; Padulosi, S.; Traore, I.; Diawara, G.; Sangaré, A.R.; Mbosso, C. Revitalizing cultivation and strengthening the seed systems of fonio and Bambara groundnut in Mali through a community biodiversity management approach. *Plant Genet. Resour.* **2020**, *19*, 31–48.

agronomy

MDPI

Article

Genetic Composition of Korean Ginseng Germplasm by Collection Area and Resource Type

Kyung Jun Lee, Raveendar Sebastin, Seong-Hoon Kim, Eunae Yoo, Sookyeong Lee, Gyu-Taek Cho, Manjung Kang and Do Yoon Hyun *

National Agrobiodiversity Center, National Institute of Agricultural Sciences (NAS), RDA, Jeonju 54874, Jeollabuk-do, Korea; lkj5214@korea.kr (K.J.L.); raveendars@gmail.com (R.S.); shkim0819@korea.kr (S.-H.K.); eung77@korea.kr (E.Y.); xsanta7@korea.kr (S.L.); gtcho@korea.kr (G.-T.C.); mjkang@korea.kr (M.K.)

* Correspondence: dyhyun@korea.kr; Tel.: +82-63-238-4912

Received: 24 September 2020; Accepted: 22 October 2020; Published: 26 October 2020

Abstract: To improve crops, it is important to secure plant genetic source material and evaluate the genetic diversity. Ginseng (*Panax ginseng* C.A. Meyer) has long been used as a medicinal herb in Korea and China. Since ginseng originated from wild ginseng with low genetic diversity, it is also expected to have low genetic diversity. In this study, the genetic diversity of 451 ginseng accessions conserved in the National Agrobiodiversity Center (NAC) at Korea was analyzed using 33 SSR markers. Another objective was to establish a strategy for NAC to manage ginseng germplasm based on these results. The 451 accessions were collected from 22 cities in six provinces in South Korea. Among the 451 ginseng accessions, 390 (86.5%) and 61 (13.5%) were landraces and breeding lines, respectively. In the STRUCTURE results for the accessions, there was no relationship between assigned genotypes and collection areas, but there was a population genetic structure. In addition, genetic differentiation within populations of each analysis was low, indicating that the ginseng accessions conserved at NAC are extensively dispersed throughout the collection areas. The results of this study suggest that NAC should increase the genetic diversity of ginseng accessions for breeding programs, and alternatives are needed for securing ginseng genetic resources.

Keywords: ginseng; genetic composition; genetic diversity; SSR

1. Introduction

Plant breeding in agriculture has decreased the diversity of many crops, which has caused bottlenecks in crop domestication, dispersal, and modernization. Researchers, in particular, have identified significant negative effects of plant breeding on diversity following the modernization bottleneck [1]. The loss of crop variation caused by the modernization of agriculture has been described as genetic erosion [2]. Since loss of genetic variation could decrease the potential of species to persist in the face of abiotic and biotic environmental changes, genetic erosion could pose a severe threat to long-term global food security [3]. Crop improvement largely depends on immediate conservation of genetic resources for their effective and sustainable utilization. For this, it is necessary to conserve and breed the vast genetic variation found in populations of the wild progenitors and landraces of cultivated plants [4,5].

Plant genetic resources are the most important components of agro-biodiversity, which encompasses primitive forms of cultivated plant species and landraces, modern and obsolete cultivars, breeding lines and genetic stocks, weedy types, and related wild species [6]. Since the International Board for Plant Genetic Resources was established in 1974, over 1750 gene banks have been activated worldwide, about 130 of which hold more than 10,000 accessions each [7,8]. In Korea, the National Agrobiodiversity Center (NAC) was established in 2008, and the center now has conserved 258,984 accessions of 3126 species [9].

The *Panax* genus of the Araliaceae family comprises 17 species, of which *P. ginseng*, *P. notoginseng*, and *P. quinquefolium* are widely cultivated for medicinal value [10,11]. The Korean ginseng, *P. ginseng* C.A. Meyer, is an important medical plant, which is well-known for its remarkable pharmacological effects [10–14]. Generally, the origination area and time of *P. ginseng* are known to be in Sangdang, China and the first century B.C. during the Han dynasty, era (Chinese origin theory of ginseng). However, it is fairly obvious that this theory does not answer appropriately to the fundamental questions of the origin of ginseng; (1) why ginseng suddenly appeared in that time, Han China, (2) how explain clearly the formation of ginseng character. Since *P. ginseng* naturally exists in only three regions: Korea (33.7°–43.1°), Manchuria (43°–47°), and the Littoral province of Siberia, it is sometimes referred that the originating place is not Shangdang of Shansi area of China, but Manchuria and Korea [15].

Many previous researchers have studied the medical and pharmacological effects and genetic diversity of *P. ginseng* [16–23]. Based on the growing environment and the method of cultivation, commercial trade ginseng is classified into three grades: cultivated, mountain cultivated, and mountain wild [24]. In general, the cultivated ginseng is known to have originated from wild ginseng, a rare, endangered plant [25]. The genetic consequence of rare or endangered species is increased genetic drift and inbreeding, and these increased phenomena contribute to a reduction in genetic diversity [26,27].

Geographic differences in the distribution of genetic diversity are extremely common. Populations can vary in all aspects of diversity such as number of alleles, identities of alleles, and the effect on the characteristics in the population. The breeding system of the species is very important in determining the differences between populations from different geographic locations [28]. In a previous study, the researchers reported on narrow genetic diversity among 1109 ginseng accessions conserved in the NAC, but there was a lack of sufficient information on the genetic diversity of ginseng accessions such as the collection site [29]. In this study, the genetic diversity of 451 Korean ginseng accessions with information on the collection area was analyzed; in particular, each accession was genetically differentiated and sorted by the collection area. Based on these results, this study includes proposals for how to efficiently manage the ginseng germplasm in NAC.

2. Materials and Methods

2.1. Plant Materials

A total of 451 ginseng accessions were obtained from the NAC at the Rural Development Administration in South Korea (Table S1).

2.2. DNA Extraction

Genomic DNA was extracted from the leaves of ginseng accessions using Qiagen DNA extraction kit (Qiagen, Hilden, Germany). DNA quality and quantity were measured using 1% (w/v) agarose gel and a spectrophotometry (Epoch, BioTek, Winooski, VT, USA). Extracted DNA was diluted to 30 ng/uL and stored at −20 °C until further PCR amplification.

2.3. SSR Genotyping

For SSR (simple sequence repeat) analysis, a total of 33 SSRs were selected previous studies [30,31] (Table S2). They were fluorescently labelled (6-FAM, HEX, and NED) and used to detect amplification products. PCR reactions were carried out in a 25 uL reaction mixture that contained 30 ng template DNA, 1.5 mM $MgCl_2$, 0.2 mM of each dNTPs, 0.5 um of each primer, 1 U *Taq* polymerase (Inclone, Yongin, Korea). The amplification was performed under the following cycling conditions: initial denaturation at 94 °C for 5 min, followed by 35 cycles of denaturation at 95 °C for 30 s, annealing at 55 °C for 30 s, extension at 72 °C for 1 min, and a final extension step at 72 °C for 10 min. Each amplicon was resolved on an ABI3500 DNA sequence (Thermo Fisher Scientific Inc., Wilmington, DE, USA) and scored using Gene Mapper Software (Version 4.0, Thermo Fisher, Wilmington, DE, USA).

2.4. Genetic Diversity Analysis

The number of alleles (Na), the number of genotypes (Ng), Shannon-Wiener index (H), Nei's genetic diversity (I), Eveness, polymorphic information content (PIC) were calculated using R software [32]. Analysis of molecular variance (AMOVA) within and between gene pools was performed using GenAlEx software v. 6.5 [33].

2.5. Population Structure Analysis

Population structure was analyzed using STRUCTURE v.2.3.4 [34]. In STRUCTURE analysis, Bayesian clustering was performed; three independent runs were tested with K from 1 to 15, each run with a burn-in period of 50,000 iterations and 500,000 Monte Carlo Markov iterations, assuming an admixture model. The output was subsequently visualized with STRUCTURE HARVESTER v.0.9.94 [35].

3. Results

3.1. Distribution of Collection Areas in Korean Ginseng Accessions

The 451 ginseng accessions in this study were collected from 22 cities in six provinces in South Korea (Figure 1 and Table S3). The collection areas were four cities in Gangwon (GW), five in Gyeonggi (GG), two in Gyeongsangbuk (GB), two in Jeollabuk-do (JB), five in Chungcheongnam (CN), and four in Chungcheongbuk (CB). Among 451 ginseng accessions, 203 (45.0%) were collected from CN, 124 (27.5%) from GW, 70 (15.5%) from GG, 24 (5.3%) from JB, 20 (4.4%) from CB, and 10 (2.2%) from GB. Of 22 cities, the largest ginseng accessions were collected from Geumsan (19.3%), followed by Hoengseong (14.4%), with less than 10% collected from other cities. Among the 451 accessions, 390 (86.5%) and 61 (13.5%) were landraces and breeding lines, respectively. All 61 breeding lines were collected from Geumsan in CN.

3.2. Profiling of SSR Markers in Korean Ginseng Accessions

A total of 226 alleles were detected among the 451 ginseng accessions (Table 1 and Table S4). Polymorphic information content (PIC) of 33 SSR markers ranged from 0.025 (GM104) to 0.770 (GM116), with an average of 0.469. In the 390 landraces and 61 breeding lines, there were 223 and 148 alleles, respectively. The number of observed alleles (Na) ranged from 2 to 24, with an average of 6.85, of which there were 6.76 and 4.49 ginseng landraces and breeding lines, respectively. The number of genotypes (Ng) ranged from 2 to 100, with an average of 17.79, of which there were 16.83 and 6.18 landraces and breeding lines, respectively. The Shannon-Wiener index (H) showed the range from 0.06 to 1.998, with an average of 0.92. Observed heterozygosity (Ho) and Nei's genetic diversity (GD) ranged from 0.02 to 1.00 (mean: 0.88) and 0.02 to 0.80 (mean: 0.52), respectively. H, Ho, and GD were similar in the ginseng landraces and breeding lines.

Table 1. Genetic diversity parameters of 33 SSR markers in 451 ginseng accessions.

	No. Acc	Total Alleles	Na [1]	Ng	H	Ho	GD
Landraces	390	223	6.76 ± 5.72	16.83 ± 23.14	0.92 ± 0.43	0.88 ± 0.27	0.52 ± 0.16
Breeding lines	61	148	4.49 ± 3.97	6.18 ± 7.02	0.93 ± 0.43	0.88 ± 0.30	0.52 ± 0.16
Total	451	226	6.85 ± 5.82	17.79 ± 24.60	0.92 ± 0.43	0.88 ± 0.27	0.52 ± 0.16

[1] Na, Number of observed alleles; Ng, Number of genotypes; H, Shannon-Wiener index; Ho, observed heterozygosity; GD, Nei's gene diversity.

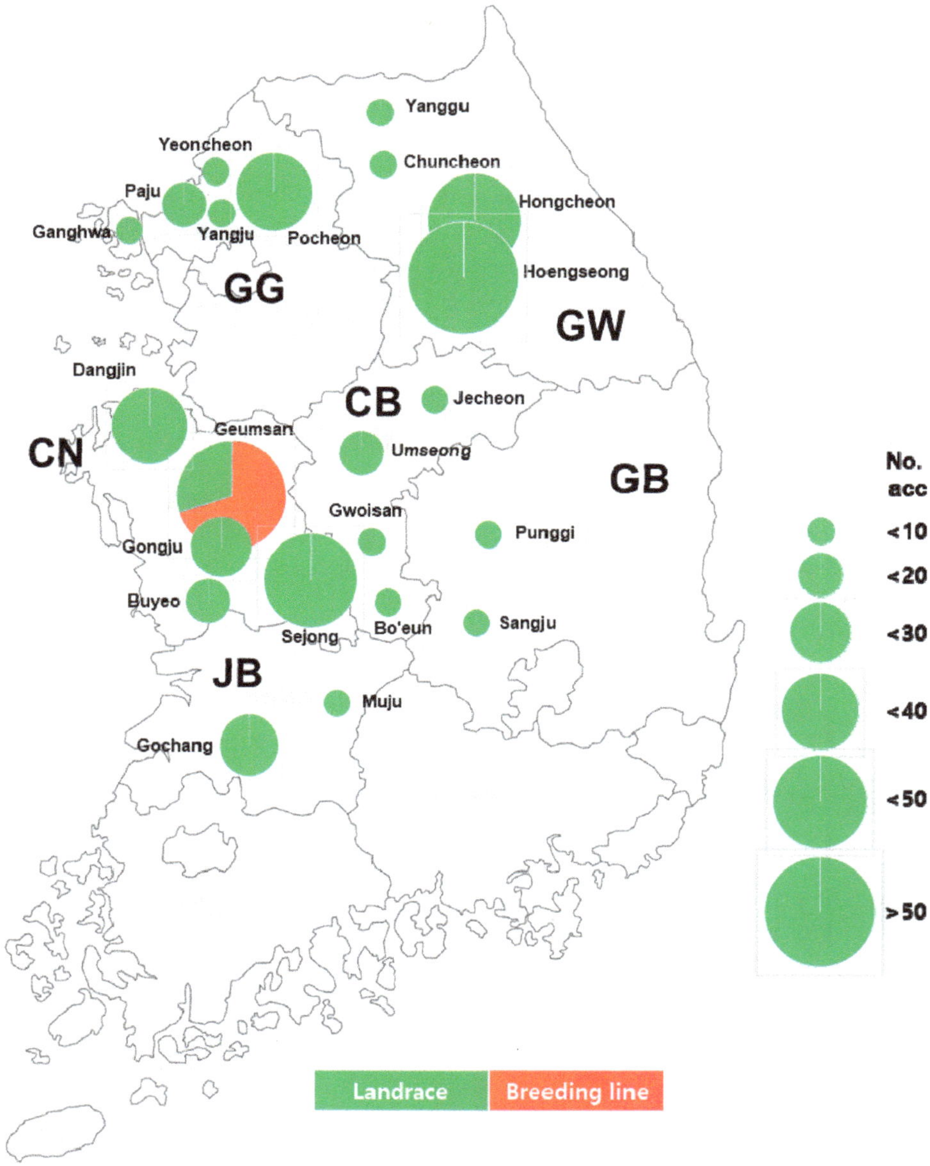

Figure 1. Distribution of collected areas in 451 ginseng accessions. CB, Chungcheongbuk; CN, Chungcheongnam; GB, Gyeongsangbuk; GG, Gyeonggi; GW, Gangwan; JB, Jeollabuk.

3.3. Population Structure of Korean Ginseng Accessions

To understand the pattern of the population structure, a Bayesian clustering analysis in STRUCTURE was performed. The STRUCTURE results suggested the best grouping number (K = 2) based on the delta K (Figure 2A). Population 1 (Pop 1) and 2 (Pop 2) consisted of 211 and 240 accessions, respectively. In pop I, 20 accessions are breeding lines and 191 are landraces. In pop II, 41 accessions are breeding lines and 199 are landraces. By collection site, the western region (GG, CN, and JB) in South Korea showed a high rate of ginseng accessions in Pop I and the eastern part (GW and GB) had a high percentage of accessions in Pop II, while CB showed the same number of accessions as Pop I and Pop II (Table S1).

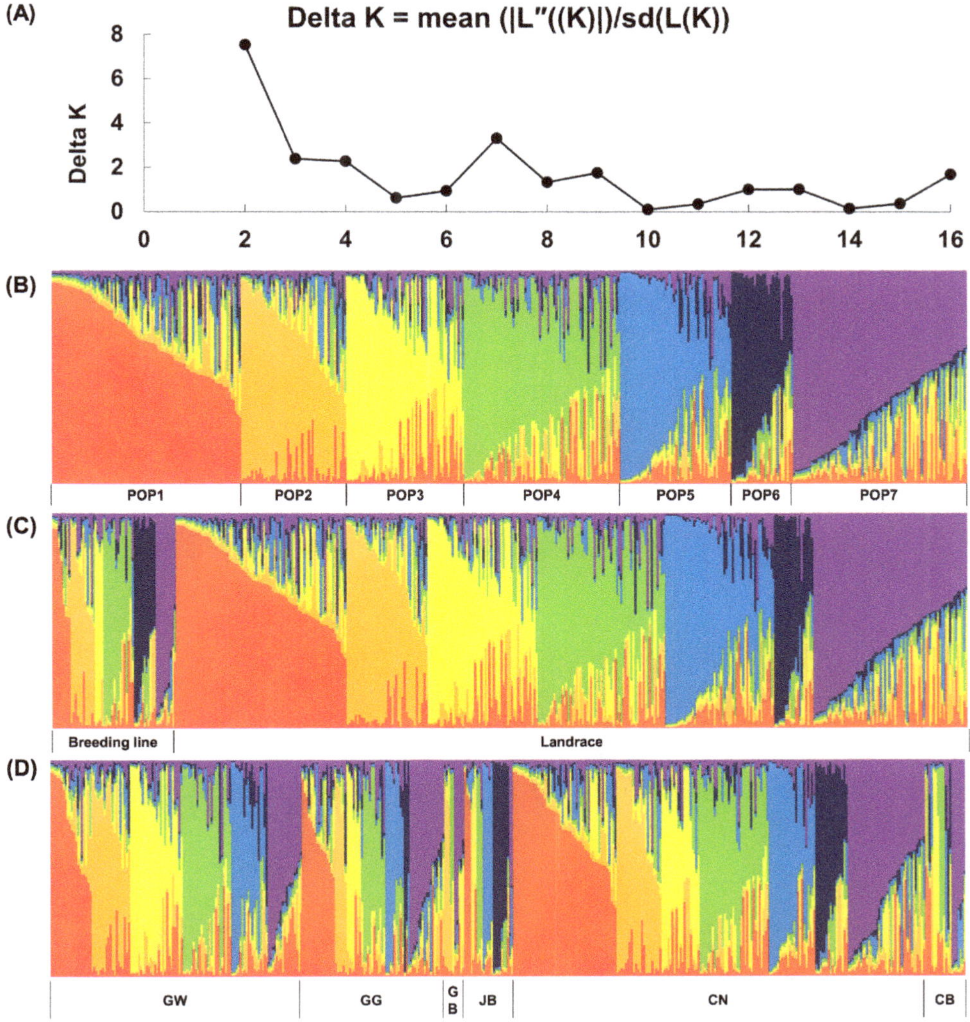

Figure 2. Population structure of 451 ginseng accessions based on 33 SSRs markers. (**A**) Estimation of population using LnP(D) derived ΔK. (**B**) Seven population sorted by Q, which represent estimated membership probabilities of the individual to inferred populations. Each accession is represented by a thin vertical line. (**C**) The population structure showed by different types of germplasms in Korea. (**D**) The population structure showed by accessions from different geographic regions in Korea. CB, Chungcheongbuk; CN, Chungcheongnam; GB, Gyeongsangbuk; GG, Gyeonggi; GW, Gangwan; JB, Jeollabuk.

Because of the further increase of LnP(D), simulation was performed, and LnP(D) was maximized at K = 7 (Figure 2A). All the ginseng accessions could be grouped into seven populations: POP1, POP2, POP3, POP4, POP5, POP6, and POP7 (Figure 2B); the clusters contained, respectively, 93, 52, 58, 77, 55, 30, and 86 accessions. The population structure was separately analyzed to reveal the differentiation between breeding lines and landraces (Figure 2C); all showed complex population structures. Among the 61 breeding lines, 14 accessions (23%) were assigned to POP4, followed by 12 (19.7%) in POP2, and 11 (18%) in POP6, while only one (IT239306) was in POP5. The 390 ginseng landraces could also be assigned into the seven population clusters: 84 accessions (21.5%) in POP1, 76 (19.5%) in POP7, and 63 (16.2%) in POP4. The population structure was also estimated to display the regional differentiation of ginseng accessions (Figure 2D). Results showed that 26 (21.0%) and 24 (19.4%) of the ginseng accessions collected from GW clustered into POP4 and POP5, respectively. Among the 70 accessions from GG, 22.9% and 24.3% were in POP1 and POP7, respectively. In GB, 50% were in POP7, 30% were POP2 and 20% were in POP4. Among the 24 ginseng accessions from JB, 33.3% were in POP6 and 20.8% were in POP5. In CN and CB, 25.1% of the accessions in POP1 and 35.0% were in POP7.

In the ginseng landraces, STRUCTURE determined 2 to be the best grouping number based on the delta K (Figure 3A). Population 1 (L-Pop 1) and 2 (L-Pop 2) consisted of 201and 189 accessions, respectively. Due to the further increase of LnP(D), simulation was performed, and LnP(D) was maximized at K = 12 (Figure 3A). All the ginseng accessions could be grouped into 12 populations, L1 to L12 (Figure 3B), ranging from 2 (L6) to 53 (L9) accessions; the distributions of the 12 groups varied by region (Figure 3C). Among 12 groups, L6 was in GG and JB only. All groups could be found in GG, there were 11 in GG and CN (excluding L6). JB contained ten groups (not L1 and L10), and there were seven in CB: L2, L3, L4, L7, L9, L10, and L11. GB showed the fewest groups, L1, L2, L3, L4, L8, and L11.

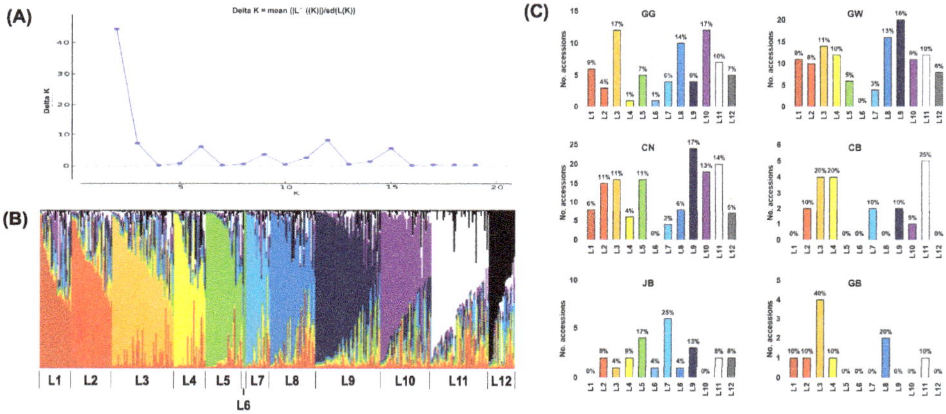

Figure 3. Population structure of 390 ginseng landraces based on 33 SSRs markers. (**A**) Estimation of population using LnP(D) derived ΔK. (**B**) Twelve populations sorted by Q, which represent estimated membership probabilities of the individual to inferred populations. Each accession is represented by a thin vertical line. (**C**) Distribution of twelve populations in each collection area.

In ginseng breeding lines, STRUCTURE proposed 2 as the best grouping number based on the delta K (Figure 4A). Breeding lines population (B) 1 and 2 consisted of 33 and 28 accessions, respectively (Figure 4B).

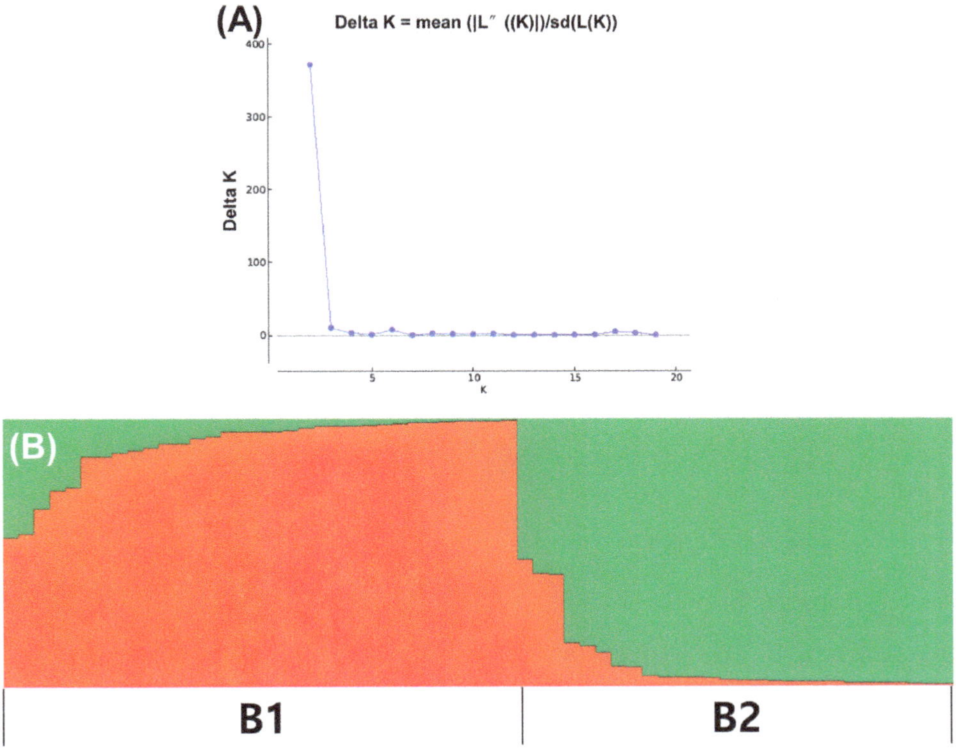

Figure 4. Population structure of 61 ginseng breeding lines based on 33 SSRs markers. (**A**) Estimation of population using LnP(D) derived ΔK. (**B**) Two populations sorted by Q, which represent estimated membership probabilities of the individual to inferred populations.

3.4. *Genetic Diversity and Population Differentiation of Korean Ginseng Accessions*

AMOVA was collected for the genetic differentiation among and within the ginseng accession populations by collection area and STRUCTURE results (Table 2). In the collection area findings, AMOVA revealed that 97% of the total genetic variation was attributable to differences within populations, which were notably and significantly greater than the 3% total variation found among populations. PhiPT and gene flow (Nm) for 451 ginseng accessions were 0.028 ($p < 0.0001$) and 17.213, respectively. Pairwise population PhiPT for four clusters ranged from 0.007 (GG–CB) to 0.085 (CB–JB) (Table S5). Pairwise population estimates of Nm for the four clusters ranged from 5.395 (CB–JB) to 67.845 (GG–CB) migrants per cluster. In the STRUCTURE analysis, 97% of the total genetic variation was again attributable to differences within populations. PhiPT and Nm for the 451 accessions in these results were 0.031 ($p < 0.0001$) and 15.431, respectively. Pairwise population PhiPT for the four clusters ranged from 0.000 (POP1–POP2) to 0.106 (POP6–POP7) (Table S6). Pairwise population estimates of Nm for the four clusters ranged from 4.223 (POP6–POP7) to 39.744 (POP1–POP3) migrants per cluster. In results of, POP1 and POP2 showed no genetic differentiation in PhiPT and Nm.

Table 2. Analysis of molecular variance (AMOVA) of ginseng accessions.

Source	df [1]	SS	MS	Est. Var.	%	PhiPT	Nm
		Collection area of total ginseng accessions					
Among Pops.	5	236.102	47.220	0.487	3%	0.028	17.213
Within Pops	445	7466.739	16.779	16.779	97%	($p < 0.001$)	
Total	450	7702.840		17.267	100%		
		STRUCTURE of total ginseng accessions					
Among Pops	6	305.185	50.864	0.54	3%	0.031	15.431
Within Pops	444	7397.655	16.661	16.661	97%	($p < 0.001$)	
Total	450	7702.84		17.201	100%		
		STRUCTURE of ginseng landraces					
Among Pops	11	385.070	35.006	0.578	3%	0.034	14.322
Within Pops	378	6255.419	16.549	16.549	97%	($p < 0.001$)	
Total	389	6640.490		17.126	100%		
		STRUCTURE of ginseng breeding lines					
Among Pops	1	19.859	19.859	0.178	1%	0.012	40.761
Within Pops	59	854.239	14.479	14.479	99%	($p < 0.028$)	
Total	60	874.098		14.656	100%		

[1] df, degree of freedom; SS, sum of squares; MS, mean squares; Est.Var, estimates of variance; %, percent of variance, PhiPT, genetic differentiation; Nm, gene flow.

In the ginseng landrace AMOVA findings, 97% of the total genetic variation was contributed by differences within populations (Table 2). PhiPT and Nm for the 390 landraces were 0.034 ($p < 0.0001$) and 14.322, respectively, and the Pairwise population PhiPT for four clusters ranged from −0.097 (L1–L6) to 0.084 (L7–L11) (Table S7). Pairwise population estimates of Nm for the four clusters ranged from 5.455 (L7–L11) to 403.3 (L1–L5) migrants per cluster.

In ginseng breeding lines, AMOVA revealed that 99% of the total genetic variation was attributable to within- rather than across-population differences (Table 2). PhiPT and Nm for 61 ginseng breeding lines were 0.012 ($p < 0.0001$) and 40.761, respectively.

4. Discussion

4.1. Distribution of Korean Ginseng Accessions

The collection areas for the ginseng accessions used in this study were mainly distributed in northern and central Korea. In South Korea, the ginseng cultivation area is from 36° to 38° N [14]. According to the 15th-century Korean history book Sejong Silokjiriji, Korean ginseng is native to all of Korea except Jeju Island. In the Joseon Dynasty, agriculturalists began cultivating ginseng in areas where native ginseng was found and gradually expanded cultivation throughout the country. However, based on differences in quality, ginseng is cultivated mainly in CN, CB, and GG [13]. The collection for the 390 ginseng landraces used in this study were distributed across six provinces (Figure 1 and Table S3); among them, the main areas were CN (36.4%), GW (31.8%), and GG (17.9%).

At the NAC, the seeds of plant germplasm are generally collected from farmers in the main cultivation areas of the target crops, except wild relatives. CN and GG are main ginseng production areas in the south, centered around Daejeon (CN), and in the north, the major area, bounded by the demilitarized zone and the two coasts (GG) [14]. GW is a mountainous region and is famous for mountain cultivated ginseng, while ginseng in GG and GN is cultivated in the field using shading systems [25–39]. Three provinces accounted for 40.3% of all ginseng production in South Korea in 2018 [40]. In traditional agriculture, farmers create and conserve new varieties, the preponderance of which are landraces. They decide which materials to conserve or discard, and they recycle seed on their farms, which are very often quite small. Such systems are highly dynamic in terms of genetic content and adaptation of the landraces [41].

4.2. Genetic Differentiation of Korean Ginseng Accessions

Although all of the ginseng accessions could be clustered into one of seven populations using STRUCTURE software, the accession clustering did not show clear separation according to geographic origin (Figure 2). In addition, AMOVA showed low genetic differentiation (PhiPT) and gene flow (Nm) between collection area (PhiPT = 0.028, Nm = 17.213) and STRUCTURE group (PhiPT = 0.031, Nm = 15.431) (Table 2). The ginseng landraces also showed low PhiPT (0.034) and high Nm (14.322), although STRUCTURE divided them into 12 groups.

Ginseng seed and seedlings necessarily originate from wild ginseng [42], and wild ginseng, which is rare, has been increasingly endangered because of high demand [43]. Natural populations of rare and threatened species are typically small and geographically isolated [44], and the growth and persistence of small populations are highly influenced by stochastic events such as genetic drift and inbreeding; such events result in reduced genetic diversity and fitness [45,46]. Karron [47,48] reported that some species with restricted distributions showed less genetic diversity than congeneric species with widespread distributions, but Cole [49] observed that it is not clear that rare plant species will necessarily have less gene flow than common species. Ellstrand [44] summarized arguments for expecting that gene flow might increase in rare plant populations, resulting in less differentiation. In Korea, farmers have been cultivating, self-seed gathering, and seed-selling the ginseng shoots themselves [50,51], and it seems that this method of ginseng cultivation is responsible for the low PhiPT and Nm in the Korean ginseng landraces at NAC. Cultivated ginseng seeds obtained from a few wild ginsengs are amplified and self-seed gathered by local farmers, and then they are sold on the ginseng market. As this cycle repeated over time, ginseng seeds of the same genetic background would have spread to other regions; it is assumed that this is the reason the ginseng landraces collected from different regions showed low PhiPT and high Nm.

The ginseng breeding lines also showed low PhiPT (0.012) and high Nm (40.761), although STRUCTURE divided the breeding lines into only two groups. Interestingly, both groups included two ginseng varieties—gumpoong and chunpoong. The two varieties, which were registered in 2002 (cv. Chunpoong) and 2003 (cv. Gumpoong), were bred by selecting individual plants from Korean landraces [11]. In Korea, 27 ginseng varieties have been registered with the Korea Seed & Variety Service, and all were bred by selecting individual plants from landraces [52]. In general, selecting individual plants is more common for developing new ginseng varieties than is crossbreeding because of the long generation period and small number of seeds with plant selection [11–53]. However, these varieties were selected for their excellent root form, but their genetic background has been little analyzed [11]. Meanwhile, the low genetic diversity in ginseng breeding lines might be because plants have been selected for morphological traits only; because morphological traits are easily affected by environmental conditions [54], their gene profiles could be similar in spite of their different phenotypes.

4.3. Increasing the Genetic Diversity of the NAC's Ginseng Germplasm

As mentioned above, the population of Korean ginseng landraces is expected to be very small, and it has been reported that small populations increase genetic drift and inbreeding [26], which could in turn influence these small populations by changing patterns of genetic diversity. One concerning drift-induced genetic change is this erosion of genetic variation; loss of genetic variation may decrease the potential for a species to persist in the face of abiotic and biotic environmental change [55,56]. The primary solution to the genetic impoverishment of crop germplasm is genetic conservation and utilization in breeding the vast genetic variation found in populations of the wild progenitors and landraces of cultivated plants [3]. However, it is very difficult to collect wild ginseng because it is such a rare plant. To overcome the shortage of genetic diversity in ginseng, researchers in Korea have attempted crossing interspecies, tissue culture, and mutation breeding [11]. Ahn et al. [57] and Kim et al. [58] attempted to cross *P. ginseng* and *P. quinquefolius*, but they did not get seeds from F2 plants. Kim et al. [56] studied the tissue culture of F1 plant propagation in a ginseng hybrid (*P. ginseng* × *P. quinquefolius*) to solve the seedless problem. In addition, some research groups have attempted mutation breeding using irradiation or chemicals, but they did not develop new ginseng varieties; they did confirm plants with high ginsenosides [59–63].

Bang et al. [11] highlighted the need to develop new varieties that are strong against abnormal weather conditions and reflect different trends. To develop such new varieties, NAC should attempt to secure the genetic diversity of ginseng. Collecting and securing *P. ginseng* and related species will be required for increasing the diversity, and the effort will also require collaboration with other ginseng research institutes to establish effective breeding methods.

Supplementary Materials: The following are available online at http://www.mdpi.com/2073-4395/10/11/1643/s1; Table S1. List of 451 ginseng accessions in this study; Table S2. List of 33 SSR markers in this study; Table S3. Distribution of collected areas in 451 ginseng accessions; Table S4. Genetic diversity parameters of 33 SSR markers; Table S5. Pairwise population PhiPT values (above diagonal) and Nm values based on 999 permutations (below diagonal) from AMOVA using collection area. All PhiPT values were significantly greater than 0 ($p < 0.0001$); Table S6. Pairwise population PhiPT values (above diagonal) and Nm values based on 999 permutations (below diagonal) from AMOVA using STRUCTURE result of 451 ginseng accessions. All PhiPT values were significantly greater than 0 ($p < 0.0001$); Table S7. Pairwise population PhiPT values (above diagonal) and Nm values based on 999 permutations (below diagonal) from AMOVA using STRUCTURE result of 39 ginseng landraces. All PhiPT values were significantly greater than 0 ($p < 0.0001$).

Author Contributions: Conceptualization, K.J.L. and D.Y.H.; data curation, K.J.L. and D.Y.H.; formal analysis, K.J.L. and R.S.; resources, M.K.; investigation, S.-H.K., E.Y., and S.L.; writing—original draft, K.J.L.; writing—review and editing, D.Y.H.; funding acquisition, G.-T.C. All authors have read and agreed to the published version of the manuscript.

Funding: This research was funded by the [Research Program for Agricultural Science and Technology Development, National Institute of Agricultural Sciences, Rural Development Administration, Republic of Korea], grant number [PJ01355701].

Conflicts of Interest: The authors declare no conflict of interest.

References

1. Louwaars, N.P. Plant breeding and diversity: A troubled relationship? *Euphytica* **2018**, *214*, 114. [CrossRef] [PubMed]
2. Van de Wouw, M.; Kik, C.; Hintum, T.; Treuren, R.; Visser, B. Genetic erosion in crops: Concept, research results and challenges. *Plant. Genet. Resour.* **2010**, *8*, 1–15. [CrossRef]
3. Hammer, K.; Teklu, Y. Plant Genetic Resources: Selected Issues from Genetic Erosion to Genetic Engineering. *J. Agr. Rural Dev. Trop.* **2008**, *109*, 15–59.
4. Tanksley, S.D.; McCouch, S.R. Seed Banks and Molecular Maps: Unlocking Genetic Potential from the Wild. *Science* **1997**, *277*, 1063–1066. [CrossRef]
5. Simmonds, N.W. *Genetic Resources in Plants—Their Exploration and Conservation*; Frankel, O.H., Bennett, E., Eds.; Blackwell Scientific Publications: Oxfordshire, UK, 1970; p. 554.
6. Upadhyaya, H.; Gowda, C.; Sastry, D.V.S.S.R. Plant genetic resources management: Collection, characterization, conservation and utilization. *J. SAT Agric. Res.* **2008**, *6*, 16.
7. Migdadi, H.; Fayad, M.; Ajloni, M.; Syouf, M.; Brake, M.; Abulila, K.; Tahebsum, Z. *The Second Report on the State of the World's Plant Genetic Resources for Food and Agriculture*; Commission on Genetic Resources for Food and Agriculture Food and Agriculture Organization of the United Nations: Rome, Italy, 2010.
8. Engelmann, F. 17—Germplasm collection, storage, and conservation. In *Plant Biotechnology and Agriculture*; Altman, A., Hasegawa, P.M., Eds.; Academic Press: San Diego, CA, USA, 2012; pp. 255–267.
9. NAC Hompage. Available online: https://genebank.rda.go.kr. (accessed on 10 September 2020).
10. Van Dan, N.; Ramchiary, N.; Choi, S.R.; Uhm, T.S.; Yang, T.-J.; Ahn, I.-O.; Lim, Y.P. Development and characterization of new microsatellite markers in *Panax ginseng* (C.A. Meyer) from BAC end sequences. *Conserv. Genet.* **2010**, *11*, 1223–1225. [CrossRef]
11. Bang, K.-H.; Kim, Y.-C.; Lee, J.-W.; Cho, I.-H.; Hong, C.-E.; Hyun, D.-Y.H.; Kim, J.-U. Major Achievement and Prospect of Ginseng Breeding in Korea. *Kor. J. Breed. Sci.* **2020**, 170–178. [CrossRef]
12. Jang, W.; Jang, Y.; Kim, N.-H.; Waminal, N.E.; Kim, Y.C.; Lee, J.W.; Yang, T.-J. Genetic diversity among cultivated and wild *Panax ginseng* populations revealed by high-resolution microsatellite markers. *J. Ginseng Res.* **2020**, *44*, 637–643. [CrossRef]

13. Ginseng, K.A.O. *Korean Ginseng*; Ministry for Food, Agriculture, Forestry and Fisheries (MIFAFF) and Korea Association of Ginseng, Ed.; Korea Association of Ginseng: Seoul, Korea, 2012.
14. Proctor, J.T.A.; Lee, J.C.; Lee, S.-S. Ginseng Production in Korea. *HortScience* **1990**, *25*, 746–750. [CrossRef]
15. Yang, J.; Yeo, I. A study on the origins of 'Korean Ginseng'. *Kor. J. Med. Hist.* **2004**, *13*, 1–19.
16. Zhuravlev, Y.; Reunova, G.D.; Kats, I.L.; Muzarok, T.I.; Bondar, A.A. Molecular variation of wild *Panax ginseng* C.A. Meyer (Araliaceae) by AFLP markers. *Chin. Med.* **2010**, *5*.
17. Zhuravlev, Y.; Koren, O.D.; Reunova, G.I.; Muzarok, T.; Gorpenchenko, T.L.; Kats, I.; Khrolenko, Y. *Panax ginseng* natural populations: Their past, current state and perspectives. *Acta Pharmacologica Sinica* **2008**, *29*, 1127–1136. [PubMed]
18. Shin, B.-K.; Kwon, S.W.; Park, J.H. Chemical diversity of ginseng saponins from *Panax ginseng*. *J. Ginseng Res.* **2015**, *39*, 287–298. [CrossRef]
19. Li, S.; Li, J.; Yang, X.-L.; Cheng, Z.; Zhang, W.-J. Genetic diversity and differentiation of cultivated ginseng (*Panax ginseng* C.A. Meyer) populations in North-east China revealed by inter-simple sequence repeat (ISSR) markers. *Genet. Resour. Crop. Evol.* **2011**, *58*, 815–824. [CrossRef]
20. Li, M.-R.; Shi, F.-X.; Li, Y.-L.; Jiang, P.; Jiao, L.; Liu, B.; Li, L.-F. Genome-Wide Variation Patterns Uncover the Origin and Selection in Cultivated Ginseng (*Panax ginseng* Meyer). *Genome Biol. Evol.* **2017**, *9*, 2159–2169. [CrossRef] [PubMed]
21. Choi, K.T. Botanical characteristics, pharmacological effects and medicinal components of Korean *Panax ginseng* C.A. Meyer. *Acta Phamacologica Sinica* **2008**, *29*, 1109–1118. [CrossRef] [PubMed]
22. Cho, C.-W.; Kim, Y.-C.; Rhee, Y.K.; Lee, Y.-C.; Kim, K.-T.; Hong, H.-D. Chemical composition characteristics of Korean straight ginseng products. *J. Ethn. Foods* **2014**, *1*, 24–28. [CrossRef]
23. Bang, K.-H.; Jo, I.-H.; Chung, J.-W.; Kim, Y.-C.; Lee, J.-W.; Seo, A.Y.; Park, J.-H.; Kim, O.; Hyun, D.-Y.; Kim, D.-H.; et al. Analysis of Genetic Polymorphism of Korean Ginseng Cultivars and Foreign Accessions using SSR Markers. *Kor. J. Med. Crop. Sci.* **2011**, *19*. [CrossRef]
24. Liu, D.; Li, Y.-G.; Xu, H.; Sun, S.-Q.; Wang, Z.-T. Differentiation of the root of Cultivated Ginseng, Mountain Cultivated Ginseng and Mountain Wild Ginseng using FT-IR and two-dimensional correlation IR spectroscopy. *J. Mol. Struct.* **2008**, 883–884. [CrossRef]
25. Suh, H.; Seo, S.M.; Woo, S.Y.; Lee, D.-S. Forest cultivated ginseng in Korea: All cure medicinal plants. *J. Med. Plants Res.* **2011**, *5*, 5331–5336.
26. Ellstrand, N.C.; Elam, D.R. Population Genetic Consequences of Small Population Size: Implications for Plant Conservation. *Annu. Rev. Ecol. Evol. Syst.* **1993**, *24*, 217–242. [CrossRef]
27. Furlan, E.; Stoklosa, J.; Griffiths, J.; Gust, N.; Ellis, R.; Huggins, R.M.; Weeks, A.R. Small population size and extremely low levels of genetic diversity in island populations of the platypus, Ornithorhynchus anatinus. *Ecol. Evol.* **2012**, *2*, 844–857. [CrossRef] [PubMed]
28. Ramanatha Rao, V.; Hodgkin, T. Genetic diversity and conservation and utilization of plant genetic resources. *Plant Cell Tissue Org. Cult.* **2002**, *68*, 1–19. [CrossRef]
29. Lee, K.J.; Lee, J.-R.; Sebastin, R.; Cho, G.-T.; Hyun, D.Y. Molecular Genetic Diversity and Population Structure of Ginseng Germplasm in RDA-Genebank: Implications for Breeding and Conservation. *Agronomy* **2020**, *10*, 68. [CrossRef]
30. Kim, N.-H.; Choi, H.-I.; Ahn, I.-O.; Yang, T.-J. EST-SSR Marker Sets for Practical Authentication of All Nine Registered Ginseng Cultivars in Korea. *J. Ginseng Res.* **2012**, *36*, 298–307. [CrossRef] [PubMed]
31. Choi, H.-I.; Kim, N.-H.; Kim, J.H.; Choi, B.S.; Ahn, I.-O.; Lee, J.-S.; Yang, T.-J. Development of Reproducible EST-derived SSR Markers and Assessment of Genetic Diversity in *Panax ginseng* Cultivars and Related Species. *J. Ginseng Res.* **2011**, *35*, 399–412. [CrossRef]
32. Kamvar, Z.N.; Tabima, J.F.; Grünwald, N.J. Poppr: An R package for genetic analysis of populations with clonal, partially clonal, and/or sexual reproduction. *PeerJ* **2014**, *2*, e281. [CrossRef] [PubMed]
33. Peakall, R.; Smouse, P.E. GenAlEx 6.5: Genetic analysis in Excel. Population genetic software for teaching and research—An update. *Bioinformatics* **2012**, *28*, 2537–2539. [CrossRef]
34. Pritchard, J.K.; Stephens, M.; Donnelly, P. Inference of population structure using multilocus genotype data. *Genetics* **2000**, *155*, 945–959.
35. Earl, D.A.; von Holdt, B.M. STRUCTURE HARVESTER: A website and program for visualizing STRUCTURE output and implementing the Evanno method. *Conserv Genet. Resour* **2012**, *4*, 359–361. [CrossRef]

36. Genebank Management System (GMS). Available online: http://genebank.rda.go.kr (accessed on 10 September 2020).
37. FAO. *Genebank Standards for Plant. Genetic Resources for Food and Agriculture*; FAO: Rome, Italy, 2013.
38. Sung, M.-H. Ginseng Industry changes and policy tasks. In *Ginseng Industry Changes and Policy Tasks*; Korea Rural Economic Institute: Jeollanam-do, Korea, 2019.
39. The Korean Society of Wild Ginseng. *A study on the Development Suggestions in Mountain Ginseng (Panax ginseng C.A. Meyer) Industry*; Forest Service: Daejeon, Korea, 2005.
40. Ministry of Agriculture Food and Rural Affairs. *Ginseng Annual Report*; Ministry of Agriculture Food and Rural Affairs: Sejong-si, Korea, 2019.
41. Berthaud, J.; Pressoir, G.; Ramírez-Corona, F.; Bellon, M. Farmers Management of Maize Landrace Diversity. A Case Study in Oaxaca and beyond. In Proceedings of the 7th International Symposium on The Biosafety of Genetically Modified Organisms, Beijing, China, 10–16 October 2002.
42. Park, H. The history of ginseng cultivation in the orient. *Acta Horticiltirae* **2003**, *620*, 453–460. [CrossRef]
43. Shahrajabian, M.H.; Sun, W.; Cheng, Q. A review of Ginseng species in different regions as a multipurpose herb in traditional Chinese medicine, modern herbology and pharmacological science. *J. Med. Plant. Res.* **2019**, *13*, 213–226.
44. Ellstrand, N.C. Gene Flow by Pollen: Implications for Plant Conservation Genetics. *Oikos* **1992**, *63*, 77–86. [CrossRef]
45. Ouborg, N.J.; Pertoldi, C.; Loeschcke, V.; Bijlsma, R.; Hedrick, P.W. Conservation genetics in transition to conservation genomics. *Trends Genet.* **2010**, *26*, 177–187. [CrossRef] [PubMed]
46. Lee, S.R.; Choi, J.E.; Lee, B.Y.; Yu, J.N.; Lim, C.E. Genetic diversity and structure of an endangered medicinal herb: Implications for conservation. *AoB PLANTS* **2018**, *10*, ply021. [CrossRef]
47. Karron, J.D. A comparison of levels of genetic polymorphism and self-compatibility in geographically restricted and widespread plant congeners. *Evol. Ecol.* **1987**, *1*, 47–58. [CrossRef]
48. Karron, J.D. Patterns of genetic variation and breeding systems in rare plant species. In *Genetics and Conservation of Rare Plants*; Falk, D.A., Holsinger, K.E., Eds.; Oxford University Press: New York, NY, USA, 1991; pp. 87–98.
49. Cole, C. Genetic Variation in Rare and Common Plants. *Annu. Rev. Ecol. Evol. Syst.* **2003**, *34*, 213–237. [CrossRef]
50. Horticulture Business Division. *2017 Ginseng Data Sheet*; Ministry of Agriculture Food and Rural Affairs: Sejong-si, Korea, 2018.
51. NIHHS. *Distribution Status of Panax ginseng in China and Comparative Analysis on the Characteristics of Korean and China ginseng*; RDA: Wansan-gu, Korea, 2016.
52. Korean Seeds and Variety Service. Available online: www.seed.go.kr (accessed on 10 September 2020).
53. Lee, S.-S.; Lee, J.-H.; Ahn, I.-O. Characteristics of new cultivars in *Panax ginseng* C.A. Meyer. In Proceedings of the Ginseng Society Conference, Daejeon, Korea, 23 May 2008.
54. Westberg, E.; Ohali, S.; Shevelevich, A.; Fine, P.; Barazani, O. Environmental effects on molecular and phenotypic variation in populations of Eruca sativa across a steep climatic gradient. *Ecol. Evol.* **2013**, *3*, 2471–2484. [CrossRef]
55. Simberloff, D. The Contribution of Population and Community Biology to Conservation Science. *Annu. Rev. Ecol. Evol. Syst.* **1988**, *19*, 473–511. [CrossRef]
56. Soule, M.E. Thresholds for survival: Maintaining fitness and evolutionary potential. In *Conservation Biology: An Evolutionary-Ecological Perspective*; Sinauer Associates: Sunderland, MA, USA, 1980.
57. Ahn, I.-O.; Lee, S.-S.; Lee, J.-H.; Lee, B.-S.; In, J.-G.; Yang, D.-C. Characteristics of Plantlets Redifferentiated from F1 Hybrid between *Panax ginseng* and *Panax quinquefolius*. *J. Plant. Biotech.* **2006**, *33*, 45–48.
58. Kim, Y.J.; Silva, J.; Zhang, D.; Shi, J.; Joo, S.C.; Jang, M.G.; Kwon, W.S.; Yang, D.C. Development of interspecies hybrids to increase ginseng biomass and ginsenoside yield. *Plant. Cell Rep.* **2016**, *35*, 779–790. [CrossRef] [PubMed]
59. Kim, J.Y.; Adhikari, P.B.; Ahn, C.H.; Kim, D.H.; Chang Kim, Y.; Han, J.Y.; Kondeti, S.; Choi, Y.E. High frequency somatic embryogenesis and plant regeneration of interspecific ginseng hybrid between *Panax ginseng* and *Panax quinquefolius*. *J. Ginseng Res.* **2019**, *43*, 38–48. [CrossRef] [PubMed]
60. Choi, K.T.; Ahn, S.D.; Park, K.J. Biological effects of ethyl methane sulphonate treatment on ginseng seeds. *J. Ginseng Res.* **1981**, *5*, 163–169.

61. Kim, D.S.; Kim, S.Y.; Jeong, I.Y.; Kim, J.B.; Lee, G.J.; Kang, S.Y.; Kim, W. Improvement of ginsenoside production by *Panax ginseng* adventitious roots induced by γ-irradiation. *Biologia Plantarum* **2009**, *53*, 408. [CrossRef]
62. Jung Woo, L.; Jo, I.-H.; Kim, J.-U.; Hong, C.E.; Bang, K.-H.; Park, Y.-D. Determination of mutagenic sensitivity to gamma rays in ginseng (*Panax ginseng*) dehiscent seeds, roots, and somatic embryos. *Hortic. Environ. Biotech.* **2019**, *60*, 721–731.
63. Zhang, J.-Y.; Bae, T.-W.; Boo, K.-H.; Sun, H.-J.; Song, I.-J.; Pham, C.-H.; Ganesan, M.; Yang, D.-H.; Kang, H.-G.; Ko, S.-M.; et al. Ginsenoside Production and Morphological Characterization of Wild Ginseng (*Panax ginseng* Meyer) Mutant Lines Induced by γ-irradiation ((60)Co) of Adventitious Roots. *J. Ginseng Res.* **2011**, *35*, 283–293. [CrossRef] [PubMed]

agronomy

MDPI

Article

Congruity of the Polymorphisms in the Expressed and Noncoding Parts of the *Gli-B1* Locus in Common Wheat

Yulia Popovych [1], Sabina Chebotar [1], Viktor Melnik [2,†], Marta Rodriguez-Quijano [2], Laura Pascual [2], William John Rogers [3] and Eugene Metakovsky [2,*]

[1] Department of Genetics and Molecular Biology, National I.I. Mechnikov University, 65058 Odessa, Ukraine; popovych1818@gmail.com (Y.P.); s.v.chebotar@onu.edu.ua (S.C.)
[2] Unit of Genetics, Department of Biotechnology—Plant Biology, Universidad Politécnica de Madrid, 28040 Madrid, Spain; meller@vigg.ru (V.M.); marta.rurquiaga@upm.es (M.R.-Q.); laura.pascual@upm.es (L.P.)
[3] Departamento de Biología Aplicada, CIISAS, CIC-BIOLAB, CONICET-INBIOTEC, CRESCA, Facultad de Agronomía, Universidad Nacional del Centro de la Provincia Buenos Aires, Av. Rep. Italia 780, 7300 Azul, Provincia de Buenos Aires, Argentina; rogers@faa.unicen.edu.ar
* Correspondence: emetakovsky@gmail.com
† Current address: Vavilov Institute of General Genetics, Russian Academy of Sciences, 119991 Moscow, Russia.

Received: 8 August 2020; Accepted: 28 September 2020; Published: 4 October 2020

Abstract: The previously defined pairs of primers GliB1.1 and GliB1.2 were found to produce three and four principal variants, respectively, of PCR sequence length for the γ-gliadin pseudogene in 46 *Triticum aestivum* L. cultivars from 15 countries carrying 19 known alleles at the *Gli-B1* locus. A congruity was established between this polymorphism, allelic sets of the *Gli-B1*-produced gliadins (especially of the electrophoretic mobility in acid gels of the encoded γ-gliadin) and the presence in the wheat genotype of the *Gli-B5b* + *Rg-1* allelic combination. Six different alleles at the *Gli-B1* locus encoding an identical γ-gliadin produced a PCR sequence of about 400 bp (GliB1.1). Nine *Gli-B1d*-carrying genotypes from four countries produced an identical sequence of about 409 bp (GliB1.2), while three cultivars with *Gli-B1h* and four with *Gli-B1b* produced three and two specific sequences, respectively, of slightly different length. Allele *Gli-B1j* might be the result of recombination between coding and noncoding DNA sequences within the *Gli-B1* locus. These observations imply that genetic diversity of the agriculturally important region of chromosome 1B marked by variants of the *Gli-B1* locus is rather limited among common wheat cultivars of the 20th century, specifically to eight principal versions. These might have been incorporated into common wheat from diverged genotypes of diploid donor(s), and, due to the scarcity of recombination, subsequently maintained relatively intact. As well as its evolutionary significance, this information is of potential use in wheat breeding and we consider it likely that novel variants of the *Gli-B1* locus will be found in hitherto unstudied germplasm.

Keywords: *Triticum aestivum* L.; γ-gliadins; *Gli-B1*; polymorphism; molecular markers; PCR analysis

---

1. Introduction

The gliadins are major wheat endosperm storage proteins [1]; the characterisation of the structure of their encoding loci and of their allelic variation is an important field of study due to (i) their being an integral constituent of gluten, the composition of which is one of the principal determinants of end-use quality; (ii) the presence within some of their amino acid sequences of epitopes that are toxic in a proportion of the population; and (iii) most importantly for crop improvement programmes, their potential use as selective markers through linkage to other genes affecting quality or other agronomic traits. In common wheat, their synthesis is mainly controlled by six major gliadin-encoding loci (*Gli-A1*, *Gli-B1*, *Gli-D1*, *Gli-A2*, *Gli-B2* and *Gli-D2*) located in the distal regions of the short arms of the homeologous chromosome groups 1 and 6 [1,2]. A single wheat genotype produces more than 50 different gliadin polypeptides [3] belonging to three major types, the α-, γ-, and ω-gliadins. The primary and domain structures of gliadins of different types is known [4,5].

The diversity of gliadin polypeptides produced by a given genotype can be readily revealed using a simple acid gel electrophoresis procedure (APAGE). Analysis of segregating progenies from intervarietal crosses by this method showed that any allele at a *Gli* locus encodes two or more gliadin polypeptides (a block of electrophoretic bands) inherited together almost as a Mendelian unit, i.e., with virtually no recombination, and showing multiple allelism [6,7]. Hence, it is firmly established that any *Gli* locus in its different allelic states harbours different sets of expressed gliadin genes encoding various gliadin polypeptides.

The recombination rate is distributed non-proportionally along wheat chromosomes [8–10] and the distance between expressed genes in high recombination areas is considerable, reaching 55–120 Kb per 1 cM of genetic distance [11]. The *Glu-3* loci, mapped in the distal region of the short arms of homoeologous group 1 close to the *Gli-1* loci [1,2], have been shown to locate in the high recombination region, but the average distance between expressed genes inherited together of one *Glu-3* locus was as large as 81 Kb, and, in some cases, even much larger [12]. The size of one storage protein gene, including its regulatory sequences, is about 0.8–1.5 Kb [4,13].

Gliadin pseudogenes and gene fragments located between expressed genes account for more than 80% of all gliadin-coding DNA sequences found at a *Gli* locus [14]. PCR sequence length polymorphism was discovered among some wheat species and a few common wheat cultivars for the γ-gliadin pseudogene *GAG56B* located within the *Gli-B1* locus [15]. γ-Gliadin SNP markers (PCR primers GliB1.1 and GliB1.2) specific for a pseudogene located inside the *Gli-B1* locus were also developed. Each of 50 Australian common wheat cultivars tested produced a PCR sequence with either primer GliB1.1 (369 bp) or GliB1.2 (397 bp), whereas three cultivars carrying the 1BL.1RS translocation (and therefore no DNA of the *Gli-B1* locus) did not result in amplification [16].

More than 20 allelic variants of the *Gli-B1* locus were revealed in about 1000 officially registered common wheat cultivars worldwide [7]. These alleles were not proportionally distributed over this germplasm; instead, significant differentiation was observed, where cultivars bred in a particular country carried only a few of the known *Gli-B1* alleles. For example, more than 70% and 20% of cultivars bred in Australia during the 20th century had alleles *Gli-B1b* and *Gli-B1i*, respectively [17], meaning there is only narrow genetic variation at the *Gli-B1* locus among Australian cultivars.

In this work, we aimed to verify whether the *Gli-B1* alleles carried by a range of wheat cultivars would produce one of the two PCR sequences observed using two primers denominated GliB1.1 or GliB1.2 [16]. Additionally, we compared the level of genetic diversity between gliadin-encoding and noncoding parts of the Gli-B1 locus (the locus marks the agriculturally important [8,11] distal region of chromosome 1B) among common wheat cultivars bred in the 20th century. For this purpose, 46 common wheat genotypes from 15 countries with different known alleles at the *Gli-B1* locus were selected.

2. Materials and Methods

2.1. Plant Materials

Most of the grain samples of common wheat (*T. aestivum* L.) cultivars were obtained from genetic and/or breeding laboratories in their countries of origin. Some samples were taken from local collections of Spain and Ukraine. In total, 43 registered and three unregistered cultivars (wheat genotypes) from 15 countries were selected, from our list of approximately 1000 cultivars studied [7], to cover the widest possible allelic variation at the *Gli-B1* locus in common wheat germplasm of the 20th century.

Names of cultivars, country of origin, and year of registration and spring/winter habit were taken from the internet site wheatpedigree.net.

2.2. Alleles at the Gli-B1 Locus and PCR Sequence Length Polymorphism in the Set of Cultivars Studied

Alleles at the *Gli-B1* locus were distinguished through the application of current knowledge of the gliadin polypeptide blocks encoded by each allele and observed in acid polyacrylamide gel electrophoregrams (APAGE) of gliadin proteins of single seeds [7].

DNA was extracted from dry single seeds using the CTAB method [18]. PCR amplification was performed in the thermocycler Analytik Jena (Flex Cycler, Germany) using the GliB1.1 and GliB1.2 primers [16]. Products of amplification for each cultivar were fractionated in 7% polyacrylamide gels (280 V) for two hours and stained as proposed by the Silver sequence TM DNA Sequencing System Technical Manual (Promega, Madison, WI, USA; https://www.promega.de/-/media/files/resources/protocols/technical-manuals/0/silver-sequence-dna-sequencing-system-protocol.pdf). The program GelAnalyzer 19.1 was used for the calculation of the approximate length of the products of amplification by comparing them with the set of standard markers pUc19/Mspl containing fragments of known length. Each fine difference in the length of the PCR sequence was registered in neighbouring lanes of a slab gel and confirmed in different runs.

3. Results

3.1. Allelic Variation at the Gli-B1 Locus and PCR Product Length Polymorphism among the Cultivars Studied

3.1.1. Allelic Variation

In 46 cultivars (wheat genotypes) studied, 19 different alleles at the *Gli-B1* locus (including the allele *Gli-B1l* represented by the 1BL.1RS translocation) were distinguished (Table 1). Gliadin protein electrophoregrams of some of the wheat genotypes studied, and blocks of gliadin electrophoretic bands controlled by each of 18 alleles (all except *Gli-B1l*), are shown in Figure 1 and Figure S1.

Table 1. Cultivars (grain samples) studied and their habit, country of origin, year of release, *Gli-B1* allele, and the approximate length of the PCR product obtained.

	Cultivar or Grain Sample [1]	Habit	Country	Year	Allele *Gli-B1*	Primer	Length [2]
1	**Bezostaya-1**	w	Russia	1959	*b*	GliB1.1	369
2	**Gabo**	s	Australia	1942	*b*	GliB1.1	369
3	**Marquis**	s	Canada	1907	*b*	GliB1.1	369
4	**Mironovskaya-808**	w	Ukraine	1963	*b*	GliB1.1	369
5	"Kavkaz" [3]				*b*	GliB1.1	369
6	Federation	s	Australia	1901	*i*	GliB1.1	400
7	**Insignia**	s	Australia	1946	*i*	GliB1.1	400
8	Mentana	s	Italy	1913	*k*	GliB1.1	400
9	Pane-247	w	Spain	1960	*k*	GliB1.1	400
10	Titien	w	France	1985	*m*	GliB1.1	400
11	**Aragon-03**	s, w	Spain	1940	*o*	GliB1.1	400
12	Potam-70	s	Mexico	1970	*p*	GliB1.1	400
13	Gazul	s	Spain	1992	*r*	GliB1.1	400
14	"Argelato" [3]				*m*	GliB1.1	400

Table 1. *Cont.*

	Cultivar or Grain Sample [1]	Habit	Country	Year	Allele *Gli-B1*	Primer	Length [2]
15	**Cluj-650**	w	Rumania	1954	*j*	GliB1.1	400
16	Intensivnaya	i	Kyrgyzstan	1978	*n*	GliB1.1	372
17	Goelent	w	France	1985	*q*	GliB1.1	372
18	**Salmone**	w	Italy	1980	*s*	GliB1.1	372
19	Cartaya	s	Spain		*l*	GliB1.1	-
20	**Chinese-Spring**	s	China	-	*a*	GliB1.2	415
21	Cajeme-71	s	Mexico, USA	1971	*d*	GliB1.2	409
22	Katepwa	s	Canada	1981	*d*	GliB1.2	409
23	**Laura**	s	Canada	1986	*d*	GliB1.2	409
24	Pavon-76	s	Mexico	1976	*d*	GliB1.2	409
25	Rinconada	s	Spain	1981	*d*	GliB1.2	409
26	Roblin	s	Canada	1986	*d*	GliB1.2	409
27	Suneca	s	Australia	1981	*d*	GliB1.2	409
28	Yecora-S	s	Mexico	1972	*d*	GliB1.2	409
29	"Inia-66" [3]				*d*	GliB1.2	409
30	Diego	s	Spain	1983	*c*	GliB1.2	397
31	Prinqual	s	France	1978	*c*	GliB1.2	397
32	**Siete-Cerros-66**	s	Mexico	1966	*c*	GliB1.2	397
33	Escualo	w	Spain	1981	*e*	GliB1.2	397
34	Glenlea	s	Canada	1972	*e*	GliB1.2	397
35	**Arminda**	w	Netherlands	1976	*f*	GliB1.2	397
36	Cappelle-Desprez	w	France	1946	*f*	GliB1.2	397
37	Darius	w	France	1974	*f*	GliB1.2	397
38	Recital	w	France	1986	*f*	GliB1.2	397
39	Sideral	w	France	1990	*f*	GliB1.2	397
40	"Libero" [3]				*f*	GliB1.2	397
41	Argelato	s, w	Italy	1964	*g*	GliB1.2	397
42	**Galahad**	w	UK	1983	*g*	GliB1.2	397
43	Ardec	s	Belgium	1979	*h*	GliB1.2	402
44	Caia	s	Portugal	-	*h*	GliB1.2	403
45	**Krasnodonka**	w	Ukraine	1969	*h*	GliB1.2	401
46	**Newcaster**	w	USA	1946	*h*	GliB1.2	403? [4]

[1] In bold, the cultivars whose gliadin electrophoregrams are shown in the Figure 1. [2] The approximate length (bp) of PCR product in cultivars with different alleles at the *Gli-B1* locus. Three or more single seeds for each grain sample giving a definite PCR product were analysed and no intra-varietal non-uniformity was noted for any of the cultivars studied. [3] The name was written on the label. As the genotype of this sample's seeds was different from that of the cultivar with this name studied earlier (Metakovsky et al. 2018), the allele at the *Gli-B1* locus was identified in this sample using gliadin electrophoresis (data not shown). [4] For the cultivar Newcaster, only a very weak PCR product was obtained; therefore, its length was not definitively established.

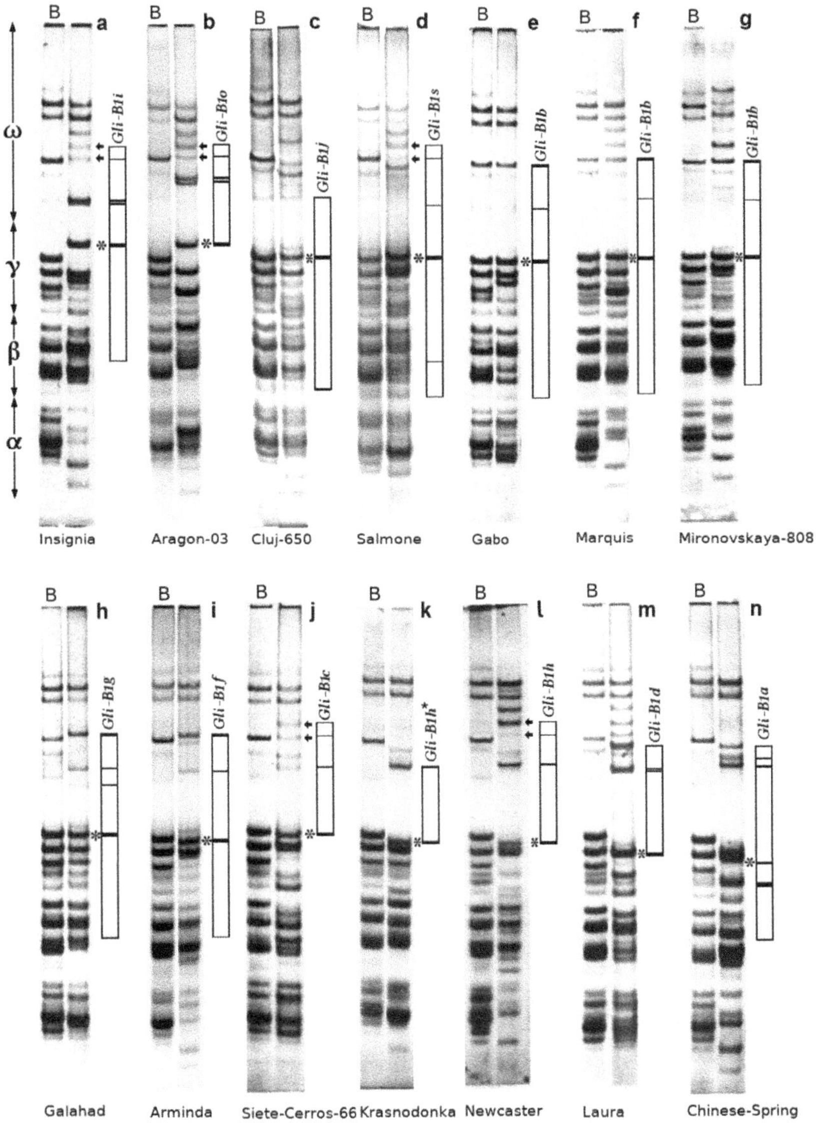

Figure 1. Gliadin electrophoregrams (APAGE) of some wheat cultivars studied. (**a**) Insignia; (**b**) Aragon-03; (**c**) Cluj-650; (**d**) Salmone; (**e**) Gabo; (**f**) Marquis; (**g**) Mironovskaya-808; (**h**) Galahad; (**i**) Arminda; (**j**) Siete-Cerros-66; (**k**) Krasnodonka; (**l**) Newcaster; (**m**) Laura; (**n**) Chinese-Spring. B, the cultivar Bezostaya-1. Blocks of jointly inherited gliadin electrophoretic bands are shown schematically. The major γ-gliadin (asterisk) and ω-gliadins encoded by the allele *Gli-B5b* (arrows) are indicated. *: see text.

Alleles at the *Gli-B1* locus differed, in particular, in the electrophoretic mobility (EM) in APAGE of their encoded major γ-gliadin polypeptide. There were six main variants of the EM of this γ-gliadin (for example, from the slowest to the fastest, Figure 1a,d,e,k,m,n).

Two variants (I and II) of relatively slow EM for this γ-gliadin were observed in several alleles per variant. Thus, variant I (the slowest γ-gliadin) occurred in alleles *Gli-B1i* (Figure 1a), *o* (Figure 1b), *k*, *m*, *p* and *r*; these alleles, with their γ-gliadin of apparently identical EM, differed among them in the composition of the encoded ω-gliadins (Figure S1). Variant II occurred in alleles *Gli-B1j* (Figure 1c), *s* (Figure 1d), *n* and *q*.

Alleles of variant III encoded this γ-gliadin of intermediate EM (Figure 1e–j), although the γ-gliadin encoded by the allele *Gli-B1f* had a slightly higher EM compared with that of other alleles of the group (for example, Figure 1h,i; Figure S1). In contrast, alleles *Gli-B1b* and *Gli-B1c* differed considerably from each other in their encoded blocks and from those produced by other alleles of variant group III, *Gli-B1e*, *g*, and *f* (Figure S1). Therefore, alleles *Gli-B1b* and *Gli-B1c* were considered as two more independent variants of the *Gli-B1* locus.

Three variants of the major γ-gliadin of greater EM each occurred in an individual allele: *Gli-B1h* (Figure 1k,l), *Gli-B1d* (Figure 1m), and *Gli-B1*a (the fastest γ-gliadin, Figure 1n).

3.1.2. PCR Product Length Polymorphism

At least seven principal variants of the PCR product length were revealed when the 46 selected cultivars from around the world were tested in our study, three variants with the GliB1.1 primers (Figure 2, zone A) and four with the GliB1.2 set (Figure 3, zone A). The approximate length of these products varied from 369 bp (GliB1.1) to 415 bp (GliB1.2) (Table 1).

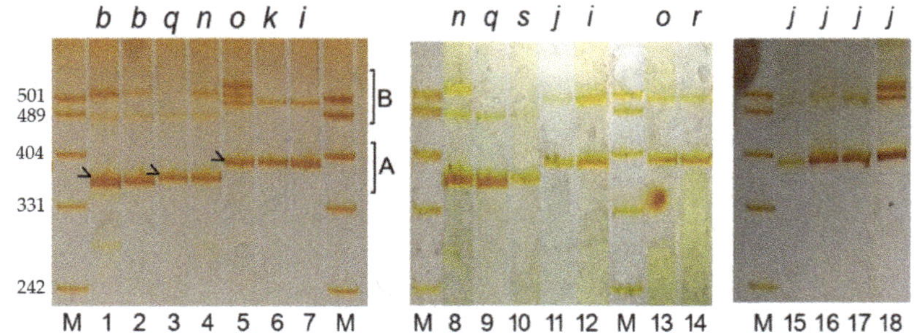

Figure 2. PCR product length polymorphism (GliB1.1 primers) among wheat cultivars. 1, Gabo; 2, Marquis; 3, Goelent; 4, Intensivnaya; 5, Aragon-03; 6, Mentana; 7, Insignia; 8, Intensivnaya; 9, Goelent; 10, Salmone; 11, Cluj-650; 12, Federation; 13, Aragon-03; 14, Gazul; 15, 16, 17, 18, Cluj-650. Alleles at the *Gli-B1* locus are indicated above lanes. M, marker pUC19/MspI, with DNA fragment lengths in bp given on the left-side of the figure. The three principal variants are signalled by > in the left-hand panel of the figure.

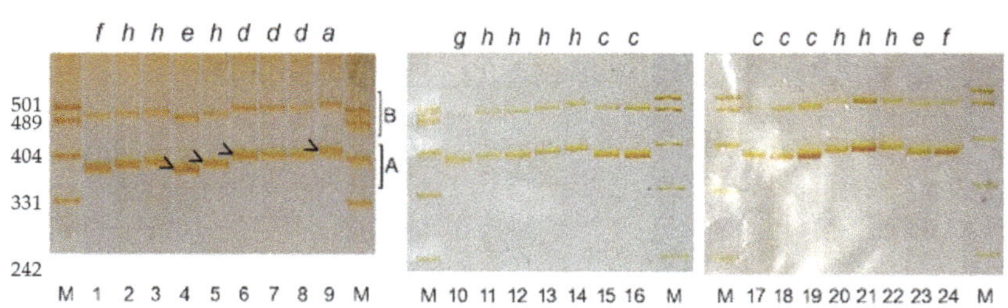

Figure 3. PCR product length polymorphism (GliB1.2 primers) among wheat cultivars. 1, Galahad; 2, Krasnodonka; 3, Caia; 4, Glenlea; 5, Ardec; 6, Laura; 7, Suneca; 8, Inia-66; 9, Chinese-Spring; 10, Argelato; 11, 12, Krasnodonka; 13, Ardec; 14, Caia; 15, Siete-Cerros-66; 16, Cappelle-Desprez; 17, Diego; 18, Prinqual; 19, Siete-Cerros-66; 20, Krasnodonka; 21, Ardec; 22, Caia; 23, Escualo; 24, Arminda. Alleles at the *Gli-B1* locus are indicated above lanes. M, marker pUC19/MspI, with DNA fragment lengths in bp given on the left-side of the figure. The four principal variants are signalled by > in the left-hand panel of the figure.

3.2. The Correspondence between PCR Product Polymorphism and Allelic Variation at the Gli-B1 Locus

The alleles tested in our work fell into two contrasting groups, in accordance with previous observations [17]. One group (alleles *Gli-B1b,i,j,k,m,n,o,p,q,r,s*) produced a PCR sequence with the GliB1.1 primers (Figure 2), while the other (*Gli-B1a,c,d,e,f,g,h*) gave a product with the GliB1.2 primers (Figure 3). Generally, the alleles encoding the slow-moving γ-gliadins (groups I and II) and the allele *Gli-B1b* produced PCR sequences with the primers GliB1.1, whereas the other alleles did so with the GliB1.2 (Table 1). Hence, variants of the *Gli-B1* locus encoding slow-moving γ-gliadins were accessible for the GliB1.1 primers and faster moving γ-gliadins were accessible for the GliB1.2 primers. As expected, the cultivar Cartaya carrying the 1BL.1RS translocation (the *Gli-B1l* allele) did not give a PCR product with either pair of primers (Table 1).

A strict congruency was discovered among the EM variants of the γ-gliadin and the PCR sequence length variants of a γ-gliadin pseudogene tested. For example, the lowest EM of the *Gli-B1* encoded γ-gliadin of variant group I corresponded to the PCR (GliB1.1) sequence of approximately 400 bp. Alleles of group II (except *Gli-B1j*) generated the PCR (GliB1.1) sequence of 372 bp. Alleles *Gli-B1e* and *Gli-B1g*, differing in the presence of one minor ω-gliadin, as well as the allele *Gli-B1f* encoding a γ-gliadin of slightly higher EM (Figure 1h,i; Figure S1), each produced the PCR (GliB1.2) sequence of 397 bp. The alleles *Gli-B1d* and *Gli-B1a* encoding different fast-moving γ-gliadins gave PCR (GliB1.2) products of 409 bp and 415 bp, respectively (Table 1; Figure 3).

The allele *Gli-B1b* (Figure 1e–g) produced a unique PCR sequence, the shortest observed amongst all alleles studied, of 369 bp (Table 1), and its controlled block differed strongly from those encoded by alleles of group III (Figure S1). Therefore, we consider the apparent equivalence of the EM of the encoded γ-gliadins of the group III and the *Gli-B1b* as being coincidental.

The allele *Gli-B1c*, while differing significantly from the other alleles of group III in its block's composition (the major ω-gliadin was absent and the major γ-gliadin was weaker) (compare Figure 1h–j), gave a PCR (GliB1.2) sequence identical in length to that of the other alleles of the group (Table 1).

3.3. The Probable Recombinational Nature of the Allele Gli-B1j

The allele *Gli-B1j* occurred in only one cultivar, Cluj-650, in more than 1000 common wheat cultivars studied to date (Table S1 in [7]). The EM of the γ-gliadin encoded by this allele was similar to those controlled by the group II alleles, for example, *Gli-B1s* (Figure 1c,d, respectively). Nevertheless, the PCR sequence produced by this cultivar was of approximately 400 bp in length, which is characteristic of the group I alleles encoding the slowest γ-gliadin (Figure 2, lanes 8–14; Table 1).

We suggest that the allele *Gli-B1j* might be the product of a rare intralocus recombination event. As a result, the *Gli-B1* locus could have generated this allele by obtaining its γ-gliadin coding DNA (tested with protein electrophoresis) from some allele belonging to group II (any of *Gli-B1s*, *q* or *n*) and its noncoding DNA (tested with the use of the GliB1.1 primers) from some allele of group I. The alternative mutational transition of the EM of the γ-gliadin from group I into the other known group II polypeptides seems much less probable. In addition, the composition (set of gliadin polypeptides) of the *Gli-B1j* encoded block was similar to that encoded by the group II alleles, not to those of group I (Figure S1).

3.4. Variation in PCR Product Length among Cultivars Having Identical Alleles at the Gli-B1 Locus

As a rule, the same allele occurring in different wheat genotypes produced a PCR sequence of the same apparent length. For example, nine genotypes with the allele *Gli-B1d* produced PCR sequences (GliB1.2) of approximately 409 bp (Table 1; Figure 3, lanes 6, 7, 8).

However, some genotypes with identical alleles at *Gli-B1* might provide PCR sequences differing slightly in length, with the allele *Gli-B1h* being the most interesting of this kind of observation. The *Gli-B1h* encoded γ-gliadin has its own unique EM in different cultivars (data not shown), but this allele produced, among three cultivars studied, three PCR (GliB1.2) sequences of slightly different length (approximately, 400–403 bp) that were not, however, produced by any other allelic variant of the locus. The shortest sequence was reproducibly produced by this allele in cultivar Krasnodonka and the longest in cultivar Caia, with the length of the PCR product of the cultivar Ardec probably intermediate between these two variants (Figure 3, lanes 2, 3, 5; 12, 14, 13; and 20, 22, 21, respectively). Two different grain samples of Krasnodonka were used with identical results (lanes 11, 12).

A further example was given by four cultivars carrying allele *Gli-B1b*, where two of the cultivars, Marquis and Gabo, produced identical PCR sequences (Figure 2, lanes 1, 2), slightly shorter (Figure 4, lanes 1, 4–9) than those in the remaining two, Bezostaya and Mironovskaya-808 (Figure 4, lanes 2, 3, 10–14).

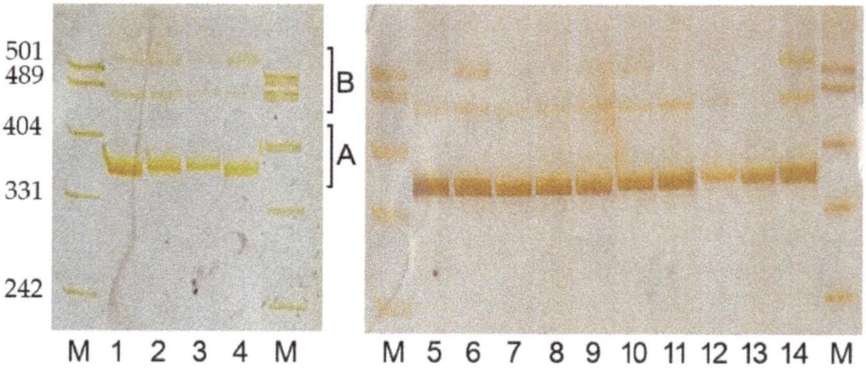

Figure 4. PCR product differences between cultivars carrying allele *Gli-B1b* in their genotypes 1, Gabo; 2, Bezostaya-1; 3, Mironovskaya-808; 4, Marquis; 5, 6, Gabo; 7, 8, 9, Marquis; 10, 11, Bezostaya-1; 12, 13, 14, Mironovskaya-808. M, marker pUC19/Mspl, with DNA fragment lengths in bp given on the left-side of the figure.

A final example was given by the allele *Gli-B1c*, which, when present in the cultivar Diego (Figure 3, lane 17), might generate a longer PCR product compared with the other cultivars carrying this allele, Prinqual and Siete-Cerros-66 (Figure 3, lanes 18, 19).

The cases of a small difference of the length of the PCR product among cultivars with the same allele help to confirm the congruity between the EM of the encoded γ-gliadin and the particular length of the PCR product.

3.5. Polymorphism of the PCR Products in "Zone B" of the Electrophoregram

In addition to the polymorphism detected in "zone A" of the DNA electrophoregram, variation was observed among PCR-derived DNA fragments of higher length ("zone B"). As a rule, the DNA sequence in this zone was about 100 bp longer compared to zone A in the same genotype. For example, the approximate length of the PCR product in zone B was about 497 bp and 495 bp for the alleles of groups I and III, respectively. Although the nature of the PCR products in zone B remains to be studied, the polymorphism in this zone is reasonably reproducible and may contribute to the polymorphism of the noncoding sequences of the *Gli-B1* locus.

For example, this polymorphism reinforced the difference between some wheat genotypes carrying different alleles at the *Gli-B1* locus, but with PCR products (in zone A) of apparently identical length. The most striking difference in zone B of two DNA electrophoregrams identical in zone A was observed between cultivar Aragon-03 (*Gli-B1o*) and other cultivars having alleles of group I, for example, Mentana (*Gli-B1k*) and Insignia (*Gli-B1i*) (Figure 2, lanes 5, 6, 7, respectively). The less prominent, but reproducible, differences in zone B were observed between some seeds of Gabo (Figure 2, lane 1; Figure 4, lane 6) and Marquis (Figure 2, lane 2; Figure 4, lane 7); between cultivars Goelent (Figure 3, lanes 3, 9) and Intensivnaya (Figure 3, lanes 4, 8); and so on.

Furthermore, some cases were observed of differences in zone B among seeds of the same cultivar, for example, in the cultivar Cluj-650 (Figure 2, lanes 15–18), in cultivars with the allele *Gli-B1b*, Gabo (Figure 4, lanes 5, 6), Bezostaya-1 (lanes 10, 11) and Mironovskaya-808 (lanes 12–14), but not in the cultivar Marquis (lanes 7–9).

3.6. The PCR Sequence Corresponds to Genotype, Not to the Label of the Grain Sample under Study

Four grain samples studied contained some seeds representing other genotypes than those identified previously in these cultivars. For example, the seeds of the grain sample labelled "Kavkaz" gave a PCR product (of length 369 bp), in spite of the fact that the Russian cultivar Kavkaz is one of the best known carriers and donors of the 1BL.1RS translocation and, therefore, lacking a true *Gli-B1* locus from its genotype (Table 1). Gliadin analysis (APAGE) of seeds of this sample showed that its genotype included allele *Gli-B1b* (data not shown), and therefore its labelling as "Kavkaz" was incorrect.

The genotype of only one of the two grain samples of the cultivar Argelato studied included the allele *Gli-B1g* characteristic of this cultivar. There were two more cases of discrepancy between genotypes of grains studied earlier [7] and in this work, Inia-66 and Libero. Thus, it is remarkable that the PCR product always corresponded to genotypes identified by APAGE in this work, not to the name of the cultivar written on the label (Table 1). The presence of admixtures of alien genotypes in grain samples of wheat and errors in labelling was noted earlier in different wheat collections [19–21].

One seed of the Canadian cultivar Katepwa successfully produced PCR sequences with both pairs of primers, 409 bp (GliB1.2) and 372 bp (GliB1.1). We suggest that this seed was heterozygous at the *Gli-B1* locus for allele *Gli-B1d* (produces the sequence of the 409 bp) characteristic for this cultivar (Table 1) and some allele of group II (for example, *Gli-B1q*, present in genotypes of some Canadian cultivars). It was thereby implied that this genotype produced PCR sequences from both homologous chromosomes. Heterozygous seeds originating from undetected cross-pollination may occur in wheat grain samples of different origin and are considered as foreign seeds or admixtures [21].

4. Discussion

All the expressed genes located at a particular *Gli* locus are known to inherit virtually as a Mendelian unit, almost without recombination [6]. The rate of recombination between two of the most remote gliadin-coding genes harboured in a *Gli-1* locus does not exceed 1% [22]. These data imply tight clustering of the expressed gliadin genes at this locus.

However, analysis of large DNA sequences of the *Gli* loci showed that the distance of 10 Kb between two gliadin genes, around ten times more than the length of a gliadin gene [4,13] itself, may be considered as unusually short and that the genes can be considered to be clustered [23]. In addition to noncoding gliadin sequences (pseudogenes, gene fragments) [14], transposable elements occupy most of the noncoding DNA between seed storage protein genes and occur in abundance at the distal end of the short arm of chromosome 1B [12,23–25].

The reason for the ability of several expressed gliadin genes, in spite of being relatively remote within the *Gli* locus, to be transmitted through intervarietal crosses as a unit with little visible signs of recombination, remains unclear. In order to understand the organization and function of the entire *Gli* locus, study and use of the polymorphism of its noncoding DNA sequences is required.

Two pairs of primers, GliB1.1 and GliB1.2, were developed for the analysis of the γ-gliadin pseudogene located within the *Gli-B1* locus [16]. Nineteen known allelic variants of the *Gli-B1* locus (revealed using APAGE of gliadin) were tested in our work using these pairs of primers, and a strict correspondence was discovered between variants of the EM of the γ-gliadin encoded by different alleles at the *Gli-B1* locus and variants of the length of the PCR-produced sequence of the γ-gliadin pseudogene tested. At least seven principal variants of the *Gli-B1* locus were distinguished. These variants are represented by four single alleles, *Gli-B1a*, *Gli-B1b*, *Gli-B1d*, *Gli-B1h*, and three allelic groups I, II, and III (without *Gli-B1b*). The seven variants differed in both their coding sequences (firstly, the EM of the major γ-gliadin, and, secondly, the composition of the entire block) and noncoding sequences (the length of the PCR-produced γ-gliadin pseudogene sequence).

The difference between variants at the distal end of chromosome 1B marked by different *Gli-B1* alleles may be reinforced by two more genetic loci situated distally from, but closely linked to [26], *Gli-B1*, namely, *Gli-B5* (where allele *Gli-B5b* encodes two faint ω-gliadins) and *Rg-1*, responsible for red glume colour. Analysis of a collection of common wheat varieties showed that the presence, in a certain wheat genotype, of any allele of groups I and II was nearly always accompanied by alleles *Gli-B5b* (Figure 1 and Figure S1) and *Rg-1*, while other alleles at the *Gli-B1* locus occurred in genotypes together with the alleles *Gli-B5a* (the absence of the two faint bands) and *rg-1* (white glume colour) [27]. This pattern was confirmed in Italian [28] and Spanish [29] wheat cultivars, although in each germplasm collection studied, single recombinant genotypes might occur. Hence, alleles of groups I and II, and the alleles *Gli-B1c* and *Gli-B1h*, occurred in wheat genotypes mainly with the allelic combination *Gli-B5b* + *Rg-1*, while the presence of any other allele was accompanied by the combination *Gli-B5a* + *rg-1*. In the set of cultivars studied in the current work, the cultivar Krasnodonka (Figure 1k) had a recombinant genotype: the "red-colour allele" at *Gli-B1* together with alleles *Gli-B5a* and *rg-1*.

The allele *Gli-B1c* (Figure 1j) produced the corresponding PCR sequence similar in length to that of the group III alleles at the *Gli-B1* locus (Table 1). However, this allele marks a recombined variant of the distal part of the chromosome 1B: it encodes the set of gliadins (block), especially of the ω-gliadins, different from that of the group III (Figure S1), and differs from other alleles of this group in being linked to the "red-colour" allelic combination, *Gli-B5b* + *Rg-1*. Therefore, the allele *Gli-B1c* may be considered as an eighth variant relatively independent from the group III alleles. In general, each homologous variant of chromosome 1B in common wheat germplasm is marked by one of the eight distinct variants of the *Gli-B1* locus.

Diploid donors of the genomes A, B and D might already have been polymorphic at the time of the origin of the tetraploid AABB and hexaploid AABBDD wheats, due to the differentiation of their genotypes at the diploid level [30]. We suggest that variants of the distal part of different homologous 1B chromosomes marked by each one of eight different variants of the *Gli-B1* locus were introduced into common wheat by considerably diverged genotypes of the donor(s) of the B genome. Due to the scarcity of recombination, each of the homologous variants of this chromosome's region may maintain intact (preserved from mixing) the precise gene (allelic) combination that came into common wheat from a particular genotype of the diploid donor. It is important to note that there are many agriculturally valuable genes in the distal part of the chromosome arms 1S [8,11].

The differences between homologous variants of common wheat chromosomes, especially of the B genome, were registered using cytological approaches [31,32]. It was assumed that these differences might cause a reduction in the number of chiasmata and chromosome pairing in intervarietal F1 hybrids compared to that of their parental inbred lines [31,33,34]. For the short arm of chromosome 1B (with the *Gli-B1* locus in its distal region) this reduction, in some crosses, might reach 40% [35].

The degree of differences between homologous chromosomes may influence the recombination rate observed in intervarietal crosses. For example, the genetic distance between *Gli-A3* and *Gli-A1* loci varied among intervarietal crosses from 12.6% to 35.3% [36] or from 7 cM to 27 cM [37]. The distance between the *Nor-B2* and *Gli-B2* loci was about 2.0 cM in a cross between the varieties Chinese Spring and Timstein, but 10.2 cM for Chinese Spring crossed to the variety Cheyenne [38]. Hence, the differences between the DNA sequences in Chinese Spring and Timstein might be greater than those between Chinese Spring and Cheyenne. The recombination frequency between *Gli-B1* and *Rg-1* loci varied from 0.0% (two different crosses) to 2.7% and 4.8% in two further crosses [26]. In each of these four crosses, one parent was the cultivar Salmone (*Gli-B1s*), and the frequency of recombination in each case might have depended upon the variant of the *Gli-B1* locus present in the second parent (alleles *Gli-B1m*, *e*, *b*, *b*, respectively).

For chromosome 3B, 82% of crossovers occurred in the distal region of chromosome 3B, representing only 19% of the chromosomal length [39]. Its "cold" areas (of low recombination rate) had a peculiar composition of transposable elements: a higher level of retrotransposons compared to DNA transposons [40]. The reduction in recombination rate in regions with a high density of retrotransposons could be due to DNA sequence variation created by their activity [41]. The negative impact of retroelements on recombination could also be attributed to their ability to condense DNA and lock the regions, preventing their accessibility to the recombination machinery [40]. The "cold" areas carry fewer and less expressed genes [40] and more single nucleotide substitutions (SNP) [10].

It is not clear if *Gli-B1* is analogous to the case of a genetic locus situated inside the "cold" region of chromosome arm 3BS. Nevertheless, eight variants of alleles at the *Gli-B1* locus may serve as markers of the distal part of homologous, but diverged, 1B chromosomes.

The primary structure of the γ-gliadin polypeptides includes, besides other sequences, variable numbers of CAA-repeats and a small microsatellite encoding poly-glutamine [5]. The number of CAA-repeats in different γ-gliadin genes varied, in 170 wheat genotypes studied, from seven to 22 [13]. The visible differences in PCR sequence lengths produced by the γ-gliadin pseudogene *GAG56B* in several common wheat cultivars were caused by variation in the number of CAA-repeats in this sequence [15]. Our preliminary results showed that common wheat genotypes with alleles *Gli-B1a* and *Gli-B1b* produced, respectively, the longest (31 repeats) and the shortest (seven repeats) PCR sequence of *GAG56B* (L. Pascual, unpublished). There is no doubt that polymorphism in the length of the γ-gliadin pseudogene tested in our work with the GliB1.1 and GliB1.2 pairs of primers was mainly due to the number of CAA-repeats in its PCR sequence.

Microsatellite sequences might frequently change their length, with the mutation rate up to 2.4×10^{-3} [42], or up to 5×10^{-3} [43] per locus, per generation. In a series of microsatellite alleles, they differed from one another, as a rule, by two or three nucleotide pairs [44]. Hence, the difference in length of the microsatellite may cause the absence of exact identity between the PCR sequences produced by different cultivars having the same allele at the *Gli-B1* locus (*Gli-B1b*, *Gli-B1h* or *Gli-B1c*) documented in our work.

The velocity of DNA divergence in noncoding regions was shown to be at least 3 to 4 times higher than in transcribed genes [45,46]. In the case of the *Gli-B1* locus, however, the variation in the length of the γ-gliadin pseudogene PCR sequence correlated strictly with the polymorphism of the EM of the encoded γ-gliadin, which merits further study.

5. Conclusions

In this work, we compared nearly all allelic variants of the *Gli-B1* locus revealed earlier in our study of gliadin genotypes in about 900 common wheat registered cultivars bred in 20th century. Analysis of allelic variation at the *Gli-B1*, *Gli-B5*, *Rg-1* loci and at the γ-gliadin pseudogene located within the *Gli-B1* locus permitted the identification of eight well-distinguished versions of the distal part of chromosome 1B in the germplasm studied. A correspondence was discovered between the length of the PCR sequences produced by different variants of the pseudogene, the variants present of the *Gli-B1*-controlled sets of gliadins (blocks), especially of the electrophoretic mobility of the encoded γ-gliadin, and the presence of the *Gli-B5b* + *Rg-1* allelic combination present in a certain wheat genotype. We suggest that distinct versions of the distal part of different homologous 1B chromosomes (each of them marked by one of eight variants of the *Gli-B1* locus), known to carry many agriculturally-important genes, were introduced into common wheat by diverged genotypes of the donor(s) of the B genome. Due to the scarcity of recombination, each of the homologous variants of this chromosome's region might maintain intact (preserving from mixing) the precise gene (allelic) combination that was incorporated into common wheat from a particular genotype of the diploid donor. Therefore, genetic polymorphism in this region of chromosome 1B in registered cultivars bred in the 20th century might be essentially limited to these eight principal variants.

As well as their evolutionary significance, these results are of potential use in wheat breeding in providing a simple means to select, via markers, for certain allelic combinations for this region across the generations of a genetic improvement programme. Furthermore, the results provide the knowledge base with which to compare the variation potentially present in hitherto unstudied collections of wheat germplasm, such as cultivars from other countries, landraces, and accessions of species related to common wheat, where we consider new and agronomically useful variants are likely to be found.

Supplementary Materials: The following is available online at http://www.mdpi.com/2073-4395/10/10/1510/s1, Figure S1: Scheme of sets of gliadin polypeptides (blocks) produced by allelic variants studied of the *Gli-B1* locus. ω-Gliadins encoded by the allele *Gli-B5b* are shown. B, the electrophoregram and scheme of its electrophoretic bands of the standard cultivar Bezostaya-1.

Author Contributions: Conceptualisation, E.M.; methodology, S.C., Y.P. and L.P.; software, V.M. and Y.P.; validation, W.J.R., L.P. and E.M.; investigation, Y.P., S.C. and E.M.; resources, M.R.-Q., S.C., L.P. and E.M.; writing—original draft preparation, E.M. and W.J.R. All authors have read and agreed to the published version of the manuscript.

Funding: This research received no external funding.

Conflicts of Interest: The authors declare no conflict of interest.

References

1. Payne, P.I. Genetics of wheat storage proteins and the effect of allelic variation on bread-making quality. *Annu. Rev. Plant Physiol.* **1987**, *38*, 141–153. [CrossRef]
2. Shewry, P.R.; Halford, N.G.; Lafiandra, D. Genetics of wheat gluten proteins. *Adv. Genet.* **2003**, *49*, 111–184. [PubMed]
3. Gobaa, S.; Bancel, E.; Branlard, G.; Kleijer, G.; Stamp, P. Proteomic analysis of wheat recombinant inbred lines: Variations in prolamin and dough rheology. *J. Cereal Sci.* **2008**, *47*, 610–619. [CrossRef]
4. Shewry, P.R.; Halford, N.G. Cereal seed storage proteins: Structures, properties and role in grain utilization. *J. Exp. Bot.* **2002**, *53*, 947–958. [CrossRef] [PubMed]
5. Barak, S.; Mudgil, D.; Khatkar, B.S. Biochemical and functional properties of wheat gliadins: A review. *Crit. Rev. Food Sci. Nutr.* **2015**, *55*, 357–368. [CrossRef]
6. Sozinov, A.A.; Poperelya, F.A. Genetic classification of prolamins and its use for plant breeding. *Ann. Technol. Agric.* **1980**, *29*, 229–245.
7. Metakovsky, E.; Melnik, V.A.; Rodriguez-Quijano, M.; Upelniek, V.P.; Carrillo, J.M. A catalog of glidin alleles: Polymorphism of 20th-century common wheat germplasm. *Crop J.* **2018**, *6*, 629–641. [CrossRef]
8. Sandhu, D.; Champoux, J.A.; Bondareva, N.; Gill, K.S. Identification and physical localization of useful genes and markers to a major gene-rich region on wheat group 1S chromosomes. *Genetics* **2001**, *157*, 1735–1747.
9. Saintenac, C.; Faure, S.; Remay, A.; Choulet, F.; Ravel, C.; Paux, E.; Balfourier, F.; Feuillet, C.; Sourdille, P. Variation in crossover rates across a 3-Mb contig of bread wheat (*Triticum aestivum*) reveals the presence of a meiotic recombination hotspot. *Chromosoma* **2011**, *120*, 185–198. [CrossRef]
10. Jordan, K.W.; Wang, S.; He, F.; Chao, S.; Lun, Y.; Paux, E.; Sourdille, P.; Sherman, J.; Akhunova, A.; Blake, N.K.; et al. The genetic architecture of genome-wide recombination rate variation in allopolyploid wheat revealed by nested association mapping. *Plant J.* **2018**, *95*, 1039–1054. [CrossRef]
11. Gill, K.S.; Gill, B.S.; Endo, T.R.; Taylor, T. Identification and high-density mapping of gene-rich regions in chromosome group 1 of wheat. *Genetics* **1996**, *144*, 1883–1891. [PubMed]
12. Huang, X.Q.; Cloutier, S. Molecular characterization and genomic organization of low molecular weight glutenin subunit genes at the *Glu-3* loci in hexaploid wheat (*Triticum aestivum* L.). *Theor. Appl. Genet.* **2008**, *116*, 953–966. [CrossRef] [PubMed]
13. Qi, P.F.; Wei, Y.M.; Ouellet, T.; Chen, Q.; Tan, X.; Zheng, Y.L. The γ-gliadin multigene family in common wheat (*Triticum aestivum*) and its closely related species. *BMC Genom.* **2009**, *10*, 168–181. [CrossRef] [PubMed]
14. van Herpen, T.W.J.M.; Goryunova, S.V.; van der Schoot, J.; Mitreva, M.; Salentijn, E.; Vorst, O.; Schenk, M.F.; van Veelen, P.A.; Koning, F.; van Soest, L.J.M.; et al. Alpha-gliadin genes from the A, B, and D genomes of wheat contain different sets of celiac disease epitopes. *BMC Genom.* **2006**, *7*, 1. [CrossRef]
15. von Büren, M. Polymorphisms in two homoeologous gamma-gliadin genes and the evolution of cultivated wheat. *Genet. Res. Crop Evol.* **2001**, *48*, 205–220. [CrossRef]

16. Zhang, W.; Gianibelli, M.C.; Ma, W.; Rampling, L.R.; Gale, K.R. Identification of SNPs and development of allele-specific PCR markers for gamma-gliadin alleles in *Triticum Aestivum*. *Theor. Appl. Genet.* **2003**, *107*, 130–138. [CrossRef]
17. Metakovsky, E.; Melnik, V.A.; Pascual, L.; Wrigley, C.W. Gliadin genotypes worldwide for spring wheats (*Triticum aestivum* L.). 2. Strong differentiation of polymorphism between countries and regions of origin. *J. Cereal Sci.* **2019**, *87*, 311–317. [CrossRef]
18. Doyle, J.J.; Doyle, J.L. Isolation of plant DNA from fresh tissue. *Focus* **1990**, *12*, 13–15.
19. Jackson, E.A.; Morel, M.H.; Sontag-Strohm, T.; Branlard, G.; Metakovsky, E.; Redaelli, R. Proposal for combining the classification systems of alleles of *Gli-1* and *Glu-3* loci in bread wheat (*Triticum aestivum* L.). *J. Genet. Breed.* **1996**, *50*, 321–336.
20. Cao, W.; Scoles, G.; Hucl, P.; Chibbar, R.N. The use of RAPD analysis to classify *Triticum* accessions. *Theor. Appl. Genet.* **1999**, *98*, 602–607. [CrossRef]
21. Metakovsky, E.; Melnik, V.A.; Pascual, L.; Romanov, G.A.; Wrigley, C.W. Types, frequences and value of intra-varietal genotypic non-uniformity in common wheat cultivars: Authentic biotypes and foreign seeds. *J. Cereal Sci.* **2019**, *89*. [CrossRef]
22. Metakovsky, E.; Branlard, G.; Chernakov, V.M.; Upelniek, V.P.; Redaelli, R.; Pogna, N.E. Recombiation mapping of some chromosome 1A-, 1B-, 1D- and 6B-controlled gliadins and low molecular weight glutenin subunits in common wheat. *Theor. Appl. Genet.* **1997**, *94*, 788–795. [CrossRef]
23. Gao, S.; Gu, Y.Q.; Wu, J.; Coleman-Derr, D.; Huo, N.; Crossman, C.; Jia, J.; Zuo, Q.; Ren, Z.; Anderson, O.D.; et al. Rapid evolution and complex structural organization in genomic regions harboring multiple prolamin genes in the polyploid wheat genome. *Plant Mol. Biol.* **2007**, *65*, 189–203. [CrossRef] [PubMed]
24. Charles, M.; Belcram, H.; Just, J.; Huneau, C.; Violett, A.; Couloux, A.; Segurens, B.; Carter, M.; Huteau, V.; Coriton, O.; et al. Dynamics and differential proliferation of transposable elements during the evolution of the B and A genomes of wheat. *Genetics* **2008**, *180*, 1071–1086. [CrossRef]
25. Wicker, T.; Mayer, K.F.X.; Gundlach, H.; Martis, M.; Steuernagel, B.; Scholz, U.; Šimková, H.; Kubaláková, M.; Choulet, F.; Taudien, S.; et al. Frequent gene movement and pseudogene evolution is common to the large and complex genomes of wheat, barley, and their relatives. *Plant Cell* **2011**, *23*, 1706–1718. [CrossRef]
26. Pogna, N.E.; Metakovsky, E.; Redaelli, R.; Raineri, F.; Dachkevitch, T. Recombination mapping of *Gli-5*, a new gliadin-coding locus on chromosome 1A and 1B in common wheat. *Theor. Appl. Genet.* **1993**, *87*, 113–121. [CrossRef]
27. Koval, S.F.; Metakovsky, E.; Kudryavtsev, A.M.; Sozinov, A.A. Non-accidental combinations of alleles at the *Hg-1* and *Rg-1* loci with definite alleles at the *Gli-A1* and *Gli-B1* loci in wheat. *Selskokhoz. Biologia* **1986**, *2*, 31–36.
28. Metakovsky, E.; Pogna, N.E.; Biancardi, A.M.; Redaelli, R. Gliadin allele composition of common wheat cultivars grown in Italy. *J. Genet. Breed.* **1994**, *48*, 55–66.
29. Metakovsky, E.; Gòmez, M.; Vàzquez, J.F.; Carrillo, J.M. High genetic diversity of Spanish common wheats as judged from gliadin alleles. *Plant Breed.* **2000**, *119*, 37–42. [CrossRef]
30. Dvorak, J.; Luo, M.C.; Akhunov, E.D. NI Vavilov theory of centres of diversity in the light of current understanding of wheat diversity, domestication and evolution. *Czech J. Genet. Plant Breed.* **2011**, *47*, S20–S27. [CrossRef]
31. Endo, T.R.; Gill, B.S. Somatic karyotype, heterochromatin distribution, and nature of chromosome differentiation in common wheat, *Triticum aestivum* L. em Thell. *Chromosome* **1984**, *89*, 361–369. [CrossRef]
32. Friebe, B.; Gill, B.S. C-band polymorphism and structural rearrangements detected in common wheat (*Triticum aestivum*). *Euphytica* **1994**, *78*, 1–5.
33. Dvořák, J.; Chen, K.C. Distribution of nonstructural variation between wheat cultivars along chromosome arm 6Bp: Evidence from the linkage map and physical map of the arm. *Genetics* **1984**, *106*, 325–333.
34. Dubcovsky, J.; Luo, M.-C.; Dvořák, J. Differentiation between homoeologous chromosomes 1A of wheat and 1Am of *Triticum monococcum* and its recognition by the wheat *Ph1* locus. *Proc. Nat. Acad. Sci. USA* **1995**, *92*, 6645–6649. [CrossRef] [PubMed]
35. Dvořák, J.; McGuire, P.E. Nonstructural chromosome differentiation among wheat cultivars, with special reference to differentiation of chromosomes in related species. *Genetics* **1981**, *97*, 391–414. [PubMed]

36. Metakovsky, E.; Chernakov, V.M.; Upelniek, V.P.; Redaelli, R.; Dardavet, M.; Branlard, G.; Pogna, N.E. Minor ω-gliadin-coding loci on chromosome 1A of common wheat: A revision. *J. Genet. Breed.* **1996**, *50*, 277–286.
37. Dubcovsky, J.; Echaide, M.; Giancola, S.; Rousset, M.; Luo, M.-C.; Joppa, L.R.; Dvořák, J. Seed-storage-protein loci in RFLP maps of diploid, tetraploid and hexaploid wheat. *Theor. Appl. Genet.* **1997**, *95*, 1169–1180. [CrossRef]
38. Dvořák, J.; Appels, R. Investigation of homologous crossing over and sister chromatid exchange in the wheat *Nor-B2* locus coding for rRNA and *Gli-B2* locus coding for gliadins. *Genetics* **1986**, *113*, 1037–1056.
39. Choulet, F.; Alberti, A.; Theil, S.; Glover, N.; Barbe, V.; Daron, J.; Pingault, L.; Sourdille, P.; Couloux, A.; Paux, E.; et al. Structural and functional partitioning of bread wheat chromosome 3B. *Science* **2014**, *345*, 1249721. [CrossRef]
40. Darrier, B.; Rimbert, H.; Balfourier, F.; Pingault, L.; Josselin, A.A.; Servin, B.; Navarro, J.; Choulet, F.; Paux, E.; Sourdille, P. High-resolution mapping of crossover events in the hexaploid wheat genome suggest a universal recombination mechanism. *Genetics* **2017**, *206*, 1373–1388. [CrossRef]
41. Dvořák, J.; Yang, Z.L.; You, F.M.; Luo, M.C. Deletion polymorphism in wheat chromosome regions with contrasting recombination rates. *Genetics* **2004**, *168*, 1665–1675. [CrossRef] [PubMed]
42. Thuillet, A.-C.; Bru, D.; David, J.; Roumet, P.; Santoni, S.; Sourdille, P.; Bataillon, T. Direct estimation of mutation rate for 10 microsatellite loci in durum wheat, *Triticum turgidum* (L.) Thell. ssp *durum* desf. *Mol. Biol. Evol.* **2002**, *19*, 122–125. [CrossRef] [PubMed]
43. Raquin, A.-L.; Depaulis, F.; Lambert, A.; Galic, N.; Brabant, P.; Goldringer, I. Experimental estimation of mutation rates in a wheat population with a gene genealogy approach. *Genetics* **2008**, *179*, 2195–2211. [CrossRef] [PubMed]
44. Huang, X.Q.; Börner, A.; Röder, M.S.; Ganal, M.W. Assessing genetic diversity of wheat (*Triticum aestivum* L.) germplasm using microsatellite markers. *Theor. Appl. Genet.* **2002**, *105*, 699–707. [CrossRef]
45. Wicker, T.; Stein, N.; Albar, L.; Feuillet, C.; Schlagenhauf, E.; Keller, B. Analysis of a contiguous 211 Kb sequence in diploid wheat (*Triticum monococcum*) reveals multiple mechanisms of genome evolution. *Plant J.* **2001**, *26*, 307–316. [CrossRef]
46. Isidore, E.; Scherrer, B.; Chalhoub, B.; Feuillet, C.; Keller, B. Ancient haplotypes resulting from extensive molecular rearrangements in the wheat A genome have been maintained in species of three different ploidy levels. *Genome Res.* **2005**, *15*, 526–536. [CrossRef]

agronomy

MDPI

Article

Salt Stress Induces Differentiated Nitrogen Uptake and Antioxidant Responses in Two Contrasting Barley Landraces from MENA Region

Fatma Ezzahra Ben Azaiez [1,†], Sawsen Ayadi [1,†], Giorgia Capasso [2], Simone Landi [2], Valeria Paradisone [2], Salma Jallouli [1], Zied Hammami [3], Zoubeir Chamekh [4], Inès Zouari [1,5], Youssef Trifa [1] and Sergio Esposito [2,*]

1. Laboratory of Genetics and Cereal Breeding, National Agronomic Institute of Tunisia, Carthage University, 3 Avenue Charles Nicole, Tunis 1082, Tunisia; zahra.azaiez@gmail.com (F.E.B.A.); sawsen.ayadi@gmail.com (S.A.); salmajallouli@gmail.com (S.J.); ines.zouari28@gmail.com (I.Z.); youssef.trifa@gmail.com (Y.T.)
2. Department of Biology, University of Naples "Federico II", Via Cinthia, 80126 Napoli, Italy; giorgia.271291@gmail.com (G.C.); simone.landi@unina.it (S.L.); valeriaparadisone@libero.it (V.P.)
3. Crop Diversification and Genetics Section, International Center for Biosaline Agriculture (ICBA), Dubai P.O. Box 14660, UAE; z.hammami@biosaline.org.ae
4. National Institute of Agronomic Research of Tunisia INRAT, Carthage University, Rue Hédi Karray CP, Ariana 2049, Tunisia; zoubeirchamek@gmail.com
5. High Institute of Agronomy of Chott Mériem, Sousse University, Chott Mériem B.P. 47, 4042 Sousse, Tunisia

* Correspondence: sergio.esposito@unina.it

† These authors equally contributed, joint first authors.

Received: 31 July 2020; Accepted: 16 September 2020; Published: 19 September 2020

Abstract: The interaction between salinity and nitrogen metabolism has been investigated in two barley landraces, one tolerant ("100/1B") and one susceptible to salinity ("Barley medenine") from the Middle East and North Africa (MENA) region. Barley plants were exposed to 50 mM NaCl for 7 days; then, salinity was increased to 150 mM NaCl in the presence (10 mM) or limitation (1 mM) of ammonium as a nitrogen source. Upon salinity, "100/1B" was shown to support N assimilation by enhancing the glutamine synthetase (GS) and glutamine oxoglutarate aminotransferase (GOGAT) cycle under high N, and the stimulation of the glutamate dehydrogenase (GDH) pathway under low N treatment. In "Barley medenine", salinity reduced the GS/GOGAT cycle, and increased GDH activity. Upon salinity, Heat Shock Proteins 70 and PEPC remained unchanged in "100/1B", while they decreased in "Barley medenine". The tolerance degree is a determining factor in enzymes' occurrence and regulation: exposed to salinity, "100/1B" rapidly increased APX and PEPC activities, while this was delayed in "Barley medenine". Salinity increased cyt-G6PDH levels in "100/1B", while "Barley medenine" showed a decrease in G6PDH isoforms. Correlation analyses confirm GOGAT was related to G6PDH; GDH and APX with PEPC in "100/1B" under moderate salinity; severe salinity correlated GDH with G6PDH and PEPC. In "Barley medenine" under salinity, GOGAT was correlated with G6PDH, while APX showed a relation with PEPC. Therefore, specific enzymatic activities and occurrence can be used to determine stress responsiveness of different landraces. We suggest that the rapid increase in G6PDH, APX, and nitrogen assimilation enzymes activities represents an index of tolerance in "100/1B" and a stress symptom in "Barley medenine".

Keywords: salt stress; nitrogen metabolism; oxidative stress response; G6PDH; GDH; GS/GOGAT

---

1. Introduction

High salinity is one of the most widespread abiotic stresses affecting plant physiology, growth, and development [1–3]. Remarkable reduction in cereals productivity was recurrently reported in recent decades by abiotic constrains, namely salinity and drought [4–6]. Exposure to an excess of salts triggers different metabolic changes by modifying the balance between nutrient availability and plant requirements, and inducing associated constraints, namely osmotic, ionic, and oxidative stresses [7]. These conditions disturb cells water potential, disrupt nutrient availability caused by competitive uptake, and generate reactive oxygen species (ROS) [8].

One of the most important mechanisms conferring acclimation and tolerance of plants to abiotic stresses and salinity is the regulation of ROS levels. This stress response is regulated by a specific scavenging complex, composed by both enzymes and non-enzymatic antioxidant compounds [9–11]. Critical enzymes involved in ROS detoxification are ascorbate peroxidase (APX, E.C. 1.11.1.11), catalase (CAT, E.C. 1.11.1.6), superoxide dismutase (SOD, E.C. 1.15.1.1), and glutathione reductase (GR, E.C. 1.8.1.7) [10,12]. Furthermore, an assisting role as reductants' supplier has been recently proposed for glucose 6-phosphate dehydrogenase (G6PDH, E.C. 1.1.1.49), which provides additional NAPDH for scavenging enzymes upon abiotic stresses such as salinity and drought [13–16].

On the other hand, salinity affects nitrogen (N) metabolism inducing nutritional imbalances [8]. N is a crucial plant nutrient required for growth and development [17]. Therefore, the interaction between salinity and N metabolism is particularly complex, because of the antagonism in plant response under these combined interactions. N supply improves nutritional plant status and alleviates toxicities of abiotic stress [18]. In this context, crucial roles were played by enzymes such as phospho*enol*pyruvate carboxylase (PEPC, E.C. 4.1.1.32), which regulate the connection between carbon and N metabolism for the replacement of carbon skeletons and the replenishment of the tricarboxylic acid cycle; G6PDH plays a pivotal role in reductants provision also for enzymes involved in N metabolism and assimilation [19]. Furthermore, N availability showed contrasting roles in abiotic stress susceptibility/tolerance: deficiency of this nutrient could reduce water loss but N transporters, namely *At*NRT1.1 (*At*1g12110), *At*NRT1.8 (*At*4g21680), and CLCA (*At*5g40890), showed positive effects on plant response to abiotic stresses [20–22].

Nowadays, the enhancement of crop yields upon high saline environments led to a renewed interest for plant physiology researchers. Mediterranean regions, especially arid and semi-arid areas from Southern Europe and Northern Africa, were usually characterized by ecosystems subjected to multiple stress conditions characterized by contemporary onset of excess of salinity and/or water scarcity, heat, and nutrients [23]. Therefore, farmers throughout centuries have developed a number of traditional varieties. These genotypes, generally described as landraces, showed the ability to tolerate environmental changes maintaining unaltered yields [3,24]. In the Middle East and North Africa (MENA) region, barley represents a critical agronomic resource in semi-desert environments, especially in developing countries where this crop is a critical component of cereal rotation, providing a stable source to sustain smallholder farmers, and often replaces wheat or other cereals in more arid areas [25]. On the other hand, barley is the fourth most produced cereal in the world [26], representing one of the main sources of carbohydrates in developing countries, due to its natural stress tolerance, thus supporting small farmers in many arid areas [25,27,28]. Furthermore, barley is an excellent model organism to investigate abiotic and biotic stress resistance, being tolerant to different environmental stress [29].

It is worth pointing out that the identification of novel alleles QTL and peculiarities from unexploited genotypes is a central challenge for researchers, in order to improve crop productivity in vulnerable environments and provide new tools to increase crop yields [24,30]. For example, genes such as *Hs*CBL8 (calcium-sensor calcineurin B-like) and *Hs*CIPKs—identified in Tibetan genotypes of *Hordeum spontaneum*—were overexpressed in rice, contributing to an increased tolerance to salinity, drought, and heavy metals [5].

The aim of this paper is to investigate the effects of the interaction between salinity and N metabolism in two contrasting barley landraces. Our strategy is finalized to understand how the selected landraces respond to the detrimental effects induced by salt stress and how salinity regulates the N metabolism. To do this, biochemical activities of enzymes involved in responses to salinity and nitrogen assimilation have been studied. Our hypothesis is that landraces showing contrasting responses to salinity and nitrogen assimilation might reflect distinct regulations in their main metabolic pathways in order to adapt to environmental conditions. Therefore, specific enzymatic activities and protein occurrence can be identified as stress responsive sensors in different cereal genotypes; this possibility will be discussed under the light of the results obtained.

2. Materials and Methods

2.1. Plant Material and Growth Conditions

Two barley landraces, "100/1B" and "Barley medenine", which differ in their response to salinity, were chosen for this study. "100/1B" is a salt-tolerant landrace from Oman; the seeds were supplied by the Laboratory of Genetics and Cereal breeding—INAT, University of Tunis. "Barley medenine" is a salt-susceptible landrace selected by International Center for Agricultural Research in the Dry Areas (ICARDA) for "PRF sud project" in 2009; the seeds were obtained from the National Research Institute for Rural Engineering, Water and Forestry, Ariana, Tunisia in collaboration with the International Center for Biosaline Agriculture, Dubai, United Arab Emirates. Soil salinity in the origin area of "100/1B" and "Barley medenine" is 2 and 1.3 (dS m^{-1}), respectively [25]. Seeds were germinated for 7 days in the dark on moistened paper, then seedlings were grown in hydroponic solution in darkened plastic bottles at 20 °C, at 60–80% relative humidity, under a 16h-light/8h-dark regime, with approximately 180 µmol photons m^{-2} s^{-1}. The composition of the medium (modified Hoagland solution), continuously bubbled with air, has been previously described [23]. Hydroponic solutions were controlled for pH (6–6.5) and adjusted daily.

2.2. Nitrogen and Salt Treatments

After 7 days of hydroponic culture and before starting any treatments, 10 seedlings from each landrace were collected for analysis (-N/-NaCl). Then, plants were separated into 4 groups: Two control groups were grown without NaCl, upon low or high N supply with 1 and 10 mM of $(NH_4)_2SO_4$, respectively. The other two groups were grown under the same concentration of $(NH_4)_2SO_4$ and subjected to moderate salt stress (50 mM NaCl), which was increased to severe stress (150 mM NaCl) after 7 days. Seedlings were collected 3 and 7 days after moderate NaCl stress (50 mM NaCl); further samples were collected after 1 and 3 days under imposing severe stress (150 mM NaCl). This experimental scheme was designed to simulate field conditions where salinity sequentially increases from low to high levels. These degrees of salinity (50 and 150 mM NaCl) allow investigating of both tolerance and adaptation to this abiotic stress. Sampling days have been selected based on previous biochemical investigation on similar enzymes upon abiotic stress [13,23], which indicated 3 and 7 as the most stressful days. The whole experimental design is depicted in Figure 1.

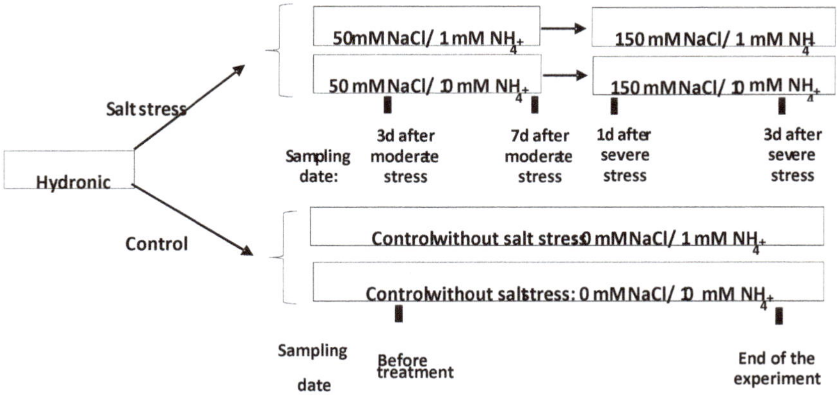

Figure 1. Scheme of the adopted experimental strategy. Stressed plants treated by low and high N were collected after 3 and 7 days of moderate stress (50 mM NaCl); then, a severe stress (150 mM NaCl) was applied and plants were collected after further 1 and 3 days. Control plants (without NaCl) were collected before any treatment, and after N application at the end of experiment.

2.3. Enzyme Extraction and Activities' Determination of NADH-GOGAT and NADH-GDH

NADH-GOGAT and NADH-GDH were extracted by grounding 300 mg of barley leaves in the same buffer containing 1 mL of 100 mM KH_2PO_4 buffer (pH 7.5), 2 mM EDTA, 2 mM dithiothreitol (DTT), and plant-specific proteases inhibitor cocktail (Sigma-Aldrich, St. Louis, USA). GOGAT and GDH assays were measured by following the oxidation of NADH at 340 nm for 10 min as described by Groat and Vance [31] and Singh and Srivastava [32], respectively. Specific activity was expressed as nmol NADH mg^{-1} protein min^{-1}. The reaction mixture for NADH-GOGAT assay contained 100 mM KH_2PO_4 buffer (pH 7.5), 18.75 mM2-oxoglutarate,15 mM aminooxyacetate, 0.15 mM NADH, 7.5 mM L-glutamine, and extract. GDH activity was measured using 100 mM KH_2PO_4 buffer (pH 7.5), 200 mM $(NH_4)_2SO_4$, 0.15 mM NADH, 2.5 mM 2-oxoglutarate, and extract. Enzymatic activities were determined spectrophotometrically using a Cary 60 spectrophotometer (Agilent Technologies, Santa Clara, CA, USA) at 25 °C. For all enzymatic assays, the protein concentration in the samples was measured using the Bradford method.

2.4. Enzyme Extraction and Activities Determination of G6PDH, APX, and PEPC

G6PDH, APX, and PEPC activities were extracted by grounding 300 mg of barley leaves in the same extraction buffer containing 50 mM Tris-HCl, a pH 8.0, 5 mM $MgCl_2$, 4 mM EDTA, 10% glycerol, and plant-specific proteases inhibitor cocktail (Sigma-Aldrich, St. Louis, USA). (1) G6PDH was assayed as previously described by Castiglia et al. [33]. $NADP^+$ was determined at 340 nm for 2–10 min. The reaction mixture contained 50 mM Tris-HCl buffer (pH 8.0), 5 mM $MgCl_2$, 0.15 mM $NADP^+$, 3 mM G6P, and 100 µL of extract. Enzyme activity was expressed as nmol-reduced $NADP^+$ min^{-1} mg^{-1} protein. (2) APX activity was determined following the oxidation of ascorbic acid at 290 nm for 5–10 min. APX assay was carried out as described by Nakano and Asada [34]; the assay mixture contained 30 mM potassium phosphate buffer (pH 7.5), 1 mM EDTA, pure hydrogen peroxide H_2O_2, 5 mM ascorbic acid, and extract. The enzyme activity was expressed as µmol of ascorbic acid min^{-1} mg^{-1} protein. (3) PEPC was assayed by following the method of Fontaine et al. [35] with some modifications. For PEPC determination, a reaction coupled to malate dehydrogenase (MDH) is required. PEPC activity was measured by following NADH oxidation at 340 nm for 10 min. The reaction was assayed at 25 °C, in a mixture containing 100 mM Tris–HCl (pH 8.2), 20 mM $MgCl_2$, 100 mM $NaHCO_3$, 0.2 mM NADH, MDH (3µL/mL Tris HCl pH 8.2); 10 mM PEP was used to start the reaction; a control without PEP was prepared for each assay. Enzymatic activities were determined spectrophotometrically using a Cary

60 spectrophotometer (Agilent Technologies, Santa Clara, USA) at 25 °C. For all enzymatic assays, the protein concentration in the samples was measured using the Bradford method.

2.5. Western Blotting

For Western blotting analysis, proteins were extracted as described previously (Section 2.4) and then, separated using electrophoresis SDS-PAGE. Polypeptides were transferred on a Hybond membrane (Ge Healthcare, Chicago, IL, USA) using a Transblot turbo transfer system (Biorad, Hercules, CA, USA). The membrane was incubated with primary G6PDH antibody raised against potato cytosolic (cyt), chloroplastic (P1), and plastidial (P2) G6PDH isoforms [36]; barley Fd-GOGAT [37]; HSP70 (cytosolic and chloroplast—Agrisera, Vännäs, Sweden) and PEPC from antiserum from *Amarantus edulis*. After incubation of the membrane with secondary antibodies, cross-reacting polypeptides were identified by enhanced chemiluminescence using the Western Bright Quantum kit (Advansta—Aurogene, Roma, Italy). Images were acquired by the BioRad Chemidoc system/Quantity One software (BioRad, Hercules, CA, USA).

2.6. Statistical Analyses

R software (R- 64 3.6.1) was used for statistical analysis. The significance of saline stress, N treatment, and duration of salt stress single effects and their interaction were analyzed using a linear variance analysis model (ANOVA $p < 0.05$). Then, means were compared using Duncan's test. Correlation analysis using Pearson's parametric correlation test was performed to determine the correlation coefficients for all possible pairs of columns in the data table.

3. Results

3.1. Contrasting Barley Landraces Showed Diversified Stress and Antioxidant Responses

The ability of two contrasting barley landraces to respond to a combination of salinity and nitrogen starvation stresses was analyzed by biochemical investigation on N metabolism and antioxidant enzymes.

The different behavior in the salt stress condition and N concentrations is clearly depicted in Figure 2, which shows leaves from these two landraces under experimental conditions. "100/1B" leaves showed a reduced damage induced by salinity, while the "Barley medenine" landrace showed strongly stressed leaves, both upon high and low N.

Figure 2. Effects of salinity and N supply on "100/1B" and "Barley medenine" leaves.

The salt tolerant landrace "100/1B" showed a reduced involvement of APX upon salinity under low N conditions (Figure 3). A rapid APX activity increase of about 50%, related to severe salt stress, was observed in "100/1B" plants grown in high N, whereas under low conditions, plants showed an APX activity essentially unchanged by salinity. The presence of an adequate source of N (10 mM NH_4) induced an increase in APX activity upon 150 mM of NaCl. The involvement of G6PDH in "100/1B" appeared to be mainly related to N assimilation by a significant increase upon the low N concentration condition compared with the high N concentration. On the other hand, susceptible "Barley medenine" showed increased activities of APX and G6PDH upon salinity both in high N and low N concentration. It should be underlined that the sensitive landrace "Barley medenine" showed higher G6PDH activities than "100/1B" when expressed on a mg^{-1} prot. The maximum activity in low N plants was observed after 24 h of 150 mM NaCl exposure, while high N induced a marked increase up to 216 nmol min^{-1} mg^{-1} prot after 3 days of severe stress.

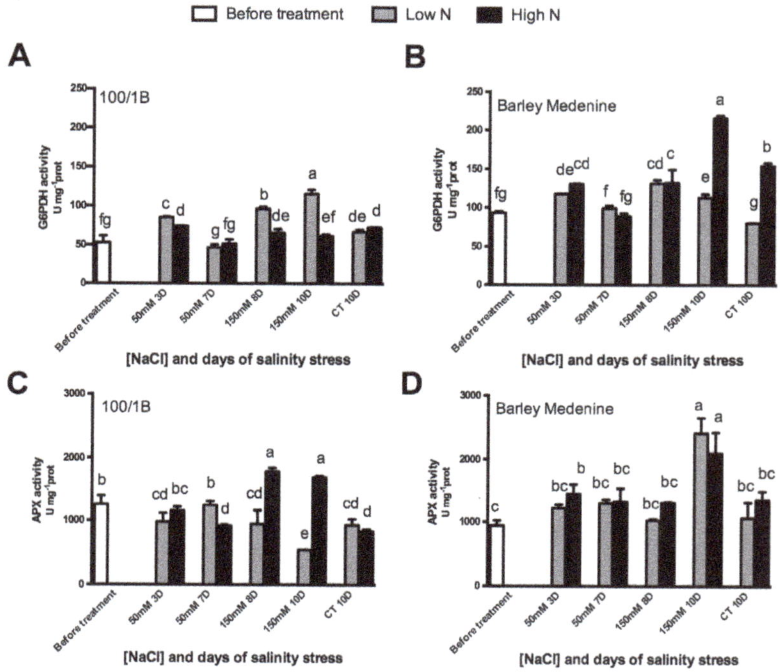

Figure 3. Effects of salinity and N concentration on G6PDH and APX enzymatic activities in barley plants growth in hydroponic system. Levels of low N concentrations are in grey bars; high N concentrations are in black bars. Legend: (BT), before treatment; (50 mM 3D), 3 days of moderate stress; (50 mM 7D), 7 days of moderate stress; (150 mM 8D), 7 days in 50 mM NaCl and 1 day in 150 mM NaCl (severe stress); (150 mM 10d) 7 days in 50 mM NaCl and 3 days in 150 mM NaCl (severe stress); (CT 10D) control treatment for 10 days. (**A**)100/1B, G6PDH activity; (**B**): Barley Medenine, G6PDH activity; (**C**) 100/1B, APX activity; (**D**): Barley Medenine, APX activity. Letters indicate significant differences between different treatments. ($p < 0.05$, ANOVA: Duncan's test).

In order to investigate the contribution of different G6PDH isoforms, antibodies constructed against cytosolic (Cyt-G6PDH), chloroplastic (P1-G6PDH), and plastidial (P2-G6PDH) were used by Western blotting (WB—Figure 4). The results showed an increase in cyt-G6PDH abundance upon 150 mM NaCl in high N "100/1B" plants, while an evident decrease was observed in low N plants (Figure 3). P1-G6PDH remained unaffected at low N regimes in "100/1B" and decreased under high N under both moderate and high salinity.

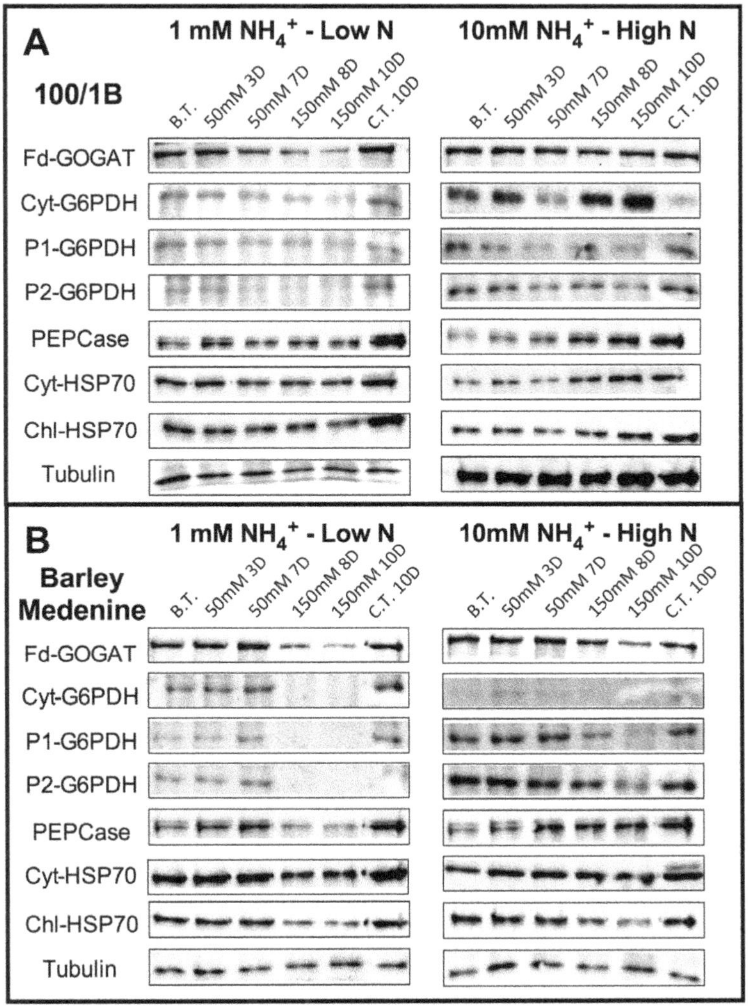

Figure 4. Western blotting of leaves of "100/1B" (**A**) and "Barley medenine" (**B**) grown under low N (1 mM NH_4^+) and high N (10 mM NH_4^+) using antisera Fd-GOGAT, cyt-G6PDH, P1-G6PDH, P2-G6PDH, Chl-HSP70, Cyt-HSP70, PEPCase, and Tubulin (as control for equal loading). Legend: (BT), before treatment; (50 mM 3D), 3 days of moderate stress; (50 mM 7D), 7 days of moderate stress; (150 mM 8D), 7 days in 50 mM NaCl and 1 day in 150 mM NaCl (severe stress); (150 mM 10d) 7 days in 50 mM NaCl and 3 days in 150 mM NaCl (severe stress); (CT 10D) control treatment for 10 days. Images are representative of two or three WB from different experiments.

Plastidic P2-G6PDH decreased in "100/1B" under low N and salinity and remained unaffected by salt stress under high N. Similar behaviors were observed in "Barley medenine". Finally, the degree of the stress conditions was monitored by the HSP70s occurrence (Figure 4). Upon moderate stress, cyt-HSP70 protein occurrence remained unchanged in low N conditions in "100/1B", while high salinity induced a sensible increase in cyt-HSP70 upon high N. By contrast, "Barley medenine" showed a high stability in cyt-HSP70 abundance upon high N independently of NaCl concentration. Chl-HSP70 showed no differences in "100/1B" in each analyzed condition with the exception of an increase upon high salinity and high N after 10D. In "Barley medenine", Chl-Hsp70 showed a decrease upon high salinity, independently by N supply.

3.2. *N Metabolism Modifications upon Salinity: The Role of GOGAT and GDH*

Salinity did not induce substantial changes in NADH-GOGAT activity both under high and low N in tolerant "100/1B" (Figure 5). The only exception was represented by the higher stress point (10 days, 150 mM NaCl, 10 mM NH_4), where the NADH-GOGAT activity showed a 22% increase. Furthermore, GDH activity showed an 24% increase upon severe salt stress in low N after 8 days. On the other hand, in "Barley medenine", NADH-GOGAT showed reduced differences between controls and salt stressed conditions both upon low and high N concentrations. Contrarily, an evident increase in GDH activity was observed in the susceptible "Barley medenine" in both low N and high N conditions; this increase was more evident under low N. Fd-GOGAT occurrence was investigated using specific antibodies. As shown in Figure 4, "100/1B" reported a decrease in Fd-GOGAT occurrence in low N grown plants under salinity, while under high N conditions, no differences were observed upon salt stress. Similar behavior in "Barley medenine", where only a more severe decrease in Fd-GOGAT occurrence, was observed under 150 mM NaCl.

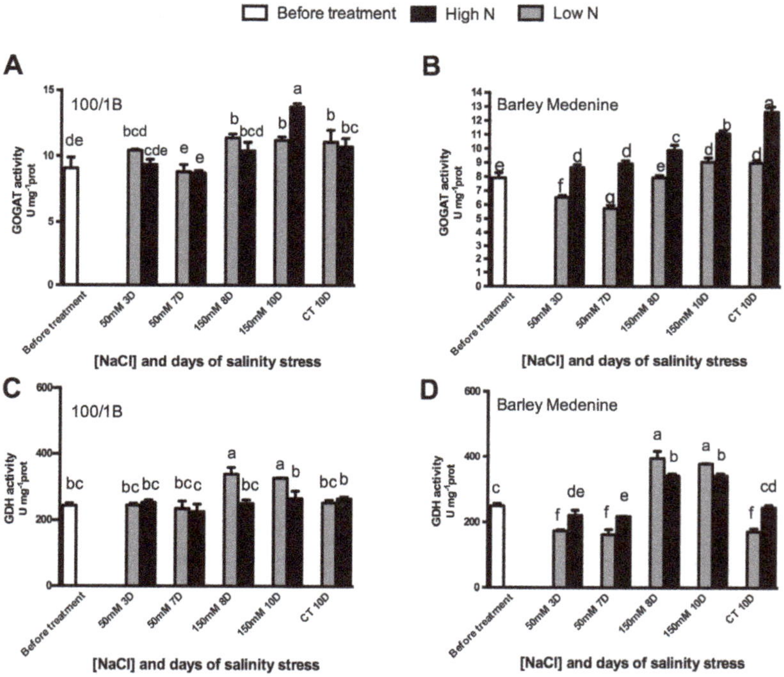

Figure 5. Effects of salinity and N concentration on NADH-GOGAT and GDH enzymatic activities in barley plants growth in hydroponic system. Levels of low N concentrations are in grey bars; high N concentrations are in black bars. Legend: (BT), before treatment; (50 mM 3D), 3 days of moderate stress; (50 mM 7D), 7 days of moderate stress; (150 mM 8D), 7 days in 50 mM NaCl and 1 day in 150 mM NaCl (severe stress); (150 mM 10d) 7 days in 50 mM NaCl and 3 days in 150 mM NaCl (severe stress); (CT 10D) control treatment for 10 days. (**A**)100/1B, GOGAT activity; (**B**): Barley Medenine, GOGAT activity; (**C**) 100/1B, GDH activity; (**D**): Barley Medenine, GDH activity.Letters indicate significant differences between different treatments. ($p < 0.05$, ANOVA: Duncan's test).

3.3. Change in the Activity of PEPcase

As shown in Figure 6, the tolerant "100/1B" had an increase in PEPC activity upon N supply and NaCl. In the susceptible "Barley medenine", PEPC activity under high N was relatively stable throughout salt stress but lower than control. A 3-d response increasing PEPC activity was observed in low N under both moderate and severe salinity. Western blotting analysis showed an unchanged PEPC occurrence in low N "100/1B" plants and an increase in high N plants under severe salinity (Figure 4). On the contrary, in low N "Barley medenine" plants, PEPC occurrence was reduced under high salinity, while a high N supply maintained a relative stability in PEPC abundance, even under severe stress (Figure 4).

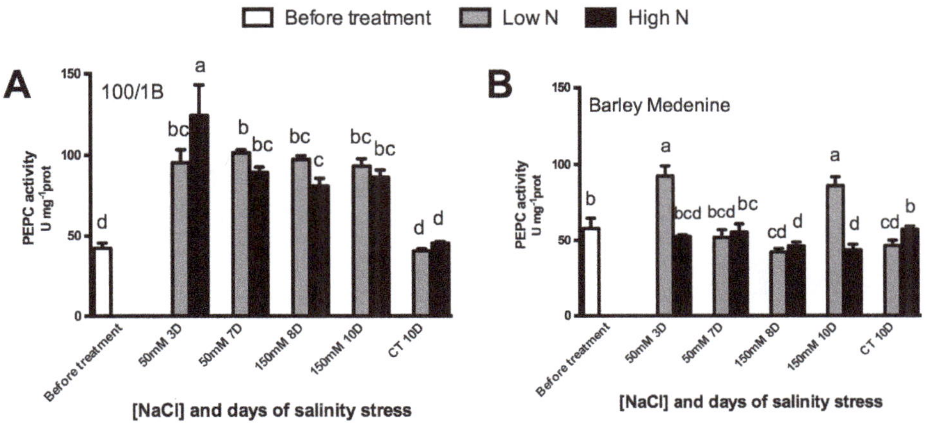

Figure 6. Effects of salinity and N concentration on PEP carboxylase (PEPC) enzymatic activity in barley plants growth in hydroponic system. Levels of low N concentrations are in grey bars; high N concentrations are in black bars. Legend: (BT), before treatment; (50 mM 3D), 3 days of moderate stress; (50 mM 7D), 7 days of moderate stress; (150 mM 8D), 7 days in 50 mM NaCl and 1 day in 150 mM NaCl (severe stress); (150 mM 10d) 7 days in 50 mM NaCl and 3 days in 150 mM NaCl (severe stress); (CT 10D) control treatment for 10 days. (**A**)100/1B, PEPC activity; (**B**): Barley Medenine, PEPC activity; Letters indicate significant differences between different treatments. ($p < 0.05$, ANOVA: Duncan's test).

3.4. Correlation between Suppliers of Reducing Power (G6PDH and PEPC) and Different Enzymes Involved in N Metabolism or Defense against Stress

Both G6PDH and PEPC participate, directly or indirectly, in N assimilation, by providing NAD(P)H and carbon skeletons supply, respectively; therefore, under stress conditions, a diversion of metabolism to ROS detoxification can be possible. To achieve a deeper understanding of G6PDH and PEPC effects on other enzymes, the correlations between different enzymes have been evaluated. In the "100/1B" landrace cultivated under 50 mM NaCl, GOGAT was strongly and positively correlated with G6PDH ($r = 0.84$, $p = 0.011$) (Table 1). GDH and APX were moderately correlated with PEPC activity ($r = 0.54$, $p = 0.176$ and $r = 0.52$, $p = 0.195$, respectively). Under severe stress, GDH was strongly and positively correlated with both G6PDH and PEPC ($r = 0.87$, $p = 0.005$ and $r = 0.9$, $p = 0.002$, respectively) and inversely correlated with APX ($r = -0.85$, $p = 0.008$) (Table 1). On the other hand, "Barley medenine" showed a positive correlation of GOGAT with G6PDH ($r = 0.76$, $p = 0.027$), and of APX with PEPC ($r = 0.7$, $p = 0.058$) upon severe salinity (Table 2).

Table 1. Coefficients of correlation (*r*) between different enzymes measured in "100/1B" cultivated under 50 mM NaCl (moderate salt stress—lower diagonal data) or 150 mM NaCl (severe salt stress—upper diagonal data). Numbers in brackets indicate *p* values.

"100/1B"	GOGAT	GDH	G6PDH	APX	PEPC
GOGAT		0.53 (0.173)	0.84 (0.011)	−0.28 (0.5)	−0.07 (0.875)
GDH	−0.11 (0.799)		0.4 (0.345)	0.21 (0.622)	0.54 (0.176)
G6PDH	−0.35 (0.389)	0.87 (0.005)		−0.24 (0.564)	0.25 (0.558)
APX	0.26(0.539)	−0.85 (0.008)	−0.97 (0.001)		0.52 (0.195)
PEPC	0.03 (0.942)	0.90 (0.002)	0.76 (0.030)	−0.78 (0.021)	

Table 2. Coefficients of correlation (*r*) between different enzymes measured in "Barley medenine" cultivated under 50 mM NaCl (moderate salt stress—lower diagonal data) or 150 mM NaCl (severe salt stress—upper diagonal data). Numbers in brackets indicate *p* values.

"Barley Medenine"	GOGAT	GDH	G6PDH	APX	PEPC
GOGAT		0.92 (0.001)	0.04 (0.916)	0.33 (0.430)	−0.35 (0.402)
GDH	−0.84 (0.009)		0.14 (0.737)	0.55 (0.155)	−0.35 (0.389)
G6PDH	0.76 (0.027)	−0.57 (0.141)		0.20 (0.642)	0.26 (0.534)
APX	0.48 (0.229)	−0.17 (0.684)	0.19 (0.645)		−0.39 (0.336)
PEPC	−0.16 (0.705)	0.23 (0.576)	−0.50 (0.206)	0.7 (0.058)	

4. Discussions

In this work, we investigated the biochemical modifications involved in the response to salt stress and nitrogen deprivation in two contrasting landraces of barley, one salt tolerant "100/1B", a landrace from Oman; and one salt susceptible, "Barley medenine", a landrace from ICARDA/Tunis [25]. Recently, these landraces—together with other 38 genotypes from North Africa—were analyzed in crop fields for their potential in salinity tolerance in the Middle East and North Africa (MENA) region. These genotypes were grouped in four different clusters, where "100/1B" belong to the best salt tolerant one. This investigation highlighted K^+, Ca^{2+} content, and relative water content as the most important traits explaining barley yield reduction ratio under salinity stress [25]. It is worth pointing out that even if agronomic data are available, no investigation about antioxidant response and N metabolism was analyzed in landraces from this germplasm collection. In this context, our results showed different behaviors in enzymatic activities and protein occurrence in the selected landraces. Heat shock proteins (HSP70s) are crucial regulators of response against abiotic stress [23]. The maintenance or the increase in HSP70s isoforms occurrence in "100/1B" and its decrease in "Barley medenine" in response to salinity could explain the tolerance of the first and susceptibility of the second.

High levels of antioxidant enzymes are generally considered as a biochemical index of plant tolerance against salinity [38]. Contrarily, a number of authors consider scavenging enzymes as a stress symptom, suggesting their increase a consequence of oxidative damage [10,39]. Our results suggested a long-term activation of APX and G6PDH in salt-susceptible "Barley medenine", thus highlighting a severe oxidative stress induced by stress. By contrast, the delayed response of "Barley medenine" in the increased APX and PEPC activities, observed only 3d after salt stress, and evidently upon low N, may suggest that high levels of APX and PEPC activities can be considered as a symptom of oxidative stress in this sensitive landrace. Similarly, Maksimović et al. [39] investigated the role as a stress sensor of APX, by analyzing the increase in enzymatic activity in barley plants under oxidative stress. Furthermore, our results confirm the role of N metabolism in maintaining metabolic functions under salt stress [40]. In accordance with our findings, previous studies showed that different plants respond to salinity by increasing APX and PEPC activities [41,42]. On the other hand, N-fed "100/1B" showed a precocious increase in APX and PEPC, indicating their central role in the ability of this landrace to tolerate high salinity. Interestingly, "100/1B" G6PDH activity showed reduced efforts upon N-fed salinity, reporting a probable preferred activation to sustain N assimilation in salt-tolerant landraces. G6PDH plays an important role in sustaining the redox state of plant cells through supplying the

NADPH reductants needed for ROS-detoxification mechanisms [19]. The G6PDH increase in both landraces reflects the oxidative damage produced by salinity and the high requirement for reductants to trigger a defense system against ROS. Therefore, the higher occurrence of cyt-G6PDH in "100/1B" in response to salinity and N availability can justify the tolerance of this landrace.

"100/1B" and "Barley medenine" behaviors suggested different N assimilation and detoxification mechanisms upon salinity. NADH-GOGAT and GDH regulations were related to N availability in the tolerant "100/1B", which was least affected by salinity compared to the sensitive "Barley medenine".

Tolerant "100/1B" stimulated the GS/GOGAT pathway under sufficient N nutrition and salinity. This result is in accordance with previous studies on chickpea (*Cicer arietinum* L.), soybean (*Glycine max* L.), and halophyte species, [43,44] thus it could allow assimilation of the excess of NH_4^+ accumulated under salinity.

"100/1B" subjected to salinity is able to switch the main GS/GOGAT cycle to GDH assimilation under low N, possibly to limit energy-consuming metabolism. In fact, despite its high efficiency, the GS/GOGAT pathway is an energy consuming process when compared to GDH assimilation, by requiring an extra ATP molecule [45]. Through this mechanism, plants can save energy for the defense system and adaptation to salinity

In "Barley medenine", GOGAT activity significantly decreased upon salinity. Previous studies on different crops showed salt stress inhibition of the GS/GOGAT cycle, induced by both salt ions damaging and/or substrate limitation [46,47].

The noticeable NADH-GDH increase after 24 h of severe salinity in both landraces reflects a rapid leaf protein catabolism induced by salt and leading to high intracellular NH_4^+ levels [8]. The alternative N assimilation pathway (GDH) activation by salinity is strongly required to protect plants against ammonium toxicity, fill the glutamate pool, and allow the synthesis of stress-protective metabolites [48]. The GDH enzyme presents low affinity for NH_4^+, and it can be stimulated to synthesize glutamate only when NH_4^+ concentration >1 mM, [46]; therefore, the fact that GDH was stimulated under low N (1 mM NH_4^+) supports the hypothesis of intracellular ammonia accumulation under salt stress.

Taken together, the tolerant "100/1B" seems to have established a strong regulation mechanism under salinity, by maintaining an efficient energy-saving N metabolism, delivering carbon sources, synthesizing oxoglutarate, incorporating NH_4^+ to synthesize amino acids required for cell osmotic adjustments, and rapidly triggering a defense system after stress exposure. The participation of G6PDH in the NADPH recycling machinery and the role of PEPC in furnishing carbon skeletons during ammonium assimilation may indicate a correlation between different enzymes, in its turn depicting different defense scenarios of the two landraces upon salinity. Under moderate stress, the positive correlation between G6PDH and GOGAT suggests that the reductants produced via G6PDH can be used in N metabolism in the GS/GOGAT cycle. The positive correlation of PEPC with both APX and GDH indicates PEPC involvement in both the ROS scavenging system and N assimilation via GDH pathway. Under severe salinity, based on the strong correlation between GDH with both G6PDH and PEPC, we suggest that severe salinity leads to a diversion of reductants supplied by G6PDH and PEPC to the alternative N assimilation pathway (GDH). Based on the observed correlation under high salinity in the sensitive "Barley medenine", we suggest the involvement of G6PDH in N assimilation via the GS/GOGAT cycle and the participation of PEPC [19] in response against salinity as well.

5. Conclusions

The results presented here support the hypothesis that activities and occurrence of specific enzymes can be used as a useful index of metabolism in landraces. Particularly, the activities of GS/GOGAT and GDH represent a fast and reliable tool to identify the rapid response of barley landraces to salinity. Furthermore, biochemical investigations on scavenging and reductants' supplier enzymes represent crucial aspects to analyze genotypes, ecotypes, and landraces with different responses in abiotic stress conditions. Finally, our findings proved that the tolerance of "100/1B" and the sensibility of "Barley medenine" to salinity, which have been previously selected on agronomic parameters [26], are based on different biochemical and metabolic regulation of enzymes of the central metabolism. This study demonstrated that specific enzymatic activities and protein occurrence can be utilized as tools and/or sensors to investigate plant salt stress response. Further investigations will be necessary to clarify the possible molecular pathways of regulation which coordinate the basal metabolism fluxes in adverse environments.

Author Contributions: F.E.B.A., Conceptualization, methodology, and writing; S.A., Conceptualization, methodology, and writing; G.C. biochemical and Western blotting analysis; S.L., biochemical analysis and wrote and amended the paper; V.P., biochemical analysis; S.J., data curation; Z.H., validation and statistical analysis; Z.C., data analysis; I.Z., review and editing; Y.T., supervision; S.E., coordinated the research project, analyzed the data, and wrote and amended the paper. All authors have read and agreed to the published version of the manuscript.

Funding: This work was supported by grants from the Tunisian Ministry of Agriculture, Water Resources and Fisheries and the Tunisian Ministry of Higher Education, Scientific Research. Giorgia Capasso acknowledges Project "Garanzia Giovani" 2018 by Regione Campania, DD566/2014/23599. Simone Landi acknowledges Project PON-AIM LINEA 1 "MOBILITA' DEI RICERCATORI" Codice Attività = AIM1823408_2.

Acknowledgments: The authors would like to thank the laboratory of Genetics and Cereal breeding (LR14GAC01)—INAT; and the National Research Institute for Rural Engineering, Water and Forestry (Ariana, Tunisia) in collaboration with the International Center for Biosaline Agriculture (Dubai, United Arab Emirates), for the generous gift of barley seeds. The authors thank Antje von Schaewen very much (Muenster—Germany) for the generous gift of potato G6PDH antibodies (Wendt et al. 2000). The authors would like to thank Robert P. Walker, (Sheffield—UK) for the kind gift of PEPCase antibodies from *Amaranthus* edulis and A. Marquez (Univ. of Sevilla, Spain) for the generous gift of barley Fd-GOGAT antibodies, respectively.

Conflicts of Interest: The authors declare no conflict of interest.

References

1. Martinez-Atienza, J.; Jiang, X.; Garciadeblas, B.; Mendoza, I.; Zhu, J.-K.; Pardo, J.M.; Quintero, F.J. Conservation of the salt overly sensitive pathway in rice. *Plant Physiol.* **2007**, *143*, 1001–1012. [CrossRef] [PubMed]
2. Wang, H.; Zhang, M.; Yang, C.; Shi, D.; Liu, B.; Lin, X.; Yang, C. Effects of salt stress on ion balance and nitrogen metabolism of old and young leaves in rice (*Oryza sativa* L.). *BMC Plant Biol.* **2012**, *12*, 194. [CrossRef] [PubMed]
3. Ruggiero, A.; Landi, S.; Punzo, P.; Possenti, M.; Van Oosten, M.J.; Costa, A.; Morelli, G.; Maggio, A.; Grillo, S.; Batelli, G. Salinity and ABA Seed Responses in Pepper: Expression and Interaction of ABA Core Signaling Components. *Front. Plant Sci.* **2019**, *10*, 304. [CrossRef]
4. Lobell, D.B.; Schlenker, W.; Costa-Roberts, J. Climate trends and global crop production since 1980. *Science* **2011**, *333*, 616–620. [CrossRef] [PubMed]
5. Guo, W.; Chen, T.; Hussain, N.; Zhang, G.; Jiang, L. Characterization of Salinity Tolerance of Transgenic Rice Lines Harboring HsCBL8 of Wild Barley (*Hordeum spontanum*) Line from Qinghai-Tibet Plateau. *Front. Plant Sci.* **2016**, *7*, 1678. [CrossRef]
6. Halford, N.G.; Curtis, T.Y.; Chen, Z.; Huang, J. Effects of abiotic stress and crop management on cereal grain composition: Implications for food quality and safety. *J. Exp. Bot.* **2015**, *66*, 1145–1156. [CrossRef]
7. Bartels, D.; Sunkar, R. Drought and salt tolerance in plants. *Crit. Rev. Plant Sci.* **2005**, *24*, 23–58. [CrossRef]
8. Ashraf, M.; Shahzad, S.M.; Imtiaz, M.; Rizwan, M.S. Salinity effects on nitrogen metabolism in plants–focusing on the activities of nitrogen metabolizing enzymes: A review. *J. Plant Nutr.* **2018**, *41*, 1065–1081. [CrossRef]
9. Mittler, R.; Vanderauwera, S.; Gollery, M.; Van Breusegem, F. Reactive oxygen gene network of plants. *Trends Plant Sci.* **2004**, *9*, 490–498. [CrossRef]
10. Gill, S.S.; Tuteja, N. Reactive oxygen species and antioxidant machinery in abiotic stress tolerance in crop plants. *Plant Physiol. Biochem.* **2010**, *48*, 909–930. [CrossRef]
11. Jallouli, S.; Ayadi, S.; Landi, S.; Capasso, G.; Santini, G.; Chamekh, Z.; Zouari, I.; Ben Azaiez, F.E.; Trifa, Y.; Esposito, S. Physiological and Molecular Osmotic Stress Responses in Three Durum Wheat (*Triticum Turgidum ssp Durum*) Genotypes. *Agronomy* **2019**, *9*, 550. [CrossRef]

12. Caverzan, A.; Passaia, G.; Rosa, S.B.; Ribeiro, C.W.; Lazzarotto, F.; Margis-Pinheiro, M. Plant responses to stresses: Role of ascorbate peroxidase in the antioxidant protection. *Genet. Mol. Biol.* **2012**, *35*, 1011–1019. [CrossRef] [PubMed]
13. Landi, S.; De Lillo, A.; Nurcato, R.; Grillo, S.; Esposito, S. In-field study on traditional Italian tomato landraces: The constitutive activation of the ROS scavenging machinery reduces effects of drought stress. *Plant Physiol. Biochem.* **2017**, *118*, 150–160. [CrossRef] [PubMed]
14. Wang, H.; Yang, L.; Li, Y.; Hou, J.; Huang, J.; Liang, W. Involvement of ABA- and H2O2-dependent cytosolic glucose-6-phosphate dehydrogenase in maintaining redox homeostasis in soybean roots under drought stress. *Plant Physiol. Biochem.* **2016**, *107*, 126–136. [CrossRef]
15. Lentini, M.; De Lillo, A.; Paradisone, V.; Liberti, D.; Landi, S.; Esposito, S. Early responses to Cadmium exposure in barley plants: Effects on biometric and physiological parameters. *Acta Physiol. Plant.* **2018**, *40*, 178. [CrossRef]
16. Zhao, C.; Wang, X.; Wang, X.; Wu, K.; Li, P.; Chang, N.; Wang, J.; Wang, F.; Li, J.; Bi, Y. Glucose-6-phosphate dehydrogenase and alternative oxidase are involved in the cross tolerance of highland barley to salt stress and UV-B radiation. *J. Plant Physiol.* **2015**, *181*, 83–95. [CrossRef]
17. Arghavani, M.; Zaeimzadeh, A.; Savadkoohi, S.; Samiei, L. Salinity tolerance of Kentucky bluegrass as affected by nitrogen fertilization. *J. Agric. Sci. Technol.* **2017**, *19*, 173–183.
18. Singh, M.; Singh, V.P.; Prasad, S.M. Responses of photosynthesis, nitrogen and proline metabolism to salinity stress in *Solanum lycopersicum* under different levels of nitrogen supplementation. *Plant Physiol. Biochem.* **2016**, *109*, 72–83. [CrossRef]
19. Esposito, S. Nitrogen assimilation, abiotic stress and glucose 6-phosphate dehydrogenase: The full circle of reductants. *Plants* **2016**, *5*, 24. [CrossRef]
20. De Angeli, A.; Monachello, D.; Ephritikhine, G.; Frachisse, J.-M.; Thomine, S.; Gambale, F.; Barbier-Brygoo, H. The nitrate/proton antiporter *At*CLCa mediates nitrate accumulation in plant vacuoles. *Nature* **2016**, *442*, 939–942. [CrossRef]
21. Álvarez-Aragón, R.; Rodríguez-Navarro, A. Nitrate-dependent shoot sodium accumulation and osmotic functions of sodium in Arabidopsis under saline conditions. *Plant J.* **2017**, *91*, 208–219. [CrossRef] [PubMed]
22. Landi, S.; Esposito, S. Nitrate Uptake Affects Cell Wall Synthesis and Modeling. *Front. Plant Sci.* **2017**, *8*, 1376. [CrossRef] [PubMed]
23. Landi, S.; Capasso, G.; Ben Azaiez, F.E.; Jallouli, S.; Ayadi, S.; Trifa, Y.; Esposito, S. Different roles of heat shock proteins (70 kDa) during abiotic stresses in barley (*Hordeum vulgare*) genotypes. *Plants* **2019**, *8*, 248. [CrossRef] [PubMed]
24. Dwivedi, S.L.; Ceccarelli, S.; Blair, M.W.; Upadhyaya, H.D.; Are, A.K.; Ortiz, R. Landrace germplasm for improving yield and abiotic stress adaptation. *Trends Plant Sci.* **2016**, *21*, 31–42. [CrossRef] [PubMed]
25. Hammami, Z.; Gauffreteau, A.; BelhajFraj, M.; Sahli, A.; Jeuffroy, M.H.; Rezgui, S.; Bergaoui, K.; McDonnell, R.; Trifa, Y. Predicting yield reduction in improved barley (*Hordeum vulgare* L.) varieties and landraces under salinity using selected tolerance traits. *Field Crop. Res.* **2017**, *211*, 10–18. [CrossRef]
26. FAOstat. Statistical Database 2017. Available online: http://www.fao.org/faostat/en/#home (accessed on 2 July 2020).
27. Shen, Q.; Fu, L.; Dai, F.; Jiang, L.; Zhang, G.; Wu, D. Multi-omics analysis reveals molecular mechanisms of shoot adaption to salt stress in Tibetan wild barley. *BMC Genom.* **2016**, *17*, 889. [CrossRef] [PubMed]
28. Lee, S.; Lee, T.; Yang, S.; Lee, Y. BarleyNet: A Network-Based Functional Omics Analysis Server for Cultivated Barley, *Hordeum vulgare* L. *Front. Plant Sci.* **2020**, *11*, 98. [CrossRef]
29. Gürel, F.; Øztürk, Z.N.; Uçarlı, C.; Rosellini, D. Barley genes as tools to confer abiotic stress tolerance in crops. *Front. Plant Sci.* **2016**, *7*, 1137. [CrossRef]
30. Tuberosa, R. Phenotyping for drought tolerance of crops in the genomics era. *Front. Physiol.* **2012**, *3*, 347. [CrossRef]
31. Groat, R.G.; Vance, C.P. Root nodule enzymes of ammonia assimilation in alfalfa (*Medicago sativa* L). *Plant Physiol.* **1981**, *67*, 1198–1203. [CrossRef]
32. Singh, R.P.; Srivastava, H.S. Increase in glutamate synthase (NADH) activity in maize seedlings in response to nitrate and ammonium nitrogen. *Physiol. Plant.* **1986**, *66*, 413–416. [CrossRef]

33. Castiglia, D.; Cardi, M.; Landi, S.; Cafasso, D.; Esposito, S. Expression and characterization of a cytosolic glucose 6 phosphate dehydrogenase isoform from barley (*Hordeum vulgare*) roots. *Protein Expr. Purif.* **2015**, *112*, 8–14. [CrossRef] [PubMed]
34. Nakano, Y.; Asada, K. Hydrogen peroxide is scavenged by ascorbate-specific peroxidase in spinach chloroplasts. *Plant Cell Physiol.* **1981**, *22*, 867–880.
35. Fontaine, V.; Pelloux, J.; Podor, M.; Afif, D.; Gérant, D.; Grieu, P.; Dizengremel, P. Carbon fixation in *Pinus halepensis* submitted to ozone. Opposite response of ribulose-1,5-bisphosphate carboxylase/oxygenase and phospho*enol*pyruvate carboxylase. *Physiol. Plant.* **1999**, *105*, 187–192. [CrossRef]
36. Wendt, U.K.; Wenderoth, I.; Tegeler, A.; Von Schaewen, A. Molecular characterization of a novel glucose-6-phosphate dehydrogenase from potato (*Solanum tuberosum* L.). *Plant J.* **2000**, *23*, 723–733. [CrossRef]
37. Pajuelo, P.; Pajuelo, E.; Márquez, A.J. Proteolysis of barley (*Hordeum vulgare* L.) T. ferredoxin-glutamate synthase affects ferredoxin-and methyl viologen-dependent enzyme activities differently. *J. Plant Physiol.* **2000**, *157*, 575–579. [CrossRef]
38. Li, Y.J.; Hai, R.L.; Du, X.H.; Jiang, X.N.; Lu, H. Over-expression of a *Populus* peroxisomal ascorbate peroxidase (PpAPX) gene in tobacco plants enhances stress tolerance. *Plant Breed.* **2009**, *128*, 404–410. [CrossRef]
39. Maksimović, J.D.; Zhang, J.; Zeng, F.; Živanović, B.D.; Shabala, L.; Zhou, M.; Shabala, S. Linking oxidative and salinity stress tolerance in barley: Can root antioxidant enzyme activity be used as a measure of stress tolerance? *Plant Soil* **2013**, *365*, 141–155. [CrossRef]
40. Chang, Z.; Liu, Y.; Dong, H.; Teng, K.; Han, L.; Zhang, X. Effects of cytokinin and nitrogen on drought tolerance of creeping bentgrass. *PLoS ONE* **2016**, *11*, e0154005. [CrossRef]
41. Sagi, M.; Dovrat, A.; Kipnis, T.; Lips, H. Nitrate reductase, phospho*enol*pyruvate carboxylase, and glutamine synthetase in annual ryegrass as affected by salinity and nitrogen. *J. Plant Nutr.* **1998**, *21*, 707–723. [CrossRef]
42. Shi, W.M.; Muramoto, Y.; Ueda, A.; Takabe, T. Cloning of peroxisomal ascorbate peroxidase gene from barley and enhanced thermotolerance by overexpressing in Arabidopsis thaliana. *Gene* **2001**, *273*, 23–27. [CrossRef]
43. Soussi, M.; Ocaña, A.; Lluch, C. Effects of salt stress on growth, photosynthesis and nitrogen fixation in chick-pea (*Cicer arietinum* L.). *J. Exp. Bot.* **1998**, *49*, 1329–1337. [CrossRef]
44. Stewart, G.R.; Rhodes, D. Nitrogen metabolism of halophytes III. Enzymes of ammonia assimilation. *New Phytol.* **1978**, *80*, 307–316. [CrossRef]
45. Abobatta, W.F. Plant Responses and Tolerance to Combined Salt and Drought Stress. In *Salt and Drought Stress Tolerance in Plants*; Hasanuzzaman, M., Tanveer, M., Eds.; Springer: Berlin/Heidelberg, Germany, 2020; pp. 17–52. [CrossRef]
46. Wang, Z.Q.; Yuan, Y.Z.; Ou, J.Q.; Lin, Q.H.; Zhang, C.F. Glutamine synthetase and glutamate dehydrogenase contribute differentially to proline accumulation in leaves of wheat (*Triticum aestivum*) seedlings exposed to different salinity. *J. Plant Physiol.* **2007**, *164*, 695–701. [CrossRef]
47. Ullah, A.; Li, M.; Noor, J.; Tariq, A.; Liu, Y.; Shi, L. Effects of salinity on photosynthetic traits, ion homeostasis and nitrogen metabolism in wild and cultivated soybean. *PeerJ* **2019**, *7*, 8191. [CrossRef]
48. Goel, P.; Singh, A.K. Abiotic stresses downregulate key genes involved in nitrogen uptake and assimilation in *Brassica juncea* L. *PLoS ONE* **2015**, *10*, e0143645. [CrossRef]

agronomy

MDPI

Article

Genetic Diversity, Population Structure and Inter-Trait Relationships of Combined Heat and Drought Tolerant Early-Maturing Maize Inbred Lines from West and Central Africa

Alimatu Sadia Osuman [1,2,3], Baffour Badu-Apraku [2,*], Beatrice E. Ifie [1], Pangirayi Tongoona [1], Ebenezer Obeng-Bio [3] and Ana Luísa Garcia-Oliveira [4]

1 West Africa Center for Crop Improvement (WACCI), University of Ghana, PMB 30 Legon, Accra, Ghana; asosuman@wacci.ug.edu.gh (A.S.O.); bifie@wacci.ug.edu.gh (B.E.I.); ptongoona@wacci.ug.edu.gh (P.T.)

2 International Institute of Tropical Agriculture (IITA), PMB 5320, Ibadan 200001, Nigeria

3 CSIR-Crops Research Institute, Ghana. P.O. Box 3785, 039-5028 Kumasi, Ghana; obengbio2000@gmail.com

4 Excellence in Breeding (EiB) Platform, International Maize and Wheat Improvement Centre (CIMMYT), ICRAF, PO Box 1041, 00621 Nairobi, Kenya; a.oliveira@cgiar.org

* Correspondence: b.badu-apraku@cgiar.org; Tel.: +234-810-848-2590

Received: 18 July 2020; Accepted: 28 August 2020; Published: 4 September 2020

Abstract: Adequate knowledge and understanding of the genetic diversity and inter-trait relationships among elite maize inbred lines are crucial for determining breeding strategies and predicting hybrid performance. The objectives of this study were to investigate the genetic diversity of 162 early maturing white and yellow tropical maize inbred lines, and to determine the population structure, heterotic groups and inter-trait relationships among the lines. Using 9684 DArT single nucleotide polymorphism (SNP) markers, a gene diversity (GD) of 0.30 was recorded for the inbred lines with polymorphic information content (PIC) ranging from 0.08 to 0.38. The genetic relatedness among the inbred lines evaluated revealed six different groups based on the history of selection, colour of endosperm and pedigree. The genotype-by-trait (GT) biplot analysis identified inbred 1 (TZEI 935) as outstanding in terms of combined heat and drought (HD) tolerance with the base index analysis identifying 15 superior inbreds in the HD environment. A wide range of genetic variability was observed among the inbred lines, indicating that they are an invaluable resource for breeding for HD tolerance in maize breeding programmes, especially in West and Central Africa.

Keywords: genetic diversity; DArT SNP markers; maize; early maturity; heat and drought tolerance

1. Introduction

Molecular analyses in combination with morphological and agronomic evaluation of germplasm are useful for providing complementary information on the genetic diversity in a breeding programme [1]. Thus, monitoring of genetic diversity in the base material helps breeders to formulate breeding strategies by selecting parental combinations for advance selection [2], determining the level of genetic variability whilst identifying the main clusters selected for specific traits [3]. Information on genetic diversity is also useful in classifying inbred lines into heterotic groups [4,5], and predicting hybrid performance [6]. By definition, a heterotic group encompasses a set of directly related inbred lines which produce high yielding and vigorous hybrids when crossed to inbred lines from different heterotic groups, but not when crossed to inbred lines from the same heterotic group [7].

Additionally, the availability of information on the genetic diversity and population structure facilitates the assessment of the likelihood of loss in genetic diversity during preservation or selection [8], as well as assessment of the relative strengths of evolutionary forces [9,10]. Molecular marker technology has been a resourceful technique for assessing genetic diversity and delimiting heterotic groups [11].

Previous studies have reported a strong association between molecular marker-based genetic distance and hybrid performance in maize [11,12]. However, these findings should be used with caution as it is not always that these findings can predict hybrid performance, heterosis and specific combining ability (SCA) of single cross hybrids [13,14]. Recent studies involving early (90–95 days to physiological maturity) and extra early (80–85 days to physiological maturity) maize inbred lines have convincingly shown that genetic variability is of utmost significance in demonstrating heterosis [15,16]. Primarily, inbred lines are developed by crossing the best lines from the same heterotic group followed by inbreeding and selection, whereas hybrids are developed by crossing lines from dissimilar heterotic groups [17]. When lines lose vigour over time as a result of inbreeding depression, there is the need to bulk-sib within them to regain an acceptable level of vigour to ensure survival. The concept of heterosis is very important in the development of climate-smart maize inbred lines and hybrids [18]. Climate-smart inbred lines are able to tolerate the major stresses (for example, drought, heat and low soil fertility) that limit their growth. Identification of stress-tolerant lines and their utilization in the development of productive hybrids has been found to be a sustainable means of developing climate-resilient maize varieties.

During the past decade, molecular markers, especially simple sequence repeat (SSR) markers, have been widely used to study the correlations among hybrid performance and genetic distance [15,19]. Recently, Richard et al. [20] successfully used single nucleotide polymorphism (SNP) markers to classify 45 maize inbred lines of temperate and tropical origin into four heterotic groups corresponding to the Salisbury white (N), Southern cross (SC), Iowa stalk synthetic (BSS) and the Lancaster heterotic groups. Based on the results, it was possible to demonstrate that the genetic divergence of the temperate inbreds was greater than that of the tropical inbreds. Similarly, the SNP marker method used for grouping inbred lines into heterotic groups has been shown to be more efficient than the SCA and general combining ability (GCA) methods, as well as the heterotic grouping based on GCA of multiple traits (HGCAMT) [21].

During the past two decades, the International Institute of Tropical Agriculture (IITA) early and extra-early maturing maize breeding programme (IITA-MBP) has focused on developing germplasm with improved resistance/tolerance to multiple stresses, including drought, and combined HD stress tolerance [22]. This strategy has resulted in the development of broad-based normal endosperm maize populations including TZE-W Pop DT STR and TZE-Y Pop DT STR in the early maturity group and TZEE-Y Pop DT STR and TZEE-W Pop DT STR within the extra-early maturity group [23,24]. The National Agricultural Research Systems (NARS) and farmers in different agro-ecological zones of West and Central Africa (WCA) have widely adopted outstanding maize hybrids and synthetic varieties developed from inbred lines with desirable agronomic characteristics that have adapted to multiple-stress environments [15]. The current resurgence of climate variability especially in WCA has made the development of drought and combined HD-tolerant maize varieties in the sub-region very critical [22]. It is anticipated that this would lead to accelerated achievement of food security and improved livelihood of resource poor farmers in WCA.

Another important focus of this study was to understand which traits have the maximum likelihood of complementing grain yield of the inbred lines. This information was necessary because high grain yield of inbred *per se* [25], coupled with other desirable agronomic characteristics, are major criteria for selecting an inbred line as a suitable seed parent. This trait could be transferred to their hybrids and such information could facilitate prediction of superior hybrid performance, including excellent plant and ear aspects which are highly heritable, and delayed stay-green characteristic under combined heat and drought stress. Therefore, analysis of inter-relationships among traits of maize genotypes is crucial to maize improvement programmes involved in genetic enhancement of grain yield under combined HD stress. This is because most economic traits of interest in maize including grain yield have low heritability because the inheritance is quantitative. In view of this, gains from direct selection for such traits are very low. However, this problem could be mitigated through indirect selection of secondary traits that have high genetic correlations with grain yield coupled with high heritability and that are easily measured. Yan and Kang [26] proposed genotype-by-trait (GT) analysis for multiple trait-based assessment of genotypes. Such statistical tools aid in the identification of genotypes with specific desirable traits that could be employed as a genetic resource in a breeding programme or released for commercialization while also presenting a graphical display of the genetic correlations among traits [27,28]. Thus, the objectives of this study were (1) to assess the genetic diversity and population structure of selected heat- and drought-tolerant early maturing white and yellow maize inbreds using DArT SNP markers and (2) examine the interrelationships among traits of the early maturing inbreds in an effort to identify those strongly associated with grain yield in maize under combined HD stress.

2. Materials and Methods

2.1. Description of Germplasm

A total of 162 early maturing maize inbred lines developed in the IITA-MBP, Ibadan, Nigeria, with tolerance to drought and combined heat and drought stress were used in the present study (Table 1). The IITA inbred lines were extracted from the broad-based populations (TZE Comp5-Y C_6, TZE-W P_{OP} STR 104 C_0, TZE-W P_{OP} STR 107 C_0, TZE-Y P_{OP} STR 106 C_0, TZE-W P_{OP} STR 108 C_0, TZE-W Pop STR C_0, TZE-W Pop × LD C_0 and TZE-Y Pop STR C_0) developed from exotic and local germplasm identified as a result of many years of extensive testing for broad adaptation to dry and hot climates of the savannas of WCA. Furthermore, some of the inbred lines were derived from crosses between the broad-based drought-tolerant population, TZE COMP 3 DT C_2 F_2, and elite *Striga* resistant inbred lines, TZEI 18, TZEI 4, TZEI 60 and TZEI 98, as well as crosses between the broad-based *Striga* resistant but drought-susceptible population, TZE COMP 5-W DT C_7, and the drought tolerant inbred lines TZEI 2, TZEI 18, TZEI 31, TZEI 56 and TZEI 65. In addition to the 34 lines derived from two IITA bi-parental populations [(TZEI 11 × TZEI 8) and (TZEI 7 × TZEI 3)], the genetic materials for this study also included three inbred lines derived from two CIMMYT broad-based populations (M37W/ZM607 and Cuba/Guad C_3).

Table 1. Source populations of 162 maize inbred lines used in the genetic diversity study.

S/N	Source of Population	Number of Extracted Inbred Lines	Grain Colour
1	(TZE COMP 3 DT C2 F2 × TZEI 18) S8	1	White
2	(TZE Comp 3 DT C2 F2 × TZEI 4) S8	4	White
3	(TZE COMP 3 DT C2 F2 × TZEI 60) S8	5	White
4	(TZE COMP 3 DT C2 F2 × TZEI 98) S8	12	White
5	(TZE COMP 5-W DT C7× TZEI 18) S8	1	White
6	(TZE COMP 5-W DT C7 × TZEI 2) S8	10	White
7	(TZE COMP 5-W DT C7 × TZEI 31) S8	4	White
8	(TZE COMP 5-W DT C7 × TZEI 56) S8	5	White
9	(TZE COMP 5-W DT C7 × TZEI 65) S8	17	White
10	(TZEI 11 × TZEI 8) S8	27	Yellow
11	(TZEI 7 × TZEI 3) S8	7	White
12	[M37W/ZM607#bF37sr-2-3sr-6-2-X]-8-2-X-1-BB-B-xP84c1 F27-4-3-3-B-1-B] F29-1-2-2 x [KILIMA ST94A]-30/MSV-03-101-08-B-B-1xP84c1 F27-4-1-4-B-3-B] F2-1-2-1-1-1-B x CML486]-1-1	1	Yellow
13	[M37W/ZM607#bF37sr-6-2-X]-8-2-X-1-BB-B-xP84c1 F27-4-3-3-B-1-B] F29-1-2-1-6 x [KILIMA ST94A]-30/MSV-03-2-10-B-1-B-B-xP84c1 F27-4-1-6-B-5-B]3-1-2-B/CML442)-1-1	1	White
14	Cuba/Guad C3 F85-3-3-1-B-B-B-B	1	White
15	TZE Comp5-Y C6 S8	2	Yellow
16	TZE-W POP STR 104 S8	2	White
17	TZE-W POP STR 107 S8	12	White
18	TZEE-Y POP STR 106 S8	3	Yellow
19	TZE-Y POP STR 106 S8	4	Yellow
20	TZE-W POP STR 104 S8	11	White
21	TZE-W POP STR 108 S8	15	White
22	TZE-W Pop STR Co S8	3	White
23	TZE-W Pop x LD S8	1	White
24	TZE-Y POP STR 106 S8	7	White
25	TZE-Y Pop STR Co S8	5	Yellow
26	WEC STR S8 Inbred	1	White

2.2. DNA Extraction and Genotyping

The 162 early maturing inbred lines were planted at the IITA research field, Ibadan, Nigeria under optimal growing conditions in 2018. At two weeks after planting, leaf samples were collected from each of the inbred lines, dried in a Labconco Freezone 2.5 L System lycophilizer (Marshall Scientific, Hampton, NH, USA) and ground using a Spex™ Sample Prep 2010 Geno/Grinder (Thomas Scientific, Swedesboro, NJ, USA).

2.3. Data Filtering and Analysis

Genomic DNA samples were extracted from freeze-dried leaf tissues of each inbred line using the DArT protocol (www.diversityarrys.com/files/DArTDNAisolation.pdf) as previously reported [18]. The quality of the DNA sample was determined using a 1% agarose gel test whilst the quantity was measured using the Nanodrop system (Thermo Scientific, Swedesboro, NJ, USA). Genotyping was performed at the IGSS-BecA-ILRI Platform, Nairobi, Kenya.

Following the methodology previously described [15,29], the DArT-SNP markers with missing data above 5% and 20% heterozygosity and minor allele frequency of below 0.05 were removed. From the initial DArTseq markers obtained from sequencing, a total of 9684 SNP markers distributed across the 10 maize chromosomes were retained for all subsequent analyses. The major allele frequency, polymorphic information content (PIC), number of alleles, heterozygosity and the gene diversity were calculated using Power Marker software V3.2.5 [30]. Analysis of the population structure was done by employing the admixture model using the STRUCTURE 2.3.4 software package [31]. The results from the model were achieved by changing the number of clusters (K) from one to twenty with ten to twenty adjustments for each K and subjecting the data to 10,000 Markov Chain Monte Carlo and 10,000 burn-in as described previously [15,25,29]. The most appropriate K value was identified by running the data obtained into the STRUCTURE HARVESTER using the Evanno method [32,33]. Each inbred was allocated to a group using a cut off threshold of 70%, while inbreds showing less than the designated threshold value were assigned to an additional group described as the mixed group.

Subsequent to verification of the number of sub-populations using the STRUCTURE, the phylogenetic relationship among the inbred lines was accessed using the neighbor-joining (NJ) method under 30,000 bootstraps in the DARwin software [34]. The genetic distance matrix was obtained using the Jaccard similarity test [35] using the formula dij = (a + b)/ (a + (a + c)), where dij is the dissimilarity between units i and j, a is the number of variables where Xi and Xj are present, b is the number of variables where Xi is present and Xj is absent, c is the number of variables where Xi is absent and Xj is present.

2.4. Assessment of the Performance of the Inbreds Under Combined Heat and Drought Stress Conditions

The set of 162 inbred lines were planted at Manga (11° 3′N, 0° 14′W, altitude 252 m, 800 mm and 1,100 mm annual rainfall) in the Upper East Region of Ghana and Kadawa (11°45′ N, 8°45′ E, altitude 468.5 m, 884 mm annual rainfall) in the Kano state of Nigeria during the dry seasons (February to May) of 2018 and 2019. The maize plants were exposed to intense DH stress for 2-3 weeks in April at Kadawa and Manga, with the day temperatures of 36 to 41.2 °C and night temperatures of 18 to 27.4 °C for Kadawa (Figure 1), while at Manga the day and night temperatures were 31.6 to 42.8 °C and 22.0 to 29 °C, respectively (Figure 2). A 9 × 18 alpha lattice design with two replications was used for the inbred experiments. Single row plots each 4 m long, with a spacing of 0.75 m between rows and 0.40 m within rows, constituted an experimental unit. Three seeds were planted per hill and later thinned to two plants per stand to obtain a population density of about 66 666 per ha. No rainfall was recorded at Kadawa during the grain filling period of April 2018 and 2019, resulting in the highest day temperature throughout the growth cycle of the plants (Table 2). Similarly, at Manga, although some amount of rainfall was recorded in the month of April during 2018 and 2019, the highest day and night temperatures recorded during the growth cycle of the plants were obtained in April

(Table 2). The maize plants were irrigated twice a week using a drip irrigation system that supplied 17 mm of water per irrigation from planting to 2–3 weeks before anthesis at Manga, while at Kadawa, a furrow irrigation system supplied the same amount of water from planting to 2–3 weeks before anthesis when the irrigation water was withdrawn at each test location. A follow-up rescue irrigation was employed 2 weeks after the withdrawal of irrigation water at both test locations and resumed 10–12 days after the "rescue irrigation" until physiological maturity. The rescue irrigation was applied based on the assessment of some physiological stress indicator traits such as severe leaf rolling during the early mornings, leaf senescence, tassel blasting and leaf wilting. Basal fertilizer rate of 60 kg each of N, P_2O_5 and K_2O ha^{-1} of N, P, K (15:15:15) was applied followed by top dressing 2 weeks after planting (WAP) using 60 kg ha^{-1} of urea. Weed management was carried out using the post emergence herbicide (Roundup) and the pre-emergence herbicides (Caliherb 720 SL and Atrazine) in Ghana, while atrazine and gramozone were applied in Nigeria at 5 l ha^{-1} followed by manual weeding to ensure weed-free trials.

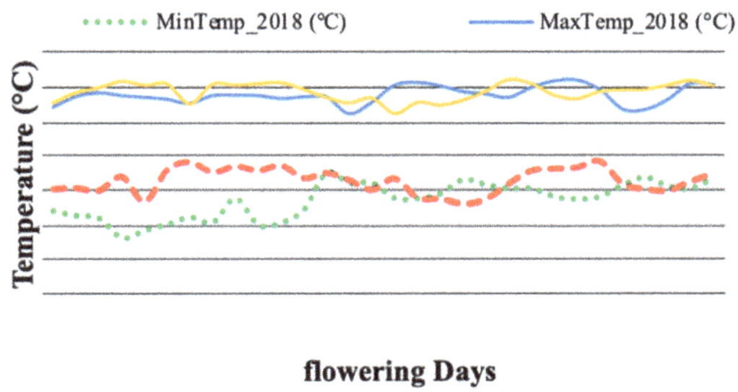

Figure 1. Mean temperature record for April 2018 and 2019 flowering days at Kadawa.

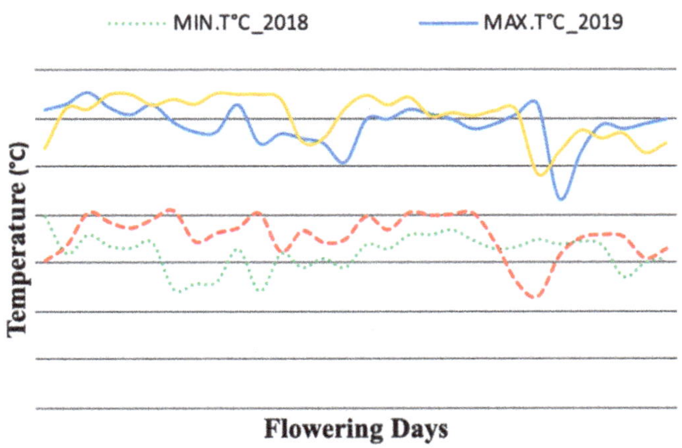

Figure 2. Mean temperature record for April 2018 and 2019 flowering days at Manga.

Table 2. Mean monthly rainfall and temperature (day & night) recorded at Manga and Kadawa during the 2018 and 2019 experimental period.

Kadawa	2018			2019		
Month	**Day °C**	**Night °C**	**Rainfall (mm)**	**Day °C**	**Night °C**	**Rainfall (mm)**
February	36.0	17.0	0.0	31.0	18.0	0.0
March	38.3	19.4	0.0	37.0	24.0	0.0
April	40.0	24.0	0.0	39.4	26.2	0.0
May	39.0	26.0	18.0	37.0	26.0	37.0
June	37.1	25.0	47.0	34.1	24.1	36.0
July	34.3	23.0	99.0	31.1	23.0	407.0
Manga	**2018**			**2019**		
Month	**Day °C**	**Night °C**	**Rainfall(mm)**	**Day °C**	**Night °C**	**Rainfall (mm)**
February	38.2	23.0	0.0	37.1	20.3	0.0
March	40.0	25.4	0.0	40.5	25.3	0.0
April	39.4	26.1	18.2	40.1	28.0	40.4
May	38.0	25.0	7.9	37.0	26.2	12.8
June	34.0	24.0	15.0	33.4	24.4	15.6
July	31.4	23.0	16.5	32.0	23.0	17.5

2.5. Data Collection

Agronomic data recorded under combined heat and drought stress at both Kadawa and Manga testing sites were similar to that described by Nelimor et al. [16]. We measured days to anthesis (AD) as the number of days from planting to when 50% of the plants in a plot had extruded pollen, and days to silking (SD) as when 50% of the plants had produced silks. The anthesis–silking interval (ASI) was recorded as the difference between SD and AD. Plant and ear heights were measured at about three weeks after flowering. Leaf death (LD) was determined by estimating the percentage of leaves dead from the base of the plant to the flag leaf using a scale of 1–9 where 1 = 1% to 10% of leaves dead and 9 = 81% to 100% of leaves dead at 70 days after planting (DAP). At about three weeks post-flowering, data were collected on husk cover (HUSKC), plant aspect (PLASP), plant height (PLTH), ear height (EHT), leaf firing and tassel blasting (TB), as described by Badu-Apraku and Fakorede [22]. A few days prior to harvesting, data were recorded on root and stalk lodging and number of plants per plot. At harvesting, records were taken on ear aspect and number of ears per plant (EPP). EPP was obtained by dividing the total number of ears per plot by the number of plants per plot. For the experiments conducted under optimal environments, the moisture content of the grains at harvest was recorded for the shelled grains of five randomly selected ears per plot using a moisture meter. Grain yield (kg ha^{-1}) was estimated from field weight of ears per plot, assuming a shelling percentage of 80, adjusted to moisture content of 15%. For the trials conducted under combined HD stress, grain weight on plot basis was obtained by shelling all harvested ears of each plot.

2.6. Statistical Analysis

The data were subjected to analysis of variance employing PROCGLM of SAS as well as the RANDOM statement with a TEST option [36] for grain yield and other measured traits. In the statistical model, environments, replicates within environments and incomplete blocks within replicates × environment interaction were considered as random factors, while the inbred lines were regarded as a fixed factor. Block effects on genotype means were adjusted using the lattice design proposed by Cochran and Cox [37]. Mean values of measured characters were separated using the standard error of difference (SED). The multiple trait selection index was adopted for the identification of inbred lines combining tolerance to drought and heat stress environments. The base index was computed using Yield, EPP, ASI, PLASP, EASP and LD [38] as follows:

$$MI = [(2 \times Yield + EPP) - ASI - PLASP - EASP - LD] \quad (1)$$

Mean values of the six measured traits were standardized prior to calculation of the MI as the traits were measured in different units. Furthermore, estimated means of the different traits of 15 best and 10 worst inbred lines were adopted for the Genotype × Trait (GT) biplot analysis [39,40] to obtain information on the secondary traits which could be relied on for indirect selection for combined HD stress. The polygon and the vector views of the GT biplot were constructed using the measured characters and the genotype-focused singular value partition ('SVP = 2'), rendering the polygon view suitable for visualizing the relationship between traits.

3. Results

3.1. Summary Statistics

The summary statistics from the 9684 DArT SNP markers is presented in Table 3. Heterozygosity values were in the range of 0.00 to 0.20 with a mean value of 0.08. The 9684 DArT SNP markers had a major allele frequency ranging from 0.50 to 0.95 with an average value of 0.79. Gene diversity (GD) values varied from 0.09 to 0.50 with a mean of 0.30. For this set of inbred lines, the polymorphic information content (PIC) varied from 0.08 to 0.38 with a mean value of 0.25 (Table 3). The minor allele frequencies were (MAF) 0.05 for minimum and 0.50 for maximum, with an average of 0.21.

Table 3. Diversity statistics of the 162 maize inbred lines based on 9684 DArT single nucleotide polymorphism (SNP) markers.

	MaF	GD	He	PIC	MAF
Minimum	0.50	0.09	0.00	0.08	0.05
Maximum	0.95	0.50	0.20	0.38	0.50
Mean	0.79	0.30	0.08	0.25	0.21

MaF = Major allele frequency, GD = Gene diversity, He = Heterozygosity, PIC = Polymorphic information content, MAF = Minor allele frequency.

3.2. Genetic Distance and Population Structure Using DArT SNP Markers

From the pairwise comparisons of the inbred lines, the genetic distance for the lines was in the range of 0.02 to 0.48, with a mean value of 0.42. About 70.32% of the genetic distance among the inbred lines fell within 0.416 and 0.461 (Figure 3). Lines TZdEI 134 and TZdEI 198 recorded the least genetic distance (Supplementary Table S1), consistent with their genetic background as they shared similar ancestry and history of selection (Supplementary Table S2). Inbred lines TZEI 500 and TZEI 7 recorded the highest (0.48) genetic distance, indicating that the two inbred lines diverged significantly. TZEI 500 was developed from the bi-parental cross involving TZEI 11 × TZEI 8 through repeated selfing up to the 8th generation of inbreeding, while TZEI 7 was developed from repeated self-pollination of WEC STR up to the 8th generation. Moreover, TZEI 500 is characterized by yellow endosperm while TZEI 7 has white endosperm (Supplementary Table S2).

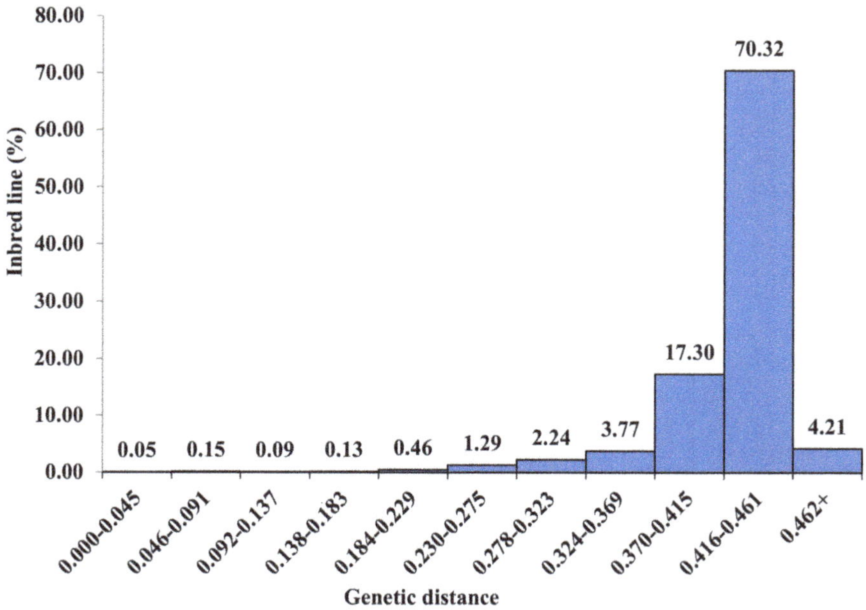

Figure 3. Frequency distribution of genetic distances based on Euclidean method for 162 early yellow and white inbred lines genotyped using 9864 DArT SNPs.

The population structure analysis of the 162 early white and yellow tropical maize inbred lines using the selected 9684 SNPs revealed an optimal k of six sub-populations (Figure 4A). Sub-population 1 comprised 8.02% (13 of the inbred lines), 16.67% (27 of the inbred lines in sub-population 2) and 8.64% (14 of the inbred lines in sub-population 3), while sub-populations 4, 5 and 6 contained 3.09% (5 lines), 6.17% (10 lines) and 4.32% (7 inbred lines), respectively (Supplementary Table S2). About 53.09% (86 of the inbreds) with the probability of association of less than 70% were grouped as mixed individuals (Supplementary Table S2). These six sub-populations were separated based largely on the pedigrees and endosperm colours. Sub-populations 1, 2, 4, 5 and 6 consisted of white endosperm maize inbred lines while sub-population 3 comprised yellow endosperm lines, with about 53% consisting of a mixture of white and yellow endosperm colour (Supplementary Table S2). The results showed that subpopulations 2, 3 and 4 contained only an inbred line, each derived from the different source populations: TZEI 11 × TZEI 8 S_8, TZE COMP 5-W DT C7 × TZEI 65 S_8 and TZE COMP 3 DT C_2 F_2 × TZEI 98 S_8, respectively. Sub-population 5 consisted of two source populations with inbred lines derived from TZE W- P_{OP} STR 108 S_8 and TZE –W P_{OP} STR 104 S_8. Sub-population 1 consisted of four dissimilar source populations, which included TZE COMP 5 W DT C7 × TZEI 2, (TZEI 7 × TZEI 3) S_8, TZE – Y P_{OP} STR 106 S_5, and TZE COMP 5 –W DT C7 × TZEI 31 S_8. Sub-population 6 had the highest genetic diversity and comprised five source populations, namely WEC STR S_8, TZE-W P_{OP} Co S_8, TZE –W P_{OP} x LD S_8, TZE COMP$_3$ DT C_2 F_2 × TZEI 60 S_8, TZE-Y P_{OP} STR 106 S_8 and TZE W P_{OP} STR 107 S_8.

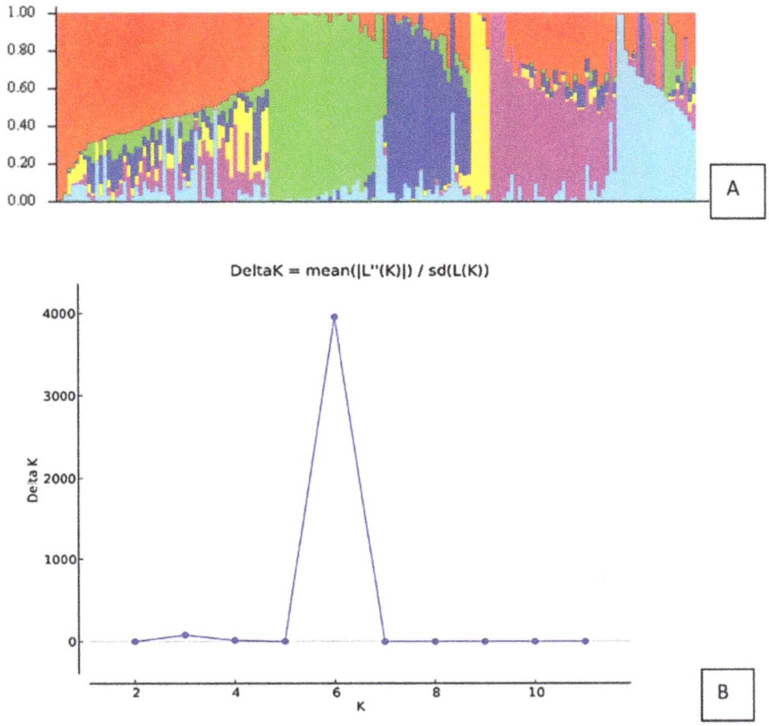

Figure 4. The six sub-populations (1 = red, 2 = green, 3i = blue, 4 = yellow, 5 = pink and 6 = teal) of the 162 early maturing yellow and white inbred lines using DArT SNP markers. (**A**) population structure bar plot of the 162 early yellow and white inbred lines as shown by 9684 DArT SNP markers for K = 6. (**B**) Assessment of the best delta K by the Evanno method.

Within the six sub-populations, the expected heterozygosity amongst the inbred lines ranged from 0.10 for sub-population 4 to 0.43 for sub-population 6 with a mean of 0.25 (Supplementary Table S3a). Overall, sub-populations 4 and 6 recorded the highest allele frequency (0.23), while sub-populations 1 and 5 recorded the least allele frequency (0.09) (Supplementary Table S3b). For each of the sub-populations, the STRUCTURE analysis estimated the fixation index (F_{ST}) and indicated significant divergence within the six sub-populations. Estimated F_{ST} values of 0.34, 0.49, 0.47, 0.81, 0.45 and 0.28 were recorded for sub-populations 1–6, respectively (Supplementary Table S3c).

3.3. Cluster Analysis and Principal Coordinate Analysis (PCoA)

The neighbour joining (NJ) phylogenetic tree constructed assigned the 162 inbred lines into three main clusters followed by several sub-clusters (Figure 5). In conformity with the biplot output of the STRUCTURE analysis (Figure 4A,B), the six clusters could be clearly depicted in the phylogenetic tree (Figure 5). Unique colours were used to represent each sub-population with red, green, blue, yellow, pink and teal denoting 1–6, respectively, and the black colour represented the mixed population when 70% probability threshold was considered.

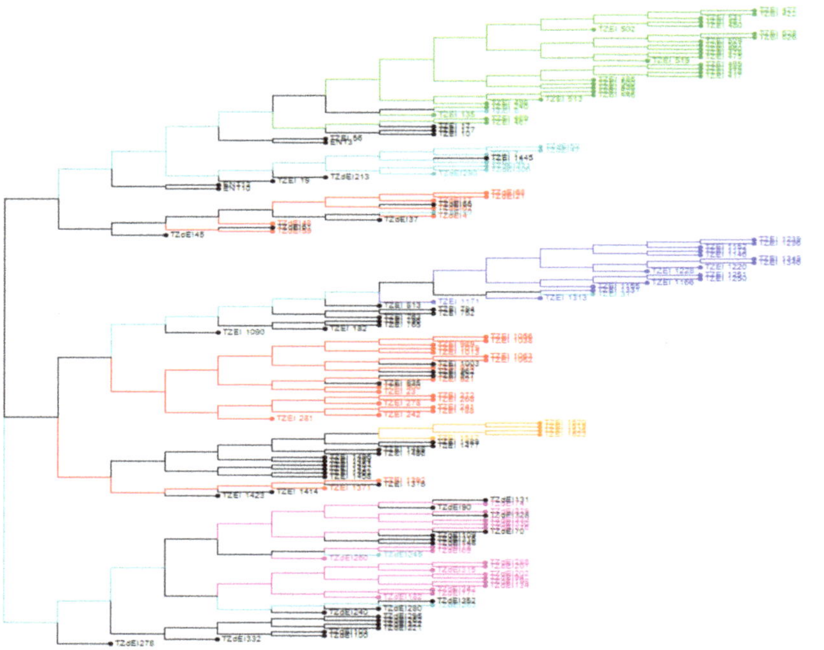

Figure 5. Clustering of 162 yellow and white early tropical maize inbred lines based on 9684 DArT SNP markers (red (cluster 1), green (cluster 2), blue (cluster 3), yellow (cluster 4), pink (cluster 5), teal (cluster 6) and black (admixture)).

Based on the phylogenetic tree, a total of 27 inbred lines were placed in cluster 1, with 13 out of the 27 inbreds derived from six different white maize backgrounds, TZE-W Pop STR Co S6, (TZEI 7 × TZEI 3) S6, (TZE COMP 5-W DT C7 × TZEI 31) S6, (TZE COMP 5-W DT C7 × TZEI 2) S6 and TZEE-Y POP STR 106 S5 189/194, representing sub-population 1 (red bars; Figure 5). The second cluster included 29 inbred lines, of which 27 clearly represented sub-population 2 (green bars). All the inbred lines in this cluster were derived from the yellow endosperm bi-parental cross (TZEI 11 × TZEI 8) S_8 (Supplementary Table S2). The third cluster consisted of 16 inbred lines with 14 being similar to those generated from the population structure analysis (sub-population 3; blue bars). All the inbred lines from this cluster were derived from one white kernel source population (TZE COMP 5-W DT C_7 X TZEI 65) S_8. Cluster 4 had 23 inbred lines out of which 5 were derived from the white endosperm population, (TZE COMP 3 DT C_2 F_2 × TZEI 98) S_8, and were similar to those generated from the population structure (sub-population 4; orange bars). The fifth cluster obtained from the phylogenetic analysis comprised a total of 18 inbred lines, with 10 out of these belonging to the white endosperm maize population, TZE-W Pop STR 104 S_5 (sub-population 5; pink colour). The results of the grouping of the 10 inbreds lines into sub-population 5 was consistent with the results obtained from the population structure analysis. The remaining eight inbred lines of the fifth cluster were derived from the source population, TZE-W Pop STR 108, and were also characterized by white endosperm kernels. The sixth cluster consisted of 12 inbred lines, with seven of them similar to those generated from the population structure analysis (indicated with teal colour lines) and represented by sub-population 6. Inbred lines placed in this cluster were derived from six different source populations with each inbred line extracted from a different source population, with the exception of the inbred lines TZdEI 47 and TZdEI 52 derived from the same source population, TZE-Y Pop STR 106 (Supplementary Table S2).

To understand and confirm the dynamics of the inbred population structure, we also adopted the principal coordinate analysis (PCoA). Based on the pairwise genetic distance matrix among the

162 inbred lines, a clear distinction among the six groups of inbred lines could be visualized, and this was in concordance with the results of the population STRUCTURE biplot (Figure 6).

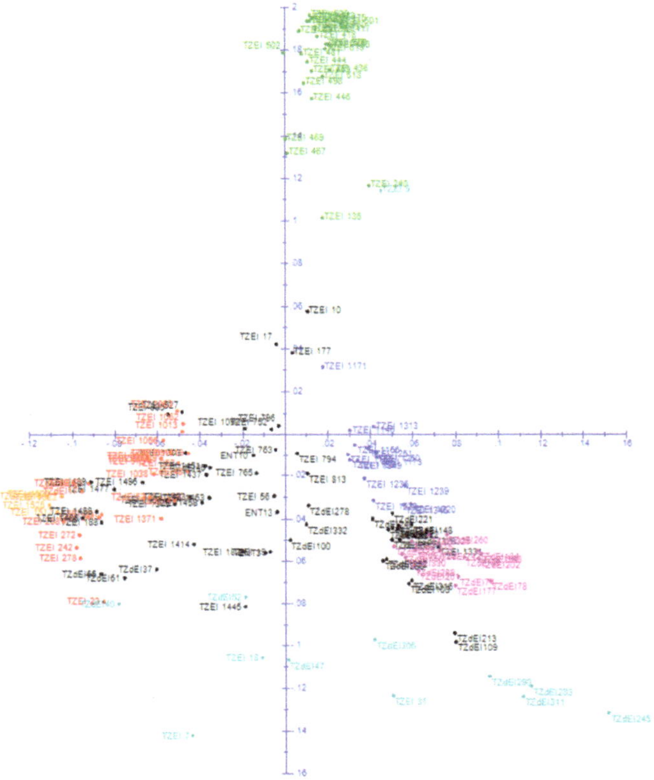

Figure 6. Principal coordinate analysis (PCoA) of 162 early yellow and white inbred lines genotyped using 9864 DArT SNPs.

3.4. Genotype-by-Trait (GT) Interaction and the Multiple Trait Selection Index

The GT biplot analysis is one of the most reliable statistical tools available for selecting inbred lines for hybrid development using secondary traits. This analysis can identify secondary traits which are directly or indirectly related to traits of major breeding interest such as grain yield and ears per plant. The inbred lines were evaluated under combined heat and drought stress in this study to select desirable seed parents that could be used in productive hybrid combinations. Moreover, the evaluation aimed to examine the inter-relationships among traits measured for the inbreds as a way of predicting outstanding hybrid performance. The present study identified 15 inbred lines with outstanding performance as well as those with poorest performance (worst 10 lines) under combined heat and drought environments using the GT biplot (Figure 7). The inbred lines were selected based on the premise that the best selected lines should not record 50% or more percentage count for leaf firing and tassel blasting under combined HD stress. Approximately 52.56% of the total variation among the measured traits of the inbred lines was accounted for by the principal component axes PCI and PC2. In the biplot view, high values of grain yield and EPP for inbreds were considered as desirable while high values for the secondary traits such as LF, TB, EASP, PLASP, LD, ASI, HUSKC, AD, SD, and SL of inbreds were considered undesirable. The inbred(s) at each vertex of the polygon (vertex inbred) had the highest values for the traits within its sector. The results revealed that the sector with inbreds 21 and

8 as the vertex inbreds contained the traits LD, TB, LF, ASI, HUSKC, SL, EASP and PLASP, indicating that these inbreds had high values for these traits and that they were the low-yielding inbreds based on the high values of these traits. Inbred lines 25 and 20 were positioned at the vertex that contained the RL traits, signifying that these inbreds recorded high values for root lodging. In contrast, inbred 1 (the best performing inbred line) was the vertex inbred in the segment containing EPP and grain yield, suggesting the superiority of the inbred line in terms of grain yield and EPP. Inbred line 7 was the vertex inbred that recorded no trait within its sector, implying that it was not outstanding in any of the measured traits. Fifteen inbred lines were identified as outstanding in performance based on the measured traits. TZEI 935 was identified as the best performing inbred line with a yield of 3461 kg/ha. Contrarily, 10 inbred lines were identified as the worst in performance based on the measured traits, with TZdEI 109 as the worst performing inbred line with a yield of 424 kg/ha (Table 4).

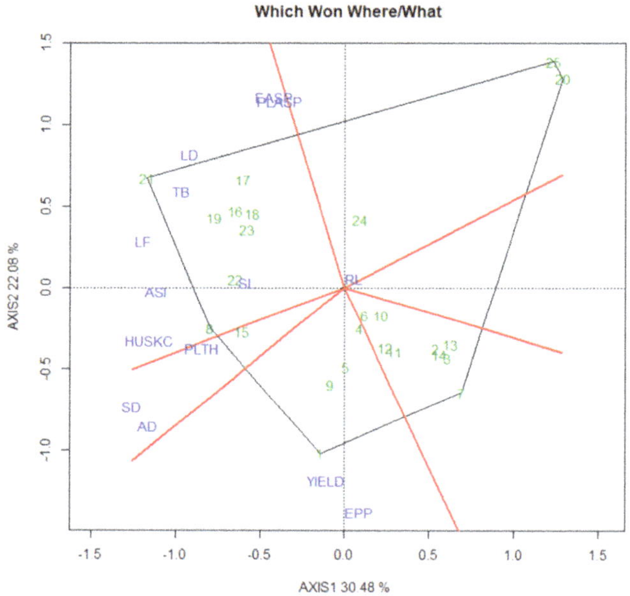

Figure 7. A "Which is best/worst for what" or "Which wins where" of GxT biplot of the 25 early maturing maize inbreds assessed under combined heat and drought conditions at Kadawa in Nigeria and Manga in Ghana during the dry seasons of 2018 and 2019. PC1 and PC2 for model 2 explained 52.56% of the variation among traits. YIELD = yield, EPP = ears per plant, RL = root lodging, SL = stalk lodging, SD = days to silking, AD= days to anthesis, ASI = anthesis silking interval, PLTH = plant height, LF = leaf firing, TB = tassel blast, LD = leaf death, HUSKC = husk cover, EASP = ear aspect, PLASP = plant aspect.

Table 4. Mean grain yield and other agronomic characters and codes of 25 selected inbred lines evaluated under combined heat and drought environments at Manga-Ghana and Kadawa-Nigeria during the 2018 and 2019 dry seasons.

Code	Entry	Yield (ka/ha)	EPP	AD	SD	ASI	RL	HUSKC	PLASP	PLTH	SL	EASP	LD	LF	TB	M.I
1	TZEI 935	3461.12	1.83	58.72	61.61	2.89	0.67	3.68	4.58	90.37	1.51	4.99	2.81	0.03	0.02	12.96
2	TZEI 1019	1447.28	1.43	54.07	56.51	2.40	0.66	2.96	4.06	92.66	1.46	3.96	2.84	0.04	0.01	9.68
3	TZEI 240	1384.51	0.90	55.53	57.71	2.14	0.34	3.00	3.75	97.93	1.14	3.92	2.25	0.03	0.00	8.82
4	TZEI 56	1833.74	0.87	57.36	59.76	2.41	0.31	3.54	4.53	94.43	0.82	4.10	2.86	0.05	0.09	7.04
5	TdZEI 295	1273.46	0.98	58.95	61.60	2.65	1.03	3.38	4.03	90.21	1.03	3.62	2.61	0.18	0.07	6.72
6	TdZEI 311	834.04	0.75	56.85	58.44	1.57	0.85	4.02	4.08	112.35	1.32	3.90	2.79	0.07	0.05	6.03
7	TZEI 1228	1002.42	0.86	55.91	58.49	2.58	0.86	3.21	2.88	85.49	0.17	3.39	2.27	0.05	0.02	5.72
8	TdZEI 278	1420.57	1.09	58.33	61.74	3.41	1.17	3.76	3.97	97.71	1.85	3.52	3.41	0.37	0.42	5.52
9	TZEI 478	1572.27	1.09	57.89	61.27	3.37	0.61	3.25	3.93	104.35	0.50	3.52	2.91	0.24	0.00	5.48
10	TZEI 475	1061.64	0.80	56.29	57.77	1.52	0.37	3.46	4.02	94.07	2.37	3.47	2.97	0.15	0.02	5.26
11	TZEI 135	1216.49	0.81	56.69	58.64	1.94	2.61	3.76	3.87	97.14	1.59	3.63	2.50	0.07	0.01	5.22
12	TZEI 763	1195.72	0.86	59.19	61.52	2.34	0.23	3.87	4.38	74.40	0.32	4.24	2.23	0.02	0.03	4.85
13	TZEI 23	854.59	0.76	55.57	57.80	2.22	0.48	3.08	3.26	78.24	1.75	3.38	2.54	0.03	0.02	4.82
14	TZEI 17	1041.86	0.98	57.51	59.17	1.67	0.84	3.59	3.80	80.81	0.33	4.48	2.04	0.07	0.01	4.73
15	TdZEI 69	1261.13	0.99	56.74	59.23	2.49	1.47	3.96	4.45	155.73	1.37	4.08	2.71	0.48	0.09	4.72
16	TZEI 278	828.89	0.54	59.30	61.68	2.42	0.35	4.73	5.01	90.07	1.48	5.49	3.57	0.17	0.01	−5.26
17	TZEI 31	699.92	0.44	57.71	60.51	2.81	0.98	3.84	4.70	85.13	2.21	4.92	3.56	0.23	0.28	−5.29
18	TdZEI 280	820.45	0.56	58.30	61.33	3.03	0.43	3.93	4.83	97.40	1.39	5.34	3.10	0.20	0.19	−5.33
19	TZEI 427	931.83	0.52	57.80	61.88	4.10	0.19	3.76	4.37	92.45	1.57	5.28	3.01	0.30	0.36	−5.42
20	TZEI 502	493.36	0.36	48.93	50.01	1.08	1.43	2.18	4.70	62.69	0.66	5.44	2.80	0.07	0.12	−5.50
21	TZEI 1003	698.00	0.45	58.36	61.88	3.51	0.28	3.65	4.74	114.21	1.85	5.10	3.10	0.51	0.57	−5.50
22	TdZEI 306	845.25	0.59	61.13	65.04	3.92	0.56	3.85	4.19	90.09	0.61	4.95	3.59	0.15	0.02	−5.66
23	TdZEI 61	722.60	0.52	58.06	62.21	4.15	2.77	3.60	4.33	106.77	0.62	5.01	3.43	0.21	0.13	−6.46
24	TZEI 1349	436.40	0.39	54.69	58.80	4.09	0.54	3.82	4.41	94.67	−0.03	5.00	2.64	0.06	0.01	−6.71
25	TdZEI 109	423.70	0.36	47.42	48.89	1.47	0.79	2.48	5.34	74.93	0.76	4.97	3.11	0.02	0.00	−7.51
	Means	1110.45	0.79	56.69	59.34	2.65	0.83	3.53	4.25	94.17	1.15	4.39	2.87	0.15	0.10	

EPP = ear per plant, RL = root lodging, SL = stalk lodging, SD = days to silking, AD = days to silking, ASI = anthesis silking interval, PLTH = plant height, LF = leaf firing, TB = tassel blast, LD = leaf death, HUSKC = husk cover, EASP = ear aspect, PLASP = plant aspect, M.I = multiple trait base index.

4. Discussion

Molecular markers have proven to be among the most reliable tools for identifying the genetic components of inbred lines [15,29]. In the present study, 9684 DArT-based SNP markers were used to assess the genetic diversity of 162 early maturing yellow and white endosperm tropical maize inbreds possessing varying degrees of tolerance to drought and/or combined HD stress. Moreover, the population structure, heterotic groups and inter-trait relationships were also examined in the studied materials.

Generally, high PIC values indicate the effectiveness of markers for linkage analysis when estimating the inheritance between parental lines and the derived hybrids [41], whereas the GD or expected heterozygosity (He) reveals gene diversity for haploid markers and gives information on the proportion of the genetic distance and average heterozygosity within a given population [41,42]. In the present study, an average PIC of 0.25 was obtained, which is comparable to that reported earlier by Dao et al. [27] and Zhang et al. [28], as well as the 0.24 of Yu et al. [43] and 0.29 of Wu et al. [44], but higher compared with the results of Adu-Boakyewaa et al. [15] and Silva et al. [45] who reported average PIC values of 0.19 and 0.17, respectively. The high mean PIC value observed implied that majority of the selected SNPs were informative and polymorphic enough to bring out the differences among the 162 inbred lines studied. The average GD of 0.30 obtained in this study is also comparable to 0.32 previously reported in maize [46,47], but higher than the value (0.22) reported by other researchers [15,44,48,49]. These results indicated a broad genetic variability among the inbred lines and the relatively lower number of pairwise individuals with low genetic distance.

Of the 9684 SNPs deployed across the 162 lines, approximately 65% had MAF of between 0.05 and 0.25 (Supplementary Table S2), which is lower compared to the range reported by Semagn et al. [17] involving 450 CIMMYT lines studied using 1065 SNPs. Nevertheless, it is comparable to the values obtained by Adewale et al. [19] in early maturing white tropical maize. Consistent with the results of Adu-Boakyewaa et al. [15] and Adewale et al. [29], the average heterozygosity of 0.08 obtained in the present study is close to zero as expected for inbred lines, thus making them a useful genetic resource for association mapping and genetic studies in maize in which uniformity of the genetic materials is invaluable.

A higher F_{ST} value for the sub-populations was reported in the present study. Together with the relatively moderate allele frequency divergence among sub-populations, the results indicated that the alleles of the inbred lines were fixed and could be categorized into groups with distinct characteristics, and this finding is in conformity with previous results [15]. The six clusters observed in this study indicated that individuals from the different groups were therefore expected to harbour different favourable alleles that could be exploited for breeding for combined HD-tolerant hybrids and synthetics. The results also demonstrated that the DArT-based SNP-derived markers were informative in providing genome profiles which were very useful for the identification of unique characteristics among the inbred lines [17,50,51].

Population structure analysis provides a guide for assigning a set of inbred lines into heterotic groups based on genotypic information of individual ancestry of the lines [52]. In this study, six sub-populations (K = 6) were established among the 162 inbred lines. The inbred lines were placed in distinct groups based on their relatedness in terms of pedigree records and selection history, such as endosperm colour, tolerance or susceptibility to drought and combined HD. This finding was affirmed by the high F_{ST} values ranging from 0.28 to 0.81 for the six sub-populations and that inbred lines from the different groups could harbour unique favourable alleles for development of productive hybrids for population improvement programmes in the tropics. The high F_{ST} values obtained for the sub-populations in the present study were comparable to the results of other researchers. For example, Kashiani et al. [53], Aci et al. [54] and Adu-Boakyewaa et al. [15] indicated F_{ST} values of as high as 0.94, 0.33 and 0.83, respectively, for the sub-populations obtained in their genetic diversity studies for different sets of tropical maize lines. In the present study, cluster 1 contained all the inbred lines with white endosperm and were generally tolerant to drought, cluster 2 had inbred lines with yellow endosperm as well as tolerance to drought, while cluster 3 had highly drought-tolerant white endosperm lines. However, all the inbred lines within these three sub-groups showed moderate tolerance to combined HD stress. This observation is not surprising because the inbred lines have long been selected under drought conditions compared to the results of the more recent evaluations under combined HD conditions.

The multivariate clustering methods showed high consistency regarding the number of groups and individuals assigned to each group, indicating the existence of six genetically distinct groups. This result corroborated the findings of Adu-Boakyewaa et al. [15] and Obeng-Bio et al. [51] who reported a high consistency among the model-based structure analysis, PCoA and the Neighbour-Joining phylogeny using the Roger's and Euclidian genetic distance methods, respectively. However, this result disagreed with the reports of Dao et al. [27] and Semagn et al. [17] who found a high consistency between the model-based structure and the PCoA but very low concordance with the Neighbour-Joining phylogeny generated using the Roger's genetic distance method. The differences in the results could be due to the variation in the inbred lines used in the different studies, the method used in deriving the genetic distances between the inbreds, as well as the different clustering algorithms used. In concordance with earlier researchers, it was evident that the clustering of tropical maize populations is largely consistent with the pedigree information [15,21,27,48,55].

Generally, it is advocated that source genotypes with favourable alleles for combined heat and drought tolerance should possess good agronomic characteristics necessary for combined HD-tolerant hybrid development without placing too much emphasis on the *per se* grain yield. The GT analysis identified ASI, EASP, PLASP, LF, TB and EPP as important secondary traits that could be utilized to accelerate genetic gains from selection for enhanced grain yield under combined HD conditions. Nelimor et al. [16] suggested that lines that scored 50% or more for LF and TB under combined HD conditions should be rejected, implying the importance of the information on inter-relationships among secondary traits. From the GT biplot view, the best 15 performing lines were placed in five different heterotic groups. As a guide, crosses among the lines from different heterotic groups should be given priority in terms of hybrid combinations towards maximization of heterosis. For example, the best performing inbred line (TZEI 935) placed in the mixed group could be crossed to inbred lines TdZEI 311 and TZEI 1228 which belonged to groups six and three, respectively, to increase the chances of developing productive hybrids. The significant and positive association observed between PLASP and EASP implied that there was a close relationship among these traits and that either of them could be eliminated from the base index without sacrificing the precision in selecting for combined HD-tolerant genotypes. Badu-Apraku et al. [38] also found a positive correlation between these traits and advocated that either PLASP or EASP could be discarded in the base index in selecting for drought- and low-N-tolerant genotypes. Similarly, there was high positive correlation between AD, SD, and ASI, but none of them were found to be redundant as the ASI is dependent on the SD and AD [38]. These three traits were very important in selecting stress-tolerant maize genotypes [56,57], as a relatively constant value (lower ASI) is desirable in selection of early maturing maize [38].

5. Conclusions

To ensure the selection of appropriate parental lines for crossing and devising strategies for significant gains from selection, adequate knowledge and understanding of the genetic diversity of available inbred lines in a breeding programme is essential. The population structure analysis, PCoA and phylogenetic clustering methods confirmed the existence of six different groups for the panel of 162 maize inbred lines studied. The lines were grouped largely based on the pedigree information, endosperm colour, selection history and similarity of ancestry. There were high genetic distances between the paired inbred lines, demonstrating the distinctiveness of the inbred lines and the presence of high genetic variability which could be exploited in tropical maize breeding programmes for hybrid development. The low heterozygosity as well as relatively low divergence among the sub-populations indicated that the present set of inbred lines had great potential for genetic studies and could serve as future sources of promising inbred lines by contributing new and favourable alleles. The GT biplot analysis revealed that EPP, PLASP, EASP, LF and TB were the most reliable traits for selecting genotypes for tolerance to combined HD stress. Inbred 1 (TZEI 935) was identified as the ideal under combined HD stress environments. Using the base index, the 15 top-yielding inbred lines were identified for use in the maize breeding programmes of SSA for the development of productive hybrids and synthetics with combined HD tolerance. Among the inbred lines studied, inbred TZEI 935 was identified as one of the ideal lines under combined HD stress environments and should be further explored. Finally, the results of the present study will serve as an important guide for appropriate decision-making for future development of productive hybrids, particularly for SSA.

Supplementary Materials: The following are available online at http://www.mdpi.com/2073-4395/10/9/1324/s1, Table S1. Genetic distance matrix of the 162 yellow and white inbred lines using Euclidean method in the present study. Table S2. Summary of the 9684 SNP markers used in the current study. Table S3. a. Expected Heterozygosity of the 162 yellow and white inbred lines; b. Allele frequency divergence among populations with 162 inbred lines assessed and c. mean values of Fixation Index (Fs_T) of 162 yellow and white inbred lines.

Author Contributions: Conceptualization, A.S.O.; B.B.-A.; methodology, A.S.O., A.L.G.-O., B.B.-A.; formal analysis, A.S.O., A.L.G.-O., E.O.-B.; genetic resources, B.B.-A.; data curation, A.S.O., A.L.G.-O., E.O.-B.; supervision, B.B.-A., B.E.I., P.T.; writing—original draft preparation, A.S.O.; writing—review & editing; funding acquisition, B.B.-A., B.E.I., P.T. and A.L.G.-O. All authors have read and agreed to the published version of the manuscript.

Funding: This research was funded by the German Academic Exchange Service (DAAD) and Bill and Melinda Gates Foundation (Grant/Award Number: OPP1134248) under the Stress Tolerant Maize for Africa (STMA) Project of IITA. Genotyping of the lines was supported by the Integrated Genotyping Service and Support (IGSS) platform grant (ref. number PJ-002507) of BecA-ILRI, Kenya.

Acknowledgments: The authors appreciate the funding support from the German Academic Exchange Service (DAAD) and Bill and Melinda Gates Foundation under the Stress Tolerant Maize for Africa (STMA) Project of IITA. Sincere appreciation also goes to the Integrated Genotyping Service and Support (IGSS) platform of BecA-ILRI, Kenya for supporting the genotyping of the inbred lines.

Conflicts of Interest: The authors declare no conflict of interest.

Data Availability: The DArTseq datasets used in the present study have been deposited at the IITA CKAN repository.

References

1. Singh, S.; Nodari, R.; Gepts., P. Genetic diversity in cultivated bean germplasm: I. Allozymes. *J. Hered.* **1991**, *76*, 447–450.
2. Barrett, B.A.; Kidwell, K.K. AFLP-based genetic diversity assessment among wheat cultivars from the Pacific Northwest. *Crop Sci.* **1998**, *38*, 1261–1271. [CrossRef]
3. Mohammadi, S.A.; Prasanna, B.M. Analysis of genetic diversity in crop plants—Salient statistical tools and considerations. *Crop Sci.* **2003**, *43*, 1235–1248. [CrossRef]
4. Flint-Garcia, S.A.; Buckler, E.S.; Tiffin, P.; Ersoz, E.; Springer, N.M. Heterosis is prevalent for multiple traits in diverse maize germplasm. *PLoS ONE* **2009**, *4*, e7433. [CrossRef]
5. Reif, J.C.; Melchinger, A.E.; Xia, X.C.; Warburton, M.L.; Hoisington, D.A.; Vasal, S.K.; Frisch, M. Genetic distance based on simple sequence repeats and heterosis in tropical maize populations. *Crop Sci.* **2003**, *43*, 1275–1282. [CrossRef]
6. Betrán, F.J.M.; Ribaut, D.; Beck, D.; Gonzalez, D.L. Genetic diversity, specific combining ability, and heterosis in tropical maize under stress and nonstress environments. *Crop Sci.* **2003**, *43*, 797–806. [CrossRef]
7. Lee, M. DNA markers and plant breeding programs. In *Advances in Agronomy*; Academic Press: Cambridge, MA, USA, 1995; Volume 55, pp. 265–344.
8. Reif, J.C.; Zhang, P.; Dreisigacker, S.; Warburton, M.L.; van Ginkel, M.; Hoisington, D.; Melchinger, A.E. Wheat genetic diversity trends during domestication and breeding. *Theor. Appl. Genet.* **2005**, *110*, 859–864. [CrossRef]
9. Ouborg, N.J.; Piquot, Y.; Van Groenendael, J.M. Population genetics, molecular markers and the study of dispersal in plants. *J. Ecol.* **1999**, *87*, 551–568. [CrossRef]
10. Falconer, D.S.; Mackay, T.F.C. *Introduction to Quantitative Genetics*; Prentice Hall: Harlow, UK, 1996.

11. Melchinger, A.E.; Coors, A.; Pandey, S. Genetic diversity and heterosis. In Proceedings of the International Symposium on the Genetics and Exploitation of Heterosis in Crops, Mexico City, Mexico, 17–22 August 1999; pp. 99–118.
12. Kiula, B.A.; Lyimo, N.G.; Botha, A.M. Association between AFLP-based genetic distance and hybrid performance in tropical maize. *Plant Breed.* **2008**, *127*, 140–144. [CrossRef]
13. Melchinger, A.E.; Lee, M.; Lamkey, K.R.; Hallauer, A.R.; Woodman, W.L. Genetic diversity for restriction fragment length polymorphisms and heterosis for two diallel sets of maize inbreds. *Theor. Appl. Genet.* **1990**, *80*, 488–496. [CrossRef]
14. Legesse, B.W.; Myburg, A.A.; Pixley, K.V.; Twumasi-Afriyie, S.; Botha, A.M. Relationship between hybrid performance and AFLP based genetic distance in highland maize inbred lines. *Euphytica* **2008**, *162*, 313–323. [CrossRef]
15. Adu-Boakyewaa, G.A.; Badu-Apraku, B.; Akromah, R.; Garcia-Oliveira, A.L.; Awuku, F.J.; Gedil, M. Genetic diversity and population structure of early-maturing tropical maize inbred lines using SNP markers. *PLoS ONE* **2019**, *14*, e0214810. [CrossRef] [PubMed]
16. Nelimor, C.; Badu-Apraku, B.; Tetteh, A.Y.; N'guetta, A.S. Assessment of genetic diversity for drought, heat and combined drought and heat stress tolerance in early maturing maize landraces. *Plants* **2019**, *8*, 518. [CrossRef] [PubMed]
17. Nyombayire, A.; Derera, J.; Sibiya, J.; Gasura, E. Genetic diversity among maize inbred lines selected for the mid-altitudes and highlands of Rwanda. *Maydica* **2018**, *61*, 7.
18. Richard, C.; Osiru, D.S.; Mwala, M.S.; Lubberstedt, T. Genetic diversity and heterotic grouping of the core set of southern African and temperate maize (*Zea mays* L) Inbred lines using SNP markers. *Maydica* **2016**, *61*, 9.
19. Badu-Apraku, B.; Annor, B.; Oyekunle, M.; Akinwale, R.O.; Fakorede, M.A.B.; Talabi, A.O.; Fasanmade, Y. Grouping of early maturing quality protein maize inbreds based on SNP markers and combining ability under multiple environments. *Field Crops Res.* **2015**, *183*, 169–183. [CrossRef]
20. Badu-Apraku, B.; Fakorede, M.A.B. *Advances in Genetic Enhancement of Early and Extra-Early Maize for Sub-Saharan Africa*; Springer: Cham, Switzerland, 2017. [CrossRef]
21. Badu-Apraku, B.; Fakorede, M.A.B.; Ouedraogo, M.; Carsky, R.J. Impact, challenges, and prospects of maize research and development in West and Central Africa. In Proceedings of the Regional Maize Workshop, Cotonou, Benin, 14–18 May 2001; pp. 4–7.
22. Estep, M.C.; Gowda, B.S.; Huang, K.; Timko, M.P.; Bennetzen, J.L. Genomic characterization for parasitic weeds of the genus by simple sequence analysis. *Plant Genome* **2012**, *5*, 30–41. [CrossRef]
23. Pswarayi, A.; Vivek, B. Combining ability amongst CIMMYT's early maturing maize (*Zea mays* L.) germplasm under stress and non-stress conditions and identification of testers. *Euphytica* **2008**, *162*, 353–362. [CrossRef]
24. Yan, W.; Kang, M.S. GGE biplot analysis: A graphical tool for breeders. In *Geneticists and Agronomist*; Kang, M.S., Ed.; CRC Press: Boca Raton, FL, USA, 2003; p. 271.
25. Ober, E.S.; Bloa, M.L.; Clark, C.J.A.; Royal, A.K.; Jaggard, W.; Pidgeon, K.W. Evaluation of physiological traits as indirect selection criteria for drought tolerance in sugar beet. *Field Crop Res.* **2005**, *91*, 231–249. [CrossRef]
26. Badu-Apraku, B.; Akinwale, R.O. Cultivar evaluation and trait analysis of tropical early maturing maize under *Striga*-infested and *Striga*-free environments. *Field Crop Res.* **2010**, *121*, 186–194. [CrossRef]
27. Dao, A.; Sanou, J.; Mitchell, S.E.; Gracen, V.; Danquah, E.Y. Genetic diversity among INERA maize inbred lines with single nucleotide polymorphism (SNP) markers and their relationship with CIMMYT, IITA, and temperate lines. *BMC Genet.* **2014**, *15*, 127. [CrossRef] [PubMed]
28. Zhang, X.; Zhang, H.; Li, L.; Lan, H.; Ren, Z.; Liu, D.; Pan, G.; Gao, S. Characterizing the population structure and genetic diversity of maize breeding germplasm in Southwest China using genome-wide SNP markers. *BMC Genom.* **2016**, *17*, 697. [CrossRef] [PubMed]
29. Semagn, K.; Magorokosho, C.; Vivek, B.S.; Makumbi, D.; Beyene, Y.; Mugo, S.; Warburton, M.L. Molecular characterization of diverse CIMMYT maize inbred lines from eastern and southern Africa using single nucleotide polymorphic markers. *BMC Genom.* **2012**, *13*, 113. [CrossRef] [PubMed]
30. Liu, K.; Muse, S.V. PowerMarker: An integrated analysis environment for genetic marker analysis. *Bioinformatics* **2005**, *21*, 2128–2129. [CrossRef] [PubMed]
31. Earl, D.A. STRUCTURE HARVESTER: A website and program for visualizing STRUCTURE output and implementing the Evanno method. *Conserv. Genet. Resour.* **2012**, *4*, 359–361. [CrossRef]

32. Evanno, G.; Regnaut, S.; Goudet, J. Detecting the number of clusters of individuals using the software STRUCTURE: A simulation study. *Mol. Ecol.* **2005**, *14*, 2611–2620. [CrossRef]
33. Falush, D.; Stephens, M.; Pritchard, J.K. Inference of population structure using multilocus genotype data: Dominant markers and null alleles. *Mol. Ecol. Notes* **2007**, *7*, 574–578. [CrossRef]
34. Perrier, X.; Jacquemoud-Collet, J.P. DARwin Software. 2006. Available online: https://darwin.cirad.fr/product.php (accessed on 23 March 2020).
35. Jaccard, P. Nouvelles recherches sur la distribution florale. *Bull. Soc. Vaud. Sci. Nat.* **1908**, *44*, 223–270.
36. SAS Institute. *SAS System for Windows. Release 9.4*; SAS Institute Inc.: Cary, NC, USA, 2012.
37. Cochran, W.G.; Cox, G.M. *Experimental Designs*; John Wiley and Sons: New York, NY, USA, 1960.
38. Badu-Apraku, B.; Akinwale, R.O.; Ajala, S.O.; Menkir, A.; Fakorede, M.A.B.; Oyekunle, M. Relationships among traits of tropical early maize cultivars in contrasting environments. *Agron. J.* **2011**, *103*, 717–729. [CrossRef]
39. Yan, W. GGEbiplot—A Windows application for graphical analysis of multi-environment trial data and other types of two-way data. *Agron. J.* **2001**, *93*, 1111–1118. [CrossRef]
40. Yan, W.; Manjit, S.; Kang. *GGE Biplot Analysis: A Graphical Tool for Breeders, Geneticists, and Agronomists*; CRC Press: Boca Raton, FL, USA, 2002.
41. Shete, S.; Tiwari, H.; Elston, R.C. On estimating the heterozygosity and polymorphism information content value. *Theoret. Populat. Biol.* **2000**, *57*, 265–271. [CrossRef] [PubMed]
42. Luo, Z.; Brocj, J.; Dyer, J.M.; Kutchan, T.; Schachtman, D.; Augustin, M.; Ge, Y.; Fahlgren, N.; Abdel-Haleem, H. Genetic diversity and population structure of a *Camelina sativa* spring panel. *Front. Plant Sci.* **2019**, *10*, 184. [CrossRef] [PubMed]
43. Yu, J.; Zhang, Z.; Zhu, C.; Tabanao, D.A.; Pressoir, G.; Tuinstra, M.R.; Buckler, E.S. Simulation appraisal of the adequacy of number of background markers for relationship estimation in association mapping. *Plant Genom.* **2009**, *2*, 63–77. [CrossRef]
44. Wu, X.; Li, Y.; Li, X.; Li, C.; Shi, Y.; Song, Y.; Wang, T. Analysis of genetic differentiation and genomic variation to reveal potential regions of importance during maize improvement. *BMC Plant Biol.* **2015**, *15*, 256. [CrossRef]
45. Silva, K.J.; Guimarães, C.T.; Guilhen, J.H.S.; Guimarães, P.E.D.O.; Parentoni, S.N.; Trindade, R.D.S.; Bernardes, C.D.O. High-density SNP-based genetic diversity and heterotic patterns of tropical maize breeding lines. *Crop Sci.* **2019**, *60*, 779–787. [CrossRef]
46. Van Inghelandt, D.; Melchinger, A.E.; Lebreton, C.; Stich, B. Population structure and genetic diversity in a commercial maize breeding program assessed with SSR and SNP markers. *Theoret. Appl. Genet.* **2010**, *120*, 1289–1299. [CrossRef]
47. Lu, Y.; Yan, J.; Guimaraes, C.T.; Taba, S.; Hao, Z.; Gao, S.; Magorokosho, C. Molecular characterization of global maize breeding germplasm based on genome-wide single nucleotide polymorphisms. *Theoret. Appl. Genet.* **2009**, *120*, 93–115. [CrossRef]
48. Obeng-Bio, E. Genetic Analysis of Grain Yield and Other Traits of Early Maturing Provitamin A-Quality Protein Maize Inbred Lines under Drought and Low Soil Nitrogen Conditions. Ph.D. Thesis, University of Ghana, Accra, Ghana, 2018.
49. Yang, X.; Gao, S.; Xu, S.; Zhang, Z.; Prasanna, B.M.; Li, L.; Yan, J. Characterization of a global germplasm collection and its potential utilization for analysis of complex quantitative traits in maize. *Mol. Breed.* **2011**, *28*, 511–526. [CrossRef]
50. Huttner, E.; Wenzl, P.; Akbari, M.; Caig, V.; Carling, J.; Cayla, C.; Uszynski, G. Diversity arrays technology: A novel tool for harnessing the genetic potential of orphan crops. In *Discovery to Delivery: BioVision Alexandria 2005, Proceedings of the 2004 Conference of The World Biological Forum*; CABI Publishing: Oxford, UK, 2005; pp. 145–155.
51. Obeng-Bio, E.; Badu-Apraku, B.; Ifie, B.E.; Danquah, A.; Blay, E.T.; Dadzie, M.A.; Noudifoulè, G.T.; Talabi, A.O. Genetic diversity among early provitamin A quality protein maize inbred lines and the performance of derived hybrids under contrasting nitrogen environments. *BMC Gen.* **2020**, *21*, 78. [CrossRef]
52. Lawson, D.J.; Falush, D. Population identification using genetic data. *Ann. Rev. Genom. Hum. Genet.* **2012**, *13*, 337–361. [CrossRef]

53. Kashiani, P.; Saleh, G.; Panandam, J.M.; Abdullah, N.A.P.; Selamat, A. Demarcation of informative chromosomes in tropical sweet corn inbred lines using microsatellite DNA markers. *Genet. Mol. Biol.* **2012**, *35*, 614–621. [CrossRef] [PubMed]
54. Aci, M.M.; Lupini, A.; Maugeri, A.; Morsli, A.; Khelifi, L.; Sunseri, F. Genetic variation and structure of maize populations from Saoura and Gourara oasis in Algerian Sahara. *BMC Genet.* **2018**, *19*, 51. [CrossRef] [PubMed]
55. Warburton, M.L.; Ribaut, J.M.; Franco, J.; Crossa, J.; Dubreuil, P.; Betran, F.J. Genetic characterization of 21 elite CIMMYT maize inbred lines using RFLP markers. *Euphytica* **2005**, *142*, 97–106. [CrossRef]
56. Bänziger, M.; Lafitte, H.R. Breeding tropical maize for low N environments: II. The values of secondary traits for improving selection gains under low N. *Crop Sci.* **1997**, *37*, 1110–1117. [CrossRef]
57. Lafitte, H.R.; Banziger, M. Maize population improvement for low soil N: Selection gains and the identification of secondary traits. In Proceedings of the Developing Drought-and Low N-Tolerant Maize Symposium, El Batan, Mexico, 25–29 March 1996.

agronomy

MDPI

Article

Assessment of Agro-Morphologic Performance, Genetic Parameters and Clustering Pattern of Newly Developed Blast Resistant Rice Lines Tested in Four Environments

Raieah Saiyedah Sabri [1], Mohd Y. Rafii [1,2,*], Mohd Razi Ismail [1,2], Oladosu Yusuff [1], Samuel C. Chukwu [1,3] and Nor'Aishah Hasan [1,4]

1 Institute of Tropical Agriculture and Food Security, Universiti Putra Malaysia, Serdang, Selangor 43400, Malaysia; raieahsaiyedah@gmail.com (R.S.S.); razi@upm.edu.my (M.R.I.); oladosuy@gmail.com (O.Y.); chukwusamuel54@yahoo.com (S.C.C.); aishahnh@ns.uitm.edu.my (N.A.H.)
2 Department of Crop Science, Faculty of Agriculture, Universiti Putra Malaysia, Serdang, Selangor 43400, Malaysia
3 Department of Crop Production and Landscape Management, Ebonyi State University, Abakaliki PBM 053, Nigeria
4 Faculty of Applied Science, Universiti Teknologi MARA, Kuala Pilah, Negeri Sembilan 72000, Malaysia
* Correspondence: mrafii@upm.edu.my; Tel.: +60397691043

Received: 30 May 2020; Accepted: 28 July 2020; Published: 29 July 2020

Abstract: Multi-environmental yield trial is very vital in assessing newly developed rice lines for its adaptability and stability across environments especially prior to release of the newly developed variety for commercial cultivation. The growth performance and phenotypic variability of these genotypes are the combination of environment, genotype and genotype by environment (G×E) interaction factors. Thus, evaluation creates an opportunity for effective selection of superior genotypes. The objectives of this study were to evaluate the newly developed blast resistant rice lines in varied environmental conditions, precisely measure the response of the advanced lines in multiple environments and classify the genotypes into groups that could serve as varieties for commercial cultivation. Genetic materials included 18 improved blast resistant rice lines and the recipient parent MR219. The total of 19 newly developed genotypes was evaluated under four varied environments in Peninsular Malaysia. The experiments were carried out using randomized complete block design (RCBD) with three replications at each environment. Data were collected on the vegetative, yield and yield component traits. Descriptive statistics (mean performance) and analysis of variance were conducted using SAS Software version 9.4. Genotypic and phenotypic coefficients, phenotypic variance component, heritability and genetic advance were also determined. Analysis of variance revealed that all traits were significantly different for genotypes except days to maturity, number of filled grains and total number of grains. Meanwhile, all the traits differed significantly for genotype × environment (G×E) except number of tillers per hill and number of panicles per hill. Low heritability (<30%) was found for all the traits. Similarly, low genetic advance was also observed for all the traits except for number of tillers per hill and number of panicles per hill. yield per hectare had significant and positive correlation with most evaluated traits except for days to flowering, days to maturity, plant height and number of unfilled grains. Cluster analysis classified the 19 evaluated genotypes into six groups. Therefore, the six clusters/groups of genotypes were recommended as varieties for commercial cultivation in Malaysia and other rice growing regions.

Keywords: rice genotypes; blast resistant genotype; genotypic coefficient of variation (GCV); phenotypic coefficient of variation (PCV); heritability values

1. Introduction

Rice (*Oryza sativa* L.) is a staple food for over three billion of the world's population—mostly in the Asian continent [1]. The present and future global food security on rice is important due to the amount of daily calories and protein intake derived from it [2]. Over the years, there is a continuous increase in human population and rice consumption. This has caused vast gap between rice demand and its production, thus creating a need to increase the current rice yield potential [3]. However, rice productivity and sustainability are constrained by biotic and abiotic stress which was further intensified by climate change, weather variability and water shortage [4]. In Malaysia, rice is completely guarded by subsidizing, price control and import tariffs because it is heavily influenced by social, economic and political sensitivity [5,6]. The current mean yield of Malaysian rice production was 4.03 t/ha with an annual production of 2.2 million tons [7]. This output can only meet 72.3% of its local demands [8]. Malaysia lacks comparative advantage in paddy production and limited available land area for rice production compared to bordering countries like Vietnam and Thailand. Consequently, the country needs to depend on imported rice to fulfill its level of self-sufficiency (SSL) [9].

One of the significant factors that is seriously threatening high production of rice is blast disease caused by the fungal pathogen *Magnaporthe oryzae* [10–12]. Blast disease accounted for 10%–20% yield loss in susceptible varieties and can be more than 80% in severe conditions [13]. Nasruddin and Amin [14] reported the use of integrated management strategy which includes resistant cultivar, suitable date of planting and fungicide to control blast disease. However, this disease occurs due to the interaction of a favorable environment, a susceptible genotype and a virulent pathogen. Hence, the utilization of resistant cultivar is a favorable way to reduce the use of destructive pesticides [15,16]. Molecular markers have been intensively used to pyramid beneficial and multiple alleles to develop new rice blast resistant varieties [17–19]. This approach is the most cost-effective and environmental safety in order to manage rice blast disease [20]. However, an evaluation on the performance of the improved genotypes is needed before the selection of ideal genotype. This is because the phenotypic performance and adaption to adverse environments are influenced by the genotype (G), environment (E) and genotype × environment (G×E) interaction [21,22]. Furthermore, variation in their performance is more significant due to higher proportion of this G×E interaction than genotypic main effects [23–25]. This G×E interaction is described as the differential response in respective genotype performance in dissimilar or unpredictable environments [26]. Genotype selection based on stability and adaptability in multiple environmental conditions is important before the recommendation for large cultivation [27–30]. Therefore, this study was conducted to evaluate the newly developed blast resistant rice lines in varied environmental conditions, precisely measure the response of these advanced lines in multiple environments and classify the genotypes into groups that could serve as varieties for commercial cultivation.

2. Materials and Methods

2.1. Planting Materials

Eighteen advance lines of BC_2 F_2 generation from backcross of MR219 and Pongsu Seribu 2 were used in this study. Introgression lines were developed from these parents (MR219 x Pongsu Seribu 2) using marker-assisted backcross selection (MABS). The donor parent, Pongsu Seribu 2 (PS2), developed by Malaysian Agricultural Research and Development Institute (MARDI) possesses broad-spectrum resistance against blast fungal isolates. MR219 has high yielding potential with a suitable grain quality and good eating quality. Unfortunately, this variety is susceptible to blast. The F_1 plant produced from the cross between MR219 and Pongsu Seribu 2 was later backcrossed with MR219 to produce BC_1F_1 and subsequently backcrossed again with MR219 to derive BC_2F_1. The BC_2F_1 was allowed to self to produce BC_2 F_2 population among which 18 advance lines with the highest genome recovery and phenotypic characteristics similar to the MR219 were selected. The 18 advance lines derived had advantage of blast resistance in the four environments as well as high yielding characteristic similar to its recurrent parent MR219 [12]. The planting materials, i.e., the 18 advance breeding genotypes and MR219 were subjected to multi-environmental field trials.

2.2. Experimental Environments

The experiment was carried out between 2015 and 2017 in four different environments that represented the major rice growing areas in Peninsular Malaysia. The environment is referred to as the combination of year and location that covered a wide range of conditions such as differing temperatures (warm to moderate climate conditions), rainfall, water regimes (full and supplementary irrigation), soil types (loam and clay loam), biotic (pests and diseases), cropping seasons (main and off season) and conditions of management (research stations and farmers' fields). The four environments tested and their site description as well as planting period for each site is as presented (Table 1).

Table 1. Environmental data and description.

Code	Planting Period	Location	Altitude (m)	Av. Temp. Min–Max (°C)	Av. Humidity (%)	Rainfall (Mean)
EN1	September 2015–January 2016	3° 25′N 101° 10′E	3	23–31	83	782.4 (195.6)
EN2	February–June 2016	3° 25′N 101° 10′E	3	25–37	65	482.7 (120.7)
EN3	December 2016–March 2017	5° 59′N 100° 24′E	18	25–38	63	486.9 (121.7)
EN4	May–September 2017	3° 02′N 101° 42′E	32	24–38	67	623.4 (115.9)

Note: EN1—TanjungKarang; Selangor; EN2—TanjungKarang; Selangor; EN3—Kota SarangSemut; Kedah; EN4—Kota SarangSemut.

2.3. Experimental Scheme and Cultural Practices

Field experiment at each environment was laid out in a randomized complete block design (RCBD) in three replications with plot size of 16.25 by 4 m. The subplot size for each replication is 4 by 4 m with planting distance of 25 cm within and between rows. The 18 advance lines together with the MR219 control was first deactivated seed dormancy by oven drying at 40 °C for 24 h. The seeds were germinated by fully soaking in water overnight in a petri dish to induce pre-germination. Later, the water was removed, and the seeds were kept moist for three days. To avoid drying out during the three days, water was added to the seeds to keep it moist. After three days, the seeds were transferred to soil-filled plastic trays that were prepared earlier. The seedlings were allowed to grow for 21 days in the nursery before transplanting it to the field. A single seedling per hill at 21 days was manually transplanted to the rice field at each of the environment. All the cultural practices from land preparation until harvesting were done following MARDI recommendations. The field was irrigated with an average of 10 cm water above ground surface level throughout the experiments. Fertilization was applied following the recommendations from MARDI where 42 kg/ha muriate of potash and 57 kg/ha triple superphosphate were applied at day 15 after transplanting. Urea was applied at 35, 55 and 75 days after transplanting in splits at 80, 12 and 20 kg/ha, respectively. The insecticides; Malathion and Lambda-cyhalothrin were applied at the recommended rate of 20 mL per 15 L knapsack sprayer when needed. Regular, hand-weeding was done to remove narrow leaf weeds. In the case of broad leaf weeds, halosulfuron-methyl was applied at the rate of 40 mL per 15 L sprayer.

2.4. Data Collection

Data were collected on 13 quantitative traits, viz; number of days to flowering (DTF) and maturity (DTM), plant height (PH cm), number of tillers per hill (TPH) and panicles per hill (PPH), panicle length (PL cm), number of filled grains (FG) and unfilled grains (UFG) per panicle, total number of grains per panicle (TG), percentage of filled grains (PFG), 1000-grain weight (THW g), total weight of grains per hill (TW g) and yield (YLD t/ha). Sampling was conducted on five plants for each genotype from each replication following the IRRI [31] procedure.

2.5. Data Analysis

Analysis of variance (ANOVA) was done for all the traits using the SAS program version 9.4 (SAS Institute, Cary, NC, USA) to determine the variation among the genotypes, environment and genotypes by environment. In addition, descriptive statistics such as mean, range, standard deviation and coefficient of variation (CV) were calculated for each trait. Mean comparisons were performed using Tukey's test. Correlation coefficients were analyzed using SAS Software (version 9.4) to study the relationship between traits. Multivariate analysis was done using SAS software to determine the pattern of interaction between the genotype and environment. The method used in this study was cluster analysis. The Euclidean distances amidst the 19 improved genotypes evaluated were observed by using the standardized morphologic data to construct an UPGMA dendrogram. The variance components were also determined from the expected mean square using proc varcomp with restricted maximum likelihood (REML) method in SAS [10,27]. However, phenotypic variance was calculated using the formula (Equation (1)):

$$\sigma^2_p = \sigma^2_g + \sigma^2_{ge} + \sigma^2_\epsilon \quad (1)$$

where, σ^2_p=Phenotypic variance, σ^2_g = genotypic variance, σ^2_{ge} =variance of G×E and σ^2_ϵ = error variance.

The percentage of GCV and PCV values were classified as low if the range is between 0% and 10%, 10%–20% as moderate and 20% and above as high according to Sivasubramaniam & Madhava [32]. Classification of heritability percentage was done according to Falconer [21] that categorized it into three; low (0%–30%), moderate (30%–60%) and high (≥ 60%). Johnson et al. [33] classified the percentage of genetic advance into low when the value is 0%–10%, moderate with value of 10%–20% and high when the value is more than 20%. The estimation of genetic parameters such as variance components, broad sense heritability and expected genetic advance was done as follows:

i.	phenotypic coefficient of variation (PCV)	$PCV = \frac{\sqrt{\sigma^2 p}}{X} \times 100$	Where: $\sigma^2 p$ = phenotypic variance; X= mean of the trait
ii.	genotypic coefficient of variation (GCV)	$GCV = \frac{\sqrt{\sigma^2 g}}{X} \times 100$	Where; $\sigma^2 g$ = genotypic variance; X = mean of the trait
iii.	broad sense heritability	$h^2B = \frac{\sigma^2 g}{\sigma^2 p}$	Where: $\sigma^2 g$ = genotypic variance; $\sigma^2 p$ = phenotypic variance
iv.	Expected genetic advance	$GA = K \times \sqrt{\sigma^2 p} \times h^2B$	Where: K = constant that represents the selection intensity (when k is 5% the value is 2.06); $\sqrt{\sigma^2 p}$= standard deviation of phenotypic variance; h^2B = heritability in a broad sense

3. Results

3.1. Agro-Morphologic Traits, Genotype and G×E Interactions

The combined analysis of variance and mean comparison for 13 agro-morphologic traits evaluated in four different environments for varietal assessment are as presented in Tables 2 and 3, respectively. days to flowering showed a significant difference ($p \leq 0.05$) among genotypes while highly significant differences ($p \leq 0.01$) were observed for G×E. Environment accounted for the highest percentage of variation of 75%, while genotypes and G×E accounted for 8.9% and 4.89% variations, respectively (Table 2). The combined analysis showed that number of days to flowering started 70.92 days after transplanting in G7 followed by G6, G10 and G15 at 71.67, 71.17 and 71.58 days, respectively. For days to maturity, no significant difference was observed for genotypes while significant ($p \leq 0.01$) G×E interaction was observed. total number of grains had highly significant difference ($p \leq 0.01$) for G×E with 14.54% of variation. Genotypes had no significant difference in total number of grains but recorded 12.10% of variation. The genotypes varied significantly ($p \leq 0.05$) in percentage of filled grains with 1.66% total variation. A high significant difference was observed for G×E with total of 0.90% variation. Highest percentage of filled grains was observed in G18 with 83.89%. thousand-grain weight recorded a significant difference for genotypes while highly significant difference was observed in G×E. Genotypes and G×E recorded 17.66% and 8.71% variation, respectively. Thousand grain weights showed all genotypes had weight of more than 21 g with an average of all genotypes at 23.31 g. The highest thousand-grain weight was recorded in G18 with 24.78 g. This genotype had nearly the same weight with G4 (24.19 g), G14 (24.04 g) and MR219 (24.42 g). Meanwhile, total grains weight revealed a high significant difference ($p \leq 0.01$) for genotypes and a significant difference ($p \leq 0.05$) for G×E. Percentage of variation was counted at 23.09% for genotypes and 8.38% for G×E. The mean of total grains weight per hill for all genotypes was 37.85 g (Table 3). Genotype G18 had highest total grains weight with 54.23 g. High significant difference ($p \leq 0.01$) for genotypes and a significant difference ($p \leq 0.05$) for G×E were observed for yield per hectare. Variation at 23.10% for genotypes and 8.38% for G×E were recorded for yield per hectare trait. The highest yield per hectare in tons (t) was recorded in G18 with the value of 8.68 t/ha. It was followed by MR219 with 7.33 t/ha and G17 with 7.03 t/ha.

3.2. Genotypic and Phenotypic Coefficient of Variability (GCV and PCV)

The level of variability among the 13 agro-morphologic traits with respect to variance components was presented in Table 4. The GCV values ranged from 0.61–11.91%. Low GCV value was recorded for all traits, having value less than 10% except for the number of tillers per hill, number of panicles per hill, number of unfilled grains, total grains weight and yield per hectare. These five traits had a moderate percentage of GCV (10–20%). As for PCV, number of tillers per hill, number of panicles per hill, number of unfilled grains, total grains weight and yield per hectare recorded a high percentage of PCV (>20%). However, moderate value was recorded for the number of filled grains and total number of grains (10–20%). Low PCV value was observed in other traits (<10%).

3.3. Heritability and Genetic Advance

The result in Table 4 shows the broad-sense heritability and genetic advance of the evaluated traits. Low heritability value was recorded for all 13 traits which range between 2.42% and 27.08%. The highest heritability value was found in number of panicles per hill (27.08%) followed by panicle length (26.42%) and number of tillers per hill (26.11%). In this study, GA value of these 13 traits ranged between 0.24–12.40%. Moderate genetic advance (GA) was found in number of tillers (12.05%) and number of panicles (12.40%). Other 11 traits showed low GA percentage.

3.4. Correlation and Cluster Analyses

Besides days to 50% flowering and number of unfilled grains per hill, all other ten traits showed a positive relationship with the final yield per hectare (Table 5). Highly significant correlation coefficients with yield were discovered in nine traits. days to maturity correlated positively with days to flowering, number of tillers per hill, number of panicles per hill, filled grain, percentage filled grain and thousand grain weight, but had negative association with plant height and unfilled grain. number of tillers per hill had a significant positive correlation with number of panicles per hill, filled grain, percentage filled grains, total weight and yield. panicle length also had a significant positive correlation with unfilled grain, total number of grains, thousand grain weight, total weight and yield, but negatively correlated with percentage filled grain. Filled grain negatively correlated with unfilled grain and thousand grain weight but had a significant positive correlation with total number of grains, percentage filled grains, total weight and yield. The cluster analysis classified the genotypes into six major groups at 0.19 dissimilarity coefficients (Figure 1). This revealed the effectiveness of quantitative/morphologic traits in grouping the rice genotypes. It also showed a high level of morphologic variations present among the assessed genotypes. The largest group was Group I that consist of eight genotypes which are G1, G5, G4, G12, G8, G14, G6 and G10. This was followed by Group V with five genotypes (G2, G3, G13, G15 and G7) and Group III with three genotypes (G17, G18 and MR219). Group II, IV and VI only consist of one genotype in each group (Figure 1).

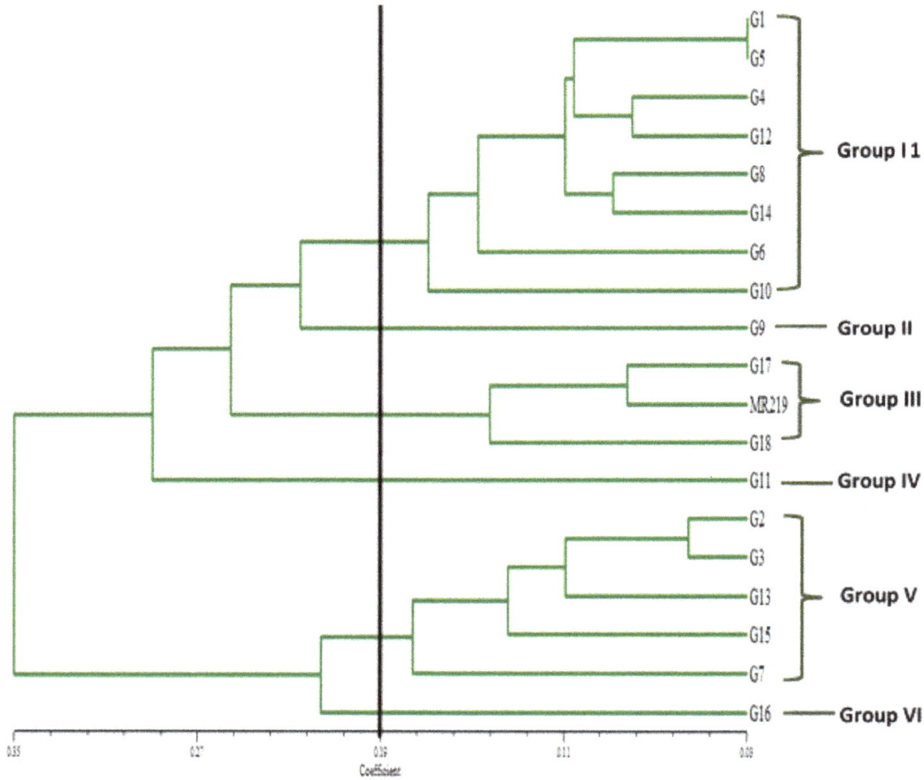

Figure 1. Clustering pattern of 19 rice genotypes evaluated in four different environments.

Table 2. mean square of combined analysis of variance for 13 traits assessed in four environments.

	SOV	Blocks (Environment)	Genotypes (G)	Environments (S)	G×S	Error
	DF	8	18	3	54	144
DTF	MS	18.34**	17.78*	149.97**	9.77**	3.94
	TSS (%)	9.18	8.9	75.06	4.89	1.97
DTM	MS	133.93**	19.96^{ns}	856.91**	15.20**	8.41
	TSS (%)	12.95	1.93	82.84	1.47	0.81
PH	MS	127.44**	60.67**	12,680.89**	22.82**	12.11
	TSS (%)	0.99	0.47	98.27	0.18	0.09
NTH	MS	95.95**	105.02**	726.42**	11.81^{ns}	23.13
	TSS (%)	9.97	10.91	75.49	1.23	2.4
NPH	MS	68.43**	78.49**	478.86**	12.40^{ns}	15.13
	TSS (%)	10.47	12.01	73.3	1.9	2.32
PL	MS	1.54**	4.67**	97.22**	1.29**	0.53
	TSS (%)	1.46	4.44	92.37	1.23	0.5
FG	MS	1791.21**	2877.37^{ns}	38,887.19**	1743.21**	608.58
	TSS (%)	3.9	6.27	84.71	3.8	1.32
UFG	MS	1979.64**	700.88*	45,613.77**	352.45**	212.92
	TSS (%)	4.05	1.43	93.36	0.72	0.44
TG	MS	1384.19*	1595.14^{ns}	5975.58**	1327.15**	686.72
	TSS (%)	12.62	14.54	54.48	12.1	6.26
PFG	MS	280.32**	145.62*	8234.20**	78.53**	34.06
	TSS (%)	3.2	1.66	93.86	0.9	0.38
TGW	MS	10.46**	6.97*	17.05**	3.44**	1.56
	TSS (%)	26.49	17.66	43.19	8.71	3.95
TW	MS	449.08**	380.11**	594.18**	138.06**	85.08
	TSS (%)	27.27	23.09	36.09	8.38	5.17
YLD	MS	11.49**	9.73**	15.19**	3.53**	2.18
	TSS (%)	27.28	23.1	36.06	8.38	5.18

Note: * significant at $p \leq 0.05$, ** highly significant at $p \leq 0.01$, ns: not significant at $p > 0.05$; SOV—source of variation; DF—degrees of freedom; MS—mean square; TSS—total sum of square; DTF—days to flowering; DTM—days to maturity; PH—plant height; NTH—number of tillers per hill; NPH—number of panicles per hill; PL—panicles length; FG—number of filled grains per panicle; UFG—number of unfilled grains per panicle; TG—total number of grains; PFG—percentage of filled grains; TGW—thousand-grain weight; TW—total weight; YLD—yield per hectare.

Table 3. Mean for 13 traits of 19 rice genotypes tested in four environments.

Genotypes	DTF (Day)	DTM (Day)	PH (cm)	NTH (no)	NPH (no)	PL (cm)	FG (no)	UFG (no)	TG (no)	PFG (%)	TGW (g)	TW (g)	YLD(t/ha)
G1	72.25 ^{ab}	104.25	98.52 ^{ab}	19.92 ^{de}	17.00 ^{de}	24.30 ^{abc}	192.33	47.91	240.25	79.35	23.21 ^{ab}	38.60 ^{ab}	6.17 ^{ab}
G2	72.00 ^{ab}	103.17	100.72 ^{ab}	19.83 ^{e}	17.50 ^{de}	24.81^{abc}	178.67	55.41	234.08	77.29	22.89 ^{ab}	33.05^{b}	5.29 ^{b}
G3	73.50 ^{ab}	105.33	100.93 ^{ab}	22.17 ^{bcde}	18.58 ^{cde}	24.85^{abc}	176.83	55.75	232.59	76.06	23.93 ^{ab}	34.30 ^{b}	5.49 ^{b}
G4	72.50 ^{ab}	104.75	100.12 ^{ab}	22.25 ^{bcde}	18.25 ^{cde}	24.86^{abc}	186.58	48.25	234.83	78.39	24.19 ^{ab}	33.83 ^{b}	5.41 ^{b}
G5	73.00 ^{ab}	105	99.32 ^{ab}	20.75 ^{cde}	18.67 ^{cde}	24.38^{abc}	194.25	49.33	243.58	79.97	23.85 ^{ab}	39.95 ^{ab}	6.39 ^{ab}
G6	71.67 ^{ab}	104.17	101.73 ^{a}	19.75 ^{e}	16.58 ^{e}	25.15 ^{a}	193.42	40.75	234.17	81.91	23.22 ^{ab}	37.80 ^{ab}	6.05 ^{ab}
G7	70.92 ^{b}	101	99.55 ^{ab}	22.75^{bcde}	19.83^{bcde}	24.53^{abc}	165.92	64.33	230.25	72.49	23.65 ^{ab}	33.27 ^{b}	5.32 ^{b}
G8	72.33 ^{ab}	103.83	98.93 ^{ab}	23.17^{bcde}	20.25^{bcde}	24.67^{abc}	185.08	44.75	229.83	80.13	23.07 ^{ab}	38.12 ^{ab}	6.10 ^{ab}
G9	74.25 ^{ab}	106.17	97.25 ^{ab}	25.17 ^{ab}	20.75^{abcde}	23.36^{abc}	169.42	48.17	217.59	79.14	22.54 ^{ab}	35.16 ^{b}	5.63 ^{b}
G10	71.17 ^{b}	102.75	94.35 ^{b}	22.08^{bcde}	19.83^{bcde}	23.53^{abc}	195.33	51.92	247.25	79.71	22.75 ^{ab}	36.82 ^{ab}	5.89 ^{ab}
G11	73.92 ^{ab}	105.58	98.61 ^{ab}	22.83^{bcde}	19.92^{bcde}	25.15 ^{a}	226.58	43.33	269.92	82.92	22.49 ^{ab}	41.00 ^{ab}	6.56 ^{ab}
G12	75.92 ^{a}	106.25	98.05 ^{ab}	21.25 ^{cde}	18.25 ^{cde}	24.36^{abc}	191.92	46.5	238.42	80.47	23.00 ^{ab}	34.21^{b}	5.47 ^{b}
G13	73.25 ^{ab}	103.83	99.04 ^{ab}	22.83^{bcde}	19.00 ^{cde}	23.37^{bc}	175.17	60.5	235.67	74.42	23.06 ^{ab}	33.44 ^{b}	5.35 ^{b}
G14	72.00 ^{ab}	103.58	94.57 ^{ab}	24.67^{abcde}	21.33^{abcde}	23.48^{abc}	178.67	47.08	225.75	79.23	24.04 ^{ab}	40.91 ^{ab}	6.55 ^{ab}
G15	71.58 ^{ab}	103.58	94.28 ^{b}	26.92 ^{ab}	23.25 ^{abc}	23.17 ^{c}	170.58	56.42	227	74.75	21.63 ^{b}	33.76 ^{b}	5.40 ^{b}
G16	73.67 ^{ab}	105.17	97.12 ^{ab}	21.17 ^{cde}	17.67 ^{de}	24.48^{abc}	156.92	58.59	215.5	72.97	23.01 ^{ab}	31.01 ^{b}	4.96 ^{b}
G17	73.83 ^{ab}	105.33	98.79 ^{ab}	25.08 ^{abcd}	22.08^{abcd}	24.62^{abc}	194.42	39.25	233.66	83.61	23.08 ^{ab}	43.95 ^{ab}	7.03 ^{ab}
G18	73.25 ^{ab}	104.75	100.48 ^{ab}	29.58 ^{a}	25.00 ^{ab}	25.06 ^{ab}	199.92	38.33	238.25	83.89	24.78 ^{a}	54.23 ^{a}	8.68 ^{a}
MR219	72.83 ^{ab}	105.5	101.10 ^{ab}	29.67 ^{a}	25.67 ^{a}	24.44^{abc}	193.17	38.83	232	82.83	24.42 ^{ab}	45.79 ^{ab}	7.33 ^{ab}
Mean	72.83	104.42	98.6	23.25	19.97	24.36	185.53	49.23	234.77	78.92	23.31	37.85	6.06
HSD(p = 0.05)	4.31	5.88	7.2	5.18	5.31	1.71	62.97	28.31	54.94	13.37	2.8	17.72	2.83
CV	4.03	3.9	5	26.93	23.76	4.34	20.54	43.32	13.94	11.98	7.73	32.64	32.64
Max	77.33	112.67	111.73	49	30.33	26.7	275	132.33	315.33	90.97	28.36	70.17	11.23
Min	65.33	95.67	88.13	12.67	10.33	22.07	108	25.33	169.33	46.92	18.32	13.78	2.2

Note: HSD—honestly significant difference by Tukey's test (means within each column with same letter are not significantly different with HSD test $p > 0.05$), CV—coefficient of variation, DTF—days to flowering, DTM—days to maturity, PH—plant height, NTH—number of tillers per hill, NPH—number of panicles per hill, PL—panicles length, FG—number of filled grains per panicle, UFG—number of unfilled grains per panicle, TG—total number of grains per panicle, PFG—percentage of fertile grains, TGW—thousand-grain weight; TG—total grains per hill weight; YLD—yield per hectare.:

Table 4. Variance components, coefficients of variation, heritability and genetic advance of the rice genotypes for 13 traits assessed in four environments.

SOV	DTF (Day)	DTM (Day)	PH (cm)	NTH (No)	NPH (No)	PL (cm)	FG (No)	UFG (No)	TG (No)	PFG (%)	TGW (g)	TW (g)	YLD (t/ha)
σ^2_g	0.67	0.40	3.15	7.08	5.34	0.28	94.51	29.04	22.33	5.59	0.29	20.17	0.52
σ^2_{gs}	1.94	2.26	3.57	0.00	0.00	0.25	378.21	46.51	213.48	14.82	0.63	17.66	0.45
σ^2_e	3.94	8.41	12.11	20.04	14.38	0.53	608.58	212.92	686.72	34.06	1.56	85.08	2.18
σ^2_p	6.55	11.07	18.83	27.12	19.72	1.06	1081.30	288.47	922.53	54.47	2.48	122.91	3.15
Mean	72.83	104.42	98.60	23.25	19.97	24.36	185.54	49.23	234.77	78.92	23.31	37.85	6.06
h^2_B (%)	10.23	3.61	16.72	26.11	27.08	26.42	8.74	10.07	2.42	10.26	11.70	16.41	16.51
GCV (%)	1.12	0.61	1.80	11.44	11.57	2.17	5.24	10.95	2.01	3.00	2.31	11.87	11.91
PCV (%)	3.51	3.19	4.40	22.40	22.24	4.23	17.72	34.50	12.94	9.35	6.76	29.29	29.30
GA (%)	0.74	0.24	1.52	12.05	12.40	2.30	3.19	7.15	0.65	1.98	1.63	9.90	9.97

Note: SOV—source of variation; DF—degrees of freedom; σ^2_g—variance of genotypic; σ^2_{gs}—variance of genotypic by environment; σ^2_e—variance of error; σ^2_p—variance of phenotypic; h^2_B—broad-sense heritability; PCV—phenotypic coefficient of variation; GCV—genotypic coefficient of variation; GA—genetic advance; DTF—days to flowering; DTM—days to maturity; PH—plant height; NTH—number of tillers per hill; NPH—number of panicles per hill; PL—panicle length; FG—number of filled grains per panicle; UFG—number of unfilled grains per panicle; TG—total number of grains per panicle; PFG—percentage of filled grains per panicle; TGW—thousand-grain weight; TW—total grains weight per hill; YLD—yield per hectare.

Table 5. Correlation coefficients among the quantitative traits evaluated.

	DTF	DTM	PH	NTH	NPH	PL	FG	UFG	TG	PFG	TGW	TW	YLD
DTF		0.43**	-0.30**	0.183**	-0.02	-0.14*	0.20**	-0.19**	0.06	0.22**	-0.12	-0.07	-0.07
DTM			-0.30**	0.28**	0.42**	-0.06	0.30**	-0.27**	0.12	0.31**	0.13*	0.122	0.128
PH				-0.49**	-0.27**	0.72**	-0.51**	0.71**	0.05	-0.73**	0.25**	0.02	0.02
NTH					0.80**	-0.31**	0.36**	-0.41**	0.05	0.45**	-0.03	0.38**	0.38**
NPH						-0.14*	0.32**	-0.28**	0.14*	0.33**	0.11	0.45**	0.45**
PL							-0.16**	0.55**	0.33**	-0.49**	0.26**	0.22**	0.22**
FG								-0.63**	0.65**	0.79**	-0.14*	0.46**	0.46**
UFG									0.18**	-0.96**	0	-0.20**	-0.20**
TG										0.06	-0.17**	0.39**	0.38**
PFG											-0.04	0.31**	0.31**
THW												0.22**	0.22**
TW													1**
YLD													

Note: *significant at 0.05 probability level, **highly significant at 0.01 probability level, DTF—days to flowering; DTM—days to maturity; PH—plant height; NTH—number of tillers per hill; NPH—number of panicle per hill; PL—panicle length; FG—number of filled grain; UFG—number of unfilled grain; TG—total number of grain per panicle; PFG—percentage of fertile grain; THW—thousand-grain weight; TW—total weight per hill; YLD—yield in t/ha.

4. Discussion

The number of tillers in rice is a significant agronomic trait that directly affects grain yield due to its positive relationship with the production of panicle that will bear the rice grains. The presence of fewer tillers produced fewer panicles, while excess tillers caused higher tillers abortions, small panicles and poor grain filling [34,35]. Ranawake and Amarasinghe [36] reported the positive relationship of tillers and panicles that caused a decrease in grain output. In terms of individual tiller development, a water deficit at these reproductive stages causes irreparable loss of potential yield. Sometimes, several tillers had two panicles where the original panicle is sterile, but the second panicle became nonfunctional. This panicle originated from the flag leaf node that was developed after the original assimilate sink. In this study, the number of tillers produced was between 19 and 29, while the number of panicles ranged from 16 to 25. These are moderate amount of tillers and panicles in rice production. Rice yields are influenced by many factors, yet number of tillers and panicles are mainly considered as key factors.

Panicle length determines the amount of spikelet and grain that can be produced in a panicle. With good environmental conditions, longer panicle length gives room for more production of spikelet and total grain, resulting in high final yield output [37]. However,, high number of total grains is meaningless if the percentage of fertile grains is low. Percentage of fertile grains is the calculation on amount of filled grains over total grains in a panicle. High percentage as recorded in this study signifies high number of fertile grains on the panicle and it is preferred by farmers since it will directly influence the final yield. As revealed by this study, number of filled grains and percentage of fertile grains were positively correlated with the yield. All genotypes evaluated recorded high number of filled grains with average of more than 150 grains. In addition, number of unfilled grains produced was recorded at low levels (38–64 grains), hence, signify higher yield of rice being produced. However, if there were limitation or ineffective translocation of nutrition from the sources, it will affect the condition of the plant parts together with the rice grain. This results in poor production of rice grains [38]; a smaller number of total grains, small size, half-filled or unfilled grains. Although there is a high production of total grains, total grain weight is an important determinant of yield in rice. The study recorded moderate (31.01 g) and high (54.23 g) total grains weight. A decrease in grain weight is expected when there is water deficit in rice during grain filling. If the current photosynthate supply is limited, the ability to mobilize and translocate reserves would be adaptive. Reduction of grain weight may be due to defective grains. This can be signified by the value of 1000-grain weight. The low weight of 1000-grain indicates grains appearance; slender, small and thick hulls. In this work, moderate and high value of 1000-grain weight was observed that revealed the good appearance of rice grains; big and fully filled (fat). Therefore, final yield also was observed with the same results.

This study revealed wide phenotypic variability for 13 traits evaluated among the improved genotypes. The result indicates the existence of sufficient variations among the assessed genotypes for those traits under consideration that can help breeders in selection of ideal genotypes. These variations of the improved lines in relation to their agro-morphologic traits may be due to the fact that these lines were evaluated in environments that differed in temperature, humidity, rainfall and soil type. This signifies the need for multi-environmental trials of rice at various locations or environments in order to see how the genotypes react in different environments due to the presence of G×E interaction. Such variations in relation to the environment that influenced rice growth performance evaluation studies have extensively been reported previously [39–41].

Several reports have also been published on significant phenotypic variation among rice accessions although evaluated in only one environment [42–44]. Islam et al. [45] observed highly significant genetic variability in 113 rice genotypes for 18 traits studied. Newest study by Hosagoudar et al. [46] reported that there were significant differences that existed for traits investigated in 18 genotypes evaluated under hilly conditions. In Malaysia area of rice study, Tuhina-Khatun et al. [47] showed a significant existence of diversity level on 43 upland rice genotypes for 22 traits evaluated. Previously, Sohrabi et al. [48] presented significant differences in 50 Malaysian upland rice germplasms evaluated for 12 traits. A significant variation, especially for G×E, signifies that the evaluated genotypes do react differently in a different environment. This fact shows that G×E interaction greatly gives influence in the selection process of superior genotype for release and this constraint needs to be solved by rice breeders

The study showed that the first environment (Tanjung Karang) planted from September 2015–January 2016 had the highest average relative humidity and rainfall although on a lower altitude of 3 m. There is no doubt that the period falls within the Malaysian heavy rain period (October–January). However, the rainfall regime and other environmental factors recorded in the other three environments were not the same. This corresponds with the significant difference observed in environment for most traits. The coefficient of variation in this study showed that PCV was relatively higher than GCV for all traits. The magnitude of differences between the value of PCV and GCV demonstrated how much environment influences the trait, where big differences referred to large environmental effect and little differences showed high genetic influence (27,39). The low differences between GCV and PCV obtained from this study implied that these traits can be used as selection criteria for further crop improvement since the variation in observed variables were mostly due to genetic factors. Previous study by Osman et al. [49] and Habib et al. [50] also found the same outcome for these traits. Anis et al. [51] and Nishanth et al. [52] also showed a difference of GCV and PCV values for several traits evaluated in 20 rice varieties and 525 germplasm lines for submergence tolerance, respectively. However, other seven traits showed high differences between PCV and GCV especially number of unfilled grains, total grain weight and yield per hectare. According to Hosagoudar and Kovi [53], the yield and yield component traits had large differences of PCV and GCV values among the 15 advanced rice genotypes evaluated for leaf blast reaction, genotypic performance and correlations. This big difference indicated large contribution of environmental factors on the phenotypic expression, hence, selection referring to these traits will not be much effective (27).

High value of GCV also indicates the existence of high genetic variation and selection using these traits to improve the genotypes could be effective. The results showed moderate value of GCV in number of tillers, number of panicles, number of unfilled grains, total grains weight and grain yield indicating moderate variability. This study was in agreement with Hasan et al. [54] who recorded moderate GCV in number of tillers, number of panicles and yield of newly developed blast resistant lines derived from crossing between MR263 and Pongsu Seribu 1. Srujuna et al. [55] also reported moderate GCV percentage for plant height, number of tillers and number of panicles on 29 evaluated rice genotypes. Value of GCV gives details on the genetic variability existing in quantitative traits, but the amount of variation that was heritable from the value of GCV is impossible to be determined. Thus, the amount of advance to be anticipated from selection is the best being visualized when using GCV value with the help of heritability estimates [56].

There were low heritability values for all traits evaluated in this study. This is an indication that direct selection using these traits would be ineffective due to high influence of environment. The results obtained in GA from this study were similar to the report by Kole et al. [57] for traits such as days to flowering, plant height and panicle length. Akinwale et al. [58] also reported days to maturity, panicle length, number of tillers and 1000-grain weight with low GA value. In contrast, Immanuel et al. [59] found high heritability value of more than 90% and high GA for most traits except for days to maturity and panicle length which had a moderate genetic advance. Work of Govintharaj et al. [60] also observed high and moderate heritability for several traits in the segregating population of blast introgressed lines.

In this study, the result showed that except days to flowering and number of unfilled grains per panicle, all other traits revealed positive relationship with yield per hectare. The association was highly significant with all traits studied except for days to maturity and plant height that were not significant with yield. The positive and significant correlation coefficients recorded between yield per hectare and other 10 quantitative traits indicated that such traits influenced the final yield of rice. These traits were suitable to be considered as yield prediction factors and deserved considerable importance during selection in further study [39,61]. Manipulation of these traits may be useful to develop high yielding genotypes with desired traits. High yield was the results from the contribution of high number of tillers and number of panicles that helped in producing high number of total grains. Furthermore, higher yield was produced when there was high number of filled grains than unfilled grains. Number of tillers is a significant agronomic trait that directly affects grain yield due to its positive relationship with the production of panicle that will bear the rice grains. The presence of fewer tillers produced fewer panicles, while, excess tillers caused higher tiller abortions, small panicles and poor grain filling [34,35,37]. Ranawake and Amarasinghe [36] showed positive relationship between tillers and panicles that resulted in a decrease in grain output. The number of tillers and panicles produced in this study were moderate. In addition, longer panicle produces more grain yield. Few studies had been done that reported positive relationship between panicle length and yield [62–64].

This study revealed the effectiveness of quantitative or morphologic traits in grouping the rice genotypes. Group III clustered the genotypes that had high yield per hectare with good performances of other traits. It revealed that not only G17 and G18 have high yield, but also their morphologic performances were almost the same as the commercialized rice variety, MR219. Consequently, these genotypes could be fully utilized in a further breeding program for genotypes improvement. Ahmadikhah et al. [65] clustered 58 inbred rice lines into three distinct classes using 18 morphologic traits. The level of genetic diversity in crops is an important matter in maintaining and increasing agricultural productivity and it needs considerable attention.

5. Conclusions

Genotypes G17 and G18 maintained superiority in yield across the four environments and this showed that they have potentials for selection in further varietal improvement. Consequently, these genotypes can be fully utilized in a further breeding program for genotypes improvement. In particular, such genotypes would be useful to breeders in selecting high-yielding blast resistant rice varieties for different environments. The wide variability obtained among the 19 genotypes evaluated across the four varied environments is a cornerstone for plant breeding and the genotypes could serve as classical model for crop improvement. There was a high influence of environment in the genotypes' performance. However, the GxE result obtained for yield components such as number of tillers and number of panicles per hill revealed the existence of genetic variation at a considerable amount. The positive and significant correlation of these traits with yield showed that an increase in selection pressure on such traits could improve their agronomic yield. The new blast disease-resistant rice lines have great potentials in high yielding, blast resistance and useful for further crop improvement. Therefore, the six clusters/groups of genotypes obtained from this study are recommended as varieties for commercial cultivation in Malaysia and other rice growing regions.

Author Contributions: Conceived and designed the experiments: M.Y.R., R.S.S and M.R.I.; Performed the experiment: R.S.S. and M.Y.R. Analyzed the data: R.S.S., M.Y.R. and O.Y. Wrote and revised the article: R.S.S., M.Y.R., Y.O., S.C.C. and N.H., All authors have read and agreed to the published version of the manuscript.

Funding: The authors are grateful to the Ministry of Education Malaysia, for adequate research Higher Institution Centers of Excellence (HICoE) funding to conduct research on improvement of crop varieties for adaption to biotic and abiotic stress with grant number 6369105.

Conflicts of Interest: The authors declare no conflicts of interest.

References

1. Chukwu, S.C.; Rafii, M.Y.; Ramlee, S.I.; Ismail, S.I.; Hasan, M.M.; Oladosu, Y.A.; Magaji, U.G.; Akos, I.; Olalekan, K.K. Bacterial leaf blight resistance in rice: A review of conventional breeding to molecular approach. *Mol. Biol. Rep.* **2019**, *46*, 1519–1532. [CrossRef]
2. Prasad, R.; Shivay, Y.S.; Kumar, D. Current Status, Challenges, and Opportunities in Rice Production. In *Rice Production Worldwide*; Chauhan, B.S., Jabran, K., Mahajan, G., Eds.; Springer International Publishing: Cham, Switzerland, 2017; pp. 1–32.

3. Conforti, P. *Looking ahead in World Food and Agriculture: Perspectives to 2050*; Food and Agriculture Organization of the United Nations (FAO): Rome, Italy, 2011.
4. Maclean, J.; Hardy, B.; Hettel, G. *Rice Almanac: Source Book for One of the Most Important Economic Activities on Earth*; IRRI: Los Baños, Philippines, 2013.
5. Vengedasalam, D.; Harris, M.; MacAulay, G. Malaysian rice trade and government interventions. In Proceedings of the 55th Annual Conference of the Australian Agricultural and Resource Economics Society, Melbourne, VI, Australia, 8–11 February 2011.
6. Ramli, N.N.; Shamsudin, M.N.; Mohamed, Z.; Radam, A. Impact of price support policy on Malaysian rice industry. In Proceedings of the International Annual Symposium on Sustainability Science and Management, University Malaysia Terengganu (UMT), Terengganu, Malaysia, 9–11 July 2012.
7. USDA. *World Agricultural Production*; Foreign Agricultural Service Circular Series WAP 08-17; USDA: Washington, DC, USA, 2017.
8. Department of Statistics Malaysia. *Supply and Utilization Accounts Selected Agricultural Commodities, Malaysia 2012–2016*; Department of Statistics Malaysia: Sandakan, Malaysia, 2017.
9. Arshad, F.M.; Alias, E.F.; Noh, K.M.; Tasrif, M. Food security: Self-sufficiency of rice in Malaysia. *Int. J. Manag. Stud.* **2011**, *18*, 83–100.
10. Chukwu, S.C.; Rafii, M.Y.; Ramlee, S.I.; Ismail, S.I.; Oladosu, Y.; Kolapo, K.; Ahmed, M. Marker-assisted introgression of multiple resistance genes confers broad spectrum resistance against bacterial leaf blight and blast diseases in Putra-1 rice variety. *Agronomy* **2020**, *10*, 42. [CrossRef]
11. Divya, B.; Biswas, A.; Robin, S.; Rabindran, R.; Joel, A.J. Gene interactions and genetics of blast resistance and yield attributes in rice (*Oryza sativa* L.). *J. Genet.* **2014**, *93*, 415–424. [CrossRef] [PubMed]
12. Tanweer, F.A.; Rafii, M.Y.; Sijam, K.; Rahim, H.A.; Ahmed, F.; Latif, M.A. Current advance methods for the identification of blast resistance genes in rice. *Comptesrendusbiologies* **2015**, *338*, 321–334. [CrossRef] [PubMed]
13. Koutroubas, S.D.; Katsantonis, D.; Ntanos, D.A.; Lupotto, E. Blast disease influence on agronomic and quality traits of rice varieties under Mediterranean conditions. *Turk. J. Agricult. For.* **2009**, *33*, 487–494.
14. Nasruddin, A.; Amin, N. Effects of Cultivar, Planting Period, and Fungicide Usage on Rice Blast Infection Levels and Crop Yield. *J. Agricult. Sci.* **2012**, *5*, 160. [CrossRef]
15. Miah, G.; Rafii, M.Y.; Ismail, M.R.; Puteh, A.B.; Rahim, H.A.; Asfaliza, R.; Latif, M.A. Blast resistance in rice: A review of conventional breeding to molecular approaches. *Mol. Biol. Rep.* **2013**, *40*, 2369–2388. [CrossRef]
16. Chukwu, S.C.; Rafii, M.Y.; Ramlee, S.I.; Ismail, S.I.; Oladosu, Y.; Okporie, E.; Jalloh, M. Marker-assisted selection and gene pyramiding for resistance to bacterial leaf blight disease of rice (*Oryza sativa* L.). *Biotechnol. Biotechnol. Equip.* **2019**, *33*, 1–16. [CrossRef]
17. Miah, G.; Rafii, M.Y.; Ismail, M.R.; Puteh, A.B.; Rahim, H.A.; Latif, M.A. Recurrent parent genome recovery analysis in a marker-assisted backcrossing program of rice (*Oryza sativa* L.). *Comptesrendusbiologies* **2015**, *338*, 83–94. [CrossRef]
18. Tanweer, F.A.; Rafii, M.Y.; Sijam, K.; Rahim, H.A.; Ahmed, F.; Ashkani, S.; Latif, M.A. Introgression of blast resistance genes (putative *Pi-b* and *Pi-kh*) into elite rice cultivar MR219 through marker-assisted selection. *Front. Plant Sci.* **2015**, *6*, 1002. [CrossRef] [PubMed]
19. Jayawardana, W.A.D.; Jayasekera, G.A.U.; Wijesundera, R.L.C.; Dissanayake, D.M.; Sooriyapathirana, S.D.S.S.; Weebadde, C.K.; Hettige, P. Evaluation of DNA markers linked to blast resistant genes, Pikh, Pit (p), and Pita, for parental selection in Sri Lankan rice breeding. *Trop. Agricult. Res.* **2015**, *26*, 82–93. [CrossRef]
20. Mackill, D.J.; Bonman, J.M. Inheritance of blast resistance in near-isogenic lines of rice. *Phytopathology* **1992**, *82*, 746–749. [CrossRef]
21. Falconer, D.S. *Introduction to Quantative Genetics*; Longman: Harlow, UK, 1975.
22. Azizi, P.; Rafii, M.Y.; Maziah, M.; Abdullah, S.N.; Hanafi, M.M.; Latif, M.A.; Rashid, A.A.; Sahebi, M. Understanding the shoot apical meristem regulation: A study of the phytohormones, auxin and cytokinin, in rice. *Mech. Dev.* **2015**, *135*, 1–5. [CrossRef] [PubMed]
23. Kang, M.S. Breeding: Genotype-by-Environment Interaction. In *Encyclopedia of Plant and Crop Science*; Goodman, R.M., Ed.; Marcel-Dekker: New York, NY, USA; Taylor & Francis: Park Drive, UK, 2004; pp. 218–221.
24. Inthapanya, P.S.; Sihavong, P.; Sihathep, V.; Chanphengsay, M.; Fukai, S.; Basnayake, J. Genotypic performance under fertilized and non-fertilised conditions in rainfed lowland rice. *Field Crops Res.* **2000**, *65*, 1–14. [CrossRef]

25. Wade, L.J.; McLaren, C.G.; Quintana, L.; Harnpichitvitaya, D.; Rajatasereekul, S.; Sarawgi, A.K.; Siopongco, J. Genotype by environment interactions across diverse rainfed lowland rice environments. *Field Crops Res.* **1999**, *64*, 35–50. [CrossRef]
26. Cooper, M.; Byth, D.E.; Hammer, G.L. Understanding plant adaptation to achieve systematic applied crop improvement—A fundamental challenge. In *Plant Adaptation and Crop Improvement*; Centre for Agriculture and Bioscience International: Wallingford, UK, 1996; pp. 5–23.
27. Oladosu, Y.; Rafii, M.Y.; Abdullah, N.; Magaji, U.; Miah, G.; Hussin, G.; Ramli, A. Genotype × Environment Interaction and Stability Analyses of Yield and Yield Components of Established and Mutant Rice Genotypes Tested in Multiple Locations in Malaysia. *Acta Agricult. Scand. Sect. B Soil Plant Sci.* **2017**, *67*, 590–606. [CrossRef]
28. Xavier, A.; Jarquin, D.; Howard, R.; Ramasubramanian, V.; Specht, J.E.; Graef, G.L.; Nelson, R. Genome-Wide Analysis of Grain Yield Stability and Environmental Interactions in a Multiparental Soybean Population. *G3 Genes Genomes Genet.* **2017**, *8*, 519–529. [CrossRef]
29. Hashemi, F.G.; Rafii, M.Y.; Ismail, M.R.; Mahmud, T.M.; Rahim, H.A.; Asfaliza, R.; Malek, M.A.; Latif, M.A. Biochemical, genetic and molecular advances of fragrance character in rice. *Crit. Rev. Plant Sci.* **2013**, *32*, 445–457. [CrossRef]
30. Tariku, S.; Lakew, T.; Bitew, M.; Asfaw, M. Genotype by Environment Interaction and Grain Yield Stability Analysis of Rice (*Oryza sativa* L.) Genotypes Evaluated in North Western Ethiopia. *Net J. Agricult. Sci.* **2013**, *1*, 10–16.
31. IRRI. *Standard Evaluation System for Rice*, 5th ed.; International Rice Research Institute: Manila, Philippines, 2013; p. 55.
32. Sivasubramanian, S.; Menon, P.M. Genotypic and phenotypic variability in rice. *Madras Agric. J.* **1973**, *60*, 1093–1096.
33. Johnson, H.W.; Robinson, H.F.; Comstock, R.E. Genotypic and Phenotypic Correlations in Soybeans and Their Implications in Selection 1. *Agron. J.* **1955**, *47*, 477–483. [CrossRef]
34. Shakeel, A.; Abid, H.; Hakoomat, A.; Ashfaq, A. Transplanted fine rice (*Oryza sativa* L.) productivity as affected by plant density and irrigation regimes. *Int. J. Agricult. Biol.* **2005**, *7*, 445–447.
35. Peng, S.; Khush, G.S.; Cassman, K.G. Evolution of the new plant ideotype for increased yield potential. In *Breaking the Yield Barrier Proceedings of a Workshop on Rice Yield Potential in Favorable Environments*; International Rice Research Institute: Los Banos, Philippines, 1994; pp. 5–20.
36. Ranawake, A.L.; Amarasinghe, U.G.S. Relationship of yield and yield related traits of some traditional rice cultivars in Sri Lanka as described by correlation analysis. *J. Sci. Res. Rep.* **2014**, *3*, 2395–2403. [CrossRef]
37. Oladosu, Y.; Rafii, M.Y.; Abdullah, N.; Abdul Malek, M.; Rahim, H.A.; Hussin, G.; Kareem, I. Genetic variability and selection criteria in rice mutant lines as revealed by quantitative traits. *Sci. World J.* **2014**, *2014*, 1–12. [CrossRef]
38. Yang, J.; Zhang, J. Grain-filling problem in 'super' rice. *J. Exp. Bot.* **2009**, *61*, 1–5. [CrossRef]
39. Oladosu, Y.; Rafii, M.Y.; Magaji, U.; Abdullah, N.; Miah, G.; Chukwu, S.C.; Kareem, I. Genotypic and phenotypic relationship among yield components in rice under tropical conditions. *Biomed. Res. Int.* **2018**, *2018*, 1–10. [CrossRef]
40. Konate, A.K.; Zongo, A.; Kam, H.; Sanni, A.; Audebert, A. Genetic variability and correlation analysis of rice (*Oryza sativa* L.) inbred lines based on agro-morphological traits. *Afr. J. Agricult. Res.* **2016**, *11*, 3340–3346.
41. Kumar, P.; Sao, A.; Salam, J.L.; Kanwar, R.R.; Poonam, K.; Gupta, A.K.; Kar, S. Formulating phenological equations for rainfed upland rice in Bastar plateau and assessment of genotype x environment interaction. *Pak. J. Bot* **2016**, *48*, 1609–1618.
42. Begum, M.; Hossain, M.A.; Hossain, F.M.M.; Hasan, A.K. Genetic variability for grain yield and yield associated traits in transplant Aman rice (*Oryza sativa* L.). *Res. Agricult. Livestock Fish.* **2015**, *2*, 207–213. [CrossRef]
43. Chouhan, S.K.; Singh, A.K.; Singh, A.; Singh, S.P.; Singh, N.K.; Singh, P.K. Agro-morphological diversity in wild rice accessions of Eastern Indo-Gangetic region of India. *Bangladesh J. Bot.* **2015**, *43*, 337–344. [CrossRef]
44. Sajid, M.; Khan, S.A.; Khurshid, H.; Iqbal, J.; Muhammad, A.; Saleem, N.; Shah, S.M.A. Characterization of rice (*Oryza Sativa* L.) germplasm through various agro-morphological traits. *Sci. Agricult.* **2015**, *9*, 83–88.

45. Islam, M.Z.; Khalequzzaman, M.; Bashar, M.K.; Ivy, N.A.; Haque, M.M.; Mian, M.A.K. Variability assessment of aromatic and fine rice germplasm in Bangladesh based on quantitative traits. *Sci. World J.* **2016**, *2016*, 1–14. [CrossRef] [PubMed]
46. Hosagoudar, G.N.; Kovi, B.S.; Sheela, H.K. Genetic Variability, Correlation and Leaf Blast Resistance Studies in Early Maturing Rice Genotypes under Hilly Conditions. *Int. J. Curr. Microbiol. App. Sci.* **2017**, *6*, 3451–3459. [CrossRef]
47. Tuhina-Khatun, M.; Hanafi, M.M.; Rafii Yusop, M.; Wong, M.Y.; Salleh, F.M.; Ferdous, J. Genetic variation, heritability, and diversity analysis of upland rice (*Oryza sativa* L.) genotypes based on quantitative traits. *BioMed Res. Int.* **2015**, *2015*, 1–7. [CrossRef]
48. Sohrabi, M.; Rafii, M.Y.; Hanafi, M.M.; Siti Nor Akmar, A.; Latif, M.A. Genetic diversity of upland rice germplasm in Malaysia based on quantitative traits. *Sci. World J.* **2012**, *2012*, 1–9. [CrossRef]
49. Osman, K.A.; Mustafa, A.M.; Ali, F.; Yonglain, Z.; Fazhan, Q. Genetic variability for yield and related attributes of upland rice genotypes in semi arid zone (Sudan). *Afr. J. Agricult. Res.* **2012**, *7*, 4613–4619. [CrossRef]
50. Habib, S.H.; Bashar, M.K.; Khalequzzaman, M.; Ahmed, M.S.; Rashid, E.S.M.H. Genetic analysis and morpho-physiological selection criteria for traditional Biroin Bangladesh rice germplasms. *J. Biol. Sci.* **2005**, *5*, 315–318.
51. Anis, G.; El-Sabagh, A.; Ghareb, A.; El-Rewainy, I. Evaluation of Promising Lines in Rice (*Oryza sativa* L.) to Agronomic and Genetic Performance under Egyptian Conditions. *Int. J. Agron. Agricult. Res.* **2016**, *8*, 52–57.
52. Nishanth, G.K.; Dushyanthakumar, B.M.; Gangaprasad, S.; Gowda, T.H.; Nataraju, S.P.; Shashidhar, H.E. Screening and Genetic Variability Studies in Submergence Tolerance Rice Germplasm Lines Under Flood Prone Lowlands of Hill Zone of Karnataka, India. *Int. J. Curr. Microbiol. Appl. Sci.* **2017**, *6*, 1254–1260. [CrossRef]
53. Hosagoudar, G.N.; Kovi, B.S. Evaluation of Advanced Varietal Rice Genotypes for Leaf Blast Reaction, Genotypic Performance and Correlation Studies. *Int. J. Curr. Microbiol. Appl. Sci.* **2018**, *7*, 75–81. [CrossRef]
54. Hasan, M.M.; Yusop, M.R.; Ismail, M.R.; Mahmood, M.; Rahim, H.A.; Latif, M.A. Performance of yield and yield contributing characteristics of BC_2F_3 population with addition of blast resistant gene. *Ciênc. Agrotecnol.* **2015**, *39*, 463–476. [CrossRef]
55. Srujana, G.; Suresh, B.G.; Lavanya, G.R.; Ram, B.J.; Sumanth, V. Studies on Genetic Variability, Heritability and Genetic Advance for Yield and Quality Components in Rice (*Oryza sativa* L.). *J. Pharmacogn. Phytochem.* **2017**, *6*, 564–566.
56. Burton, G.W.; Devane, E.H. Estimating heritability in tall fescue (*Festuca arundinacea*) from replicated clonal material 1. *Agron. J.* **1953**, *45*, 478–481. [CrossRef]
57. Kole, P.C.; Chakraborty, N.R.; Bhat, J.S. Analysis of variability, correlation and path coefficients in induced mutants of aromatic non-basmati rice. *Trop. Agricult. Res. Ext.* **2010**, *11*, 60. [CrossRef]
58. Akinwale, M.G.; Gregorio, G.; Nwilene, F.; Akinyele, B.O.; Ogunbayo, S.A.; Odiyi, A.C. Heritability and Correlation Coefficient Analysis for Yield and Its Components in Rice (*Oryza sativa* L.). *Afr. J. Plant Sci.* **2011**, *5*, 207–212.
59. Immanuel, S.C.; Pothiraj, N.; Thiyagarajan, K.; Bharathi, M.; Rabindran, R. Genetic parameters of variability, correlation and pathcoefficient studies for grain yield and other yield attributes among rice blast disease resistant genotypes of rice (*Oryza sativa* L.). *Afr. J. Biotechnol.* **2011**, *10*, 3322–3334. [CrossRef]
60. Govintharaj, P.; Shalini, T.; Manonmani, S.; Robin, S. Estimates of genetic variability, heritability and genetic advance for blast resistance gene introgressed segregating population in rice. *Int. J. Curr. Microbiol. App. Sci.* **2016**, *5*, 672–677. [CrossRef]
61. Chukwu, S.C.; Ekwu, L.G.; Onyishi, G.C.; Okporie, E.O.; Obi, I.U. Correlation between agronomic and chemical characteristics of maize (*Zea mays* L.) genotypes after two years of mass selection. *Int. J. Sci. Res.* **2015**, *4*, 1708–1712.
62. Kumar, V.; Singh, D. Association Studies for Yield and Its Component Traits in Basmati Genotypes of Himachal Pradesh, India. *Int. J. Curr. Microbiol. App. Sci.* **2018**, *7*, 1243–1249. [CrossRef]
63. Sowmiya, C.A.; Venkatesan, M. Studies on Correlation and Path Coefficient Analysis in Rice (*Oryza sativa* L.). *Int. J. Curr. Microbiol. App. Sci.* **2017**, *6*, 1757–1763. [CrossRef]

64. Sarawagi, A.K.; Ojha, G.C.; Koshta, N.; Pachauri, A. Genetic Divergence and Association Study for Grain Yield in Rice (*Oryza sativa* L.) Germplasm Accessions. *Ecoscan* **2015**, *9*, 217–223.
65. Ahmadikhah, A.; Nasrollanejad, S.; Alishah, O. Quantitative studies for investigating variation and its effect on heterosis of rice. *Int. J. Plant Prod.* **2008**, *2*, 297–308.

agronomy

MDPI

Article

Isolate-Dependent Inheritance of Resistance Against *Pseudoperonospora cubensis* in Cucumber

Tomer Chen, Daniel Katz, Yariv Ben Naim, Rivka Hammer, Bat Hen Ben Daniel, Avia E. Rubin and Yigal Cohen *

Faculty of Life Sciences, Bar Ilan University, Ramat-Gan 529000, Israel; tomerlivechen@gmail.com (T.C.); 1danielkatz@gmail.com (D.K.); yar2710@gmail.com (Y.B.N.); rivka.sh.h@gmail.com (R.H.); bathen.b.d@gmail.com (B.H.B.D.); evgenia.rubin@gmail.com (A.E.R.)

* Correspondence: yigal.cohen1@gmail.com; Tel.: +972-3-5318251; Fax: +972-3-5354133

Received: 9 July 2020; Accepted: 23 July 2020; Published: 27 July 2020

Abstract: Six wild accessions of *Cucumis sativum* were evaluated for resistance against each of the 23 isolates of the downy mildew oomycete *Pseudoperonospora cubensis*. The isolates originated from Israel, Europe, USA, and Asia. *C. sativum* PI 197088 (India) and PI 330628 (Pakistan) exhibited the highest level of resistance against multiple isolates of *P. cubensis*. Resistance was manifested as reduced lesion number, lesion size, sporangiophores and sporangia per lesion and enhanced encasement of haustoria with callose and intensive accumulation of lignin in lesions of both Plant Introductions (PIs) compared to the susceptible *C. sativum* SMR-18. In the field, much smaller AUDPC (Area Under Disease Progress Curve) values were recorded in PI 197088 or PI 330628 as compared to SMR-18. Each PI was crossed with SMR-18 and offspring progeny plants were exposed to inoculation with each of several isolates of *P. cubensis* in growth chambers and the field during six growing seasons. F1 plants showed partial resistance. F2 plants showed multiple phenotypes ranging from highly susceptible (S) to highly resistant (R, no symptoms) including moderately resistant (MR) phenotypes. The segregation ratio between phenotypes in growth chambers ranged from 3:1 to 1:15, depending on the isolate used for inoculation, suggesting that the number of genes, dominant, partially dominant, or recessive are responsible for resistance. In the field, the segregation ratio of 1:15, 1:14:1, or 1:9:6 was observed. F2 progeny plants of the cross between the two resistant PI's were resistant, except for a few plants that were partially susceptible, suggesting that some of the resistance genes in PI 197088 and PI 330328 are not allelic.

Keywords: *Cucumis sativum*; downy mildew; genetics; inheritance; oomycetes; resistance

1. Introduction

Downy mildew (DM) is a devastating foliar disease of cucurbits with a global distribution. The causal agent, *Pseudoperonospora cubensis* (Berk. and Curt.) Rost. (Oomycota, Peronosporaceae), is an obligate biotrophic oomycete pathogen that attacks over 40 host plant species belonging to 20 genera of the *Cucurbitaceae* [1,2]. Typical symptoms in cucumber consist of chlorotic irregular lesions with sporulation on the lower leaf surface. Several review articles provide basic information on the biology, epidemiology, and control of the disease [1–5]. Infection may take place if free leaf moisture is available for ≥2 h at an appropriate temperature [6]. The sporangia release biflagellate zoospores that swim towards the stomata where they encyst, germinate, and penetrate. Hyphae grow into the intercellular space, colonize the mesophyll tissue, and establish intracellular haustoria for nutrient uptake. Haustoria also deliver effector proteins to facilitate the establishment and/or combat the host plant's defense response system [3,7]. At ≥4 days post-inoculation, hyaline sporangiophores emerge from stomata bearing dark sporangia at their tip. Sporangia are dispersed by wind or rain and continue the asexual disease cycle. Lesion development is strongly affected by temperature and light.

Sporulation occurred in darkness at a moisture-saturated atmosphere. Sporangial yield is strongly influenced by the availability of photosynthate. Labeled C^{14}-CO_2 it was supplied to infected cucumber plants during the day and found in carbohydrates in the sporangia that were produced the following night [8].

Current control measures of downy mildew in cucurbits rely on fungicide applications. However, the frequent appearance of insensitive isolates to the current chemistries rendered them ineffective. The novel fungicide oxathiapiprolin that targets oxysterol binding proteins provides a relief due to its excellent systemic activity against DM in cucumber [9,10].

In the past two decades, major changes in the population structure of *P. cubensis* occurred: new genotypes, races, pathotypes, and mating types were reported from around the globe [4,5,11–13]. Possible mechanisms may involve the cultivation of large acreages of monocultures, the introduction of new cultivars, changes in climatic conditions (e.g., global warming), and the migration, mutation, and sexual recombination of the pathogen. By using ISSR (Inter-Simple Sequence Repeats) and SRAP (Sequence Related Amplified Polymorphism) markers, Polat et al. [14] discovered remarkable genetic diversity within and among isolates of *P. cubensis* in Europe and Asia. While isolates from Turkey and the Czech Republic exhibited uniform genetic background, the isolates from Israel were clearly distinguished from the others, probably due to migration and/or frequent sexual reproduction of the pathogen in Israel. Wallace et al. [15] used 10 SSR (Single Sequence Repeats) markers to show that in the USA, *P. cubensis* has two distinct, host-adapted clades at the cucurbit species level. Clade 1 isolates preferentially infect *Cucurbita pepo, Cucurbita maxima, Cucurbita moschata, Citrullus lanatus,* and wild hosts *Momordica charantia,* and *Momordica balsamina,* while clade 2 isolates preferentially infect *Cucumis sativus, Cucumis melo,* and the wild host *Lagenaria siceraria.* Clade 1 showed random mating and evidence of recombination and clade 2 non-random mating and no evidence of recombination. In Israel, the A1 isolates preferably infect *Cucumis sativus, Cucumis melo,* and *Lagenaria* sp. while the A2 isolates preferably infect *Cucurbita pepo, Cucurbita maxima, and Cucurbita moschata* [16].

Several introductions of wild cucumber were reported to carry resistance genes/QTLs (Quantitative Trait Loci) against *P. cubensis*, including PI 197085, PI 197087, PI 197088, PI 330628, Chinese Long, TH118FLM, and *Cucumis hystrix* [17]. Overall, many QTLs associated with resistance to DM have been identified across seven chromosomes [18]. However, researchers in different countries reported on a different number of genes or QTLs that confer resistance against the disease, probably because they worked in different environments, used different isolates of the pathogen, and/or used different evaluation methods. PI 197087 was reported to carry one, two, or three recessive genes, or two or three partially dominant genes for resistance. PI 197088 was reported to have 3–14 QTLs residing on chromosomes 1–7. Table 1 summarizes the genetic data available in the literature on genes/QTLs conferring resistance of cucumber against DM caused by *P. cubensis*.

Table 1. A literature survey showing the reported genes and QTLs responsible for the resistance of cucumber genotypes against downy mildew caused by *Pseudoperonospora cubensis*. (Abbreviations: RAPD—Random Amplification of Polymorphic DNA; SCAR—Sequence Characterized Amplified Regions; SNP—Single nucleotide polymorphism; RIL—Recombinant Inbred Line; SSR—Single Sequence Repeats).

Genotype	Country	Method	Genes/QTLs	Chromosome	Reference
PI 197085	EU	F2, RAPD, SCAR	3	5	[19]
PI 197087	EU		1 recessive		[20]
PI 197087	USA		3		[21]
PI 197087	Russia		3 partially dominant		[22]

Table 1. *Cont.*

Genotype	Country	Method	Genes/QTLs	Chromosome	Reference
PI 197087	EU		3 recessive		[23]
PI 197087	Israel		2 partially dominant		[24]
PI 197088	EU		2 recessive		[25]
PI 197088	USA	SNP	3	2, 4, 5	[26]
PI 197088	Japan	RIL	14	1, 3, 5, 6, 7	[27]
PI 197088	USA	RIL/SNP	11	1, 2, 3, 4, 5, 6	[28]
PI 197088	China	F2:F3/SSR	5	1, 3, 4, 5	[18]
PI 197088	Israel	F2/SNP	9	1, 2, 3, 4, 5, 6, 7	[29]
PI 330628	USA, EU	F2:F3/SNP	5	2, 4, 5, 6	[30]
Chinese Long	China	F2:F3/SSR	5	1, 5, 6	[31]
TH118FLM	Korea	F2:F3/SNP	5	2, 4, 5, 6	[32]

Despite the extensive screening and breeding efforts that were done to identify sources of resistance and to incorporate them into commercial cultivars [9], no cultivars currently offer a high level of resistance to the populations of *P. cubensis* that occur in different parts of the world. The reasons for the lack of resistant cultivars may derive from the heterozygosity of the resistant sources used for breeding, the continuous changes in the population structure of the pathogen, and the difficulty to pyramid the number of genes/QTLs in one cultivar.

The most promising current sources of resistance are PI 197088 and PI 330628 [30]. However, no data are available on the magnitude of their resistance against different isolates of *P. cubensis* from different parts of the world.

The objectives of this study were to: (i) stabilize PI 197088 and PI 330628 for resistance against multiple isolates of *P. cubensis* from different parts of the world. (ii) study the mechanism of resistance of PI 197088 and PI 330628 against *P. cubensis*. (iii) determine the mode of inheritance of resistance in PI 197088 and PI 330628 against multiple isolates of *P. cubensis* from different parts of the world.

2. Methods and Materials

2.1. Cucumber Accessions

Nine accessions of cucumber (*Cucumis sativus* L) were obtained from Todd Whener, NCSU Charleston, NC, USA. Three accessions M-21, SMR-18, and Sumter are commercial cultivars whereas six Plant Introduction (PI) accessions PI 197085, PI 197087, PI 197088, PI 606015, PI 605996 and PI 432875 are wild *C. sativum*.

2.2. Pathogen

Forty-four field isolates of *P. cubensis* that were collected during 1980–2018 from 13 countries were available to perform the different experiments in the present study (Table 2), including two F1 hybrid isolates that were produced in our laboratory by crossing A1 and A2 field isolates from different hosts as described before [2]. A subset of 23 isolates was used to screen the resistance of PI 197088 and PI 330628 while other subsets of isolates were used to determine the resistance of F2 and F3 populations. The isolates were maintained by repeated inoculation of detached cucumber leaves of the universal susceptible cucumber line Nadiojny (own bred). Long-term maintenance of the isolates was done by storing freshly-sporulating leaves in dry paper bags at −80 °C.

Table 2. Isolates of *Pseudoperonospora cubensis* used in this study. (For pathotype description see Table 1 in [4].

	Isolate	Year	Country	Host	Mating Type	Pathotype
1	PCHS	1980	Japan	Unknown	A1	3
2	C1	1982	South Carolina	*C. melo*	A1	3
3	62	1995	Czech Republic	*C. sativum*	A1	3
4	66	2000	France	*C. sativum*	A1	3
5	23 C	2008	Israel	*C. sativum*	A2	4
6	US-299	2008	Michigan	*C. sativum*	A1	3
7	US-163	2008	Florida	*C. lanatus*	A1	5
8	5	2010	Bulgaria	*C. sativum*	A1	3
9	7	2010	Bulgaria	*C. sativum*	A1	3
10	17	2010	Turkey	*C. sativum*	A1	3
11	21	2010	Turkey	*C. sativum*	A1	3
12	56	2010	Turkey	*C. sativum*	A1	3
13	81 C	2011	Spain	*C. sativum*	A0	2
14	83 C	2011	Spain	*C. sativum*	A1	3
15	84 C	2011	Spain	*C. sativum*	A1	3
16	88 P	2011	Israel	*C. maxima*	A2	6
17	90 p	2012	Israel	*C. maxima*	A2	7
18	98 P	2011	Israel	*C. maxima*	A1	6
19	101 D	2011	Israel	*C. maxima*	A2	6
20	109	2011	Ukraine	*C. sativum*	A1	3
21	148	2011	Israel	*C. sativum*	A1	3
22	US-504	2011	New York	*C. sativum*		
23	US-506	2011	Ohio	*C. sativum*	A1	3
24	Noam 19P	2011	Kenya	*C. maxima*	A2	6
25	Noam C	2011	Israel	*C. sativum*	A1	3
26	TW-01	2011	Taiwan	*C. sativum*	A1	3
27	151/17	2012	Israel	*C. sativum*	A1	3
28	185	2012	Israel	*C. melo*	A1	3
29	171	2012	Israel	*C. sativum*	A1	3
30	182 D	2012	Israel	*C. sativum*	A1	3
31	183/2	2012	Israel	*C. sativum*	A1	3
32	184	2012	Israel	*C. moschata*	A2	6
33	C-29	2012	China	*C. sativum*	A1	3
34	Harbin 10	2012	China	*C. sativum*	A1	3
35	SG-11	2012	China	*C. sativum*	A1	3
36	Petiole 1	2012	Israel	*C. moschata*	A2	6
37	172 B × 183 C, F1	2012	Israel	*C. melo*	Hybrid	
38	83 C x 98 P, F1	2012	Israel	*C. melo*	Hybrid	

Table 2. *Cont.*

	Isolate	Year	Country	Host	Mating Type	Pathotype
39	197 C	2013	Israel	*C. sativum*	A1	3
40	198 B	2013	Israel	*C. pepo*	A2	6
41	245	2015	Israel	*C. sativum*	A1	3
42	Pol 1	2016	Poland	*C. sativum*	A1	3
43	Pol 4	2016	Poland	*C. sativum*	A1	3
44	260	2018	Israel	*C. sativum*	A1	3

2.3. Crosses

The susceptible SMR-18 and the resistant PI 197088 and PI 330628 were self-pollinated for three generations to ensure homozygosity (see below). Crosses were made between the susceptible SMR-18 and each of the resistant PI 197088 or PI 330628. Another cross was done between these two resistant accessions. F1 plants were grown in a net-house (insect-proof) and self-pollinated to obtain F2 populations. A single fruit was harvested from each F1 plant and the F2 seeds were grown the following season in net houses and self-pollinated to obtain F3 plants.

2.4. Inoculation of Detached Leaves

Parents, F1, F2, and F3 plants were grown erect in net-houses during 2014–2019. The third leaf from the top of 15–20-leaf plants were excised, placed on a wet filter paper in flat plastic trays (60 × 40 × 5 cm), lower surface uppermost, and spray-inoculated with a sporangial suspension of *P. cubensis* (2000 sporangia per mL). Trays were covered with transparent plastic bags and kept for 16 h in a dew chamber at 18 °C in the dark and thereafter in a growth chamber at 20 °C (14 h light/day, 100 $\mu mole \cdot s^{-1} \cdot m^2$) for 7 days.

2.5. Disease Assessment in Detached Leaves

Two readings were taken from each detached leaf at 7 days post-inoculation (dpi): the proportion of leaf area (0–100%) occupied with downy mildew lesions and the intensity of sporulation (0–3 scale) as visualized with a ×10 magnifying lens. The two values were multiplied to obtain a disease scoring scale of 0–300 (Figure 1). Leaves showing a score of 0–20; 21–200; 201–300 were considered resistant (R); moderately resistant (MR) and susceptible (S), respectively.

Figure 1. Phenotypic score panel of downy mildew in detached leaves of cucumber infected with *P. cubensis*. A score composes of percent leaf area infected multiplied by sporulation intensity (at ×10 magnification on a scale of 0–3). The scores given to the leaves in the upper row are: 0, 0.5, 20, and 80, and in the lower row are: 150, 180, 237.5, and 285. Plants showing a score of 0–20, 21–100, and 101–300 were considered resistant, moderately resistant, and susceptible, respectively.

2.6. Inheritance of Resistance in Detached Leaves Taken from Adult Plants

Leaves (3rd leaf from the top) were detached daily from adult F2 plants growing in the field (summer 2014), placed on moistened filter paper inside sealed transparent trays and spray-inoculated with sporangia of each of the 22 isolates (PI 197088 × SMR-18) or 14 isolates (PI 330628 × SMR-18) of *P. cubensis*. The percentage of leaf area infected and sporulation intensity at ×10 magnification on a scale of 0–3 were recorded at 7 dpi.

2.7. Inheritance of Resistance in Intact Field-Grown Plants

Two-leaf plants of parents of F1 and F2 families were transplanted to the field in mid-March. Plants were grown erect in net houses (50 × 6 m) during the spring seasons of 2013, 2014, 2016, 2017, and 2019. In 2016, plants were also grown in the autumn season. F3 plants were similarly grown in the field in 2014. Natural infection with downy mildew initiated at 2–3 weeks after planting. Percentage of leaf area occupied by downy mildew lesions was visually assessed in each plant once or twice a week for 5–6 weeks after planting (unless stated otherwise). AUDPC (area under disease progress curve) was used to categorize plants as resistant, moderately resistant, or susceptible to the disease. A plant showing an AUDPC value of ≤5% of the maximal AUDPC value in that season was considered resistant whereas a plant showing an AUDPC value of ≥80% of that maximal value was considered susceptible. All other plants were considered moderately resistant.

2.8. Microscopy

The methods used by Cohen et al. [33] were used. Briefly, healthy and infected leaf discs were clarified in boiling ethanol, placed in aniline blue solution (0.05% aniline blue in 0.05 M K_2HPO_4, pH 8.9) at 4 °C for 24 h, stained with 0.01% calcofluor (Sigma) and examined with an Olympus A70 epi-fluorescent microscope. Sporangiophores on leaf surface fluoresced blue and sporangia looked dark. Fungal structures inside the leaf showed green-yellow fluorescence. Callose-encased haustoria were seen yellow. Staining for lignin was done with ethanol-clarified leaf discs. They were placed on microscope slides, treated with 2% phloroglucinol in methanol, and then with 0.25% HCl. A red color was visible in the lignified mesophyll cell.

3. Results

3.1. Resistance of C. sativus to Multiple Isolates of P. cubensis

The fruits of nine accessions of cucumber used in this study: SMR-18, M-21, Sumter, PI 606015, PI 432875, PI 605996, PI 197085, PI 197088, and PI 330628 are shown in Figure 2. Also shown are the F1 fruits of PI 197088 × SMR-18 and PI 330628 × SMR-18.

Figure 2. *Cont.*

Figure 2. Fruits of three cultivated and six wild *Cucumis sativus* L. and F1s of two wild types with SMR-18.

The resistance of nine accessions of *Cucumis sativus* (15–28 plants per accession, detached leaf bioassay) against each of five Israeli isolates of *P. cubensis* is shown in Figure 3. Most resistant accessions were PI 197088 and PI 330628. To test the homozygosity of the two most resistant accessions, detached leaves taken from 19 individual plants of PI 197088 and 19 individual plants of PI 330628 were each inoculated with each of the 23 isolates of *P. cubensis* from different parts of the world. The resistance profile of each plant is shown in Table 3. Nine PI 197088 plants (Figure 4A) and three PI 330628 plants (Figure 4B) were resistant to all 23 isolates of the pathogen, suggesting heterozygosity of the original accessions. One resistant plant of each PI was self-pollinated for two more generations and its offspring plants were all found resistant to these isolates of the pathogen. These two plants were used to study the resistance mechanisms and the mode of inheritance of resistance.

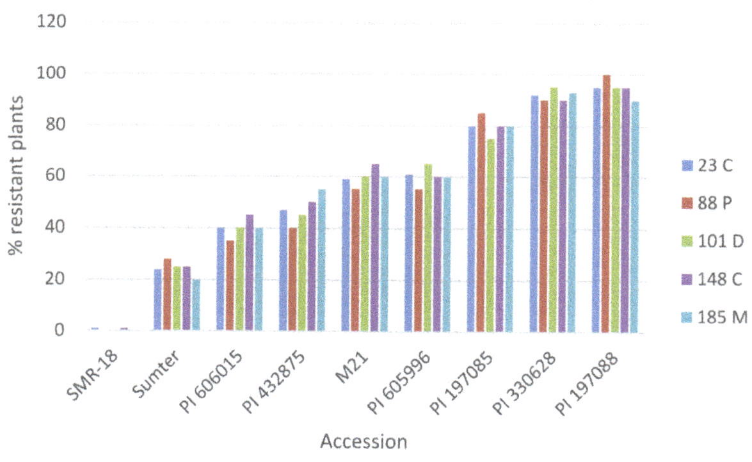

Figure 3. Resistance of nine accessions of *Cucumis sativus* (15–28 plants per accession, detached leaf bioassay) against each of five Israeli isolates of *P. cubensis*. The origin of isolates is shown in Table 2.

Table 3. Response of detached leaves of 19 individual plants of PI 197088 and 19 individual plants of PI 330628 to inoculation with each of the 23 isolates of *P. cubensis*. Isolates originated from Israel, Europe, USA, and Asia. White R = resistant (score 0–20). Blue S = susceptible (score 21–300). Green * = not determined. Disease records were taken at 7 dpi (20 °C, 14 h light/day).

	Isolate																						
Plant Number	**88**	**98**	**101D**	**148**	**151/17**	**171**	**Noam C**	**17**	**21**	**56**	**66**	**84C**	**7**	**5**	**109**	**62**	**C1**	**US-163**	**US-504**	**US-506**	**US-299**	**TW-01**	**PCHS**
PI 197088 (1)	S	R	R	R	R	R	R	R	R	R	R	R	R	S	R	R	R	R	R	R	R	R	R
PI 197088 (2)	R	R	R	R	R	R	R	R	R	R	R	R	R	R	R	R	R	R	R	R	R	R	R
PI 197088 (3)	S	R	R	R	R	R	R	R	R	R	S	R	R	R	R	S	S	R	R	R	R	R	R
PI 197088 (4)	R	R	R	R	R	R	R	R	R	R	R	R	R	R	R	R	R	R	R	R	R	R	R
PI 197088 (5)	R	R	R	R	R	R	R	R	R	R	R	R	R	R	R	R	R	R	R	R	R	R	R
PI 197088 (6)	R	R	R	R	R	R	R	R	R	R	R	R	R	R	R	R	R	R	R	R	R	R	R
PI 197088 (7)	R	R	R	R	R	R	R	R	R	R	R	R	R	R	R	R	R	R	R	R	R	R	R
PI 197088 (8)	R	R	R	R	S	S	R	R	R	S	R	R	R	R	R	R	R	S	R	R	R	R	R
PI 197088 (9)	R	R	R	R	R	R	R	R	R	R	R	R	R	R	R	R	R	R	R	R	R	R	R
PI 197088 (10)	R	R	R	R	R	R	R	R	R	R	R	R	R	R	R	R	R	R	R	R	R	R	R
PI 197088 (11)	R	R	R	R	R	R	R	R	R	R	R	R	R	R	R	R	S	R	R	R	R	R	S
PI 197088 (12)	R	R	R	R	R	R	R	R	R	R	R	R	R	R	R	R	R	R	R	R	R	R	R
PI 197088 (13)	R	R	R	R	S	S	R	S	R	S	S	R	R	R	R	R	S	R	R	R	R	R	R
PI 197088 (14)	R	R	R	R	S	S	R	R	R	R	R	R	R	R	R	R	R	R	R	R	R	R	R
PI 197088 (15)	R	R	R	R	R	R	R	R	R	R	R	R	R	R	R	R	R	S	R	R	R	R	S
PI 197088 (16)	R	R	R	R	R	R	R	R	R	R	R	R	R	R	R	R	R	R	R	R	R	R	R
PI 197088 (17)	S	R	R	R	R	R	R	R	R	R	R	R	R	R	R	R	R	R	R	R	R	R	S
PI 197088 (18)	R	R	R	R	R	S	R	R	R	R	R	R	R	R	S	R	R	R	R	R	R	R	R
PI 197088 (19)	R	R	R	R	R	S	R	R	R	R	R	R	R	R	R	R	R	R	R	R	R	R	R
PI 330628 (1)	S	R	R	R	R	S	R	R	S	S	S	R	R	S	S	S	S	R	R	R	R	R	S
PI 330628 (2)	R	R	R	R	R	R	R	R	R	R	R	R	R	R	S	R	R	R	R	R	R	R	R

Table 3. *Cont.*

Plant Number	Isolate																					
	88	98	101D	148	151/17	171	Noam C	17	21	56	66	84C	7	5	109	62	C1	US-163	US-504	US-506	US-299	TW-01	PCHS
PI 330628 (3)	R	R	R	R	R	R	R	R	R	S	S	R	R	R	R	R	S	R	R	R	R	R	R
PI 330628 (4)	R	R	R	R	R	R	R	R	R	R	R	R	R	R	R	R	R	R	R	R	R	R	R
PI 330628 (5)	R	R	R	R	R	R	R	R	R	S	R	R	R	R	R	R	R	R	R	R	R	R	R
PI 330628 (6)	R	R	R	R	S	S	R	S	R	S	S	R	S	R	S	R	R	R	R	R	R	R	R
PI 330628 (7)	R	R	R	R	R	R	R	R	R	R	R	R	R	S	R	R	R	R	R	R	R	R	R
PI 330628 (8)	R	R	R	R	R	R	R	R	R	R	R	R	R	R	R	R	R	R	R	R	R	R	R
PI 330628 (9)	S	R	R	R	R	R	R	R	S	R	S	R	R	R	R	R	R	R	R	R	R	R	S
PI 330628 (10)	R	R	R	R	R	R	R	R	R	R	R	R	R	R	R	R	R	R	R	R	R	R	S
PI 330628 (11)	R	R	R	R	R	R	R	R	R	R	R	R	R	S	R	R	R	R	R	R	R	R	R
PI 330628 (12)	S	R	R	R	R	R	R	S	R	S	S	R	R	R	R	S	R	R	R	R	R	R	R
PI 330628 (13)	R	R	R	R	R	R	R	R	R	R	R	R	R	R	R	R	R	R	R	R	R	R	R
PI 330628 (14)	R	R	R	R	R	R	R	R	R	R	R	*	R	R	R	R	R	R	R	R	R	R	R
PI 330628 (15)	R	R	R	R	R	S	R	R	R	S	R	R	R	R	R	R	R	R	R	R	R	R	S
PI 330628 (16)	R	R	R	R	R	R	R	R	R	R	R	*	R	R	R	R	*	R	R	R	R	R	R
PI 330628 (17)	R	R	R	R	R	R	R	R	R	R	R	R	R	R	R	R	*	R	R	R	R	R	R
PI 330628 (18)	R	R	R	*	R	R	R	R	R	R	R	R	R	R	R	R	*	R	R	R	R	*	R
PI 330628 (19)	R	R	R	*	R	S	R	*	R	*	*	*	*	R	*	R	*	R	R	R	R	*	R

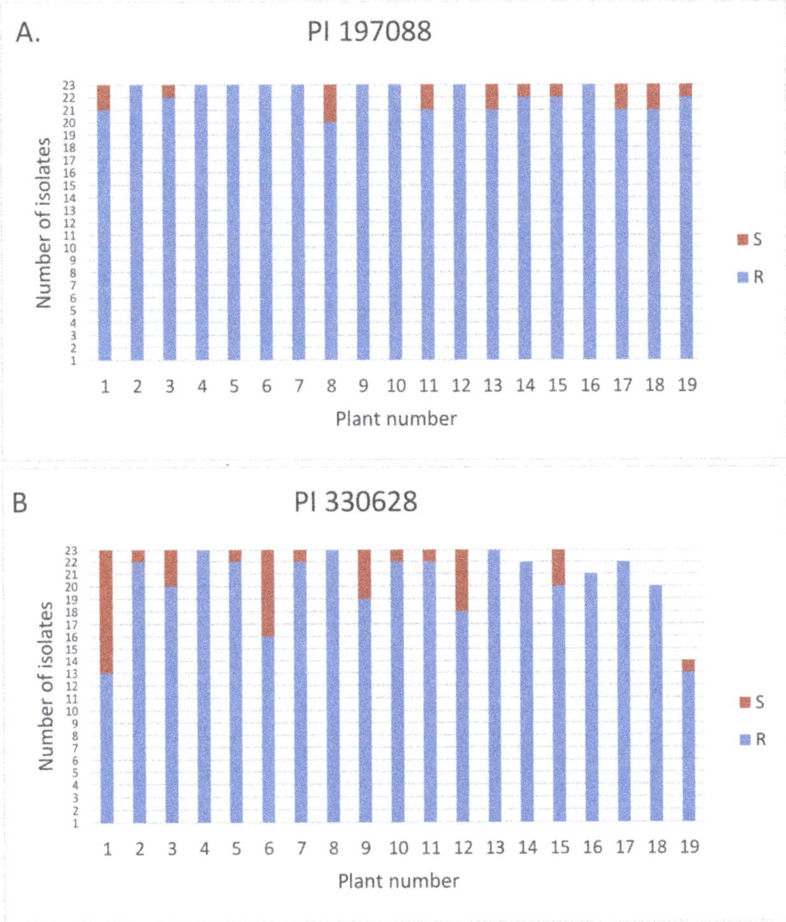

Figure 4. Response of 19 individual plants of *Cucumis sativus* PI 197088 (**A**) and 19 plants of PI 330628 (**B**) to inoculation with 23 isolates of *P. cubensis*. A detached leaf bioassay. S = susceptible. R = resistant.

3.2. Microscopy of Resistance in PI 197088 and PI 330628

Leaf discs (15 mm diameter) were removed at 6 dpi from detached infected leaves (isolate 83C) of SMR-18, PI 197088, and PI 330628 and examined under UV illumination. Abundant sporangiophores with sporangia were seen in the susceptible SMR-18 (Figure 5A–C). In contrast, a few mycelium runners bearing callose-encased haustoria were seen in PI 197088 (Figure 5E,G). Similar callose depositions were seen in PI 330628 (not shown). Sporangiophores were partially branched with no sporangia seen in the resistant plants (Figure 5F). Bright-field microscopy of phloroglucinol-stained infected leaf discs showed no lignin staining in SMR-18 (Figure 5D) but heavy lignin accumulation in the resistant PI 197088 (Figure 5H). Similar lignin accumulation was seen in the resistant PI 330628 (not shown). Resistance to downy mildew in PI 197088 and PI 330628 was stable at a colonization temperature of 14 °C (not shown).

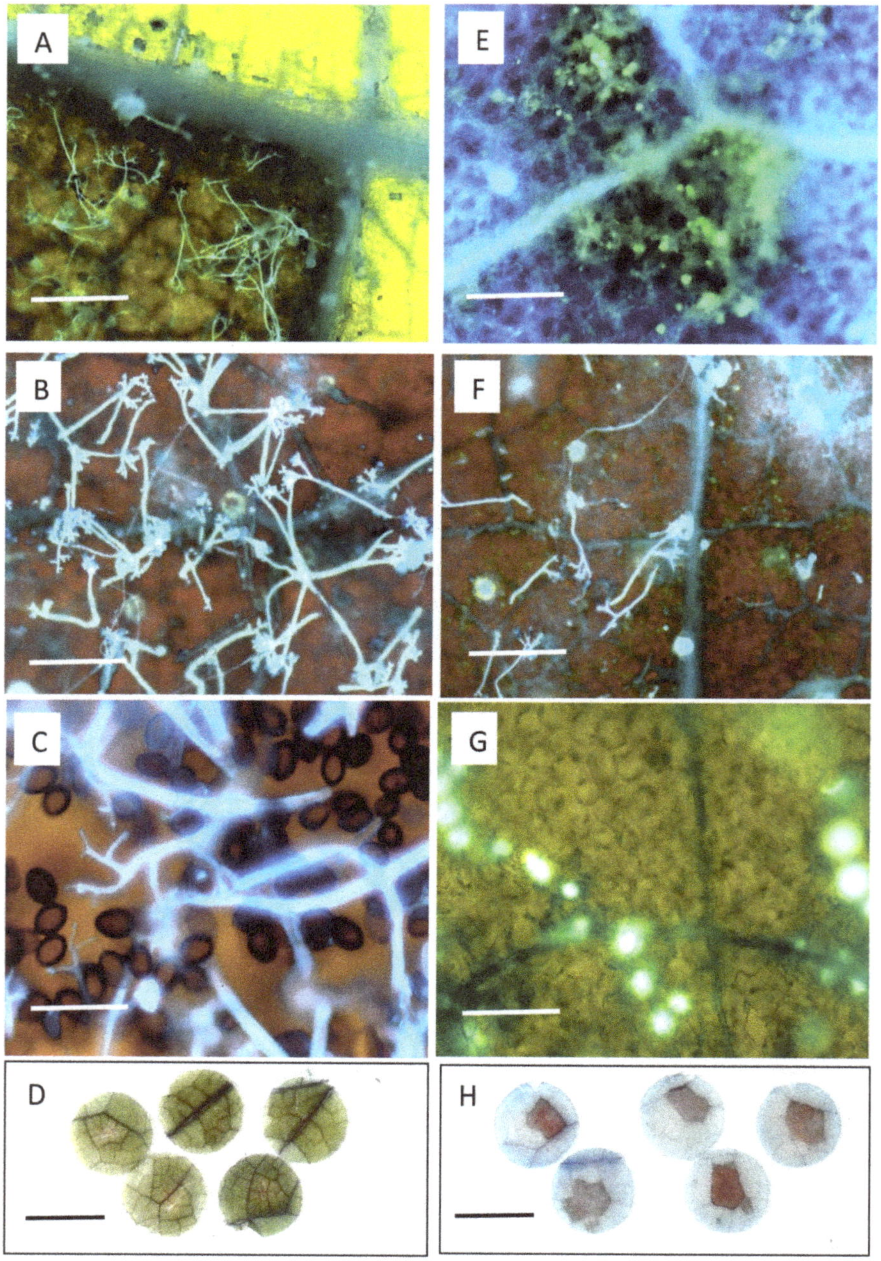

Figure 5. Development of *P. cubensis* (isolate 83C) in detached leaves of cucumber at 6 dpi in a growth chamber at 20 °C. (**A–D**), SMR-18, susceptible. Note in (**A**) that the lesion is confined between the veins. Note in (**B**) the dichotomously branched sporangiophore in blue. Note in (**C**) the dark oval sporangia. (**E–H**), PI 197088 or PI 330628, resistant. Note in (**E**) and (**G**) the yellow dotes which are callose-encased haustoria. Note in (**F**) that sporangiophores are branched only once. Scale bar: (**A**) and (**E**) = 200 μm. (**B**) and (**F**) = 100 μm. (**C**) and (**G**) = 50 μm. (**D–H**), 15 mm. (**A–C**) and (**E–G**) are UV micrographs after calcofluor staining; (**D–H**), bright field micrographs, phloroglucinol staining of ethanol-clarified leaf discs.

3.3. Quantification of Resistance in Adult Plants

The third leaf from the top of adult SMR-18, PI 197088, and PI 330628 plants (n = 10) grown in a net house was detached and drop-inoculated on the lower leaf surface with isolate 260. Fifteen mm leaf discs were sampled at 7 dpi for microscopic examination of sporulation. UV-epifluorescence microscopy of calcofluor-stained leaf discs revealed 650 ± 123 sporangiophores/cm^2 in SMR-18 as against 11 ± 8 and 18 ± 12 sporangiophores/cm^2 in PI 197088 and PI 330628, respectively (Figure 6A). The sporangiophores in SMR-18 were branched dichotomously three times, twice in PI 330628 and only once in PI 197088. The number of sporangia produced in SMR-18, PI 197088, and PI 330628 was 91 ± 5, 1 ± 0.3, and 6 ± 1 thousands of sporangia per cm^2, respectively (Figure 6B). Similar results were obtained with other isolates of the pathogen (not shown). The data indicated that PI 197088 is slightly more resistant to downy mildew than PI 330628.

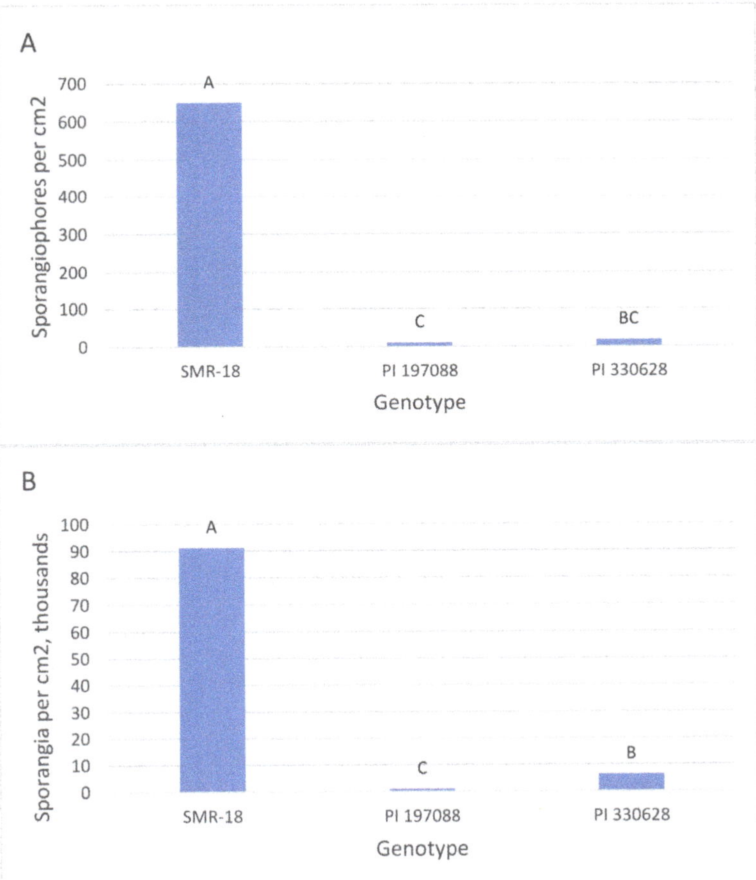

Figure 6. Quantification of resistance against isolate 260 of *P. cubensis* in detached leaves of PI 197088 and PI 330628 (F3) in growth chambers. (**A**) Formation of sporangiophores at 7 dpi. (**B**) Sporangial production at 7 dpi. Different letters on bars indicate a significant difference at α = 0.05 (t-test).

Artificial inoculation of intact plants (10-leaf stage) in the field with isolate 260 resulted at 9 dpi with the production of chlorotic lesions in SMR-18 and minute necrotic lesions in PI 197088 and PI 330628 (Figure 7A–C). Some infected leaves were detached, placed on wet filter paper and incubated in a growth chamber at 20 °C for three days (14 h light a day, 100 μmole·s^{-1}·m^2). At 13 dpi, SMR-18 produced the largest lesions (17.5 ± 3.5 mm) and the highest number of sporangia 71.9 ± 6.2 × 10^3

per lesion whereas PI 197088 produced the smallest lesions (1.5 ± 0.7 mm) with the lowest number of sporangia per lesion ($0.4 \pm 0.1 \times 10^3$) (Figure 7G,H). The F1 hybrid plants produced intermediate-size lesions with a moderate number of sporangia per lesion (Figure 7G,H). Percent leaf area infected at 27 dpi in fully-grown, blooming plants in the field is shown in Figure 7I. While SMR-18 exhibited 72% infected leaf area, PI197088 and PI330628 showed 1.2 and 4.4% infected leaf area, respectively. Their F1 plants were moderately infected (Figure 7I). AUDPC values at the end of the season are shown in Figure 7J. These results indicated that resistance of 197088 and 330628 is controlled by partially dominant gene(s).

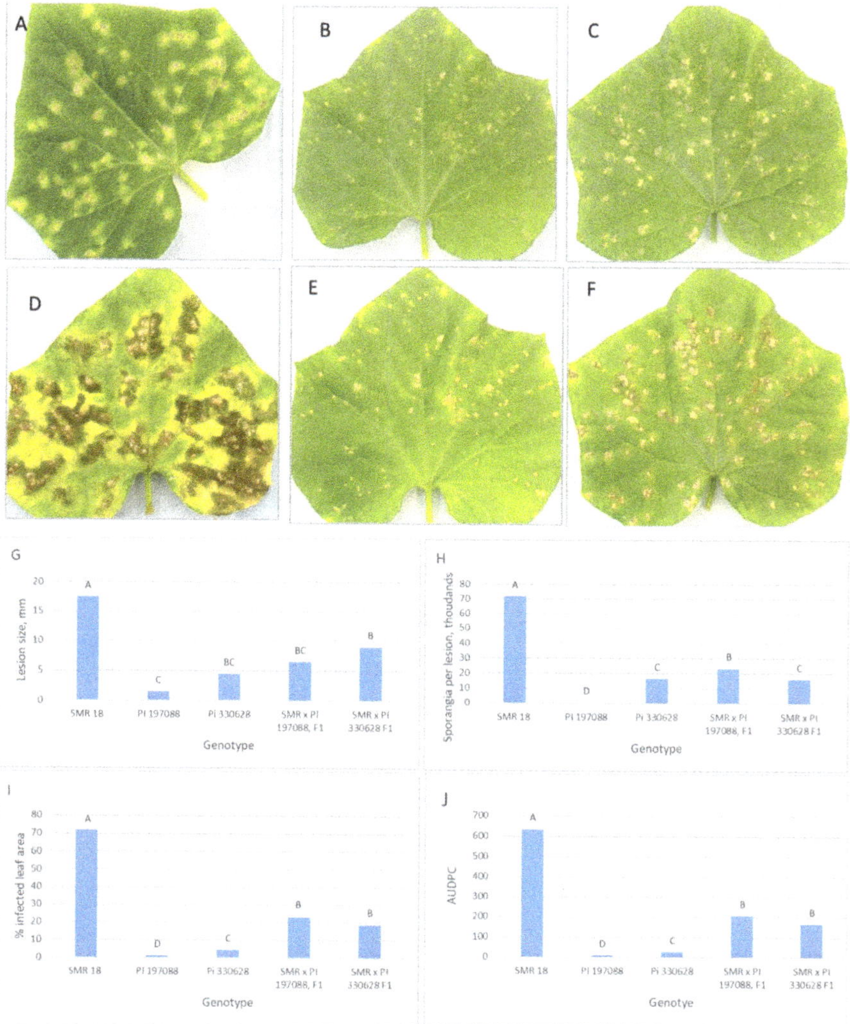

Figure 7. Phenology of downy mildew symptoms in leaves of field-grown intact cucumber plants (n = 10) after artificial spray inoculation with sporangia of isolate 260 of *P. cubensis*. (**A–C**), symptoms at 9 dpi. (**D**,**E**) symptoms after an additional 4 days of incubation in a growth chamber at 20 °C (13 dpi). (**A**,**D**)—SMR-18. (**B**,**E**)—PI 197088. (**C**,**F**)—PI 330628. (**G**,**H**), lesion size, and sporulation at 13 dpi of parental genotypes and their F1 progeny plants. (**I**)—Percentage infected leaf area in the field at 27 dpi. (**J**) AUDPC in the field at 56 dpi. Different letters on bars indicate a significant difference at $\alpha = 0.05$ (*t*-test).

When SMR-18, PI 197088, and PI 330628 (n = 10) were planted in a net house and thus exposed to natural infection, severe downy mildew developed in SMR-18 within four weeks after planting whereas no disease symptoms were seen in PI 197088 or PI 330628 (Figure 8). The disease was not visible in these two PI's even at four months after planting when they carried mature fruits (produced by hand pollination), suggesting that homozygosity may avoid the appearance of the susceptible reaction reported to occur in these accessions at an advanced stage of growth in the field [28].

Figure 8. The appearance of downy mildew caused by *P. cubensis* in field-grown cucumber plants at 4 weeks (**left**) and 12 weeks (**right**) after plating. Note extensive disease symptoms in the susceptible SMR-18 but no disease symptoms in the resistant PI 197088 nor in the resistant PI 330628. BIU Farm, summer 2019 (**left**), spring 2020 (**right**).

Resistance to downy mildew in melon *Cucumis melo* PI 124111F was shown to be active at colonization temperatures of >15 °C [34] due to the expression of *eR* genes [35]. Here, we observed that both PI 197088 and PI 330628 sustained full resistance to multiple isolates of *P. cubensis* when incubated after inoculation at either 14 °C or 20 °C (detached leaf bioassay), suggesting that unlike melon, expression of resistance against downy mildew in cucumber occurs at a low temperature of 14 °C.

3.4. Inheritance of Resistance in Detached Leaves of Adult Plants

A set of 22 isolates was used to inoculate F2 plants of the cross PI 197088 × SMR 18 and a set of 14 isolates (all included in the former set) was used to inoculate F2 plants of the cross PI 330628 ×

SMR 18. Graphical illustrations of the disease scores are given for only 10 and 8 isolates of the above respective crosses (Figures 9 and 10). Full numerical scores for all 22 and 14 isolates are given in Table 4. Data in Figures 9 and 10 show a unique segregation pattern for each isolate. With some isolates, a similar pattern was seen for F2 of both crosses PI 197088 × SMR 18 and PI 330628 × SMR 18. Large differences in the response to inoculation were observed between plants, depending on the isolate used for inoculation. A single plant could react with different scores when inoculated with different isolates. The Mendelian analyses of the data are presented in Table 4. When plants' responses were classified into two categories, F2 plants of the cross PI 197088 × SMR-18 showed five segregation ratios, depending on the isolate used for inoculation: 3:1 (2 isolates), 1:3 (11 isolates), 1:15 (6 isolates), 9:7 (2 isolates), and 13:3 (one isolate) (Table 4A). When three categories were used for classification, only 2 isolates out of 22, obeyed the Mendelian segregation of 9:6:1 (Table 4A). No inheritance model could be assigned to the results obtained with the other 19 isolates. Five segregation ratios were observed when two categories were applied for classification of the F2 plants of the cross PI 330628 × SMR-18: 3:1 (1 isolate), 1:3 (8 isolates), 1: 15 (2 isolates), 7:9 (2 isolates), and 9:7 (1 isolate) (Table 4B). When three categories were used for classification, only one isolate out of 14 obeyed a segregation ratio of 9:7. No inheritance model could be assigned to the results obtained for the other isolates (Table 4B).

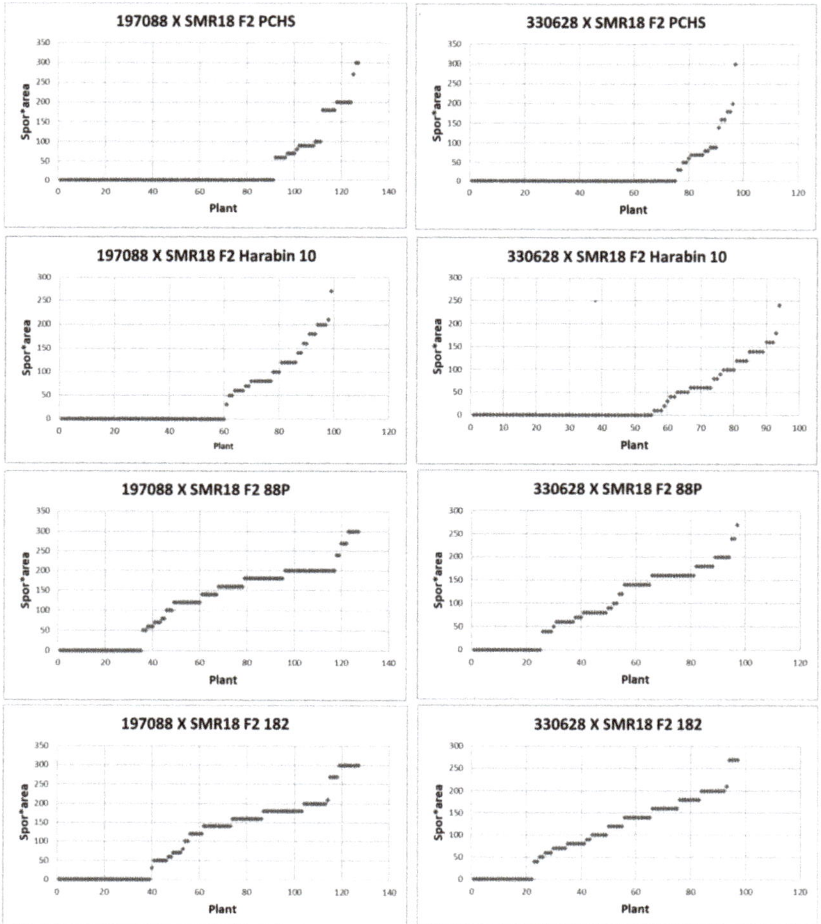

Figure 9. Scores of the response of F2 plants derived from the cross PI 197088 × SMR-18 or the cross PI 330628 × SMR-18 to inoculation with five isolates of *P. cubensis*. A detached leaf bioassay. Leaves were taken from field-grown plants.

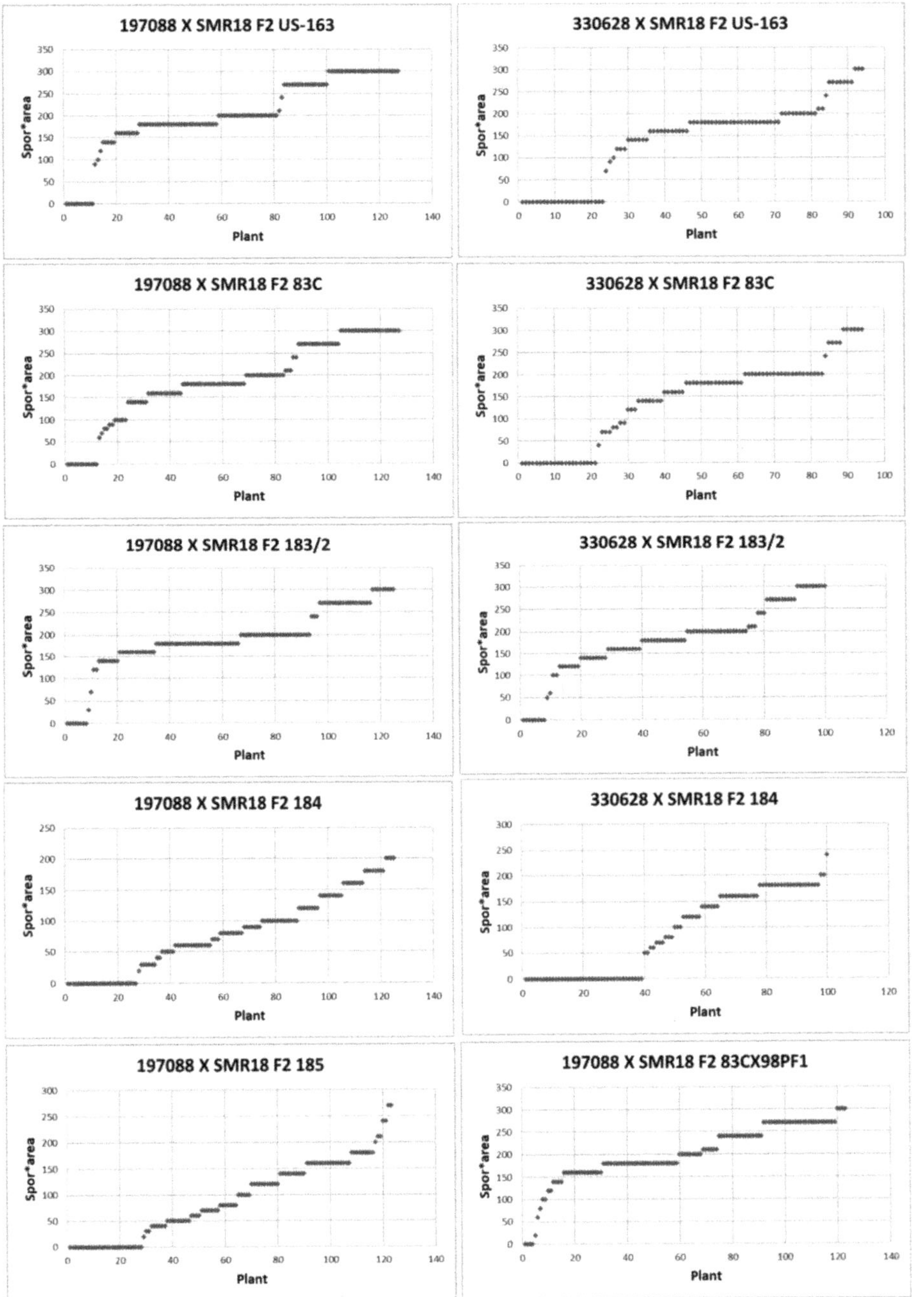

Figure 10. Scores of the response of F2 plants derived from the cross PI 197088 × SMR-18 or the cross PI 330628 × SMR-18 to inoculation with six (or five) isolates of *P. cubensis*. A detached leaf bioassay. Leaves were taken from field-grown plants.

Table 4. Inheritance of resistance against multiple isolates of *P. cubensis* in detached leaves of cucumber. R = resistant (score 0–20). MR—moderately resistant (score 21–200). S = susceptible (score 201–300). * = model not accepted. ** = no Mendelian model.

A. 197088 × SMR-18, F2													
	Two categories						Three categories						
	Number of plants						Number of plants						
Isolate	**Total**	**R**	**S**	**Ratio**	p	X^2	**Total**	**R**	**MR**	**S**	**Ratio**	p	X^2
PCHS	127	91	36	3:1	0.384	0.759	127	91	33	3	9:6:1	*	*
Harbin 10	100	61	39	9:7	0.338	0.917	100	61	33	6	9:6:1	0.622	0.951
88P	127	35	92	1:3	0.505	0.444	127	35	84	8	1:2:1	*	*
182D	127	39	88	1:3	0.137	2.207	127	39	75	13	1:2:1	*	*
US-163	127	11	116	1:15	0.262	1.260	127	11	72	44	**	**	**
83C	127	12	115	1:15	0.136	2.218	127	12	76	39	**	**	**
183/2	125	8	117	1:15	0.945	0.005	125	8	88	29	**	**	**
184	123	25	98	1:3	0.231	1.434	123	25	98	0	**	**	**
185	123	29	94	1:3	0.716	0.133	123	29	92	2	**	**	**
83CX98P, F1	123	4	119	1:15	0.170	1.887	123	4	87	32	**	**	**
172BX183C, F1	118	24	94	1:3	0.242	1.367	118	24	91	3	**	**	**
US-504	123	70	53	9:7	0.883	0.022	123	70	47	6	9:6:1	0.820	0.397
SG-11	104	7	97	1:15	0.839	0.041	104	7	32	65	**	**	**
Noam 19P	118	41	77	1:3	*	*	118	41	75	2	**	**	**
Petiole 1	104	5	99	1:15	0.543	0.369	104	5	38	61	**	**	**
23 C	155	36	119	1:3	0.610	0.260	155	36	111	8	**	**	**
Pol 1	155	47	108	1:3	0.126	2.342	155	47	89	19	1:2:1	*	*
Pol 4	158	134	24	13:3	0.252	1.315	158	134	24	0	**	**	**
US-506	155	114	41	3:1	0.676	0.174	155	114	41	0	**	**	**
81 C	151	30	121	1:3	0.145	2.121	151	30	121	0	**	**	**

Table 4. *Cont.*

88 C	148	35	113	1:3	0.704	0.144	148	35	113	0	**	**	**
90 P	156	46	110	1:3	0.196	1.675	156	46	110	0	**	**	**
B. 330628 × SMR-18, F2													
	Number of plants						Number of plants						
Isolate	**Total**	**R**	**S**	**Ratio**	p	X^2	**Total**	**R**	**MR**	**S**	**Ratio**	P	X^2
PCHS	97	75	22	3:1	0.598	0.278	97	75	21	1	12:3:1	*	*
Harbin 10	94	55	39	9:7	0.659	0.195	94	55	39	0	9:7	0.659	0.195
88P	97	25	72	1:3	0.860	0.031	97	25	69	3	**	**	**
182D	97	22	75	1:3	0.598	0.278	97	22	71	4	**	**	**
US-163	94	23	71	1:3	0.905	0.014	94	23	58	13	1:2:1	*	*
83C	93	20	73	1:3	0.436	0.606	93	20	62	11	**	**	**
183/2	99	7	92	1:15	0.736	0.114	99	7	72	20	**	**	**
184	99	38	61	1:3	*	*	99	38	60	1	**	**	**
185	85	28	57	1:3	0.091	2.859	85	28	43	14	1:2:1	*	*
83CX98P, F1	86	7	79	1:15	0.469	0.524	86	7	54	25	**	**	**
172BX183C, F1	68	12	56	1:3	*	*	68	12	45	11	1:2:1	*	*
US-504	129	50	79	7:9	0.253	1.305	129	50	58	21	**	**	**
101D	125	55	70	7:9	0.955	0.003	125	55	65	5	**	**	**
Noam 19P	67	16	51	1:3	0.832	0.045	67	16	44	7	1:2:1	*	*

3.5. Inheritance of Resistance in Intact Field-Grown Plants

The response of the parents F1 and F2 plants to downy mildew in the field in 2019 is shown in Figure 11. A continuous response pattern to the disease was observed in F2 plants of PI 197088 × SMR-18 (Figure 11A,B) and of PI 330628 × SMR-18 (Figure 11D,E). Both resistant parents were completely resistant all along the season (until fruit maturity) whereas F1 plants were moderately resistant (Figure 11C,F).

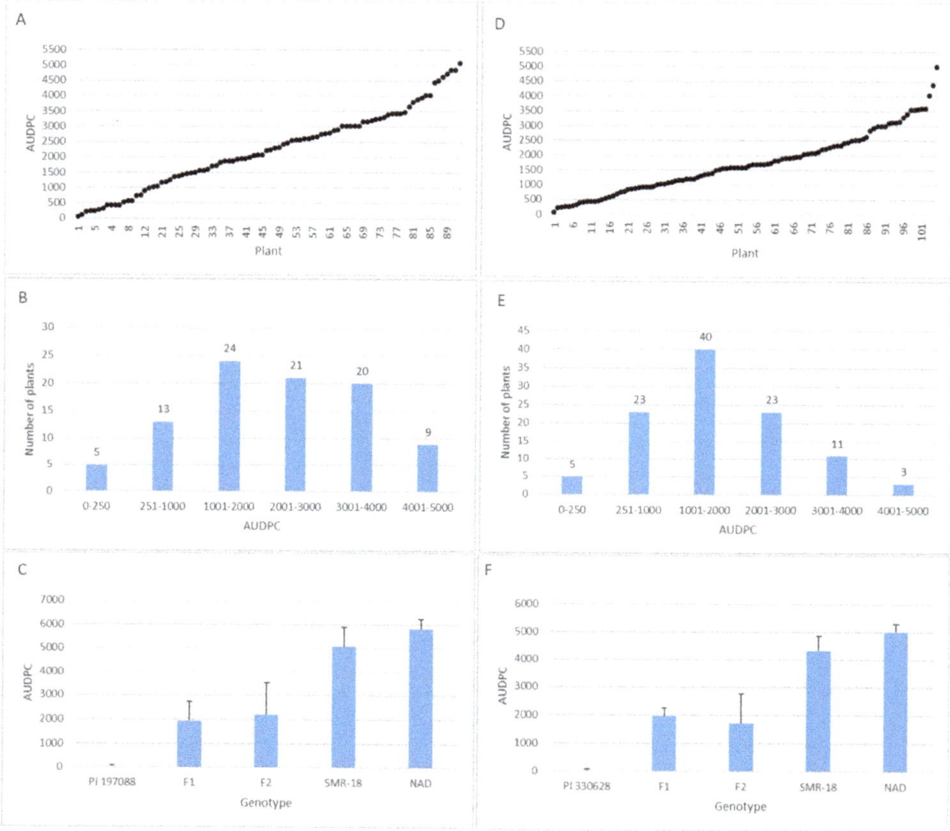

Figure 11. Development of downy mildew in cucumber plants during 52 days under field conditions in autumn 2019. (**A**) The area under disease progress curve (AUDPC) of 92 F2 plants of the cross PI 197088 × SMR-18. (**B**) Categorical distribution of the data presented in (**A**). (**C**) Mean AUDPC of the resistant parent PI 197088, the susceptible parent SMR-18, their F1 and F2 progeny plants. NAD is another susceptible line. (**D**) The area under disease progress curve (AUDPC) of 105 F2 plants of the cross PI 330628 × SMR-18. (**E**) Categorical distribution of the data presented in (**D**). (**F**) Mean AUDPC of the resistant parent PI 330628, the susceptible parent SMR-18, their F1 and F2 progeny plants. NAD is another susceptible line.

Data in Table 5 summarize the segregation for the resistance of F2 plants in the field during 2013–2019. F2 plants of the cross PI 197088 × SMR-18 were tested in six seasons whereas F2 plants of the cross PI 330628 × SMR-18 were tested in two seasons. When two categories were used to classify the response of the plants to the disease (R and S),, two segregation ratios were observed, 1:15 or 1:63. When three categories were used for classification (R, MR, and S), two segregation ratios were observed, 1:14:1 or 1:9:6. The data suggest that genetic control of resistance in F2 plants varies between seasons, probably depending on the isolate prevailing in the field at each season.

Table 5. Inheritance of resistance against *P. cubensis* in intact cucumber plants in multiple years under field conditions at BIU Farm, Israel. R = resistant (AUDPC—≤5% of maximal value). MR—moderately resistant (AUDPC—6–79% of maximal value). S = susceptible (AUDPC—≥80% of maximal value).

A. 197088 × SMR-18, F2														
			Number of plants							**Number of plants**				
Year	**Isolate**	**Total**	**R**	**S**	**Ratio**	p	X^2	**Total**	**R**	**MR**	**S**	**Ratio**	p	X^2
2013	natural	75	5	70	1:15	0.881	0.022	75	5	68	2	1:14:1	0.439	1.648
2014	artificial, 83C	180	11	169	1:15	0.939	0.006	180	11	158	11	1:14:1	0.994	0.013
2016	artificial, 245C	95	3	86	1:15	0.212	1.559	95	3	89	3	1:14:1	0.190	3.322
2016	natural	156	3	153	1:63	0.717	0.132	156	3	95	58	1:9:6	*	*
2017	natural	103	5	98	1:15	0.558	0.342	103	5	57	41	1:9:6	*	*
2019	artificial, 260	92	5	87	1:15	0.747	0.104	92	5	78	9	1:14:1	0.366	2.012
B. 330628 × SMR-18, F2														
2013	natural	88	6	82	1:15	0.826	0.048	88	6	75	7	1:14:1	0.776	0.506
2019	artificial, 260	105	5	100	1:15	0.529	0.397	105	5	97	3	1:14:1	0.274	2.592

* = model not accepted.

3.6. Resistance in F2: F3 Plants

Two field-grown resistant F2 plants (107 and 111) of the cross PI 197088 × SMR-18, were tested (detached leaf bioassay) for resistance against 14 isolates of *P. cubensis*. Plant 107 was resistant to all isolates, except to the hybrid isolate 83C × 98P, while plant 111 was resistant to all isolates except to the hybrid isolate 172 B × 183 C (Table 6A), suggesting enhanced virulence of hybrid isolates. Each plant was self-pollinated and detached leaves from the F3 plants growing in the field were inoculated with each of the four isolates of the pathogen. The segregation data are shown in Table 6B. F3 plants segregated into R: S at a ratio of 3:1, 1:3, 1:15, or 7:9, depending on the isolate used for inoculation. No genetic model fits the segregation data when three response categories were used because no MR-scored plants were detected in the progenies (Table 6B).

Table 6. Segregation of resistance to downy mildew in F2 and F3 plants derived from the cross PI 197088 × SMR-18. A—Response of two F2 plants (107 and 111) to inoculation with 14 isolates of *P. cubensis*. R = resistant (score 0–20). S = susceptible (score 21–300). B—Response and segregation for the resistance of F3 plants derived from plants 107 and 111 after inoculation with four isolates of *P. cubensis*. R = resistant (score 0–20). MR- moderately resistant (score 21–200). S = susceptible (score 201–300). Plants were grown in the field and used for detached leaf bioassays.

A. 197088 × SMR-18, F2		
Isolate	**Plant 107**	**Plant 111**
PCHS	R	R
Harbin 10	R	R
88P	R	R
182D	R	R
US-163	R	R
83C	R	R
183/2	R	R
184	R	R
185	R	R
83C × 98P, F1	S	R
172B × 183C, F1	R	S
US-504	R	R
SG-11	R	R
Petiole 1	R	R

B. 197088 × SMR-18, F3						
		Number of plants				
Plant/Isolate	**Total**	**R**	**S**	**Ratio**	X^2	p
Plant 107						
182 D	40	30	10	3:1	1	0
83C × 98P, F1	42	14	28	1:3	0.212	1.556
83C	41	5	36	1:15	0.116	2.473
US-504	40	11	29	1:3	0.715	0.133
Plant 111						
182 D	34	11	23	1:3	0.322	0.980
83C × 98P, F1	36	8	28	1:3	0.700	0.148

Table 6. *Cont.*

83C	35	7	28		1:3	0.495	0.467
US-504	35	13	22		7:9	0.097	2.752
		Number of plants					
Plant/Isolate	**Total**	**R**	**MR**	**S**	**Ratio**	X^2	p
Plant 107							
182 D	40	30	10	0	3:1:0	**	**
83C × 98P, F1	42	14	28	0	1:3:0	**	**
83 C	41	5	36	0	1:3:0	**	**
US-504	40	11	29	0	1:3:0	**	**
Plant 111							
182 D	34	11	23	0	1:2:0	**	**
83C × 98P, F1	36	8	28	0	1:3:0	**	**
83 C	35	7	28	0	1:3:0	**	**
US-504	35	13	21	1	*	**	**

* = model not accepted. ** = no Mendelian models.

3.7. Resistance of F3 Plants in the Field

Seventy-nine F2 plants of the cross PI 197088 × SMR-18, six resistant (score 0), and 73 susceptible (score ≥ 200) (Figure 10) were self-pollinated to produce F3 seeds. Ten plants of each F3 entry were transplanted to the field and exposed to natural infection. Disease records were taken at weekly intervals for 22 days after the onset of the disease. Mean AUDPC and SD values for each F3 entry (n = 10) are shown in Figure 12. Mean AUDPC of SMR-18, PI 197088, and their F1 were 635, 10, and 206, respectively. F3 entries derived from resistant F2 plants showed AUDPC values ranging from 62 to 293 and those derived from susceptible F2 plants showed AUDPC values ranging from 31–545 (Figure 12). The results indicate that F3 plants are heterozygous for resistance, regardless of whether they were derived from a resistant or a susceptible F2 plant.

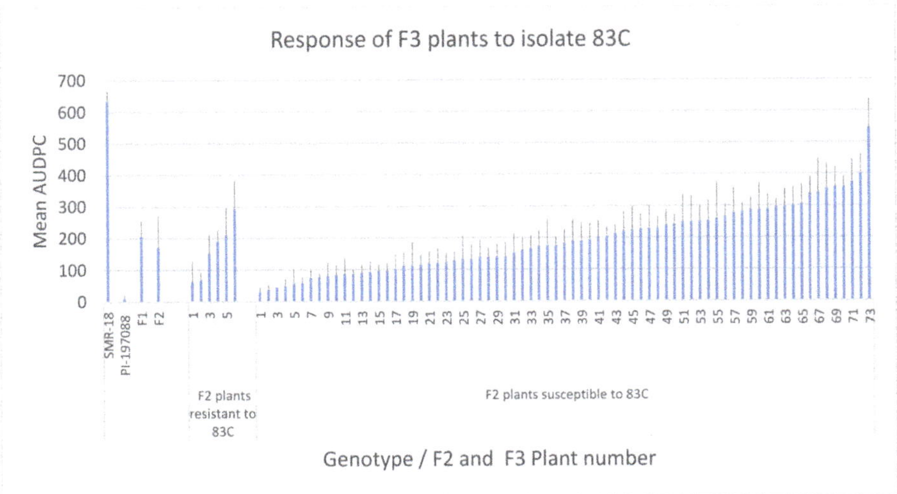

Figure 12. The response of F3 plants (derived from 6 F2 resistant plants and 73 F2 susceptible plants) to isolate 83C of *P. cubensis* in the field. The F3 data represent the mean and SD of 10 plants per entry.

4. Discussion

The population of *P. cubensis* in the field may consist of many isolates, pathotypes, or races with varying degrees of pathogenicity or virulence thus rendering host resistance ineffective. Indeed, combating downy mildew (DM) in cucumber through host plant resistance or fungicide applications has become more complex in the past two decades due to the emergence of new pathotypes, races, and mating types of the causal agent *P. cubensis*. Old cucumber cultivars resistant to DM succumbed to the new pathotypes, and the old fungicidal chemistries lost activity due to the prevalence of resistant isolates of the pathogen [4,13]. Breeding cucumber for DM resistance is a long and laborious task due to the lack of stable, multi-race resistant sources and the complex mode of resistance inheritance.

Here we identified two sources of wild cucumber with multi-race/pathotype resistance. We characterized the mechanism of their resistance and determined the way they inherit resistance to their progeny plants. Because the resistance of accession to a local isolate of *P. cubensis* does not necessarily mean that it will be resistant to isolates that prevail in other locations, we used a large collection of isolates from different parts of the world to screen resistance. We developed a detached leaf bioassay in which we could determine the resistance of a single plant to multiple isolates of *P. cubensis*. We thus were able, for the first time, to study the mode of inheritance of resistance to multiple isolates and predict the performance of the resistant pedigrees in other countries.

Of the six *Cucumis sativum* genotypes known to exhibit resistance against *P. cubensis* [4], only PI 197088 and PI 330628 [28,30] exhibited multiple-isolate resistance. They were self-pollinated for three generations to bring their multiple-isolate resistance to homozygosity. The stabilized lines were used for the inheritance studies reported here.

When grown in the field under natural epiphytotic conditions, no disease was observed on the leaves of PI 197088 or PI 330628. However, when artificially inoculated in the field or in growth chambers, a few necrotic lesions did appear. Microscopic observations revealed that PI 197088 and PI 330628 exhibit similar responses to artificial inoculation with *P. cubensis*. The pathogen ceased developing at a relatively late stage after penetration and developed some initial hyphae and haustoria. The haustoria formed were encased with callose, which probably inhibits the intake of nutrients into the mycelium, while the infected cells accumulated lignin-like, phloroglucinol-positive materials. A similar structural mode of resistance was observed in melons resistant to *P. cubensis* [33,34]. These defense compounds still allowed some deteriorated sporangiophores to emerge from the stoma but almost totally prevented sporangial production. We show here that unlike the resistance in melon which breaks down at 14 °C [34], the resistance in PI 197088 and PI 330628 remained effective at a low colonization temperature of 14 °C.

We used a double visual scoring system (percent infected leaf area and sporulation intensity) to determine the level of resistance to DM in detached leaves (Figure 1). We observed that leaves taken from F2 plants of the cross PI 197088 × SMR-18 or 330628 × SMR-18 segregated in their phenotypic responses to infection with *P. cubensis* ranging from complete resistance to high susceptibility. The pattern of segregation depended on the isolate used for inoculation. The differential pattern of response to different isolates indicated that a number of genes might be involved in resistance. The Mendelian analysis was employed to the segregated populations after categorical classification into S: R or R: MR: S. The analysis of two categories R and S indicated that resistance in PI 197088 or PI 330628 is controlled by either 1 dominant, 1 recessive, or 2 recessive genes, depending on the isolate used for inoculation. Analysis with three categories of S, MR, and R did not fit, in most cases, any Mendelian model of segregation.

F3 plants, derived from either susceptible or resistant F2 plants of the cross PI 197088 × SMR-18, showed a continuous pattern of variable resistance to DM in the field. When detached leaves were inoculated with different isolates, F3 plants segregated R:S 3:1 or 1:3, depending on the isolate used for inoculation, reaffirming that inheritance of resistance to DM is isolate-dependent.

Interestingly, field-grown F2 plants of the cross between the two resistant genotypes PI 197088 × PI 330628 were all fully resistant at the end of the season (no DM symptoms). However, two out of

75 plants in 2017 and one out of 130 plants in 2019 showed a few DM lesions, consisting of about 5% infected leaf area. Neighboring SMR-18 plants showed about 90% leaf area infected. This suggests that PI 197088 and PI 330628 differ in at least one gene for resistance. On the other hand, they share one QTL dm4.1 as suggested by Wang et al. [28].

Isolate-dependent inheritance of disease resistance is a rarely reported phenomenon [36]. Lapin et al. [37] showed that unlike most natural *Arabidopsis thaliana* accessions that are susceptible to one or more isolates of the downy mildew pathogen *Hyaloperonospora arabidopsidis*, accession C24 is resistant to all isolates tested. The resistance of C24 was found to be a multigenic trait with complex inheritance. Many identified resistance loci were isolate-specific and located on different chromosomes. Among the C24 resistance QTLs, there were dominant, codominant, and recessive loci. Interestingly, none of the identified loci significantly contributed to resistance against all three tested isolates.

Unlike wild cucumbers, resistance of the wild melon (*Cucumis melo* L) PI 124111F against *P. cubensis* is broad-spectrum but not isolate-specific [38]. That resistance was controlled genetically by two partially dominant, complementary loci [39]. Unlike other plant disease resistance genes, which confer an ability to resist infection by pathogens expressing corresponding avirulence genes, the resistance of PI 124111F to *P. cubensis* is controlled by enhanced expression of the enzymatic resistance (eR) genes *At1* and *At2*. These constitutively expressed genes encode the photorespiratory peroxisomal enzyme proteins glyoxylate aminotransferases. The low expression of *At1* and *At2* in susceptible melon lines is regulated mainly at the transcriptional level. This regulation is independent of infection with the pathogen. Transgenic melon plants overexpressing either of these eR genes displayed the enhanced activity of glyoxylate aminotransferases and remarkable resistance against *P. cubensis* [35,40]. Our attempts to transfer *At1* and *At2* to cucumber did not succeed (Cohen, *unpublished data*).

The results presented here corroborate with other studies in which multiple QTLs for resistance against *P. cubensis* were identified in PI 197088 and PI 330628. (Table 1). Wang et al. [30] reported QTL mapping results for DM resistance with F2:3 families from the cross between DM-resistant inbred line PI 330628 (WI7120) and susceptible '9930'. Four QTLs, *dm2.1*, *dm4.1*, *dm5.1*, and *dm6.1* were consistently and reliably detected across at least three of the four environments which together could explain 62–76% phenotypic variations. Among them, *dm4.1* and *dm5.1* were major effect QTL and *dm2.1* and *dm6.1* had moderate and minor effects, respectively.

Wang et al. [28] used recombinant inbred lines from a cross between PI 197088 and the susceptible line 'Coolgreen'. Phenotypic data on responses to natural DM infection were collected in three years and five locations from replicated field trials in North Carolina. The observed ratings followed a normal distribution that covered a large range of ratings at each environment and date. The interaction effects of genotype-by-location and genotype-by-year were significant at all ratings. QTL analysis identified 11 QTL for DM resistance harbored on chromosomes 1–6, accounting for more than 73.5% total phenotypic variance. Among the 11 DM resistance QTLs, *dm5.1*, *dm5.2*, and *dm5.3* were major effect contributing QTL whereas *dm1.1*, *dm2.1*, and *dm6.2* conferred susceptibility. The QTL *dm4.1* which had a moderate effect was likely the same as the major-effect QTL *dm4.1* detected in PI 330628 [30]. Three DM QTLs *dm2.1*, *dm5.2*, and *dm6.1*, were co-localized with powdery mildew (PM) QTLs, *pm2.1*, *pm5.1*, and *pm6.1*, respectively, which was consistent with the observed linkage of PM and DM resistances in PI 197088.

Katz et al. [29] reported on nine QTLs associated with resistance of PI 197088 against each of the seven isolates of *P. cubensis*. They examined for two years the response of a segregating F2 family (PI-197088 × SMR-18, n = 170) to seven isolates in growth chambers and the field. NGS (Next-Generation Sequencing) was performed for genotyping, and polymorphic SNPs were obtained from the same populations in both years. QTLs obtained for isolate 23C- resided on chromosomes 4 and 5; for isolate Pol.1- on chromosomes 1, 4, and 5; for isolate Pol.4- on chromosome 7; for isolate US-506- on chromosomes 1 and 2; for isolate 81C- on chromosomes 4 and 5; for isolate 88C- on chromosomes 3 and 6; for isolate 90C- on chromosomes 1, 4, and 6; for field isolate 2016, on chromosomes 3 and 5, and for field isolate 2017- on chromosomes 4 and 5. These authors concluded that the inheritance of resistance against DM in PI 197088 was isolate-dependent.

Tian et al. [41] sequenced 14% of the genome of one isolate of *P. cubensis* and identified 32 putative RXLR effector proteins and 29 secreted peptides with high similarity to RXLR effectors. They suggested that these effectors might play pivotal roles in pathogen fitness and pathogenicity. Sexual reproduction of the pathogen [12,42] may result in recombinant isolates which carry various combinations of effector proteins. It might, therefore, occur that isolates evolved in different parts of the world and therefore belong to different races, pathotypes, and mating types, each carries a unique set of effectors. Of this set of effectors, some might be secreted while others may not. Of the secreted effectors, some may recognize certain R genes in the host while others will not. This will make some host genotypes resistant to some genotypes of the pathogen.

The isolate-dependent inheritance of resistance of cucumber against *P. cubensis* may indicate that each isolate secretes a different battery of effectors that ignite a unique set of R genes in the host, thus making the inheritance of resistance isolate-dependent.

Author Contributions: T.C., D.K., Y.B.N., R.H., and B.H.B.D. did the investigations. A.E.R.—curation. Y.C. supervision and writing of the manuscript. All authors have read and agreed to the published version of the manuscript.

Funding: This research received no external funding

Conflicts of Interest: The authors declare no conflict of interest.

References

1. Palti, J.; Cohen, Y. Downy mildew of cucurbits (Pseudoperonospora cubensis)—The fungus and its hosts, distribution, epidemiology and control. *Phytoparasitica* **1980**, *8*, 109–147. [CrossRef]
2. Lebeda, A.; Cohen, Y. Cucurbit downy mildew (Pseudoperonospora cubensis)—Biology, ecology, epidemiology, host-pathogen interaction and control. *Eur. J. Plant Pathol.* **2011**, *129*, 157–192. [CrossRef]
3. Savory, E.; Granke, L.L.; Quesada-Ocampo, L.M.; Varbanova, M.; Hausbeck, M.K.; Day, B. The cucurbit downy mildew pathogen Pseudoperonospora cubensis. *Mol. Plant Pathol.* **2011**, *12*, 217–226. [CrossRef]
4. Cohen, Y.; Langenberg, K.M.V.D.; Wehner, T.C.; Ojiambo, P.S.; Hausbeck, M.; Quesada-Ocampo, L.M.; Lebeda, A.; Sierotzki, H.; Gisi, U.; Whener, T.C. Resurgence of Pseudoperonospora cubensis—The causal agent of cucurbit downy mildew. *Phytopathology* **2015**, *105*, 1–15. [CrossRef]
5. Ojiambo, P.; Gent, D.H.; Quesada-Ocampo, L.M.; Hausbeck, M.K.; Holmes, G.J. Epidemiology and population biology of Pseudoperonospora cubensis: A model system for management of downy mildews. *Annu. Rev. Phytopathol.* **2015**, *53*, 223–246. [CrossRef]
6. Cohen, Y. The combined effects of temperature, leaf wetness and inoculum concentration on infection of cucumbers with Pseudoperonospora cubensis. *Can. J. Bot.* **1977**, *55*, 1478–1487. [CrossRef]
7. Whisson, S.C.; Boevink, P.C.; Moleleki, L.N.; Avrova, A.O.; Morales, J.G.; Gilroy, E.M.; Armstrong, M.R.; Grouffaud, S.; Van West, P.; Chapman, S.; et al. A translocation signal for delivery of oomycete effector proteins into host plant cells. *Nature* **2007**, *450*, 115–118. [CrossRef]
8. Perl, M.; Cohen, Y.; Rotem, J. The effect of humidity during darkness on the transfer of assimilates from cucumber leaves to sporangia of Pseudoperonospora cubensis. *Physiol. Plant Pathol.* **1972**, *2*, 113–122. [CrossRef]
9. Cohen, Y. The novel oomycide oxathiapiprolin inhibits all stages in the asexual life cycle of Pseudoperonospora cubensis—Causal agent of cucurbit downy mildew. *PLoS ONE* **2015**, *10*, e0140015. [CrossRef]
10. Cohen, Y.; Rubin, A.E.; Galperin, M. Control of cucumber downy mildew with novel fungicidal mixtures of oxathiapiprolin. *Phytoparasitica* **2018**, *46*, 689–704. [CrossRef]
11. Call, A.D.; Criswell, A.D.; Wehner, T.C.; Ando, K.; Grumet, R. Resistance of cucumber cultivars to a new strain of cucurbit downy mildew. *HortScience* **2012**, *47*, 171–178. [CrossRef]
12. Cohen, Y.; Rubin, A.E. Mating type and sexual reproduction of Pseudoperonospora cubensis, the downy mildew agent of cucurbits. *Eur. J. Plant Pathol.* **2012**, *132*, 577–592. [CrossRef]
13. Thomas, A.; Carbone, I.; Lebeda, A.; Ojiambo, P. Virulence structure within populations of Pseudoperonospora cubensis in the United States. *Phytopathology* **2017**, *107*, 777–785. [CrossRef]

14. Polat, I.; Baysal, O.; Mercati, F.; Kitner, M.; Cohen, Y.; Lebeda, A.; Carimi, F. Characterization of Pseudoperonospora cubensis isolates from Europe and Asia using ISSR and SRAP molecular markers. *Eur. J. Plant Pathol.* **2014**, *139*, 641–653. [CrossRef]
15. Wallace, E.C.; D'Arcangelo, K.N.; Quesada-Ocampo, L.M. Population analyses reveal two host-adapted clades of Pseudoperonospora cubensis, the causal agent of cucurbit downy mildew, on commercial and wild cucurbits. *Phytopathology* **2020**. [CrossRef]
16. Cohen, Y.; Rubin, A.E.; Galperin, M. Host preference of mating type in Pseudoperonospora cubensis, the downy mildew causal agent of cucurbits. *Plant Dis.* **2013**, *97*, 292. [CrossRef]
17. Pang, X.; Zhou, X.; Wan, H.; Chen, J. QTL mapping of downy mildew resistance in an introgression line derived from interspecific hybridization between cucumber and Cucumis hystrix. *J. Phytopathol.* **2013**, *161*, 536–543. [CrossRef]
18. Li, L.; He, H.; Zou, Z.; Li, Y. QTL analysis for downy mildew resistance in cucumber inbred line PI 197088. *Plant Dis.* **2018**, *102*, 1240–1245. [CrossRef]
19. Szczechura, W.; Staniaszek, M.; Kłosińska, U.; Kozik, E.U. Molecular Analysis of New Sources of Resistance to Pseudoperonospora cubensis (Berk. et Curt.) Rostovzev in Cucumber. *Russ. J. Genet.* **2015**, *51*, 974–979. [CrossRef]
20. Van Vliet, G.J.A.; Meysing, W.D. Inheritance of resistance to Pseudoperonospora cubensis Rost. in cucumber (Cucumis sativus L.). *Euphytica* **1974**, *23*, 251–255. [CrossRef]
21. McFerson, J.R. A Screening Procedure for Determining Levels of Resistance in Cucumber. Master's Thesis, Texas A&M University, College Station, TX, USA, 1978.
22. Pershin, A.F.; Medvedeva, N.I.; Medvedev, A.V. Quantitative approach to studying the genetics of disease resistance. IV. Interaction of the genetic systems for resistance to powdery and downy mildews in cucumber. *Genet. USSR.* **1988**, *24*, 484–493.
23. Doruchowski, R.W.E.; Lakowska-Ryk, E. Inheritance of Resistance to Downy Mildew (Pseudoperonospora cubensis Berk & Curt) in Cucumis sativus. In *Proceedings of the 5th EUCARPIA Symposium, Warsaw, Poland, 27–31 July 1992*; Doruchowski, R.W., Kozik, E., Niemirowicz Szczytt, K., Eds.; Research Institute of Vegetable Crops and Warsaw Agricultural University: Warsaw, Poland, 1992; pp. 132–138.
24. Petrov, L.; Boogert, K.; Sheck, L.; Baider, A.; Rubin, E.; Cohen, Y. Resistance to downy mildew, Pseudoperonospora cubensis, in cucumbers. *Acta Hortic.* **2000**, *510*, 203–209. [CrossRef]
25. Angelov, D. Inheritance of resistance to downy mildew, Pseudoperonospora cubensis (Berk. & Curt.) Rostow. In Proceedings of the Report of the 2nd National Symposium of Plant Immunity, Plovdiv, Bulgaria, 1994; Volume 3, pp. 99–105.
26. Caldwell, D.; Chan, E.; de Vries, J.; Joobeur, T.; King, J.; Reina, A.; Shetty, N. Methods and Compositions for Identifying Downy Mildew Resistant Cucumber Plants. U.S. Patent Application No. 2011/0126309 A1, April 2011.
27. Yoshioka, Y.; Sakata, Y.; Sugiyama, M.; Fukino, N. Identification of quantitative trait loci for downy mildew resistance in cucumber (Cucumis sativus L.). *Euphytica* **2014**, *198*, 265–276. [CrossRef]
28. Wang, Y.; Vandenlangenberg, K.; Wen, C.; Wehner, T.C.; Weng, Y. QTL mapping of downy and powdery mildew resistances in PI 197088 cucumber with genotyping-by-sequencing in RIL population. *Theor. Appl. Genet.* **2018**, *131*, 597–611. [CrossRef]
29. Katz, D.; Hammer, R.; Ben Naim, Y.; Cohen, Y. Resistance to downy mildew in cucumber breeding lines. *Phytoparasitica* **2018**, *46*, 320.
30. Wang, Y.; Vandenlangenberg, K.; Wehner, T.C.; Kraan, P.A.G.; Suelmann, J.; Zheng, X.; Owens, K.; Weng, Y. QTL mapping for downy mildew resistance in cucumber inbred line WI7120 (PI 330628). *Theor. Appl. Genet.* **2016**, *1298*, 1493–1505. [CrossRef]
31. Zhang, S.P.; Liu, M.M.; Miao, H.; Zhang, S.Q.; Yang, Y.H.; Xie, B.Y.; Wehner, T.C.; Gu, X.F. Chromosomal mapping and QTL analysis of resistance to downy mildew in Cucumis sativus. *Plant Dis.* **2013**, *97*, 245–251. [CrossRef]
32. Win, K.T.; Vegas, J.; Zhang, C.; Song, K.; Lee, S. QTL mapping for downy mildew resistance in cucumber via bulked segregant analysis using next-generation sequencing and conventional methods. *Theor. Appl. Genet.* **2017**, *130*, 199–211. [CrossRef]
33. Cohen, Y.; Eyal, H.; Hanania, J.; Malik, Z. Ultrastructure of Pseudoperonospora cubensis in muskmelon genotype susceptible and resistant to downy mildew. *Physiol. Mol. Plant Pathol.* **1989**, *34*, 27–40. [CrossRef]

34. Balass, M.; Cohen, Y.; Bar-Joseph, M. Temperature-dependent resistance to downy mildew in muskmelon: Structural responses. *Physiol. Mol. Plant Pathol.* **1993**, *43*, 11–20. [CrossRef]
35. Taler, D.; Galperin, M.; Benjamin, I.; Cohen, Y.; Kenigsbuch, D. Plant eR genes encoding for photorespiratory enzymes confer resistance against disease. *Plant Cell* **2004**, *16*, 172–184. [CrossRef]
36. Chen, T.; Cohen, Y. Isolate-dependent inheritance of resistance to downy mildew in cucumber. *Phytoparasitica* **2013**, *41*, 462.
37. Lapin, D.; Meyer, R.C.; Takahashi, H.; Bechtold, U.; Ackerveken, G.V.D. Broad-spectrum resistance of Arabidopsis C24 to downy mildew is mediated by different combinations of isolate-specific loci. *New Phytol.* **2012**, *196*, 1171–1181. [CrossRef]
38. Cohen, Y.; Eyal, H. Downy mildew-, powdery mildew- and fusarium wilt-resistant muskmelon breeding line PI-124111F. *Phytoparasitica* **1987**, *15*, 187–195. [CrossRef]
39. Kenigsbuch, D.; Cohen, Y. Inheritance of resistance to downy mildew in Cucumis melo PI 124112 and commonalty of resistance genes with C. melo PI 124111F. *Plant Dis.* **1992**, *76*, 615–617. [CrossRef]
40. Benjamin, I.; Kenigsbuch, D.; Galperin, M.; Abrameto, J.A.; Cohen, Y. Cisgenic melons over expressing glyoxylate-aminotransferase are resistant to downy mildew. *Eur. J. Plant Pathol.* **2009**, *125*, 355–365. [CrossRef]
41. Tian, M.; Win, J.; Savory, E.; Burkhardt, A.K.; Held, M.A.; Brandizzi, F.; Day, B. 454 Genome sequencing of Pseudoperonospora cubensis reveals effector proteins with a QXLR translocation motif. *Mol. Plant Microbe Interact.* **2011**, *24*, 543–553. [CrossRef]
42. Thomas, A.; Carbon, I.; Cohen, Y.; Ojiambo, P. Occurrence and distribution of mating types of Pseudoperonospora cubensis in the USA. *Phytopathology* **2017**, *107*, 313–321. [CrossRef]

agronomy

MDPI

Article

The Variability of Puroindoline-Encoding Alleles and Their Influence on Grain Hardness in Modern Wheat Cultivars Cultivated in Poland, Breeding Lines and Polish Old Landraces (*Triticum aestivum* L.)

Mateusz Przyborowski [1,*], Sebastian Gasparis [1], Maciej Kała [1], Wacław Orczyk [2] and Anna Nadolska-Orczyk [1]

[1] Department of Functional Genomics, Plant Breeding and Acclimatization Institute–National Research Institute, Radzików, 05-870 Błonie, Poland; s.gasparis@ihar.edu.pl (S.G.); maciek.kala@gmail.com (M.K.); a.orczyk@ihar.edu.pl (A.N.-O.)

[2] Department of Genetic Engineering, Plant Breeding and Acclimatization Institute–National Research Institute, Radzików, 05-870 Błonie, Poland; w.orczyk@ihar.edu.pl

* Correspondence: m.przyborowski@ihar.edu.pl

Received: 19 June 2020; Accepted: 22 July 2020; Published: 25 July 2020

Abstract: Wheat (*Triticum aestivum* L.) grain hardness is determined mainly by variations in puroindoline genes (*Pina-D1* and *Pinb-D1*), which are located on the short arm of chromosome 5D. This trait has a direct effect on the technological properties of the flour and the final product quality. The objective of the study was to analyze the mutation frequency in both *Pin* genes and their influence on grain hardness in 118 modern bread wheat cultivars and breeding lines cultivated in Poland, and 80 landraces from Poland. The PCR products containing the *Pin* gene coding sequences were sequenced by the Sanger method. Based on detected the SNPs (single-nucleotide polymorphisms) we designed CAPS (cleaved amplified polymorphic sequence) markers for the fast screening of *Pinb* alleles in a large number of genotypes. All analyzed cultivars, breeding lines, and landraces possess the wild-type *Pina-D1a* allele. Allelic variation was observed within the *Pinb* gene. The most frequently occurring allele in modern wheat cultivars and breeding lines (over 50%) was *Pinb-D1b*. The contribution of the remaining alleles (*Pinb-D1a*, *Pinb-D1c*, and *Pinb-D1d*) was much less (approx. 15% each). In landraces, the most frequent allele was *Pinb-D1a* (over 70%), followed by *Pinb-D1b* (21% frequency). *Pinb-D1c* and *Pinb-D1g* were found in individual varieties. SKCS (single-kernel characterization system) analysis revealed that grain hardness was strictly connected with *Pinb* gene allelic variation in most tested cultivars. The mean grain hardness values were significantly greater in cultivars with mutant *Pinb* variants as compared to those with the wild-type *Pinb-D1a* allele. Based on grain hardness measured by SKCS, we classified the analyzed cultivars and lines into different classes according to a previously proposed classification system.

Keywords: grain hardness; PIN; kernel texture; triticum; wheat; SKCS

---

1. Introduction

In Europe, common wheat (*Triticum aestivum* L.) is cultivated on almost 61 million hectares of land, and, in Poland, on 2.416 million hectares, which accounts for about 31% of all cereals cultivated in Poland [1]. Wheat is an important component of food and feed. Wheat grains, depending on their technological properties, are used for various food products [2]. Grain hardness, one of the most important technological traits, is considered as significant as gluten protein composition [3]. It has a direct effect on the grinding and baking characteristics, flour particle sizes, and water absorption ability, which represent crucial technological flour properties [3,4]. In common wheat, grain hardness is primarily controlled by two puroindoline genes, *Pina* and *Pinb*. Both genes are intronless, with a 447 bp coding sequence and are located in locus *Ha* on the short arm of chromosome 5D [5]. The *Pin* gene coding sequences are 70.2% similar. In the course of wheat evolution, puroindoline genes located on chromosomes 5A and 5B were lost after wheat ancestor polyploidization [6]. *Pina* and *Pinb* encode the following puroindoline proteins: puroindoline a (PINA), and puroindoline b (PINB). Puroindolines are small (~13 kDa), cysteine-rich, lipid-binding proteins consisting of 148 amino acids [7,8]. PIN proteins exhibit a tryptophan-rich domain, which consists of five tryptophan residues in PINA and three residues in PINB [5]. Increased grain hardness is the result of mutations within one or both wild-type *Pin* alleles, which cause amino acid sequence changes in puroindoline proteins [9]. Thus far, 30 *Pina* alleles and 26 *Pinb* alleles have been described in different wheat species [10–13]. In most cases, the differences between wild-type and mutant alleles are single nucleotide polymorphisms (SNPs), such as in *Pinb-D1b* and *Pinb-D1c*, or one base deletion, like in *Pinb-D1p* and *Pinb-D1u*. Allele *Pinb-D1b*, originally described by Giroux and Morris [14], is characterized by guanine to adenine substitution at position 223, resulting in glycine to serine substitution in PINB at position 46 and changes in the tryptophan-rich domain (TRD). It is the most frequent *Pin* gene mutation among all domesticated wheat cultivars. *Pinb-D1c* and *Pinb-D1d* are the second most frequent alleles found in common wheat cultivars [11]. *Pinb-D1c* is characterized by thymine to cytosine substitution at position 266, leading to leucine to proline change at position 60 [15]. In *Pinb-D1d*, thymine to adenine substitution at position 217 causes tryptophan to arginine change at position 44 and leads to changes in the TRD [15]. An example of the change in reading frame is the allele *Pinb-D1g*, where a single nucleotide change at position 255 (cytosine to adenine substitution) causes a cysteine to stop codon change at position 56 [16]. An example of a SNP which does not change the amino acid sequence is the *Pinb-D1ae* allele, in which thymine is substituted by adenine at position 93 [17]. Less common alleles possess more than one SNP, e.g., two SNPs in *Pinb-D1v* [18] or one SNP and one base deletion in *Pinb-D1s* [19]. *Pina-D1b* is a null mutation resulting from the deletion of a 15.3 Kb chromosome fragment [20,21].

Genetic diversity and grain hardness allelic variation were studied in Australian commercial wheat cultivars [22], the Watkins collection [23], and bread wheat landraces from Andalusia [24]. Additionally, the trait is also easily genetically engineered [25–27]. Both *Pin* genes are considered the main grain hardness determinants. TRD is the most important domain of PIN proteins because it is responsible for their lipid-binding properties. Changes in TRD caused by SNP in *Pinb-D1b* and *Pinb-D1d* alleles result in decreased affinity to anionic phospholipids and altered interaction between PINA and PINB proteins [28,29]. TRD is comprised of five tryptophan (Trp) residues in PINA and three Trp residues in PINB [5]. Apart from puroindoline-encoding genes, very similar genes showing over 70% similarity with the *Pinb* nucleotide sequence have been identified. They are categorized as *Pinb-2* variants and are located on chromosomes 7A, 7B, and 7C [30–33]. However, further studies clearly demonstrated that the *Pinb-2* variants expression levels were very low and these genes had no significant effect on grain hardness [31,34].

In this study, we analyzed allelic variation in *Pin* genes in collection of hexaploid wheat cultivated in Poland, including Polish and European cultivars, as well as in Polish breeding lines and landraces, and we determined how this variation affected grain hardness. The cleaved amplified polymorphic sequence (CAPS) markers adapted to the screening of *Pinb* alleles proved to be useful.

2. Materials and Methods

2.1. Plant Material

The plant material consisted of 67 wheat cultivars cultivated in Poland in 2015–2018, and 51 breeding lines from Polish breeding companies, and all 80 landraces collected in Poland between 1960 and 1976, and deposited in the National Centre for Plant Genetic Resources (KCRZG) of the Plant Breeding and Acclimatization Institute – National Research Institute. Plants were grown during 2018–19 in a greenhouse, with eight plants in a twelve liter pot using standard agronomic practices including preventative spraying with plant protection products. The full list of wheat genotypes is presented in Supplementary Tables S1 and S2.

2.2. DNA Extraction

The seedlings were grown in beakers with 4 mm glass balls and autoclaved tap water, in darkness at 23 °C. Genomic DNA was extracted from coleoptile fragments and leaves of six-day old seedlings using a modified cetrimonium bromide (CTAB) method devised by Murray and Thompson [35]. DNA was extracted from the pooled tissue samples of at least six individuals from the same variety. The DNA purity and concentration were checked using a Nanodrop 1000 (NanoDrop Technologies) and diluted to 50 ng/μL for further PCR analysis.

2.3. *Pina* and *Pinb* PCR Amplification and Sanger Sequencing

Both alleles of the *Pin* genes were sequenced in all examined (198) wheat genotypes. Two primer pairs were used for PCR amplification and sequencing: *Pina* forward: 5′-CATCTATTCATCTCCA CCTGC-3′; reverse: 5′-GTGACAGTTTATTAGCTAGTC-3′ (product length—524 bp); *Pinb* forward: 5′-GAGCCTCAACCCATCTATTCATC-3′; reverse: 5′-CAAGGGTGA TTTTATTCATAG-3′ (product length—595 bp) [36]. The PCR mix contained: 100 ng DNA, 0.5 μM of each primer, 200 μM deoxynucleotide triphosphates (dNTPs) mix, 0.4 U Phusion Hot Start II DNA Polymerase (Thermo Scientific), 4 μL 5× Phusion HF buffer, and PCR-grade water up for a 20 μL final volume. PCR was performed under the following conditions: initial denaturation at 98 °C for 30 s, followed by 36 cycles consisting of denaturation at 98 °C for 10 s, annealing at 58 °C (for both primer pairs) for 30 s, DNA elongation at 72 °C for 15 s, and, after 36 cycles, the final extension step at 72 °C for 8 min.

The amplification products were purified using the GeneJET PCR Purification Kit (Thermo Scientific), according to the manufacturer's instructions, and then sequenced by the Sanger method using the same primers that were used for amplification. The samples were sequenced by Genomed S.A. Sequences, which were aligned using MegAlgin (DNASTAR) software.

2.4. CAPS Analysis

The *Pinb* gene primer sequences for amplification were as follows: forward: 5′-ATGAAGACCT TATTCC TCCTA-3′; reverse: 5′-AGTAATAGCCACTAGGGAACTT-3′ (product length—442 bp) [5]. The PCR mix was composed of: 100 ng DNA, 0.5 μM of each primer, 200 μM dNTP mix, 0.4 U Phusion Hot Start II DNA Polymerase (Thermo Scientific), 4 μL 5× Phusion HF buffer, and PCR-grade water up to a final volume of 20 μL. PCR was performed under the following conditions: initial denaturation at 98 °C for 30 s followed by 36 cycles consisting of denaturation at 98 °C for 10 s, annealing at 53 °C for 30 s, DNA elongation at 72 °C for 15 s, and, after 36 cycles, the final extension step at 72 °C for 8 min. PCR product was digested with restriction enzymes selected using the SNP2CAPS program [37] to confirm the sequencing results. The digest reaction was composed of: 5 μL PCR product, 0.7 μL restriction enzyme, 2 μL 10× compatible buffer, and PCR grade water up to a final reaction volume of 20 μL. Samples were incubated at 37 °C in a water bath for 3 h. Digested fragments were separated on a 2.5% agarose gel by electrophoresis in SB buffer (Brody and Kern, 2004). *Mbi*I (Thermo Scientific) was used to confirm *Pinb-D1b*, *Pvu*II (Thermo Scientific) for *Pinb-D1c*, and *Mnl*I (Thermo Scientific) for *Pinb-D1d*. The amplicons containing *Pinb-D1a, Pinb-D1c* and *Pinb-D1d* alleles cut by *MbiI* gave two fragments, which were—318 and 129 bp long. The amplicon with *Pinb-D1b* cut with the same enzyme gave three fragments, of 223, 129 and 95 bp. Amplicons with *Pinb-D1a*, *Pinb-D1b* and *Pinb-D1d* digested with *PvuII* gave two fragments, which were—264 and 183 bp long, and the amplicon with *Pinb-D1c*, that was not cut with this enzyme had a fragment that was 442 bp long. Digestion of amplicons with *Pinb-D1a*, *Pinb-D1b*, and *Pinb-D1c* with *MnlI* led to nine fragments, which had the following lengths—137, 112, 66, 45, 29, 25, 20, 5 and 3 bp. Digestion with the same enzyme of *Pinb-D1d* amplicon gave ten fragments, which had the following lengths—115,112, 66, 45, 29, 25, 22, 20, 5, and 3 bp.

2.5. Grain Hardness Measurement

The grain hardness was measured using the Single Kernel Characterization System (SKCS) 4100 (Perten Instruments). Single kernel characterization was measured according to Approved Method 55–31.01 [38]. The hardness of each wheat kernel was determined by the instrumental measurement of the force required to crush the kernel. The SKCS instrument was calibrated to compute the hardness index including the weight, moisture, and diameter of each kernel. For each sample, 300 individual kernels were analyzed.

2.6. Percentage of Protein, Starch, and Wet Gluten

The percentage of protein, starch, and wet gluten was measured using the Infratec Nova Grain Analyser (Foss). Analyses were performed for landraces in two independent replicates. Samples of 30 g grains were used.

2.7. Statistical Analysis

Statistical analysis was performed using the R v. 3.6.2 and the FSA v. 0.8.27 package [39]. The normality distribution was verified by the Shapiro—Wilk test. The Kruskal—Wallis test was performed to compare variable distributions of *Pinb* allele groups and hardness index. The Dunn test was performed in order to establish which groups were different from each other. Pearson correlationa between hardness and protein or starch or gluten content were calculated separately for landraces with the wild-type *Pinb* alleles and mutant alleles, as well as for all landraces, regardless of their allelic status. The chart of hardness index ranges was performed using gglot2 v. 3.2.1 package [40].

3. Results

The allelic variation of *Pina* and *Pinb* was determined by Sanger sequencing. All tested cultivars, breeding lines, and landraces had the wild-type allele of the *Pina* gene (*Pina-D1a*). The frequencies of *Pinb-D1* alleles are shown in Table 1. The most frequent allele among modern cultivars and breeding lines was *Pinb-D1b* (57%). The wild type *Pinb-D1a* allele occurred in 19 lines (16%) and the frequencies of the remaining alleles, *Pinb-D1c* and *Pinb-D1d*, were at a similar level (12% and 15%, respectively).

Table 1. The *Pinb* allele frequency in the tested genotypes and mean hardness index (HI) in each allele group.

Pinb Allele	No. Cultivars (%)	Mean HI (SD)
	Cultivars and Breeding Lines	
Pinb-D1a	19 (16%)	31.31 (±19.50)
Pinb-D1b	67 (57%)	59.36 (±11.73)
Pinb-D1c	14 (12%)	63.70 (±10.43)
Pinb-D1d	18 (15%)	59.08 (±13.80)
	Landraces	
Pinb-D1a	57 (71%)	23.34 (±6.09)
Pinb-D1b	21 (26%)	49.23 (±10.97)
Pinb-D1c	1 (1%)	65.80
Pinb-D1g	1 (1%)	60.79

In landraces, the *Pinb* frequencies were different to modern cultivars and breeding lines. Over 71% had the *Pinb-D1a* allele, while 26% (21) had the *Pinb-D1b* allele. *Pinb-D1c* and *Pinb-D1g* were each found in one tested landrace. Alignment of the nucleotide and amino acid sequences of *Pinb* alleles with the marked mutations is shown in Figure 1, and chromatogram fragments of five alleles found in tested wheat genotypes are shown in Supplementary Figure S1.

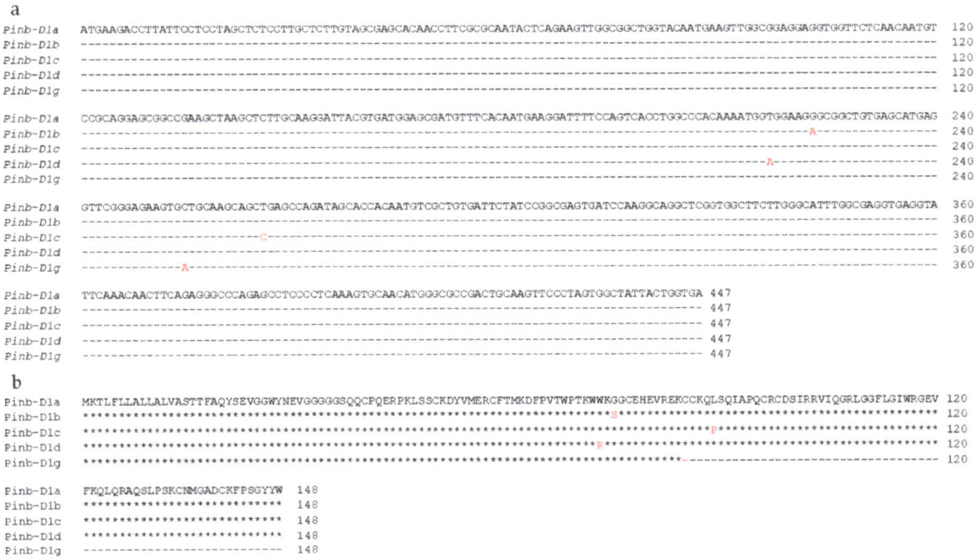

Figure 1. (**a**) Alignment of the nucleotide and (**b**) amino acid sequences of *Pinb* alleles detected in the tested cultivars.

Next, we examined whether the mutations in the *Pinb* coding sequence could be detected by CAPS analysis. Restriction enzyme digestion of the amplified full length *Pinb* coding sequences produced allele-specific fragment patterns, which were unique for each enzyme. The *Pinb-D1b*, *Pinb-D1c*, and *Pinb-D1d* alleles can be distinguished from the wild-type *Pinb-D1a* allele by restriction analysis with *Mbi*I, *Pvu*II, and *Mnl*I, respectively (Figure 2). The SKCS hardness values were grouped by each *Pinb* allele and by genotype (modern vs. landraces) (Figure 3). In modern cultivars and breeding lines the hardness index for each allele group ranged as follows: for *Pinb-D1a*, from 6 to 57; for *Pinb-D1b*, from 35 to 80; for *Pinb-D1c*, from 42 to 77; for *Pinb-D1d* from 40 to 82. In landraces, the hardness index for the *Pinb-D1a* group ranged from 9 to 37 and for *Pinb-D1b* from 18 to 63. The landrace with *Pinb-D1c* had a hardness index of 66, and another with *Pinb-D1g* had a hardness index of 61 (Supplementary Tables S1 and S2).

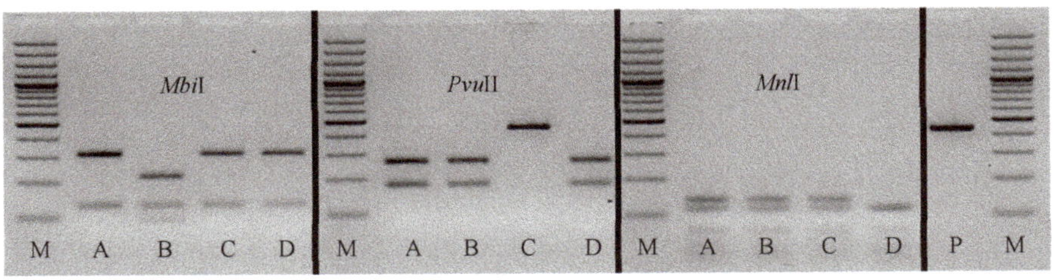

Figure 2. Examples of cleaved amplified polymorphic sequence (CAPS) analysis for detection of different *Pinb* alleles. M—molecular weight marker; A—*Pinb-D1a*; B—*Pinb-D1b*; C—*Pinb-D1c*; D—*Pinb-D1d*; P—undigested 447 bp PCR product.

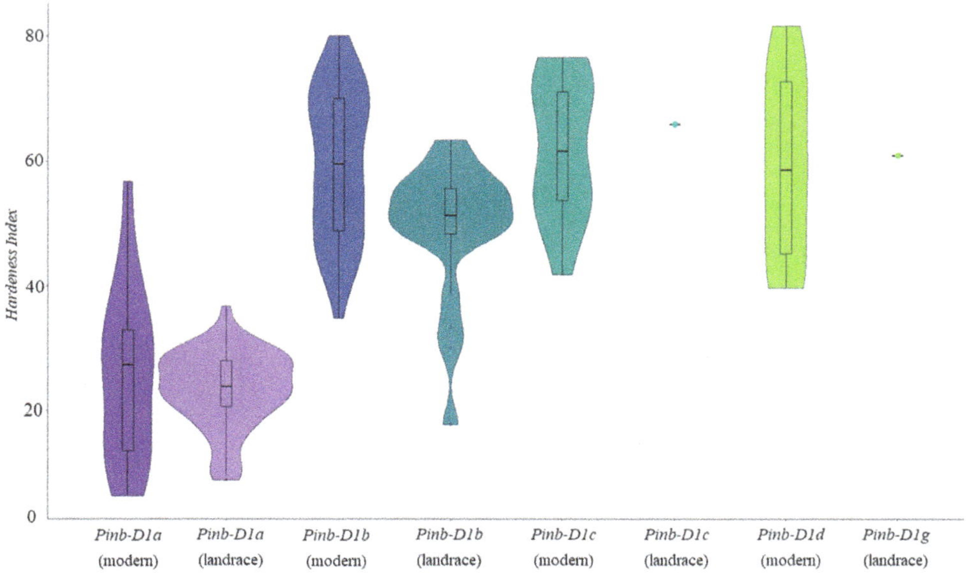

Figure 3. Hardness index ranges for individual *Pinb* alleles divided into modern wheats and landraces. The widths of the figures indicates the frequency of occurrence of a given hardness index.

The tested genotypes were classified, as proposed by Corona et al. [41] into three classes, namely soft, medium hard, and hard types with the SKCS values in the range of 15–40, 41–70, and 71–95 respectively. In the modern cultivars group, 20 genotypes were classified as soft type, 74 as medium hard, and 24 as hard type. In the group of landraces, 61 genotypes were classified as soft and 19 as medium hard.

The protein content in the tested landraces ranged from 13.10% to 23.15%. The starch content ranged from 55.85% to 67.70%, and the wet gluten content ranged from 28.05% to 53.15% (Supplementary Table S3). None of the above values were significantly correlated at $p < 0.05$ with the hardness measured by SKCS. The correlation coefficient between grain hardness and protein, starch and wet gluten content calculated for all landraces was −0.18, 0.19, and −0.16, respectively. The same correlation coefficients calculated for the wild-type *Pinb* landraces were −0.18 (SKCS vs. protein content), 0.15 (SKCS vs. starch content), and −0.03 (SKCS vs. wet gluten content), and for the mutant landraces, these values were −0.41, 0.36, and −0.45, respectively. The normality distribution of hardness values in each allele group was verified by the Shapiro–Wilk test. The test was positive only for the *Pinb-D1b* group in modern wheats. The Kruskal–Wallis test was performed to compare variable distributions between *Pinb* allele groups with a breakdown into modern wheat and landraces, and hardness, and excluding single observations (*Pinb-D1c* and *Pinb-D1g* in landraces). The results proved to be statistically significant (p-value $< 2.2 \times 10^{-16}$). In the next step, the Dunn test was performed in order to establish which groups were different from each other. The SKCS values of modern wheats and landraces with *Pinb-D1b*, *Pinb-D1c*, or *Pinb-D1d* were significantly different compared to genotypes with wild-type *Pinb* allele (Table 2). There were no significant differences between genotypes carrying the *Pinb-D1b*, *Pinb-D1c*, or *Pinb-D1d* alleles (Table 2).

Table 2. Dunn test for multiple comparisons of SKCS hardness in six allele groups. (L)—landraces, (M)—modern cultivars and breeding lines.

Tested Pairs	Z	P.unadj	P.adj	Difference
Pinb-D1a (L)–Pinb-D1b (L)	−4.931	0.000	0.000	True
Pinb-D1a (L)–Pinb-D1b (M)	−9.762	0.000	0.000	True
Pinb-D1a (L)–Pinb-D1c (M)	−6.448	0.000	0.000	True
Pinb-D1a (L)–Pinb-D1d (M)	−6.873	0.000	0.000	True
Pinb-D1a (M)–Pinb-D1a (L)	0.158	0.875	1.000	False
Pinb-D1a (M)–Pinb-D1b (L)	−3.588	0.000	0.005	True
Pinb-D1a (M)–Pinb-D1b (M)	−6.025	0.000	0.000	True
Pinb-D1a (M)–Pinb-D1d (M)	−5.096	0.000	0.000	True
Pinb-D1a (M)–Pinb-D1c (M)	−5.053	0.000	0.000	True
Pinb-D1b (L)–Pinb-D1c (M)	−1.926	0.054	0.812	False
Pinb-D1b (L)–Pinb-D1d (M)	−1.688	0.091	1.000	False
Pinb-D1b (M)–Pinb-D1b (L)	2.042	0.041	0.617	False
Pinb-D1b (M)–Pinb-D1c (M)	−0.516	0.606	1.000	False
Pinb-D1b (M)–Pinb-D1d (M)	−0.058	0.954	1.000	False
Pinb-D1c (M)–Pinb-D1d (M)	0.394	0.694	1.000	False

4. Discussion

We analyzed grain hardness within a collection of various wheat cultivars, breeding lines currently applied in wheat breeding programs and old landraces collected during multiple expeditions between 1960 and 1976 in Poland. Cultivars were tested both at the molecular level (by determining *Pin* gene allelic status) and at the phenotype level (by measuring wheat kernel SKCS hardness index).

Surprisingly, no allelic variability within the *Pina* gene was observed, although mutant alleles of this gene were detected in many other European cultivars [15,24,42]. All tested cultivars, breeding lines, and landraces had the wild-type *Pina-D1a* allele. This allele, originally described by Gautier et al. [5], was proven to be the most common *Pina* allele among wheat genotypes [10,12]. The second *Pina* allele found in European cultivars was *Pina-D1b*, discovered by Giroux and Morris [9] which is extremely rare. Other *Pina* alleles mainly occurred in landraces from China or India [17,18]. The lack of *Pina* variability in the European modern wheat cultivars and breeding lines tested here indicates negative mutant gene selection during the breeding process.

More common allelic variability was observed for the *Pinb* gene. The *Pinb-D1b* allele, detected in 57% of tested modern cultivars and breeding lines, was the most frequent. The remaining *Pinb* alleles were the wild-type *Pinb-D1a* and the mutant *Pinb-D1c* and *Pinb-D1d*. They were represented by a similar variant number (11–16%). Similar distribution of the same *Pinb* alleles was observed previously in European wheat populations [15,24,43]. As for the landraces, the *Pinb* allele distributions were different. The main, wild-type *Pinb-D1a* allele was found in over 70% of tested landraces. The next most frequent was *Pinb-D1b*, detected in 26% of the tested landraces. *Pinb-D1c* was found only in one variety, and none contained the *Pinb-D1d* allele. It was surprising for us to find a landrace with the *Pinb-D1g* allele. To date, this allele has not been observed in European wheat cultivars. It was reported for the first time in historical varieties from North America [16]. We did not find such high variability among *Pina* and *Pinb* alleles, which were observed in hard wheat cultivars originating from other continents. For example, among those originating from Asia, seven various *Pinb* alleles, including four newly described, such as *Pinb-D1ad*, *Pinb-D1ae*, *Pinb-D1af* and *Pinb-D1ag*, have been found [17]. In another study, nine *Pinb* alleles were described [44]. Therefore, our observation confirms that these four alleles found among modern wheats cultivated in Poland dominate in modern European varieties. This is probably caused by the narrow genetic variability of breeding lines as well as being the result of breeding pressure.

The most commonly used methods of determination of grain hardness include the particle size index (PSI), near-infrared reflectance (NIR) with a properly prepared library, and the single kernel characterization system (SKCS). We decided to use SKCS due to the high reliability and accuracy of this method, and the possibility of analyzing 300 individual grains in a single run. Based on the SKCS value, wheats can be classified into different categories. The first classification was proposed by Morris et al. [16], who distinguished four groups according to the SKCS hardness index: (I) from 0 to 33, (II) from 34 to 46, (III) from 47 to 59, and (IV) above 60. Slightly different categories were proposed by Corona et al. [41] who divided wheat into three main groups: soft (SKCS 15–40), medium hard (41–70), and hard (71–95). The second wheat classification system, used also by us, is the most common. However, in many regions, other classes are used as well, depending on local needs or regulations. For example, Sharma et al. [45] proposed five classes for a more precise classification of soft and hard cultivars from India. They included very soft (SKCS 34 or less), soft (35–54), medium hard (SKCS 55–74), and hard (90 or more).

It was noted that some of the tested cultivars were classified into groups contrary to ourexpectations based on their allelic status. For example, in case of the *Pinb-D1a*, a group of six cultivars and breeding lines had the SKCS indexes ranging from 31 to 56, with a mean value of 46, which was higher than usual for wild-type alleles. In previous reports, cultivars with both wild-type *Pina* and *Pinb* alleles had low grain hardness values ranging from 7 to 40, with a mean value of 27 [41], or from 15 to 29, with a mean value of 24.5 [16]. Moreover, some cultivars, breeding lines, and landraces with *Pinb-D1b* or *Pinb-D1d* mutant alleles, which were expected to have higher grain hardness values, were classified as soft, with a hardness index of 15 to 40. Similarly, one cultivar and one breeding line carrying *Pinb-D1d* were classified as soft.

Particularly interesting is the landrace with the accession number I31930, (*Pinb-D1b*) of which the hardness index is 17.43 (±18.59). Such a low value has never been reported in the case of the *Pinb-D1b* allele. In five breeding lines and seven landraces, the standard deviation of SKCS hardness was above 20, which might suggest that the lines represented a mixture of hard and soft genotypes. This is possible because landraces have been sourced from farmland or local markets and may be contaminated by seeds from other genotypes. In turn, the breeding lines may not yet be genetically homogenous.

Statistical analyses do not show differences in grain hardness between genotypes with the mutant alleles of *Pinb* in both groups. Statistically significant differences occurred only between genotypes carrying the wild-type allele *Pinb-D1a* and genotypes with mutated alleles, which proved that the mutation causing the amino acid sequence change leads to increased grain hardness.

Differences in grain hardness between the same alleles may be caused by other biochemical components of the grain [11,43,46]. Therefore, we also investigated the total protein content, starch, and wet gluten measured by the NIR method. Unfortunately, no correlation between any of these factors was found. This contradicts the results reported by Salmanowicz et al. [47], who found a strong correlation between grain hardness and protein content. It should be noted however, that in the aforementioned work the grain hardness was measured using the NIR and the PSI methods, whereas in the current study the SKCS was applied. As mentioned earlier, the method of measurement influences the results because each method defines the grain hardness in a slightly different way [46,48,49]. The influence of protein content on grain hardness was also reported in other papers [11,50,51]. Moreover grain hardness might be affected by other grain components such as pentosans, primarily arabinoxylans and arabinogalactans [11,52], and gliadin compositions [50], as well as gluten proteins [53]. Other reasons for differences in grain hardness may be the result of differences in genetic background other than variability in *Pin* genes.

Nirmal et al. [34] described significant differences between the expression of *Pin* genes in hard and soft wheat and concluded that higher expression at earlier stages of grain development causes a softer grain, but this does not fully explain entirety of the variability between the genotypes studied. On the other hand, increased expression of *Pin* genes in rice, durum wheat, or in common wheat caused a significant decrease in grain hardness [27,54,55]. The opposed relationship was shown by the silencing of the *Pin* genes, which resulted in increased grain hardness [25,26]. Another known genetic factor, which might affect grain hardness is nucleotide polymorphisms of the transcription activator SPA (storage protein activator) [56]. In addition many QTLs (quantitative trait locus) have been described that affect the grain hardness. The QTLs related to grain hardness are located on chromosomes 1A, 1B, 2B, 2D, 3B, 4A, 4B, 5AL, 5BL, 5D, 6A, 6BL, 7A, and 7B [57–62]. It has also been proven that grain hardness is affected by environmental factors [49,50,63]. However, in our investigations all cultivars, breeding lines and landraces grew under the same optimal conditions, so we assume that this factor was eliminated. Another factor, which was proven to influence kernel texture is the grain moisture [11,51]. However, in our research, this parameter was almost identical in all tested lines, with the standard deviation values below 1%.

5. Conclusions

In conclusion, no new *Pin* alleles have been discovered however, the *Pinb-D1g* allele, unique to European wheat, has been found among landraces from Poland. In addition, we found many interesting objects whose grain hardness seemed to be determined not only by *Pina* and *Pinb* alleles but also by other genetic or biochemical factors. In further studies, we aim to unravel the genetic background of these inconsistences.

Supplementary Materials: The following are available online at http://www.mdpi.com/2073-4395/10/8/1075/s1, Table S1: Means for kernel hardness, weight, moisture and diameter of modern wheat. Table S2: Means for kernel hardness, weight, moisture and diameter of Polish landraces. Table S3: Means for protein, starch, and wet gluten contents of Polish landraces. Figure S1: Chromatogram fragments of five alleles found in tested wheat genotypes.

Author Contributions: Conceptualization, A.N.-O. and M.P.; methodology, M.P. and S.G.; formal analysis, M.P. and S.G.; investigation, M.P., S.G.; and M.K.; Data curation, M.P. and M.K.; resources, M.P.; writing—original draft preparation, M.P.; writing—review and editing, A.N.-O., S.G., M.P., and W.O.; visualization, M.P. and M.K.; supervision, A.N.-O.; project administration A.N.-O.; funding acquisition A.N.-O., All authors have read and agreed to the published version of the manuscript.

Funding: Ministry of Agriculture and Rural Development Grant PW (2015–2020), Task 2.1.

Acknowledgments: This research was supported by the Ministry of Agriculture and Rural Development Grant PW (2015–2020), Task 2.1. We thank Agnieszka Głowacka, Izabela Skuza and Małgorzata Wojciechowska for their technical support.

Conflicts of Interest: The authors declare no conflict of interest.

References

1. FAO, S. Food and Agriculture Organization of the United Nations. *Retrieved* **2010**, *3*, 2012.
2. Igrejas, G.; Gaborit, T.; Oury, F.X.; Chiron, H.; Marion, D.; Branlard, G. Genetic and environmental effects on puroindoline-a and puroindoline-b content and their relationship to technological properties in French bread wheats. *J. Cereal Sci.* **2001**, *34*, 37–47. [CrossRef]
3. Ribeiro, M.; Rodriguez-Quijano, M.; Giraldo, P.; Pinto, L.; Vazquez, J.F.; Carrillo, J.M.; Igrejas, G. Effect of allelic variation at glutenin and puroindoline loci on bread-making quality: Favorable combinations occur in less toxic varieties of wheat for celiac patients. *Eur. Food Res. Technol.* **2017**, *243*, 743–752. [CrossRef]
4. Morris, C.F. Puroindolines: The molecular genetic basis of wheat grain hardness. *Plant Mol. Biol.* **2002**, *48*, 633–647. [CrossRef] [PubMed]
5. Gautier, M.F.; Aleman, M.E.; Guirao, A.; Marion, D.; Joudrier, P. Triticum-aestivum puroindolines, 2 basic cystine-rich seed rroteins-cDNA sequence-analysis and developmental gene-expression. *Plant Mol. Biol* **1994**, *25*, 43–57. [CrossRef] [PubMed]
6. Chantret, N.; Salse, J.; Sabot, F.; Rahman, S.; Bellec, A.; Laubin, B.; Dubois, I.; Dossat, C.; Sourdille, P.; Joudrier, P.; et al. Molecular basis of evolutionary events that shaped the hardness locus in diploid and polyploid wheat species (Triticum and aegilops). *Plant Cell* **2005**, *17*, 1033–1045. [CrossRef]
7. Blochet, J.E.; Chevalier, C.; Forest, E.; Pebaypeyroula, E.; Gautier, M.F.; Joudrier, P.; Pezolet, M.; Marion, D. Complete amino-acid-sequence of puroindoline, a new basic and cystine-rich protein with a unique tryptophan-rich domain, isolated from wheat endosperm by triton X-114 phase partitioning. *FEBS Lett.* **1993**, *329*, 336–340. [CrossRef]
8. Day, L.; Bhandari, D.G.; Greenwell, P.; Leonard, S.A.; Schofield, J.D. Characterization of wheat puroindoline proteins. *FEBS J.* **2006**, *273*, 5358–5373. [CrossRef]

9. Giroux, M.J.; Morris, C.F. Wheat grain hardness results from highly conserved mutations in the friabilin components puroindoline a and b. *Proc. Natl. Acad. Sci. USA* **1998**, *95*, 6262–6266. [CrossRef]
10. Iftikhar, A.; Ali, I. Kernel softness in wheat is determined by starch granule bound Puroindoline proteins. *J. Plant Biochem. Biotechnol.* **2017**, *26*, 247–262. [CrossRef]
11. Bhave, M.; Morris, C.F. Molecular genetics of puroindolines and related genes: Regulation of expression, membrane binding properties and applications. *Plant Mol. Biol.* **2008**, *66*, 221–231. [CrossRef] [PubMed]
12. Nadolska-Orczyk, A.; Gasparis, S.; Orczyk, W. The determinants of grain texture in cereals. *J. Appl. Genet.* **2009**, *50*, 185–197. [CrossRef]
13. Morris, C.F.; Bhave, M. Reconciliation of D-genome puroindoline allele designations with current DNA sequence data. *J. Cereal Sci.* **2008**, *48*, 277–287. [CrossRef]
14. Giroux, M.J.; Morris, C.F. A glycine to serine change in puroindoline b is associated with wheat grain hardness and low levels of starch-surface friabilin. *Theor. Appl. Genet.* **1997**, *95*, 857–864. [CrossRef]
15. Lillemo, M.; Morris, C.F. A leucine to proline mutation in puroindoline b is frequently present in hard wheats from Northern Europe. *Theor. Appl. Genet.* **2000**, *100*, 1100–1107. [CrossRef]
16. Morris, C.F.; Lillemo, M.; Simeone, M.C.; Giroux, M.J.; Babb, S.L.; Kidwell, K.K. Prevalence of puroindoline grain hardness genotypes among historically significant North American spring and winter wheats. *Crop Sci.* **2001**, *41*, 218–228. [CrossRef]
17. Kumar, R.; Arora, S.; Singh, K.; Garg, M. Puroindoline allelic diversity in Indian wheat germplasm and identification of new allelic variants. *Breed. Sci* **2015**, *65*, 319–326. [CrossRef]
18. Chang, C.; Zhang, H.P.; Xu, J.; Li, W.H.; Liu, G.T.; You, M.S.; Li, B.Y. Identification of allelic variations of puroindoline genes controlling grain hardness in wheat using a modified denaturing PAGE. *Euphytica* **2006**, *152*, 225–234. [CrossRef]
19. Ram, S.; Jain, N.; Shoran, J.; Singh, R. New frame shift mutation in puroindoline b in Indian wheat cultivars Hyb65 and NI5439. *J. Plant. Biochem. Biotechnol.* **2005**, *14*, 45–48. [CrossRef]
20. Ikeda, T.M.; Ohnishi, N.; Nagamine, T.; Oda, S.; Hisatomi, T.; Yano, H. Identification of new puroindoline genotypes and their relationship to flour texture among wheat cultivars. *J. Cereal Sci.* **2005**, *41*, 1–6. [CrossRef]
21. Tranquilli, G.; Heaton, J.; Chicaiza, O.; Dubcovsky, J. Substitutions and deletions of genes related to grain hardness in wheat and their effect on grain texture. *Crop Sci.* **2002**, *42*, 1812–1817. [CrossRef]
22. Pickering, P.A.; Bhave, M. Comprehensive analysis of Australian hard wheat cultivars shows limited puroindoline allele diversity. *Plant Sci.* **2007**, *172*, 371–379. [CrossRef]
23. Qamar, Z.U.; Bansal, U.K.; Dong, C.M.; Alfred, R.L.; Bhave, M.; Bariana, H.S. Detection of puroindoline (Pina-D1 and Pinb-D1) allelic variation in wheat landraces. *J. Cereal Sci.* **2014**, *60*, 610–616. [CrossRef]
24. Ayala, M.; Guzman, C.; Pena, R.J.; Alvarez, J.B. Genetic diversity and molecular characterization of puroindoline genes (Pina-D1 and Pinb-D1) in bread wheat landraces from Andalusia (Southern Spain). *J. Cereal Sci.* **2016**, *71*, 61–65. [CrossRef]
25. Gasparis, S.; Orczyk, W.; Zalewski, W.; Nadolska-Orczyk, A. The RNA-mediated silencing of one of the Pin genes in allohexaploid wheat simultaneously decreases the expression of the other, and increases grain hardness. *J. Exp. Bot.* **2011**, *62*, 4025–4036. [CrossRef] [PubMed]
26. Gasparis, S.; Kała, M.; Przyborowski, M.; Orczyk, W.; Nadolska-Orczyk, A. Artificial MicroRNA-based specific gene silencing of grain hardness genes in polyploid cereals appeared to be not stable over transgenic plant generations. *Front. Plant Sci.* **2017**, *7*, 1–13. [CrossRef] [PubMed]
27. Li, Y.; Mao, X.; Wang, Q.; Zhang, J.; Li, X.; Ma, F.; Sun, F.; Chang, J.; Chen, M.; Wang, Y. Overexpression of Puroindoline a gene in transgenic durum wheat (Triticum turgidum ssp. durum) leads to a medium–hard kernel texture. *Mol. Breed.* **2014**, *33*, 545–554. [CrossRef]
28. Alfred, R.L.; Palombo, E.A.; Panozzo, J.F.; Bhave, M. The co-operative interaction of puroindolines in wheat grain texture may involve the hydrophobic domain. *J. Cereal Sci.* **2014**, *60*, 323–330. [CrossRef]
29. Clifton, L.A.; Green, R.J.; Frazier, R.A. Puroindoline-b mutations control the lipid binding interactions in mixed puroindoline-a: Puroindoline-b systems. *Biochemistry* **2007**, *46*, 13929–13937. [CrossRef]
30. Chen, F.; Beecher, B.S.; Morris, C.F. Physical mapping and a new variant of Puroindoline b-2 genes in wheat. *Theor. Appl. Genet.* **2010**, *120*, 745–751. [CrossRef]
31. Giroux, M.J.; Kim, K.H.; Hogg, A.C.; Martin, J.M.; Beecher, B. The puroindoline b-2 variants are expressed at low levels relative to the puroindoline D1 genes in wheat seeds. *Crop Sci.* **2013**, *53*, 833–841. [CrossRef]

32. Ramalingam, A.; Palombo, E.A.; Bhave, M. The Pinb-2 genes in wheat comprise a multigene family with great sequence diversity and important variants. *J. Cereal Sci.* **2012**, *56*, 171–180. [CrossRef]
33. Wilkinson, M.; Wan, Y.F.; Tosi, P.; Leverington, M.; Snape, J.; Mitchell, R.A.C.; Shewry, P.R. Identification and genetic mapping of variant forms of puroindoline b expressed in developing wheat grain. *J. Cereal Sci.* **2008**, *48*, 722–728. [CrossRef]
34. Nirmal, R.C.; Furtado, A.; Wrigley, C.; Henry, R.J. Influence of gene expression on hardness in wheat. *PLoS ONE* **2016**, *11*, e0164746. [CrossRef]
35. Murray, M.G.; Thompson, W.F. Rapid isolation of high molecular-weight plant DNA. *Nucleic Acids Res.* **1980**, *8*, 4321–4325. [CrossRef] [PubMed]
36. Lillemo, M.; Chen, F.; Xia, X.C.; William, M.; Pena, R.J.; Trethowan, R.; He, Z.H. Puroindoline grain hardness alleles in CIMMYT bread wheat germplasm. *J. Cereal Sci.* **2006**, *44*, 86–92. [CrossRef]
37. Thiel, T.; Kota, R.; Grosse, I.; Stein, N.; Graner, A. SNP2CAPS: A SNP and INDEL analysis tool for CAPS marker development. *Nucleic Acids Res.* **2004**, *32*, e5. [CrossRef]
38. *AACC Approved Methods of Analysis*, 11th ed.; Method 55-31.01. Single-kernel characterization system for wheat kernel texture; Cereals & Grains Association: St. Paul, MN, USA, 1999. [CrossRef]
39. Ogle, D.H. FSA: Fisheries stock analysis. *R Package Vers. 0.8* **2017**, *17*, 636.
40. Wickham, H. *ggplot2: Elegant Graphics for Data Analysis*; Springer: New York, NY, USA, 2016.
41. Corona, V.; Gazza, L.; Boggini, G.; Pogna, N.E. Variation in friabilin composition as determined by A-PAGE fractionation and PCR amplification, and its relationship to grain hardness in bread wheat. *J. Cereal Sci.* **2001**, *34*, 243–250. [CrossRef]
42. Mohler, V.; Schmolke, M.; Paladey, E.; Seling, S.; Hartl, L. Association analysis of Puroindoline-D1 and Puroindoline b-2 loci with 13 quality traits in European winter wheat (*Triticum aestivum* L.). *J. Cereal Sci.* **2012**, *56*, 623–628. [CrossRef]
43. Huang, X.-Q.; Röder, M.S. Development of SNP assays for genotyping the puroindoline b gene for grain hardness in wheat using pyrosequencing. *J. Agric. Food Chem.* **2005**, *53*, 2070–2075. [CrossRef] [PubMed]
44. Li, X.; Li, Y.; Zhang, M.; Yu, X.; Hu, R.; Chang, J.; Yang, G.; Wang, Y.; He, G. Diversity of Puroindoline genes and their association with kernel hardness in Chinese wheat cultivars and landraces. *Mol. Breed.* **2019**, *39*, 61. [CrossRef]
45. Sharma, R.; Rawat, A.; Misra, B.; Nagarajan, S. Distribution of grain hardness in Indian wheat varieties and landraces. *Wheat Inf. Serv.* **2012**, *114*, 1–8.
46. Dobraszczyk, B.; Whitworth, M.; Vincent, J.; Khan, A. Single kernel wheat hardness and fracture properties in relation to density and the modelling of fracture in wheat endosperm. *J. Cereal Sci.* **2002**, *35*, 245–263. [CrossRef]
47. Salmanowicz, B.P.; Adamski, T.; Surma, M.; Kaczmarek, Z.; Karolina, K.; Kuczyńska, A.; Banaszak, Z.; Ługowska, B.; Majcher, M.; Obuchowski, W. The relationship between grain hardness, dough mixing parameters and bread-making quality in winter wheat. *Int. J. Mol. Sci.* **2012**, *13*, 4186–4201. [CrossRef]
48. Morris, C.F.; Massa, A.N. Puroindoline genotype of the US national institute of standards & technology reference material 8441, wheat hardness. *Cereal Chem.* **2003**, *80*, 674–678.
49. Oury, F.-X.; Lasme, P.; Michelet, C.; Rousset, M.; Abecassis, J.; Lullien-Pellerin, V. Relationships between wheat grain physical characteristics studied through near-isogenic lines with distinct puroindoline-b allele. *Theor. Appl. Genet.* **2015**, *128*, 913–929. [CrossRef]
50. Huebner, F.R.; Gaines, C.S. Relation between Wheat Kernel Hardness, Environment, and Gliadin Composition. *Cereal Chem.* **1992**, *69*, 148–151.
51. Turnbull, K.M.; Rahman, S. Endosperm texture in wheat. *J. Cereal Sci.* **2002**, *36*, 327–337. [CrossRef]
52. Lillemo, M.; Ringlund, K. Impact of puroindoline b alleles on the genetic variation for hardness in soft x hard wheat crosses. *Plant Breed.* **2002**, *121*, 210–217. [CrossRef]
53. Quayson, E.T.; Marti, A.; Morris, C.F.; Marengo, M.; Bonomi, F.; Seetharaman, K.; Iametti, S. Structural consequences of the interaction of puroindolines with gluten proteins. *Food Chem.* **2018**, *253*, 255–261. [CrossRef] [PubMed]
54. See, D.; Giroux, M.; Gill, B. Effect of multiple copies of puroindoline genes on grain softness. *Crop Sci.* **2004**, *44*, 1248–1253. [CrossRef]
55. Krishnamurthy, K.; Giroux, M.J. Expression of wheat puroindoline genes in transgenic rice enhances grain softness. *Nat. Biotechnol.* **2001**, *19*, 162–166. [CrossRef] [PubMed]

56. Ravel, C.; Martre, P.; Romeuf, I.; Dardevet, M.; El-Malki, R.; Bordes, J.; Duchateau, N.; Brunel, D.; Balfourier, F.; Charmet, G. Nucleotide polymorphism in the wheat transcriptional activator SPA influences its pattern of expression and has pleiotropic effects on grain protein composition, dough viscoelasticity, and grain hardness. *Plant Physiol.* **2009**, *151*, 2133–2144. [CrossRef] [PubMed]
57. Boehm, J.D.; Ibba, M.I.; Kiszonas, A.M.; See, D.R.; Skinner, D.Z.; Morris, C.F. Genetic analysis of kernel texture (grain hardness) in a hard red spring wheat (*Triticum aestivum* L.) bi-parental population. *J. Cereal Sci.* **2018**, *79*, 57–65. [CrossRef]
58. Breseghello, F.; Finney, P.L.; Gaines, C.; Andrews, L.; Tanaka, J.; Penner, G.; Sorrells, M.E. Genetic loci related to kernel quality differences between a soft and a hard wheat cultivar. *Crop Sci.* **2005**, *45*, 1685–1695. [CrossRef]
59. Campbell, K.G.; Bergman, C.J.; Gualberto, D.G.; Anderson, J.A.; Giroux, M.J.; Hareland, G.; Fulcher, R.G.; Sorrells, M.E.; Finney, P.L. Quantitative trait loci associated with kernel traits in a soft x hard wheat cross. *Crop Sci.* **1999**, *39*, 1184–1195. [CrossRef]
60. Carter, A.H.; Garland-Campbell, K.; Morris, C.F.; Kidwell, K.K. Chromosomes 3B and 4D are associated with several milling and baking quality traits in a soft white spring wheat (*Triticum aestivum* L.) population. *Theor. Appl. Genet.* **2012**, *124*, 1079–1096. [CrossRef]
61. Osborne, B.G.; Turnbull, K.M.; Anderssen, R.S.; Rahman, S.; Sharp, P.J.; Appels, R. The hardness locus in Australian wheat lines. *Aust. J. Agric. Res.* **2001**, *52*, 1275–1286. [CrossRef]
62. Igrejas, G.; Leroy, P.; Charmet, G.; Gaborit, T.; Marion, D.; Branlard, G. Mapping QTLs for grain hardness and puroindoline content in wheat (*Triticum aestivum* L.). *Theor. Appl. Genet.* **2002**, *106*, 19–27. [CrossRef]
63. Gazza, L.; Taddei, F.; Corbellini, M.; Cacciatori, P.; Pogna, N. Genetic and environmental factors affecting grain texture in common wheat. *J. Cereal Sci.* **2008**, *47*, 52–58. [CrossRef]

agronomy

MDPI

Article

Analysis of Proteomic Profile of Contrasting Phosphorus Responsive Rice Cultivars Grown under Phosphorus Deficiency

Aadil Yousuf Tantray [1], Hayssam M. Ali [2] and Altaf Ahmad [1,*]

[1] Department of Botany, Aligarh Muslim University, Aligarh 202002, India; tantray.rs@amu.ac.in
[2] Botany and Microbiology Department, College of Science, King Saud University, P.O. Box. 2455, Riyadh 11451, Saudi Arabia; hayhassan@ksu.edu.sa
* Correspondence: aahmad.bo@amu.ac.in

Received: 15 June 2020; Accepted: 14 July 2020; Published: 16 July 2020

Abstract: Phosphorus (P) deficiency is one of the major limiting factors for crop productivity. The yield of rice (*Oryza sativa* L.) is severely limited by phosphorus deficiency. An attempt has been made in this study to identify P deficiency responsive differentially expressed proteins of rice through analysis of leaf proteome of contrasting P-responsive rice cultivars under P deficiency conditions because genetic variability has been found in the rice cultivars for adaptive response to P deficiency and a controlled regulatory system is involved in the P deficiency adaptation response. Phosphorus-efficient (cv. Panvel) and P-inefficient (cv. Nagina 22) rice cultivars were hydroponically grown in the nutrient medium under control environmental conditions at low-P level (2.0 μM) and optimum-P level (320 μM) treatments. Expression patterns of the proteins of the leaves of both the cultivars were analyzed in 30-day-old plants. The identification of these proteins through mass spectrometry and MASCOT software (Matrix Science Inc., Boston, USA) revealed that these differentially expressed proteins were homologous to known functional proteins involved in energy metabolism, biosynthesis, photosynthesis, signaling, protein synthesis, protein folding, phospholipid metabolism, oxidative stress, transcription factors, and phosphorus metabolism. It has been observed that rice cultivars responded differently to low-P treatment through modification in protein expressions pattern to maintain the growth of the plants. Therefore, the expression patterns of proteins were different in both of the cultivars under low-P treatment. Higher potential of protein stability, stress tolerance, osmo-protection, and regulation of phosphorus uptake was observed in cv. Panvel than cv. Nagina 22. This study could help to unravel the complex regulatory process that is involved in adaptation to P deficiency in rice.

Keywords: phosphorus use efficiency; phosphorus; proteomics; rice

1. Introduction

The food crisis in developing countries is due to a decrease in crop productivity, on the one hand, and increased the global human population, on the other side [1]. Phosphorus (P) is macro-nutrients and an essential component of phospholipids, proteins, and nucleotide-containing molecules, like RNA and DNA. It is one of the major limiting factors that influence crop production. Estimation shows that almost 50% of the world's agricultural lands are P-deficient [2]. Phosphorus fertilizers are applied in the agricultural field in order to overcome this deficiency and maintain crop production. However, the phosphorus use efficiency (PUE) of crop plants is only 10–20% [3]. This low PUE has increased the global demand of P fertilizers that reached 47 million tonnes [4] and created severe environmental problems. like eutrophication, because of unutilized P in the agricultural field. Because the only sources of P fertilizers are phosphate rocks, and the accessibility of phosphate rocks is estimated to last in the next 300–400 years according to the International Fertilizer Development Centre [5], the improvement in the P efficiency of crop plants would be an important contribution to the sustainability of agroecosystems.

Rice is a staple food for more than 50% of the world's population [6]. It has been found that there is a gradual decline in rice productivity. Estimations show that it will reach 40% of the present productivity by the end of the 21st century [7]. P deficiency is one of the reasons for the decline in the production of rice. Plants respond to P deficiency stress at morphological, physiological, and molecular levels. Leaf growth inhibition in response to low-P supply has been reported that results in the reduction of plant growth [8]. Phosphorus deficiency affected many metabolic processes that led to a decrease in the biomass accumulation and overall growth of the plants. A decrease in the photosynthesis of rice by P deficiency has been observed through the reduction in the activities of the enzyme of the Calvin cycle [9]. It has been reported that the sufficient supply of P improved the activities of sucrose phosphate synthase, glutamine synthase, and PEP carboxylase in leaves of rice [10]. The exposure of plants to stresses leads to the identification of stress-responsive genes, numerous downstream effectors, and signaling factors/components through analysis of the expression profiling of genes. It is suggested that there is a well sophisticated regulatory network of genes in plants to access and control the nutrient homeostasis [11–15]. Therefore, there is a need to explore the research on the identification of the regulatory gene(s) of rice that are involved in the mechanism of P-deficiency tolerance. Kim et al. [16] have shown that proteins that are involved in the metabolism and defense/stress response, like putative glyceraldehydes-3-phosphate dehydrogenase, S-adenosyl-L-methionine synthetase, ATP synthase subunit alpha, and root-specific pathogenesis-related protein ten, were differentially regulated in the roots of rice under P-starvation. In another study on rice, proteins that are involved in signal transduction and plant defense responses, like abscisic ASR1, superoxide dismutase, glutathione S-transferase, salt stress-induced proteins, and putative r40c1 protein, were found to be differentially expressed under P-starvation conditions [17]. These studies are primarily limited to the proteome profiling of the roots of rice under P-starvation conditions and have not provided information about the proteome profiling in leaves. Since the leaves are the sites of assimilation and utilization of P, the present study was conducted to study the leaf proteome profiling of P-efficient and P-inefficient cultivars of rice to identify differential expression of P-starvation responsive proteins, and to determine the mechanism of rice adaptation to P-starvation. A basis of investigation of the function and regulation of protein involved in low-P tolerance will be provided by this study in order to help develop rice with high PUE through genetic manipulation.

2. Materials and Methods

2.1. Plant Material and Growth Conditions

Seeds of P-inefficient (cv. Nagina 22) and P-efficient (cv. Panvel) rice cultivars were obtained from the Indian Agricultural Research Institute, New Delhi, India. The cv. Nagina 22 is tall, susceptible to blast, resistant to drought, and cultivated in the Uttar Pradesh state of India. The grains of this variety are short, bold, and white. The yield is 20–25 Q/ha. The cv. Panvel is semi-dwarf, resistant to neck blast, and cultivated in the Maharashtra state of India. The grains are short, bold, and yield is 40–43 Q/ha. The cv. Nagina 22 is an *Aus*-type and cv. Panvel is an *Indica* subspecies. The seeds of cv. Nagina 22 and cv. Panvel rice cultivars were surface sterilized with 75% alcohol and 0.1% mercuric chloride for 5 min and thoroughly rinsed four times with distilled water. Sterilized seeds were germinated in distilled water for six days in a hydroponic culture system (90 seedlings in 5 L tank). After six days, the distilled water was replaced by Yoshida nutrient solution without P. The P was supplied in the form of phosphoric acid (H_3PO_4) with two concentrations, i.e., low (2 μM) and optimum (320 μM, according to Yoshida) [18]. Each treatment consisted of three replications. The medium was replaced after every three days and properly aerated with an aquarium air-filter pump (aerated solution increases P accumulation in root and shoot [19]), and pH was maintained at 5.6 ± 0.2.

The experiment was conducted in controlled conditions of a plant growth chamber, 14:10 h light/dark period, the relative humidity of 60%, 28/22 °C day/night temperature, and a photosynthetic photon flux density of 430 μmol m^{-2} s^{-1}. The initiation of primary tillering in rice starts from the fourth week after sowing. Top leaf samples (third and fourth leaf) were collected with three biological and three technical replicates after 30 days of sowing (Figure S1). For proteomic analysis, the sampled leaves were frozen in liquid nitrogen before their storage at −80 °C. Morphological parameters, such as length, fresh weight and dry weight of shoot and root, and phosphorus content of leaf and root, were measured. Shoot and root length were measured while using a measuring scale and represented in centimeter (cm). The phosphorus content of leaf and root samples was measured by digesting samples in a di-acid mixture (HNO_3:$HClO_4$ = 9:4 v/v) and spectrophotometrically determined at 625 nm after adding using vanadate-molybdate reagent [20]. The fresh weight of shoot and root samples were measured at the time of harvesting, and the dry weights were measured after drying the shoot and root samples in an oven at 65 °C for 72 h and represented as g $plant^{-1}$. The PUE was calculated using Syers et al. [21] formula, the total P uptake divided by the amount of P applied and multiplied by 100.

2.2. Protein Extraction

The stored leaves were used for protein extraction by the modified phenol method of Isaacson et al. [22]. Fine powder of two grams of leaf material was prepared in liquid nitrogen and homogenized in 10 mL extraction buffer (700 mM sucrose, 50 mM HEPES, 100 mM KCl, 50 mM EDTA, and 2% SDS), with pH 7.5 and on homogenizing, 2% β-mercaptoethanol and 1 mM PMSF were also added. Then 15 mL of molecular grade phenol (Sigma–Aldrich, St. Louis, Missouri, USA) was added, and the solution was mixed for 30 min on ice-rocker and centrifugation at 3200× *g* for 10 min at 4 °C to recovered top phenolic layer. The protein was precipitated in ice-cold 0.1 M ammonium acetate solution at −20 °C overnight. The protein pellet was collected by centrifugation at 6500× *g* (15 min at 4 °C) and then washed two times with cold acetone at the same speed and then lyophilized (Labconco, Kansas City, USA) at −50 °C. The pellet was solubilized in a cocktail buffer (7 M urea, 2 M thiourea, 50 mM DTT, and 4% 3-[(3-cholamidopropyl) dimethylammonio]-1-propanesulfonate). Using a 2D Quant Kit (GE Healthcare, Munich, Germany), protein concentration was estimated with bovine serum albumin (BSA) used as a reference.

2.3. Two-Dimensional Gel Electrophoresis

Bio-Rad's apparatus and chemical grades were used for two-dimensional electrophoresis (2-DE) and performed by the method of O'Farrell [23]. An immobilized dry strip gel (11 cm, linear-gradient, pH 4–7; Bio-Rad, Hercules, California, USA) was rehydrated at 20 °C for 12 h in 180 μL of a sample containing 350 μg proteins. The first dimension, i.e., isoelectric focusing was carried on the following program; 250 V for 30 min, 500 V for 30 min, 1000 V for 1 h, 2000 V for 1 h followed by a linear increase of 6000 V to a total of 65.00 kV, and completed in 16:30 h. The strips were then subjected to reduction by equilibration buffer (containing 8 M urea, 20% glycerol, 50 mM Tris, 2% SDS, and 130 mM DTT) at pH 8.8 and, followed by alkylation using same equilibration buffer with 135 mM iodoacetamide instead of DTT. The second dimension, SDS-PAGE was carried out for the separation of proteins, using 12% SDS in large vertical format electrophoresis cell (PROTEAN® II Xi Cell, Bio-Rad, Hercules, California, USA) at a constant voltage of 150 V. The gels were stained overnight with Coomassie Brilliant Blue G-250 dye and then destained with sterilized MilliQ water (Milli-Q®, Merck KGaA, Darmstadt, Germany) by washing several times.

2.4. Gel Analysis

Digital imaging of the gel was captured by the gel documentation system (GS-900™ Calibrated Densitometer, Bio-Rad, Hercules, California, USA) for further analysis of set parameters (relative abundance, spot density, and location by comparing pH and mass). The image master PDQuest software (version 8.0, Bio-Rad, Hercules, California, USA) was used for gel image analysis. Optimized parameters were considered as partial threshold 4, saliency 2.0, and minimum area 50. Each spot was normalized with the total volume percentage of all of the spots for rectification of unevenness due to quantitative diversity in spot intensities. The quantification of spots was held based on their relative volume and quality, which was concluded by the ratio of the single spot volume to the whole comparative set of the spots.

2.5. In-Gel Digestion and MALDI-TOF/TOF-MS Analysis

Differential responsive protein spots were excised from the gels and, to remove the excess SDS, the gels were washed three times with sterilized MilliQ water. Destaining of excised gel pieces was done using 50 mM ammonium bicarbonate and dehydrated with 100% acetonitrile (ACN). 15 mM DTT was used for the reduction of dehydrated protein spots at 60 °C for 1 h, and protein spots were subjected to alkylation in the dark for 15 min by using 100 mM iodoacetamide. Before drying in a speed vacuum, these protein spots were rehydrated with 50 mM ammonium bicarbonate. Dried gel pieces were digested with 15 µL of working trypsin (10 ng/µL) (Promega, Madison, Wisconsin, USA) overnight at 37 °C. The supernatant was taken in a fresh centrifuge tube, and 20:1 percentage of acetonitrile and the formic acid solution were added for further extraction of remaining gel pieces. To reduce the volume of the final supernatant to 25–50 µL, it was dried in a speed vacuum. The final volume was subjected to a mass spectrometer (Applied Biosystems SCIEX 4800 MALDI TOF/TOF™, Foster City, CA, USA), which was controlled by the 4000 Series Explorer™ software (Applied Biosystems, Foster City, CA, USA) using set parameters with the peptide charge of 1+ and peptide tolerance of 150 ppm. A mass ranges from 800 to 4000 with a focus mass of 2000 was used to record the mass spectrometer (MS) spectra in reflector mode. The instrument was calibrated by a CalMix5 standard (ABI 4700 Calibration Mixture, Foster City, CA, USA), and 25 sub-spectra with 125 shots per sub spectrum were amassed while using a random search pattern for each MS spectrum. Trypsin autolysis peaks were used for MS calibration, and MS/MS acquisition of precursors was selected up to 10 of the most intense ion signals, excluding the trypsin autolysis peaks and the matrix ion signals. For one main MS spectrum, 50 sub spectra with 50 shots per sub-spectrum were accumulated in MS/MS positive ion mode, while using a random search pattern. Air was used as collision gas, 2 kV collision energy, and Glu1-Fibrinopeptide B spotted onto the Cal 7 positions of the MALDI target were used as in default calibration.

2.6. Protein Identification and Database Searching

Following the calibration, the data were picked while using GPS Explorer (Applied Biosystems 2006, Foster City, CA, USA) for monoisotopic peak analysis, and the MASCOT program (http://www.matrixscience.com) was used for analyzing the monoisotopic peak lists, but only significant peak hits with probability analysis ($p < 0.05$), were accepted. The peptides were searched with Protein-NCBI database [24] allowing for single trypsin missed cleavage, partial modification of cysteine carbamidomethylated, and methionine oxidized with the pyro-Glu formation of N-terminal. To achieve identification results with high confidence (≥95%), the protein should have a valuable MOWSE (Molecular Weight Search) score, sequence coverage greater than 15%, and at least six peptides matched. The functional information of identified proteins was assembled with the help of NCBI [24] and Uniprot [25] databases. The subcellular location of identified proteins was also assimilated by pTARGET [26] and Uniprot databases in order to understand the function of the identified proteins.

2.7. Statistical Analysis

In the present study, the statistical analysis of physiological traits was carried by two-ANOVA analysis, for each response, treatment and cultivars were selected as factors at $p < 0.05$ (Minitab 17.0). The significance between treatments of each rice cultivar was determined at * $p < 0.05$, ** $p < 0.01$, and *** $p < 0.001$. The protein samples were taken in three biological and three technical replicates. For the normalization of protein, spots paired Student's t-test with the significance of 95% was performed, with the help of SPSS software (SPSS for Windows, Version 16.0. SPSS Inc., Chicago, IL, USA). The spots in analyzer manager with greater than two-fold change intensity in volume during the comparison or with significant variation between the control and other treatments decisive by the paired Student's *t*-test ($p \leq 0.05$), were regarded as the treatment-responsive proteins. The protein spots with increased >2-fold charge intensity in volume at low-P treatment than optimum-P were considered to be up-regulated proteins. The protein spots with decreased >2-fold charge intensity in volume at low-P treatment than optimum-P were considered as down-regulated proteins.

3. Results

3.1. Morphological Traits as Influenced by Low-P and Optimum-P Treatments

The morphological traits, like the length of root and shoot, plant biomass, P concentration of leaf and root, and PUE of cv. Panvel and cv. Nagina 22, were measured under the treatments of low-P and optimum-P. All of the traits were significantly decreased under low P treatment, except root length, which significantly increased under low-P treatment in both of the cultivars as compared with the optimum-P treatment (Table 1). The increase in the root length was 23.24% and 16.34% in cv. Nagina 22 and cv. Panvel, respectively, under low P treatment, when compared with the optimum-P treatment. Low-P treatment reduced the shoot length of cv. Nagina 22 and cv. Panvel by 25.2% and 12.9%, respectively, when compared with the optimum P treatment. Plant dry weight was also reduced by low P treatment. However, the reduction of dry weight in both the cultivars varied. The percent reduction in the dry weight of the cv. Nagina 22 and cv. Panvel were 39.2% and 17.3%, respectively. The plant dry weight of cv. Nagina 22 was low than cv. Panvel even at optimum-P. Phosphorus concentration of leaf and root was lesser in cv. Nagina 22 than cv. Panvel under both the treatments. Low P treatment decreased the concentration of the P also in both the cultivars. The decrease in P concentration was significantly more in cv. Nagina than cv. Panvel. The PUE of cv. Panvel was higher than cv. Nagina 22 at both the treatments of P.

Table 1. Morphological traits and phosphorus concentration and phosphorus use efficiency of cv. Nagina 22 and cv. Panvel rice under optimum-P and low-P treatments.

Physiological Traits	Cv. Nagina 22		Cv. Panvel		Statistical Analysis ($p < 0.05$)		
	Optimum-P	Low-P	Optimum-P	Low-P	C	T	C × T
Root length (cm)	14.2 ± 2.55	18.5 ± 2.78 **	12.8 ± 1.98	15.3 ± 1.67 *	0.031	0.012	0.033
Shoot length (cm)	32.5 ± 3.66	24.3 ± 3.11 *	36.4 ± 4.23	31.6 ± 0.34 *	0.033	0.021	0.035
Plant dry weight (g plant^{-1})	2.04 ± 0.47	1.24 ± 0.35 *	2.59 ± 0.61	2.14 ± 0.37 *	0.007	0.015	0.026
Leaf P concentration (mg g^{-1} DW)	1.63 ± 0.24	1.04 ± 0.21 ***	2.05 ± 0.32	1.57 ± 0.22 **	0.013	0.003	0.007
Root P concentration (mg g^{-1} DW)	1.78 ± 0.23	1.22 ± 0.20 **	2.14 ± 0.33	1.72 ± 0.25 *	0.021	0.005	0.011
Phosphorus use efficiency (%)	12.96 ± 1.24	8.61 ± 1.03 **	15.97 ± 2.01	12.54 ± 1.73 *	0.002	0.000	0.006

Data are represented as mean ± standard error (n = 3). The ANOVA analysis was carried at $p < 0.05$ for cultivars (C), treatments (T), and their interaction (C × T). The asterisks in the table represent the significance of the treatment in each cultivar at * $p < 0.05$, ** $p < 0.01$ and *** $p < 0.001$.

3.2. Number of Differentially Expressed Proteins

Figure 1 shows the distribution of protein spots in the proteomic maps of Nagina 22 and Panvel rice cultivars, grown under optimum-P and low-P conditions. The reproducible rate of protein spots was reasonable, and a total of 512 protein spots were detected in each gel. After analysis, 63 (12.3%) protein spots were differentially expressed in rice cultivars at low-P treatment. Out of the 63 differentially expressed protein (DEP) spots, 42 (66.67%) spots were up-regulated and 21 (34.92%) spots were down-regulated (Table 2). In cv. Nagina 22, 21 proteins were up-regulated and 17 proteins were down-regulated at low-P treatment. The numbers of upregulated proteins were 35, and down-regulated proteins were 11 in cv. Panvel under the low-P treatment (Table 2). These results suggested that P-deficiency resulted in major changes in the leaf proteome of rice.

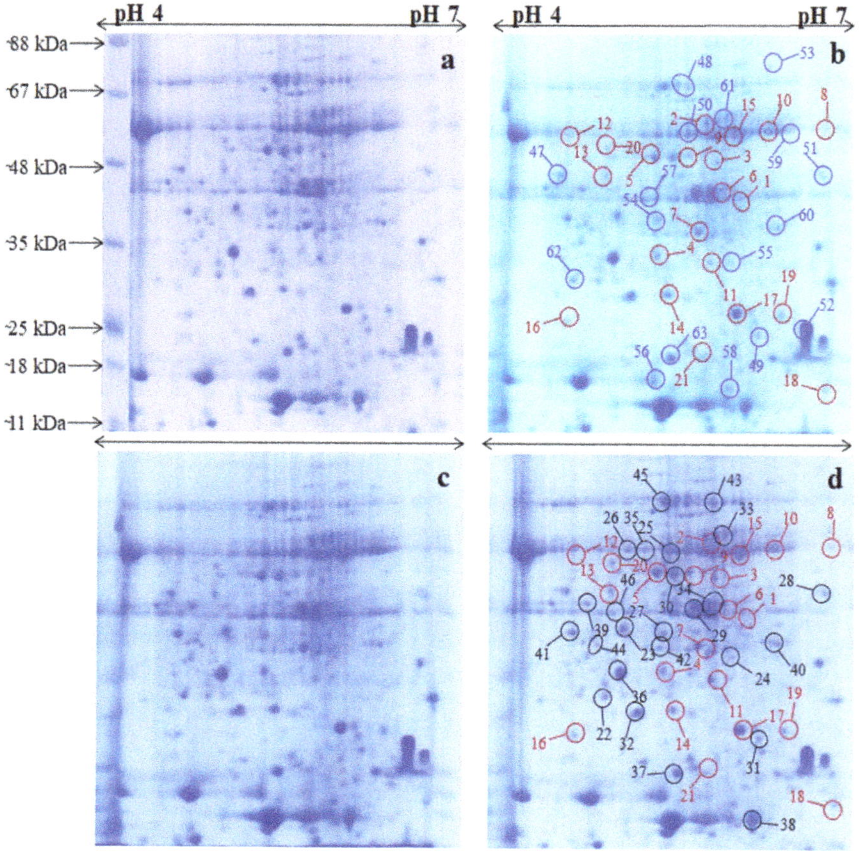

Figure 1. 2-DE gel maps of total leaf proteome of rice varieties. (**a**) cv. Nagina 22 under optimum-P condition; (**b**) cv. Nagina 22 under low-P condition; (**c**) cv. Panvel under optimum-P condition; (**d**) cv. Panvel under low-P condition. Differentially expressed proteins that were common in both cv. Nagina 22 and cv. Panvel under low-P condition are marked with red circle. The differentially expressed proteins of cv. Nagina 22 only under low-P condition are marked with blue circle. The differentially expressed proteins of cv. Panvel only under low-P condition are marked with are black circle.

Table 2. Number of differentially expressed proteins in rice cultivars under low phosphorus level. The expression pattern was compared with the optimum P level.

Distribution of DEPs	Cv. Panvel	Cv. Nagina 22	Cv. Panvel + Cv. Nagina 22	Total DEPs
Up regulated	21	7	14	42
Down regulated	4	10	7	21

DEPs–differentially expressed proteins

3.3. Spatial and Functional Categorization of Differentially Expressed Proteins

The differentially expressed proteins were identified of different sub-cellular sites (Figure 2a). These belonged to chloroplast (36%), ribosome (12%), nucleus (16%), mitochondrion (9%), and cytosol (19%). Some of the low range proteins belonged to the endoplasmic reticulum (1%), cell wall (3%), cell membrane (2%), and peroxisome (2%). According to the exhibited homology known function of the proteins, 59 (93.65%) among the 63 identified proteins were justified, whereas the rest 4 (6.35%) was unknown. Functional categorization of these differentially expressed proteins showed that these proteins belonged to eleven groups (Figure 2b), viz., metabolism (10%), protein synthesis (8%), transcription/signal transduction mechanisms (14%), cell rescue/defense (12%), photosynthesis (16%), energy (6%), cell cycle/transport (3%), oxidative stress (14%), protein fate/stabilization (9%), and unclassified function (8%).

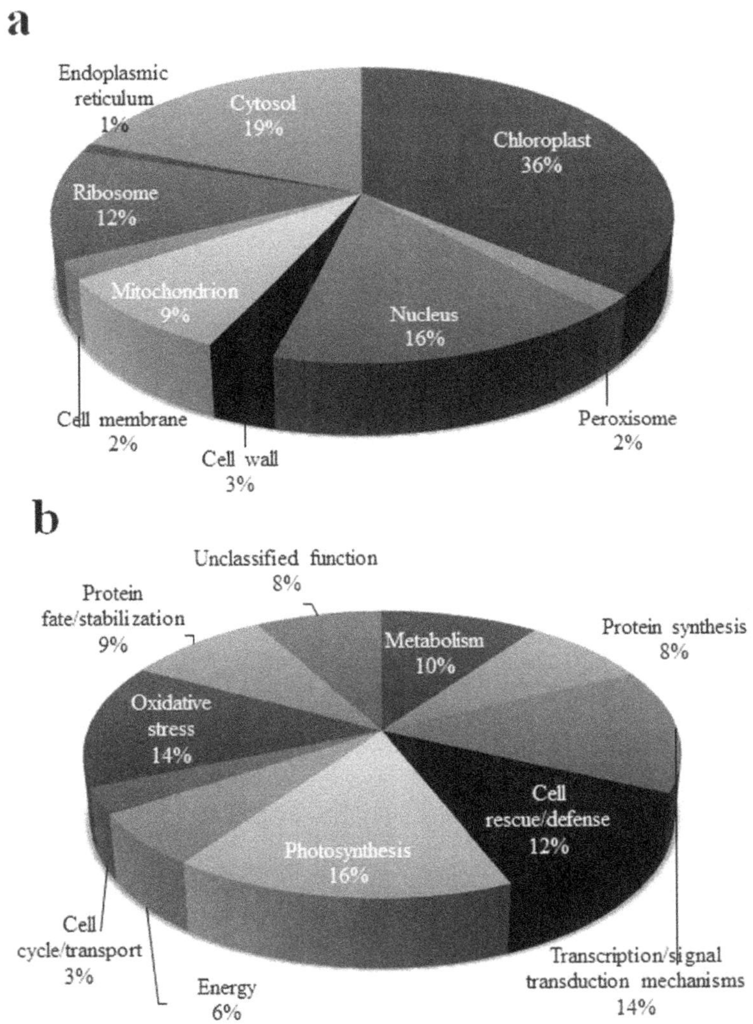

Figure 2. Percentage of spatial (**a**) and functional (**b**) distribution of differentially expressed proteins in rice leaf at low phosphorus conditions.

3.4. *Differentially Expressed Proteins of Rice Cultivars under Low P Condition*

The P-efficient (cv. Panvel) and P-inefficient (cv. Nagina) cultivars of rice both showed differential expression (up-regulation/down-regulation) of proteins under low P treatment when compared to optimum-P condition. While some differentially expressed proteins (DEPs) are common in both of the cultivars (Table 3), some DEPs are cultivar-specific (Tables 4 and 5). The identified upregulated and down-regulated proteins are described in the following sub-sections.

3.4.1. Common Differentially Expressed Proteins of Cv. Nagina and Cv. Panvel under Low P Condition

The expression patterns of twenty-one proteins of leaves was common in both the cultivars of rice under low P condition. Fourteen proteins were upregulated in both of the cultivars by low P treatment when compared to optimum P treatment (Table 3). However, the level of upregulation of these proteins differed in these rice cultivars. Similarly, the expressions of seven proteins were downregulated in both of the cultivars by low P treatment. However, the level of down-regulation differed in both of the cultivars. Fructose-bisphosphate aldolase was upregulated in both of the cultivars under low P treatment, when compared to the optimum P treatment. However, upregulation level was more in cv. Nagina than cv. Panvel. Maturase K, NADPH-dependent FMN reductase, ABA-responsive element-binding protein 3, methionine synthase, enolase 1 were identified as upregulated proteins in both of the cultivars at the same levels under low P treatment, when compared with optimum P treatment. The expression of succinate dehydrogenase flavoprotein subunit, glutathione S-transferase, photosystem II oxygen-evolving complex protein 1, glyoxylase I 7, chaperonin GroEL, DNA binding transcription factors, and protein disulfide isomerase were upregulated in both of the cultivars by low P treatment, but the markedly higher expression was found in cv. Panvel than cv. Nagina. Ribulose-1,5-bisphosphate carboxylase/oxygenase large subunit, NADP-dependent malate dehydrogenase, Small ribosomal protein 4, Ribulose-1,5-bisphosphate carboxylase/oxygenase small subunit, and Elongation factor Tu were identified as down-regulated proteins under low P treatment in both the cultivars. The level of down-regulation of Elongation factor Tu was higher in cv. Panvel than cv. Nagina. However, in all the other down-regulated proteins, the level of down-regulation was more in cv. Nagina than cv. Panvel (Table 3).

3.4.2. Differentially Expressed Proteins of Cv. Panvel only under low P Condition

Twenty-five proteins were differentially expressed in cv. Panvel only under low P condition, when compared to optimum P condition (Table 4). Out of these proteins, 21 were upregulated and four were down-regulated proteins. The upregulated proteins belonged to the category of photosynthesis (Phosphoribulokinase, NADP-dependent oxidoreductase P1), energy (chloroplastic Glyceraldehyde-3-phosphate dehydrogenase A, ATPase), metabolism (Succinyl-CoA synthetase beta subunit, 6-Phosphogluconolactonase, phosphogluconate dehydrogenase, inorganic pyrophosphatase family protein), protein synthesis (RNA-binding protein), protein stabilization (Chaperonin 60 β precursor, Heat shock protein 40, Heat shock responsive transcription factor), oxidative stress (59-epimerase, Flavonol synthase), signal transduction (14¨C3¨C3 protein), and cell defence (Pyrroline-5-carboxylate synthetase). Down-regulated proteins belonged to the category of energy (Sedoheptulose-1,7-bisphosphatase), photosynthesis (Ribosomal protein S19), and metabolism (Malate dehydrogenase, pyruvate orthophosphate dikinase).

3.4.3. Differentially Expressed Proteins of Cv. Nagina only under Low P Condition

Seventeen proteins were differentially expressed in cv. Nagina under low P treatment, when compared with the optimum P treatment (Table 5). Out of these proteins, seven proteins were upregulated and ten were down regulated under low P condition. The down-regulated proteins belong to the category of Photosynthesis (Rubisco activase chloroplast precursor, ferredoxin, light-harvesting complex I protein precursor LHCA3, chloroplastic Transketolase) and metabolism (Phosphoenolpyruvate carboxylase, Triosephosphate isomerase) and transcription/STMs (RNA polymerase β chain, Integrin-linked protein kinase family protein). The upregulated proteins belonged to oxidative stress, protein stabilization, and energy.

Table 3. Identification, subcellular localization, and quantitative analysis of differentially expressed leaf proteins of both the rice cultivars, cv. Nagina 22 and cv. Panvel, under low level of phosphorus.

S. N.	Accession No.	Name of Protein	Exp. MW (kDa)	Exp. Pi	M.S.	No. of Matched Peptides	Location	Process	Mode of Regulation	Relative spot intensity (Optimum P:Low P)	
										Nagina 22	Panvel
1	ABL74560	Fructose-bisphosphate aldolase	45.6	6.45	82	12	Cytosol	Energy	Upregulated	1.00:3.13	1.00:2.91
2	XP_015646992	Succinate dehydrogenase flavoprotein subunit	65.8	6.12	134	15	Mitochondrion	Energy	Upregulated	1.00:3.22	1.00:3.50
3	BAA00147	Ribulose-1,5-bisphosphate carboxylase/oxygenase large subunit	54.1	6.31	348	15	Chloroplast	Photosynthesis	Downregulated	1.00: −3.51	1.00: −3.23
4	AAS93256	Glutathione S-transferase	34.2	5.49	94	11	Cytoplasm	Oxidative stress	Upregulated	1.00:2.40	1.00:2.91
5	XP_015643741	Enolase 1	57.8	5.59	107	11	Chloroplast	Cell rescue/defense	Upregulated	1.00:3.00	1.00:3.02
6	BAT16624	NADP-dependant malate dehydrogenase	49.3	6.14	114	13	Mitochondrion, chloroplast, cytosol	Metabolism	Downregulated	1.00: −3.32	1.00: −3.10
7	2002393A	Photosystem II oxygen-evolving complex protein 1	37.2	6.87	159	13	Chloroplast	Photosynthesis	Upregulated	1.00:2.54	1.00:3.00
8	BAV53208	Small ribosomal protein 4	53.2	6.93	89	9	Ribosome, mitochondrion	Protein synthesis	Downregulated	1.00: −3.43	1.00: −3.11
9	AEP20544	Ribulose-1,5-bisphosphate carboxylase/oxygenase small subunit	53.3	6.19	96	11	Chloroplast	Photosynthesis	Downregulated	1.00: −4.35	1.00: −3.72
10	AAP12937	Transposon protein, putative, CACTA, En/Spmsub-class	56.2	6.8	86	16	Nucleus	-	Upregulated	1.00:2.97	1.00:3.68
11	Q948T6	Glyoxylase I 7	33.4	6.3	86	14	Peroxisome	Oxidative stress	Upregulated	1.00:4.20	1.00:5.20
12	ABR25753	Chaperonin GroEL	65.3	4.91	87	10	Cytoplasm	Protein fate/stabilization	Upregulated	1.00:3.53	1.00:5.11
13	AAM74563	Elongation factor Tu	52.1	4.98	123	16	Cytosol, plastid, mitochondrion	Protein synthesis	Downregulated	1.00: −2.33	1.00: −2.72
14	NP_973937	DNA binding transcription factors	27.8	5.7	96	8	Nucleus	Transcription/STMs	Upregulated	1.00:3.22	1.00:4.61
15	Q2QLY4	Methionine synthase	55.1	6.5	112	10	Cytosol	Metabolism	Upregulated	1.00:4.21	1.00:4.52
16	CAD41255	OSJNBa0067K08.7	33.4	5.33	141	8	-	-	Downregulated	1.00: −3.84	1.00: −3.51
17	D7LVK3	ABA-responsive element binding protein 3	32.8	5.93	123	11	Nucleus	Transcription/STMs	Upregulated	1.00:4.42	1.00:4.61

Table 3. *Cont.*

S. N.	Accession No.	Name of Protein	Exp. MW (kDa)	Exp. Pi	M.S.	No. of Matched Peptides	Location	Process	Mode of Regulation	Relative spot intensity (Optimum P:Low P)	
										Nagina 22	Panvel
18	P69667	Ribosomal protein L23	10.13	7.02	81	8	Mitochondrion	Protein synthesis	Downregulated	1.00: −2.11	1.00: −2.02
19	Q6NPS8	NADPH-dependent FMN reductase	32.1	6.55	82	8	Nucleus, cytoplasm	Cell cycle/transport	Upregulated	1.00:3.62	1.00:3.64
20	AAX85991	Protein disulfide isomerase	57.1	5.3	152	6	Endoplasmic reticulum	Protein fate/stabilization	Upregulated	1.00:2.71	1.00:3.22
21	ANG44638	Maturase K	20.9	6.36	87	7	Nucleus	Cell rescue/defense	Upregulated	1.00:4.84	1.00:4.80

S.N. = Spot number, Exp. Mw = Experimental molecular weight; Exp. Pi = Experimental isoelectric point; M.S. = Mascot score. PDQuest software was used for spot volume analysis. The fold change of up-regulated protein spot volumes was calculated by treatment/optimum, whereas the change fold of downregulated protein spot volumes was calculated by optimum/treatment. The relative spot intensity of proteins was shown in fold changes between optimum-P and low-P. Kd = Kilo-Dolton; STMs = Signal transduction mechanisms.

Table 4. Identification, sub-cellular localization, and quantitative analysis of differentially expressed leaf proteins of cv. Panvel only under low P condition.

S. N.	Accession No.	Name of Protein	Exp. Mw (kDa)	Exp. Pi	M. S.	No. of Matched Peptides	Location	Process	Mode of Regulation	Relative Spot Intensity (Optimum P: Low P)
22	ACA50522	14˙C3˙C3 protein	34.2	4.93	112	9	Nucleus	Transcription/STMs	Upregulated	1.00:3.60
23	BAD07865	Phosphoribulokinase	46.1	5.45	91	14	Chloroplast, cytosol	Photosynthesis	Upregulated	1.00:3.22
24	Q0DYB1	Inorganic pyrophosphatase family protein	33.2	6.1	110	12	Cytoplasm	Metabolism	Upregulated	1.00:4.63
25	BAD67774	Phosphogluconate dehydrogenase	54.2	5.89	104	14	Cytoplasm	Metabolism	Upregulated	1.00:2.92
26	BAF92702	Chaperonin 60 β precursor	55.1	5.4	128	8	Mitochondrion	Protein fate/stabilization	Upregulated	1.00:2.51
27	AAB33001	Sedoheptulose-1,7-bisphosphatase	41.9	5.81	143	14	Chloroplast	Metabolism	Downregulated	1.00: −2.11
28	3E5R_A	Glyceraldehyde-3-phosphate dehydrogenase A, chloroplastic precursor	55.2	6.47	94	13	Chloroplast	Energy	Upregulated	1.00:3.52

Table 4. *Cont.*

S. N.	Accession No.	Name of Protein	Exp. Mw (kDa)	Exp. Pi	M. S.	No. of Matched Peptides	Location	Process	Mode of Regulation	Relative Spot Intensity (Optimum P: Low P)
29	ABA18619	59-epimerase	43.5	5.89	104	8	Cytosol	Oxidative stress	Upregulated	1.00:3.23
30	BAG24017	RNA-binding protein	49.7	5.30	106	9	Cytosol, nucleus	Transcription/STMs	Upregulated	1.00:4.50
31	M1NZ56	Pyrroline-5-carboxylate synthetase	28.1	6.7	114	13	Cytoplasm	Cell rescue/defense	Upregulated	1.00:3.11
32	ACJ54888	Heat shock protein 40	28.7	5.8	96	8	Nucleus, mitochondrion, ER	Protein fate/stabilization	Upregulated	1.00:3.50
33	AAF85973	PR-10b protein	17.2	6.7	23	9	Nucleus	-	Upregulated	1.00:2.55
34	ABA39947	UDP-glucose epimerase	42.5	5.78	102	8	Cytosol	Oxidative stress	Upregulated	1.00:1.63
35	BAS78758	Os02g0493300	56.4	6.15	87	8	-	-	Upregulated	1.00:1.80
36	AJB98433	6-Phosphogluconolactonase	34.2	5.46	87	10	Cytosol	Metabolism	Upregulated	1.00:3.41
37	Q40693	Heat shock protein 70	27.5	5.3	92	9	Nucleus, mitochondrion	Protein fate/stabilization	Upregulated	1.00:1.50
38	BAA11351	Ribosomal protein S19	12.6	6.8	58	13	Ribosome, mitochondrion	Protein synthesis	Downregulated	1.00: −3.22
39	ABI74568	Phosphoglycerate kinase	46.2	5.48	167	14	Cytosol	Energy	Upregulated	1.00:3.53
40	ABA92415	NADP-dependent oxidoreductase P1	40.3	6.31	119	13	Chloroplast	Photosynthesis	Upregulated	1.00:3.28
41	AAQ23061	Heat shock responsive transcription factor	37.9	5.1	54	9	Nucleus, cytoplasm	Protein fate/stabilization	Upregulated	1.00:3.97
42	BAC00625	Malate dehydrogenase	37.7	5.41	86	9	Mitochondrion	Metabolism	Downregulated	1.00: −3.24
43	AAK92626	ATPase	74.8	6.48	157	11	Plasma membrane	Energy	Upregulated	1.00:3.32
44	BAD17324	Flavonol synthase	35.9	5.43	104	10	Cytoplasm, nucleus	Oxidative stress	Upregulated	1.00:4.61
45	Q6AVA8	Pyruvate orthophosphate dikinase	73.3	5.15	128	22	Chloroplast, cytoplasm	Metabolism	Downregulated	1.00: −3.00
46	Q6K9N6	Succinyl-CoA synthetase beta subunit	44.2	5.64	63	10	Mitochondrion	Metabolism	Upregulated	1.00:3.62

S.N. = Spot number, Exp. Mw = Experimental molecular weight; Exp. Pi = Experimental isoelectric point; M.S. = Mascot score. PDQuest software was used for spot volume analysis. The fold change of up-regulated protein spot volumes was calculated by treatment/optimum, whereas the change fold of downregulated protein spot volumes was calculated by optimum/treatment. The relative spot intensity of proteins was shown in fold changes between optimum-P and low-P. Kd = Kilo-Dolton; STMs = Signal transduction mechanisms.

Table 5. Identification, subcellular localization, and quantitative analysis of differentially expressed leaf proteins of cultivar cv. Nagina 22 only under low P condition.

S. N.	Accession No.	Name of Protein	Exp. Mw (kDa)	Exp. Pi	M. S.	No. of Matched Peptides	Location	Process	Mode of Regulation	Relative Spot Intensity (Optimum P: Low P)
47	P93431	Rubisco activase chloroplast precursor	52.21	4.93	148	6	Chloroplast	Photosynthesis	Downregulated	1.00: −3.26
48	XP_015641702	Ferredoxin-nitrite reductase, chloroplastic	71.3	6.15	94	11	Chloroplast	Photosynthesis	Downregulated	1.00: −4.43
49	BAP76084	Phosphoenolpyruvate carboxylase	25.9	6.87	98	7	Chloroplast, mitochondrion	Metabolism	Downregulated	1.00: −3.26
50	Q01859	F1-ATP synthase, beta subunit	66.4	5.39	165	18	Mitochondrion	Energy	Upregulated	1.00:4.70
51	A3C4S4	GDP-D-mannose-3′,5′-epimerase	54.3	6.63	102	11	Cytosol	Oxidative stress	Upregulated	1.00:3.32
52	ABA96472	FKBP-type peptidyl-prolyl cis-trans isomerase	27.2	6.84	86	8	Nucleus	Cell rescue/defense	Upregulated	1.00:3.83
53	AAS46111	RNA polymerase β chain	81.1	6.93	215	10	Chloroplast	Transcription/STMs	Downregulated	1.00: −3.90
54	AAB65699	Ferredoxin	44.5	6.35	147	16	Chloroplast	Photosynthesis	Downregulated	1.00: −3.44
55	A2Y7D9	Chloroplast light-harvesting complex I protein LHCA3	31.1	6.46	85	7	Chloroplast	Photosynthesis	Downregulated	1.00: −3.37
56	P48494	Triosephosphate isomerase	21.2	5.5	151	13	Cytoplasm	Metabolism	Downregulated	1.00: −3.70
57	-	Unnamed protein product	46.3	5.49	97	14	-	-	Upregulated	1.00:3.80
58	XP_015643207	Transketolase chloroplastic	17.8	6.12	84	16	Chloroplast	Photosynthesis	Downregulated	1.00: −3.52
59	ABL74559	Glyceraldehyde 3-phosphate dehydrogenase	53.8	6.72	84	8	Cytoplasm	Energy	Upregulated	1.00:3.45
60	AET04420	Integrin-linked protein kinase family protein	34.2	6.11	104	12	Ribosome	Transcription/STMs	Downregulated	1.00: −3.11
61	XP_015621402	Asparaginyl-tRNA synthetase, cytoplasmic 3	65.1	6.13	173	23	Cytoplasm	Protein synthesis	Downregulated	1.00: −3.24
62	AAP13093	Ascorbate peroxidase cytoplasmic	31.4	5.75	92	12	Cytoplasm	Oxidative stress	Upregulated	1.00:4.42
63	AAB63591	Chaperonin 10 Kd subunit	25.8	5.66	82	8	Cytoplasm	Protein stabilization	Upregulated	1.00:3.54

S.N. = Spot number, Exp. Mw = Experimental molecular weight; Exp. Pi = Experimental isoelectric point; M.S. = Mascot score. PDQuest software was used for spot volume analysis. The fold change of up-regulated protein spot volumes was calculated by treatment/optimum, whereas the change fold of downregulated protein spot volumes was calculated by optimum/treatment. The relative spot intensity of proteins was shown in fold changes between optimum-P and low-P. Kd = Kilo-Dolton; STMs = Signal transduction mechanisms.

4. Discussion

The performance of P-inefficient cultivar (cv. Nagina 22) and P-efficient cultivar (cv. Panvel) in terms of morphological traits, like plant height, plant biomass, and P concentration in shoot and root, showed that cv. Nagina 22 was more affected than cv. Panvel by low P treatment. The plant biomass was low in the cv. Nagina 22 than the cv. Panvel even at optimum-P, which may be due to the expression differences of some proteins. The PUE of the latter is higher than the former cultivar. Based on these observations, it can be revealed that there is genetic variability in these rice cultivars in response to low P condition. The cv. Panvel has a better adaptive response to P deficiency than cv. Nagina 22. Earlier studies have also reported the genetic variability in the rice genotypes in response to P deficiency [27,28]. Phosphorus deficiency affected many metabolic processes that led to a decrease in the biomass accumulation and overall growth of the plants. Some rice genotypes develop an adaptive mechanism to tolerate P deficiency stress to some extent through modification in the root architecture [29], and changes in the expression level of enzymes of key metabolic pathways. The reduction in the activities of the enzyme of the Calvin cycle in the leaves of rice by P deficiency has been reported [9]. Changed activities of the enzymes of carbon metabolism, nitrogen metabolism, and energy metabolism have been observed in the leaves of rice when the supply of P was altered [10]. It is suggested that a well sophisticated regulatory network of genes in plants control the nutrient homeostasis through differential expression of proteins. A proteomics approach was used in the present study to investigate the proteins involved in the adaptive response of the P-efficient and P-inefficient rice cultivars to P deficiency. Previous studies on the proteomics of rice, maize, oilseed rape, and *Arabidopsis* under P deficiency are limited to the expression pattern of the proteins of root only [12,15,17,30]. However, the assimilation and utilization of P in the key metabolic processes take place in the leaves of the plants. Therefore, the leaf proteome profiles of contrasting P-efficient rice cultivars were analyzed in this study under low P and optimum P conditions. The cv. Nagina 22 is fully sequenced and cv. Panvel is not fully sequenced at the genomic level, and of both the cultivars belong to two different subspecies (*Aus* and *Indica*) of rice due to which there was little difference in sequences of the identified proteins. Comparative analysis of the leaf proteome of the P-efficient and P-inefficient rice cultivars provided a better way for the identification of the P deficiency adaptive proteins in rice. The identified differentially expressed proteins (DEPs) of this study were discussed with their functions in their respective pathways. Most of the DEPs were involved in photosynthesis, where these proteins affect light and dark reactions of the photosynthesis process. Some DEPs were involved in oxidative stress and defense pathways, which help plants to tolerate P stress, and few DEPs were related to protein synthesis, where they either have an impact on translation or protein stabilization. Other DEPs were involved in Kreb's cycle, sucrose metabolism, Embden-Mayerhoff and Hexose Monophosphate pathways. Figure 3 shows the schematic model of organized mechanisms of adaptation to P deficiency in P-efficient rice cultivar (cv. Panvel).

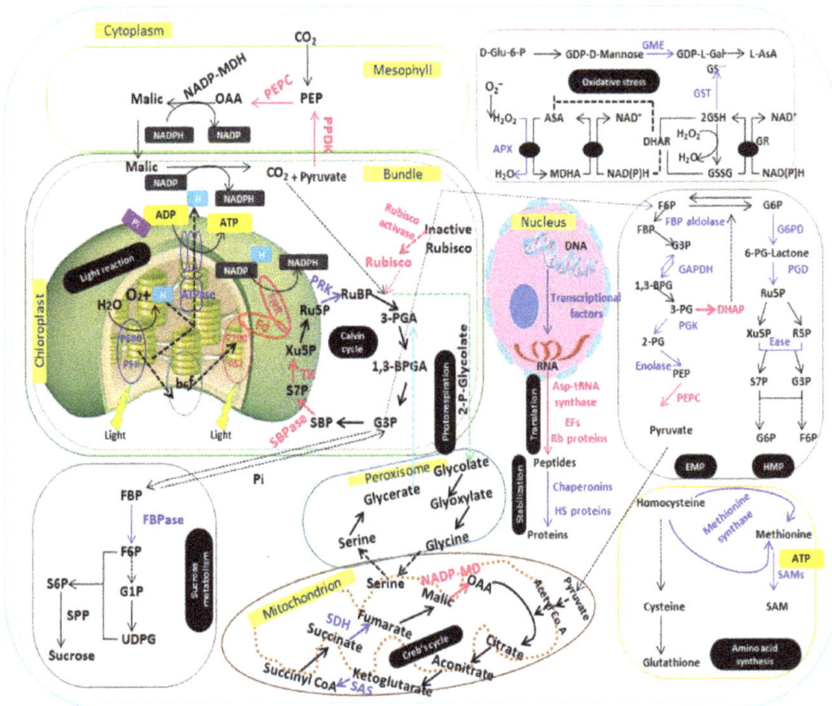

Figure 3. Schematic model of organized mechanisms of adaptation to P deficiency in P-efficient rice cultivar. Identified proteins were portrayed into subcellular location according to their molecular and metabolic pathways. Protein expression patterns were indicated by marking protein names and arrows in blue (upregulated) or pink (downregulated). 1,3-BPG: 1,3-bisphosphoglycerate; 1,3-BPGA: 1,3-bisphosphoglycerate; 2PG: 2-phosphoglycerate; 3PG: 3-phosphoglycerate; 3-PGA: 3-phosphoglycerate; 6-PG-lactone: 6-phosphoglucono-lactone; ADP: adenosine diphosphate; APX: ascorbate peroxidase; ASA: ascorbic acid; ASP: aspartic acid; ATP: adenosine triphosphate; D-Glu6P: D-Glucose 6-phosphate; DHAP: dihydroacetone phosphate; DHAR: dehydroascorbate reductase; EF: elongation factors; EMP: Embden-Meyerhof-Parnas pathway; F6P: fructose-6-phosphate; FBP aldose: fructose-1,6-bisphosphate aldose; FBP: fructose-1,6-bisphosphate; FBPase: fructose-1,6-bisphosphatase; Fd: ferredoxin; FNR: ferredoxin-NADP reductase; G1P: glucose-1-phosphate; G3P: glyceraldehyde-3-phosphate; G6P: glucose-6-phosphate; G6PD: glucose-6-phosphate dehydrogenase; GAPDH: glyceraldehyde-3-phosphate dehydrogenase; GDP-L-Gal: GDP-L-galactose; GME: GDP-D-mannose-3,5-epimerase; GR: glutathione reductase; GSH: reduced glutathione; GSSG: oxidized glutathione; GST: glutathione-S-transferase; HMP: hexose-monophophate-pathway; HS proteins: heat shock proteins; MDHA: monodehydroasorbate; NADP-MDH: NADP-malate dehydrogenase; OAA: oxalacetic acid; PEP: phosphoenolpyruvate; PEPC: phosphoenolpyruvate carboxylase; PEPC: phosphoenolpyruvate carboxylase;PGD: phosphogluconate dehydrogenase; PGK: phosphoglycerate kinase; PPDK: pyruvate orthophosphate dikinase; PRK: phosphoribolokinase; PS I: photosystem I; PS II: photosystem II; R5P: ribose-5-phosphate; RB proteins: ribosomal proteins; Ru5P: ribulose-5-phosphate; Rubisco: ribose 1,5-bisphosphate carboxylase/oxygenase; RuBP: ribulose-1,5-diphosphate; S6P: sucrose-6-phosphate; S7P: 7-phosphosedoheptose; SAMs: S-adenosymethionine synthase; SAS: succinyl-CoA-synthetase; SBP: sedoheptulose-1,7-bisphosphate; SDH: succinate dehydrogenase; SPP: sucrose-6-phosphate phosphohydrolase; TCA cycle: tricarboxylic acid cycle; TK: transketolase; UDPG: uridine diphosphoglucose; Xu5P: xylulose-5-phosphate.The source of the pathways, Calvin cycle, protein synthesis, oxidative stress and EMP pathway [15], and for light reaction, Creb's cycle, sucrose metabolism and HMP pathway [30].

4.1. Expression Pattern of the Proteins of Energy Metabolism under P-deprivation

Modifications in P and energy metabolism under P-starvation greatly affect the plant physiology. Phosphorus deficiency affects the proteins that are involved in electron transport and glycolytic enzymes. Remobilization of inorganic phosphate from different substrates is due to the scavenging allocation of enzymes by internal P sources, such as phospholipases, nucleases, and phosphatases [31]. Plants need to maintain the P homeostasis to release the stress under low-P conditions. In this experiment, the PUE was low in both the cultivars under low-P conditions by the difference in expression of some proteins which may be directly or indirectly involved in PUE in the plants. Our study depicted diversified expression results concerning PUE and energy metabolism under low-P conditions. Phosphogluconate dehydrogenase (PGDH, five protein families) in cv. Panvel may regulate carboxylation reactions by functioning on the precursor of ribulose 1,5-bisphosphate (RuBP, 11 proteins, and seven protein family) as a carboxylation/Rubisco substrate and, hence, may replace Rubisco with its reversible reductive carboxylase activity [32]. While the upregulation of fructose-1,6-bisphosphate in both of the cultivars showed that there is an enhancement in carbon fixation, the upregulation of sedoheptulose-1,7-bisphosphatase (BiBPase) in cv. Panvel only revealed that it redirects the carbon in coordination with aldolase that functions as the rate-limiting enzyme in the Calvin cycle [33]. The upregulation of phosphoglycerate kinases and 6-phosphogluconolactonase in cv. Panvel suppressed more CO_2 assimilation and starch degradation under P-deficient conditions in a process to save energy [34]. These proteins can be assumed to be candidate P deficiency adaptive proteins of rice. Succinate dehydrogenase (SDH) and succinyl-CoA synthetase function in the tricarboxylic acid cycle and their upregulation in cultivar cv. Panvel under P deficient conditions suggested that there is an enhancement in the fumarate formation from succinate, and the production of GTPs (Guanosine-5′-triphosphate) [35,36]. Glyceraldehyde-3-phosphate dehydrogenase (GAPDH, 20 proteins, and 16 protein family) was upregulated in cv. Panvel. It plays an active part by redirecting energy fluxes and improving biomass production according to nutrient availability [37]. An intermediate enzyme of the glycolytic pathway enolase 1 was upregulated in both the cultivars. It catalyzes the conversion of 2-phosphoglycerate (2-PGA) to phosphoenolpyruvate (PEP) in plastid [38]. The differential expression pattern of proteins of energy metabolism and P homeostasis in cv. Panvel and cv. Nagina 22 under low P conditions showed that the P-efficient cultivar of rice managed the expression of these proteins in a more efficient way than the P-inefficient cultivar in order to adapt to P deficiency to some extent. The phenotypic traits, like plant dry weight and PUE, have shown strong cultivar × treatment interaction, which also suggests that there is a difference in protein expression between the rice cultivars.

4.2. Expression Pattern of the Proteins Involved in Transcription and Translation

The down-regulation of RNA polymerase β chain and Integrin-linked protein kinase family protein in cv. Nagina 22 suggested that there was a reduction in the translation process under low-P condition. The expression of these transcriptional factors has a significant impact on biomass and P content in rice, as reported earlier, the role of OsPTF1 under P-deficient condition in rice [39]. The 14¨C3¨C3 protein is involved in cellular regulatory pathways for signal transduction, metabolism. Its differential expression functions as an activator/repressor with phosphorylation of target proteins [40]. This upregulation of this protein in cv. Panvel suggested that this protein can help in the regulation of P deficiency stress stimuli and defense signaling factors in P-efficient cultivar.

The involvement of P in energy metabolism and transcription process, and as a component of nucleic acids, suggested that P can affect some steps of the protein synthesis [41]. One of the limiting factors of protein synthesis is elongation factor-Tu (EF-Tu, seven proteins). It was downregulated in both the rice cultivars under low P condition. The EF-Tu helps in the binding of aminoacyl-tRNAs to the ribosome sub-unit and refolding of denatured proteins [42], thus playing a role of chaperone activity. Asparaginyl-tRNA synthetase (APRS) is the class II enzyme of aminoacyl-tRNA synthetases (AARS, only one protein) enzyme family with ubiquitous function and helps in the translation of mRNA codons into their corresponding amino acids [43]. It was down regulated in cultivar cv. Nagina 22 in low P condition. A group of ribosomal proteins was downregulated in both rice cultivars under low P condition. These are ribosomal protein L23 and ribosomal protein S4. Interestingly, the ribosomal protein S19 was downregulated in cv. Panvel only. The differential expression of ribosomal proteins was reported in response to abiotic stress [44]. It has been suggested that these proteins regulate the mechanism of protein synthesis by recognition of mRNA [45]. The differential expression pattern of the proteins that are involved in the transcription and translation processes under low P conditions in rice cultivars showed that there was a higher down-regulation of these proteins in cv. Nagina 22 than cv. Panvel, suggesting that cv. Panvel developed an adaptive mechanism of maintaining the protein synthesis at a normal level to some extent under low P stress conditions.

4.3. Photosynthesis and CO_2 Regulation under P-deficiency

The photosynthesis process comprises two parts, light reactions to produce ATP and NADPH in the thylakoid membrane and fixing of CO_2 into organic molecules with the use of ATP and NADPH in the light-independent carbon reactions [46]. The downregulation of light-harvesting complex I (LHC, 29 proteins, and 19 protein family) in cv. Nagina 22 under low P condition showed the lesser capacity of this cultivar for low light absorption and light utilization by photosystem I (PSI) than cv. Panvel. Similar findings of reduction in light absorption along with oxygen-evolving complex proteins with a decrease of LHCs and chlorophyll content were observed under other stresses earlier [47]. Moreover, proteins that are involved in light energy transfer in photosynthesis, like ferredoxin, ferredoxin reductase (FNR, 12 proteins, and nine protein family), and beta and gamma subunit of ATP synthase (27 proteins and six protein family) were also down-regulated in cultivar cv. Nagina 22 under low P condition. The above-mentioned proteins are helpful in the production in NADPH and ATP [48,49]. The down-regulation of these proteins reduced the photosynthetic efficiency of cv. Nagina 22 under low P condition.

Expressions of the proteins that were involved in the dark reaction of photosynthesis were also down-regulated in P-inefficient rice cultivar (cv. Nagina 22). These proteins are ribulose-1, 5-bisphosphate carboxylase/oxygenase (Rubisco, five-member family, *OsRBCS1–OsRBCS5*), Rubisco activase (RCA, three proteins belong to AAA+ superfamily), phosphoglycerate kinase (PGK, 20 proteins and seven protein family)), and transketolase. Rubisco is a rate-limiting enzyme of CO_2 fixation [50], which is activated by an ATP-dependent enzyme Rubisco activase [51]. The downregulation of transketolase coincided with previous findings under stress conditions in tobacco and rice [52,53]. These findings demonstrated the repression of ribulose-1, 5-biphosphate regeneration, and photosynthesis (Figure 4). Other proteins that were involved in CO_2 fixation were also differentially expressed under low P conditions in rice cultivars. These were phosphoenolpyruvate carboxylase (PEPC), NADP-malate dehydrogenase (NADP-MDH), and pyruvate orthophosphate dikinase (PPDK). The PEPC helps in the CO_2 fixation by bicarbonate formation and it regulates the photo-respiratory pathway. This protein was downregulated in cv. Nagina 22 under low P conditions, but upregulated in cv. Panvel, which suggested P deficiency adaptive mechanism and involves the photosystem II functioning and its structural stability [54,55].

4.4. Expression Proteins of Antioxidant Defense System under P-deficiency

The generation of ROS (reactive oxygen species) is due to abiotic stress that has a negative impact on plants, as it oxidizes life-sustaining biomolecules, like proteins, lipids, carbohydrates, and nucleic acids. The upregulation of glutathione S-transferases (GSTs) in both the rice cultivars at low P conditions suggested that low P condition causes oxidative stress also through the generation of ROS. Ascorbate peroxidase was upregulated in cultivar cv. Nagina 22, which has multifunction towards ROS depletion, like scavenging H_2O_2. Its overexpression enhances stress tolerance in plants [56]. The upregulation of glyoxylase I7 in both the cultivars and upregulation of 59-epimerase only in cv. Panvel under low-P condition suggested that the expression antioxidative defense system of cv. Panvel is better than cv. Nagina 22. These proteins are involved in the biosynthesis of ascorbic acid, which enhances stress tolerance in plants [57]. The chaperones (Chaperonin 60 β precursor and heat shock protein) have functions in intracellular protein folding and act as intercellular signals with a wide variety of biological effects. The Chaperonin 60 β has been found to be a pathogenic factor in a wide range of diseases [58]. These proteins were upregulated in cv. Panvel. A stress response protein, peptidyl-prolyl cis, trans-isomerase was upregulated in cv. Nagina 22. This protein has an important role in better survival and tolerance to stress conditions [59]. The PR-10 protein (pathogenesis-related protein) was upregulated in cultivar cv. Panvel and plays a role in the defense mechanisms for plant tolerance against pathogen attack and abiotic stimuli [60]. Differential expression patterns of the proteins of the antioxidant defense system in cv. Panvel and cv. Nagina 22 suggested that this system is also involved in the P deficiency adaptation mechanism of rice. Plants undertake some changes to tolerate the stress conditions, like decreasing ROS production by enhancing antioxidant defense or change in root architecture, like increasing root length, which was revealed under the low-P condition in the study. Most of the DEPs are responsive to P-deficiency because of their expression changes under low-P conditions, and some of the responsive proteins, like transcriptional factors and antioxidants, are directly involved in stability and tolerance mechanisms, which enable plants to tolerant P-deficiency.

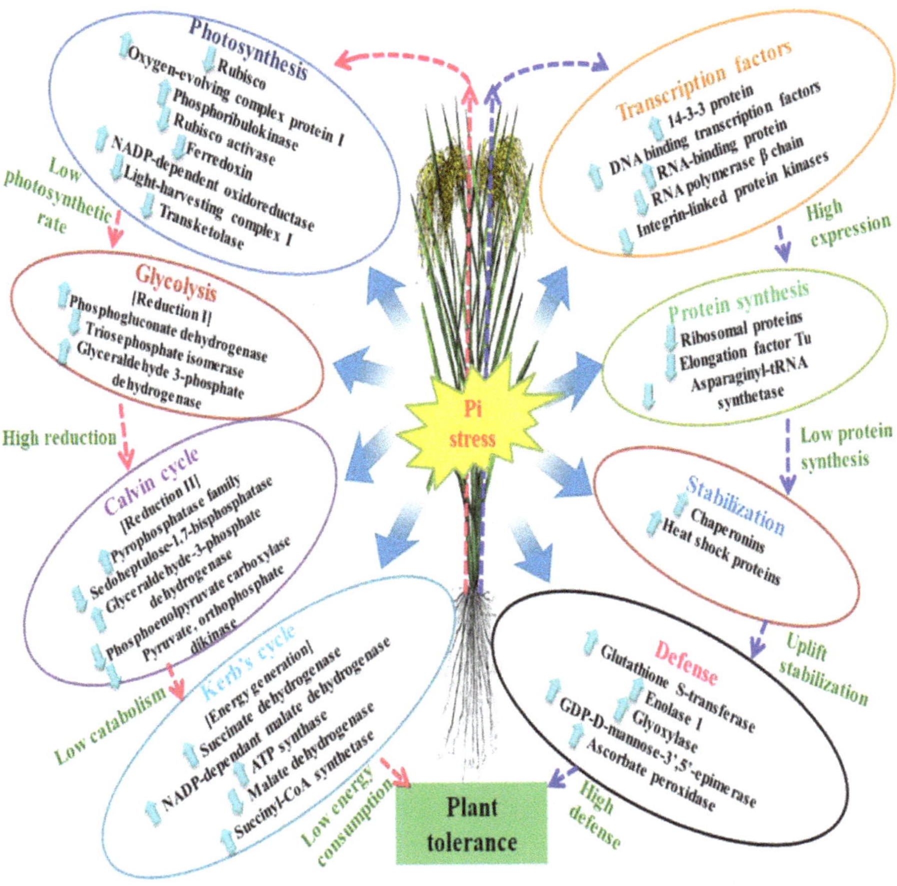

Figure 4. Representation of differentially expressed proteins and their alliance with the biological processes under low phosphorus. In the right side of the figure, the arrows with blue colour depict modulation of molecular factors to raise the defense level. Whereas, left side arrows with pink colour shows the regulation of low energy consumption under phosphorus stress. Pi = Inorganic Phosphorus

5. Conclusions

The leaf proteome profile of the P-efficient and P-inefficient cultivars of rice suggested that P-efficient rice cultivar developed a P-deficiency adaptive mechanism through the change in the expression pattern of the proteins that are involved in energy metabolism, photosynthesis, and CO_2 assimilation. Rubisco activase, phosphoenolpyruvate carboxylase, F1-ATP synthase, chloroplast light-harvesting complex I protein, and glyceraldehyde 3-phosphate dehydrogenase are potential P deficiency adaptive protein candidates. The regulation of these proteins can improve the tolerance of rice under P starvation. The upregulation of antioxidant enzymes (like glyoxylase I7, 5′-epimerase and Ascorbate peroxidase) and defense proteins (like pyrroline-5-carboxylate synthetase) suggested that P deficiency creates oxidative stress, and the P-efficient cultivar of rice developed stronger protection from ROS accumulation damage than P-inefficient cultivar under low P condition. The information on the differential expression pattern of proteins of P-efficient and P-inefficient rice cultivars under low P condition will help develop strategies for generating rice with high PUE.

Supplementary Materials: The following are available online at http://www.mdpi.com/2073-4395/10/7/1028/s1, Figure S1. Plants of 30-days-old rice cultivars, Nagina 22 and Panvel, grown under optimum-P (A, C) and low-P (B, D) conditions.

Author Contributions: Conceptualization, A.Y.T. and A.A.; methodology, A.Y.T. and A.A.; software, A.Y.T., H.M.A. and A.A.; validation, A.Y.T., H.M.A. and A.A.; formal analysis, A.Y.T. and A.A.; investigation, A.Y.T., H.M.A. and A.A.; resources, H.M.A. and A.A.; data analysis, A.Y.T., H.M.A.; writing—original draft preparation, A.Y.T., and A.A.; writing—review and editing, A.Y.T., H.M.A. and A.A.; supervision, A.A.; project administration, H.M.A. and A.A.; funding acquisition, H.M.A. and A.A. All authors have read and agreed to the published version of the manuscript.

Funding: This research received no external funding.

Acknowledgments: The authors would like to extend their sincere appreciation to the King Saud University, Riyadh, Saudi Arabia for its support to the Researchers Supporting Project number (RSP-2020/123).

Conflicts of Interest: The authors declare no conflict of interest.

Abbreviations

PUE	Phosphate utilization efficiency
2DE	Two-dimensional gel electrophoresis
DEP	Differentially expressed proteins
MALDI-TOF	Matrix-assisted laser desorption/ionization-time of flight

References

1. Fischer, R.A.; Edmeades, G.O. Breeding and Cereal Yield Progress. *Crop. Sci.* **2010**, *50*, S-85–S-98. [CrossRef]
2. Lynch, J.P. Root Phenes for Enhanced Soil Exploration and Phosphorus Acquisition: Tools for Future Crops. *Plant Physiol.* **2011**, *156*, 1041–1049. [CrossRef] [PubMed]
3. Doberman, A.; Fairhurst, T. *Rice: Nutrient Disorders and Nutrient Management*; IRRI & PPI & PPIC: Makati, Philippines; Singapore, 2000.
4. Heuer, S.; Gaxiola, R.; Schilling, R.; Herrera-Estrella, L.; López-Arredondo, D.; Wissuwa, M.; Delhaize, E.; Rouached, H. Improving phosphorus use efficiency: A complex trait with emerging opportunities. *Plant J.* **2017**, *90*, 868–885. [CrossRef] [PubMed]
5. Van Kauwenbergh, S.J. *World Phosphate Rock Reserves and Resources*; IFDC: Muscle Shoals, AL, USA, 2010.
6. Dawe, D. The potential role of biological nitrogen fixation in meeting future demand for rice and fertilizer. In *The Quest for Nitrogen Fixation in Rice*; Ladha, J.K., Reddy, P.M., Eds.; International Rice Research Institute: Los Banos, Philippines, 2000; pp. 93–118.
7. Van Der Velde, M.; Folberth, C.; Balkovič, J.; Ciais, P.; Fritz, S.; Janssens, I.A.; Obersteiner, M.; See, L.; Skalský, R.; Xiong, W.; et al. African crop yield reductions due to increasingly unbalanced Nitrogen and Phosphorus consumption. *Glob. Chang. Boil.* **2014**, *20*, 1278–1288. [CrossRef]
8. Chiera, J.; Thomas, J.; Rufty, T.W. Leaf initiation and development in soybean under phosphorus stress. *J. Exp. Bot.* **2002**, *53*, 473–481. [CrossRef]

9. Brooks, A. Effects of Phosphorus Nutrition on Ribulose-1, 5-Bisphosphate Carboxylase Activation, Photosynthetic Quantum Yield and Amounts of Some Calvin-Cycle Metabolites in Spinach Leaves. *Funct. Plant Boil.* **1986**, *13*, 221. [CrossRef]
10. Tang, X.R.; Yu, T.Q. Effects and mechanisms of P and K nutrients on yield and protein content of fodder rice. *Agric. Sci. China* **2002**, *1*, 432–437.
11. Qiu, H.; Liu, C.; Yu, T.; Mei, X.; Wang, G.; Wang, J.; Cai, Y. Identification of QTL for acid phosphatase activity in root and rhizosphere soil of maize under low phosphorus stress. *Euphytica* **2014**, *197*, 133–143. [CrossRef]
12. Jiang, H.; Zhang, J.; Han, Z.; Yang, J.; Ge, C.; Wu, Q. Revealing new insights into different phosphorus-starving responses between two maize (*Zea mays*) inbred lines by transcriptomic and proteomic studies. *Sci. Rep.* **2017**, *7*, 44294. [CrossRef]
13. Ganie, A.H.; Ahmad, A.; Pandey, R.; Aref, I.M.; Yousuf, P.Y.; Ahmad, S.; Iqbal, M. Metabolite Profiling of Low-P Tolerant and Low-P Sensitive Maize Genotypes under Phosphorus Starvation and Restoration Conditions. *PLoS ONE* **2015**, *10*, e0129520. [CrossRef] [PubMed]
14. Huang, G.; Liang, W.; Sturrock, C.J.; Pandey, B.K.; Giri, J.; Mairhofer, S.; Wang, D.; Müller, L.; Tan, H.; York, L.M.; et al. Rice actin binding protein RMD controls crown root angle in response to external phosphate. *Nat. Commun.* **2018**, *9*, 2346. [CrossRef]
15. Alexova, R.; Millar, A.H. Proteomics of phosphate use and deprivation in plants. *Proteomics* **2013**, *13*, 609–623. [CrossRef] [PubMed]
16. Kim, S.G.; Wang, Y.; Lee, C.H.; Mun, B.G.; Kim, P.J.; Lee, S.Y.; Kim, Y.C.; Kang, K.Y.; Rakwal, R.; Agrawal, G.K.; et al. A comparative proteomics survey of proteins responsive to phosphorous starvation in roots of hydroponically-grown rice seedlings. *J. Korean Soc. Appl. Biol.* **2011**, *54*, 667–677. [CrossRef]
17. Torabi, S.; Wissuwa, M.; Heidari, M.; Naghavi, M.R.; Gilany, K.; Hajirezaei, M.-R.; Omidi, M.; Yazdi-Samadi, B.; Ismail, A.M.; Salekdeh, G.H. A comparative proteome approach to decipher the mechanism of rice adaptation to phosphorous deficiency. *Proteomics* **2009**, *9*, 159–170. [CrossRef] [PubMed]
18. Yoshida, S. Routine Procedure for Growing Rice Plants in Culture Solution. In *Laboratory Manual for Physiological Studies of Rice*; Yoshida, S., Forno, D.A., Cock, J.H., Eds.; International Rice Research Institute: Los Banos, Philippines, 1976; pp. 61–66.
19. Wu, C.; Wang, Q.; Xue, S.; Pan, W.; Lou, L.; Li, D.; Hartley, W. Do aeration conditions affect arsenic and phosphate accumulation and phosphate transporter expression in rice (*Oryza sativa* L.)? *Environ. Sci. Pollut. Res.* **2016**, *25*, 43–51. [CrossRef] [PubMed]
20. Mahajan, G.; Pandey, R.N.; Sahoo, R.N.; Gupta, V.K.; Datta, S.C.; Kumar, D. Monitoring nitrogen, phosphorus and sulphur in hybrid rice (*Oryza sativa* L.) using hyperspectral remote sensing. *Precis. Agric.* **2016**, *18*, 736–761. [CrossRef]
21. Syers, J.K.; Johnston, A.E.; Curtin, D. Efficiency of soil and fertilizer phosphorus use. *FAO Fertil. Plant Nutr. Bull.* **2008**, *18*, 108.
22. Isaacson, T.; Damasceno, C.M.B.; Saravanan, R.S.; He, Y.; Catalá, C.; Saladié, M.; Rose, J.K.C. Sample extraction techniques for enhanced proteomic analysis of plant tissues. *Nat. Protoc.* **2006**, *1*, 769–774. [CrossRef]
23. O'Farrell, P.H. High resolution two-dimensional electrophoresis of proteins. *J. Boil. Chem.* **1975**, *250*, 4007–4021.
24. Wheeler, D.L.; Barrett, T.; Benson, D.A.; Bryant, S.H.; Canese, K.; Chetvernin, V.; Church, D.M.; DiCuccio, M.; Edgar, R.; Federhen, S.; et al. Database resources of the national center for biotechnology information. *Nucleic Acids Res.* **2006**, *35*, 5–12. [CrossRef]
25. Wu, C.H. The Universal Protein Resource (UniProt): An expanding universe of protein information. *Nucleic Acids Res.* **2006**, *34*, D187–D191. [CrossRef]
26. Guda, C. pTARGET: A web server for predicting protein subcellular localization. *Nucleic Acids Res.* **2006**, *34*, W210–W213. [CrossRef] [PubMed]
27. Chankaew, S.; Monkham, T.; Pinta, W.; Sanitchon, J.; Kaewpradit, W.; Srinives, P. Screening tolerance to phosphorus deficiency and validation of phosphorus pptake 1 (Pup1) gene linked markers in Thai indigenous upland rice germplasm. *Agronomy* **2019**, *9*, 81. [CrossRef]
28. Wissuwa, M.; Ae, N. Genotypic variation for tolerance to phosphorus deficiency in rice and the potential for its exploitation in rice improvement. *Plant Breed.* **2001**, *120*, 43–48. [CrossRef]
29. Vejchasarn, P.; Lynch, J.P.; Brown, K.M. Genetic Variability in Phosphorus Responses of Rice Root Phenotypes. *Rice* **2016**, *9*, 29. [CrossRef]

30. Zhang, K.; Liu, H.; Tao, P.; Chen, H. Comparative Proteomic Analyses Provide New Insights into Low Phosphorus Stress Responses in Maize Leaves. *PLoS ONE* **2014**, *9*, e98215. [CrossRef]
31. Zhang, W.; Gruszewski, H.A.; Chevone, B.I.; Nessler, C.L. An *Arabidopsis* Purple Acid Phosphatase with Phytase Activity Increases Foliar Ascorbate1 [OA]. *Plant Physiol.* **2007**, *146*, 431–440. [CrossRef]
32. Bar-Even, A. Daring metabolic designs for enhanced plant carbon fixation. *Plant Sci.* **2018**, *273*, 71–83. [CrossRef] [PubMed]
33. De Porcellinis, A.J.; Nørgaard, H.; Brey, L.M.F.; Erstad, S.M.; Jones, P.R.; Heazlewood, J.L.; Sakuragi, Y. Overexpression of bifunctional fructose-1,6-bisphosphatase/sedoheptulose-1,7-bisphosphatase leads to enhanced photosynthesis and global reprogramming of carbon metabolism in *Synechococcus* sp. PCC 7002. *Metab. Eng.* **2018**, *47*, 170–183. [CrossRef] [PubMed]
34. Wang, Y.; Lu, Y.; Chang, Z.; Wang, S.; Ding, Y.; Ding, C. Transcriptomic analysis of field-grown rice (*Oryza sativa* L.) reveals responses to shade stress in reproductive stage. *Plant Growth Regul.* **2018**, *84*, 583–592. [CrossRef]
35. Huang, S.; Millar, A.H. Succinate dehydrogenase: The complex roles of a simple enzyme. *Curr. Opin. Plant Boil.* **2013**, *16*, 344–349. [CrossRef] [PubMed]
36. Belt, K.; Van Aken, O.; Murcha, M.; Millar, A.H.; Huang, S. An Assembly Factor Promotes Assembly of Flavinated SDH1 into the Succinate Dehydrogenase Complex. *Plant Physiol.* **2018**, *177*, 1439–1452. [CrossRef] [PubMed]
37. Scheibe, R. Maintaining homeostasis by controlled alternatives for energy distribution in plant cells under changing conditions of supply and demand. *Photosynth. Res.* **2018**, *139*, 81–91. [CrossRef] [PubMed]
38. Troncoso-Ponce, M.; Rivoal, J.; Dorion, S.; Sánchez, R.; Venegas-Calerón, M.; Moreno-Pérez, A.J.; Baud, S.; Garces, R.; Martínez-Force, E. Molecular and biochemical characterization of the sunflower (*Helianthus annuus* L.) cytosolic and plastidial enolases in relation to seed development. *Plant Sci.* **2018**, *272*, 117–130. [CrossRef] [PubMed]
39. Yi, K.; Wu, Z.; Zhou, J.; Du, L.; Guo, L.; Wu, Y.; Wu, P. OsPTF1, a Novel Transcription Factor Involved in Tolerance to Phosphate Starvation in Rice1 [w]. *Plant Physiol.* **2005**, *138*, 2087–2096. [CrossRef] [PubMed]
40. Chen, F.; Li, Q.; Sun, L.; He, Z. The Rice 14-3-3 Gene Family and its Involvement in Responses to Biotic and Abiotic Stress. *DNA Res.* **2006**, *13*, 53–63. [CrossRef] [PubMed]
41. Zörb, C.; Ludewig, U.; Hawkesford, M.J. Perspective on Wheat Yield and Quality with Reduced Nitrogen Supply. *Trends Plant Sci.* **2018**, *23*, 1029–1037. [CrossRef]
42. Kudlicki, W.; Coffman, A.; Kramer, G.; Hardesty, B. Renaturation of rhodanese by translational elongation factor (EF) Tu. Protein refolding by EF-Tu flexing. *J. Boil. Chem.* **1997**, *272*, 32206–32210. [CrossRef]
43. Yakobov, N.; Debard, S.; Fischer, F.; Senger, B.; Becker, H. Cytosolic aminoacyl-tRNA synthetases: Unanticipated relocations for unexpected functions. *Biochim. Biophys. Acta (BBA)-Gene Regul. Mech.* **2018**, *1861*, 387–400. [CrossRef]
44. Hummel, M.; Cordewener, J.H.; de Groot, J.C.; Smeekens, S.; America, A.H.; Hanson, J. Dynamic protein composition of *Arabidopsis thaliana* cytosolic ribosomes in response to sucrose feeding as revealed by label free MS E proteomics. *Proteomics* **2012**, *12*, 1024–1038. [CrossRef]
45. Graifer, D.M.; Karpova, G. Roles of ribosomal proteins in the functioning of translational machinery of eukaryotes. *Biochimie* **2015**, *109*, 1–17. [CrossRef] [PubMed]
46. Cardona, T.; Shao, S.; Nixon, P.J. Enhancing photosynthesis in plants: The light reactions. *Essays Biochem.* **2018**, *62*, 85–94. [CrossRef] [PubMed]
47. Dalal, V.K.; Tripathy, B.C. Water-stress induced downsizing of light-harvesting antenna complex protects developing rice seedlings from photo-oxidative damage. *Sci. Rep.* **2018**, *8*, 5955. [CrossRef] [PubMed]
48. Hashida, S.-N.; Miyagi, A.; Nishiyama, M.; Yoshida, K.; Hisabori, T.; Kawai-Yamada, M. Ferredoxin/thioredoxin system plays an important role in the chloroplastic NADP status of Arabidopsis. *Plant J.* **2018**, *95*, 947–960. [CrossRef] [PubMed]
49. Medrano, H.; Parry, M.A.J.; Socias, X.; Lawlor, D.W. Long term water stress inactivates Rubisco in subterranean clover. *Ann. Appl. Boil.* **1997**, *131*, 491–501. [CrossRef]
50. Yousuf, P.Y.; Abd Allah, E.F.; Nauman, M.; Asif, A.; Hashem, A.; Alqarawi, A.A.; Ahmad, A. Responsive Proteins in Wheat Cultivars with Contrasting Nitrogen Efficiencies under the Combined Stress of High Temperature and Low Nitrogen. *Genes* **2017**, *8*, 356. [CrossRef]

51. Henkes, S.; Sonnewald, U.; Badur, R.; Flachmann, R.; Stitt, M. A small decrease of plastid transketolase activity in antisense tobacco transformants has dramatic effects on photosynthesis and phenylpropanoid metabolism. *Plant Cell* **2001**, *13*, 535–551. [CrossRef]
52. Kim, D.-W.; Rakwal, R.; Agrawal, G.K.; Jung, Y.-H.; Shibato, J.; Jwa, N.-S.; Iwahashi, Y.; Iwahashi, H.; Kim, D.H.; Shim, I.-S.; et al. A hydroponic rice seedling culture model system for investigating proteome of salt stress in rice leaf. *Electrophoresis* **2005**, *26*, 4521–4539. [CrossRef] [PubMed]
53. Yang, X.; Liu, D.; Tschaplinski, T.J.; Tuskan, G.A. Comparative genomics can provide new insights into the evolutionary mechanisms and gene function in CAM plants. *J. Exp. Bot.* **2019**, *70*, 6539–6547. [CrossRef]
54. Liu, X.; Dai, C.; Zhou, J.; Ren, C.; Li, X.; Zhang, C.; Zhang, J. Phosphoenolpyruvate carboxylase regulation in C4-PEPC -expressing transgenic rice during early responses to drought stress. *Physiol. Plant.* **2016**, *159*, 178–200. [CrossRef]
55. Zhang, Y.; Giuliani, R.; Zhang, Y.; Zhang, Y.; Araújo, W.L.; Wang, B.; Liu, P.; Sun, Q.; Cousins, A.; Edwards, G.; et al. Characterization of maize leaf pyruvate orthophosphate dikinase using high throughput sequencing. *J. Integr. Plant Boil.* **2018**, *60*, 670–690. [CrossRef]
56. Liu, F.; Huang, N.; Wang, L.; Ling, H.; Sun, T.; Ahmad, W.; Muhammad, K.; Guo, J.; Xu, L.; Gao, S.; et al. A Novel L-ascorbate Peroxidase 6 Gene, ScAPX6, Plays an Important Role in the Regulation of Response to Biotic and Abiotic Stresses in Sugarcane. *Front. Plant Sci.* **2018**, *8*, 2262. [CrossRef]
57. Tao, J.; Wu, H.; Li, Z.; Huang, C.; Xu, X. Molecular Evolution of GDP-D-Mannose Epimerase (GME), a Key Gene in Plant Ascorbic Acid Biosynthesis. *Front. Plant Sci.* **2018**, *9*, 9. [CrossRef] [PubMed]
58. Maguire, M.; Coates, A.R.M.; Henderson, B. Chaperonin 60 unfolds its secrets of cellular communication. *Cell Stress Chaperones* **2002**, *7*, 317–329. [CrossRef]
59. Alavilli, H.; Lee, H.; Park, M.; Yun, D.-J.; Lee, B.-H. Enhanced multiple stress tolerance in Arabidopsis by overexpression of the polar moss peptidyl prolyl isomerase FKBP12 gene. *Plant Cell Rep.* **2017**, *37*, 453–465. [CrossRef] [PubMed]
60. Jain, S.; Kumar, D.; Jain, M.; Chaudhary, P.; Deswal, R.; Sarin, N.B. Ectopic overexpression of a salt stress-induced pathogenesis-related class 10 protein (PR10) gene from peanut (*Arachis hypogaea* L.) affords broad spectrum abiotic stress tolerance in transgenic tobacco. *Plant Cell Tissue Organ Cult. (PCTOC)* **2011**, *109*, 19–31. [CrossRef]

agronomy

MDPI

Article

Genetic Variation and Sequence Diversity of Starch Biosynthesis and Sucrose Metabolism Genes in Sweet Potato

Kai Zhang [1,2,3], Kai Luo [1,4], Shixi Li [1,2], Deliang Peng [1], Daobin Tang [1,2,3], Huixiang Lu [1], Yong Zhao [1,2,3], Changwen Lv [1,2,3] and Jichun Wang [1,2,3,*]

1 College of Agronomy and Biotechnology, Southwest University, Beibei, Chongqing 400715, China; kaizhang2013@gmail.com (K.Z.); luokai1112@126.com (K.L.); swulsx@163.com (S.L.); pdl20170608@163.com (D.P.); tdbin@swu.edu.cn (D.T.); lhx1991@swu.edu.cn (H.L.); zhyzhj35@swu.edu.cn (Y.Z.); lvcgwn@163.com (C.L.)

2 Key Laboratory of Biology and Genetic Breeding for Tuber and Root Crops in Chongqing, Beibei, Chongqing 400715, China

3 State Cultivation Base of Crop Stress Biology for Southern Mountainous Land of Southwest University, Beibei, Chongqing 400715, China

4 The Agricultural Science Research Institute of Liupanshui, Guizhou 553001, China

* Correspondence: wjchun@swu.edu.cn; Tel.: +86-23-68251264

Received: 27 March 2020; Accepted: 23 April 2020; Published: 29 April 2020

Abstract: Knowledge of genetic variations can provide clues into the molecular mechanisms regulating key crop traits. Sweet potato (*Ipomoea batatas* (L.) Lam.) is an important starch-producing crop, but little is known about the genetic variations in starch biosynthesis and sucrose metabolism genes. Here, we used high-throughput sequencing of pooled amplicons of target genes to identify sequence variations in 20 genes encoding key enzymes involved in starch biosynthesis and sucrose metabolism in 507 sweet potato germplasms. After filtering potential variations between gene copies within the genome, we identified 622 potential allelic single nucleotide polymorphisms (SNPs) and 85 insertions/deletions (InDels), including 50 non-synonymous SNPs (nsSNPs) and 12 frameshift InDels. Three nsSNPs were confirmed to be present in eight sweet potato varieties with various starch properties using cleaved amplified polymorphic sequence (CAPS) markers. Gene copy with loss of the fifth intron was detected in *IbAGPb3* genes, and loss of multiple introns were observed in *IbGBSS1-1* genes and various among germplasms based on intron length polymorphism (ILP) markers. Thus, we identified sequence variations between germplasms in 20 genes involved in starch biosynthesis and sucrose metabolism, and demonstrated the diversity in intron-loss alleles among sweet potato germplasms. These findings provide critical genetic information and useful molecular markers for revealing regulatory mechanism of starch properties.

Keywords: starch; SNP; InDel; CAPS; intron-loss; NGS; sweet potato

1. Introduction

The discovery of genetic variation is essential for revealing the molecular mechanism controlling important traits. Many sequence variations, such as single nucleotide polymorphisms (SNPs), insertions/deletions (InDels), and intron length polymorphisms (ILPs), have been identified in crops [1–4], providing comprehensive tools for analyzing the genome and identifying genes and genomic regions that contribute to phenotypes of interest. Many variations associated with disease resistance or important agronomic traits have been identified in crops [5,6]. Genetic markers developed from sequence variations have been used extensively for diverse genetic analyses, including genetic diversity assessment, trait association mapping, and fine-mapping of QTLs regulating important agronomic traits [7].

Sweet potato (*Ipomoea batatas* (L.) Lam.) is widely grown throughout the world and is critical for food security and nutrition due to its high yield, rich nutrient content, low input requirements, multiple uses, and adaptability under a range of environmental conditions [8–10]. Sweet potato yields a large amount of energy per unit area per unit time [11]. Starch levels in the storage root are 20% to 30% of the wet weight [12] and 50% to 80% of the dry weight [10]. The high starch content and reliable starch yield of sweet potato make it a good source of carbohydrates and, thus, an excellent raw material for starch-based industries and environmentally friendly ethanol biofuel production. Indeed, sweet potato may have an even greater potential than maize (*Zea mays*) as an ethanol source [11–13].

The quality and yield of carbohydrates and ethanol from sweet potato depend on starch properties such as yield, content, and composition. The starch content of the storage root influences post-harvest processing and ethanol yield [11,14]. The starch composition, particularly the amylose–amylopectin ratio, affects the starch structure and physicochemical properties of the starch [15] and also the ethanol yield [11]. A higher proportion of short amylopectin branched chains may lower the gelatinization temperature of starch and thereby help reduce energy consumption and CO_2 emission in ethanol biofuel production [12]. Thus, improving the starch content and starch quality remains an important goal, especially in the field of biotechnology [16]. Starch content and composition vary greatly among different sweet potato varieties [14,15,17], but the genetic basis and regulatory mechanism of these important traits is insufficiently understood.

Starch is synthesized through a complex biosynthetic pathway, and the properties of the starch are determined by the activities of several key starch biosynthetic and metabolic enzymes [18]. Natural genetic variations of genes encoding these key enzymes are associated with starch properties in crops. For example, Kharabian-Masouleh et al. [19] identified two SNPs in the Granule bound starch synthase I (GBSSI) gene—a G/T SNP at the exon1/intron1 boundary and a C/T SNP in exon 10—with a significant association with amylose content in rice (*Oryza sativa*). Candidate gene association mapping showed that rice amylose content was also associated with SNPs in starch synthases (SS). SNPs in sucrose synthase, starch-branching enzyme (SBE), adenosine diphosphate (ADP)-glucose pyrophosphorylase (AGPase) large subunit, α-amylase, and β-amylase (BMY) genes were associated with starch properties of maize (*Zea may*) kernels [20]. SNPs in the BMY gene *BMY-8/2*, the starch phosphorylase gene *PHO1b*, and the AGPase large (β) subunit gene *AGPaseS* were associated with tuber starch content and starch yield in potato (*Solanum tuberosum*) [21]. These sequence variations will provide a basis for further study of the regulation of starch-related traits.

In sweet potato, the key enzymes involved in starch biosynthesis and metabolism, and their associated genes, have been isolated and studied [10,18]. RNA interference studies confirmed that several of these genes directly regulate starch properties. The starch content of storage roots was reduced in transgenic sweet potato plants in which the expression of the starch branching enzyme II gene (*IbSBEII*) [22] or *GBSSI* [23] was suppressed by RNA interference. Overexpression of the soluble starch synthase I gene *IbSSI* resulted in a significant increase in starch content and granule size, as well as in the proportion of amylopectin [24]. Using CRISPR/Cas9 technology, Wang et al. [25] knocked out *IbGBSSI* and *IbSBEII* in sweet potato and, thereby, altered the starch content, amylopectin chain length distribution, and amylose percentage. These results demonstrated that starch biosynthesis genes could be exploited to improve the starch properties of sweet potato through biotechnology. However, the genetic variation in these genes, which may provide critical clues into the genetic basis of starch properties, has rarely been investigated.

Sequence variation in candidate genes can be mined by sequencing PCR amplicons from bulk DNA from a population or germplasm collection. The development of next-generation sequencing (NGS) technology and improved computational algorithms for sequence analysis, along with the availability of increasing sequence information in public databases, make it possible to identify all existing genetic variation in an entire genome and in targeted sequences of interest. The major advance offered by NGS methods is the ability to cheaply and reliably sequence DNA on a large scale and generate large volumes of sequence data [26,27]. Many genetic variations have been identified from amplicon, transcriptome, exome, and genome resequencing [1–3]. Su et al. [28] developed 795,794 SNP sites in sweet potato based on specific length amplification fragment sequencing (SLAF-seq) technology. Based on transcriptome data, SNPs were detected between two sweet potato varieties, Xushu 18 and Xu 781, and 32 SNP markers were verified in these two varieties using Tetra-primer Amplification Refractory Mutation System PCR (ARMS-PCR) [29]. Using double-digest restriction site-associated DNA sequencing (ddRAD-Seq), based on NGS technology, 94,361 SNPs were identified in an S1 population generated through self-pollination of Xushu 18, and a high-density genetic map of sweet potato was constructed [30].

Despite this progress, it remains a challenge to discover allelic variations in sweet potato, which is an allohexaploid and has a large, heterozygous genome [31]. Variations may arise both between allelic (homologous) sequences within individual subgenomes and between homoeologous sequences among subgenomes, in addition to paralogous variation between duplicated gene copies [32]. Deep sequencing of target genes or gene regions is an effective method to discover homologous variations, but potential paralogous variations between duplicated gene copies need to be eliminated.

In the present study, to discover sequence variations in genes encoding key enzymes regulating starch biosynthesis and metabolism in sweet potato, we performed deep sequencing of target sequences/genes in a large natural population with wide genetic and phenotypic diversity. SNPs, InDels, and ILPs were detected by aligning the reads to the reference sequences. Through alignment of all the gene copies explored in the hexaploid sweet potato genome, potential variations within the genome were filtered, and SNPs and InDels between germplasms were identified. CAPS and ILP markers were developed and verified in sweet potato germplasms. The sequence variations detected in this work will provide powerful tools for association analysis, marker-assisted selection, high-resolution genetic mapping, and research into regulatory mechanisms in sweet potato.

2. Materials and Methods

2.1. Plant Materials

To capture the widest possible variability in the gene sequences, 507 sweet potato germplasms, including varieties, breeding lines, wild varieties, farmer varieties, and landraces, were used (Supplementary Materials Table S1). These germplasms have a wide range of morphological types, and are derived from prominent sweet potato production regions in China and other countries, including Japan, Nigeria, Brazil, and the USA. These sweet potato germplasms were conversed by our lab and collected from other four institutions, and have been identified based on morphological characteristics, quality traits and molecular markers. Most of these germplasms have been reported in our previous genetic diversity and association analysis studies [14,33,34]. The sweet potato germplasms were planted in the experimental field of the Key Laboratory of Biology and Genetic Breeding for Tuber and Root Crops in Chongqing, China.

2.2. Sample Preparation and DNA Extraction

The fresh young leaves of each sweet potato germplasm were harvested from the experimental field and immediately stored in liquid nitrogen. Genomic DNA was extracted following the CTAB protocol [35]. The resulting DNA samples were examined by agarose gel electrophoresis, and the concentration and quality of RNA were determined with a NanoDrop ND-2000 Spectrophotometer (Thermo Scientific, Waltham, MA, USA) following the manufacturer's protocol. All DNA samples were normalized to 50 ng/μL concentration.

2.3. Genetic Diversity Analysis, Population Structure Analysis, and Starch Properties Evaluation

The genetic diversity and population structure of the 507 germplasms were analyzed based on inter-simple sequence repeat (ISSR) markers selected in our previous study [33]. Genomic DNA of each germplasm was used as template for PCR amplification using previously described method [33], and ISSR bands were used to assign loci for each primer and scored as present (1) or absent (0). The band presence/absence data matrix was analyzed using NTSYS pc2.10 [36] to estimate the Nei's standard genetic distance [37] between the tested germplasm. The genetic distance matrix was computed and clustered using the neighbor-joining (NJ) method [38] using MEGA X [39].

Population structure was assessed using the model-based method implemented in STRUCTURE v2.3.3 [40]. The number of subgroups (*K*) was set from 1 to 20 based on models characterized by admixture and correlated allele frequencies. For each K, five runs were performed separately, with 100,000 iterations carried out for each run after a burn-in period of 10,000 iterations. A *K* value was selected when the estimate of LnPr (*X*|*K*) peaked in the range of 1 to 20 subpopulations. Since the distribution of LnP (D) did not show a clear cut-off point for the true *K* value, an ad hoc measure, ΔK, was used to determine the numbers of subpopulations [41]. The run with the maximum likelihood was applied to subdivide the accessions into different subpopulations using a membership probability threshold of 0.55 as well as the maximum membership probability among subgroups. Those accessions with a membership probability of less than 0.55 were retained in the admixed group (AD). The results from STRUCTURE were displayed by DISTRUCT 1.1 software [42].

The starch properties, including the storage root starch content and amylose–amylopectin ratio, of these germplasms were evaluated using previously described methods [14].

2.4. Candidate Gene Selection

Genes encoding key enzymes involved in starch biosynthesis and metabolism, which have been confirmed to show different expression patterns in sweet potato varieties with different starch properties in our previous study [10], were considered as candidate genes.

2.5. Sequence Analysis and Primer Design

Candidate sequences of these genes were explored against National Center for Biotechnology Information (NCBI) (http://www.ncbi.nlm.nih.gov/) and previously reported sweet potato transcriptome sequencing data. The nucleotide sequences of each target gene were analyzed on Geneious 4.8.5 to determine the open reading frame (ORF), untranslated region (UTR), and exon and intron regions. The sequences, including complete coding sequences, were used for primer design. Forward and reverse primer sequences were up- and downstream of the ORF sequence, respectively, to ensure that the entire ORF was obtained in the PCR products. Primers were designed using Primer Premier 6 according to the manual.

2.6. Gene Cloning and Reference Sequence Determination

The candidate genes were cloned from 10 sweet potato germplasms, i.e., Yushu No.2, Xushu22, Shangqiu52-7, D01414, Yushu33, Chaoshu No.1, Xinxiang, Suyu No.1, Mianfen No.1, and S1-5, that had diverse starch content and quality properties, or from the other germplasms listed in Supplementary Materials Table S1, using PCR. TransStart FastPfu Fly DNA polymerase (TransGen Biotech, Beijing) was used to amplify sequences in PCR reactions, containing 50 ng genomic DNA, 1 μL of each forward and reverse primer (10 μm), 10 μL of 5×TransStart FastPfu Fly Buffer (50 μm), 4 μL dNTPs (10 mM), 1 μL TransStart FastPfu DNA polymerase, and ddH_2O to a final volume of 50 μL. The PCR reactions were conducted using a 9700 Thermal Cycler (ABI, USA) under the following cycle profile: 1 cycle of 95 °C for 2 min, followed by 35 cycles of 95 °C for 1 min, annealing temperature based on the Tm of the primers (Tm–5 °C) for 30 s, and 72 °C for 1–4 min. The final extension was performed at 72 °C for 10 min. The amplicons were cloned into pMD19-T vectors (TARAKA) and Sanger sequencing (Invitrogen, Shanghai) was performed to ensure that the target gene sequences were obtained.

2.7. DNA Equivalent Pooling

To obtain the gene sequences from 507 sweet potato germplasms, a uniform pooling strategy was applied for all samples. The genomic DNA of 507 sweet potato germplasms was divided into 25 individual pools based on starch content, containing equally mixed genomic DNA of 20–21 germplasms with various (high, medium, and low) levels of starch content each. The starch contents of germplasms in each pool were listed in Supplementary Materials Table S1. For each gene, PCR was carried out using each genomic DNA pool as template, respectively, as above. The concentration of PCR products from these pools was measured using a NanoDrop 2000c Spectrophotometer, and the final mega pool containing 5 μg of amplicons from each pool was prepared according to the sequencing manufacturer's protocol (Illumina, San Diego, CA).

2.8. Amplicon Sequencing

The pooled amplicons were subjected to Illumina paired-end (PE125) sequencing using the Hiseq2000 platform at Novogene Biotech, Beijing, China. Quality control, such as filtering and trimming, was performed using the DNASeq-QC software package, which includes DNASeq-*PrimerFilter*, DNASeq-*LowqFilter*, and DNASeq-*KHHNBaseFilter*, established by Novogene Bioinformatics Technology Co. Ltd, Beijing, China.

2.9. Genotype Calling and Variation Filtering

Data analysis, including reads assembly, filtering, trimming, and mapping to the reference sequences, was performed using CLC genomic workbench 7.5.1. The CLC genomic workbench general parameters were set as follows: the conflict resolution was changed into all four nucleotides (select A, C, G, and T), and nonspecific and masking references were ignored. The mismatch cost, insertion cost, deletion cost and length fraction for all of the paired end reads were 2, 3, and 3, respectively. The sequence data were assembled de novo with a sequence similarity of 0.8 over 0.5 of the read length. To minimize the bias introduced by PCR amplification, and errors caused by sequencing and read alignment, the variations were called with a minimum coverage of 10 and a minimum variant frequency of 20%. Thus, sequences that varied by less than 20% were excluded in this test, but the authenticity of the identified variations was ensured.

The reference sequences were aligned to the hexaploid sweet potato genome (http://public-genomes-ngs.molgen.mpg.de/SweetPotato/) [31] using BLASTN (E value cut-off of 1E-5) in Geneious Prime, and potential homologous, homoeologous, or paralogous genes of target genes were explored using an E value of 1e-5 and identity of above 85%. The sequence variations between these genes or gene copies were filtered.

2.10. Total Polymorphism Rate Calculation and Non-Synonymous SNP (nsSNP) Detection

The SNPs/InDels identified in the coding regions were detected, and SNPs/InDels that caused amino acid changes or reading frame shifts were analyzed. The total polymorphism rate and nsSNP rate were calculated as described by Kharabian-Masouleh et al. [1]. To predict the impact of the detected SNPs/InDels to enzyme activity or protein function, the conserved domain and site of each proteins encoded by the candidate genes were analyzed on InterProscan (http://www.ebi.ac.uk/interpro/search/sequence/) [43], and the likelihood of nsSNP to cause a functional impact on the protein was estimated using PANTHER (http://www.pantherdb.org/tools/csnpScoreForm.jsp) [44].

2.11. Marker Development and Identification

CAPS markers were developed using the online restriction/SNP-RFLP analysis tool Watcut (http://watcut.uwaterloo.ca). Eight sweet potato germplasms with various starch properties were used as templates and PCRs were performed using CAPS marker primers. The PCR products were digested using the corresponding restriction endonucleases and the products were compared. For InDel, primers were designed based on the Insert/Deletion region using Primer Premier 6. DNA extracted from the sweet potato germplasms was used as templates for PCRs using the InDel marker primers, and the length of PCR amplicons was detected using agarose or PAGE electrophoresis. Association analysis between marker and starch properties was performed as described [14].

2.12. ILP Marker Development and Identification

The sequences of each candidate gene and the corresponding mRNA sequences were aligned to examine the gene structure (number of introns and positions of splice sites). Putative ILPs among the obtained sequences were identified by aligning the entire sequences of available genes (gDNA) and their corresponding cDNA using Geneious Prime. ILP markers were developed and exon-primed intron crossing PCR (EPIC-PCR) primers were designed using Primer Premier 6. ILP markers were identified by PCR amplification on a population of 192 sweet potato germplasms, and the amplified products were isolated on an 8% non-denaturing PAGE gel. Silver staining was used to visualize DNA bands.

3. Results

3.1. The Sweet Potato Germplasms Exhibited High Genetic and Phenotypic Diversity

To evaluate the genetic diversity and structure of 507 sweet potato germplasms, the genetic distances between germplasms were calculated and the population structure was analyzed using ISSR markers. Based on the 216 polymorphic bands generated using 17 ISSR markers, the calculated Nei's genetic distances [37] among 507 sweet potato germplasms ranged from 0.1286 to 1.9869. Cluster analysis revealed eight subgroups and high genetic diversity in the tested sweet potato population (Figure 1a, Supplementary Materials Table S2).

Population structure analysis revealed the presence of ten main subpopulations within the 507 germplasms (Figure 1b). With membership probabilities of 0.55, 66, 44, 8, 31, 9, 54, 8, 23, 72, and 76 accessions were assigned into the ten subpopulations, respectively, and 116 germplasms exhibited various membership probabilities between the two subpopulations and were assigned to the admixed group (AD) (Figure 1c, Supplementary Materials Table S3).

The starch properties of the 507 sweet potato germplasms were tested; the storage root starch content ranged from 4.480% to 29.131% and the amylose–amylopectin ratio ranged from 0.247 to 0.429. However, no relationship was detected between starch properties and subgroup assignment in the cluster analysis, or subpopulation assignment in the population structure analysis on the 507 germplasms, based on the 17 ISSR markers. The storage root starch content of germplasms assigned to each subgroup and each subpopulation were listed in Supplementary Materials Table S2 and Table S3, respectively.

In summary, these results showed the 507 germplasms exhibited high levels of genetic diversity, complex population structure, and various starch properties, and sufficient genetic variations could be explored from this set of germplasms.

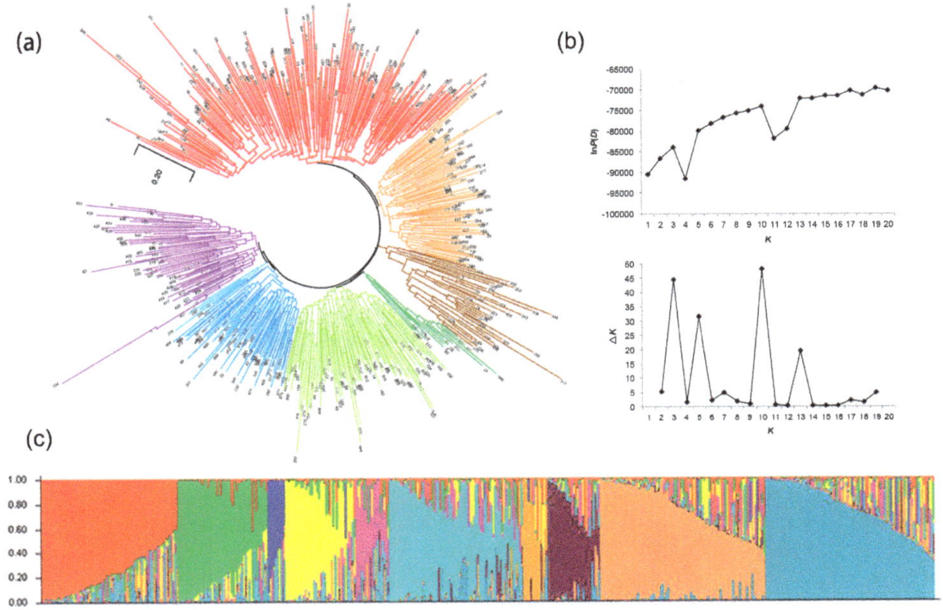

Figure 1. Genetic diversity and population structure analysis of 507 sweet potato germplasms. (**a**) The 507 germplasms could be divided into eight subgroups based on neighbor-joining (NJ) clustering. The different colored lines represent the subgroups inferred by NJ cluster analysis. (**b**) STRUCTURE estimation of the number of subpopulations for *K* ranging from 1 to 20 by LnP (D) did not show a clear cut-off point for the true *K* value, and delta *K* values (ΔK) indicating the presence of ten subpopulations; (**c**) Ten subpopulations in the 507 germplasms inferred from population structure analysis. The vertical coordinate of each subgroup indicates the membership coefficient of each individual, and the colors of the bar indicate the ten subpopulations identified using STRUCTURE.

3.2. Twenty Candidate Genes Were Captured for Variation Detection

By searching against the NCBI database and sweet potato transcriptome sequencing data, we collected sequences of genes that had previously been reported to encode key enzymes involved in starch biosynthesis and metabolism, and were differentially expressed in sweet potato germplasms with various starch properties [10]. The primers were designed based on the downloaded sequences and candidate genes were amplified using DNA extracted from 10 sweet potato varieties as templates. The primers used for PCR amplification of each gene are listed in Supplementary Materials Table S4.

Twenty genes were successfully amplified, and no unspecific PCR product was detected in the sequenced clones. The cloned sequences showed polymorphisms among different sweet potato germplasms. These genes encode key enzymes involved in starch granule formation, starch degradation, and starch and sucrose metabolism: (1) AGPase (EC 2.7.7.27) large (β) subunit 1 gene *IbAGPb1A*; (2) AGPase large (β) subunit 2 gene *IbAGPb1B*; (3) AGPase large (β) subunit 3 gene *IbAGPb2*; (4) AGPase large (β) subunit 4 gene *IbAGPb3*; (5) AGPase small (α) subunit 1 gene *IbAGPa1*; (6) AGPase small (α) subunit 2 gene *IbAGPa2*; GBSSI (EC 2.4.1.242) genes (7) *IbGBSS1-1*, (8) *IbGBSS1-2*, and (9) *IbSPSS67*; (10) granule-bound starch synthase 2 (GBSS2, EC 2.4.1.21) gene *IbGBSS2*; (11) soluble starch synthase (SSS, EC 2.4.1.21) gene *IbSSS1*; (12) starch-branching enzyme (SBE, EC 2.4.1.18) gene *IbSBE1*; (13) isoamylase (ISA, EC 3.2.1.68) gene *IbIsal*; (14) starch phosphorylase/α-1,4 glucan phosphorylase L isozyme (EC 2.4.1.1) gene *IbSP*; sucrose synthase (EC 2.4.1.13) genes (15) *IbSuSy1*, (16) *IbSuSy2*, and (17) *IbSuSy3*; and uridine diphosphate glucose dehydrogenase/UDP-glucose 6-dehydrogenase (UDPGH, EC 1.1.1.22) genes (18) *IbUDPGH3*, (19) *IbUDPGH*, and (20) *IbUDPGH13*.

IbSSS, which encodes soluble starch synthase in sweet potato, was first isolated based on our sweet potato transcriptome data [10], and we used the clone from sweet potato variety Shangqiu52-7 as the reference gene in this study. The gene names, accession numbers, and reference gene lengths are listed in Supplementary Materials Table S5 and the reference gene sequences are provided in Supplementary Materials File S1.

3.3. Number of Reads and Average Coverage Obtained from NGS

To detect sequence polymorphisms in sweet potato, the candidate genes were amplified from pooled DNA and NGS was performed on the pooled PCR products. Two pools, Mix1 and Mix2, were subjected to NGS; genes that generated insufficient reads were amplified again, pooled (Mix3), and sequenced. A total of 5.594, 5.038, and 1.715 Gb of raw data was obtained for Mix 1, 2, and 3, respectively, which yielded 5.496, 4.954, and 1.704 Gb of clean data. The sequencing data are deposited in the BIG Data Center under BioProject accession code PRJCA002386.

Sequencing of pooled amplicons of 20 candidate genes generated approximately 90,206,712 reads of clean data, of which 71,696,295 (79.48%) mapped to the reference sequences (Supplementary Materials Table S6). The highest and lowest numbers of reads were obtained for *IbAGPa1* and *IbSBE1*, with 13,734,507 and 912 reads, respectively. The average coverage of genes ranged from 28.56 to 1,458,236.09. The average coverage was above 6,000 for all target genes except *IbSBE1*. Thus, sufficient sequencing data were obtained for variation exploration.

3.4. Detection of SNPs and InDels

SNPs and single/multi-base InDels were detected in the target gene sequence by alignment with the reference genes. In total, we detected 1,113 SNPs and 85 InDels across the 20 studied genes. To further select allelic variation between germplasms, all of the potential homologous genes or gene copies of target genes were explored in the hexaploid sweet potato genome and aligned. For each gene, 1 to 37 potential homologous genes or gene copies were identified (Supplementary Materials Table S7), and 350 SNPs and 17 InDels, which were identified as being variations between genomic sequences, were filtered. Surprisingly, 145 variants were detected only in our cloned sequences but not in the genomic sequences. These variants were also filtered. Furthermore, six SNPs were further identified as PSVs by alignment of the gene sequences cloned from the same germplasm and filtered. A total of 44.115% and 31.765% of the SNPs and InDels were filtered from the variations identified from pooled gene amplicons, respectively. Finally, a total of 622 SNPs and 58 InDels were detected in the 20 gene sequences. The position and characteristics of the SNPs and InDels are listed in Supplementary Materials Table S8.

The SNPs and InDels rates in the detected gene sequences were 10.568 and 0.985/kb, respectively. SNP rates in intron, UTR, and coding sequences (CDS) were 7.136, 0.578, and 2.871 /kb, respectively, and the corresponding InDel rates were 0.731, 0.034, and 0.204/kb.

The number and distribution of SNPs and InDels in each gene are listed in Supplementary Materials Table S5. On average, the SNP rate ranged from 0 SNPs/kb (*IbAGPb1A*) to 26.792 SNPs/kb (*IbAGPa2*) for these candidate genes within this set of germplasms. Of the 622 SNPs identified, 169 (27.170% of the total) were located in coding regions, and of these, 50 SNPs were recognized as non-synonymous SNPs (nsSNPs) predicted to cause a change in amino acid sequence or ORF pre-termination. Most of the SNPs and InDels were detected in intron regions of the candidate gene sequences. We detected fewer polymorphisms in the 5′ and 3′ untranslated regions (UTRs) than in the ORF regions (Figure 2a, Supplementary Materials Table S5).

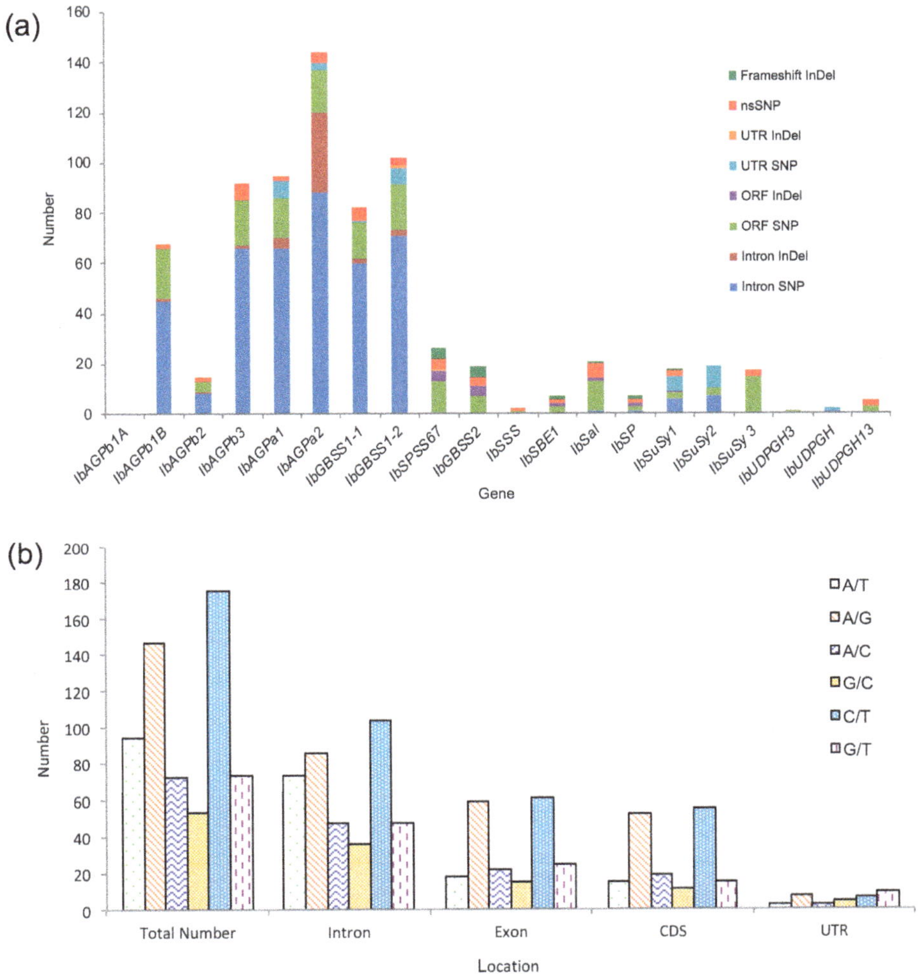

Figure 2. Summary of the single nucleotide polymorphisms (SNPs) and insertions/deletions (InDels) identified in this study. (**a**) Summary of the number of SNPs and InDels detected in the 20 genes. ORF, open reading frame; UTR, untranslated region. (**b**) Number of substitutions in the detected SNPs. CDS, coding sequence;

The C/T and A/G transitions were the top two frequent substitutions, which accounted for 28.502% and 23.941% of the detected SNPs, respectively. Except for transitions, a high proportion of A/C transversions was detected in CDS regions, which accounted for 11.377% of the SNPs detected in the CDS; and the G/T transversion accounted for 28.125% of detected substitutions in the CDS (Figure 2b).

The number of polymorphisms varied among the individual genes in our study. The AGPase large subunit genes *IbAGPb1B* and *IbAGPb3*, the small subunit genes *IbAGPa1* and *IbAGPa2*, and two *IbGBSS1* genes showed a high degree of polymorphism in the germplasm collection, with more than 100 SNPs and InDels detected in their sequences. By contrast, we detected zero SNPs in *IbAGPb1A*, three SNPs in *IbSP*, and one SNP in *IbSSS*, even though we obtained high sequence coverage for these genes (101,057.31, 137,132.69, and 104,531.88 respectively, Supplementary Materials Table S6).

Low sequence coverage might have contributed to the low level of variation detected in the *IbSBE1* and *IbSuSy3* gene sequences. The three UDP-glucose 6-dehydrogenase genes also showed a low level of sequence polymorphism in this study, although an average coverage of 42,870.36 to 72,770.73 was obtained for SNP calling in the three genes.

3.5. Non-Synonymous Substitutions were Identified in Starch Biosynthesis and Metabolism Genes

We next analyzed the non-synonymous substitutions we detected. Synonymous substitutions outnumbered non-synonymous substitutions in the ORF regions (Supplementary Materials Table S8). In the set of germplasms we analyzed, we detected 50 nsSNPs in the 20 candidate gene sequences (Table 1), 16 of which would cause a change in polarity at the corresponding amino acid position and thus might affect the structure or activity of the resulting protein.

Nineteen nsSNPs would cause amino acid change in conserved domain of the protein, but 16 of them would not cause functional impact on protein as estimated using PANTHER (Table 1). Three nsSNPs, the SNP detected in *IbAGPb2* (1,572 bp), in *IbSPSS67* (421 bp) *and in IbUDPGH13* (140 bp) would cause amino acid change in conserved domain of protein and possibly damage the function of enzyme as estimated using PANTHER. Two nsSNPs detected in *IbSP* (3,584 bp and 3,743 bp) would not cause amino acid change in the conserved domain or site, but possibly damage the function of enzyme as estimated using PANTHER.

Four SNPs, detected in the reference sequences of *IbGBSS1-1* (4,276 bp), *IbSBE1* (658bp), *IbSal1* (1,467 bp), and *IbUDPGH13* (1,318 bp), would cause premature termination of translation. SNP in *IbSBE1* (658bp) would result in a 170 aa amino acid sequence and loss of the conserved domains of SBE. The SNP detected in *IbSal1* (1,467 bp) would change the length of translated amino acid sequence from 786 to 484 aa, and lost the glycosyl hydrolase domain. The SNP in *IbUDPGH13* (1,318 bp) would change the sequence of a conserved domain. Furthermore, the SNP detected in *IbAGPb3* (2 bp) would cause loss of the first 10 aa in amino acid sequence, but would not affect the conserved domain sequences.

Table 1. Non-synonymous SNPs (nsSNPs) detected in the candidate gene sequences.

No.	Gene	Position	Nucleotide Substitution	Amino acid Change	Polarity Changed	Amino Acid Change in Conserved Domain
1	*IbAGPb1B*	2,992 bp/457 aa	C/G	His to Gln	No	
2		3,304 bp/507 aa	T/C	Val to Ala	No	
3	*IbAGPb2*	1,572 bp/205 aa	G/A	Asp to Asn	Yes	Yes
4		2,454 bp/340 aa	A/G	Try to His	No	Yes
5	*IbAGPb3*	2 bp/1 aa	T/A	Start codon		
6		29 bp/10 aa	C/G	Ala to Gly	No	
7		160 bp/54 aa	G/A	Gly to Ser	Yes	
8		163 bp/55 aa	G/A	Thr to Ala	Yes	
9		166 bp/56 aa	A/G	Lys to Glu	No	
10		1,526 bp/294 aa	C/A	Pro to Gln	Yes	Yes
11		2,511 bp/445 aa	T/A	Phe to Try	Yes	
12	*IbAGPa1*	125 bp/19 aa	G/T	Glu to Thr	Yes	
13		2,263 bp/310 aa	C/T	Ala to Val	No	
14	*IbAGPa2*	2,140 bp/286 aa	C/A	Phe to Leu	No	Yes
15		2,538 bp/342 aa	A/C	Gln to Pro	Yes	Yes
16		2,717 bp/376 aa	G/C	Phe to Leu	No	
17		2,971 bp/423 aa	G/C	Val to Leu	No[a]	
18	*IbGBSS1-1*	2,296 bp/217 aa	A/C	Lys to Asn	Yes	
19		4,276 bp	G/T	pre-termination		
20		4,280 bp/585 aa	T/G	Val to Gly	Yes	
21		4,283 bp/586 aa	C/A	Cys to Asp	Yes	
22		4,311 pb/595 aa	C/G	Asp to Glu	No	
23	*IbGBSS1-2*	1,372 bp/137aa	A/G	Ile to Val	No	Yes
24		1,892 bp/216aa	A/G	Ser to Gly	No	Yes
25		2,736 bp/376 aa	T/G	Val to Gly	Yes	

Table 1. *Cont.*

No.	Gene	Position	Nucleotide Substitution	Amino acid Change	Polarity Changed	Amino Acid Change in Conserved Domain
26	*IbSPSS67*	198 bp/59 aa	T/C	Leu to Pro	No	
27		421 bp/133 aaD	A/T	Glu to Asp	No	Yes
28		441 bp/140 aa	C/T	Pro to Leu	No	Yes
29		1,779 bp/586 aa	C/A	Ala to Asp	No	
30	*IbGBSS2*	779 bp/209 aa	T/C	Phe to Leu	No	Yes
31		797–798 bp/238 aa	AA/GG	Asn to Gly	Yes	Yes
32		915 bp /277 aa	T/A	Val to Glu	No	Yes
33		1,985 bp/634 aa	A/G	Ser to Gly	Yes	
34	*IbSSS*	691 bp/111 aa	G/A	Glu to Lys	No	
35	*IbSBE1*	658 bp	G/T	per-termination		
36		1,654 bp/493 aa	T/C	Ser to Pro	Yes	Yes
37	*IbSal*	139 bp/42 aa	A/G	Lys to Arg	No	
38		453 bp/147 aa	C/A	Gln to Lys	Yes	Yes
39		641–642 bp/210 aa	TA/GT	Thr to Ser	No	
40		898 bp/295 aa	A/G	Lys to Arg	No	Yes
41		1467 bp	G/T	per-termination		
42		1472 bp/486 aa	G/T	Trp to Cys	No	Yes
43	*IbSP*	3,584 bp/458 aa	T/C	Ser to Pro	Yes	
44		3,743 bp/511 aa	A/G	Lys to Glu	No	
45	*IbSuSy1*	94 bp/9 aa	C/A	Thr to Asn	No	
46		110 bp/12 aa	A/C	Gln to Pro	Yes	
47	*IbSuSy3*	911–912 bp/289 aa	GG/CC	Gly to Ala	No	Yes
48		1,280–1,281 bp/412 aa	GT/AC	Ser to Asn	No	Yes
49	*IbUDPGH13*	140 bp/38 aa	G/A	Arg to Gln	Yes	Yes
50		1,318 bp/431 aa	A/T	Lys to stop codon		

3.6. Development of CAPS Markers and Verification of SNPs

To further test the authenticity of these nsSNPs and to develop markers that could be used for further study, six of the nsSNPs were converted to CAPS markers (Table 2) and tested in eight sweet potato varieties with various starch properties. DNA of these sweet potato varieties was used as template for PCR amplification using the marker primers, and the PCR products were purified and digested with the appropriate restriction endonucleases.

Table 2. Cleaved amplified polymorphic sequence (CAPS) markers developed based on nsSNPs and tested in sweet potato germplasms.

Marker	Gene	Position	Base Change	Restriction Endonuclease	Primer Position	Primer Sequence
CAPS1	*IbAGPa1*	2263	C/T	*Btg*ZI	1875 2511	CGCTGGAGATCACCTATACCGAATGG CAGTGAGACTTCACATAGAGCTACTG
CAPS2	*IbAGPa2*	2538	A/C	*Hind*III	2242 2794	CCCTGGAGCCAATGACTTTGGAAGTG CACTGTCCGTGACATCAGCATCAAGC
CAPS3	*IbAGPa2*	2971	G/C	*Dde*I	2730 3274	GCTCCAATCTACACTCAGCCTCGATA GCCAATGCCAATTGGGATGCTGCCC
CAPS4	*IbGBSS1-1*	4283	C/A	*Bse*YI	3878 4327	GTTGCTTGTGCTCAGTGTGAAACTG CAAGTGGTGCAATTTCGTCTCCTTC
CAPS5	*IbSP*	3584	T/C	*Rsa*I	3407 3937	TGGAGTTATGAGCTGATGGAGAAGC ATGAATCTCGGCAACTCCATTTACA
CAPS6	*IbSuSy1*	110	A/C	*Eco*RII	6 470	TGTGACACCCGGGGAGCCTTCGTTCA GCCACGTGTACCTATTTCAAGCAAAC

As shown in Figure 3, no digestion products or polymorphic bands were produced when marker CAPS1 (Figure 3a,b), CAPS5 (Figure 3c,d), and CAPS6 (Figure 3e,f) were tested in the eight sweet potato varieties. There would be digestion products if the minor alleles existed, thus the results indicate there were no minor alleles, but only major alleles of *IbAGPa1*-T_{2263}, *IbSP*-T_{3584}, and *IbSuSy1*-A_{110} in the genomes of eight sweet potato varieties. For CAPS3 (Figure 3g,h), CAPS2 (Figure 3i,j), and CAPS4

(Figure 3k,l), we obtained PCR products and digestion products of the predicted size, and the bands were polymorphic among the eight sweet potato varieties. However, the DNA of some varieties yielded both digested and undigested products, indicating either incomplete digestion of the PCR products or the presence of both major and minor alleles in the same genome.

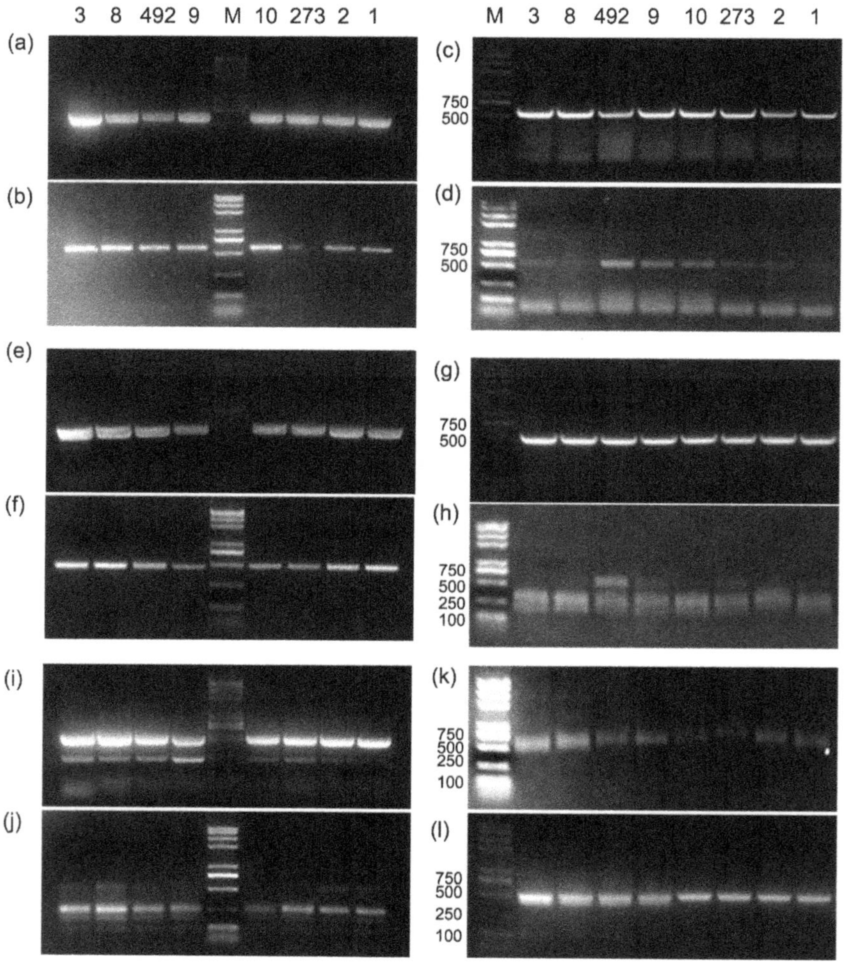

Figure 3. Test of CAPS markers in eight sweet potato germplasms. The numbers 3, 8, 492, 9, 10, 273, 2, and 1 represent the sweet potato germplasms shown in Supplementary Materials Table S1, which were Shangqiu52-7, Yushu33, Yanshu No.5, Chaoshu No.1, Xinxiang, Xiaohuaye, Mianfen No.1, and Suyu No.1, respectively. The measured starch contents of the eight germplasms were 4.476%, 25.611%, 11.036%, 10.206%, 19.584%, 14.728%, 29.386%, and 16.853%, respectively. The measured amylose–amylopectin ratio of the eight germplasms was 0.300, 0.308, 0.277, 0.309, 0.290, 0.282, 0.315, and 0.282, respectively. (**a**), (**c**), (**e**), (**g**), (**i**), and (**k**), PCR products obtained using CAPS1, CAPS5, CAPS6, CAPS3, CAPS2, and CAPS4 primers, respectively. (**b**), (**d**), (**f**), (**h**), (**j**) and (**l**), PCR products were digested using the corresponding restriction endonucleases. M, marker (Trans2K Plus II DNA marker, TransGen Biotech), the band size from top to bottom is 8000; 5000; 3000; 2000; 1000; 750; 500; 250; and 100 bp, respectively.

The result of CAPS3 detection confirmed the SNP in *IbAGPa2*-G_{2971}. InterProscan analysis predicted that the G2971/C2971 transversion in *IbAGPa2* would change Leu_{423} to Val_{423} in the

N-terminal domain interface of the protein. Even though this substitution results in no change in polarity, it could affect enzyme activity because of its location. However, no functional impact of this single amino acid polymorphism on the protein was detected using PANTHER, and the biological role of this variation need to be further investigated.

3.7. Frameshift InDels were Detected

InDels detected in the sequences of the 20 candidate genes mainly ranged from 1 to 10 bp in size. Most of the InDels were located in introns and would not be expected to significantly affect protein activity or function (Supplementary Materials Table S8). Twelve of the InDels were located in ORFs and could lead to frameshifts, premature termination of translation, or amino acid changes (Table 3) and thus to loss of function of the gene or protein.

Table 3. Frameshift InDels detected in the 20 gene sequences.

No.	Gene	Position (bp)	Deletion(D)/ Insertion(I)	Deletion/Insertion nucleotide	Influence on protein
1	*IbSPSS67*	188–189	I	C	Reading frame shift, and translation pre-terminate 64 bp downstream of the insertion site
2		234	D	A	Reading frame shift, and translation pre-terminate 32 bp downstream of the deletion site
3		437–438	I	C	Glu to Ala (139 aa), reading frame shift and translation pre-terminate 22 bp downstream of the insertion site
4		1776	D	T	Reading frame shift, amino acid changed and ORF become longer
5	*IbGBSS2*	681–682	I	A	Met to Ile (176 aa), and reading frame shift, and translation pre-terminate 30 bp downstream of the insertion site
6		873	D	A	Asp to Ala (240 aa), reading frame shift and translation pre-terminate 5 bp downstream of the deletion site
7		914	D	G	Val to Trp (254 aa), and reading frame shift, translation pre-terminate 51 bp downstream of the deletion site
8		915	D	T	Val to Gly (254 aa), and reading frame shift and translation pre-terminate 50 bp downstream of the deletion site
9	*IbSBE1*	1835	D	C	Ala to Val (553 aa), translation pre-terminate 32 bp downstream of the deletion site
10	*IbSal*	1981–1982	I	G	His to Gln, reading frame shift and translation pre-termination
11	*IbSP*	3580–3581	I	T	Amino acid changed from the insertion site and translation pre-terminate 26 bp downstream of the insertion site
12	*IbSuSy1*	96	I	T	End coding and translation pre-termination

The insertion at position 1982 bp in *Ibsal* causes a premature termination of translation and result in a protein without a glycosyl hydrolase domain, and the insertion at position 3581 bp in *IbSP* causes the change of amino acid sequence and would affect the conserved site. The deletion at position 96 bp in *IbSuSy1* result in a protein without conserved domain. Interestingly, a deletion at position 1,776 bp in *IbSPSS67* causes a shift in reading frame that would extend the ORF from 1827 to 2025 bp and the encoded protein from 608 to 674 amino acids. Although this frameshift would also change the amino acid sequence downstream of the deletion site, bioinformatics analysis showed that this sequence change is not located in conserved domains. Therefore, further research is needed to determine if this mutation would alter the activity or function of the enzyme.

3.8. Two Gene Forms were Detected in IbAGPb3

We detected a 79-bp InDel in the fifth exon of the *IbAGPb3* reference gene (Figure 4) that could potentially change the gene's structure. Gene structure analysis suggested that the 79-bp fragment is an intron; if indeed so, insertion of this fragment would change the number of exons in *IbAGPb3* from 13 to 14, and the number of introns from 12 to 13 (Figure 4a).

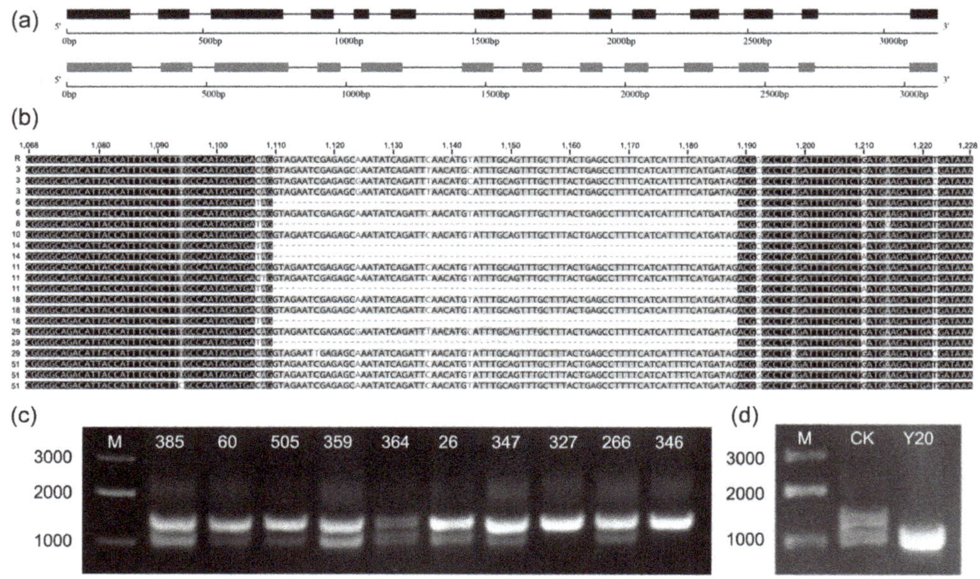

Figure 4. The two gene forms detected in the *IbAGPb3* gene sequence. (**a**) Gene structure analysis of a 79-bp insertion and deletion form of *IbAGPb3*. (**b**) Cloning and sequencing of the *IbAGPb3* sequences in sweet potato germplasms. R: reference sequence; the numbers 3, 6, 10, 11, 14, 18, 29, and 51 represent the sweet potato germplasms shown in Supplementary Materials Table S1. (**c**) and (**d**) Detection of the two gene forms in sweet potato germplasms. M: marker; the numbers represent the sweet potato germplasms shown in Supplementary Materials Table S1; Y20, sweet potato variety Yanshu No.20.

To verify the InDel, we designed primers to amplify the *IbAGPb3* gene sequence containing 79-bp InDel (Supplementary Materials Table S4), and cloned and sequenced this sequence from several sweet potato germplasms. We detected both 79-bp fragment inserted form and deleted form of *IbAGPb3* (Figure 4b). However, alignment of the cloned sequences showed that the two forms were two alleles. Except for the difference in the 79-bp sequences, the two alleles showed 95.679%–96.752% of sequence identities, indicating that the "delete" form might be the intron-loss gene copy with loss of the fifth intron of "inserted" form.

We next used PCR to screen the presence of the two gene forms in 126 sweet potato germplasms (Figure 4c shows the PCR products amplified from 10 of the germplasms). DNA from 81 germplasms showed both forms of the gene, 44 showed only the "inserted" form, and 1 (Yanshu No.20, Figure 4d) only the "deleted" form. We performed an association analysis of the presence/absence of the gene forms with starch content and composition, but detected no significant association.

3.9. Intron Loss in the *IbGBSS1-1* Genes

Abundant ILPs were detected in the 20 genes, especially in *IbGBSS1* genes. Surprisingly, we cloned a *IbGBSS1-1* gene with a length of 1893-bp that lacked all 13 introns from the genomic DNA of sweet potato variety Suyu No.1. To confirm that this allele actually exists in the plant, and did not occur because of errors in cloning or sequencing, we developed ILP markers to detect the intron-loss form in *IbGBSS1-1* genes (Table 4). Alleles lacking introns 1–4 and 7–13 could be detected in the Shangqiu 52-7 and 0929-106 genomes, but not in the D01414 and Sanheshu genomes (Figure 5 and Supplementary Materials Figure S1). Because we did not successfully develop markers to detect introns 5 and 6, the sequences containing the two introns were cloned from 23 sweet potato germplasms using primers FIbGBSS1-B and RIbGBSS1-B (Supplementary Materials Table S4). The sequences with intron loss were obtained in 15 of the germplasms (Supplementary Materials Figure S2). We further tested these ILP markers in 192 of the 507 sweet potato germplasms. Intron-loss alleles could be detected in 86–92 of the tested germplasms, indicating the intron-loss allele exists in the genomes of approximately half of the germplasms. Thus, there were intron-loss *IbGBSS1-1* genes in the sweet potato genome, and these intron losses were variations between germplasms.

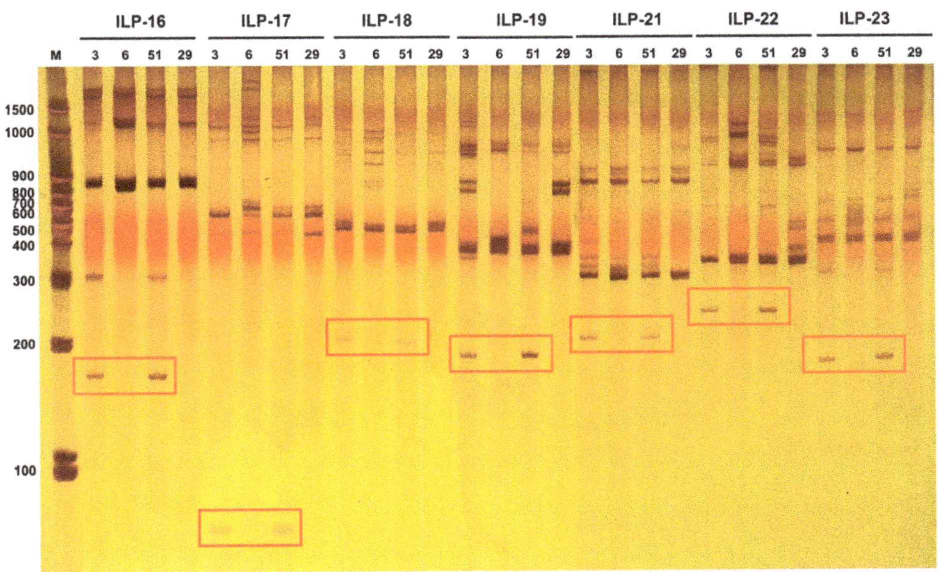

Figure 5. Detection of intron losses in *IbGBSS1-1*. The intron losses in *IbGBSS1-1* genes were detected in four sweet potato germplasms using the ILP markers shown in Table 4. M, marker. The numbers 3, 6, 51, and 29 represent the sweet potato germplasms shown in Supplementary Materials Table S1, namely, Shangqiu 52-7, D01414, 0929-106, and Sanheshu, respectively. The measured starch content of the four germplasms was 4.476%, 26.637%, 10.934%, and 22.360%, respectively. The measured amylose–amylopectin ratio of the four germplasms was 0.300, 0.291, 0.300, and 0.302, respectively. The bands in red boxes are the PCR products of intron-loss *IbGBSS1-1* genes.

Table 4. Intron length polymorphism (ILP) markers developed based on ILPs detected in *IbGBSS1-1*.

No.	Primer Name	Primer Sequence	Primer Length	Tm (°C)	Intron No. Detected	Start/Stop Site in Reference Sequence	Length of PCR Products without and with Intron Loss
1	FIbILP-16	CGTGCTTCCACACTCTTGCAGTAGCTG	27	69.16	1	412	662/166
	RIbILP-16	CCCCACTTTTGATTCTCCAGAAGTGGCA	27	67.54		1074	
2	FIbILP-17	GGACTTGGAGATGTTCTTGGAGGATTGCC	29	68.87	2	1294	444/77
	RIbILP-17	CGGGGACACACTGTCATAACTCTATGCC	28	69.01		1737	
3	FIbILP-18	GTACAAAGATGCTTGGGATACCTGTGTG	28	66.08	3 and 4	1747	392/196
	RIbILP-18	GTCCTTGTAATCTTTCCCAGCCTTGGG	27	67.64		2139	
4	FIbILP-19	CAACCAGTTGCGGTTCAGTTTGTTGTGCC	29	68.87	5 and 6	2139	357/173
	RIbILP-19	TCATGTAGATTCCTCTCGACTGGTACATGG	30	67.37		2496	
5	FIbILP-21	CACGGGTACTTGTAATGGAATGGATACCCA	30	67.37	9	2985	287/206
	RIbILP-21	TGTCTGAGCCTTTCTGCTCTTCAAGTCTG	29	67.45		3272	
6	FIbILP-22	GCAGAAAGGCTCAGACATTCTTTATGCTGC	30	67.37	10	3256	325/235
	RIbILP-22	GAGACCACACGGCTCAAATCTGCTCG	26	69.32		3581	
7	FIbILP-23	GAGCCGTGTGGTCTCTTTCAGTTGCA	26	67.75	11 and 12	3567	371/178
	RIbILP-23	CAGTGGTTATCACCTTCAGCACGTCCT	27	67.64		3938	
8	FIbILP-24	CACTGAAATGATCAAGAACTGCATGTCAC	29	64.62	13	3976	351/137
	RIbILP-24	CAAGTGGTGCAATTTCGTCTCCTTCAA	27	64.6		4327	

4. Discussion

4.1. Effective Strategies for Capturing and Identifying Allelic Variations in Hexaploid Sweet Potato

The large polyploid genome of sweet potato presents a challenge for genetic variation discovery, because variations are present within and between germplasms, and the potential presence of multiple homoeologous sequence variants (HSVs) and paralogous sequence variants (PSVs) will hinder the discovery of allelic variations [32]. In this study, we used several strategies to discover allelic variations. First, when we prepared the amplicons for sequencing, we normalized the PCR system to confirm that only target gene sequences were obtained. No non-specific PCR products were obtained in the clones from 10 sweet potato varieties, and the high mapping rate of reads to target genes also exhibited the specificity of amplicons used for NGS. Thus, the sequence variations were called from pooled amplicons of each target gene in 507 germplasms, and deep sequencing and alignments were performed only in target regions using a candidate gene sequence as reference. Second, except for strict quality control and criteria setting, additional filtering was performed based on alignment of all the potential gene copies explored from genome data. The within-genomic sequence variations, which were shown as variations between gene copies, were eliminated. Over 50% of variations were filtered at this step, indicating the high level of variations between gene copies in the individual genome.

However, due to the complexity and heterozygosity of the allohexaploid (B1B1B2B2B2B2) genome [31], it is challenging to distinguish homologous, homoeologous, and paralogous genes of target genes in the genome, and thus to discriminate all the HSVs and PSVs. Deeply re-sequencing of a hexaploid genome in multiple germplasms will help uncover additional sequence information and DNA sequence polymorphisms. Detailed information of each component of the allohexaploid genome would facilitate the identification of homozygous variations in sweet potato [2,45].

NGS data can have high error rates due to multiple factors, including base-calling and alignment errors [27]. In this study, several strategies were used to ensure the efficiency and accuracy of sequence variation detection. Pooling influences SNP discovery, because not each sample will be amplified with the same efficiency, even if the DNA samples are extracted using the same method, which may introduce bias when the pooling DNA is prepared [46]. We used small pools of DNA isolated from 20–21 germplasms each for the PCRs to minimize the chance of biased amplification of target genes and to improve the chance of generating amplicons from each germplasm. As each pool contained DNA isolated from germplasms with various starch contents, the chance of capturing variations between germplasms with different starch properties were also improved. Furthermore, the genes that did not generate enough amplicons were amplified and sequenced again. These approaches would produce lower false-positive rates in SNP screens [27]. Thus, these pooling, amplification, sequencing, calling, and filtering strategies might ensure the identification of allelic variations in sweet potato, and the variations identified in this study might serve as yardsticks in further variation discovery in sweet potato through genome re-sequencing.,

4.2. Characteristics of Gene Sequence Variation in Target Genes

Of the 622 SNPs and 58 InDels detected in this study, 169 SNPs (27.170%) and 12 InDels (20.690%) were located in coding regions. The density of SNPs and InDels was higher in introns than in UTRs and CDS in the 20 genes. All of the coding region InDels we detected cause frameshifts or premature termination. Of the coding region SNPs, 70.414% were synonymous and 29.586% were nsSNPs. Although the nsSNPs are clearly of interest because of their potential to change protein function or structure, increasing numbers of non-coding SNPs and synonymous SNPs are being identified as functionally critical in humans and plants [5,47,48]. Thus, the other SNPs we detected are also worth functional studies to further elucidate their effects on phenotype. Furthermore, we detected fewer polymorphisms in the 5′ and 3′ UTRs than in the CDS, perhaps because short sequences of UTRs were amplified in this study, and more SNPs and InDels might be detected in the entire UTRs sequences.

In this study, we obtained an SNP rate of 10.568/kb and an InDel rate of 0.985/kb in the 20 target genes, which are higher than the 4.31 SNPs/kb and 0.97 Indels/kb reported in starch-related genes of diploid rice using a similar method [1]. The high SNP rate for sweet potato could be attributed to multiple factors. First, we used 507 germplasms with high genetic and phenotypic diversity, which would be expected to provide abundant polymorphisms. Second, the high coverage ensured effective variation discovery. Furthermore, the high heterozygosity of sweet potato might contribute to an actual high SNP frequency, or to a high error rate in SNP calling, which would alter the SNP frequency [49]. Considering the large number of germplasms and high read coverage, we set the SNP parameters minimum counts and minimum frequency to 2 and 20%, respectively. Thus, alleles that were considered SNPs were represented in at least two independent sequences and had a frequency of or above 20%, and SNPs with a variant frequency of less than 20% were excluded, but the accuracy of SNP calling was high.

AGPase and GBSSI are critical enzymes in starch and sucrose metabolism in plants [21]. High levels of variations were detected in their encoding genes in this study, possibly due to the higher read coverage when compared with other key enzyme encoding genes. The difference among average coverage obtained for each target gene might be attributed to a difference in amplification efficiency of these genes. Interestingly, the subunit-encoding genes of AGPase showed different levels of sequence polymorphisms. The large subunit genes *IbAGPb1B* and *IbAGPb3* and the small subunit genes *IbAGPa1* and *IbAGPa2* showed high levels of sequence variation in the germplasm collection, and *IbAGPa2* and *IbAGPb3* had the highest SNP frequencies (26.792 and 26.266 SNPs/kb, respectively) and high number of nsSNPs among the 20 genes. By contrast, only 0 and 12 SNPs were detected in *IbAGPb1A* and *IbAGPb2*, respectively, although both genes were sequenced at high coverage (101,057.31 for *IbAGPb1A* and 61,443.71 for *IbAGPb2*, respectively, Supplementary Materials Table S6), indicating actual low sequence variation in these subunit genes. Although small number of SNPs was detected in *IbAGPb2*, the two nsSNPs detected in this gene would cause single amino acid change in the conserved domain, and one of them would influence protein function, indicating the importance of detected variations in *IbAGPb2*. As the subunits of AGPase had individual functions and expression patterns [50], they might be under different selection pressures, which could contribute to different degrees of sequence variation [51]. Our results provide a basis for further investigation and utilization of each subunit of AGPase in sweet potato.

4.3. CAPS Markers are An Effective Tool for SNP Genotyping in Sweet Potato

SNP genotyping is a major limitation for the comprehensive utilization of SNPs in crops. To address this problem, an efficient, low-cost, and versatile SNP genotyping method must be developed. We tried to genotype the detected SNPs using qPCR and MALDI-TOF mass spectrometry, but found these methods were not available for SNP genotyping in our sweet potato population, due to the polyploidy and high degree of heterozygosity of the sweet potato genome.

Since we did not find a suitable high-throughput SNP genotyping method for the allohexaploid sweet potato, we developed CAPS markers to examine the SNPs identified in this study. CAPS markers are based on PCR amplifications of DNA fragments with specific primers, followed by digestion with restriction endonucleases and separation of the products in an agarose gel. When compared with other PCR-based SNP genotyping methods, the restriction enzyme digestion step would eliminate errors caused by PCR amplification efficiency, and the products would be easier to separate through electrophoresis. As restriction endonucleases only recognize and digest specific sequences, the accuracy of genotyping would be markedly improved. Furthermore, the most important characteristics of CAPS marker is both homo- and heterozygotes could be easily recognized during genotyping [52], possibly rendering CAPS markers the most accurate and suitable method for SNP or InDel genotyping in sweet potato.

In this study, using six CAPS markers developed based on nsSNPs, genotyping was performed and SNP genotypes were easily distinguished among the different sweet potato germplasms. However, considering that not all of the SNPs could be converted into CAPS or derived CAPS (dCAPS) markers, and that it is time-consuming and challenging to perform high-throughput genotyping of the association population [52], using quantitative liquid chromatography-tandem mass spectrometry (LC-MSMS) method [53], or establishing an SNP chip/array through detecting and collecting a set of accurate SNPs might be the ideal strategies to perform high-throughput SNP genotyping of sweet potato.

Three developed CAPS in sweet potato germplasms did not show polymorphisms, possibly because of the minor alleles present in other germplasms used for polymorphism detection, but were not present in the eight germplasms used in the marker test. However, these SNPs were called with a minimum variant frequency of 20%, meaning that these variations should be present in no less than 20% of the 507 sweet potato germplasms. Thus, these markers should be polymorphic in a larger set of germplasms.

4.4. Creation of Intron-Loss Alleles Might Be a Characteristic Mechanism of Regulating Gene Expression in Sweet Potato

Intron losses were detected in *IbGBSS1-1* genes. An 1893-bp *IbGBSS1-1* gene without any of the 13 introns was isolated from sweet potato variety Suyu No.1 during cloning of reference sequences and was identified as an experimental error. However, a number of intron losses were detected in the sequencing data. We then cloned gene fragments from various germplasms, and designed ILP markers to identify intron-loss *IbGBSS1-1* alleles in various sweet potato germplasms. The results confirmed that intron loss might occur in each of the 13 introns, and that the intron-loss alleles were present in some germplasms but absent in others.

We also detected a potential intron-loss gene copy of *IbAGPb3*, and this gene copy was also shown to be present in some germplasms but absent in others. However, different with that in *IbGBSS1-1* genes, the intron loss only occurred in the fifth intron but not exhibited in other intron regions. Because association analysis showed no significant association between the presence/absence of this gene copy and starch properties, the biological relevance of loss of the fifth intron and the diversity of gene copies among sweet potato germplasms remains to be determined.

Intron use is an important component of genome adaptation, and intron gain and loss are the results of genome responses to strong selective pressures. Introns could be lost by exact genomic deletion, or by gene double recombination with a reverse-transcribed copy [54]. The all intron-loss allele might be a new gene created by reverse transcription of mRNA followed by insertion of this cDNA into the genome [54]. Introns have been shown to increase the transcriptional efficiency of genes and enhance gene expression [55,56], but increase the time needed for transcription, and intron-less alleles could be transcribed faster [56].

Genes that are strongly expressed and expressed in all tissues tend to have short introns in humans [56], and intron losses occur preferentially in highly expressed housekeeping genes in mammals [57]. Although *IbGBSS1-1* is not a housekeeping gene, it plays key roles in starch biosynthesis in plants, and our previous transcriptome and qRT-PCR analysis demonstrated that *IbGBSS1-1* (the unigene comp84815_c0_seq1 shown in Zhang et al., [10]) was expressed at very high levels in sweet potato. We speculate that intron loss is an important mechanism affecting *IbGBSS1-1* gene expression through regulating the efficiency, expense, and time of transcription.

However, we did not identify multiple intron losses in other target genes, which also showed high levels of expression in our previous study [10], the function of all intron-loss *IbGBSS1-1* alleles and the biological meaning of multiple intron losses in *IbGBSS1-1* genes require further investigation. Nevertheless, these discoveries indicated that the creation of intron-loss alleles might be a characteristic mechanism of regulating gene expression in sweet potato genome and should be emphasized in the further studies.

4.5. The Impact of Genetic Variations to Phenotype in Allohexaploid Sweet Potato

In this study, we detected genetic variations in 20 genes. It should be mentioned that there might be approximate six copies of each gene in the allohexaploid (B1B1B2B2B2B2) sweet potato genomes [31,58], and our results also showed 1–37 potential homologs or paralogs of each studied gene in the genome database. During the formation and evolution of polyploid genomes, the duplicated genes might exhibit expressional, regulatory or functional divergence [53,59]. Thus, the impact of variations in a specific gene copy on the phenotype might be determined by the contribution of this gene copy to the phenotype. The SNP or InDel detected in this study might directly affect starch properties of germplasms, if the gene copy is the major one controlling the phenotype, and this variation cause a change in the gene expression or function. Otherwise, if the gene copy is not the major loci controlling starch properties, or its function can be compensated by other gene copy [60], then, the variations detected in this gene copy probably not impact the starch properties. To reveal the effect of single gene copy and single genetic variation to the phenotype will help to elucidate the genetic basis and regulatory mechanism of starch properties in sweet potato.

Furthermore, our results showed that there were a number of multiple copy genes in the sweet potato genome, and some are unusual alleles, such as the intron-loss alleles. Given the potential functional divergence among genes, the orthologous and paralogous genes, and also unusual alleles, should be considered in gene function studies or genetic engineering efforts in sweet potato.

Supplementary Materials: The following are available online at http://www.mdpi.com/2073-4395/10/5/627/s1, Figure S1: Detection of the 13th intron loss in *IbGBSS1-1* genes in four sweet potato germplasms using the ILP marker ILP-24 shown in Table 4. M, marker. The numbers 3, 6, 51, and 29 in the figure represent the sweet potato germplasms shown in Table S1, namely, Shangqiu 52-7, D01414, 0929-106, and Sanheshu, respectively. Figure S2: The gene sequences cloned from 23 sweet potato germplasms exhibited the 5th and 6th intron losses of *IbGBSS1-1* genes in some of the germplasms. R, the reference gene sequence. The numbers present the sweet potato germplasms shown in Table S1. Table S1: The sweet potato germplasms used in this study. Table S2: Cluster analysis of 507 sweet potato germplasms. Table S3: Subpopulation assignment of the sweet potato germplasms in the population structure analysis. Table S4: Primers used for gene amplification. Table S5: Variation analysis of 20 genes. Table S6. Summary of statistics of reads mapping to the reference sequences. Table S7: Number of potential homologous gene copies of 20 genes identified on each pseudochromosome in the sweet potato genome. Table S8: SNPs and InDels detected in this study. File S1: Reference gene sequences.

Author Contributions: Methodology, K.Z.; software, K.Z., K.L., and H.L.; validation, K.L., S.L., and D.P.; formal analysis, K.Z. and K.L.; investigation, K.Z.; resources, D.T., C.L., and Y.Z.; data curation, K.Z.; writing—original draft preparation, K.Z.; writing—review and editing, K.Z.; visualization, K.Z.; supervision, J.W.; project administration, K.Z.; funding acquisition, C.L. and J.W. All authors have read and agreed to the published version of the manuscript.

Funding: This research was funded by National Key Research and Development Plan (2018YFD1000705, 2018YFD1000700), Fundamental Research Funds for the Central Universities (XDJK2020B032) and the Technology Innovation Fund of Chongqing (cstc2019jscx-msxmX0326).

Acknowledgments: We thank Nanchong Institute of Agricultural Sciences, Chongqing Three Gorges Academy of Agricultural Sciences, Jingsu Xuzhou Sweet potato Research Center and Chongqing Sweet Potato Engineering and Technology Research Center for plant materials.

Conflicts of Interest: The authors declare no conflict of interest. The funders had no role in the design of the study; in the collection, analyses, or interpretation of data; in the writing of the manuscript, or in the decision to publish the results.

References

1. Kharabian-Masouleh, A.; Waters, D.L.E.; Reinke, R.F.; Henry, R.J. Discovery of polymorphisms in starch-related genes in rice germplasm by amplification of pooled DNA and deeply parallel sequencing. *Plant Biotech. J.* **2011**, *9*, 1074–1085. [CrossRef] [PubMed]
2. Lai, K.; Duran, C.; Berkman, P.J.; Lorenc, M.T.; Stiller, J.; Manoli, S.; Hayden, M.J.; Forrest, K.L.; Fleury, D.; Bumann, U.; et al. Single nucleotide polymorphism discovery from wheat next-generation sequence data. *Plant Biotechnol. J.* **2012**, *10*, 743–749. [CrossRef] [PubMed]

3. Lu, K.; Wei, L.; Li, X.; Wang, Y.; Wu, J.; Liu, M.; Zhng, C.; Chen, Z.; Xiao, Z.; Jian, H.; et al. Whole-genome resequencing reveals *Brassica napus* origin and genetic loci involved in its improvement. *Nat. Commun.* **2019**, *10*, 1154. [CrossRef] [PubMed]
4. Jayaswall, K.; Sharma, H.; Bhandawat, A.; Sagar, R.; Yadav, V.K.; Sharma, V.; Mahajan, V.; Roy, J.; Singh, M. Development of intron length polymorphic (ILP) markers in onion (*Allium cepa* L.), and their cross-species transferability in garlic (*A. sativum* L.) and wild relatives. *Genet. Resour. Crop. Evol.* **2019**, *66*, 1379–1388. [CrossRef]
5. Li, W.; Zhu, Z.; Chern, M.; Yin, J.; Yang, C.; Ran, L.; Cheng, M.; He, M.; Wang, K.; Wang, J.; et al. A natural allele of a transcription factor in rice confers broad-spectrum blast resistance. *Cell* **2017**, *170*, 114–126. [CrossRef] [PubMed]
6. Li, F.; Chen, B.; Xu, K.; Wu, J.; Song, W.; Bancroft, I.; Harper, A.L.; Trick, M.; Liu, S.; Gao, G.; et al. Genome-wide association study dissects the genetic architecture of seed weight and seed quality in rapeseed (*Brassica napus* L.). *DNA Res.* **2014**, *21*, 355–367. [CrossRef]
7. Mammadov, J.; Aggarwal, R.; Buyyarapu, R.; Kumpatla, S. SNP markers and their impact on plant breeding. *Int. J. Plant Genom.* **2012**, *2012*, 728398. [CrossRef]
8. Burri, B.J. Evaluating sweet potato as an intervention food to prevent vitamin A deficiency. *Compr. Rev. Food Sci. Food Saf.* **2011**, *10*, 118–130. [CrossRef]
9. Mitra, S. Nutritional status of orange-fleshed sweet potatoes in alleviating Vitamin A malnutrition through a food-based approach. *J. Nutr. Food Sci.* **2012**, *2*, 8. [CrossRef]
10. Zhang, K.; Wu, Z.; Tang, D.; Luo, K.; Lu, H.; Liu, Y.; Dong, J.; Wang, X.; Lv, C.; Wang, J.; et al. Comparative transcriptome analysis reveals critical function of sucrose metabolism related-enzymes in starch accumulation in the storage root of sweet potato. *Front. Plant Sci.* **2017**, *8*, 914. [CrossRef]
11. Nedunchezhiyan, M.; Byju, G.; Jata, S.K. Sweet Potato Agronomy. *Fruit Veg. Cereal. Sci. Biotech.* **2012**, *6*, 1–10.
12. Srichuwong, S.; Orikasa, T.; Matsuki, J.; Shiina, T.; Kobayashi, T.; Tokuyasu, K. Sweet potato having a low temperature-gelatinizing starch as a promising feedstock for bioethanol production. *Biomass Bioenergy* **2012**, *39*, 120–127. [CrossRef]
13. Koçar, G.; Civaş, N. An overview of biofuels from energy crops: Current status and future prospects. *Renew. Sustain. Energy Rev.* **2013**, *28*, 900–916. [CrossRef]
14. Zhang, K.; Wu, Z.; Tang, D.; Lv, C.; Luo, K.; Zhao, Y.; Liu, X.; Huang, Y.; Wang, J. Development and identification of SSR markers associated with starch properties and β-Carotene content in the storage root of sweet potato (*Ipomoea batatas* L.). *Front. Plant Sci.* **2016**, *7*, 223. [CrossRef] [PubMed]
15. Zhou, W.; Yang, J.; Hong, Y.; Liu, G.; Zheng, J.; Gu, Z.; Zhang, P. Impact of amylose content on starch physicochemical properties in transgenic sweet potato. *Carbohydr. Polym.* **2015**, *122*, 417–427. [CrossRef] [PubMed]
16. Ren, Z.; He, S.; Zhao, N.; Zhai, H.; Liu, Q. A sucrose non-fermenting-1-related protein kinase-1 gene, IbSnRK1, improves starch content, composition, granule size, degree of crystallinity and gelatinization in transgenic sweet potato. *Plant Biotechnol. J.* **2019**, *17*, 21–32. [CrossRef]
17. Tumwegamire, S.; Kapinga, R.; Rubaihayo, P.R.; Labonte, D.R.; Grüneberg, W.J.; Burgos, G.; Felde, T.Z.; Carpio, R.; Pawelzik, E.; Mwanga, R.O.M. Evaluation of dry Matter, protein, starch, sucrose, β-carotene, iron, zinc, calcium, and magnesium in East African sweetpotato [*Ipomoea batatas* (L.) Lam] germplasm. *HortScience* **2011**, *46*, 348–357. [CrossRef]
18. Lai, Y.C.; Wang, S.Y.; Gao, H.Y.; Nguyen, K.M.; Nguyen, C.H.; Shih, M.C.; Lin, K.H. Physicochemical properties of starches and expression and activity of starch biosynthesis-related genes in sweet potatoes. *Food Chem.* **2016**, *199*, 556–564. [CrossRef]
19. Kharabian-Masouleh, A.; Waters, D.L.E.; Reinke, R.F.; Ward, R.; Henry, R.J. SNP in starch biosynthesis genes associated with nutritional and functional properties of rice. *Sci. Rep.* **2012**, *2*, 557. [CrossRef]
20. Cook, J.P.; McMullen, M.D.; Holland, J.B.; Tian, F.; Bradbury, P.; Ross-Ibarra, J.; Buckler, E.S.; Flint-Garcia, S.A. Genetic architecture of maize kernel composition in the nested association mapping and inbred association panels. *Plant Physiology* **2012**, *158*, 824–834. [CrossRef]
21. Schreiber, L.; Nader-Nieto, A.C.; Schönhals, E.M.; Walkemeier, B.; Gebhardt, C. SNPs in genes functional in starch-sugar interconversion associate with natural variation of tuber starch and sugar content of potato (*Solanum tuberosum* L.). *G3 Genes Genomes Genet.* **2014**, *4*, 1797–1811. [CrossRef] [PubMed]

22. Shimada, T.; Otani, M.; Hamada, T.; Kim, S.H. Increase of amylose content of sweetpotato starch by RNA interference of the starch branching enzyme II gene (*IbSBEII*). *Plant Biotech.* **2006**, *23*, 85–90. [CrossRef]
23. Otani, M.; Hamada, T.; Katayama, K.; Kitahara, K.; Kim, S.H.; Takahata, Y.; Kim, S.-H.; Yasuhiro Takahata, Y.; Suganuma, T.; Shimada, T. Inhibition of the gene expression for granule-bound starch synthase I by RNA interference in sweet potato plants. *Plant Cell Rep.* **2007**, *26*, 1801–1807. [CrossRef] [PubMed]
24. Wang, Y.; Li, Y.; Zhang, H.; Zhai, H.; Liu, Q.; He, S. A soluble starch synthase I gene, *IbSSI*, alters the content, composition, granule size and structure of starch in transgenic sweet potato. *Sci. Rep.* **2017**, *7*, 2315. [CrossRef]
25. Wang, H.X.; Wu, Y.L.; Zhang, Y.D.; Yang, J.; Fan, W.J.; Zhang, H.; Zhao, S.S.; Yuan, L.; Zhang, P. CRISPR/Cas9-based mutagenesis of starch biosynthetic genes in sweet potato (*Ipomoea Batatas*) for the improvement of starch quality. *Int. J. Mol. Sci.* **2019**, *20*, 4702. [CrossRef]
26. Metzker, M.L. Sequencing technologies—The next generation. *Nat. Rev. Genet.* **2010**, *11*, 31–46. [CrossRef]
27. Nielsen, R.; Paul, J.S.; Albrechtsen, A.; Song, Y.S. Genotype and SNP calling from next-generation sequencing data. *Nat. Rev. Genet.* **2011**, *12*, 443–451. [CrossRef]
28. Su, W.J.; Zhao, N.; Lei, J.; Wang, L.J.; Chai, S.S.; Yang, X.S. SNP sites developed by specific length amplification fragment sequencing (SLAF-seq) in Sweetpotato. *Sci. Agric. Sinica* **2016**, *49*, 27–34.
29. Kou, M.; Xu, J.; Li, Q.; Liu, Y.; Wang, X.; Tang, W.; Yan, H.; Zhang, Y.G.; Ma, D.F. Development of SNP markers using RNA-seq technology and tetra-primer ARMS-PCR in sweetpotato. *J. Integr. Agric.* **2017**, *16*, 464–470. [CrossRef]
30. Shirasawa, K.; Tanaka, M.; Takahata, Y.; Ma, D.F.; Cao, Q.H.; Liu, Q.C.; Zhai, H.; Kwak, S.-S.; Jeong, J.C.; Cheol, J.; et al. A high-density SNP genetic map consisting of a complete set of homologous groups in autohexaploid sweetpotato (*Ipomoea batatas*). *Sci. Rep.* **2017**, *7*, 44207. [CrossRef]
31. Yang, J.; Moeinzadeh, M.H.; Kuhl, H.; Helmuth, J.; Xiao, P.; Haas, S.; Liu, G.; Zheng, J.L.; Sun, Z.; Weijuan Fan, W.J.; et al. Haplotype-resolved sweet potato genome traces back its hexaploidization history. *Nat. Plants* **2017**, *3*, 696–703. [CrossRef] [PubMed]
32. Kaur, S.; Francki, M.G.; Forster, J.W. Identification, characterization and interpretation of single-nucleotide sequence variation in allopolyploid crop species. *Plant Biotech. J.* **2012**, *10*, 125–138. [CrossRef] [PubMed]
33. Zhang, K.; Wu, Z.D.; Li, Y.H.; Zhang, H.; Wang, L.P.; Zhou, Q.L.; Tang, D.B.; Fu, Y.F.; He, F.F.; Jiang, Y.C.; et al. ISSR-based molecular characterization of an elite germplasm collection of sweet potato (*Ipomoea batatas* L.) in China. *J. Integr. Agric.* **2014**, *13*, 2346–2361. [CrossRef]
34. Luo, K.; Lu, H.X.; Wu, Z.D.; Wu, X.L.; Yin, W.; Tang, D.B.; Wang, J.C.; Zhang, K. Genetic diversity and population structure analysis of main sweet potato breeding parents in southwest China. *Sci. Agric. Sinica* **2016**, *49*, 593–608.
35. Kim, S.H.; Hamada, T. Rapid and reliable method of extracting DNA and RNA from sweetpotato, *Ipomoea batatas* (L). Lam. *Biotechnol. Lett.* **2005**, *27*, 1841–1845. [CrossRef]
36. Rohlf, F.J. *NTSYS-pc: Numerical Taxonomy and Multivariate Analysis System, Version 2.1. Exeter Software*; Setauket: New York, NY, USA, 2002.
37. Nei, M. Genetic distance between populations. *Am. Nat.* **1972**, *106*, 283–292. [CrossRef]
38. Saitou, N.; Nei, M. The neighbor-joining method: A new method for reconstructing phylogenyetic trees. *Mol. Biol. Evol.* **1987**, *4*, 406–425.
39. Kumar, S.; Stecher, G.; Li, M.; Knyaz, C.; Tamura, K.; Notes, A. MEGA X: Molecular evolutionary genetics analysis across computing platforms. *Mol. Biol. Evol.* **2018**, *35*, 1547–1549. [CrossRef]
40. Hubisz, M.J.; Falush, D.; Stephens, M.; Pritchard, J.K. Inferring weak population structure with the assistance of sample group information. *Mol. Ecol. Resour.* **2009**, *9*, 1322–1332. [CrossRef]
41. Evanno, G.; Regnaut, S.; Goudet, J. Detecting the number of clusters of individuals using the software STRUCTURE: A simulation study. *Mol. Ecol.* **2005**, *14*, 2611–2620. [CrossRef]
42. Rosenberg, N.A. DISTRUCT: A program for the graphical display of population structure. *Mol. Ecol Notes* **2004**, *4*, 137–138. [CrossRef]
43. Jones, P.; Binns, D.; Chang, H.Y.; Fraser, M.; Li, W.; McAnulla, C.; McWilliam, H.; Maslen, J.; Mitchell, A.; Nuka, G.; et al. InterProScan 5: Genome-scale protein function classification. *Bioinformatics* **2014**, *30*, 1236–1240. [CrossRef] [PubMed]
44. Tang, H.; Thomas, P.D. PANTHER-PSEP: Predicting disease-causing genetic variants using position-specific evolutionary preservation. *Bioinformatics* **2016**, *32*, 2230–2232. [CrossRef] [PubMed]

45. Isobe, S.; Shirasawa, K.; Hirakawa, H. Current status in whole genome sequencing and analysis of *Ipomoea* spp. *Plant Cell Rep.* **2019**, *38*, 1365–1371. [CrossRef] [PubMed]
46. Ingman, M.; Gyllensten, U. SNP frequency estimation using massively parallel sequencing of pooled DNA. *Eur. J. Hum. Genet.* **2009**, *17*, 383–386. [CrossRef] [PubMed]
47. Madelaine, R.; Notwell, J.H.; Skariah, G.; Halluin, C.; Chen, C.C.; Bejerano, G.; Mourrain, P. A screen for deeply conserved non-coding GWAS SNPs uncovers a MIR-9-2 functional mutation associated to retinal vasculature defects in human. *Nucleic Acids Res.* **2018**, *46*, 3517–3531. [CrossRef]
48. Deng, N.; Zhou, H.; Fan, H.; Yuan, Y. Single nucleotide polymorphisms and cancer susceptibility. *Oncotarget* **2017**, *8*, 110635–110649. [CrossRef]
49. Hyma, K.E.; Barba, P.; Wang, M.; Londo, J.P.; Acharya, C.B.; Mitchell, S.E.; Sun, Q.; Reisch, B.; Cadle-Davidson, L. Heterozygous mapping strategy (HetMappS) for high resolution genotyping-by-sequencing markers: A case study in grapevine. *PLoS ONE* **2015**, *10*, 0134880. [CrossRef]
50. Mugford, S.T.; Fernandez, O.; Brinton, J.; Flis, A.; Krohn, N.; Encke, B.; Regina Feil, R.; Sulpice, R.; Lunn, J.E.; Stitt, M.; et al. Regulatory properties of ADP glucose pyrophosphorylase are required for adjustment of leaf starch synthesis in different photoperiods. *Plant Physiol.* **2014**, *166*, 1733–1747. [CrossRef]
51. Henry, R.J.; Nevo, E. Exploring natural selection to guide breeding for agriculture. *Plant Biotechnol. J.* **2014**, *12*, 655–662. [CrossRef]
52. Shavrukov, Y.N. CAPS markers in plant biology. *Russ. J. Genet. Appl. Res.* **2016**, *6*, 279–287. [CrossRef]
53. van Wesemael, J.; Hueber, Y.; Kissel, E.; Campos, N.; Swennen, R.; Carpentier, S. Homeolog expression analysis in an allotriploid non-model crop via integration of transcriptomics and proteomics. *Sci. Rep.* **2018**, *8*, 1353. [CrossRef] [PubMed]
54. Deshmukh, R.K.; Sonah, H.; Singh, N.K. Intron gain, a dominant evolutionary process supporting high levels of gene expression in rice. *J. Plant Biochem. Biotechnol.* **2016**, *25*, 142–146. [CrossRef]
55. Hir, H.L.; Nott, A.; Moore, M.J. How introns influence and enhance eukaryotic gene expression. *TRENDS Biochem. Sci.* **2003**, *28*, 215–220. [CrossRef]
56. Jeffares, D.C.; Mourier, T.; Penny, D. The biology of intron gain and loss. *TRENDS Genet.* **2006**, *22*, 16–22. [CrossRef] [PubMed]
57. Mourier, T.; Jeffares, D.C. Eukaryotic Intron Loss. *Science* **2003**, *300*, 1393. [CrossRef]
58. Gao, M.; Soriano, S.F.; Cao, Q.H.; Yang, X.S.; Lu, G.Q. Hexaploid sweetpotato (*Ipomoea batatas* (L.) Lam.) may not be a true type to either auto- or allopolyploid. *PLoS ONE* **2020**, *15*, e0229624. [CrossRef]
59. Fan, Y.H.; Yu, M.N.; Liu, M.; Zhang, R.; Sun, W.; Qian, M.C.; Duan, H.C.; Chang, W.; Ma, J.Q.; Qu, C.M.; et al. Genome-wide identification, evolutionary and expression analyses of the *GALACTINOL SYNTHASE* gene family in rapeseed and tobacco. *Int. J. Mol. Sci.* **2017**, *18*, 2768. [CrossRef]
60. Gu, Z.; Steinmetz, L.M.; Gu, X.; Scharfe, C.; Davis, R.W.; Li, W.-H. Role of duplicate genes in genetic robustness against null mutations. *Nature* **2003**, *421*, 63–66. [CrossRef]

agronomy

MDPI

Article

Development of a Multipurpose Core Collection of Bread Wheat Based on High-Throughput Genotyping Data

Laura Pascual [1], Mario Fernández [1], Nieves Aparicio [2], Matilde López-Fernández [1], Rosario Fité [3], Patricia Giraldo [1] and Magdalena Ruiz [3,*]

[1] Department of Biotechnology-Plant Biology, School of Agricultural, Food and Biosystems Engineering, Universidad Politécnica de Madrid, 28040 Madrid, Spain; laura.pascual@upm.es (L.P.); mario.fernandez.gonzalez@alumnos.upm.es (M.F.); matilde.lopez@upm.es (M.L.-F.); patricia.giraldo@upm.es (P.G.)

[2] Instituto Tecnológico Agrario de Castilla y León (ITACyL), 47071 Valladolid, Spain; apagutni@itacyl.es

[3] Plant Genetic Resources Centre, National Institute for Agricultural and Food Research and Technology, 28800 Alcalá de Henares, Spain; rfite@inia.es

* Correspondence: mruiz@inia.es; Tel.: +34-915026681

Received: 12 March 2020; Accepted: 4 April 2020; Published: 8 April 2020

Abstract: Modern plant breeding practices have narrowed the genetic base of wheat throughout the world, increasing crop vulnerability. Therefore, there is clearly a need for introducing new germplasm in breeding programs to search for variability related to traits of agronomic interest for wheat improvement. The existence of subsets of accessions (core collections) that represent the diversity conserved in germplasm collections is a favored approach for breeders to explore novel variation and enhance the use of germplasm. In this study, a core collection of Spanish landraces of bread wheat has been created using high-throughput genotyping technologies (DArTseq), which yielded more than 50 K molecular markers. This marker system not only provides a robust estimate of the diversity, but also information about its distribution in the genome. Two core collections of 94 entries were created by using two common sampling strategies: the maximization strategy and the population structure-based method. Both core collections showed high geographic, phenotypic and genetic representativeness, but the collection obtained with the maximization strategy captured better the diversity displayed by the initial collection. This core collection, which includes a broad range of adapted genotypes, can be efficiently utilized for mining new alleles for useful traits in wheat breeding.

Keywords: DArTseq markers; GBS; genetic diversity; *Triticum aestivum*

1. Introduction

Bread wheat (*Triticum aestivum* L.) is a major staple food crop that is widely grown throughout the world. Currently, in order to meet the increasing requirements of a growing population and tackle the challenges of global climate change, the genetic improvement of this crop must achieve several goals, including higher yield, adaptation to specific environments, tolerance to biotic stresses and quality enhancement. Modern plant breeding practices, in which only a small number of elite cultivars are included in breeding programs, have narrowed the genetic base of wheat throughout the world, increasing crop vulnerability. Therefore, there is clearly a need for introducing new germplasm in breeding programs so as to broaden the gene pool in which to search for new traits of agronomic interest necessary for wheat improvement. Wheat landraces are among the most suitable germplasm resource where the genetic variation required to that end can be searched [1] and utilized through a pre-breeding process. These locally adapted varieties, traditionally grown with less artificial resource inputs, are genetically diverse repositories of unique traits that have evolved in local environments, which cover a wide range of biotic and abiotic conditions [2,3]. Different studies have shown that Mediterranean wheat landraces represent a particularly important group of genetic resources, where extensive genetic variability as well as tolerance to drought, resistance to diseases and adaptability to low-input farming systems have been documented (see review [3]).

However, the use of wheat landraces in breeding is limited because the accessions preserved in Genebanks have not been globally characterized, leading to a scarcity of genotypic and phenotypic information. In particular, the lack of genome-wide genotypic information due to the characteristics of the wheat genome (large size and polyploidy-related complexity), along with phenotyping costs for specific traits, represent the main limiting factors for the use of such germplasm in breeding programs. It is, therefore, essential to generate subsets of accessions of a suitable size to represent the diversity conserved in germplasm collections, facilitating the availability of fine phenotyping for breeders. This approach is currently being pursued in some international breeding programs, as in the CIMMYT Seeds of Discovery program [4]. In this context, core collections, defined as a limited set of accessions chosen to represent at least 70% of the genetic variation of an entire collection with minimal redundancy, can be a powerful tool for increasing the efficiency of utilization of the germplasm stored in Genebanks [5,6].

Although conservation of allelic variation is important, the challenge of preserving quantitative genetic variation in conjunction with marker variation should be considered. Indeed, the final overall objective of a core collection is its effective use in breeding programs. The evaluation of the quality of the collection with phenotypic and genotypic data is, therefore, essential to confirm that prevalent variation types are preserved in the core collection [7]. There are two basic requirements for setting up useful core collections. The first concerns the use of efficient marker systems for unravelling the diversity of the collection. The second relates to appropriate sampling strategies to retain maximum diversity. To estimate the diversity of a collection, genotypic values are preferred over phenotypic traits to minimize genotype × environmental (GE) interactions [8]. The assessment of genome-wide diversity by genotyping by sequencing (GBS) methods provides a robust estimate of diversity and has been increasingly adopted as a fast, high-throughput cost-effective tool for whole-genome genetic diversity analysis in large germplasm sets [9]. Moreover, this approach may reveal new alleles in wheat germplasm that might exhibit a high value for prebreeding [10–12]. DArTseq markers, based on GBS [13], efficiently target low-copy-number sequences via a complexity reduction method and provide data at a more affordable cost, especially in complex polyploid species such as wheat [14] where they have been extensively used [15].

Regarding sampling strategies, the most common approaches using molecular markers are the M (maximization) strategy [16], and stratified sampling [6,16,17]. In the M-strategy, accessions are directly selected from the whole collection by maximizing the probability of retaining all observed alleles in order to construct cores with high allelic richness [16]. This strategy, based on the existence of correlations (shared coancestry) among marker and target loci, reduces the degree of redundancy in the core collection and leads to a more effective capture of localized, high frequency alleles [18]. On the other hand, stratified sampling requires a previous knowledge of the genetic structure of the collection; all genetic groups should contribute to the core collection with the goal of optimizing the representativeness of the genetic diversity in the core collection. Different allocation strategies can be implemented to decide the number of accessions to be selected per group. The H methodology determines the size of the sample per group in proportion to their within group genetic diversity, whereas the D method determines the size in proportion to a genetic distance and/or allele diversity index within the group [19]. The use of a diversity index seems to be more effective in maximizing allele richness, especially the expected heterozygosity, which leads to core subsets less likely to be homozygous for a number of different loci [20,21]. The studies that compare the maximization and the stratified sampling strategies have reported that stratified sampling used in conjunction with the D method formed core collections that had higher average genetic distances between genotypes, whereas the M-strategy captured more allelic diversity [20,21].

There are several computer programs to aid in core collection design. The Core Hunter algorithm has proved to be a fast and powerful method for designating core collections with increased genetic diversity, which can be applied with or without stratification of the whole collection [21–23]. Core Hunter has additional advantages: repeating the selection process produces a consistent solution, it is freely available and it is less time-consuming than other algorithms [21,22].

The objectives of the research were (a) to apply DArTseq GBS technology to provide a molecular basis for the design of a core collection of Spanish wheat landraces using the two most common approaches, M-strategy and stratified sampling, and (b) to evaluate the quality of the resulting core collections in order to select the most appropriate for a more efficient use of this germplasm in breeding. The use of the GBS technology has allowed the selection of a core collection based on more than 50K molecular markers distributed along the whole genome, and the detection of the presence of genomic regions where the different sampling methods employed performed differently. The results showed that the core collection created with the M-strategy using the Core Hunter algorithm performed better at retaining the diversity available in the initial collection.

2. Materials and Methods

2.1. Materials

The Spanish National Plant Genetic Resources Centre, CRF-INIA (Centro de Recursos Fitogenéticos, INIA, Madrid), maintains the national collection composed of 522 Spanish landraces of *Triticum aestivum* subsp. *vulgare* (Vill.). From this collection, a total of 189 genotypes were selected based on their collection site data (altitude, longitude, latitude [24]) and morphological spike traits (see Supplementary Tables S1 and S2) to represent the available diversity [25]. Homozygous lines were derived from these selected genotypes by collecting single bagged spikes from single selected plants during three generations. These 189 genotypes constituted the primary subset collection (PS) from which the entries for the final core collection were selected.

In the present study, the term "accessions" refer to genotypes that constitute the PS and "entries" are genotypes of the core collection [26].

2.2. Genetic and Phenotypic Characterization

High-throughput genotyping data for the PS accessions were obtained by DArTseq GBS technology at SAGA (Genetic Analysis Service for Agriculture, Mexico City, Mexico) as described in Pascual et al. [27]. This genotyping technology produces two different sets of markers: SNPs (Single Nucleotide Polymorphisms) and PAVs (Presence Absence Variants), from now on referred as DArTs (Diversity Arrays Technology markers). For this study, we selected a total of 59,276 DArTs and 14,830 SNPs, which were obtained after filtering out the markers that presented the same allelic profile or more than 10% missing data, as described by [27]. In that study, the genetic structure of the 189 accessions of the present research were analysed based on DArT markers, and the allelic profiles for the vernalization gene *Vrn-A1* and the *Glu-1* homoeoloci, determinants of wheat quality [28], were obtained. Both winter and spring landraces are included in the PS [27].

For phenotypic characterization, the accessions were sown in an augmented design during the season 2016–2017 at Alcala de Henares (Madrid). Seven qualitative (growth habit, awnedness, awn color, spike density, glume hairiness, glume color and seed color) and five quantitative agromorphological traits (days to heading and to maturity, plant height, spike length and spikelets per spike) were recorded according to the International Board of Plant Genetic Resources (IBPGR) [29] from five different plants in each accession.

2.3. Creation of the Core Collections

In order to determine the optimal collection size, simulations for sizes ranging from 5% to 100% of accessions included in the PS were performed with the DArT markers in the software Bio-R [30], which provides a graphical interface for the Core Hunter algorithm [21]. For simulations, the heterozygosity of the selected collections (HE = 1) was maximized, while the default values for the rest of parameters were maintained. Finally, the relationship between collection size and genetic diversity, quantified as the number of polymorphic markers retained, was examined. The optimal size was established as the point where the number of polymorphic markers increased asymptotically.

Two core collections were constructed based on the DArT markers. The maximization core collection (MCC) was obtained with the M-strategy using the Core Hunter algorithm and the expected heterozygosity (HE = 1) as the criteria of maximization (as described for the simulations). The stratified core collection (SCC) was created using the stratified sampling strategy. First, accessions were grouped based on populations. Then, inside each population, a number of accessions proportional to the genetic diversity (H_s), calculated with the DArTs as described by Nei [31], was selected to maximize the expected heterozygosity (HE = 1). Finally, a random core collection (RCC), where accessions were sampled randomly from the PS, was created to serve as reference.

2.4. Evaluation of the Core Collections

The quality of the different core collections created was evaluated using geographic, agromorphological and genetic data. Statistical analyses were performed with the software R version 3.5.2 [32].

For qualitative agromorphological traits and allelic profiles for the *Vrn-A1* and *Glu-1* loci, significant differences between the frequencies in the core collections and the PS were checked by Fisher's Exact Test (p-value < 0.05) [33]. For quantitative characters, the mean, variance, range and coefficient of variation were calculated for the PS, and for each one of the core collections. A homogeneity test (F-test) for variances and a t-test for means (p-value < 0.05) were used to compare the core collections and PS. The following evaluation parameters were calculated as described by Hu et al. [8]: mean difference percentage (MD), variance difference percentage (VD), coincidence rate of range (CR) and variable rate of coefficient of variation (VR). According to these parameters, a core collection can be considered representative if the percentage of traits with significant differences in their means is less than 20% (MD ≤ 20) and the coincidence rate of the range retained by the core collection is greater than 80% (CR ≥ 80%) [8].

The genetic diversity captured in each core collection was assessed with SNP markers. Different approaches were followed to evaluate the collections. First, the genetic diversity (H_S; [31]) was calculated for the PS and each of the core collections. Second, SNPs markers were classified according to their *MAF* (Minimum Allele Frequency) in: >0.1 (present in at least 19 accessions), ≥0.05 (9 accessions), ≥0.03 (6 accessions), >0.01 (2 accessions) and ≤0.01 (only in one accession). The number of markers fixed for each category in the core collections was calculated. Third, accessions selected and non-selected in each core collection were plotted in the Principal Coordinate Analysis (PCoA) performed by [27] to detect areas not sufficiently covered by the different core collections. Fourth, H_s along the bread wheat genome in the PS and different core collections was calculated based on SNP markers located in the bread wheat genome as described by [27]. Results were analyzed in order to detect genomic regions in which the core collections failed to retain the available diversity.

Finally, in order to quantify the degree of dissimilarity, Gower's genetic distances [34] between accessions were computed using the agromorphological data, SNP markers and *Glu-1* and *Vrn-A1* alleles. For each core collection, entry to entry mean-distance (E-E), the distance between each accession in the PS and the nearest entry in the core collection (A-NE) and the distance between each entry in the core collection and the nearest neighboring entry (E-NE) were calculated and averaged over all entries as described in [26]. A–NE represents the selection of entries close to each accession in the PS; thus, lower values are obtained for greater representativeness in the core entries. The E–NE distance indicates the presence of groups of similar entries in the core; thus, both the mean and minimum E-NE reach a maximum when all the entries are far apart.

3. Results

High-throughput genotyping provides genetic information that can guarantee the full inclusion of the available genetic diversity when creating a core collection. In order to avoid constraints in terms of budget and time, a primary set of accessions with low genetic redundancy, representing the entire collection, was selected before genotyping. From the full collection of 522 *T. aestivum* subsp. *vulgare* accessions, we selected a PS of 189 landraces covering the full collection range for latitude, longitude and altitude (Supplementary Table S1). This PS included landraces collected in the first half of the 20th century from all Spanish regions (including the two Spanish archipelagos), in which nine agroecological growing zones have been described [35]. The PS also covers the variability for six traits currently used in wheat accession characterization [29] (Supplementary Table S2). According to a previous study [27], the PS was subdivided into 4 populations, with Pop 2 having the highest number of accessions. The genetic diversity values (H_s) in each population ranged from 0.13 to 0.32 (Table 1).

3.1. Creation of the Core Collections

The final size of the core collection was determined based on simulations using the 59,276 polymorphic DArT markers present in the PS. For sizes larger than 94 entries, the genetic gain (estimated as the number of polymorphic markers) increased asymptotically (Figure S1). Thus, the size of the core collection was established in 94 genotypes, which captured 56,451 of the polymorphic DArT markers.

DArT markers were also used to select entries of the core collections. The MCC was obtained with the M-strategy by maximizing the expected heterozygosity. The SCC was created using the stratified sampling strategy based on the genetic structure of the PS and the genetic diversity within each population. Thus, those populations with higher diversity contributed a higher proportion of entries to the core collection. Finally, the RCC collection was established by randomly selecting entries from the PS. The final number of entries from each population included in the MCC, SCC and RCC are indicated in Table 1.

Table 1. Genetic diversity (H_s) of DArT (Diversity Array Technology) markers in each population (Pop) of the primary set and final number of accessions from each population in the core collections.

	Pop 1	Pop 2	Pop 3	Pop 4
H_S	0.21	0.32	0.13	0.23
PS	25	112	16	36
MCC	13	50	8	23
SCC	21	37	10	26
RCC	10	55	8	21

PS, primary set; MCC, core collection generated with maximization strategy; SCC, core collection generated with stratified sampling; RCC, core collection generated with random sampling.

3.2. Evaluation of Core Collections

A core collection of germplasm is a useful approach for breeders only when it includes most of the available variation (both genotypic and phenotypic). In this study, each core collection was evaluated considering genotypic and phenotypic data not used for the selection of entries. We evaluated the quality of the three core collections (CCs) at different levels: (1) representativeness of the CC for geographic, phenotypic and allelic (*Glu-1* and *Vrn-A1* loci) variability present in the PS; (2) allelic richness estimated from SNPs; (3) degree of dissimilarity and redundancy according to distances between accessions; and (4) distribution of the genetic variability included in each CC with respect to the PS and along the full genome.

3.2.1. Representativeness of the Core Collections

The three CCs represented adequately the geographic diversity and variability of qualitative morphological traits present in the PS (Table 2, Figure 1).

Table 2. Latitude, longitude and elevation ranges covered by the primary set and the core collections.

	Latitude	Longitude	Elevation (m)
PS	433310N–281820N	0174928W–0041559E	10–1610
MCC	433310N–281820N	0174928W–0041559E	22–1540
SCC	433310N–281820N	0174928W–0041559E	35–1540
RCC	433310N–281820N	0162421W–0031238E	63–1540

PS, primary set; MCC, core collection generated with maximization strategy; SCC, core collection generated with stratified sampling; RCC, core collection generated with random sampling.

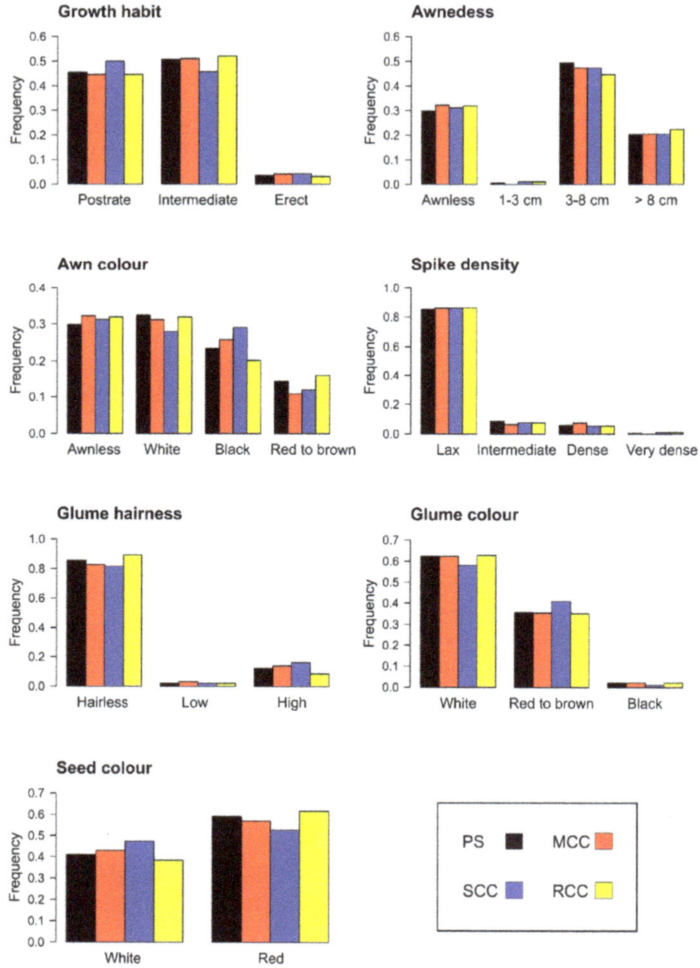

Figure 1. Relative frequencies of qualitative agromorphological traits in the primary set (PS) and core collections generated with the maximization strategy (MCC), stratified sampling (SCC) and random sampling (RCC).

The results also showed that the frequency distributions of the qualitative agromorphological traits in the CCs were not significantly different from the PS (p-values for Fisher tests ranging from 0.37 to 1). For the quantitative traits, no significant differences among means were detected in the CCs (Table 3). Moreover, the null values for the evaluation parameters MD and VD indicated that the PS was properly represented in the core collections (Table 3), and the high VR and CR values in the SCC and MCC revealed the prevalence of diverse entries in these subsets. The MCC generally had the highest coefficient of variation values, higher than those of the PS for some traits.

Table 3. Summary statistics for quantitative agromorphological traits in the primary set and core collections.

		Days to Heading (Days)	Days to Maturity (Days)	Plant Height (cm)	Spike Length (mm)	Spikelets Per Spike (Number)	Evaluation Parameter	
Mean	PS	171.23	206.86	88.27	117.03	19.11	MD	-
	MCC	171.48	206.81	87.89	118.36	19.31		0
	SCC	172.08	206.9	85.97	115.98	19.02		0
	RCC	171.19	207	88.19	119	19.28		0
Variance	PS	49.16	11.34	137.87	365.54	4.16	VD	-
	MCC	47.48	10.65	160.14	403.72	4.52		0
	SCC	48.83	9.59	137.73	340.78	4.22		0
	RCC	48.97	10.54	159.7	385.03	3.51		0
Coefficient of Variation	PS	4.09	1.63	13.3	16.34	10.68	VR	-
	MCC	4.02	1.58	14.4	16.98	11.01		102.06
	SCC	4.04	1.49	13.58	15.83	10.74		97.92
	RCC	4.09	1.57	14.33	16.49	9.72		99.18
Range	PS	155–188	199–216	53–119	59–168	14–24	CR	-
	MCC	157–188	199–216	53–115	67–168	14–24		91.53
	SCC	157–188	199–215	53–114	59–168	14–24		91.64
	RCC	155–188	200–216	53–119	59–153	14–23		89.46

PS, primary set; MCC, core collection generated with maximization strategy; SCC, core collection generated with stratified sampling; RCC, core collection generated with random sampling; MD, mean difference percentage; VD, variance difference percentage; VR, variable rate of coefficient of variation; CR, coincidence rate of range.

Regarding the genotypic data, all *Glu-1* alleles were included in the MCC and SCC (Table 4), whereas the RCC failed to capture one allele at each of the *Glu-B1* and *Glu-D1* loci. The three CCs included the three *Vrn-A1* alleles identified in the Spanish landraces in a similar proportion, especially in the MMC (Table 4).

Table 4. Alleles at the *Glu-1* homoeoloci and gene *Vrn-A1* in the primary set and the core collections.

	***Glu-1* Homoeoloci**			***Vrn-A1* Alleles (%)**		
	Glu-A1	***Glu-B1***	***Glu-D1***	***Vrn-A1***	***Vrn-A1a***	***Vrn-A1b***
PS	*a,b,c,y*	*a,al,am,aq,d,e,f,h,i,u,n2,n3,n4,n5,n6*	*a,c,d,h,j,l,n6*	18.52	52.38	29.10
MCC	*a,b,c,y*	*a,al,am,aq,d,e,f,h,i,u,n2,n3,n4,n5,n6*	*a,c,d,h,j,l,n6*	18.09	48.94	32.98
SCC	*a,b,c,y*	*a,al,am,aq,d,e,f,h,i,u,n2,n3,n4,n5,n6*	*a,c,d,h,j,l,n6*	12.77	55.32	31.91
RCC	*a,b,c,y*	*a,al,am,aq,d,e,f,h,i,u,n3,n4,n5,n6*	*a,c,d,h,j,l*	15.96	50.00	34.04

PS, primary set; MCC, core collection generated with maximization strategy; SCC, core collection generated with stratified sampling; RCC, core collection generated with random sampling.

3.2.2. Representativeness of the Core Collections

The allelic richness of the CCs was evaluated with 14,830 polymorphic SNPs in the PS. To analyze the degree of allele fixation in the three subsets, we studied the presence of monomorphic markers (Table 5). All the SNPs with predominant alleles (MAF > 0.1) were polymorphic in the three CCs. For the rest of the markers, the MCC had the lowest number of fixed markers, whereas the SCC showed the highest values. Taking into account the number of accessions in the PS, the least frequent allele in markers with MAF ≤ 0.01 was present in only one accession. Thus, it was expected that some of them would not be included. Overall, the MCC possessed the highest gene diversity value, equal to that in the PS.

Table 5. Genetic diversity (H_s) and distribution of minor allele frequency (MAF) for the SNP (Single Nucleotide Polymorphisms) markers in the primary set and those fixed in the core collections.

	SNP Markers (Number)	Fixed SNP Markers (Number)		
MAF	PS	MCC	SCC	RCC
>0.1	6376	0	0	0
≥0.05	8394	0	6	2
≥0.03	10,092	3	43	8
>0.01	14,002	534	685	586
≤0.01	828	387	419	453
Total	14,830	921	1104	1049
H_S	0.20	0.20	0.19	0.19

PS, primary set; MCC, core collection generated with maximization strategy; SCC, core collection generated with stratified sampling; RCC, core collection generated with random sampling.

3.2.3. Distances between Entries

The mean Gower's distance between entries (E-E) was higher in the CCs than in the PS, thereby providing a gain of 1% in the RCC and SCC, and 2% in the MCC (Table 6). The three CCs showed higher values for the minimum distances among the entries (mean E-NE distance and minimum E-NE distance) than the PS, especially the MCC. This last subset and the RCC also showed the lowest A-NE distances.

Table 6. Gower's genetic distance values between accessions in the primary set and core collections.

	E-E	A-NE	E-NE	Min E-NE
PS	0.189	-	0.073	0.0061
MCC	0.192	0.043	0.090	0.0067
SCC	0.190	0.051	0.081	0.0065
RCC	0.191	0.043	0.081	0.0067

PS, primary set; MCC, core collection generated with maximization strategy; SCC, core collection generated with stratified sampling; RCC, core collection generated with random sampling; E-E, mean distance between entries; A-NE, average distance between each accession in the PS and the nearest entry in the core collection; E-NE, average distance between each entry in the core collection and the nearest neighboring entry; E-NE min, minimum distance between each entry in the core collection and the nearest neighboring entry.

3.2.4. Distribution of Genetic Variability

The capture of the available genetic diversity in the three CCs was analyzed by representing the selected accessions in each one of them with respect to the PS by a PCoA analysis explaining 19.2% of the available SNPs diversity (Figure 2). The three subsets well covered the genetic variation, capturing accessions from the four populations of the PS. The RCC, however, failed to include some accessions from Pop 2 placed in the central part of the PCoA graph (Figure 2).

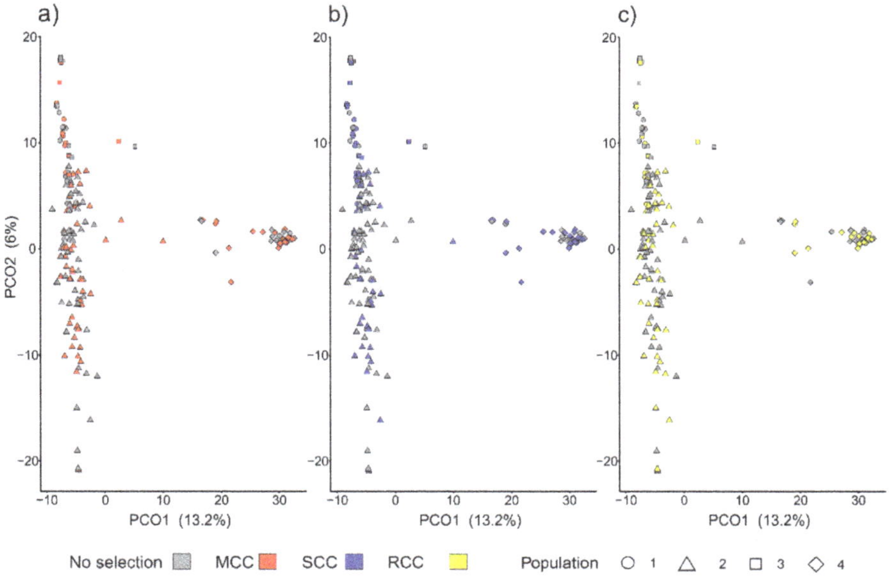

Figure 2. Biplot of the first two axes of the principal coordinate analysis according to Pascual et al. [27] showing the relative distribution of the primary set and core collections generated with (**a**) maximization strategy (MCC), (**b**) stratified sampling (SCC) and (**c**) random sampling (RCC). The genetic population is also indicated in the representation of the accessions.

Finally, we analyzed the genetic diversity along the bread wheat genome in the PS and the three CCs (Figure 3). In this case, the SCC failed to capture all the available diversity in the centromeric regions of chromosomes 2A, 4A and 2D. The RCC included less diversity in the 7D, and the MCC captured most of the diversity present in the PS along all the chromosomes. Considering both analyses, the MCC better represented the distribution of genetic variability in the PS.

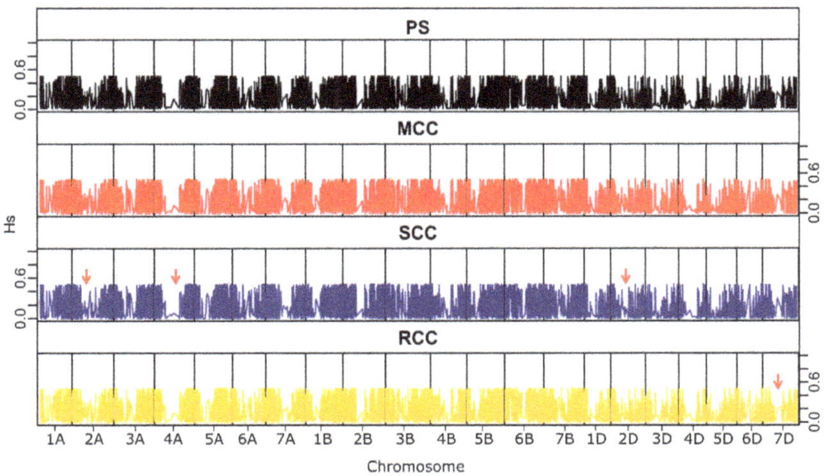

Figure 3. *Cont.*

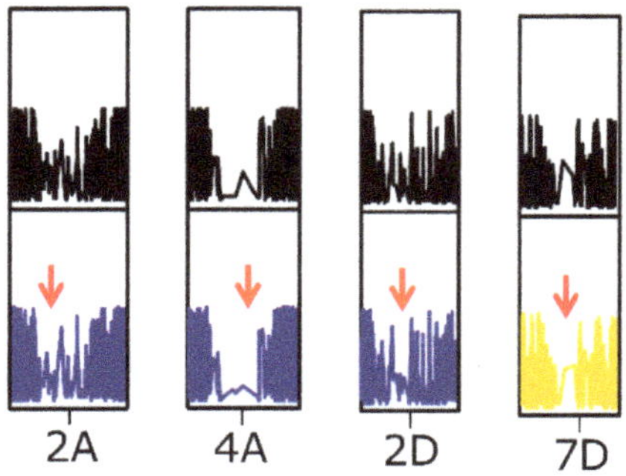

Figure 3. Genetic diversity (H_s) distribution across the genome in the primary set (PS) and the core collections generated with the maximization strategy (MCC), stratified sampling (SCC) and random sampling (RCC). Red arrows indicate lower genetic diversity regions when compared with the PS. Expanded subsections of these regions are shown.

4. Discussion

Several studies have shown the considerable variability among the Spanish wheat landraces compared to other germplasm collections [36,37]. These materials can be an important source of genes for wheat improvement, such as rust resistance [38] and quality traits [37,39]. However, even after the removal of redundant accessions, the collection maintained at CRF-INIA comprising more than 500 landraces is too large for evaluation and use in most breeding programs. Defining a core collection is, therefore, a pre-requisite and valuable tool for utilizing this germplasm, particularly for more complicated phenotypic screens, as has been demonstrated by the Spanish core collections of barley and durum wheat [40,41]. Both core subsets have facilitated detailed study of some difficult to analyze traits such as yield performance, root architecture, disease resistance or characterization of low molecular glutenin subunits [42–46].

On the other hand, the high genetic variability of the Spanish collection complicated the designing of a core collection of suitable size and able to capture the diversity present in the entire collection. To determine the optimal number of entries needed to retain an acceptable proportion of alleles present in the primary set, we tried to find a "point of compromise" between gain of genetic variation and elimination of genetic redundancy in the core collection. The final size of 94 entries captured 96% of polymorphic DArT markers present in the PS, which is in agreement with other studies that have reported values between 70 and 98% using SNPs [47,48] and SSR [40,49,50]. This size represents 18% of the entire collection, and is within the range from 5 to 30% recommended for retaining a great part of the genetic variability with a manageable number of accessions (e.g., [5,40,51]).

Preserving maximum genetic variation with a small number of accessions is a challenging task, and several improvements in sampling strategies have been devised over the last two decades. In order to create a multipurpose collection, we constructed two core collections using the most common approaches: M-strategy [16] and stratified sampling [6,16,17]. According to the objectives of these methodologies, we expected that the MCC maximized the total allelic diversity, and selected more diverse entries, whereas the SCC optimized the representativeness of the genetic diversity, including more representative entries. However, the effective utilization of the resulting core collections in breeding programs depends directly on their quality, which should be correctly evaluated.

Depending on the purpose of a core collection, a variety of metrics can be used to evaluate its quality. In the present study, considering that the aim is a multipurpose collection, the quality was evaluated using different types of variables such as geographic, phenotypic (discrete and continuous) and genotypic data that were not used in the core selection [26,51]. Also, a random core collection was generated to serve as a reference. Both the M-strategy and the stratified sampling selected entries that represented the geographic, phenotypic and genotypic variability of the PS, which validated both sampling strategies. However, the MCC included higher variation for quantitative agromorphological characters, indicating that this subset maximized the representativeness of the pattern of variation of these traits in the PS [26]. The allelic richness of the collections was analyzed with 14,830 SNPs covering all chromosomes. This genome-wide assay has proved to be a very convenient method for the analysis of the variability in germplasm collections [52]. On average, the CCs captured between 92.6 (in the SCC) and 93.8% (in the MCC) of SNPs present in the PS, and more than 90% of common alleles ($0.1 \geq MAF > 0.01$). Common-localized alleles may be biologically specialized alleles that enhanced adaptation to different agroecological conditions, and is often the class of alleles most interesting to breeders. Other studies have reported that crop improvement is accompanied by a selective advantage of rare alleles present at low frequency ($MAF < 0.05$) [53,54]. The percentage of rare allele recovery was 82.9% in the SCC and 85.7% in MCC, which was in agreement with that obtained in other core collections with SNPs [55]. The core collections performed worse in preserving very rare-localized alleles ($MAF \leq 0.01$) that were present in only one accession, which is in agreement with Wingen et al. [49]. Some studies reported that this type of allele, likely to be maintained by deleterious mutation-selection balance, would be of less interest; they seldom contribute to the improvement of elite varieties and, therefore, their inclusion in the CC might not be worthwhile [7,56,57]. In contrast, other authors have more recently proposed core subsets focused on preserving rare accessions and uncommon alleles, which may have unique genetic potentials for plant breeding [58]. In our case, the high number of accessions possessing specific rare alleles makes it difficult to retain a greater number of very rare alleles without increasing the sample size. Nevertheless, the MCC also maximized the coverage of this type of alleles and included an increased number of the most divergent accessions (56 in MCC, 52 in SCC and 50 in RCC out of the 94 accessions with the highest mean E-E distance in the PS). Moreover, the screening of the genetic diversity in each CC along the genome revealed that the MCC was able to better capture the available diversity from the PS in chromosomes 2A, 4A and 2D than the SCC.

Allelic richness is an evaluation criterion that ensures the inclusion of restricted alleles, whereas genetic distances between accessions is an evaluation criterion related to the concept of the maximization of the representation of genetic diversity in the whole collection [19,26,57]. In the present study, both phenotypic and genotypic variables were combined to calculate the distance among accessions since they provided complementary information thereby maximizing overall diversity for analysis [55,59]. The mean distance between accessions (E-E) reaches a maximum when diverse entries are sampled, but the presence of similar entries at the extreme ends of the distributions cannot be distinguished by this criterion. Therefore, two additional distances (A-NE and E-NE) recommended to evaluate the quality of multipurpose CCs were calculated [26]. The A-NE distance is a good criterion to evaluate the representativeness of genetic diversity of the PS, whereas the E-NE distance allows evaluation of whether the core collection has entries that are as different as possible from each other [26]. The lower A-NE and the higher E-NE distances in the MCC indicated that this subset maximized the representativeness of the genetic diversity of the PS and that all the entries were far apart genetically. The PCoA of the MCC also demonstrated that this subset was well distributed within the PS, covering the four genetic populations, even though the information on the genetic structure of the collection was not used to constrain the subset extraction. To some extent, this latter result could be related to the small number of populations identified in the collection [60].

Other studies have shown that the CC created based on the M-strategy maximizes the genetic variability index [23,40], whereas the structured method yields subsets that better represent the distribution of the genetic variability of the initial collections [7,20,40,57]. Also, worse values for E-NE distances have been reported in collections developed using the M-strategy [23,40]. In the present study, however, the M-strategy performed better than the stratified method, by increasing genetic diversity and reducing redundancy [26]. The higher quality of the MCC could be due to the lower degree of stratification of our collection. Only Pop 4 was clearly separated, while some overlap was shown among the other three populations, especially within Pop 2, which was the largest population [27]. In contrast, the stratified sampling performed better than the M-strategy in the Spanish durum wheat collection [40]. Comparisons of the genetic structure of the two wheat collections revealed that bread wheat populations exhibited a higher level of admixture and less genetic differentiation (Population differentiation index Dest = 0.22 in durum and 0.17 in bread wheat) [27]. This finding may explain why population-based sampling did not optimize the representativeness of the genetic diversity in bread wheat. Furthermore, the little gain by minimizing A–NE in the MCC compared to RCC could be also caused by the weaker structure of our bread wheat collection [26]. The results presented here have shown that the core collection designed with the M-strategy had superior performance; thus, this subset was selected as the Spanish core collection.

The Spanish wheat core collection constructed in the present study contains genotypes collected from every region in Spain where bread wheat is cultivated. Such coverage is essential because growing regions possess very diverse environmental conditions in terms of climate, altitude and soil characteristics. Wheat is grown from cold sub-humid areas in the northern parts of Spain to warm semi-arid regimes in the southeast [61], in basic or neutral soils in the Centre and East, and acid soils in the western regions [62]. Our core collection also includes all the *Vrn-A1* alleles for the vernalization response identified in Spanish landraces. The *Vrn-A1* gene is one of the most determinant loci involved in the transition from vegetative to reproductive growth [63], and thus for wheat adaptability. Such adaptation of Spanish landraces to different agroecological conditions has resulted in the accumulation of favorable alleles, including for stress tolerance, which can be incorporated into breeding programs [44,64–66]. Furthermore, in the case of wheat, functional quality requirements must also be kept in mind in order to have a multipurpose collection useful for wheat improvement. Our collection covers the allelic variation for 30 alleles at the *Glu-1* loci, the main genetic determinants of gluten quality [28,37].

5. Conclusions

The use of high-throughput genotyping technologies has allowed the selection of a core collection based on more than 50K molecular markers distributed across the whole genome. In addition, this approach has enabled us to detect the presence of genomic regions where different sampling methods employed performed differently. The M-method using the Core Hunter algorithm has demonstrated to be a fast and powerful method for designating core collections, especially for non-highly structured collections or in the absence of knowledge of a clear genetic structure in the whole collection. The core collection of Spanish landraces of bread wheat designed in the present study includes a broad range of adapted genotypes, and maximizes the representativeness of the genetic and phenotypic diversity in the initial collection of 522 landraces. This wheat core collection can be efficiently utilized in mining new alleles for useful traits and in broadening the genetic base in the cultivated wheat germplasm pool.

Supplementary Materials: The following are available online at http://www.mdpi.com/2073-4395/10/4/534/s1, Figure S1: Relationship between number of polymorphic DArT markers and sample size. The trend line is shown in black. Table S1: Latitude, longitude and elevation ranges covered by the initial collection of 522 accessions and the primary set. Table S2: Qualitative agromorphological traits classes included in the initial 522 accessions and primary set.

Author Contributions: Conceptualization, M.R. and L.P.; data curation, L.P., M.F., R.F., M.L.-F. and N.A.; formal analysis, L.P., M.F., N.A. and M.R.; funding acquisition, P.G.; investigation, L.P., M.F., M.L.-F., P.G. and M.R.; methodology, L.P., M.F., M.L.-F., R.F., P.G. and M.R.; project administration, P.G. and L.P.; resources, P.G., L.P., M.R.; supervision, L.P. and M.R.; validation, L.P. and M.R.; visualization, L.P., M.F., M.L.-F and M.R; writing—original draft, M.R. and L.P.; writing—review and editing, M.R., L.P. and P.G. All authors have read and agreed to the published version of the manuscript.

Funding: This research was funded by the Spanish Autonomous Community of Madrid under Grants No. AGRISOST-CM S2018/BAA-4330, the National Institute for Agricultural and Food Research and Technology under Grants No. AT2016-006 and RFP2015-00008-C04-01, the Universidad Politécnica de Madrid under Grant No. VJIDOCUPM18LPB, Ministry of Economy, Industry and Competitiveness under Grants No. AGL2016-77149-C2-1P, and Structural Funds 2014–2020 (ERDF and ESF). M.L.-F is a recipient of a predoctoral fellowship from the Programa Propio of the Universidad Politécnica de Madrid.

Acknowledgments: We thank M.J. Tomás for technical assistance.

Conflicts of Interest: The authors declare no conflicts of interest. The funders had no role in the design of the study; in the collection, analyses, or interpretation of data; in the writing of the manuscript, or in the decision to publish the results.

References

1. FAO. Harvesting Nature's Diversity. Available online: http://www.fao.org/3/v1430e/V1430E04.htm (accessed on 2 March 2020).
2. Ficiciyan, A.; Loos, J.; Sievers-Glotzbach, S.; Tscharntke, T. More than yield: Ecosystem services of traditional versus modern crop varieties revisited. *Sustainability* **2018**, *10*, 2834. [CrossRef]
3. Lopes, M.S.; El-Basyoni, I.; Baenziger, P.S.; Singh, S.; Royo, C.; Ozbek, K.; Aktas, H.; Ozer, E.; Ozdemir, F.; Manickavelu, A.; et al. Exploiting genetic diversity from landraces in wheat breeding for adaptation to climate change. *J. Exp. Bot.* **2015**, *66*, 3477–3486. [CrossRef] [PubMed]
4. CIMMYT. Seeds of Discovery. Available online: http://seedsofdiscovery.org/en/ (accessed on 5 March 2020).
5. Brown, A. The case for core collections. In *The Use of Plant Genetic Resources*; Brown, A., Frankel, O., Marshall, D., Williams, J., Eds.; Cambridge University Press: Cambridge, UK, 1989; pp. 136–156.
6. Brown, A. Core collections: A practical approach to genetic resources management. *Genome* **1989**, *31*, 818–824. [CrossRef]
7. Zhang, H.; Zhang, D.; Wang, M.; Sun, J.; Qi, Y.; Li, J.; Han, L.; Qiu, Z.; Tang, S.; Li, Z. A core collection and mini core collection of Oryza sativa L in China. *Appl. Genet.* **2011**, *122*, 49–61. [CrossRef] [PubMed]
8. Hu, J.; Zhu, J.; Xu, H. Methods of constructing core collections by stepwise clustering with three sampling strategies based on the genotypic values of crops. *Appl. Genet.* **2000**, *101*, 264–268. [CrossRef]
9. Holtz, Y.; Ardisson, M.; Ranwez, V.; Besnard, A.; Leroy, P.; Poux, G.; Roumet, P.; Viader, V.; Santoni, S.; David, J. Genotyping by sequencing using specific allelic capture to build a high-density genetic map of durum wheat. *PLoS ONE* **2016**, *11*, e0154609. [CrossRef]
10. Kilian, B.; Graner, A. NGS technologies for analyzing germplasm diversity in genebanks. *Brief. Funct. Genom.* **2012**, *11*, 38–50. [CrossRef]
11. Heslot, N.; Rutkoski, J.; Poland, J.; Jannink, J.-L.; Sorrells, M.E. Impact of marker ascertainment bias on genomic selection accuracy and estimates of genetic diversity. *PLoS ONE* **2013**, *8*, e74612. [CrossRef]
12. Manickavelu, A.; Jighly, A.; Ban, T. Molecular evaluation of orphan afghan common wheat (*Triticum aestivum* L.) landraces collected by Dr. Kihara using single nucleotide polymorphic markers. *BMC Plant Biol.* **2014**, *14*, 320. [CrossRef]
13. Sansaloni, C.; Petroli, C.; Jaccoud, D.; Carling, J.; Detering, F.; Grattapaglia, D.; Kilian, A. Diversity arrays technology (DArT) and next-generation sequencing combined: Genome-wide, high throughput, highly informative genotyping for molecular breeding of Eucalyptus. *BMC Proc.* **2011**, *5*, 54. [CrossRef]
14. Diversity Arrays Technology. Available online: https://www.diversityarrays.com (accessed on 5 March 2020).
15. Rasheed, A.; Mujeeb-Kazi, A.; Ogbonnaya, F.C.; He, Z.; Rajaram, S. Wheat genetic resources in the post-genomics era: Promise and challenges. *Ann. Bot.* **2018**, *121*, 603–616. [CrossRef] [PubMed]
16. Schoen, D.J.; Brown, A. Conservation of allelic richness in wild crop relatives is aided by assessment of genetic markers. *Proc. Natl. Acad. Sci. USA* **1993**, *90*, 10623–10627. [CrossRef] [PubMed]
17. Erskine, W.; Muehlbauer, F. Allozyme and morphological variability, outcrossing rate and core collection formation in lentil germplasm. *Appl. Genet.* **1991**, *83*, 119–125. [CrossRef]
18. Bataillon, T.M.; David, J.L.; Schoen, D.J. Neutral genetic markers and conservation genetics: Simulated germplasm collections. *Genetics* **1996**, *144*, 409–417.
19. Franco, J.; Crossa, J.; Taba, S.; Shands, H. A sampling strategy for conserving genetic diversity when forming core subsets. *Crop Sci.* **2005**, *45*, 1035–1044. [CrossRef]
20. Franco, J.; Crossa, J.; Warburton, M.L.; Taba, S. Sampling strategies for conserving maize diversity when forming core subsets using genetic markers. *Crop Sci.* **2006**, *46*, 854–864. [CrossRef]
21. Thachuk, C.; Crossa, J.; Franco, J.; Dreisigacker, S.; Warburton, M.; Davenport, G.F. Core hunter: An algorithm for sampling genetic resources based on multiple genetic measures. *BMC Bioinform.* **2009**, *10*, 243. [CrossRef]
22. Díez, C.M.; Imperato, A.; Rallo, L.; Barranco, D.; Trujillo, I. Worldwide core collection of olive cultivars based on simple sequence repeat and morphological markers. *Crop Sci.* **2012**, *52*, 211–221. [CrossRef]

23. Krishnan, R.R.; Sumathy, R.; Ramesh, S.; Bindroo, B.; Naik, G.V. SimEli: Similarity elimination method for sampling distant entries in development of core collections. *Crop Sci.* **2014**, *54*, 1070–1078. [CrossRef]
24. INIA. National Inventory of Plant Genetic Resources. Available online: http://webx.inia.es/web_inventario_nacional/Introduccioneng.asp (accessed on 5 March 2020).
25. Aparicio, N.; Alvaro, F.; Sillero, J.C.; Ruiz, M.; López, P.; Cátedra, M.; Codesal, P. Bread Wheat (*Triticum aestivum*, L.) Core Collection based in Spanish landraces. In Proceedings of the 8th International Wheat Conference, St. Petersburg, Russia, 1–4 June 2010; Dzyubenko, N.I., Ed.; NI Vavilov Research Institute of Plant Industry (VIR): Saint Petersburg, Russia, 2005; p. 85.
26. Odong, T.; Jansen, J.; Van Eeuwijk, F.; van Hintum, T.J. Quality of core collections for effective utilisation of genetic resources review, discussion and interpretation. *Appl. Genet.* **2013**, *126*, 289–305. [CrossRef]
27. Pascual, L.; Ruiz, M.; López-Fernández, M.; Pérez-Peña, H.; Benavente, E.; Vázquez, J.F.; Sansaloni, C.; Giraldo, P. Genomic analysis of Spanish wheat landraces reveals their variability and potential for breeding. *BMC Genom.* **2020**, *21*, 122. [CrossRef]
28. Payne, P.I.; Nightingale, M.A.; Krattiger, A.F.; Holt, L.M. The relationship between HMW glutenin subunit composition and the bread-making quality of British-grown wheat varieties. *J. Sci. Food Agric.* **1987**, *40*, 51–65. [CrossRef]
29. IBPGR. *Revised Descriptor List for Wheat (Triticum ssp)*; International Board for Plant Genetic Resources: Rome, Italy, 1985; p. 12.
30. Pacheco, A.; Alvarado, G.; Rodríguez, F.; Burgueño, J. *BIO-R (Biodiversity Analysis with R for Windows) Version 2.0*; Cimmyt: Sonora, Mexico, 2016.
31. Nei, M. Analysis of gene diversity in subdivided populations. *Proc. Natl. Acad. Sci. USA* **1973**, *70*, 3321–3323. [CrossRef]
32. R Core Team R. *A Language and Environment for Statistical Computing*; R Foundation for Statistical Computing: Vienna, Austria, 2018.
33. Fisher, R. *Statistical Methods for Research Workers*; Oliver and Boyd: Edinburgh, Scotland, 1954.
34. Gower, J.C. A general coefficient of similarity and some of its properties. *Biometrics* **1971**, 857–871. [CrossRef]
35. Ruiz, M.; Giraldo, P.; Royo, C.; Villegas, D.; Jose Aranzana, M.; Carrillo, J.M. Diversity and genetic structure of a collection of spanish durum wheat landraces. *Crop Sci.* **2012**, *52*, 2262–2275. [CrossRef]
36. Ruiz, M.; Rodriguez-Quijano, M.; Metakovsky, E.V.; Vazquez, J.F.; Carrillo, J.M. Polymorphism, variation and genetic identity of Spanish common wheat germplasm based on gliadin alleles. *Field Crop. Res.* **2002**, *79*, 185–196. [CrossRef]
37. Giraldo, P.; Rodriguez-Quijano, M.; Simon, C.; Vazquez, J.F.; Carrillo, J.M. Allelic variation in HMW glutenins in Spanish wheat landraces and their relationship with bread quality. *Span. J. Agric. Res.* **2010**, *8*, 1012–1023. [CrossRef]
38. Martinez, F.; Niks, R.E.; Moral, A.; Urbano, J.M.; Rubiales, D. Search for partial resistance to leaf rust in a collection of ancient Spanish wheats. *Hereditas* **2001**, *135*, 193–197. [CrossRef]
39. Vázquez, J.F.; Chacón, E.A.; Carrillo, J.M.; Benavente, E. Grain mineral density of bread and durum wheat landraces from geochemically diverse native soils. *Crop Pasture Sci.* **2018**, *69*, 335–346. [CrossRef]
40. Ruiz, M.; Giraldo, P.; Royo, C.; Carrillo, J.M. Creation and validation of the spanish durum wheat core collection. *Crop Sci.* **2013**, *53*, 2530–2537. [CrossRef]
41. Igartua, E.; Gracia, M.; Lasa, J.; Medina, B.; Molina-Cano, J.; Montoya, J.; Romagosa, I. The Spanish barley core collection. *Genet. Resour. Crop Evol.* **1998**, *45*, 475–481. [CrossRef]
42. Silvar, C.; Flath, K.; Kopahnke, D.; Gracia, M.; Lasa, J.; Casas, A.; Igartua, E.; Ordon, F. Analysis of powdery mildew resistance in the Spanish barley core collection. *Plant Breed.* **2011**, *130*, 195–202. [CrossRef]
43. Yahiaoui, S.; Cuesta-Marcos, A.; Gracia, M.P.; Medina, B.; Lasa, J.M.; Casas, A.M.; Ciudad, F.J.; Montoya, J.L.; Moralejo, M.; Molina-Cano, J.L. Spanish barley landraces outperform modern cultivars at low-productivity sites. *Plant Breed.* **2014**, *133*, 218–226. [CrossRef]
44. Ruiz, M.; Giraldo, P.; González, J.M. Phenotypic variation in root architecture traits and their relationship with eco-geographical and agronomic features in a core collection of tetraploid wheat landraces (*Triticum turgidum* L.). *Euphytica* **2018**, *214*, 54. [CrossRef]
45. Ruiz, M.; Bernal, G.; Giraldo, P. An update of low molecular weight glutenin subunits in durum wheat relevant to breeding for quality. *J. Cereal Sci.* **2018**, *83*, 236–244. [CrossRef]

46. Martínez-Moreno, F.; Ruiz, M.; Blandón, M.; Giraldo, P. Resistance to leaf rust in a core collection of ancient Spanish tetraploid wheats. In Proceedings of the I Spanish Symposium on Cereal Physiology and Breeding, Zaragoza, Spain, 9–10 April 2018; p. 34.
47. Acuña-Matamoros, C.L.; Reyes-Valdés, M.H. Comparison of optimization methods for core subset selection from a large collection of Mexican wheat landraces characterized by SNP markers. *Plant Genet. Resour.* **2018**, *16*, 228–236. [CrossRef]
48. Kobayashi, F.; Tanaka, T.; Kanamori, H.; Wu, J.; Katayose, Y.; Handa, H. Characterization of a mini core collection of Japanese wheat varieties using single-nucleotide polymorphisms generated by genotyping-by-sequencing. *Breed. Sci.* **2016**, *66*, 213–225. [CrossRef]
49. Wingen, L.U.; Orford, S.; Goram, R.; Leverington-Waite, M.; Bilham, L.; Patsiou, T.S.; Ambrose, M.; Dicks, J.; Griffiths, S. Establishing the AE Watkins landrace cultivar collection as a resource for systematic gene discovery in bread wheat. *Appl. Genet.* **2014**, *127*, 1831–1842. [CrossRef]
50. Balfourier, F.; Roussel, V.; Strelchenko, P.; Exbrayat-Vinson, F.; Sourdille, P.; Boutet, G.; Koenig, J.; Ravel, C.; Mitrofanova, O.; Beckert, M. A worldwide bread wheat core collection arrayed in a 384-well plate. *Appl. Genet.* **2007**, *114*, 1265–1275. [CrossRef]
51. Van Hintum, T.J.; Brown, A.; Spillane, C. *Core Collections of Plant Genetic Resources*; Bioversity International: Rome, Italy, 2000; p. 48.
52. Varshney, R.K.; Nayak, S.N.; May, G.D.; Jackson, S.A. Next-generation sequencing technologies and their implications for crop genetics and breeding. *Trends Biotechnol.* **2009**, *27*, 522–530. [CrossRef]
53. Gale, M.; Marshall, G.A. The chromosomal location of Gai 1 and Rht 1, genes for gibberellin insensitivity and semi-dwarfism, in a derivative of Norin 10 wheat. *Heredity* **1976**, *37*, 283–289. [CrossRef]
54. Murai, M.; Takamure, I.; Sato, S.; Tokutome, T.; Sato, Y. Effects of the dwarfing gene originating from 'Dee-geo-woo-gen'on yield and its related traits in rice. *Breed. Sci.* **2002**, *52*, 95–100. [CrossRef]
55. Vikram, P.; Franco, J.; Burgueño-Ferreira, J.; Li, H.; Sehgal, D.; Saint Pierre, C.; Ortiz, C.; Sneller, C.; Tattaris, M.; Guzman, C. Unlocking the genetic diversity of Creole wheats. *Sci. Rep.* **2016**, *6*, 23092. [CrossRef]
56. Marshall, D.; Brown, A. Optimum sampling strategies in genetic conservation. In *Crop Genetic Resources for Today and Tomorrow*; Frankel, O.H., Hawkes, J.G., Eds.; Cambridge University Press: Cambridge, UK, 1975; pp. 53–70.
57. Odong, T.; van Heerwaarden, J.; Jansen, J.; van Hintum, T.J.; van Eeuwijk, F. Statistical techniques for defining reference sets of accessions and microsatellite markers. *Crop Sci.* **2011**, *51*, 2401–2411. [CrossRef]
58. Reyes-Valdés, M.H.; Burgueño, J.; Singh, S.; Martínez, O.; Sansaloni, C.P. An informational view of accession rarity and allele specificity in germplasm banks for management and conservation. *PLoS ONE* **2018**, *13*, e0193346. [CrossRef]
59. Wen, W.; Franco, J.; Chavez-Tovar, V.H.; Yan, J.; Taba, S. Genetic characterization of a core set of a tropical maize race Tuxpeño for further use in maize improvement. *PLoS ONE* **2012**, *7*, e32626. [CrossRef]
60. Corrado, G.; Caramante, M.; Piffanelli, P.; Rao, R. Genetic diversity in Italian tomato landraces: Implications for the development of a core collection. *Sci. Hortic.* **2014**, *168*, 138–144. [CrossRef]
61. Informe Cereal de Invierno GENVCE 2018. Available online: www.genvce.org/repositorio/informe-cereal-de-invierno-genvce-2018 (accessed on 23 February 2020).
62. Map of Soil pH in Europe. Available online: https://esdac.jrc.ec.europa.eu/content/soil-ph-europe (accessed on 23 February 2020).
63. Zhang, X.K.; Xiao, Y.G.; Zhang, Y.; Xia, X.C.; Dubcovsky, J.; He, Z.H. Allelic variation at the vernalization genes Vrn-A1, Vrn-B1, Vrn-D1, and Vrn-B3 in Chinese wheat cultivars and their association with growth habit. *Crop Sci.* **2008**, *48*, 458–470. [CrossRef]
64. Cantalapiedra, C.P.; García-Pereira, M.J.; Gracia, M.P.; Igartua, E.; Casas, A.M.; Contreras-Moreira, B. Large differences in gene expression responses to drought and heat stress between elite barley cultivar scarlett and a spanish landrace. *Front. Plant Sci.* **2017**, *8*, 647. [CrossRef]

65. Monteagudo Gálvez, A.; Casas Cendoya, A.M.; Pérez Cantalapiedra, C.; Contreras-Moreira, B.; Gracia Gimeno, M.P.; Igartua Arregui, E. Harnessing novel diversity from landraces to improve an elite barley variety. *Front. Plant Sci.* **2019**, *10*, 434. [CrossRef]
66. Ruiz, M.; Zambrana, E.; Fite, R.; Sole, A.; Tenorio, J.L.; Benavente, E. Yield and quality performance of traditional and improved bread and durum wheat varieties under two conservation tillage systems. *Sustainability* **2019**, *11*, 4522. [CrossRef]

agronomy

MDPI

Article

Biomorphological Characterization of Brazilian Capsicum Chinense Jacq. Germplasm

Paola Alvares Bianchi [1], Lígia Renata Almeida da Silva [2], Antônio André da Silva Alencar [1], Pedro Henrique Araújo Diniz Santos [1], Samy Pimenta [3], Cláudia Pombo Sudré [1], Lígia Erpen-Dalla Corte [4], Leandro Simões Azeredo Gonçalves [4] and Rosana Rodrigues [1,*]

[1] Plant Breeding Department, Universidade Estadual do Norte Fluminense Darcy Ribeiro (UENF), Campos dos Goytacazes RJ 28013-602, Brazil; paolaalvaresb@gmail.com (P.A.B.); antonioandre14@hotmail.com (A.A.d.S.A.); phsantos2004@yahoo.com.br (P.H.A.D.S.); claudia.pombo@yahoo.com.br (C.P.S.)

[2] National Navel Orange Engineering Research Center (NORC), Gannan Normal University, Shiyuan Road, Rongjiang Zone, Ganzhou 341000, China; ligiarasilva@gmail.com

[3] Department of Agrarian Sciences, Universidade Estadual de Montes Claros, Av. Reinaldo Viana, 2630, Morada do Sol, Janaúba MG 39440-00, Brazil; pimentasamy@gmail.com

[4] Agronomy Department, Universidade Estadual de Londrina (UEL), Rodovia Celso Garcia Cid, Km 380, Londrina, PR 86051-900 Brazil; ligia_erpen@yahoo.com.br (L.E.-D.C.); leandrosag@uel.br (L.S.A.G.)

* Correspondence: rosana@uenf.br

Received: 7 March 2020; Accepted: 20 March 2020; Published: 24 March 2020

Abstract: Loss of biodiversity and ecosystem degradation become major concerns worldwide, making the conservation process an important strategy for maintaining biodiversity. *Capsicum chinense* Jacq. is the most Brazilian species of the genus, with representatives in different biomes. Anthropic pressure, such as burns, real estate speculation, and changing cultivation habit has led to risks of genetic erosion. Conservation and characterization of conserved accessions are paramount to ensure genetic diversity, useful for the bioeconomy and for genetic improvement. We report the characterization of 55 *C. chinense* accessions from four different regions of Brazil and one accession from Peru. The accessions were characterized based on 37 morpho-agronomic variables, Inter Simple Sequence Repeats—ISSR and Simple Sequence Repeat—SSR. Qualitative descriptors were analyzed using a descriptive statistical, while the quantitative descriptors were analyzed via F test and significant differences in mean values were separated using Scott-Knott test. The relative contribution of each quantitative trait was determined. A correlation between morphological and molecular distances was calculated. Color of ripe fruit and fruit shape had the largest number of observed classes. Six distinct groups and the joint analysis presented an entanglement rate of 0.58, evidencing the divergence of accessions between the groups of both dendrograms. Bayesian analysis allowed the distinction of two clusters for SSR. A significant variability was observed among accession with potential to integrate several breeding programs.

Keywords: chili pepper; gene bank; molecular markers; morphological descriptor

1. Introduction

The *Capsicum* genus (Solanaceae), commonly known as bell and chili pepper has been very important in many cultures worldwide for spices, medicines, ornamentals and vegetables [1–4]. The genus comprises approximately 42 described species [5], with wide range of morphological variability, mainly in different shapes, sizes, colors and sensory attributes of its fruits [6]. Among these species, five are considered domesticated: *C. annuum* L., *C. frutescens* L., *C. chinense* Jacq., *C. baccatum* L. (var. *pendulum*), and *C. pubescens* Ruiz & Pav. [7].

In Brazil, chili peppers are part of the culture and are widely cultivated through the country, in most cases by small farmers. In addition, Brazil is considered a valuable source of variability for this horticultural crop, harboring domesticated species as well as semi-domesticated and wild species of *Capsicum* spp. [7,8]. Among the domesticated species, *C. chinense* is considered the most Brazilian of the genus *Capsicum*. This specie is easily found in the Amazonia region, a probable center of domestication showing a wide array of fruit size, shape, color and pungency [9–11]. It is widely used in Brazilian cuisine, especially the types "pimenta-de-cheiro", "pimenta-de-bode", "cumari-do-Pará", "murupi", "habanero", and "biquinho" [12]. However, many accessions that compose this germplasm have been threatened by genetic erosion due to the destruction of natural habitats or replacement of local varieties by modern cultivars [13,14].

Preservation of such plant genetic resources is extremely important for plant breeding as well as for society as a whole [15]. In view of the cultural and economic importance of the *Capsicum* in Brazil, the development of new and improved cultivars carrying characteristics that meet the needs of farmers and the consumers is primordial. To achieve this goal, plant breeders are dependent on plant genetic resources and need access to the widest genetic diversity available. For these reasons, the preservation of wild species, local varieties and traditional genotypes in collections or germplasm banks is very important [16–18]. In addition, the characterization of these materials is essential information for the conservation and use in plant breeding programs [19].

The development and population growth are related to the human capacity to inhabit and explore different areas and geographic locations [20]. This appropriation and exploitation of nature implies on the transformation of the environment through anthropic actions that influence the loss of biodiversity, such as habitat destruction, pollution, deforestation, degradation and intensive land use [21,22]. This makes the conservation process an important strategy for maintaining biodiversity and food security, as stated in United Nations 2030 Agenda, Objective 15, that prioritizes protect, restore and promote sustainable use of terrestrial ecosystems, sustainably manage forests, combat desertification, and halt and reverse land degradation and halt biodiversity loss [23].

The assessment of genetic diversity in chili pepper collections is traditionally performed using phenotypic descriptors such as flower and fruit attributes, plant architecture, growth habits and others [19,24–26]. Germplasm resources in a gene banks can be more detailed and reliable if morphological traits are associated with the molecular markers which can often lead to more accurate genetic information and better distinction between accessions since they are stable and detectable in all tissues regardless the plant development stage. Moreover, they do not consider environmental, pleiotropic, and epistatic effects [27,28]. Various types of molecular markers are available for identification and characterization of genetic diversity. Among them, PCR-based molecular markers such as RAPD (Random Amplification Polymorphism DNA) [10,29], AFLP (Amplified Fragment Length Polymorphism) [13,30], ISSR (Inter Simple Sequence Repeats) [12,31] and microsatellites or SSR (Simple Sequence Repeat) [32–35] have been widely used to characterize genetic diversity in several chili pepper collections. The knowledge of genetic variability estimated by these markers complements phenotypic characterization and it is essential to find additional sources of genetic diversity present in germplasm [12,14].

Several studies have addressed the characterization of the genetic diversity of *C. chinense* based on phenotypic and molecular descriptors [10,12–14,24,29,35–37]. However, most of them focused in fruit morphological descriptors and fewer studies have addressed a broad range of morpho-agronomic descriptors including both qualitative and quantitative traits. Unravelling genetic variability of the *C. chinense* germplasm restricts its utilization as donor species for interspecific hybridization and consequently limits its use in other *Capsicum* spp. breeding programs. In this study, we describe the characterization of 55 *C. chinense* accessions from different regions of Brazil and Peru using qualitative and quantitative descriptors and SSR and ISSR molecular markers.

2. Materials and Methods

2.1. Phenotyping

The 55 *C. chinense* accessions from four Brazilian region and one accession from Peru (Supplementary Table S1) are part of the gene bank of *Capsicum* spp. from the *Universidade Estadual do Norte Fluminense Darcy Ribeiro*—UENF, located in Campos dos Goytacazes, RJ, Brazil. Seeds from each accession were sown in 128-cell polystyrene trays containing the commercial substrate. Three seeds were placed per cell and after germination and growth of seedlings, two seedlings from each repetition were individually transferred to five-liters pots containing a mixture of soil, sand, and manure (1: 1: 1 ratio) with subsequent thinning. The seeds used had an average of 4 years of storage and were all cultivated varieties. The plants were kept in greenhouse conditions located in the experimental area of the same institution. There was no humidity and temperature control inside the greenhouse and the irrigation was done once a day. The experiment was conducted in completely randomized design with five repetitions, each repetition was composed of a plant and they were grown following practices recommended for chili pepper cultivation. A digital thermometer recorded the temperature and humidity conditions of the greenhouse. The irrigation shifts vary according to the plants needs at different stages of development.

The accessions were characterized based on 37 morpho-agronomic variables (Supplementary Tables S2 and S3), sixteen quantitative and twenty-one qualitative, following the Bioversity International (formerly International Plant Genetic Resources Institute, IPGRI, 1995) [38]. Using this approach, other researchers and gene bank curators can compare and easily identify traits of common interest in our collection. All phenotypic information that we generate on this study is being used to fill a platform that share genebank information with other researchers.

2.2. Genotyping

Genomic DNA was isolated from pools of fresh young leaves from five individuals accessions according to the Doyle and Doyle (1990) [39] protocol with modifications [40]. The genomic quality and integrity DNA were checked by electrophoresis for ISSR and SSR analysis using 1% agarose gel.

For ISSR analysis 35 ISSR primers were preliminarily evaluated to determine optimal amplification reaction conditions (annealing temperature and cycling time). Eighteen primers out of the thirty-five resulted in profiles of well-separated fragments and were selected for molecular characterization of the *C. chinense* accessions. A list of DNA sequence and melting temperatures (Tm) associated with ISSR markers can be found in Supplementary Table S4. PCR amplification of 10 ng DNA (2 μL) was performed in 13 μL reaction containing 1.5 μL dNTP mix (0.2 mM), 1.0 μL $MgCl_2$ (1.9 mM), 1.3 μL PCR Buffer 1X, 0.12 μL Taq DNA polymerase (0.6 U) and 1 μL of each primer (5 μM) forward and reverse. The cycling conditions were 5 min at 94 °C, followed by 35 cycles of 1 min at 94 °C, 1 min at 48–52 °C (temperature previously determined for each primer), 3 min at 72 °C, and a final extension of 7 min at 72 °C. PCR products were separated by electrophoresis in agarose gel (2%), stained with 6 μL of a mix with gel red and blue juice (1:1), visualized on a transilluminator using UV illumination and photographed using documentation system MiniBis Pro (Bio-Imaging Systems). The ISSR results were interpreted as presence or absence of bands and expressed in a binary matrix.

For SSR analysis 17 SSR primers were selected from a set of 47 SSR primers previously developed and optimized for *C. annuum* [41]. The screening was based on the amplification of well-separated fragments at different annealing temperatures (56–66 °C). A list of DNA sequence and melting temperatures (Tm) associated with SSR markers can be found in Supplementary Table S5. The PCR analysis were performed as described above. The amplified products were separated by electrophoresis in MetaPhor agarose gel (4%), stained with 6 μL of a mix with gel red and blue juice (1:1), visualized on a transilluminator using UV illumination and photographed using documentation system MiniBis Pro (Bio-Imaging Systems). MetaPhor agarose gel which produces a high resolution gel that allows the visualization of fragments between 20 bp–800 bp. According to manufacturer, "MetaPhor agarose

(FMC or Cambrex Corporation, East Rutherford, NJ, USA) is an intermediate melting temperature agarose (75 °C) that provides twice the resolution capabilities of the finest sieving agarose products. Using submarine gel electrophoresis, MetaPhor agarose gives high resolution separation of 20 to 800 bp DNA fragments that differ in size by 2%, which approximates the resolution of polyacrylamide gels". The SSR results were converted into numeric code per locus for each allele. With the polymorphic loci, an array of numerical data was constructed, assigning values from 1 to the maximum number of alleles per locus.

2.3. Data Analysis

Qualitative descriptors were analyzed using a descriptive statistical, while the quantitative descriptors were analyzed via the F test in the analysis of variance (ANOVA) and significant differences in mean values were separated using Scott-Knott test at $\alpha = 0.05$. The relative contribution of each quantitative traits on the phenotypic divergence was calculated using the method proposed by Singh [42]. The estimates of Pearson's correlation coefficients were obtained based on the average of repetitions between the characters combined two by two. Statistical analyzes were performed with GENES software [43]. The values of the 37 morpho-agronomic variables were used for Ward clustering analysis based on the Gower distance [44]. For this analyze we used R software [45].

The DNA samples of *C. chinense* accessions were amplified with ISSR primers and analyzed by visual assessment of the most consistent bands (Supplementary Figure S1). The matrix of genetic dissimilarities was obtained by arithmetic complement of the Jaccard index.

Amplification data from the SSR primers were converted to numeric code per locus for each allele (Supplementary Figure S2). The polymorphic loci were used to generate a numerical matrix. Genetic distances among *C. chinense* accessions was evaluated by Index Weighted performed with the GENES software [43] and clustered with hierarchical algorithms, such as the unweighted pair-group method by arithmetic averages (UPGMA) performed with the R program [45]. The number of alleles (Na), effective number of alleles (Ne), expected heterozygosity (He), observed heterozygosity (Ho), diversity index (I), and fixation index (F) were performed with GENES program [43].

A Bayesian-based cluster analysis was performed to determine the optimal number of genetic clusters using Structure 2.3.4 software [46] according to the method described by Evanno et al. (2005) [47], with 10,000 repetitions. K values between 1 and 10 were tested, with 10 independent iterations for each group.

Analyses of the ISSR and SSR data were performed using the Gower distance matrix and Ward clustering analysis. The correlation between molecular and morphological distances was determined using the distance matrices relative to the 37 morpho-agronomic descriptors and ISSR and SSR markers. For these analyses, we used the Dendextend package in the R program [45].

3. Results

3.1. Morpho-Agronomic Characterization Data

The qualitative traits stem shape (SS), stem pubescence (SP), leaf shape (LS), corolla shape (CS), cotyledon leaf color (CLC), and cotyledon leaf shape (CLS) were monomorphic, described as angular, sparse, oval, round, green and lanceolate, respectively. The hypocotyl color was green in 78.2% of the accessions and purple in 21.8% of the remaining accessions. For the attribute stem color (SC), three colors were observed, ranging from green (85.45%), green with purple stripes (12.7%) and purple (1.85%). Anthocyanin color of the node (ACN) was absent in 73.5% of the accessions and in those in which it was observed it was purple (10.1%), light purple (9.1%), and dark purple (7.3%).

Most accessions exhibited plant growth habit (PGH) described as erect (80%) and the remainder (20%) was compact. The flower position (FP) intermediate was observed in 50.9% of the accessions, while standing position and pendent position were observed in 38.2 and 10.9% of the accessions, respectively. For corolla color (CC), only one accession presented a pale-yellow corolla and the others were white-green. For the anther color (AC) the purple color was predominant (56.36%) followed by blue color (38.2%) and yellow color (5.5%). In relation to the number of flowers per axilla (NFA), 74.54% of the accessions presented two flowers per node and 25.45% three flowers per node.

The attributes related to fruits present wide variability among the accessions. The fruit color at intermediate stage (FCIS) varied between orange (72.7%), green (21.8%) and purple (5.5%). On the mature stage (FCMS), eight fruit color were found ranging from red (50.9 %), dark red (12.8%), light red (9.1%), orange (10.9%), light orange (10.9%), orange with purple ends (1.8%), light yellow (1.8%), and brown (1.8%). About fruit shape (FS), triangular shape (45.45%) was predominant, followed by elongated (16.4%), square (16.4%), rounded (12.7%), campanulate (5.45%), bell (1.8%), and oval (1.8%). Regarding fruit surface (FS), 41.8% presented semi-wrinkled surface, 40% smooth surface, and 18.2% wrinkled surface. The neck at the base of the fruit (NBF) was observed in only 16.4% of the accessions. The number of locules (NL) observed in the fruits were three, four and two in 80%, 10.9%, and 9.1% of the accessions, respectively. Most fruits (85.45%) presented capsaicin.

Wide variability was observed for quantitative descriptors mean values. These data are described in Supplementary Table S6 wherein Scott-Knott (1974) analysis revealed the arrangement of distinct groups for each attribute. The plant height (PH) ranged from 49 to 156 cm and the accessions were settled in seven distinct groups. The canopy diameter (CD) and stem diameter (SD) allowed the establishment of four groups and ranged from 44.18 to 123.25 cm and 0.96 to 2.08 cm, respectively.

The time of the plant development stages was different among the accessions. Days for germination (DG) ranged from 5 to 11 days, days for flowering (DFL) ranged from 62 to 92 days, days for fruiting (DFR) ranged from 62 to 117 days and days for fruit maturation (DM) ranged from 135 to 217 days. Based on four attributes, accessions were classified in seven, four, five, and six distinct groups, respectively.

Fruits attribute such as fruit weight (FW) ranged from 1.04 to 18.61g, fruit length (FL) ranged from 7.85 to 84.93 mm and fruit diameter (FD) ranged from 8.99 a 34.44 mm and according to these attributes the accessions were classified into nine, eight and five distinct groups, respectively. Cotyledon leaf length (CLL) ranged from 12.81 to 25.96 mm and cotyledon leaf diameter (CLD) ranged from 6.04 to 8.05 mm, and the accessions were arranged in seven and four distinct groups, respectively. The leaf length (LL) and leaf width (LW) traits also varied among the accessions, presenting values from 6.5 to 9.5 and 3.8 to 5.9 cm, respectively. It was possible to classify them in four and five different groups according to these two attributes. The trait peduncle length (PL) ranged from 19 to 43 mm and the accessions were separated into seven groups. The pericarp thickness (PT) varied from 1.38 a 3.08 mm and four groups were formed.

The relative contribution of each of these quantitative traits on the phenotypic divergence of the *C. chinense* accessions was estimated using the method proposed by Singh (1981) [42] (Table 1). It was possible to observe that the fruit diameter (FD) (9.83%) was the attribute that most contributed to the discrimination of the genotypes, followed by days for germination (DG) (9.1%) and fruit weight (FW) (8.84%). In contrast, cotyledon leaf length (CLL) was the characteristic that presented less relative importance for the phenotypic divergence (2.85%).

The presence of correlation between variables was analyzed using Pearson's correlation coefficient (Supplementary Table S7). High correlations were found for the characteristic plant height and plant canopy width (75%), fruit diameter and pericarp thickness (71%), days for fruiting and maturation (70%), and moderate correlation for fruit diameter and fruit weight (69%), pericarp thickness and fruit weight (66%), and leaf length and leaf diameter (64%). Moderate negative correlations were found for days for germination and fruit diameter (−39%), days for germination and cotyledonous leaf diameter (−38%), and leaf width and canopy diameter (−31%).

Table 1. Relative contribution of quantitative traits for *C. chinense* accessions genetic divergence.

Quantitative Traits	Relative Contribution (%)	Minimum Value	Maximum Value	Average Value
Fruit diameter (mm)	9.83%	8.99	34.44	21.22
Days for germination	9.1%	5	11	8
Fruit weight (g)	8.84%	1.04	18.61	6.93
Pericarp thickness (mm)	7.76%	1.38	3.08	2.33
Cotyledon leaf diameter (mm)	7.38%	6.04	8.05	6.97
Fruit length (mm)	6.77%	7.85	84.93	34.32
Leaf length (cm)	6.67%	6.5	9.5	7.57
Plant height (cm)	6.59%	49	156	84.85
Stem diameter (cm)	5.77%	0.96	2.08	1.35
Leaf width (cm)	5.52%	3.8	5.9	4.18
Days for maturation	5.46%	135	217	158
Canopy diameter (cm)	5.12%	44.18	123.25	70.46
Days for flowering	4.27%	62	92	75
Peduncle length (mm)	4.22%	19	43	24.37
Days for fruiting	3.79%	62	117	92
Cotyledon leaf length (mm)	2.85%	12.81	25.96	16.82

3.2. Morpho-Agronomic Diversity

The identification of homogeneous groups on the 55 accessions of *C. chinense* was performed by Ward clustering analysis. The dendrogram obtained shows the distinction of six groups (Figures 1a and 2). Group I was composed of four accessions, which presented fruits with four locules, semi-rough surface and orange color when immature. Group II was composed of six accessions, which presented flowers in the upright position and purple-colored cotyledon and stem. Group III was composed of 21 accessions. All with purple anthers and fruits with smooth surface and pungents. Group IV was composed of nine accessions in which the following descriptors were predominant: days for germination (DG) ranged from 7 to 9 days; stem diameter (SD) varying from 0.96 to 1.33 mm; growth habit (PGH) described as compact, fruit color at intermediate stage was green and fruit color at mature stage was orange, except for UENF 1730, the only Peruvian accession, that fruit color on mature stage was brown. Group V comprised seven accessions. Red fruit color at mature stage, triangular and elongated shape, presence of a neck at the base of the fruit are some of the characteristics of this group. These accessions also presented fruits with a small variation in length (35.53 to 44.46 mm). Group VI is characterized by clustering plants with compact growth habit, lack of a neck at the base of the fruit and fruits with three locules, and plant height (PH) ranged from 58.62 to 86.22 cm.

3.3. Molecular Characterization Data

For ISSR analysis, 15 primers out of the eighteen detected polymorphisms among accessions. These primers produced 97 fragments, of which 46 were polymorphic, representing 47% polymorphism.

For SSR analysis, from 47 pairs of microsatellite primers, previously developed and optimized for *C. annuum* [41] only 17 resulted in amplification and nine primers out of the 17 detected polymorphisms among *C. chinense* accessions.

The values for the major allele frequency (MAF) ranged from 0.5 to 0.94, with an average of 0.72 (Table 2). The diversity index (I) ranged from 0.10 to 0.50, with an average of 0.38. The observed heterozygosity (Ho) ranged from zero to 0.44, with a mean of 0.10, and the fixation index ranged from −0.14 to 1.00, with a mean of 0.77.

Following Bayesian analysis, the highest ΔK value was obtained when two clusters were formed, obtaining groups I and II (Figure 3). Group I gathered 41 accessions, while group II gathered 14 accessions.

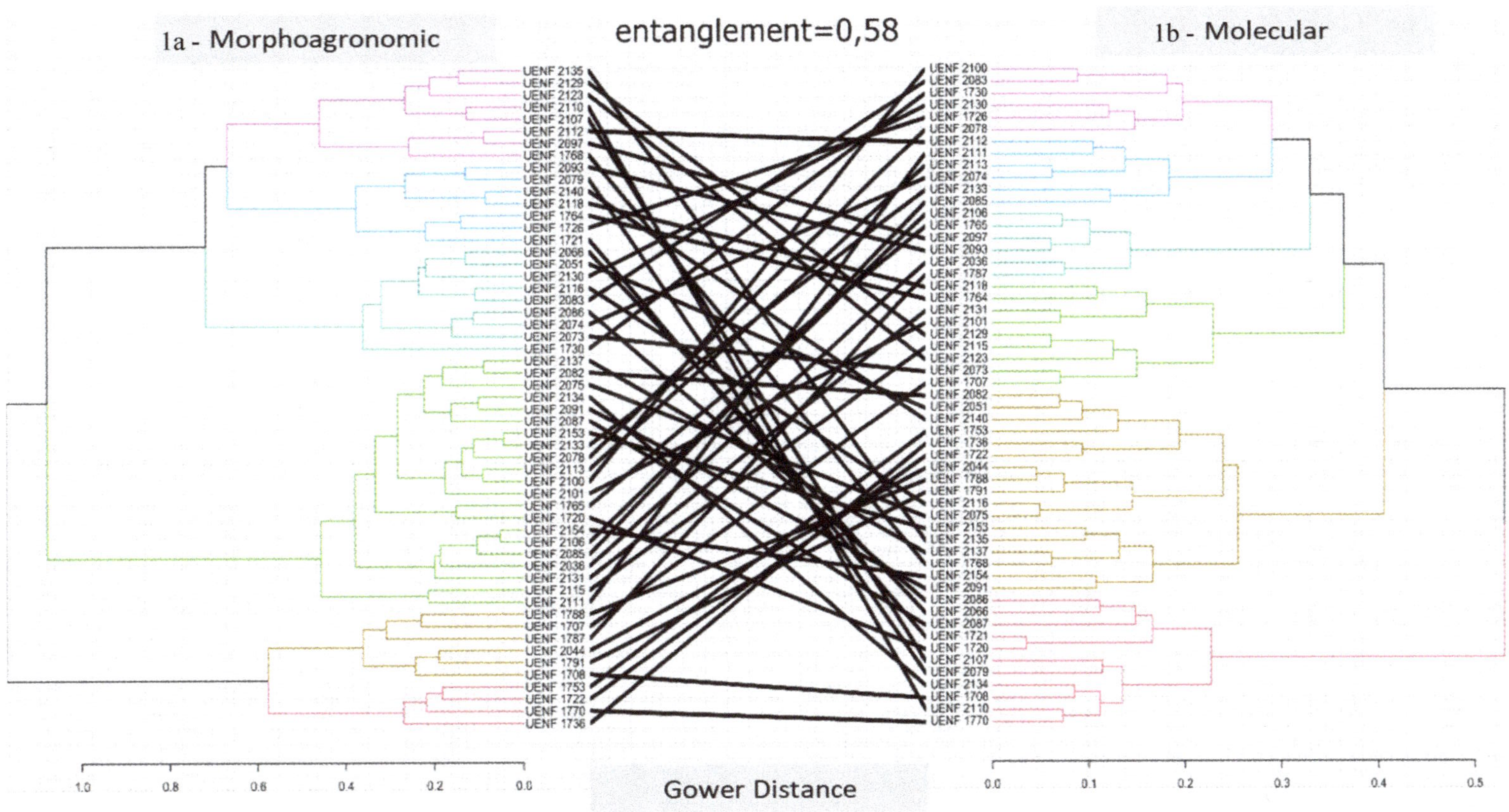

Figure 1. Grouping between 55 accessions of *C. chinense*, obtained by the Dendextend package, based on Gower Distance. **1a.** Morpho-agronomic traits; **1b.** Molecular variables.

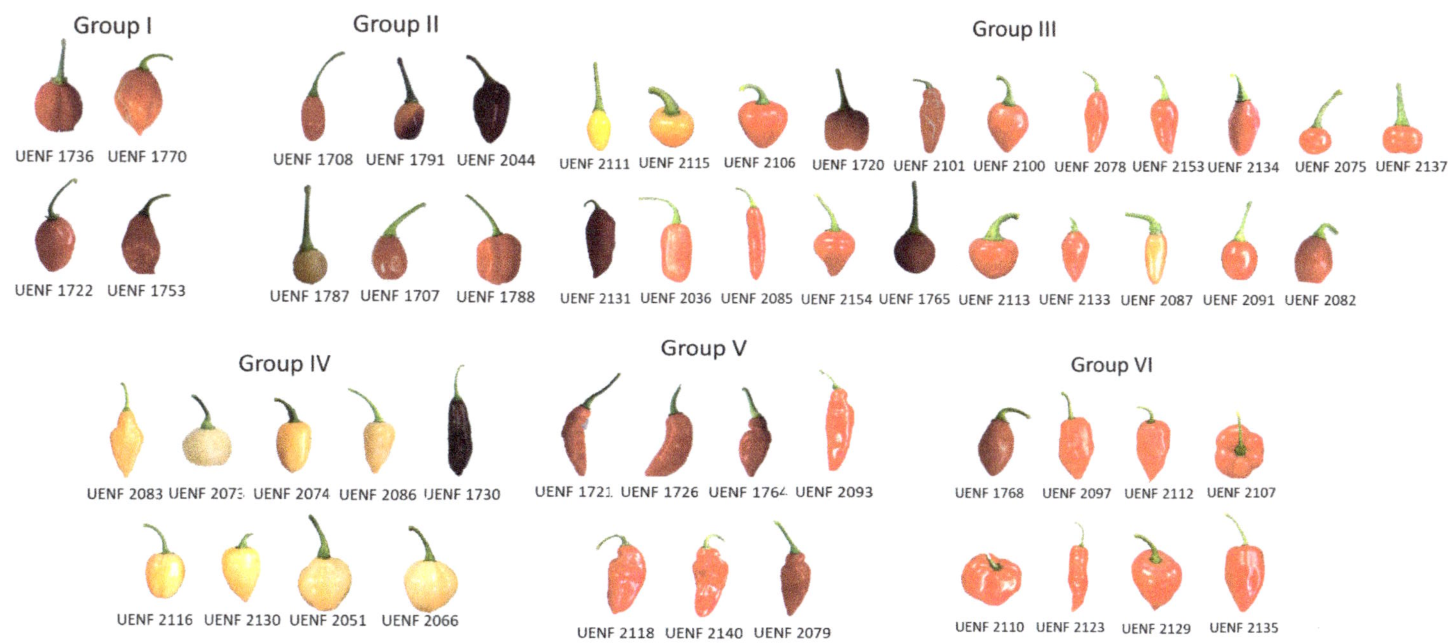

Figure 2. Phenotypic representation of fruits allocated in the dendrogram groups referring to the morpho-agronomic characterization.

Table 2. Characterization of *C. chinense* accessions obtained by molecular analysis with microsatellite markers.

Genotype	MAF	Na	I	H_o	F
UENF 1707	0.72	2	0.40	0.11	0.75
UENF 1708	0.50	2	0.50	0.33	0.38
UENF 1720	0.72	2	0.40	0.11	0.75
UENF 1721	0.72	2	0.40	0.11	0.75
UENF 1722	0.67	2	0.44	0	1.00
UENF 1726	0.78	2	0.35	0.22	0.41
UENF 1730	0.83	2	0.28	0.11	0.64
UENF 1736	0.78	2	0.35	0	1.00
UENF 1753	0.56	2	0.49	0.44	0.16
UENF 1764	0.78	2	0.35	0	1.00
UENF 1765	0.67	2	0.44	0	1.00
UENF 1768	0.83	2	0.28	0.11	0.64
UENF 1770	0.50	2	0.50	0.11	0.80
UENF 1787	0.61	2	0.48	0.11	0.79
UENF 1788	0.56	2	0.49	0	1.00
UENF 1791	0.72	2	0.40	0.11	0.75
UENF 2036	0.94	2	0.10	0.11	0
UENF 2044	0.78	2	0.35	0	1.00
UENF 2051	0.89	2	0.20	0	1.00
UENF 2066	0.81	2	0.30	0.13	0.63
UENF 2073	0.61	2	0.48	0.11	0.79
UENF 2074	0.78	2	0.35	0	1.00
UENF 2075	0.72	2	0.40	0.11	0.75
UENF 2078	0.72	2	0.40	0.11	0.75
UENF 2079	0.67	2	0.44	0.22	0.54
UENF 2082	0.83	2	0.28	0.11	0.64
UENF 2083	0.56	2	0.49	0	1.00
UENF 2085	0.72	2	0.40	0.11	0.75
UENF 2086	0.83	2	0.28	0.11	0.64
UENF 2087	0.83	2	0.28	0.11	0.64
UENF 2091	0.89	2	0.20	0	1.00
UENF 2093	0.56	2	0.49	0.13	0.77
UENF 2097	0.71	2	0.41	0	1.00
UENF 2100	0.78	2	0.35	0.22	0.41
UENF 2101	0.67	2	0.44	0	1.00
UENF 2106	0.83	2	0.28	0.33	−0.14
UENF 2107	0.67	2	0.44	0	1.00
UENF 2110	0.72	2	0.40	0.11	0.75
UENF 2111	0.56	2	0.49	0	1.00
UENF 2112	0.81	2	0.30	0.13	0.63
UENF 2113	0.83	2	0.28	0.11	0.64
UENF 2115	0.72	2	0.40	0.11	0.75
UENF 2116	0.61	2	0.48	0.11	0.79
UENF 2118	0.94	2	0.10	0.11	0
UENF 2123	0.67	2	0.44	0	1.00
UENF 2129	0.83	2	0.28	0.11	0.64
UENF 2130	0.61	2	0.48	0.11	0.79
UENF 2131	0.72	2	0.40	0.11	0.75
UENF 2133	0.67	2	0.44	0	1.00
UENF 2134	0.56	2	0.49	0	1.00
UENF 2135	0.67	2	0.44	0.22	0.54
UENF 2137	0.67	2	0.44	0	1.00
UENF 2140	0.78	2	0.35	0.22	0.41
UENF 2153	0.63	2	0.47	0	1.00
UENF 2154	0.61	2	0.48	0.11	0.79
Mean	0.72	2	0.38	0.10	0.77

MAF—Major Allele Frequency; NA—Number of alleles; I—Diversity index; Ho—Observed heterozygosity; F—Fixation index.

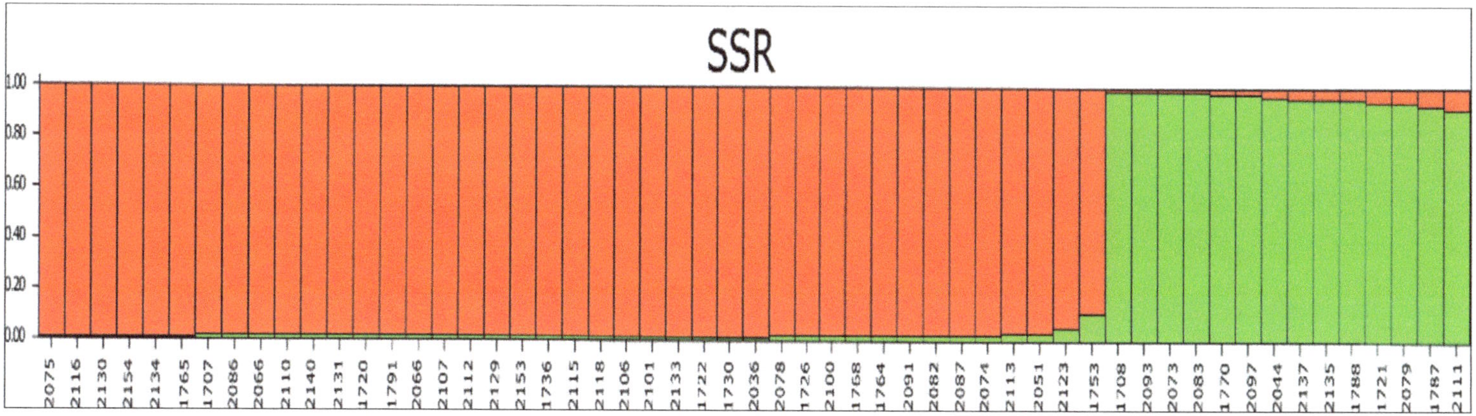

Figure 3. Genetic structure estimate for *C. chinense* accessions by Simple Sequence Repeat (SSR) primers. Group I is indicated by the color red and group II is indicated by the color green.

3.4. Molecular Diversity

A combined grouping was obtained for the ISSR and SSR markers by the Ward clustering analysis. The dendrogram obtained shows the distinction of six groups (Figure 1b). Group I was composed of 11 accessions, which presented predominance of red fruit color on the mature stage. Group II was composed of seventeen accessions which presented similarity regarding their growth habit. Group III was composed of nine accessions, with predominance of flowers in intermediate position. Group IV was composed of six accessions in which presence of capsaicin were predominant. Group V also comprised six accessions with triangular shape as a common characteristic of this group. Group VI is characterized by clustering six accessions with two flowers per node.

The entanglement obtained was 0.58, evidencing the divergence in the distribution of the accessions in molecular and morpho-agronomic dendrograms. Most accessions were allocated in different groups in both dendrograms, with the exception of UENF 1736, UENF 1722, UENF 1753, UENF 1791, and UENF 2044 accessions, which were allocated to group VI in both molecular and morpho-agronomic dendrograms.

4. Discussion

The characterization of genetic diversity among *Capsicum* genus has become essential for the breeding programs of pepper species, important as a vegetable and spice crop world-wide. The investment in screening of this diversity has the potential to reveal many traditional varieties with distinguished values. In order to do this, we have described in this study the characterization of 55 *C. chinense* accessions from different regions of Brazil. The characterization was performed using qualitative and quantitative descriptors.

The consumer market for fresh peppers has good acceptance for different pepper sizes [48]. In agreement with our description, similar fruit weight was found by Castro and Dávila (2008) [49] (1.8 to 22.2 g), Lannes et al. (2007) [50] (0.9 to 19.1 g), and Moreira et al. (2018) [14] (5.34 to 19.05 g). The variables diameter and length of fruits also have a wide variation among the accessions. Close results were obtained by Domenico et al. (2012) [51] for accessions of "pimenta de cheiro", which presented variation from 2.1 to 7.7 cm for fruit length and 1.1 to 2.5 cm for diameter. Moreira et al. (2018) [14], evaluating *C. chinense* accessions, reported 3.5 to 9.94 cm for fruit length and 1.73 to 4.10 cm for fruit diameter. Some accessions presented high values or the thickness of the pericarp, ranging from 1.38 to 3.08mm. Rêgo et al. (2011) [52] evaluating different species of the genus *Capsicum* found values between 0.04 and 1.3 cm for this trait.

A thicker pericarp is an important aspect in fruit quality as it may increase the degree of resistance to pathogens and parasites during post-harvest and give a better appearance for consumers than fruits with a thinner pericarp [53]. In addition, among the attributes that most contributed to the discrimination of the genotypes are the fruit diameter (FD) (9.83%) and fruit weight (FW) (8.84%). This abundant phenotypic variation, including traits such as fruit shapes, color, and sizes are very interesting for current pepper breeding programs, focused on meet consumer preferences and product differentiation. The characterization of variability for fruit-related characteristics to select promising accessions has been also shown in other *Capsicum* sp. studies [10,12–14,24,29,35–37].

Pearson's correlation coefficient is a measure of linear association between two quantitative variables, in which one characteristic can be selected based on another. Pessoa et al. (2019) [54] evaluating the inheritance of seedling and plant traits in ornamental pepper, identified positive correlations for the traits plant height, first bifurcation height and canopy width, 96%, 88%, and 65%, respectively. They also affirmed that the positive correlation for most of the traits indicates that the recessive alleles were generally responsible for the increase in these traits.

Agyare et al. (2016) [55] assessing the genetic diversity of agro-morphological characters of pepper (*Capsicum* sp.) and evaluating characteristics similar to those evaluated in this work identified positive correlation between cotyledon leaf length and fruiting (30.50%) and negative correlation with cotyledon leaf width (−37.50). The cotyledon leaf width had a negative association with most morpho-agronomic characteristics in *Capsicum* and tomato [55,56]. Andrade Junior et al. (2018) [57] evaluating *C. annuum* and *C. chinense* also found a positive correlation between the pericarp thickness and the fruit characteristic (fruit length) of 83% and negatively with the plant height (−85%). Peña-Yam et al. (2019) [58] in *C. chinense* identified 49.2% entre pericarp thickness and fruit weight. The fruit shape and the pericarp thickness are important traits for the classification of accessions, varieties and cultivars of peppers. The positive correlation between these characteristics can directly affect the flavoring compounds essential to the *Capsicum* fruits [59].

Accessions sampling was determined in order to contemplate the diversity of Brazilian biomes. All regions of Brazil, except the South, and four biomes were sampled (Pantanal, Cerrado, Amazon and Atlantic Forest). Accessions from Caceres, MT, corresponded to 42% of the total analyzed, due to the high number of accession found per property at the time of collection, on average seven types of pepper per property. The municipality of Caceres-MT shares a border with Bolivia, the origin center of *Capsicum* genus. This municipality has an area of 29,031 km^2 and three biomes are present in its territory: Pantanal, Cerrado, and Amazon. *Capsicum* accessions collected in this region are expected to have a high adaptive capacity, since they are present in places that annually register the occurrence of flood and fires, in a natural way, in addition to anthropic actions (including fires and deforestation). The collections were carried out after a very strong rainy period, and plants above two meters in height were observed with a stem up to 10 cm in diameter, with bush size. At the time of collection, it was explained by locals that *C. chinense* plants lose their leaves during the time they are flooded and when the soil dries, there is a regrowth, characterizing them as perennials, which differentiates them from peppers from other regions of Brazil. These features raise the hypothesis that these accessions have great potential for use in order to mitigate the effects of ongoing climate changes, whether in breeding programs or for immediate use.

The use of molecular markers is an effective strategy to complementing phenotypic characterization in detecting additional sources of genetic diversity present within the gene pool. In our study, the ISSR markers detected 47% polymorphism among accessions of *C. chinense* or 45 polymorphic loci. Different from our results, Moulin et al. (2015) [60], using 35 ISSR primers in *C. baccatum* var. *pendulum* observed a total of 201 polymorphic loci. The characterization of 81 accessions of *Capsicum* spp. with 13 ISSR markers resulted in a total of 88 amplified loci where 80 of them were polymorphic [61]. ISSR markers provide a large amount of information and enable the identification of polymorphic loci, making the correct differentiation among the accessions. The SSR markers are highly polymorphic and widely distributed in the pepper genome [62]. In addition, they have been widely used for genetic diversity assessment of germplasm because of their ability to detect multi-allelic forms of variation and being co-dominant, are able to distinguish genetic relationships between genotypes. The SSR analysis showed a transfer rate of approximately 36.17%.

The values for the major allele frequency (MAF) ranged from 0.5 to 0.94, with an average of 0.72. The effective number of alleles per locus was two alleles for all genotypes, which is in agreement with other pepper studies [63]. The diversity index (I) indicating the presence of diversity among accessions. Very close to our results, Meng et al. (2017) [64], studying the genetic diversity in the genus *Capsicum*, found values between 0.384 and 1.379, with an average of 0.508. The observed heterozygosity (Ho) suggesting the presence of homozygotes among individuals and the fixation index indicate the presence of inbreeding in the samples. Pepper is traditionally a cross-pollinated crop with its bisexual flower. The domesticated peppers are diploid and predominantly perform self-pollination, contributing to the high homozygous accessions.

The groups obtained by Bayesian analysis were not formed exclusively by accessions from the same geographical origin. This can be attributed to seed exchange between farmers and free fruit transport between the different regions of Brazil, also indicated by Cardoso et al. (2018) [12]. In contrast, Moses et al. (2014) [11], studying the genetic diversity of *C. chinense* accessions using microsatellite molecular markers, observed the formation of two distinct genetic clusters, corresponding to the upper and lower Amazon regions, suggesting two independent domestication events or two centers of diversity in these regions.

The description of diversity for phenotypic traits was established using Ward hierarchical clustering analysis allowing the identification of six groups. In general, the accessions could not be grouped based on the geographic origin, since the accessions of same geographic areas were classified in different groups. The same observation was presented by Baba et al. (2016) [13] and Moreira et al. (2018) [14]. The Ward hierarchical clustering analysis of molecular data also allowed the identification of six groups. However, when comparing clusters formed by morpho-agronomic and molecular data, an agreement of 42%, considered low, was observed. Low correlation or even no correlation between morpho-agronomic and molecular data was reported in other studies [13,65–67]. Molecular markers not related to the morphoagronomic traits evaluated used in the study should provide a plausible explanation for the absence or low correlation between formed clusters. Microsatellites are present in both coding and non-coding regions and are therefore not necessarily linked to the expression of morphological traits. We can assume that both characterization stages are important for an understanding of the genetic variability of a population and for developing effective strategies for germplasm conservation and breeding purposes. Similar to phenotypic characterization, no association with the geographical origin was observed.

5. Conclusion

A high level of variation was found among the most qualitative and quantitative traits evaluated. Pearson's correlation showed a high correlation between a few pairs of phenotypic traits: Plant height and plant canopy width, fruit diameter and pericarp thickness, and days for fruiting and maturation. Based on this genetic diversity, the clustering analysis formed a dendrogram with these distinct groups with the 55 accessions of *C. chinense*, showed the divergence between the molecular and morpho-agronomic analyzes of the accessions.

In the molecular characterization, of the 18 ISSR, 15 of those primers detected polymorphism and of the 47 SSR evaluated, only 9 were polymorphic between accessions. There was a significant variation for the Major Allele Frequency, Number of alleles, Diversity index, Observed heterozygosity and Fixation index.

There was divergence among accessions considering morpho-agronomic and molecular analyzes, with no duplicate detected. The variability observed reflects accessions adaptation to different ecogeographic and cultural conditions in which they were collected. These accessions have the potential to be used in several breeding programs, including tolerance to abiotic stresses.

Supplementary Materials: The following are available online at http://www.mdpi.com/2073-4395/10/3/447/s1, Figure S1: Representative sample of 2% agarose gel. DNA amplified fragments using ISSR marker 23. M – 100 bp DNA Ladder, Figure S2: Representative sample of 4% Metaphor gel. DNA amplified fragments using the microsatellite loci CB-164897. Table S1: Accession Identification, origin, latitude, longitude (collection place) and pollination type of 55 *C. chinense* accessions from *Universidade Estadual do Norte Fluminense Darcy Ribeiro* germplasm bank, Table S2: Descriptors used to characterize quantitative traits in 55 *C. chinense* accessions (54 from four different geographic Brazilian regions + one Peruvian) according to IPGRI (1995), Table S3: Descriptors for qualitative traits and phenotypes used to characterize 55 *C. chinense* accessions (54 from four different geographic Brazilian regions + one Peruvian) according to IPGRI (1995), Table S4: ISSR primers sequences used for genotyping 55 *C. chinense* accessions (54 from four different geographic Brazilian regions + one Peruvian), Table S5: Microsatellite (SSR) primers sequences used for genotyping 55 *C. chinense* accessions (54 from four different geographic Brazilian regions + one Peruvian), Table S6: Clustering of 16 quantitative traits means of 55 Capsicum chinense accessions by Scott-Knott method at 5% probability, Table S7: Correlations between the independent variables obtained by Pearson's correlation coefficient considering 16 descriptors used to characterize 55 *C. chinense* accessions (54 from four different geographic Brazilian regions + one Peruvian).

Author Contributions: P.A.B.: obtaining data and writing the original draft; L.R.A.d.S.: assistance in obtaining the data; A.A.d.S.A.: assistance in conducting the experiment; P.H.A.D.S.: data analysis; S.P.: data analysis; C.P.S.: assistance in obtaining the data; L.E.-D.C.: writing—proofreading and editing; L.S.A.G.: writing—proofreading and editing; R.R.: project coordinator and student supervision. All authors have read and agreed to the published version of the manuscript.

Funding: "This study was financed in part by the Coordenação de Aperfeiçoamento de Pessoal de Nível Superior—Brasil (CAPES)—Finance Code 001". FAPERJ Grant 202.985/2017. CNPq 307.569/2017-9.

Acknowledgments: The authors thank CAPES for granting the doctoral scholarship to the first author and *Universidade Estadual do Norte Fluminense Darcy Ribeiro, Conselho Nacional de Desenvolvimento Científico e Tecnológico* (CNPq), and *Fundação Carlos Chagas Filho de Amparo à Pesquisa do Estado do Rio de Janeiro (FAPERJ)* for support to research.

Conflicts of Interest: The authors declare no conflict of interest.

References

1. Hulse-Kemp, A.M.; Ashrafi, H.; Plieske, J.; Lemm, J.; Stoffel, K.; Hill, T.; Luerssen, H.; Pethiyagoda, C.L.; Lawley, C.T.; Ganal, M.W. A HapMap leads to a *Capsicum annuum* SNP infinium array: A new tool for pepper breeding. *Hortic. Res.* **2016**, *3*, 1–10. [CrossRef] [PubMed]
2. Bosland, P.W. Capsicum: Innovative uses of an ancient crop. In *Progress in New Crops*; Janick, J., Ed.; ASHS Press: Arlington, VA, USA, 1996; pp. 479–487.
3. Neitzke, R.S. *Genetic Resources of Capsicum Peppers—Exploring the Multiplicity of Uses. Tese (Doutorado em Agronomia)*; Universidade Federal de Pelotas: Pelotas, Brazil, 2012.
4. Neitzke, R.S.; Fischer, S.Z.; Vasconcelos, C.S.; Barbieri, R.L.; Treptow, R.O. Pimentas ornamentais: Aceitação e preferências do público consumidor. *Hortic. Bras.* **2016**, *34*, 102–109. [CrossRef]
5. Barboza, G.E.; Garcia, C.C.; Gonzalez, S.L.; Scaldaferro, M.; Reyes, X. Four new species of Capsicum (Solanaceae) from the tropical Andes and an update on the phylogeny of the genus. *PLoS ONE* **2019**, *14*. [CrossRef] [PubMed]
6. The Plant List. Available online: http://www.theplantlist.org (accessed on 10 December 2019).
7. DeWitt, D.; Bosland, P.W. *The Complete Chilli Pepper Book: A Gardener's Guide to Choosing, Growing, Preserving and Cooking*; Timber Press: Portland, OR, USA, 2009.
8. Barboza, G.E.; Bianchetti, L.D.B. Three New Species of *Capsicum* (Solanaceae) and a Key to the Wild Species from Brazil. *Am. Soc. Plant Taxon.* **2005**, *30*, 863–871. [CrossRef]
9. Moscone, E.A.; Scaldaferro, M.A.; Gabriele, M.; Cecchini, N.M.; García, Y.S.; Jarret, R.; Daviña, J.R.; Ducasse, D.A.; Barboza, G.E.; Ehrendorfer, F. The Evolution of Chili Peppers (Capsicum—Solanaceae): A Cytogenetic Perspective. In Proceedings of the 6th International Solanaceae Conference, Madison, WI, USA, 23–27 July 2006; Spooner, D.M., Ed.; Volume 745, pp. 137–170.
10. Pickersgill, B. Cytogenetics and evolution of *Capsicum* L. In *Chromosome Engineering in Plants: Genetics, Breeding, Evolution, Part B*; Tsuchiya, T., Gupta, P.K., Eds.; Elsevier: Amsterdam, The Netherlands, 1991; pp. 139–160.
11. Moses, M.; Umaharan, P. Genetic structure and phylogenetic relationships of *Capsicum Chinese*. *J. Am. Soc. Hortic. Sci.* **2012**, *137*, 250–262. [CrossRef]
12. Moses, M.; Umaharan, P.; Dayanandan, S. Microsatellite based analysis of the genetic structure and diversity of Capsicum chinense in the Neotropics. *Genet. Resour. Crop. Evolut.* **2014**, *61*, 741–755. [CrossRef]
13. Carvalho, S.I.C.; Ragassi, C.F.; Bianchetti, L.B.; Reifschneider, F.J.B.; Buso, G.S.C.; Faleiro, F.G. Morphological and genetic relationships between wild and domesticated forms of peppers (*Capsicum frutescens* L. and *C. chinense* Jacquin). *Genet. Mol. Res.* **2014**, *13*, 7447–7464. [CrossRef]
14. Baba, V.Y.; Rocha, K.R.; Gomes, G.P.; Ruas, C.F.; Ruas, P.M.; Rodrigues, R.; Gonçalves, L.S.A. Genetic diversity of *Capsicum chinense* accessions based on fruit morphological characterization and AFLP markers. *Genet. Resour. Crop Evolut.* **2016**, *63*, 1371–1381. [CrossRef]
15. Moreira, A.F.P.; Ruas, P.M.; Ruas, C.F.; Baba, V.Y.; Giordani, W.; Arruda, I.M.; Rodrigues, R.; Gonçalves, L.S.A. Genetic diversity, population structure and genetic parameters of fruit traits in *Capsicum Chinese*. *Sci. Hortic.* **2018**, *236*, 1–9. [CrossRef]
16. Nass, L.L.; Sigrist, M.S.; Ribeiro, C.S.C.; Reifschneider, F.J.B. Genetic resources: The basis for sustainable and competitive plant breeding. *Crop Breed. Appl. Biotechnol.* **2012**, *12*, 75–86. [CrossRef]
17. Ballina-Gomez, H.; Ruiz-Sanchez, E.; Chan-Cupul, W.; Latournerie-Moreno, L.; Hernandez-Alvarado, L.; Islas-Flores, I.; Zuñiga-Aguilar, J.J. Response of Bemisia tabaci Genn. (Hemiptera: Aleyrodidae) Biotype B to Genotypes of Pepper *Capsicum annuum* (Solanales: Solanaceae). *Neotrop. Entomol.* **2013**, *42*, 205. [CrossRef]

18. Nicolai, M.; Cantet, M.; Lefebvre, V.; Sage-Palloix, A.M.; Palloix, A. Genotyping a large collection of pepper (*Capsicum* spp.) with SSR loci brings new evidence for the wild origin of cultivated *C. annuum* and the structuring of genetic diversity by human selection of cultivar types. *Genet. Resour. Crop Evolut.* **2013**, *60*, 2375–2390. [CrossRef]
19. Quang, D.V.; Schreinemachers, P.; Berger, T. Ex-ante assessment of soil conservation methods in the uplands of Vietnam: An agent-based modeling approach. *Agric. Syst.* **2015**, *123*, 108–119. [CrossRef]
20. Sudré, C.P.; Gonçalves, L.S.A.; Rodrigues, R.; Amaral Júnior, A.T.; Riva-Souza, E.M.; Bento, C.S. Genetic variability in domesticated *Capsicum* spp. as assessed by morphological and agronomic data in mixed statistical analysis. *Genet. Mol. Res.* **2010**, *9*, 283–294. [CrossRef] [PubMed]
21. Santos, M. *Metamorfoses do Espaço Habitado*; EDUSP: São Paulo, Brazil, 2014.
22. Newman, P. The environmental impact of cities. *Environ. Urban.* **2006**, *18*, 275–295. [CrossRef]
23. Ribeiro, L.S. *A Relevância Dos Patrimônios Naturais Da Humanidade E Das Reservas Da Biosfera Para a Conservação Da Biodiversidade—O Caso Do Parque Nacional Da Chapada Dos Veadeiros (PNCV) Monografia De Especialização—Universidade De Brasília*; Centro UnB Cerrado: Alto Paraíso, Brazil, 2018.
24. Organização das Nações Unidas. Available online: https://nacoesunidas.org/pos2015/agenda2030/ (accessed on 14 December 2019).
25. Araújo, R.F.; Abud, H.F.; Silva, L.J.; Araújo, E.F.; Pinto, C.M.F.; Silva, F.W.S. Alterações fisiológicas e atividade de enzimas antioxidantes em sementes de pimentas *Capsicum chinense* Jacq e *Capsicum frutescens* L. durante o processo de maturação. *Rev. Ceres* **2018**, *65*, 534–545. [CrossRef]
26. Zonneveld, M.; Ramirez, M.; Williams, D.E.; Petz, M.; Meckelmann, S.; Avila, T.; Bejarano, C.; Ros, L.; Pea, K.; Jger, M. Screening Genetic Resources of *Capsicum* Peppers in Their Primary Center of Diversity in Bolivia and Peru. *PLoS ONE* **2015**, *10*, e0134663. [CrossRef] [PubMed]
27. Bianchi, P.A.; Dutra, I.P.; Moulin, M.M.; Santos, J.O.; Júnior, A.C.S. Morphological characterization and analysis of genetic variability among pepper accessions. *Ciência Rural* **2016**, *46*, 1151–1157. [CrossRef]
28. Agarwal, M.; Shrivastava, N.; Padh, H. Advances in molecular marker techniques and their applications in plant sciences. *Plant Cell Rep.* **2008**, *27*, 617–631. [CrossRef]
29. Grover, A.; Sharma, P. Development and use of molecular markers: Past and present. *Crit. Rev. Biotechnol.* **2014**, *36*, 1–13. [CrossRef]
30. Finger, F.L.; Lannes, S.D.; Schuelter, A.R.; Doege, J.; Comerlato, A.P.; Gonçalves, L.S.A.; Ferreira, F.R.A.; Clovis, L.R.; Scapim, C.A. Genetic diversity of *Capsicum chinense* (Solanaceae) accessions based on molecular markers and morphological and agronomic traits. *Genet. Mol. Res.* **2010**, *9*, 1852–1864. [CrossRef]
31. Wahyuni, Y.; Ballester, A.R.; Tikunov, Y.; de Vos, R.C.H.; Pelgrom, K.T.B.; Maharijaya, A.; Sudarmonowati, E.; Bino, R.J.; Bovy, A.G. Metabolomics and molecular marker analysis to explore pepper (*Capsicum* sp.) biodiversity. *Metabolomics* **2013**, *9*, 130. [CrossRef]
32. Thul, S.; Darokar, M.; Shasany, A.; Khanuja, S. Molecular Profiling for Genetic Variability in *Capsicum* Species Based on ISSR and RAPD Markers. *Mol. Biotechnol.* **2011**, *51*, 137–147. [CrossRef]
33. Pacheco-Olvera, A.; Hernandez-Verdugo, S.; Rocha-Ramirez, V.; Gonzalez-Rodriguez, A.; Oyama, K. Genetic Diversity and Structure of Pepper (*Capsicum Annuum* L.) from Northwestern Mexico Analyzed by Microsatellite Markers. *Crop Sci.* **2012**, *52*, 231–241. [CrossRef]
34. Rai, V.P.; Kumar, R.; Kumar, S.; Rai, A.; Kumar, S.; Singh, M.; Singh, S.P.; Rai, A.B.; Paliwal, R. Genetic diversity in *Capsicum* germplasm based on microsatellite and random amplified microsatellite polymorphism markers. *Physiol. Mol. Biol. Plants* **2013**, *19*, 575–586. [CrossRef]
35. Carvalho, S.I.C.; Ragassi, C.F.; Oliveira, I.B.; Amaral, Z.P.S.; Reifschneider, F.J.B.; Faleiro, F.G.; Buso, G.S.C. Transferability of microsatellite markers of *Capsicum annuum* L. to *C. frutescens* L. and *C. chinense* Jacq. *Genet. Mol. Res.* **2015**, *14*, 7937–7946. [CrossRef]
36. Cardoso, R.; Ruas, C.F.; Giacomin, R.M.; Ruas, P.M.; Ruas, E.A.; Barbieri, R.L.; Rodrigues, R.; Gonçalves, L.S.A. Genetic variability in Brazilian *Capsicum baccatum* germplasm collection assessed by morphological fruit traits and AFLP markers. *PLoS ONE* **2018**, *13*, e0196468. [CrossRef]
37. Jarret, R.L.; Berke, T. Variation for Fruit Morphological Characteristics in a *Capsicum chinense* Jacq. Germplasm Collection. *HortScience* **2008**, *43*, 1694–1697. [CrossRef]
38. Bharath, S.M.; Cilas, C.; Umaharan, P. Fruit Trait Variation in a Caribbean Germplasm Collection of Aromatic Hot Peppers (*Capsicum chinense* Jacq.). *HortScience* **2013**, *48*, 531–538. [CrossRef]

39. IPGRI. *Descriptors for Capsicum (Capsicum spp.)*; International Plant Genetic Resources Institute: Rome, Italy, 1995; 49p.
40. Doyle, J.J.; Doyle, J.L. Isolation of plant DNA from fresh tissue. *Focus* **1987**, *12*, 13.
41. Daher, R.F.; Pereira, M.G.; Tupinamba, E.A.; Amaral Júnior, A.T.; Aragão, W.N.; Ribeiro, F.E.; Oliveira, L.O.; Sakiyama, N.S. Assessment of coconut tree genetic divergence by compound sample RAPD marker analysis. *Crop Breed. Appl. Biotechnol.* **2002**, *3*, 431–438. [CrossRef]
42. Minamiyama, Y.; Tsuro, M.; Hirai, M. An SSR-basead linkage map of *Capsicum annuum*. *Mol. Breed.* **2006**, *18*, 157–169. [CrossRef]
43. Singh, D. The relative importance of characters affecting genetic divergence. *Indian J. Genet. Plant Breed.* **1981**, *41*, 237–245.
44. Cruz, C.D. GENES—A software package for analysis in experimental statistics and quantitative genetics. *Acta Sci.* **2013**, *35*, 271–276.
45. Gower, J.C. A general coefficient of similarity and some of its properties. *Biometrics* **1971**, *27*, 857–874. [CrossRef]
46. R Core Team. *R: A Language and ENVIRONMENT for statistical Computing*; R Foundation for Statistical Computing: Vienna, Austria, 2017; ISBN 3-900051-07-0. Available online: http://www.R-project.org/ (accessed on 9 May 2019).
47. Pritchard, J.K.; Stephens, M.; Donnelly, P. Inference of population structure using multilocus genotype data. *Genetics* **2000**, *155*, 945–959.
48. Evanno, G.; Regnaut, S.; Goudet, J. Detecting the number of clusters of individuals using the software STRUCTURE: A simulation study. *Mol. Ecol.* **2005**, *14*, 2611–2620. [CrossRef]
49. Singh, P.; Cheema, D.S.; Dhaliwa, M.S.; Garg, N. Heterosis and 51 combining ability for earliness, plant growth, yield and fruit attributes in hot pepper (*Capsicum annuum* L.) Involving genetic and cytoplasmic-genetic male sterile lines. *Sci. Hortic.* **2014**, *168*, 175–188. [CrossRef]
50. Castro, S.P.; Dávila, A.M.A.G. Caracterización morfológica de 93 accesiones de *Capsicum* spp del banco de germoplasma de la Universidad Nacional de Colombia—Sede Palmira. *Acta Agron.* **2008**, *57*, 247–252.
51. Lannes, S.D.; Finger, F.L.; SchuelteR, A.R.; Casali, V.W.D. Growth and quality of Brazilian accessions of *Capsicum chinense* fruits. *Sci. Hortic.* **2007**, *112*, 266–270. [CrossRef]
52. Domenico, C.I.; Coutinho, J.P.; Godoy, H.T.; Melo, A.M.T. Caracterização agronômica e pungência em pimenta de cheiro. *Hortic. Bras.* **2012**, *30*, 466–472. [CrossRef]
53. Rêgo, E.R.; Rêgo, M.M.; Matos, I.W.F.; Barbosa, L.A. Morphological and chemical characterization of fruits of *Capsicum* spp. accessions. *Hortic. Bras.* **2011**, *29*, 3. [CrossRef]
54. Ferrão, L.F.V.; Cecon, P.R.; Finger, F.L.; Silva, F.F.; Puiatti, M. Divergência genética entre genótipos de pimenta com base em caracteres morfo-agrônomicos. *Hortic. Bras.* **2011**, *29*, 354–358. [CrossRef]
55. Pessoa, A.M.S.; Rêgo, E.R.; Santos, C.A.P.; Carvalho, M.G.; Rêgo, M.M. Inheritance of seedling and plant traits in ornamental pepper (*Capsicum annuum*). *Genet. Mol. Res.* **2019**, *18*, 18120. [CrossRef]
56. Agyare, R.Y.; Akromah, R.; Abdulai, M.S. Assesment of genetic diversity in pepper (*Capsicum* spp.) landraces from Ghana using agro-morphological characters. *Am. J. Exp. Agric.* **2016**, *12*, 1–16.
57. Ghosh, K.P.; Islam, A.K.M.A.; Mian, M.A.K.; Hossain, M.M. Variability and character association in F_2 segregating population of different commercial hybrids of tomato (*Solanum lycopersicum* L.). *J. Appl. Sci. Environ. Manag.* **2010**, *14*, 91–95. [CrossRef]
58. Andrade Júnior, V.C.; Pedrosa, C.E.; Miranda, T.G.; Valadares, N.R.; Pereira, S.L.; Azevedo, A.M. Biometric evaluation of morpho-agronomic traits in pepper lines and hybrids. *Hortic. Bras.* **2018**, *36*, 357–361. [CrossRef]
59. Peña-Yam, L.P.; Muñoz-Ramírez, L.S.; Avilés-Viñas, S.A.; Canto-Flick, A.; Pérez-Pastrana, J.; Guzmán-Antonio, A.; Santana-Buzzy, N.; Aguilera-Cauich, E.A.; Mijangos-Cortes, J.O. Analysis of Genetic Parameters of Habanero Pepper (*Capsicum chinense* Jacq.) in the Yucatan, Mexico. *HortScience* **2019**, *54*, 429–433. [CrossRef]
60. Naegele, R.P.; Mitcheil, J.; Hausbeck, M.K. Genetic diversity, population structure, and heritability of fruits traits in *Capsicum annuum*. *PLoS ONE* **2016**, *11*, e0156969. [CrossRef]
61. Moulin, M.M.; Rodrigues, R.; Ramos, H.C.C.; Bento, C.S.; Sudré, C.P.; Gonçalves, L.S.A.; Viana, A.P. Construction of an integrated genetic map for *Capsicum baccatum* L. *Genet. Mol. Res.* **2015**, *14*, 6683–6694. [CrossRef]

62. Campos, A.L. Coleta e Caracterização de Acessos de *Capsicum* spp. Da Agrobiodiversidade Da Região Sudoeste Do Mato Grosso. Master's Thesis, Universidade do Estado do Mato Grosso, Mato Grosso, Brazil, 2014.
63. Mimura, Y.; Inoue, T.; Minamiyama, Y.; Kubo, N. An SSR-based genetic map of pepper (*Capsicum annuum* L.) serves as an anchor for the alignment of major pepper maps. *Breed. Sci.* **2012**, *62*, 93–98. [CrossRef]
64. Dhaliwal, M.S.; Yadav, A.; Jindal, S.K. Molecular Characterization and Diversity Analysis in Chilli Pepper Using Simple Sequence Repeats (SSR) Markers. *Afr. J. Biotechnol.* **2014**, *13*, 3137–3143.
65. Meng, C.Y.; Wei, X.C.; Zhao, Y.Y.; Yuan, Y.X.; Yang, S.J.; Wang, Z.Y.; Zhang, X.W.; Sun, J.W.; Yang, L.; Yao, Q.J.; et al. Genetic diversity analysis of *Capsicum* genus by SSR markers. *Mol. Plant Breed.* **2017**, *8*, 70–78. [CrossRef]
66. Costa, F.R.; Pereira, T.N.S.; Vitória, A.P.; Campos, K.P.; Rodrigues, R.; Silva, D.H.; Pereira, M.G. Genetic diversity among *Capsicum* accessions using RAPD markers. *Crop Breed. Appl. Biotechnol.* **2006**, *6*, 18–23. [CrossRef]
67. Leite, P.S.S.; Rodrigues, R.; Silva, R.N.O.; Pimenta, S.; Medeiros, A.M.; Bento, C.S.; Gonçalves, L.S.A. Molecular and agronomic analysis of intraspecific variability in *Capsicum baccatum* var. *pendulum* accessions. *Genet. Mol. Res.* **2016**, *15*, 4. [CrossRef]

agronomy

MDPI

Article

Genetic Variability of Eggplant Germplasm Evaluated under Open Field and Glasshouse Cropping Conditions

Nur Nadzirah Mat Sulaiman [1], Mohd Y. Rafii [1,2,*], Janejira Duangjit [3,*], Shairul Izan Ramlee [1,2], Chalermpol Phumichai [4], Yusuff Oladosu [1], Debi Rani Datta [1] and Ibrahim Musa [1]

1 Laboratory of Climate-Smart Food Crop Production, Institute of Tropical Agriculture and Food Security, Universiti Putra Malaysia (UPM), Serdang 43400, Selangor, Malaysia
2 Department of Crop Science, Faculty of Agriculture, University Putra Malaysia, Serdang 43400 UPM, Selangor, Malaysia
3 Department of Horticulture, Faculty of Agriculture, Kasetsart University, 50 Ngam Wong Wan Road Lat Yao, Chatuchak, Bangkok 10900, Thailand
4 Department of Agronomy, Faculty of Agriculture, Kasetsart University, 50 Ngam Wong Wan Road Lat Yao, Chatuchak, Bangkok 10900, Thailand
* Correspondence: mrafii@upm.edu.my (M.Y.R.); janejira.d@ku.th (J.D.); Tel.: +603-9769-1043 (M.Y.R.); +662-579-0308 (J.D.)

Received: 19 February 2020; Accepted: 19 March 2020; Published: 23 March 2020

Abstract: Knowledge of agro-morphological genetic variation and cropping conditions on vegetative and yield-related traits plays a significant role in varietal improvement and production of eggplant (*Solanum melongena* L.). Following this premise, the current study was conducted to critically asses the genetic variation of 29 eggplant accessions by using agro-morphological characterization evaluated under two cropping conditions, namely, glasshouse and open field. The experiments were laid out in randomized complete block design (RCBD) with three replications. Data on vegetative and yield characteristics were collected and subjected to analysis of variance (ANOVA) using SAS 9.4, while variance components were estimated manually. The results obtained from the analysis of variance indicated a highly significant difference ($p \leq 0.01$) for all characteristics studied in both cropping conditions. The evaluated accessions were grouped into six major clusters based on agro-morphological traits using Unweighted Pair Group Method with Arithmetic mean (UPGMA) dendrogram. Hence, crosses between group I with VI or V could be used to attain higher heterosis and vigor among the accessions. Also, this evaluation could be used as a selection criterion for important yield agronomic traits in eggplant. The methodology and the approaches used may provide a model for the enhancement of other vegetable crop diversity towards adaptability to the cropping condition decision. This result displayed importance for preserving eggplant germplasm for future varietal development and revealed that open field cropping condition is more suitable under Malaysia's agroecology.

Keywords: genetic variation; eggplant; cropping condition; yield; agro-morphological characterization

1. Introduction

Eggplant (*Solanum melongena* L.) is one of the important vegetables belonging to the family Solanaceae, which comprises other significant crop species including chilli (*Capsicum annuum* L.), tomato (*Solanum lycopersicum* L.), tobacco (*Nicotiana tabacum* L.), and potato (*Solanum tuberosum* L.). Eggplant is an old world crop species, unlike other members of the Solanaceae family [1]. According to [1], eggplant's ancestor, *Solanum incanum* or eggplant bitter apple, pre-domesticated in the subtropical species and is a native of West Asia and North Africa that is used as the source of resistance to drought and variation for phenolics content in eggplant breeding programs [2]. The global eggplant production statistic in 2017 was 52.31 million tons [3], going up by 2.18% against the previous year. This global eggplant production peaked in 2017, and the growth trend pattern is likely to be on a continuous increase [3]. China and India are the top eggplant producing countries in the world followed by Egypt, Turkey, Indonesia, Iran, Philippines, Spain, Mexico, Japan, Italy, and Syria (http://www.fao.org/faostat/en/#data/QC). In the context of Malaysia, vegetable production—especially eggplant—has been low with only 39,311.5 metric tons [4], and one factor identified was due to lack of genetic resources. This has forced the country to rely on importation of vegetable seeds to meet 265 tons annually, but locally they can only meet 13 tons [5], leading to a deficit of 252 tons/year, which is being imported from overseas [6]. With limited eggplant cultivation and studies, there is a duty call for awareness of the importance of utilization and development of our available varieties gearing towards food security and the high-value vegetable market of Malaysia. The local production could improve through the exploration of germplasm. The existence of compelling high genetic variation is critical for expanding the stricken eggplant genetic base and advancing current germplasm, whether it is local or commercial germplasm.

Generally, morphological characterization is the first step in exploring eggplant genetic variation. Hence, there is a wide variation in eggplant habitats as well as vegetative and agro-morphological characteristics [7]. Besides, the evaluation of genetic variation and adaptation to climatic conditions using agronomic traits has been the focus of research in the last decade [8]. Recently, taste, texture, and appearance are among the considered factors alongside nutrient compositions [9] that are being emphasized to meet consumer demand. In conjunction with that, eggplant has been bred for improved fruit quality, fruit yield, disease resistance, and adaptation with stable, high yielding performance across heterogeneous growing areas. Eventually, this marker was found interesting and was best applied by plant breeders due to easy scoring, low cost, and rapid method and evergreen evaluations. Also, investigation of these qualities required non-complex tools and equipment, and it could be acquired without explicit biochemical or molecular methods. Specifically, it could be aided with competent multivariate tools such as principal component analysis (PCA) [10], clustering, and discriminate analysis for assessing the genetic diversity of germplasm introduced in heterogeneous crops [11]. In addition, clustering analysis [12] was utilized together with pattern analysis for grouping prior scattered materials in which a collective use of methods was ordinated and classified for investigating the structure of the fundamental basis among germplasm [13]. We need a precise and practical classification of the eggplant germplasm genetic pool in different cultivar groups, which is vital to promoting their use of crop improvement.

A *priori*, eggplant is one of the model crops that can be grown in a heterogenous macro-environment [14]. However, knowledge of suitable cropping conditions and plant conditions is a prerequisite for cost-effective production. In any geographical area, the prevalent cropping condition is the aggregate results of the previous decisions by individuals, communities, or governments and their agencies. Hence, crop adaptability of the growing conditions such as raining seasons, species, and land use efficiency together with plant growth resources such as irrigation, climate, tradition, and experiences are among the determinant factors for efficient production. In the context of Malaysia, conventional open fields and glasshouses are the two most widely used cropping conditions for eggplant cultivation. None of the studies focused on eggplant genetic variation to simultaneously compare yield performance between these cropping conditions. Therefore, this study was conducted to evaluate genetic variation and establish relationships between vegetative, yield, and yield components using agro-morphological characterization among 29 eggplant accessions from Malaysia, Thailand, and China under two cropping conditions, namely, a fertigation system in the glasshouse and an open field condition.

2. Materials and Methods

2.1. Planting Materials and Agronomic Practice

Twenty-nine eggplants accessions, which form three main populations from Malaysia, Thailand, and China were used in this study, as presented in Table 1. The evaluation was conducted at Field 10 (S8) at Universiti Putra Malaysia (UPM), Serdang, Selangor, Malaysia, which is geographically located between 2°59′ north latitude and 101°42′ east longitude, with 45 m above sea level altitude. The other experiment was conducted in Fertigation Unit, Ladang 15, Universiti Putra Malaysia (UPM), Serdang, Selangor, Malaysia, which is between 2°59′ north latitude and 101°43′ east longitude, with an altitude of 55 m. The experiment was laid out in randomized complete block design (RCBD) with three replications. The environment is hot humid tropics with high humidity and adequate rainfall throughout the year. The seeds were sown in 104-holes seed germinating trays with 1–2 seeds per cell on peat moss growing medium. After 27 days of sowing, the seedlings were transplanted to mixed soil and peat moss with a ratio of 2:1 in polybags for the hardening phase for another 25 days before being transferred to the field and the glasshouse conditions. In each replication, the eggplant was planted with 50 cm spacing between the five plants of each accession and 1 m between the rows. Following the standard cultural practices, agronomic routines and plant maintenance such as fertilizer application, pest and disease management, and weeding were carried out. On a daily basis for the glasshouse fertigation system, the plants were supplied with modified copper formulation fertilizer consisting of N (200 mg L^{-1}), P (60 mg L^{-1}), K (300 mg L^{-1}), Ca (170 mg L^{-1}), Mg (50 mg L^{-1}), Fe (12 mg L^{-1}), Mn (2 mg L^{-1}), B (1.5 mg L^{-1}), Zn (0.1 mg L^{-1}), Cu (0.1 mg L^{-1}), and Mo (0.2 mg L^{-1}) [15] while the electron conductivity (EC) reading increased in succession according to the growing phase (0.5–3.0). In the aspect of pest and disease mitigation strategies, several pesticides were applied as recommended by the Department of Agriculture, Malaysia (http://jpn.penang.gov.my/index.php/perkhidmatan/teknologi-tanaman/sayur-sayuran/78-terung-sp-3424).

Table 1. List of 29 eggplants accessions.

No.	Accessions Code	Accessions Name	Origin Country	Fruit Type	Collection Source
1	1TR	CT Round Eggplant 01450	Thailand	Round	Village Market
2	2TL	CT Long Green Eggplant 01166	Thailand	Long	Village Market
3	3TR	CT Round Eggplant 01388	Thailand	Round	Village Market
4	4TR	CT El Ryu Eggplant 636	Thailand	Round	Village Market
5	5TR	MOP Eggplant 548	Thailand	Round	Village Market
6	6TL	MOP Eggplant 969	Thailand	Long	Village Market
7	7TR	MOP Eggplant 762	Thailand	Round	Village Market
8	8ML	Purple Dream 302	Malaysia	Long	Commercial market
9	9ML	Eggplant B. VE-023 F_1 Hybrid Long	Malaysia	Long	Commercial market
10	10ML	White Shining Eggplant 330 F_1 Hybrid	Malaysia	Long	Commercial market
11	13MR	Round Eggplant MTe2	Malaysia	Round	MARDI Gene Bank
12	14ML	Terung Belacan D/No 04-1272	Malaysia	Long	MARDI Gene Bank
13	15ML	Terung Rapuh Unggu 76	Malaysia	Round	MARDI Gene Bank
14	16ML	L. Little Nyonya 313	Malaysia	Long	Commercial market
15	17ML	L. Super Naga 312	Malaysia	Long	Commercial market
16	18ML	A. Nyonya Eggplant F_1 428	Malaysia	Long	Commercial market
17	19ML	A. Purple King F_1 418	Malaysia	Long	Commercial market
18	20ML	Pahuja	Malaysia	Long	Commercial market
19	21ML	MChina-3	China	Round	Commercial market
20	22ML	Mukta Keshi	Malaysia	Long	Commercial market
21	23ML	Makra Begun	Malaysia	Round	Commercial market
22	25ML	Brinjal Bhagan	Malaysia	Long	Commercial market
23	26CL	China 1	China	Long	Commercial market
24	27CL	TESH Eggplant 204	Malaysia	Long	Commercial market
25	29MN	NTH 08-0031	Malaysia	Long	MARDI Gene Bank
26	30MN	NTH 08-0077	Malaysia	Round	MARDI Gene Bank
27	32MN	NTH 08-0131	Malaysia	Long	MARDI Gene Bank
28	34CL	China 3	China	Long	Commercial market
29	35CL	China 2	China	Long	Commercial market

Note: MARDI = Malaysian Agricultural Research & Development Institute, TR = Thailand Round, TL = Thailand Long, ML = Malaysian Long, MR = Malaysian Round, MN = Malaysian Native, CL = China Long.

2.2. Data Collection

Thirteen sets of agro-morphological data were collected and measured from the two planting conditions following the description of the International Boards for Plant Genetic Resources [16] and European Eggplant Genetic Resources Network [17]. To obtain the means of the variables in each plot, five fruits were chosen at random from each of the three tagged plants in the middle of each replication during the harvest. The harvest was carried out five times subject to the productivity of plants that might vary between the accessions. Harvesting frequency and respective number of fruits taken were recorded (data not shown). Plants were evaluated on the yield components fruit girth (FGI), diameter of fruit (DFR), fruit length (FLE), fruit length to width ratio (FLW), total number of fruit (TNF), number of fruit per bunch (NFB), average fruit weight (AFW), and fruit yield per plant (FYP). These also included vegetative parameters such as number of primary branches (NPB), plant height (PHE), stem diameter (SDM), plant spread (PSP), and days to first flowering (DFF). All data measurement and observations were accomplished on the same day to reduce variation in plant growth developmental stage or environmental changes.

2.3. Statistical Analysis

All vegetative, yield, and yield-related data in collection were subjected to analysis of variance (ANOVA) using the SAS version 9.4 (SAS Institute Inc., Cary, NC, USA), while means comparisons were separated with least significant difference (LSD) at 5% level of significance. Additionally, grand mean, standard deviation, and coefficient of variation (CV) were recorded for each trait measured. Together with the analysis of variance (ANOVA), nested design analysis was a hierarchical design plotted with fruit type and the whole set of attributes of eggplant as an interest of evaluation in eggplant accession. Among the 29 accessions subjected in this agro-morphological analysis, ten accessions were the round type of eggplant, and 19 accessions were long shape type. Both long and round fruit types of specifically assigned eggplants were subsampled under accession and eventually made the fruit type within accession. Accession has a higher level as compared to fruit type.

Genetic relationships among the eggplant germplasm were determined using the unweighted pair group method with arithmetic mean (UPGMA algorithm) and sequential agglomerative hierarchical non-overlapping (SAHN) methods. Cluster trees [18] are important multivariate tools to assess genetic variation among the eggplant's germplasm under two cropping conditions. Utilizing the basis of comparable vegetative and yield components traits, the grouping of individual accessions was exposed by cluster analysis according to similarity and relatedness of eggplants. Other than that, restricted maximum likelihood (REML) using PROC VARCOMP in SAS version 9.4 (SAS Institute Inc., Cary, NC, USA) was used for variance components estimations. Genetic parameters such as genotypic coefficient variance (GCV), phenotypic coefficient variance (PCV), heritability in the broad sense ($H^2{}_B$), and genetic advance (GA) were calculated using the following equation [19].

$$\mathbf{PCV} = \frac{\sqrt{\sigma_P^2}}{\overline{X}} \times 100 \quad (1)$$

$$\mathbf{GCV} = \frac{\sqrt{\sigma^2{}_g}}{\overline{X}} \times 100 \quad (2)$$

where σ_P^2 is the phenotypic variance, $\sigma^2{}_g$ is the genotypic variance, and $\overline{X}$ is the traits mean. PCV and GCV were classified as low (<10%), moderate (10–20%), and high (>20%), as described by [20].

Percentage of broad-sense heritability was estimated as the proportion of phenotypic variance (σ^2_P) to genotypic variance (σ^2_g) as indicated below:

$$h^2_B = \frac{\sigma^2_g}{\sigma^2_P} \times 100 \quad (3)$$

where σ^2_P is phenotypic variance, and σ^2_g is the genotypic variance. Heritability values were estimated and standardized as low (0–30%), moderate (30–60%), and high (>60%) in accordance with Robinson et al. and Assefa et al. [21,22].

Expected genetic advance (GA) percentage was estimated following the method described by Johnson et al. [23].

$$\mathbf{GA}\% = \mathrm{K} \times \frac{\sqrt{\sigma^2_P}}{\overline{\mathrm{X}}} \times h^2_B \times 100 \quad (4)$$

where K is a constant that represents the selection intensity. At the value when k is at 5%, the rate is 2.06. $\frac{\sqrt{\sigma^2_P}}{\overline{\mathrm{X}}}$ is the phenotypic standard deviation, and h^2_B is the broad-sense heritability value. GA values of 0–10%, 10–20%, and >20% are low, intermediate, and high, respectively [21,24].

3. Results and Discussion

3.1. Yield and Yield Components Across Two Cropping Conditions

The pooled analyses of variance for yield and yield traits from two cropping conditions are presented in Table 2. Highly significant differences ($p \leq 0.01$) were observed among the accessions and the fruit types within the accessions for all yield and yield parameters measured. Similarly, highly significant differences ($p \leq 0.01$) were recorded in the cropping conditions for all yield and yield-related traits except for the number of fruits per bunch (NFB). Next, highly significant differences ($p \leq 0.01$) were observed in fruit type for all the yield and the yield-related parameters except for average fruit weight (AFW) and fruit yield per plant (FYP), where non-significant differences were observed. On the other hand, the interaction between cropping conditions with the accession showed highly significant differences ($p \leq 0.01$) for the most important yield and yield component traits, which were fruit yield per plant (FYP), fruit length (FLE), total number of fruit (TNF), and average fruit weight (AFW). This strongly signified that cropping conditions and agronomic practices play pivotal roles in affecting eggplant varietal aspects in terms of yield and yield components. The result also revealed that there was no sign of replication effect within the cropping condition except for total number of fruits (TNF), average fruit weight (AFW), and fruit yield per plant (FYP). A high coefficient of variation (CV) of more than 40% was applied to average fruit weight (AFW) and fruit yield per plant (FYP), indicating they were the most diverse quantitative agro-morphological characteristics observed in Table 2. In general, all accessions were strongly varied from each other in terms of yield characteristics. The reason for the significant difference is apparently the differences in their origins that cause the existence of variation in a population [25]. Similarly, several studies have been carried out on phenotypic variation among eggplant accessions. The outcome of this research is in agreement with the findings of Caguiat and Hautea [26]. Hence, this strongly supported a postulation by Naujeer [27] that enhanced yield and improved fruit quality are defined as the main objectives in eggplant breeding program.

Table 2. Mean square of yield and yield component of 29 eggplant accessions.

Sources of Variation	DF	FGI	DFR	FLE	FLW	TNF	NFB	AFW	FYP
Cropping condition (C)	1	123.51 **	55.48 **	483.34 **	6.08 **	28,678.25 **	0.02 ns	10,836,424.27 **	295,844,792 **
Replications within C (R/C)	4	5.24 ns	0.81 ns	6.14 ns	0.20 ns	107.24 **	0.05 ns	143,806.97 **	3,823,329.5 **
Accessions (A)	28	95.28 **	12.37 **	137.69 **	10.88 **	1017.08 **	0.59 **	261,459.39 **	3,065,055.9 **
Fruit types (F)	(1)	259.02 **	23.52 **	2613.71 **	230.36 **	3867.31 **	0.83 **	53.20 ns	1,811,853.7 ns
A/F	(27)	89.35 **	12.60 **	52.71 **	2.79 **	1043.78 **	0.58 **	271,141.10 **	3,111,470.8 **
C × A	28	5.27 ns	4.10 *	23.68 **	1.92 **	954.65 **	0.02ns	157,755.84 **	2,816,928.8 **
Error	113	4.18	2.39	5.19	0.71	26.45	0.03	37,725.85	937,710.7
CV (%)		15.9	34.95	22.22	31.44	22.52	17.42	47.28	62.64
Mean		12.87	4.38	10.36	2.65	23.68	1.07	433.92	1603.36
σ^2_g		15.85	1.52	20.65	1.57	12.57	0.09	17,283.90	41,354.50
σ^2_{gc}		0.46	0.69	6.53	0.44	328.04	0.00	40,010	626,406
σ^2_e		4.14	2.31	5.17	0.68	26.42	0.03	37,725.80	937,710.70
σ^2_p		20.45	4.52	32.35	2.69	367.03	0.12	95,019.70	1,605,471.20

Note = *, **, ns: significant at $p \leq 0.05$, $p \leq 0.01$ and non-significant, respectively, DF = degree of freedom at 0.05, CV(%) = coefficient of variation (%), σ^2_g = genotypic variance, σ^2_{gc} = genotype x cropping condition variance, σ^2_e = error variance, σ^2_p = phenotypic variance, FGI = fruit girth (cm), DFR = diameter of fruit (cm), FLE = fruits length (cm), FLW = fruits length to width ratio (no.), TNF = total number of fruits (no.), NFB = number of fruits per bunch (no.), AFW = average fruits weight (g), FYP = fruit yield per plant (g).

The yield traits' least significant difference (LSD) mean performances of 29 accessions are presented in Table 3. The fruit girth (FGI) ranged from 22.37 cm to 5.71 cm with an average value of 12.84 cm. Accession 13MR (Malaysian Round) had the longest fruit girth (FGI), while accession 6TL (Thailand Long) showed the smallest fruit girth (FGI). Concurrently, the same accession pattern was observed for the diameter of fruit (DFR) that showed an average diameter of 4.36 cm. Accession 6TL recorded the smallest diameter of 1.99 cm, and the widest diameter of fruit (DFR) was observed in 13MR at 7.31 cm. The fruit length (FLE) varied from 2.10 cm to 17.70 cm. The longest fruit length (FLE) was observed in 2TL, while accession 5TR (Thailand Round) had the shortest fruit length (FLE). The mean fruit length (FLE) among accessions was 10.31 cm. For fruit length to width ratio (FLW), accession 1TR recorded the lowest with ratio (0.73), while a ratio of 5.15 was observed in 10ML (Malaysian Long) with an average mean of 2.65. The highest total number of fruits (TNF) was produced by 15ML (63.17), and accession 4TR produced the lowest number of fruits (TNF) at 4.50 fruits. The average total number of fruits (TNF) produced was 23.19 among the accessions. The average number of fruits per bunch (NFB) was 1.07 with 6TL producing the highest (2.67), and the remaining accession had the lowest number of fruits per bunch (NFB) (1.00) except for 26CL (China Long), which had an intermediate (1.40) total number of fruits per bunch (NFB). The average fruit weight (AFW) was 433.92 g. The fruit weight ranged from 142.50 g (5TR) to 962.90 g (21ML). The overall yield per plant (FYP) means was 1603.36 g. The yield per plant (FYP) ranged from 323.90 g (4TR) to 2932.20 g (13MR). Generally, accession 13 MR had the best performance in fruit yield per plant (FYP) in both cropping conditions with slightly lower yield in glasshouse cropping conditions. Generally, mean comparisons of accession performances between two cropping condition portrayed a higher mean value (as indicated in bold) in the open field except for the number of fruits per bunch (NFB). Indeed, the yield is evergreen major parameters for evaluating cropping conditions. Pollination in the open field is more frequent, as it is aided with natural pollinators such as bees and wind flow to help dissemination and distribution of pollen. Meanwhile, in glasshouse conditions, limited aeration and higher temperature due to cladding materials in this microclimate [28] eventually can reduce the fruit set. Alternatively, hand pollination assistance through shaking flowers is seen as the savior for promoting the set of the first blossoms of the flowers, and this implicitly causes more labor work and requires higher costs. Other than that, eggplant is more susceptible to whiteflies family species (*Aleyrodidae* sp.) in a glasshouse compared to the open field. This might be due to higher temperatures causing an outbreak of whiteflies due to a thermal tolerance up to 40–45 °C in the glasshouse [29].

Table 3. Means yield and yield traits studied in 29 eggplant accessions across two cropping conditions.

Accessions	FGI (cm)			DFR (cm)			FLE (cm)			FLW (ratio)		
	OP	GH	Pooled	OP	GH	Pooled	OP	GH	Pooled	OP	GH	Pooled
1 TR	**13.41**$_{d-g}$	10.54$_{c-e}$	11.97$_{d-h}$	**9.40**$_{ab}$	3.38$_{f-i}$	6.39$_{abc}$	**4.09**$_{i-k}$	3.18$_{kl}$	3.63 $_{l}$	0.51$_{k}$	**0.94**$_{kl}$	0.73$_{l}$
2 TL	**11.41**$_{f-h}$	10.67$_{c-e}$	11.04$_{e-j}$	**6.89**$_{a-g}$	3.32$_{f-i}$	5.11$_{b-e}$	**19.04**$_{ab}$	16.37$_{a}$	17.70 $_{a}$	3.66$_{c-g}$	**4.95**$_{ab}$	4.31$_{ab}$
3 TR	**16.09**$_{cd}$	11.49$_{cd}$	13.79$_{d}$	**5.02**$_{d-i}$	3.69$_{f-h}$	4.36$_{def}$	**5.23**$_{g-k}$	3.76$_{j-l}$	4.49 $_{kl}$	**1.04**$_{i-k}$	1.02$_{kl}$	1.03$_{kl}$
4 TR	16.22$_{cd}$	**16.44**$_{ab}$	16.33$_{c}$	**5.10**$_{d-i}$	5.09$_{b-e}$	5.10$_{cde}$	6.17$_{f-j}$	**7.40**$_{g-j}$	6.79 $_{jk}$	1.22$_{i-k}$	**1.47**$_{i-l}$	1.34$_{jkl}$
5 TR	6.87$_{ij}$	**7.17**$_{ef}$	7.02$_{kl}$	2.11$_{i}$	**2.21**$_{ij}$	2.16$_{gh}$	2.06$_{k}$	**2.13**$_{l}$	2.10 $_{l}$	**0.98**$_{i-k}$	0.96$_{kl}$	0.97$_{kl}$
6 TL	**5.93**$_{j}$	5.49$_{f}$	5.71$_{l}$	**2.30**$_{i}$	1.67$_{j}$	1.99$_{h}$	4.55$_{h-k}$	**4.63**$_{i-l}$	4.59 $_{kl}$	2.18$_{g-j}$	**2.79**$_{d-h}$	2.49$_{f-i}$
7 TR	**14.56**$_{c-e}$	7.99$_{d-f}$	11.28$_{e-i}$	**4.53**$_{e-i}$	3.50$_{f-h}$	4.02$_{def}$	**4.24**$_{h-k}$	3.21$_{kl}$	3.72 $_{l}$	**0.94**$_{i-k}$	0.92$_{kl}$	0.93$_{kl}$
8 ML	**13.92**$_{c-f}$	11.60$_{cd}$	13.34$_{de}$	**8.52**$_{a-c}$	3.55$_{f-h}$	7.28$_{a}$	**18.43**$_{ab}$	13.20$_{a-d}$	17.12 $_{ab}$	3.21$_{d-h}$	**3.72**$_{b-e}$	3.34$_{b-f}$
9 ML	**13.25**$_{d-g}$	9.65$_{c-e}$	12.05$_{d-g}$	**4.41**$_{e-i}$	2.99$_{f-h}$	3.94$_{d-g}$	**20.11**$_{a}$	11.24$_{b-g}$	17.15 $_{ab}$	**4.58**$_{a-e}$	3.76$_{a-d}$	4.30$_{ab}$
10 ML	**12.52**$_{e-h}$	9.02$_{d-f}$	10.77$_{f-j}$	**3.82**$_{g-i}$	2.83$_{g-j}$	3.32$_{e-h}$	**20.25**$_{a}$	14.14$_{ab}$	17.19 $_{ab}$	**5.32**$_{ab}$	4.98$_{a}$	5.15$_{a}$
13 MR	**24.39**$_{a}$	20.36$_{a}$	22.37$_{a}$	**8.12**$_{a-d}$	6.50$_{a}$	7.31$_{a}$	**7.94**$_{f-h}$	6.69$_{h-k}$	7.31 $_{j}$	0.98$_{i-k}$	**1.03**$_{kl}$	1.00$_{kl}$
14 ML	**10.39**$_{gh}$	7.91$_{d-f}$	9.15$_{ijk}$	**3.47**$_{hi}$	2.48$_{h-j}$	2.98$_{fgh}$	**16.03**$_{bc}$	8.03$_{f-i}$	12.03 $_{fgh}$	**4.65**$_{a-e}$	3.28$_{c-g}$	3.96$_{bcd}$
15 ML	**11.82**$_{e-h}$	10.78$_{c-e}$	11.30$_{e-i}$	**3.77**$_{g-i}$	3.45$_{f-h}$	3.61$_{d-h}$	**3.94**$_{jk}$	2.62$_{kl}$	3.28 $_{l}$	**1.07**$_{i-k}$	0.76$_{l}$	0.91$_{kl}$
16 ML	**10.38**$_{gh}$	8.73$_{d-f}$	9.56$_{h-j}$	**3.60**$_{g-i}$	2.73$_{h-j}$	3.17$_{fgh}$	**17.09**$_{ab}$	6.66$_{h-k}$	11.88 $_{fgh}$	**4.79**$_{a-d}$	2.51$_{e-i}$	3.65$_{b-e}$
17 ML	12.87$_{e-g}$	**13.15**$_{bc}$	12.98$_{def}$	3.89$_{g-i}$	**4.08**$_{d-f}$	3.97$_{d-g}$	**17.27**$_{ab}$	11.00$_{b-g}$	14.76 $_{b-e}$	**4.45**$_{a-e}$	2.53$_{d-i}$	3.68$_{b-e}$
18 ML	**13.29**$_{d-g}$	13.25$_{bc}$	13.27$_{de}$	3.94$_{g-i}$	**4.13**$_{c-f}$	4.03$_{def}$	**18.09**$_{ab}$	9.53$_{c-h}$	13.81 $_{c-f}$	**4.53**$_{a-e}$	2.29$_{f-j}$	3.41$_{b-f}$
19 ML	**13.30**$_{d-g}$	11.75$_{cd}$	12.53$_{d-g}$	**4.18**$_{e-i}$	3.65$_{f-h}$	3.92$_{d-g}$	12.72$_{c-e}$	**17.07**$_{a}$	14.89 $_{b-e}$	3.37$_{d-g}$	**4.74**$_{ab}$	4.05$_{bc}$
20 ML	**11.48**$_{f-h}$	11.33$_{cd}$	11.40$_{d-i}$	**3.54**$_{g-i}$	3.47$_{f-h}$	3.50$_{d-h}$	**16.08**$_{bc}$	10.86$_{b-g}$	13.47 $_{d-g}$	**4.60**$_{a-e}$	3.16$_{c-g}$	3.88 $_{bcd}$
21 ML	**24.00**$_{a}$	17.71$_{a}$	20.86$_{ab}$	**7.35**$_{a-f}$	5.32$_{a-c}$	6.34$_{abc}$	8.49$_{f-g}$	**9.25**$_{d-h}$	8.87 $_{ij}$	1.16$_{i-k}$	**2.12**$_{g-k}$	1.64 $_{i-l}$
22 ML	**23.69**$_{a}$	20.20$_{a}$	22.29$_{a}$	**7.40**$_{a-e}$	6.08$_{ab}$	6.87$_{abc}$	**13.09**$_{cd}$	11.33$_{b-g}$	12.38 $_{efg}$	1.78$_{h-k}$	**1.86**$_{h-l}$	1.81 $_{h-k}$
23 ML	**19.85**$_{b}$	19.43$_{a}$	19.64$_{b}$	6.27$_{b-h}$	**6.29**$_{ab}$	6.28$_{abc}$	**7.68**$_{f-i}$	6.60$_{h-k}$	7.14 $_{jk}$	**1.22**$_{i-k}$	1.05$_{j-l}$	1.14 $_{jkl}$
25 ML	16.59$_{c}$	**16.72**$_{ab}$	16.66$_{c}$	5.18$_{c-i}$	5.18$_{b-d}$	5.18$_{bcd}$	**12.58**$_{c-e}$	9.42$_{d-h}$	11.00 $_{ghi}$	**2.42**$_{f-i}$	1.83$_{h-l}$	2.12 $_{g-j}$
26 CL	**9.52**$_{hi}$	8.06$_{d-f}$	8.64$_{jk}$	**3.00**$_{hi}$	2.56$_{h-j}$	2.73$_{fgh}$	**9.32**$_{ef}$	7.50$_{f-j}$	8.23 $_{j}$	**3.11**$_{e-h}$	2.89$_{c-h}$	2.97$_{d-g}$
27 CL	**13.02**$_{e-g}$	11.52$_{cd}$	12.27$_{d-g}$	**4.01**$_{f-i}$	3.50$_{f-h}$	3.75$_{d-h}$	**20.61**$_{a}$	11.64$_{b-f}$	16.13 $_{a-d}$	**5.14**$_{a-c}$	3.49$_{c-f}$	4.32$_{ab}$
29 MN	**11.57**$_{e-h}$	10.69$_{c-e}$	11.13$_{e-i}$	**3.61**$_{g-i}$	3.35$_{f-i}$	3.48$_{d-h}$	12.93$_{c-e}$	**13.65**$_{a-c}$	13.29 $_{efg}$	3.56$_{c-g}$	**4.09**$_{a-c}$	3.82$_{bcd}$
30 MN	**11.52**$_{e-h}$	10.91$_{c-e}$	11.27$_{e-i}$	**3.59**$_{g-i}$	3.44$_{f-h}$	3.53$_{d-h}$	**2.82**$_{jk}$	2.60$_{kl}$	2.73 $_{l}$	**0.79**$_{jk}$	0.76$_{l}$	0.77$_{l}$
32 MN	10.46$_{gh}$	**10.62**$_{c-e}$	10.54$_{g-j}$	**3.09**$_{hi}$	3.04$_{f-i}$	3.07$_{fgh}$	**12.23**$_{de}$	6.41$_{h-k}$	9.32 $_{hij}$	**3.99**$_{b-f}$	2.13$_{g-k}$	3.06$_{c-g}$
34 CL	**11.79**$_{e-h}$	11.10$_{c-e}$	11.45$_{d-i}$	**3.58**$_{g-i}$	3.33$_{f-i}$	3.45$_{d-h}$	**20.18**$_{a}$	8.50$_{e-i}$	14.34 $_{c-f}$	**5.70**$_{a}$	2.56$_{d-i}$	4.13$_{b}$
35 CL	**14.33**$_{c-f}$	11.83$_{cd}$	13.08$_{def}$	**9.93**$_{a}$	3.95$_{e-g}$	6.94$_{ab}$	**20.40**$_{a}$	12.19$_{b-e}$	16.29 $_{abc}$	2.34$_{g-j}$	**3.20**$_{c-g}$	2.77$_{e-h}$
Mean	**13.79**	11.89	12.84	**4.98**	3.74	4.36	**12.13**	8.48	10.31	**2.85**	2.44	2.65
LSD (p = 0.05)	3.06	4.03	2.43	3.39	1.22	1.84	3.72	4.16	2.71	1.59	1.24	1.00
SEM	0.06	0.05	0.06	0.03	0.02	0.03	0.08	0.06	0.07	0.02	0.02	0.02

Table 3. *Cont.*

Accessions	TNF (no.)			NFB (no.)			AFW (g)			FYP (g)		
	OP	GH	Pooled	OP	GH	Pooled	OP	GH	Pooled	OP	GH	Pooled
1 TR	**19.00**$_{m-p}$	10.67$_{ef}$	14.83$_{klm}$	1.00$_{b}$	1.00$_{c}$	1.00$_{c}$	**315.90**$_{j-m}$	87.76$_{bc}$	201.80$_{i}$	**974.13**$_{i-k}$	190.57$_{c-h}$	582.30$_{hij}$
2 TL	**37.33**$_{g-i}$	4.00$_{i-m}$	20.67$_{h-k}$	1.00$_{b}$	1.00$_{c}$	1.00$_{c}$	**767.70**$_{d-h}$	181.90$_{bc}$	474.80$_{d-g}$	**3317.28**$_{a-g}$	207.23$_{c-h}$	1762.30$_{b-f}$
3 TR	**71.00**$_{c}$	20.33$_{c}$	45.67$_{b}$	1.00$_{b}$	1.00$_{c}$	1.00$_{c}$	**944.58**$_{c-e}$	249.08$_{bc}$	596.80$_{b-e}$	**4556.22**$_{a-d}$	498.15$_{b-d}$	2527.20$_{abc}$
4 TR	3.33$_{q}$	**5.67**$_{g-l}$	4.50$_{o}$	1.00$_{b}$	1.00$_{c}$	1.00$_{c}$	116.89$_{m}$	**214.50**$_{bc}$	165.70$_{i}$	218.70$_{k}$	**429.10**$_{b-f}$	323.90$_{j}$
5 TR	27.00$_{i-n}$	**45.67**$_{a}$	36.33$_{cd}$	2.67$_{a}$	2.67$_{a}$	2.67$_{a}$	**185.88**$_{lm}$	99.21$_{bc}$	142.50$_{i}$	**811.47**$_{jk}$	198.40$_{c-h}$	504.90$_{ij}$
6 TL	**29.33**$_{h-m}$	26.33$_{b}$	27.83$_{ef}$	1.00$_{b}$	1.00$_{c}$	1.00$_{c}$	**212.78**$_{lm}$	101.83$_{bc}$	157.30$_{i}$	**937.06**$_{i-k}$	136.35$_{d-h}$	536.70$_{ij}$
7 TR	**35.33**$_{g-k}$	20.67$_{c}$	28.00$_{e}$	1.00$_{b}$	1.00$_{c}$	1.00$_{c}$	**370.56**$_{i-m}$	201.57$_{bc}$	286.10 ghi	**1812.47**$_{g-k}$	387.6$_{b-g}$	1100.00$_{f-j}$
8 ML	**17.67**$_{n-p}$	1.00$_{m}$	13.50$_{lmn}$	1.00$_{b}$	1.00$_{c}$	1.00$_{c}$	**379.53**$_{i-m}$	26.00$_{c}$	202.80$_{i}$	**2996.47**$_{b-j}$	26.00$_{gh}$	1511.20$_{c-i}$
9 ML	**21.00**$_{m-p}$	5.00$_{h-m}$	15.67$_{j-m}$	1.00$_{b}$	1.00$_{c}$	1.00$_{c}$	**862.45**$_{d-f}$	108.25$_{bc}$	485.40$_{d-g}$	**2300.37**$_{e-k}$	108.20$_{e-h}$	1204.30$_{g-j}$
10 ML	**40.00**$_{f-h}$	16.67$_{cd}$	28.33$_{e}$	1.00$_{b}$	1.00$_{c}$	1.00$_{c}$	**1248.01**$_{ab}$	338.89$_{bc}$	793.40$_{ab}$	**5049.57**$_{ab}$	677.77$_{b}$	2863.70$_{ab}$
13 MR	**32.00**$_{g-l}$	5.00$_{h-m}$	18.50$_{i-l}$	1.00$_{b}$	1.00$_{c}$	1.00$_{c}$	**1027.82**$_{b-d}$	295.43$_{bc}$	661.60$_{bcd}$	**5342.90**$_{a}$	521.59$_{bc}$	2932.20$_{a}$
14 ML	**64.00**$_{cd}$	10.67$_{ef}$	37.33$_{c}$	1.00$_{b}$	1.00$_{c}$	1.00$_{c}$	**926.75**$_{c-e}$	127.43$_{bc}$	527.10$_{def}$	**4467.10**$_{a-e}$	254.86$_{c-h}$	2361.00$_{a-d}$
15 ML	**116.33**$_{a}$	10.00$_{e-g}$	63.17$_{a}$	1.00$_{b}$	1.00$_{c}$	1.00$_{c}$	**483.21**$_{h-l}$	101.36$_{bc}$	292.30 ghi	**2462.54**$_{d-j}$	180.66$_{c-h}$	1321.60$_{d-j}$
16 ML	**60.00**$_{de}$	2.00$_{k-m}$	31.00$_{de}$	1.00$_{b}$	1.00$_{c}$	1.00$_{c}$	**933.33**$_{c-e}$	61.35$_{bc}$	497.30$_{d-g}$	**4666.66**$_{a-d}$	61.35$_{f-h}$	2364.00$_{a-d}$
17 ML	**25.00**$_{k-o}$	3.00$_{j-m}$	16.20$_{i-l}$	1.00$_{b}$	1.00$_{c}$	1.00$_{c}$	**821.03**$_{d-g}$	111.47$_{bc}$	466.30$_{d-g}$	**3750.81**$_{a-g}$	111.47$_{e-h}$	1931.10$_{a-f}$
18 ML	**27.67**$_{i-n}$	8.33$_{f-i}$	18.00$_{i-l}$	1.00$_{b}$	1.00$_{c}$	1.00$_{c}$	**762.35**$_{d-h}$	206.19$_{bc}$	484.30$_{d-g}$	**3742.40**$_{a-g}$	348.32$_{b-h}$	2045.40$_{a-f}$
19 ML	**12.67**$_{pq}$	2.33$_{k-m}$	7.50$_{mno}$	1.00$_{b}$	1.00$_{c}$	1.00$_{c}$	**476.10**$_{h-l}$	111.29$_{bc}$	293.70$_{ghi}$	**1785.35**$_{g-k}$	222.58$_{c-h}$	1004.00$_{f-j}$
20 ML	**40.67**$_{fg}$	13.00$_{de}$	26.83$_{efg}$	1.00$_{b}$	1.00$_{c}$	1.00$_{c}$	**891.35**$_{de}$	242.33$_{bc}$	566.80$_{c-f}$	**3779.90**$_{a-g}$	484.66$_{b-e}$	2132.30$_{a-e}$
21 ML	**25.67**$_{j-n}$	6.00$_{g-l}$	15.83$_{i-m}$	1.00$_{b}$	1.00$_{c}$	1.00$_{c}$	**981.66**$_{b-d}$	944.21$_{a}$	962.90$_{ab}$	**2225.11**$_{f-k}$	1208.66$_{a}$	1716.90$_{c-g}$
22 ML	**22.67**$_{l-p}$	4.33$_{h-m}$	13.50$_{lmn}$	1.00$_{b}$	1.00$_{c}$	1.00$_{c}$	**1194.09**$_{a-c}$	99.27$_{bc}$	646.70$_{b-e}$	**4205.55**$_{a-f}$	198.54$_{c-h}$	2202.00$_{a-e}$
23 ML	**24.00**$_{l-o}$	1.67$_{lm}$	12.83$_{lmn}$	1.00$_{b}$	1.00$_{c}$	1.00$_{c}$	**1397.98**$_{a}$	163.49$_{bc}$	780.70$_{abc}$	**4674.73**$_{a-c}$	209.87$_{c-h}$	2442.30$_{abc}$
25 ML	**14.67**$_{op}$	8.67$_{e-h}$	11.67$_{mn}$	1.00$_{b}$	1.00$_{c}$	1.00$_{c}$	**587.80**$_{f-j}$	3.72$_{c}$	295.80 ghi	**2284.97**$_{e-k}$	5.84$_{h}$	1145.40$_{f-j}$
26 CL	**49.33**$_{ef}$	3.67$_{j-m}$	26.50$_{e-h}$	1.00$_{b}$	1.67$_{b}$	1.40$_{b}$	**270.86**$_{k-m}$	165.8$_{bc}$	218.30$_{hi}$	**992.46**$_{h-k}$	283.74$_{c-h}$	638.10$_{g-j}$
27 CL	**18.67**$_{m-p}$	7.00$_{f-j}$	12.83$_{lmn}$	1.00$_{b}$	1.00$_{c}$	1.00$_{c}$	**572.50**$_{f-j}$	377.64$_{b}$	475.10$_{d-g}$	**3199.05**$_{a-h}$	377.64$_{b-h}$	1788.30$_{b-f}$
29 MN	**36.00**$_{g-j}$	6.33$_{f-k}$	21.17$_{g-j}$	1.00$_{b}$	1.00$_{c}$	1.00$_{c}$	**546.08**$_{g-k}$	167.38$_{bc}$	356.70$_{f-i}$	**2736.63**$_{c-j}$	290.92$_{c-h}$	1513.80$_{c-i}$
30 MN	**96.00**$_{b}$	4.00$_{i-k}$	59.20$_{a}$	1.00$_{b}$	1.00$_{c}$	1.00$_{c}$	**462.39**$_{i-l}$	19.66$_{c}$	241.00$_{h-i}$	**2311.94**$_{e-k}$	39.32$_{gh}$	1175.60$_{f-j}$
32 MN	**37.33**$_{g-i}$	6.33$_{f-k}$	21.83$_{f-i}$	1.00$_{b}$	1.00$_{c}$	1.00$_{c}$	**582.65**$_{f-j}$	121.85$_{bc}$	352.30$_{f-i}$	**3098.90**$_{b-i}$	201.09$_{c-h}$	1650.00$_{c-h}$
34 CL	**33.33**$_{g-l}$	9.67$_{e-g}$	21.50$_{g-j}$	1.00$_{b}$	1.00$_{c}$	1.00$_{c}$	**656.63**$_{e-i}$	197.53$_{bc}$	427.10 e–h	**2959.85**$_{b-j}$	382.49$_{b-h}$	1671.20$_{c-h}$
35 CL	**19.67**$_{m-p}$	7.00$_{f-j}$	13.33$_{lmn}$	1.00$_{b}$	1.00$_{c}$	1.00$_{c}$	**841.95**$_{d-g}$	220.15$_{bc}$	531.10$_{def}$	**2651.15**$_{c-j}$	440.30$_{b-e}$	1545.70$_{c-h}$
Mean	**36.44**	9.95	23.19	1.06	1.08	1.07	**683.48**	184.36	433.92	**2907.30**	299.42	1603.36
LSD ($p = 0.05$)	10.99	4.61	6.08	0.18	0.4	0.22	298.96	335.38	222.19	2207.90	377.7	1107.70
SEM	0.29	0.12	0.21	0	0	0	4.19	2.81	3.50	20.62	3.57	12.10

Note: OP = open field, GH = glasshouse, FGI = fruit girth (cm), DFR = diameter of fruit (cm), FLE = fruits length (cm), FLW = fruits length to width (ratio), TNF = total no of fruits (no.), NFB = number of fruits per bunch (no.), AFW = average fruits weight (g), FYP = fruit yield per plant (g), LSD = least significant difference, SEM = standard error of mean, n. s=not significant at $p > 0.05$ and means with the same letter in each column also not significantly different at 5% probability level.

3.1.1. Vegetative Traits Across Two Cropping Conditions

The combined analysis of variance for vegetative traits is presented in Table 4. Highly significant differences ($p \leq 0.01$) were observed among the accessions for number of primary branches (NPB), plant height (PHE), stem diameter (SDM), and days to first flowering (DFF), while non-significant differences were observed in plant spread (PSP). To emphasize, plant height is the most critical vegetative indicator of high yield, as postulated by [25]. Together with other vegetative parameters, these traits contributed by genetic makeup were implicitly influenced by the environment, especially cropping conditions. This was mainly due to limited sources of photosynthates partitioning to meet vigorous sink competition. Hence, the quota for yield was unfairly used by somatic cells growth, which resulted in luxurious vegetative development and obvious height. For fruit type, all vegetative parameters showed no significant differences except for the number of primary branches (NPB), where a highly significant difference was observed, and plant spread (PSP), which showed a significant difference at $p \leq 0.05$. The fruit type within accessions showed highly significant differences for all vegetative parameters except for plant spread (PSP), which illustrated no significant difference. While cropping conditions also indicated high significance ($p \leq 0.01$) for number of primary branches (NPB), plant height (PHE) and plant spread (PSP) were vegetative components that had an impact on genetic variation; stem diameter (SDM) and days to first flowering (DFF) indicated no significant difference. On the other hand, the interaction between cropping conditions with the accession showed a highly significant difference ($p \leq 0.01$) for the stem diameter (SDM) trait only, and the remaining number of primary branches (NPB), plant height (PHE), days to first flowering (DFF), and plant spread (PSP) eventually showed no significant difference. Moreover, the result implied that there was a highly significant difference ($p \leq 0.01$) in replication effect within the cropping conditions, which were number of primary branches (NPB), plant height (PHE), and plant spread (PSP). However, days to first flowering (DFF) and stem diameter (SDM) showed no significant difference.

Table 4. Mean square of vegetative parameters of 29 eggplant accessions.

Sources of Variation	DF	NPB	PHE	SDM	PSP	DFF
Cropping condition (C)	1	357.57 **	76049.97 **	0.00 ns	36858.45 **	175.06 ns
Replications within C (R/C)	4	27.59 **	473.55 **	0.05 ns	1490.84 **	294.64 ns
Accessions (A)	28	7.49 **	368.68 **	0.49 **	356.81 ns	369.05 **
Fruit types (F)	(1)	58.58 **	1.67 ns	0.00 ns	1349.57 *	8.65 ns
A/F	(27)	5.46 **	399.06 **	0.50 **	307.33 ns	381.71 **
C × A	28.	2.99 ns	187.36 ns	0.32 **	381.72 ns	136.42 ns
Error	113	2.60	143.59	0.03	280.45	158.85
CV (%)		24.59	13.52	10.13	18.52	14.25
Mean		7.58	92.09	1.68	96.75	88.81
σ^2_g		0.79	32.06	0.03	0.00	37.81
σ^2_{gc}		0.09	17.83	0.09	23.13	0.00
σ^2_e		2.62	143.04	0.03	285.71	153.80
σ^2_p		3.50	192.93	0.15	308.84	191.61

Note = *, **, ns: significant at $p \leq 0.05$, $p \leq 0.01$ and non-significant respectively, DF = degree of freedom at 0.05, CV (%) = coefficient of variation (%), σ^2_g = genotype variance, σ^2_{gc} = genotype × cropping condition variance, σ^2_e = error variance, σ^2_p = phenotypic variance, NPB = number of primary branches, PHE = plant height 90 days after tra nsplant (cm), SDM = stem diameter 90 days after tra nsplant (cm), PSP = plant spread 90 days after tra nsplant (cm), DFF = days to first flowering (day).

The mean comparison for vegetative traits is presented in Table 5. The average number of primary branches (NPB) was 7.54 and ranged from 5.67 (19ML) to 10.50 (30MN) (Malaysian Native). Plant height (PHE) varied from 73.80 cm (26CL) to 105.60 cm (4TR) with an average of 91.72 cm. The average stem diameter (SDM) for all accessions was 1.68 cm; the smallest stem diameter was observed in 14ML at 1.31 cm, while the largest was recorded by 25ML at 2.83 cm. For plant spread (PSP), 14ML had the smallest with 82.81 cm, and 3TR recorded the largest plant spread at 116.14 cm. The average plant spread (PSP) length was 96.50 cm. For days to flowering (DFF), 3TR recorded the longest days to flowering with 100.67 days, while 35CL recorded the earliest at 71.17 days. The average mean of days to flowering (DFF) was 88.81 days. Given the resulting comparison of accession mean performance for two cropping conditions, as shown in Table 5, the means for all vegetative traits were comparatively higher in the greenhouse compared to the open field cropping conditions except for day for first flowering (DFF). This showed robust growth of vegetative yield in the greenhouse cropping condition that may be affected by a significant and continuous supply of fertilizer using irrigation. The variation of vegetative growth among eggplant accessions was wide enough to indicate the perspective glass view for improving accessions studied for all characteristics that eventually support and prepare the reproductive phase of eggplant. It was evidenced that this might be due to the association of genetic composition together with the environment factor applied.

Table 5. Mea ns for vegetative characteristics studied in 29 accessio ns of eggplant across two cropping conditio ns.

Accession	NPB (no.)			PHE (cm)			SDM (cm)			PSP (cm)			DFF (day)		
	OP	**GH**	**Pooled**	**OP**	**GH**	**Pooled**	**OP**	**GH**	**Pooled**	**OP**	**GH**	**Pooled**	**OP**	**GH**	**Pooled**
1 TR	5.67	**9.00**$_{b-g}$	7.33$_{c-g}$	62	**115.00**$_{a-c}$	88.50$_{c-h}$	1.36$_{j}$	**1.57**$_{c-f}$	1.47$_{k-o}$	75.56	**117.67**$_{a-f}$	96.61	90.33$_{c-g}$	**93**	91.40$_{abc}$
2 TL	5.67	**7.33**$_{e-h}$	6.50$_{efg}$	80.1	**105.17**$_{b-e}$	92.63$_{b-g}$	**1.83**$_{b-e}$	1.67$_{c-e}$	1.75$_{d-i}$	90	**97.17**$_{f-h}$	93.58	93.00$_{a-f}$	**108.33**	100.67$_{a}$
3 TR	6.67	**10.33**$_{bc}$	8.50$_{bcd}$	71.44	**129.00**$_{a}$	100.22$_{a-d}$	**1.76**$_{d-g}$	1.67$_{c-e}$	1.72$_{e-j}$	94.78	**137.50**$_{a}$	116.14	**94.00**$_{a-f}$	86.33	90.17$_{abc}$
4 TR	5.5	**9.33**$_{b-f}$	7.80$_{b-f}$	75.25	**125.83**$_{ab}$	105.60$_{ab}$	1.53$_{g-j}$	**1.97**$_{b}$	1.75$_{d-i}$	61	**125.83**$_{a-d}$	99.9	94.33$_{a-f}$	**95.67**	95.00$_{ab}$
5 TR	8	7.67$_{d-h}$	7.80$_{b-f}$	67.11	**88.00**$_{de}$	77.56$_{hi}$	**1.79**$_{c-g}$	1.35$_{fg}$	1.57$_{h-n}$	86.56	**98.00**$_{e-h}$	92.28	**87.00**$_{f-h}$	69.67	78.33$_{cde}$
6 TL	7	**11.00**$_{ab}$	9.00$_{abc}$	52.89	**106.00**$_{b-e}$	79.45$_{ghi}$	1.42$_{h-j}$	**1.55**$_{d-g}$	1.49$_{k-o}$	70.11	**106.00**$_{c-h}$	88.06	**91.67**$_{b-g}$	86	88.83$_{abc}$
7 TR	7	**11.00**$_{ab}$	9.00$_{abc}$	57.67	**112.67**$_{a-c}$	85.17$_{e-i}$	1.34$_{j}$	**1.64**$_{c-e}$	1.49$_{k-o}$	79.33	**124.33**$_{a-d}$	101.83	**91.33**$_{b-g}$	88.33	89.83$_{abc}$
8 ML	5.67	**7.67**$_{e-h}$	6.67$_{d-g}$	72.5	**108.33**$_{a-d}$	90.42$_{c-h}$	**1.80**$_{c-f}$	1.72$_{b-e}$	1.76$_{d-g}$	81	**110.17**$_{b-h}$	95.58	94.00$_{a-f}$	**97.67**	95.83$_{ab}$
9 ML	5.67	**6.50**$_{h}$	6.00$_{fg}$	73.44	**109.00**$_{a-d}$	87.67$_{d-i}$	**1.55**$_{f-j}$	1.35$_{fg}$	1.45$_{l-o}$	74.61	**128.00**$_{a-c}$	95.97	96.50$_{a-e}$	**107**	100.00$_{a}$
10 ML	3.67	**9.00**$_{b-g}$	6.33$_{efg}$	64.11	**114.50**$_{a-c}$	89.31$_{c-h}$	1.58$_{e-j}$	1.58$_{c-f}$	1.58$_{g-n}$	74.22	**104.83**$_{d-h}$	89.53	92.00$_{a-g}$	**98**	95.00$_{ab}$
13 MR	6.33	**9.67**$_{b-e}$	8.00$_{b-e}$	70.33	**106.33**$_{b-e}$	88.33$_{c-h}$	**1.84**$_{b-e}$	1.45$_{e-g}$	1.64$_{f-k}$	89.55	**104.33**$_{d-h}$	96.94	**100.67**$_{a}$	74.33	87.50$_{a-d}$
14 ML	6	**7.00**$_{f-h}$	6.50$_{efg}$	63.89	**96.50**$_{c-e}$	80.20$_{f-i}$	**1.34**$_{j}$	1.29$_{g}$	1.31$_{o}$	77.78	**87.83**$_{h}$	82.81	88.67$_{c-h}$	**94**	91.33$_{abc}$
15 ML	8.33	**10.67**$_{a-c}$	9.50$_{ab}$	75.11	**119.50**$_{ab}$	97.31$_{a-e}$	**2.05**$_{a-c}$	1.76$_{b-d}$	1.91$_{bcd}$	85.55	**114.83**$_{a-g}$	100.19	**88.67**$_{c-h}$	83.33	86.00$_{a-e}$
16 ML	5	**10.33**$_{bc}$	7.67$_{b-f}$	60.66	**125.33**$_{ab}$	93.00$_{b-g}$	1.39$_{ij}$	**1.46**$_{e-g}$	1.43$_{mno}$	88.44	**113.83**$_{b-g}$	101.14	**89.00**$_{c-h}$	87.33	88.17$_{a-d}$
17 ML	6.67	**9.67**$_{b-e}$	8.17$_{b-e}$	76.22	**121.17**$_{ab}$	98.69$_{a-e}$	**1.65**$_{e-i}$	1.54$_{d-g}$	1.60$_{f-m}$	70.11	**111.83**$_{b-g}$	90.97	**97.33**$_{a-c}$	84	92.00$_{abc}$
18 ML	6.33	**9.33**$_{b-f}$	7.83$_{b-f}$	63.39	**125.50**$_{ab}$	94.45$_{a-e}$	1.44$_{h-j}$	**1.74**$_{b-d}$	1.59$_{g-m}$	66.72	**115.33**$_{a-g}$	91.03	97.00$_{a-d}$	**98.33**	97.67$_{a}$
19 ML	4.67	**6.67**$_{gh}$	5.67$_{g}$	76.11	**110.17**$_{a-c}$	93.14$_{b-g}$	**1.58**$_{e-j}$	1.33$_{fg}$	1.45$_{k-o}$	76.89	**106.17**$_{c-h}$	91.53	91.67$_{b-g}$	91.67	91.67$_{abc}$
20 ML	5.33	**9.00**$_{b-g}$	7.17$_{c-g}$	76.67	**122.17**$_{ab}$	99.42$_{a-d}$	**2.08**$_{ab}$	1.84$_{bc}$	1.96$_{bc}$	85.67	**123.50**$_{a-d}$	104.58	**96.67**$_{a-d}$	92.67	94.67$_{ab}$
21 ML	4.33	**9.33**$_{b-f}$	6.83$_{d-g}$	66.89	**109.50**$_{a-c}$	88.19$_{c-h}$	**2.02**$_{a-d}$	1.65$_{c-e}$	1.84$_{b-e}$	79.11	**103.83**$_{d-h}$	91.47	100.00$_{ab}$	**100.33**	100.17$_{a}$
22 ML	5.33	**8.33**$_{c-h}$	6.83$_{d-g}$	77.66	**113.33**$_{a-c}$	95.50$_{a-e}$	**2.25**$_{a}$	1.58$_{c-f}$	1.92$_{bcd}$	100.89	**106.83**$_{c-h}$	103.86	88.33$_{d-h}$	**88.5**	88.40$_{abc}$
23 ML	7.67	**10.00**$_{b-d}$	8.83$_{abc}$	71.45	**109.50**$_{a-c}$	90.47$_{c-h}$	1.67$_{e-h}$	**1.85**$_{bc}$	1.76$_{d-h}$	94.39	**129.83**$_{ab}$	112.11	**95.33**$_{a-f}$	90.67	93.00$_{abc}$
25 ML	7	**10.00**$_{b-d}$	8.80$_{abc}$	72.83	**97.50**$_{c-e}$	87.63$_{d-i}$	1.83$_{b-e}$	**3.83**$_{a}$	2.83$_{a}$	96.84	**103.17**$_{d-h}$	100.63	**90.67**$_{c-g}$	87.67	89.17$_{abc}$
26 CL	5.5	**9.67**$_{b-e}$	8.00$_{b-e}$	54.5	**86.67**$_{e}$	73.80$_{i}$	1.42$_{h-j}$	**1.64**$_{c-e}$	1.53$_{j-n}$	55.84	**104.67**$_{d-h}$	85.13	77.00$_{ij}$	**80**	78.50$_{cde}$
27 CL	5.67	**8.67**$_{b-h}$	7.17$_{c-g}$	70.54	**116.67**$_{a-c}$	93.61$_{a-f}$	1.45$_{h-j}$	**1.66**$_{c-e}$	1.56$_{j-n}$	70	**110.17**$_{b-h}$	90.08	**92.00**$_{a-g}$	90	91.00$_{abc}$
29 MN	6	**9.00**$_{b-g}$	7.50$_{c-g}$	73.95	**129.33**$_{a}$	101.64$_{abc}$	**2.17**$_{a}$	1.82$_{b-d}$	1.99$_{b}$	98.45	**107.17**$_{b-h}$	102.81	**87.67**$_{e-h}$	59	73.33$_{de}$
30 MN	8	**13.00**$_{a}$	10.50$_{a}$	71.56	**122.83**$_{ab}$	97.20$_{a-e}$	1.42$_{h-j}$	**1.70**$_{b-e}$	1.56$_{i-n}$	85.22	**109.83**$_{b-h}$	97.53	81.00$_{hi}$	**82.5**	81.60$_{b-e}$
32 MN	6	**7.33**$_{e-h}$	6.67$_{d-g}$	85.33	**129.67**$_{a}$	107.50$_{a}$	**1.68**$_{e-h}$	1.57$_{c-f}$	1.63$_{f-l}$	94.55	**105.67**$_{c-h}$	100.11	84.00$_{g-i}$	**94**	89.00$_{abc}$
34 CL	7	6.67$_{gh}$	6.80$_{d-g}$	86.17	**112.17**$_{a-c}$	101.77$_{abc}$	**1.44**$_{h-j}$	1.34$_{fg}$	1.39$_{no}$	91.67	**94.50**$_{gh}$	93.37	69.00$_{j}$	**73.67**	71.33$_{e}$
35 CL	5	**7.67**$_{d-h}$	6.33$_{efg}$	74.89	**108.50**$_{a-d}$	91.70$_{b-g}$	**1.83**$_{b-e}$	1.74$_{b-d}$	1.78$_{c-f}$	76.22	**120.83**$_{a-e}$	98.53	70.00$_{j}$	**72.33**	71.17$_{e}$
Mean	6.06	**9.02**	7.54	70.42	**113.01**	91.72	1.67	**1.68**	1.68	82.02	**110.97**	96.5	**89.96**	87.67	88.81
LSD (p= 0.05)	n. s	2.38	1.88	n. s	21.43	13.95	0.27	0.28	0.19	n. s	22.98	n. s	8.93	n. s	14.87
SEM	0.03	0.02	0.03	0.15	0.18	0.17	0	0.01	0	0.24	0.19	0.22	0.1	0.22	0.16

Note: OP = open field, GH = glasshouse, NPB = number of primary branches (no.), PHE = plant height 90 days after tra nsplant (cm), PSP = plant spread 90 days after tra nsplant (cm), LSD = least significant difference, SEM = standard error of mean, n.s = not significant at $p > 0.05$ and mea ns with the same letter in each column also not significantly different at 5% probability level.

3.1.2. Heritability and Genetic Parameters

Broad-sense heritability, phenotypic coefficient variation, genotypic coefficient variation, and genetic advance are presented in Table 6. Heritability is a dimension of physical appearance (phenotypic traits) or total variance that is handed down from the parents [25]. We could identify a bigger range of low to high broad-sense heritabilities observed for most of the yield component traits, while low broad-sense heritabilities were evidenced for all vegetative traits. Estimation of broad-sense heritability showed the highest value for the trait fruit girth (FGI) with 77.50% and the lowest for plant spread (PSP) with 0.00. Number of fruits per bunch (NFB) and fruit length gave the values of 74.98% and 63.84% heritability estimations, respectively. Moderate values (30–60%) were observed in fruit length to width ratio (FLW) and diameter of fruit (DFR), while the lowest heritability values were illustrated in number of primary branches (NPB), plant height (PHE), stem diameter (SDM), plant spread (PSP), days to first flowering (DFF), total number of fruits (TNF), average fruit weight (AFW), and fruit yield per plant (FYP). In general, for high heritability (>60%), values together with high genetic advance (>20%) were observed for fruit girth (FGI), fruit length (FLE), and number of fruits per bunch (NFB). These parameters are mainly controlled by the additive type of genes and can be used as selection criteria for significant improvement in fruit yield production of eggplant. The results obtained are in agreement with previous research [30–32]. Nevertheless, moderate heritability values but high genetic advance were observed in diameter of fruit (DFR) and fruit length to width ratio (FLW). Both lower heritability values and genetic advance were respectively indicated by plant height (PHE), stem diameter (SDM), plant spread (PSP), and days to first flowering (DFF). This explanation of the function of non-additive genes in the traits could be corrected by heterosis breeding [33,34].

Next, estimation of the phenotypic coefficient of variance (PCV) and the genotypic coefficient of variance (GCV) ranged from zero to 47.34%, portrayed by plant spread (PSP) and fruit length to width ratio (FLW), respectively. As in overall characteristics of vegetative and yield and their components traits, the highest GCV (>20%) was evidenced by FGI (30.92%), FLW (47.34%), FLE (43.85%), NFB (28.72%), and AFW (30.30%), while moderate GCV (10–20%) was observed in NPB (11.69%), SDM (10.70%), DFR (28.18%), TNF (14.97%), and FYP (12.69%). The lowest GCV (<10%) was indicated by the remaining PHE (6.15%), PSP (0.00%), and DFF (6.92%), in which their phenotypic expressions were strongly affected by the environment. Hence, the limited selection was found on these traits. Ranges for PCV values from moderate to high were observed as 15.08% in plant height (PHE) to 80.89% in the total number of fruit (TNF). High PCV was indicated in TNF (80.89%) followed by FYP (79.04%), AFW (71.04%), FLW (61.97%), FLE (54.89%), DFR (48.57%), FGI (35.12%), NFB (33.69%), NPB (24.70%), and SDM (23.18%), whereas moderate values were recorded for the rest—PHE (15.08%), PSP (18.16%), and DFF (15.59%). PCV values have undertaken GCV values for all vegetative and yield component traits in correspondence illustrating the prevalence of environmental influence—especially cropping conditions—on traits expressions. This finding is in harmony with previous work of [32,35–39]. Nonetheless, the difference between them proposing the governance of genetics and hence the selection on a phenotypic basis would remain reliable as influenced by environmental factors. This also implicitly shows the importance of germplasm adaptive capacity with the environment used in upcoming crop breeding selection. Higher genotypic coefficient of variation together with high heritability and high genetic advance provide superior indication rather than individual parameters [40]. Fruit girth (FGI), fruit length (FLE), and number of fruits per bunch (NFB) were highlighted trait candidates in meeting these criteria. Thus, it is pivotal to select one trait that gives positive manipulation to the other traits. In addition to the performance response in both vegetative and yield parameters, this study also highlighted the considerably significant degree of genetic variation with the evidence among the accession for some traits that could be further explored for the breeding program. This finding is also in agreement with other researchers, such as [35,41,42]. Additionally, accession selection would be resourceful for an eggplant improvement program established based on yield and yield component traits in particular.

Table 6. Heritability and genetic variances for quantitative traits.

Traits	H^2_b (%)	GCV (%)	PCV (%)	GA (%)
NPB	22.41	11.69	24.70	11.41
PHE	16.62	6.15	15.08	5.16
SDM	19.24	10.70	23.18	9.19
PSP	0.00	0.00	18.16	0.00
DFF	19.73	6.92	15.59	6.34
FGI	77.50	30.92	35.12	56.08
DFR	33.66	28.18	48.57	33.68
FLE	63.84	43.85	54.89	72.18
FLW	58.36	47.34	61.97	74.50
TNF	3.42	14.97	80.89	5.70
NFB	74.98	28.72	33.69	51.22
AFW	18.19	30.30	71.04	26.62
FYP	2.58	12.69	79.04	4.19

Note: H^2b (%) = broad se nse heritability, GCV (%) = genotypic coefficient variance, PCV (%) = phenotypic coefficient variance, GA = genetic advance, NPB = number of primary branches (no.), PHE = plant height 90 days after tra nsplant (cm), SDM = stem diameter 90 days after tra nsplant (cm), PSP = plant spread 90 days after tra nsplant (cm), DFF = days to first flowering(day), FGI = fruit girth (cm), DFR = diameter of fruit (cm), FLE = fruits length (cm), FLW = fruits length to width (ratio), TNF = total number of fruits (no.), NFB = number of fruits per bunch (no.), AFW = average fruits weight (g), FYP = fruit yield per plant (g)

3.1.3. Cluster Analysis of Agro-Morphological Traits

Agro-morphological characteristics including vegetative, yield, and yield components parameters were adopted based on the Euclidean distances among the 29 accessions of eggplant to construct a UPGMA dendrogram as in Figure 1. This dendrogram revealed 29 eggplant accessions clustered into six groups with a similarity coefficient of 0.35 and which were the best fit for convenience discussion, and this implied a high level of agro-morphological variation of eggplant accessions. As presented in Table 7, cluster I had four admixed accessions from Thailand and China, namely 1TR, 26CL, 5TR, and 6TL, while cluster II had the largest group of 19 accessions (2TL, 27CL, 20ML, 34CL, 10ML, 22ML, 3TR, 17ML, 18ML, 13MR, 14ML, 16ML, 29MN, 32MN, 7TR, 25ML, 19ML, 15ML, and 30MN). Meanwhile, cluster III had 9ML, 23ML, and 35CL—three admixed accessions from Malaysia and China—and the remaining clusters, IV, V, and VI, had one accession each, 8ML, 21ML, and 4TR, respectively (two accessions from Malaysia and the latter from Thailand). Indeed, there are diverse eggplant accessions commercially domesticated between these three origins, and agro-morphological traits are reliable to classify different accessions in a pool of germplasm [25,43]. Regarding morphological traits' mean performances between clusters, as shown in Table 8, cluster II achieved the best in terms of yield and yield component traits. It portrayed a fruit yield per plant of 1867.95 g/plant, which was mainly due to the highest total number of fruits (TNF) of 26.47. Utilizing the basis of comparable vegetative and yield components traits, the grouping of individual accessions was exposed by cluster analysis according to similarity and relatedness of eggplants. Hence, the large difference of each accession attribute performance proposed in the crosses between group I and IV or V could be used to attain higher heterosis and vigor among the accessions.

Table 7. List of clusters of 29 eggplant accessio ns according to cluster analysis.

Cluster	No. of Accessio ns	Accessio ns	Origin
I	4	1TR, 26CL, 5TR, 6TL	Thailand, China
II	19	2TL, 27CL, 20ML, 34CL, 10ML, 22ML, 3TR, 17ML, 18ML, 13MR, 14ML, 16ML, 29MN, 32MN, 7TR, 25ML, 19ML, 15ML, 30MN	Malaysia, Thailand, China
III	3	9ML, 23ML, 35CL	Malaysia, China
IV	1	8ML	Malaysia
V	1	21ML	Malaysia
VI	1	4TR	Thailand

Note: accession code abbreviation in Table 1.

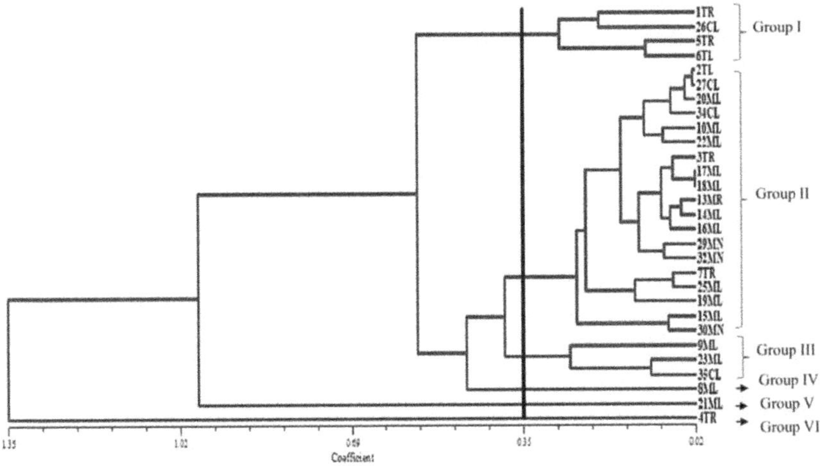

Figure 1. Dendrogram of 29 eggplant accessions based on quantitative traits generated by unweighted pair group method with arithmetic mean (UPGMA) at a 0.35 similarity coefficient.

Table 8. Cluster group and quantitative traits mean.

CLUSTER	NPB	PHE	SDM	PSP	DFF	FGI	DFR	FLE	FLW	TNF	NFB	AFW	FYP
I	8.03	79.83	1.52	90.52	84.27	8.34	3.32	4.64	1.79	26.37	1.52	179.98	565.50
II	7.64	94.56	1.70	97.30	88.87	12.90	4.14	11.25	2.95	26.47	1.00	459.75	1867.95
III	7.05	89.95	1.66	102.20	88.06	14.92	5.72	13.53	2.74	13.94	1.00	599.07	1730.77
IV	6.67	90.42	1.76	95.58	95.83	13.34	7.28	17.12	3.34	13.50	1.00	202.80	1511.20
V	6.83	88.19	1.84	91.47	100.17	20.86	6.34	8.87	1.64	15.83	1.00	962.90	1716.90
VI	7.80	105.60	1.75	99.90	95.00	16.33	5.10	6.79	1.34	4.50	1.00	165.70	323.90

Note: NPB = number of primary branches (no.), PHE = plant height 90 days after tra nsplant(cm), SDM = stem diameter 90 days after tra nsplant (cm), PSP = plant spread 90 days after tra nsplant (cm), DFF = days to first flowering (day), FGI = fruit girth (cm), DFR = diameter of fruit (cm), FLE = fruits length (cm), FLW = fruits length to width (ratio), TNF = total number of fruits (no.), NFB = number of fruits per bunch (no.), AFW = average fruits weight (g), FYP = fruit yield per plant (g).

4. Conclusions and Recommendation

This research revealed that eggplant germplasm had ample genetic variation portrayed through agro-morphological characterization via ANOVA and multivariate analysis. The pivotal understanding of agro-morphological evaluation of genetic variation on eggplant germplasm synergized with cropping condition practices leads to the finding of higher eggplant production with preferable cropping conditions. In Malaysia, it was found that the open field is more suitable for eggplant production with better efficiency of agronomic management together with sustainable production systems. Nevertheless, future work is suggested to explain the molecular approach of genetic variation together with a comprehensive validation of a few seasonal and site trials.

Author Contributions: Conceived and designed the experiments: M.Y.R., J.D., S.I.R., N.N.M.S. and C.P.; Performed the experiment: N.N.M.S. and M.Y.R., Analyzed the data: N.N.M.S., D.R.D. and I.M. Wrote and revised the article: N.N.M.S., M.Y.R., Y.O., J.D., S.I.R. and J.D., Funding acquisition: M.Y.R. and J.D. All authors have read and agreed to the published version of the manuscript.

Funding: This research was funded by the Malaysian Higher Institution Centre of Excellence (HICOE) and Thailand Horticulture Innovation Lab.

Acknowledgments: This project would be impossible without supported Master of Food Security and Climate Change (MSFSCC) project support. Also, the Malaysian Agriculture Research and Development Institute (MARDI) and Research Management Centre (RMC) of Universiti Putra Malaysia (UPM). This may include administrative and technical support of the Institute of Tropical Agriculture and Food Security (ITAFoS), UPM.

Conflicts of Interest: The authors declare no conflict of interest.

References

1. Lester, R.N. Origin and domestication of the brinjal eggplant, Solanum melongena, from *Solanum incanum*, in Africa and Asia. In *Solanaceae III: Taxonomy, Chemistry, Evolution*; The Linnean Society of London: London, UK, 1991; pp. 369–387.
2. Knapp, S.; Vorontsova, M.S.; Prohens, J. Wild relatives of the eggplant (*Solanum melongena* L.: Solanaceae): New understanding of species names in a complex group. *PLoS ONE* **2013**, *8*, e57039. [CrossRef] [PubMed]
3. FAOSTAT: Agriculture Production Data 2016. Available online: http//www.fao.org/ (accessed on 7 April 2018).
4. DOA (Department of Agriculture Peninsular Malaysia). Vegetables and Cash Crops Statistic, Malaysia. 2018. Available online: http://www.doa.gov.my/index/resources/aktiviti_sumber/sumber_awam/maklumat_pertanian/perangkaan_tanaman/perangkaan_sayur_tnmn_ladang_2018.pdf (accessed on 21 January 2019).
5. Zainol, A.A.; Izham, A.; Leong, A.C.; Shahadan, M. Production of quality seed and planting materials by MARDI. In Proceedings of the 3rd National Seed Symposium: Prospects and Opportunities, Selangor, Malaysia, 8–9 April 2003; Malaysian Association of Seed Technologies: Selangor, Malaysia, 2003; pp. 32–34.
6. Mahmood, W.J.W. Developing Malaysian seed industry: Prospects and challenges. *Econ. Technol. Manag. Rev.* **2006**, *1*, 51–59.
7. Ullah, S.; Ijaz, U.; Shah, T.I.; Najeebullah, M.; Niaz, S. Association and genetic assessment in brinjal. *Eur. J. Biotechnol. Biosci.* **2014**, *2*, 41–45.
8. Daunay, M.C.; Lester, R.; Ano, G. *Cultivated Eggplants in Tropical Plant Breeding*; Charrier, A., Iacquot, M., Hamon, S., Nicolas, D., Enfield, N.H., Eds.; Oxford University Press: Oxford, UK, 2001; pp. 220–225.
9. Causse, M.; Friguet, C.; Coiret, C.; Lépicier, M.; Navez, B.; Lee, M.; Holthuysen, N.; Sinesio, F.; Moneta, E.; Grandillo, S. Consumer preferences for fresh tomato at the European scale: A common segmentation on taste and firmness. *J. Food Sci.* **2010**, *75*, S531–S541. [CrossRef]
10. Nwangburuka, C.C.; Kehinde, O.B.; Ojo, D.K.; Denton, O.A.; Popoola, A.R. Morphological classification of genetic diversity in cultivated okra, *Abelmoschus esculentus*. L Moench using principal component analysis (PCA) and single linkage cluster analysis (SLCA). *Afr. J. Biotechnol.* **2011**, *10*, 11165–11172.
11. Oyelola, B.A. The Nigerian Statistical Association Preconference Workshop. In Proceedings of the 2004 Conference Centre, University of Ibadan, Ibadan, Nigeria, 20–21 September 2004.
12. Ulaganathan, V.; Nirmalakumari, A. Finger millet germplasm characterization and evaluation using principal component analysis. *SABRAO J. Breed. Genet.* **2015**, *47*, 79–88.
13. Oladosu, Y.; Rafii, M.Y.; Abdullah, N.; Magaji, U.; Miah, G.; Hussin, G.; Ramli, A. Genotype× Environment interaction and stability analyses of yield and yield components of established and mutant rice genotypes tested in multiple locations in Malaysia. *Acta Agric. Scand.* **2017**, *67*, 590–606. [CrossRef]
14. Kumar, S.R.; Arumugam, T.; Ulaganathan, V. Genetic diversity in eggplant germplasm by principal component analysis. *SABRAO J. Breed. Genet.* **2016**, *48*, 162–171.
15. A'fifah, A.R.; Ismail, M.R.; Puteri, E.M.W.; Abdullah, S.N.A.; Berahim, Z.; Bakhtiar, R.; Kausar, H. Optimum fertigation requirement and crop coefficients of chilli (*Capsicum annuum*) grown in soilless medium in the tropic climate. *Int. J. Agric. Biol.* **2015**, *17*, 80–88.
16. IBPGR. *Descriptors for Eggplant*; International Board for Plant Genetic Resources: Rome, Italy, 1990.
17. Van der Weerden, G.M.; Barendse, G.W. A web-based searchable database developed for the EGGNET project and applied to the Radboud University Solanaceae database. In Proceedings of the VI International Solanaceae Conference: Genomics Meets Biodiversity, Madison, WI, USA, 23–27 July 2006; pp. 503–506.
18. Rohlf, F.J. *NTSYS-Pc: Numerical Taxonomy and Multivariate Analysis System*, version 2.1 Exeter Software; Setauket: New York, NY, USA, 2000.
19. Singh, R.K.; Chaudhary, B.D. *Biometrical Methods in Quantitative Genetic Analysis*; Kalyani Publishers: New Delhi, India, 1979.
20. Sivasubramanian, S.; Madhavamenon, P. Genotypic and phenotypic variability in rice. *Madras Agric. J.* **1973**, *60*, 1093–1096.
21. Robinson, H.F.; Comstock, R.E.; Harvey, P.H. Genotypic and phenotypic correlation in corn and their implications in selection. *Agron. J.* **1949**, *43*, 282–287. [CrossRef]
22. Assefa, K.; Ketema, S.; Tefera, H.; Nguyen, H.T.; Blum, A.; Ayele, M.; Bai, G.; Simane, B.; Kefyalew, T. Diversity among germplasm lines of the Ethiopian cereal tef [Eragrostis tef (Zucc.) Trotter]. *Euphytica* **1999**, *106*, 87–97. [CrossRef]

23. Johnson, H.W.; Robinson, H.F.; Cormstock, R.E. Estimation of genetic and environmental variability in soybeans. *Agron. J.* **1955**, *47*, 314–318. [CrossRef]
24. Juangsamoot, J.; Ruangviriyachai, C.; Techawongstien, S.; Chanthai, S. Determination of capsaicin and dihydrocapsaicin in some hot chillies varieties by RP-HPLC-PDA after magnetic stirring extraction and clean up with C18 cartridge. *Int. Food Res. J.* **2012**, *19*, 1217–1226.
25. Oladosu, Y.; Rafii, M.Y.; Abdullah, N.; Abdul Malek, M.; Rahim, H.A.; Hussin, G.; Kareem, I. Genetic variability and selection criteria in rice mutant lines as revealed by quantitative traits. *Sci. World J.* **2014**, *2014*. [CrossRef]
26. Caguiat, X.G.I.; Hautea, D.M. Genetic diversity analysis of eggplant (Solanum melongena. L.) and related wild species in the Philippines using morphological and SSR markers. *SABRAO J. Breed. Genet.* **2014**, *46*, 183–201.
27. Naujeer, H.B. Morphological Diversity in Eggplant (*Solanum melongena* L.), Their Related Species and Wild Types Conserved at the National Gene Bank in Mauritius. Master's Thesis, Swedish University of Agriculture Sciences, Uppsala, Sweden, 2009.
28. Katsoulas, N.; Stanghellini, C. Modelling crop transpiration in greenhouses: Different models for different applications. *Agronomy* **2019**, *9*, 392. [CrossRef]
29. Xiao, N.; Pan, L.L.; Zhang, C.R.; Shan, H.W.; Liu, S.S. Differential tolerance capacity to unfavourable low and high temperatures between two invasive whiteflies. *Sci. Rep.* **2016**, *6*, 1–10. [CrossRef]
30. Pathania, N.K.; Rajeev Katoch, D.R.; Chaudhary, C.K.S. Genetic variability and association studies in eggplant. In Proceedings of the International Conference on Vegetables 2002, Bangalore, India, 11–14 November 2002; p. 89.
31. Chung-Won Bok, J.S.; Oh-Jusung, H.W.; Chung, W.B.; Jeong, P.S. Hwang (Genetic analysis of F_1 generation in eggplant. *J. Korean Soc. Hort. Sci.* **2003**, *44*, 44–48.
32. Kumar, S.R.; Arumugam, T.; An, C.R.; Premalakshmi, V. Genetic variability for quantitative and qualitative characters in Brinjal (*Solanum melongena* L.). *Afr. J. Agric. Res.* **2013**, *8*, 4956–4959.
33. Ridzuan, R.; Rafii, M.Y.; Mohammad Yusoff, M.; Ismail, S.I.; Miah, G.; Usman, M. Genetic diversity analysis of selected *Capsicum annuum* l. Genotypes based on morphophysiological, yield characteristics and their biochemical properties. *J. Sci. Food Agric.* **2019**, *99*, 269–280. [CrossRef]
34. Priyanka, B.; Naidu, M. Genetic variability, heritability and genetic advance of growth and yield components of chilli (*Capsicum annuum* L.) genotypes. *Int. J. Sci. Res.* **2016**, *5*, 1305–1307.
35. Akpan, N.M.; Ogbonna, P.E.; Onyia, V.N.; Okechukwu, E.C.; Atugwu, I.A. Variability studies on ten genotypes of eggplant for growth and yield performance in south eastern Nigeria. *J. Anim. Plant Sci.* **2016**, *26*, 1034–1041.
36. Mohammad, R.N.R.; Mahdiyeh, P.; Abdolrahim, G.; Javad, A. Variability, heritability and association analysis in eggplant (*Solanum melongena* L.). *ARPN J. Agric. Biol. Sci.* **2015**, *10*, 464–468.
37. Mili, C.; Bora, G.C.; Das, B.J.; Paul, S.K. Studies on variability, heritability and genetic advance in *Solanum melongena* l. (brinjal) genotypes. *Direct Res. J. Agric. Food Sci.* **2014**, *2*, 192–194.
38. Babu, S.; Patil, R. Genetic variability and correlation studies in eggplant (*Solanum melongena* L.). *Madras J. Aric Res.* **2005**, *95*, 18–23.
39. Singh, O.; Kumar, J. Variability, heritability and genetic advance in brinjal. *Indian J. Hort.* **2005**, *62*, 265–267.
40. Shabanimofrad, M.; Rafii, M.Y.; Wahab, P.M.; Biabani, A.R.; Latif, M.A. Phenotypic, genotypic and genetic divergence found in 48 newly collected Malaysian accessions of *Jatropha curcas* L. *Ind. Crop. Prod.* **2013**, *42*, 543–551. [CrossRef]
41. Senapati, N.; Mishra, H.N.; Bhoi, M.K.; Dash, S.K.; Prasad, G. Genetic variability and divergence studies in brinjal (*Solanum melongena* L.). *Veg. Sci.* **2009**, *36*, 150–154.
42. Kushwah, S.; Bandyopadhyay, B.B. Variability and correlation studies in brinjal. *Indian J. Hortic.* **2005**, *62*, 210–212.
43. Franco, J.; Crossa, J.; Ribaut, J.M.; Bertran, J.; Warburton, M.L.; Khairallah, M. A method for combining molecular markers and phenotypic attributes for classifying plant genotypes. *Theor. Appl. Genet.* **2001**, *103*, 944–952. [CrossRef]

agronomy

MDPI

Article

Genetic Basis Dissection for Eating and Cooking Qualities of *Japonica* Rice in Northeast China

Yaolong Yang †, Xin Xu †, Mengchen Zhang, Qun Xu, Yue Feng, Xiaoping Yuan, Hanyong Yu, Yiping Wang and Xinghua Wei *

State Key Laboratory of Rice Biology, China National Rice Research Institute, 310006, China; yangxiao182@126.com (Y.Y.); 82101171083@caas.cn (X.X.); thejust@126.com (M.Z.); xuqun37@hotmail.com (Q.X.); fy_555500@163.com (Y.F.); yxp641110@163.com (X.Y.); yuhy540@163.com (H.Y.); cnwangyp@126.com (Y.W.)

* Correspondence: weixinghua@caas.cn; Tel.: +86-0571-63370366

† Authors who contributed equally to this work.

Received: 19 January 2020; Accepted: 16 March 2020; Published: 20 March 2020

Abstract: The *japonica* rice in Northeast China is famous because of its high quality. Eating and cooking qualities (ECQs) are the most important factors that determine cooked rice quality. However, the genetic basis of ECQ of *japonica* varieties in Northeast China needs further study. In this study, 200 *japonica* varieties that are widely distributed in Northeast China were collected to evaluate the physicochemical indices of grain ECQs. The distribution of each trait was concentrated without large variations. Correlation analysis indicated that gel consistency (GC) had a significantly negative correlation with gelatinization temperature (GT). By integrating various analyses including kinship calculation, principal component analysis (PCA), linkage disequilibrium (LD) analysis, and original parent investigation, we found that the *japonica* varieties in Northeast China exhibited a narrow genetic basis. An association study for grain ECQs was performed and eight quantitative trait loci (QTLs) were detected. *ALK* was the major locus that regulated GT and also significantly affecting GC. Through the linkage disequilibrium (LD) and expression pattern analysis, one possible candidate gene (LOC_Os02g29980) was predicted and required further research for validation. Additionally, a different allele of *Wx* was identified in the variety CH4126, and *ALK* was not fixed in these *japonica* varieties. These results further elucidate the genetic basis of ECQs of *japonica* varieties in Northeast China and provide local breeders some assistance for improving ECQs of rice grain in rice breeding.

Keywords: genetic basis; GWAS; eating and cooking qualities; rice

---

1. Introduction

Rice (*Oryza sativa* L.) is a staple food and feeds more than half of the world's population [1]. In China, rice production has increased almost every year. Because of the increasing rice yield and living standard for customers, more and more rice breeders and consumers have been focusing on rice grain quality, which includes grain milling, appearance, eating and cooking qualities, and nutritional quality. Eating and cooking qualities (ECQs) are the most important factors that determine cooked rice quality. Amylose content (AC), gel consistency (GC), and gelatinization temperature (GT) are the key physicochemical properties of ECQs.

Starch is the major component of rice grain which usually contains two types of starch, i.e., amylose and amylopectin [2]. AC is considered to be the most important factor among rice ECQ. The rice grains with high AC are usually dry, separate, and firm when cooked, whereas rice grains with low AC are soft, glossy, and cohesive [3]. Rice grains with diverse AC meet various demands of consumers [4]. GC is the measure of rice starch gel strength and is another property that reflects firmness and stickiness of cooled cooked rice. Therefore, it has been used as a key character to determine the texture property of cooked rice [5]. It has been demonstrated that cooked rice with high GC is softer than that with low

GC. Alternatively, GT is the temperature at which the starch surface changes into a soluble state, and it is partly responsible for the rice cooking time [6]. GT is usually estimated using the alkali spreading value (ASV). Rice grains with low ASV require a short cooking time, which is a good indicator of rice grain quality.

Many studies about ECQs have been previously reported, and ECQs are considered to be the most complex quantitative traits in rice. Waxy gene (*Wx*) was the first gene identified, which is responsible for controlling AC [7]. *Wx* encodes a granule-bound starch synthase and has also been reported to be the main controller of GC and a minor controller of GT [8–10]. Generally, AC in *japonica* rice grain is lower than that in *indica* rice grain, due to the two *Wx* alleles, Wx^a and Wx^b, respectively [11]. The Wx^a allele is widely distributed in most *indica* rice, whereas the Wx^b allele is found in *japonica* rice. *ALK*, encoding the soluble starch synthase II (SSSII), is currently the only major gene known to be associated with GT [12]. *ALK* also has a minor effect on GC, which is similar to *Wx* [8]. Actually, there are close relationships among these three physiochemical properties. Tian et al. [13] found that AC negatively correlated with GC and GT, and GC positively correlated with GT. In addition, other genes, especially the starch synthesis-related genes, had minor effects on rice ECQ [14].

To facilitate rice breeding with high ECQ, genetic dissection of rice grain ECQ is essential. Quantitative trait locus (QTL) mapping is an effective method for clarifying the molecular and genetic basis of rice grain ECQ. Sun et al. [15] detected eight major QTLs and 27 epistatic QTL pairs for AC, GC, GT, and protein content. Eight physiochemical properties related to ECQs were studied using 190 RILs population with the parents of *japonica* Tainung 78 and *indica* Taichung Sen 17. As a result, 24 QTLs were detected and most of them overlapped the starch synthesis-related genes [16]. Guo et al. [17] used the doubled haploid population derived from a cross between two *japonica* varieties to analyze the grain cooking and nutrient qualities and found 20 QTLs for AC, GC, GT, and protein content. Additionally, Leng et al. [18] also used the DH population derived from a *japonica*/*indica* cross and detected 18 QTLs for ECQ-related traits.

Recently, with the development of association mapping and the reduced cost of genotyping, it has become easy to obtain the high-density single nucleotide polymorphism (SNP) markers by using next-generation sequencing or SNP chip methods. Many associated mapping studies have previously reported on rice yield, disease resistant, rice quality, and plant architecture [19,20]. Xu et al. [21] used resequencing data and performed an associated study for 10 ECQs-related traits. Many loci were detected and some of them were close to the starch synthesis-related genes. Wang et al. [22] used 258 accessions selected from the 3K Rice Genome Project and included measurements of their physicochemical quality traits to study the genetic mechanism of rice ECQs. The results showed that 19 QTLs were detected for rice ECQs, and some new genes were predicted using gene-based association mapping.

The rice grown in Northeast China is famous for its high quality. Previous reports have shown that high rice quality benefited from the excellent light and temperature conditions in Northeast China [23]. In addition to the environmental factors, the genetic basis of these varieties is also crucial for determining rice quality. However, the genetic basis of the *japonica* varieties in Northeast China is poorly understood. In this study, 200 *japonica* varieties widely distributed in Northeast China were collected to evaluate the physicochemical indices of grain ECQs. We performed an association study, in Northeast China, on grain ECQs to characterize the genetic basis of high grain quality and provide breeders, in the area, some assistance for rice breeding.

2. Methods

2.1. Plant Materials

The population comprised of 200 *japonica* varieties which are widely distributed in Northeast China was used in this study (Supplementary Table S1). All of these varieties were conserved in the China National mid-term Genebank for rice and previously described by Ye et al. [24].The rice varieties were planted at the experimental farm of Shenyang Agricultural University. Every variety of the population was planted in four rows with six plants in each row, and the planting destiny was 20 × 20 cm. At harvest time, the seeds were obtained from each line for further phenotypic identification.

2.2. AC, GC, and GT Evaluation in Rice Grain

The rice grains were harvested, dried, and stored at room temperature for at least three months before milling. The rice grains of each variety were dehulled using an electrical dehuller (Xinfeng JLGJ4.5, China), and then milled using the sample miller (Model JB-20, China). A total of 24 hulled grains was selected for the alkali spreading value (ASV) measurement, as in the previous study [12]. Approximately 30 g of grains were ground to flour in order to evaluate the AC and GC, as previously described in [18]. The measurements of AC and GC were performed with three biological replicates, respectively.

2.3. Genotyping, Kinship, and LD Decay Analysis

All 200 *japonica* rice varieties were genotyped for 90K SNPs using a high-density rice array [24]. Only 21,308 SNPs were selected for further study due to the following filter criterion: missing data ratio (MDR) <0.2 and markers with frequency of minor allele (MAF) >0.05 (Supplementary Table S2). The kinship matrix was calculated using TASSEL software [25]. The linkage disequilibrium (LD) coefficient was calculated using the PLINK software and the LD decay plot was drawn using Microsoft Excel software.

2.4. Genome-Wide Associated Mapping

The genome-wide associated study (GWAS) was performed to detect the association between phenotype and genotype using TASSEL software [25]. A mixed linear model was used for the associated mapping. Kinship and principal components analysis (PCA) were used to correct the GWAS results. Among these, the PCA was calculated using EIGENSOFT [26]. To obtain independent association signals, multiple SNPs passing the threshold on the same chromosome were clustered as one association locus, and the SNP with the minimum P-value in a cluster was considered as the lead SNP. The threshold value was determined as follows:

$$P_{threshold} = 1/N \quad (1)$$

where N indicates the number of SNPs used in GWAS and $P_{threshold}$ indicates the threshold of the P-value.

In our study, to simplicity, the $-\log_{10}(P) = 4$ was used for the threshold value Equation (1).

2.5. Candidate Gene Analysis

According to the GWAS results, we calculated the LD decay to discover the candidate regions of the significant loci using the Haploview software (Broad Institute of MIT and Harvard, Cambridge, Massachusetts, USA) [27]. The expression pattern of the genes in the candidate region was obtained from the RNA-seq database in the Rice Genome Annotation Project (http://rice.plantbiology.msu.edu/).

3. Results

3.1. Distributions of AC, GC, and GT in the Japonica Population

Except for GT, all traits showed continuous distributions and large variations in the *japonica* population (Figure 1). For GC, the value of every variety ranged from 45 to 80 mm, and mostly focused on the length between 50 and 65 mm (Figure 1B). The AC ranged from 14% to 20% and more than half of varieties harbored the AC of 16%~18% (Figure 1C). Therefore, most of the varieties exhibited low AC (12% to 20%) and medium GC (41 to 60 mm). In addition, the distribution of GT was discontinuous.

The value of GT was focused on the degree 4~5 and 6~7, with the largest portion focused on degree 6~7 (Figure 1A).

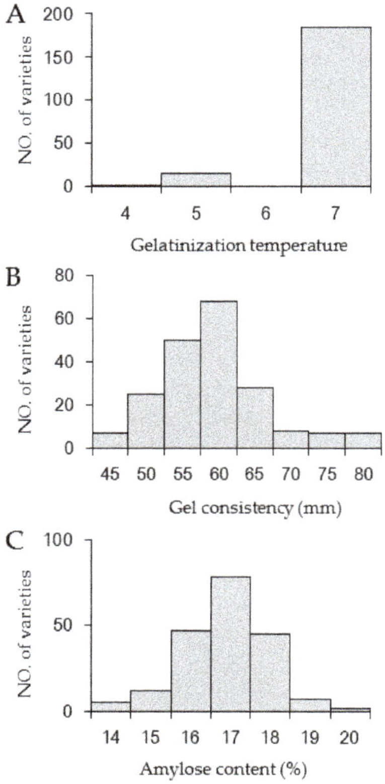

Figure 1. Distribution of eating and cooking qualities (ECQs) in *japonica* varieties of Northeast China. (**A**) Gelatinization temperature (GT); (**B**) Gel consistency (GC); (**C**) Amylose content (AC).

3.2. Geographical Distributions of Rice Varieties

We found that more than half of the varieties were from Liaoning province (LN), whereas only 52 and 20 varieties were from Heilongjiang province (HLJ) and Jilin province (JL), respectively (Figure 2A). To investigate the regional differences of these traits, we compared every trait among the three provinces and found that the AC in the varieties planted in the north was lower than that of those varieties grown in the south. Moreover, the GC of the varieties from LN was softer than that of those varieties from HLJ and JL. In addition, there was no regularity in the GT (Figure 2B–D).

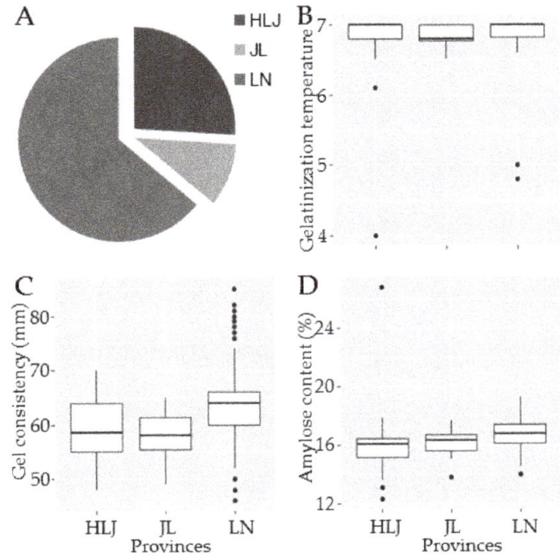

Figure 2. Geographical distributions of rice varieties. (**A**) Number of varieties in the three provinces; (**B**) Boxplot of GT in different province; (**C**) Boxplot of GC in different province; (**D**) Boxplot of AC in different province.

3.3. Correlation Analysis of AC, GC, and GT

AC, GC, and GT collectively affect the ECQ in rice grain. The correlation analysis showed that AC exhibited a significant positive correlation to GC with the correlation coefficient of 0.3. The highest negative correlation was observed between GT and GC, and the correlation coefficient reached up to -0.42. In addition, there was no significant correlation between GT and AC, and the correlation coefficient was -0.1 (Figure 3).

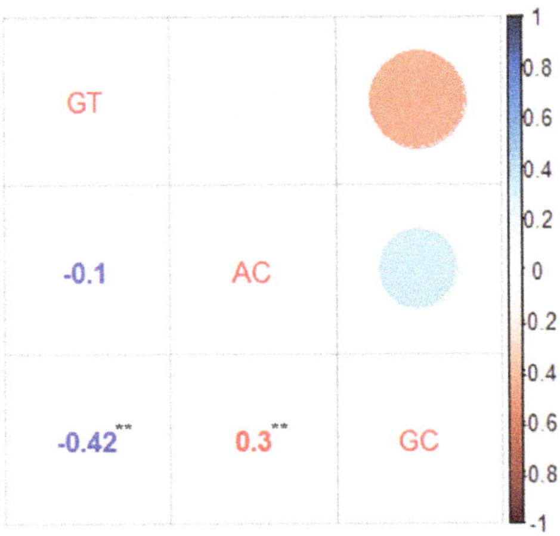

Figure 3. Correlation analysis of AC, GC, and GT. The color and areas of the circle represent the value of the correlation coefficient. Red and blue indicate negative and positive correlations, respectively. ** Represents significant difference at 1% level.

3.4. The Japonica Varieties in Northeast China Displayed Narrow Genetic Basis

On the basis of the genotype of the total 200 varieties, the pairwise relative kinship value was analyzed (Figure 4A). Only 0.6% of the values were less than 0.1, and 22.1% of the values ranged from 0.1 to 0.5. Moreover, 77.3% of the values were larger than 0.5 (Figure 4B). Together, these results suggest that there was high relatedness among the *japonica* varieties in Northeast China. Thus, we consequently investigated the genealogy of all varieties, and the original parent of these varieties focused on four varieties which included Liaogeng5, Liaogeng326, Songgeng3, and Hejiang20 (Figure 4C). Furthermore, Liaogeng5 was widely used as a parent for 79% of all the varieties, which was the possible reason for the high kinship between each of the two varieties. In addition, the linkage disequilibrium (LD), for all the *japonica* varieties, was then analyzed. The LD decay rate was measured as the chromosomal distance at which the average pairwise correlation coefficient dropped to half the value of the maximum r^2 (Figure 4D). The result showed that the genome-wide LD decay rate was approximately estimated at 1 Mb where the r^2 dropped to 0.34. The PCA analysis was also performed to study the population structure. We were not able to divide the population into some obvious subgroups with the different distributions along the two eigenvectors. PC1 and PC2 accounted for 10.3% and 10.0% of the genetic variation, respectively. The top ten PCs only represented 48.9% of the genetic variation (Supplementary Figure S1). In total, it indicated that the *japonica* varieties in Northeast China displayed a narrow genetic basis and resulted in long-range LD and high relatedness.

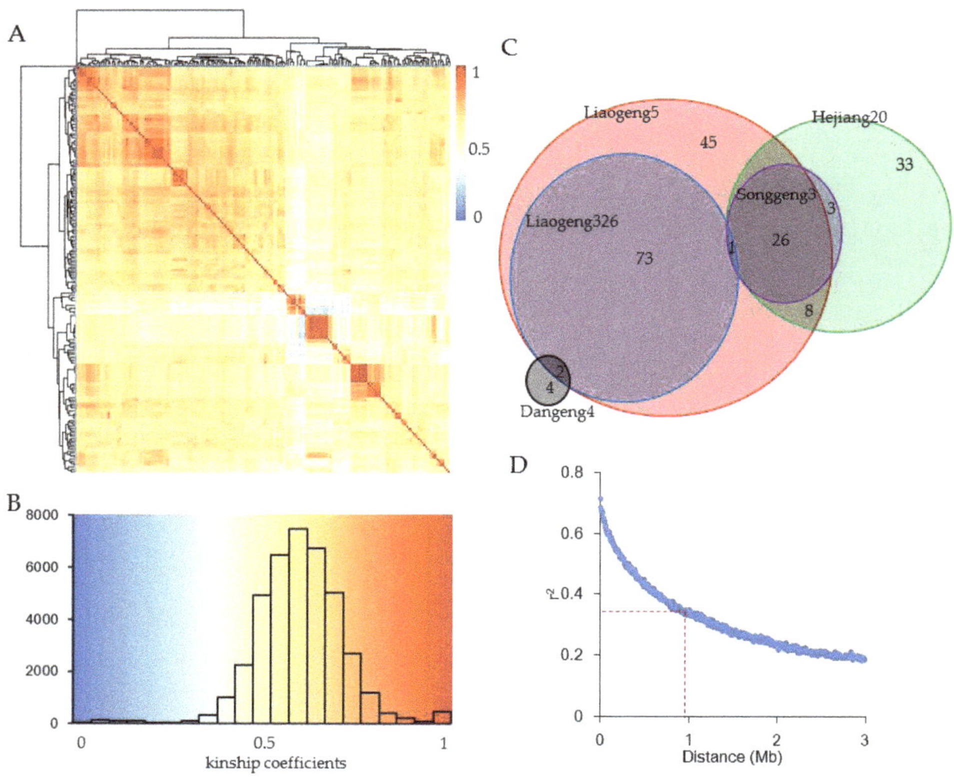

Figure 4. Genetic basis analyses of *japonica* varieties in Northeast China. (**A**) Pairwise relative kinship analysis of *japonica* varieties; (**B**) The distribution of kinship coefficients; (**C**) Common original parent analysis; (**D**) Genome-wide average linkage disequilibrium (LD) decay estimated from the *japonica* varieties.

3.5. GWAS for AC, GC, and GT

Association mapping was performed under a mixed linear model with the kinship matrix and the top three PCs as covariates. In total, 130 significant SNPs containing only eight regions associated with the three traits were detected at the threshold of four across all 12 chromosomes (Figure 5 and Table 1). For GC, eight SNP loci were detected on chromosome 6. These eight loci were close to each other, therefore, only one region, named *qGC6*, was found near the position 6.7 Mb, in the associated analysis for GC, which accounted for 21% of the phenotypic variance (Figure 5B). Two significant SNPs for AC were found at the position of 8.30 Mb on chromosome 11 and 6.84 Mb on chromosome 7, respectively. The two SNP loci, namely *qAC11* and *qAC7*, accounted for 7.38% and 12.9% of the phenotypic variance, respectively (Figure 5C). In addition, 120 significant SNPs were detected in the association mapping for GT. Most SNPs were clustered together and only five regions, namely *qGT1*, *qGT2-1*, *qGT2-2*, *qGT6-1*, and *qGT6-2*, were obtained. *qGT1*, located on chromosome 1 at the position of 15.2 Mb, accounted for 13.7% of the phenotypic variance with the $-\log_{10}(P)$ value of 5.38. *qGT2-1* and *qGT2-2* were the two loci on chromosome 2, which explained 18.9% and 9.7% of the phenotypic variance with the $-\log_{10}(P)$ value of 8.77 and 4.03, respectively. *qGT6-2* was at the position of 20.1 Mb on chromosome 6 with the lower value in the associated mapping for GT, explaining 9% of the phenotypic variance. *qGT6-1*, which had a highest-peak SNP, located on chromosome 6 near the position 6.7 Mb, explained 29.7% of the phenotypic variance (Figure 5A). Interestingly, *qGT6-1* was overlapped with the region of *qGC6*.

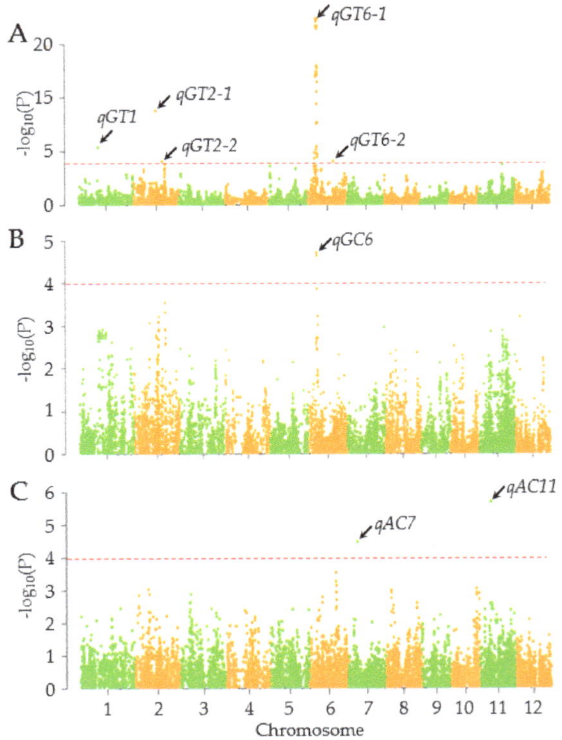

Figure 5. Genome-wide association studies of ECQs. (**A**) GT; (**B**) GC; (**C**) AC the arrow represents the position of the peak single nucleotide polymorphism (SNP). The red dotted line represents the threshold of $-\log_{10}(P)$ value; green and orange represents the different chromosome.

Table 1. Summary of the significant SNPs detected by GWAS and the overlapped QTLs reported previously.

Traits	Name	Peak SNPs	Chr	Position	P-Value	$-\log_{10}P$	R^2	Reported gene/QTL
GT	*qGT1*	seq1_15261369	1	15261369	4.13E-06	5.38	0.14	
	qGT2-1	seq2_17671386	2	17671386	1.68E-09	8.78	−0.19	
	qGT2-2	seq2_22314159	2	22314159	9.33E-05	4.03	0.10	
	qGT6-1	seq6_6726252	6	6726252	5.25E-18	17.28	0.30	*ALK*
	qGT6-2	seq6_20151786	6	20151786	8.94E-05	4.05	0.09	
GC	*qGC6*	seq6_6720486	6	6720486	1.90E-05	4.72	−0.21	*ALK*
AC	*qAC7*	seq7_6847358	7	6847358	3.36E-05	4.47	−0.20	
	qAC11	seq11_8300677	11	8300677	2.04E-06	5.69	0.07	

We compared the significant SNP loci detected in this study with the quantitative trait loci (QTLs)/genes reported previously (Table 1). *Wx* was the major QTL for AC, but in our study, it was not detected in the GWAS for AC. However, *ALK*, the major gene for gelatinization temperature, overlapped the region of *qGC6*/*qGT6-1*, which regulated both the GC and GT in rice grain. In the intervals of the other QTLs, there was no overlap as compared with the QTLs previously reported to be associated with rice ECQs.

3.6. Candidate Gene Analysis

In our study, only eight QTLs were detected, and candidate gene analysis was performed on the two major important QTLs, *qAC11* and *qGT2-1* (Figure 6). For the region of *qAC11*, according to the LD decay analysis, a total 1.02 Mb region ranging from 7.78 to 8.80 Mb on chromosome 11 (Figure 6B) was identified as the candidate region which contained more than 150 genes. It seems that it is difficult to determine which genes are useful from the large number of genes in the candidate region. A similar result was found for *qGT2-1* in the region of 17.51 to 18.44 Mb on chromosome 2, and an approximately 928 kb region was considered to be a candidate region using the LD decay analysis (Figure 6A).

Mostly, the genes regulate the ECQ in rice grain which is always highly expressed in rice endosperm, such as *Wx* and *ALK* (Figure 7E–F). We screened all of the genes located in the candidate regions using the RNA-seq data published by the Rice Genome Annotation Project and selected 33 genes that were expressed in rice endosperm (Supplementary Table S3). On the basis of the expression pattern analysis, four genes including LOC_Os02g29980, LOC_Os02g30300, LOC_Os02g30470, and LOC_Os02g30600 were relatively higher expressed in rice seed and endosperm and lower expressed in shoots and leaf (Figure 7A–D). However, there was no possible gene selected in the candidate region for *qAC11*. Except for LOC_Os02g30600, the other three genes harbored the highest expression in rice grain endosperm. Additionally, according to the gene function annotation, both LOC_Os02g30470, and LOC_Os02g30600 encode the expressed proteins with unknown function, LOC_Os02g29980 encodes an X8 domain containing protein, and LOC_Os02g30300 encodes a MYND finger family protein. Further analysis of sequences of the four genes using BLAST, revealed that the amino acid sequence of LOC_Os02g29980 was highly similar to the 1,3-beta-glucosidase, which could affect the starch synthesis, and LOC_Os02g30300 was homologous to F-box protein. Previous reports have demonstrated that the F-box protein was involved in various physiological and biochemical pathways including panicle and flower development, heading data, plant hormone, and abiotic stress [28]; however, none were directly related to the starch synthesis and rice quality. Additionally, the expression pattern of LOC_Os02g29980 was most similar to those of *Wx* and *ALK*. Therefore, it is possible that LOC_Os02g29980 can regulate GT in rice, but further research is required for validation.

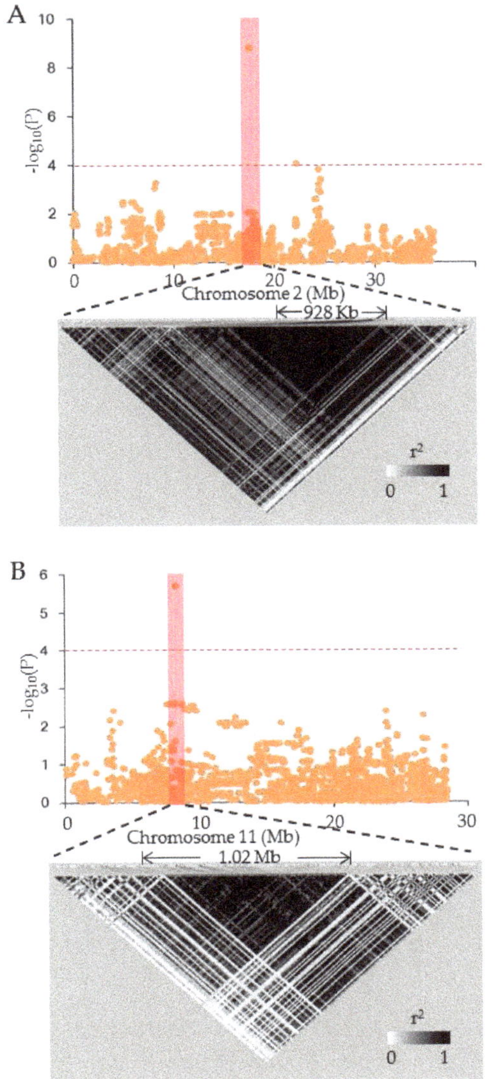

Figure 6. Candidate region estimation of major QTL for GT and AC. (**A**) Candidate region estimation of *qGT2-1* on chromosome 2 and (**B**) Candidate region estimation of *qAC11* on chromosome 11.

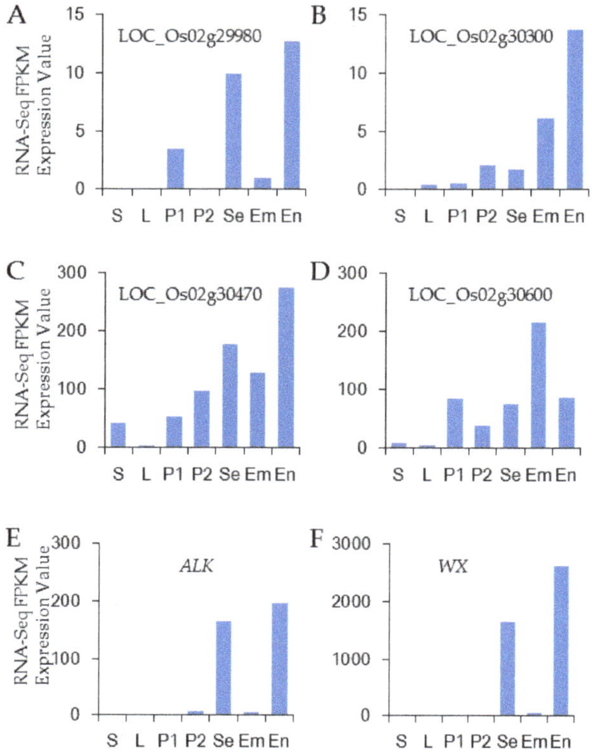

Figure 7. Expression patterns of the candidate genes. (**A**) LOC_Os02g29980; (**B**) LOC_Os02g30300; (**C**) LOC_Os02g30470; (**D**) LOC_Os02g30600; (**E**) *ALK*; and (**F**) *Wx*. S, shoots; L, leaves; P1, pre-emergence inflorescence; P2, post-emergence inflorescence; Se, seed; Em, embryo; En, endosperm.

3.7. Analysis of ALK and Wx in the Japonica Varieties

According to our GWAS results, the major gene for AC, *Wx*, was not detected. It was abnormal and strange in the QTL analysis or GWAS for AC, on the basis of many previous reports. Twenty varieties with highest AC and lower AC were selected, respectively, and their *Wx* gene was sequenced using the primer in Supplementary Table S4. The results showed that all varieties, except CH4126, exhibited the same sequence in *Wx* gene. And the sequence analysis revealed that the *Wx* gene in all sequenced varieties, including CH4126, harbored the allele of Wx^b [29]. Interestingly, the variety CH4126 had the largest AC value which reached up to 26.8%, whereas the values of AC in all other varieties were less than 20%. The sequencing analysis in CH4126 showed that it had 21 variations in the *Wx* as compared with other sequenced varieties (Figure 8A, Supplementary Table S5). Among these variations, only two variations (snp3 and snp4) were located on the exon. Snp4 was a nonsynonymous mutation at the position of 3377 bp that resulted in amino acid change from proline to serine, which could be the reason for the high AC in CH4126.

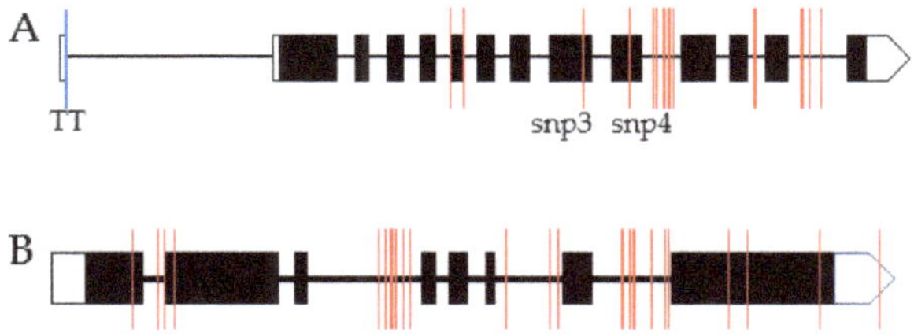

Figure 8. Sequence analysis of *Wx* and *ALK*. (**A**) *Wx*; (**B**) *ALK*. Solid box and empty box indicate the exon and untranslated region, respectively, the red lines indicate the differences, the blue line indicates the functional locus of Wx^b.

qGT6-1/*qGC6* was the largest significant locus in our GWAS results, overlapping the previously reported gene, *ALK*, which was the major gene controlling the gelatinization temperature. Sequencing analysis demonstrated that there were many variations between varieties with high GT and low GT (Figure 8B, Supplementary Table S6). Five variations were located on the exon of *ALK*, and three of them could lead to the protein change. These variations in *ALK* were similar to a previous study. Together with the results of GWAS and the sequencing analysis of *ALK*, it suggested that the gelatinization temperature of *japonica* varieties in Northeast China was mainly controlled by *ALK*.

4. Discussion

The Northeast Plain is the largest plain in China, and the rice grown in this area accounts for a significant portion of the total rice production in China [30]. High-quality rice is conducive to the commercialization of rice. Hence, with increasing rice yield, excellent rice quality has been one of the aims of breeders in Northeast China. Moreover, breeding is the process of artificial selection, which leads to low genetic diversity in the varieties planted in one region. Generally, ECQs are invisible characters and not easily selected as compared with other visible traits. In our study, we statistically analyzed the AC, GC, and GT of the 200 varieties in Northeast China. The distribution of these traits showed that the values of each trait assembled in a narrow range without a large variation. Among these traits, GC relatively displayed the largest variation in the 200 *japonica* varieties. We suggest that GC is potentially difficult to select in Northern China which could be due to the deficiency of the major control gene. In addition, a modern breeding program always involves genetically improving the elite varieties used before. In total, 98% of the varieties used one or more of the four varieties including Liaogeng5, Liaogeng326, Songgeng3, and Hejiang20 as their original parents. Because of this, it is easy to explain the close kinship and long-range LD among these varieties. Therefore, the genetic basis of the varieties in Northeast China is narrow, which could be the reason for the production bottleneck in rice breeding. In order to resolve this bottleneck, we suggest introducing other excellent genetic germplasm and broadening the genetic basis of the *japonica* varieties in Northeast China.

A previous study showed that *japonica* had the Wx^b allele, which is consistent with our results [29]. *Wx* was the only major effect gene in the rice germplasm controlling the AC in rice grain. It leads to the similar AC in most of the varieties. Similar to *Wx*, *ALK* is the only major effect gene regulating GT. In our study, about 11% of the varieties exhibited low GT, which means that *ALK* was not fixed in the *japonica* varieties in Northeast China. The process of *indica-japonica* differentiation could result in differentiation of *Wx* in *japonica* and *indica*, while *ALK* had no obvious difference between the two subspecies. Thus, *ALK* was subjected to human selection for rice breeding in Northeast China, due to the high GT in most *japonica* varieties there.

Fortunately, we found that the variety CH4126 had the highest AC of 26.8%. By sequencing *Wx*, we discovered that the functional locus of the Wx^b allele still existed. In addition, many other variations, including a nonsynonymous mutation, were detected. Interestingly, the nonsynonymous mutation in CH4126 was the rare allele of *Wx* and it had been reported that the mutation was not significantly associate with AC [31]. However, in our study, the variation in *Wx* could be the main reason for the high AC in CH4126. Of course, the results need further research for validation. These results could provide a new genetic resource to dissect the genetic mechanism of *Wx* regulating the AC in rice grain.

The GWAS results detected eight QTLs of ECQs; only one QTL for GC that overlapped with *qGT6-1* was detected, the functional gene in the region was *ALK*. Thus, *ALK* is an important gene regulating the GC in the *japonica* varieties in Northeast China. Correlation analysis showed that the highest correlation was detected between GC and GT, which verified the pleiotropism of *ALK* to some extent. The result was different from a previous report [8], which found that *ALK* had a minor effect on GC. In our study, *ALK* had a significant effect on GC, possibly due to the different genetic population. Except for *ALK*, other QTLs were new and did not overlap with the previously reported genes and QTLs. Due to the minor effects of these QTLs and long-range of LD, it was difficult to find the candidate gene. However, the genes regulating the ECQs are often highly expressed in endosperm, such as *Wx* and *ALK*, and the number of these genes is not very high. Therefore, screening the expression pattern of all genes in the candidate region is a good way for candidate gene determination.

GWAS is useful for genetic dissection of complex traits in rice and has been widely proven by many studies [19,32,33]. Furthermore, many novel genes have been identified using GWAS [34–36]. However, GWAS still has some deficiencies. Population structure is a very important factor that affects GWAS results and can produce false positive results. Its influence can be weakened by using the mixed linear model method in GWAS. In our study, 200 *japonica* varieties were collected from Northeast China. These collections exhibited a low-level population structure which was suitable for GWAS and reduced false positive results, but the close kinship among the population lead to long range LD, which made it difficult to determine the key genes.

The results in our study should be helpful not only for understanding the genetic basis of *japonica* varieties in Northeast China, but also for providing a theoretical basis of ECQs. This is a benefit to ECQ improvement for rice breeding in Northeast China.

5. Conclusions

In our study, 200 *japonica* varieties widely distributed in Northeast China were collected to evaluate the physicochemical indices of grain ECQs. We found that the *japonica* varieties displayed a narrow genetic basis in Northeast China. Through the association study, we found that *ALK* was the major locus that regulating GT, and also significantly affecting GC, which was consistent with the results in the correlation analysis. In addition, one possible candidate gene (LOC_Os02g29980) was finally predicted. Additionally, two major genes for rice ECQs (*ALK* and *Wx*) in the *japonica* population were also studied. These results further contribute to the understanding of the genetic basis of ECQs of *japonica* varieties in Northeast China and provide breeders, in the area, some assistance for improving ECQs of rice grain in rice breeding.

Supplementary Materials: The following are available online http://www.mdpi.com/2073-4395/10/3/423/s1, Table S1: Information for 200 *japonica* varieties from Northeast China, Table S2: The genotype of 200 *japonica* varieties in Northeast China, Table S3: Presentation of the candidate genes that expressed in rice endosperm, Table S4: Primers for sequencing in this study, Table S5: Gene diversity in *Wx* between CH4126 and other sequenced varieties, Table S6: Gene diversity in *ALK* between low and high GT varieties, Figure S1: Principal component analysis of 200 *japonica* varieties.

Author Contributions: Y.Y. and X.W. conceived and designed research; Y.Y. and X.X. conducted experiments; M.Z., Q.X., Y.F., and X.Y. performed the phenotypic identification; Y.Y., H.Y., and Y.W. analyzed the data; Y.Y. and X.X. wrote the manuscript; X.W. and M.Z. helped to revise the manuscript. All authors have read and agreed to the published version of the manuscript.

Funding: We sincerely thank all the participants. The present investigation is financially supported by the National Key Research and Development Program of China (2016YFD0100902-07), the National Natural Science Foundation of China (31600999), and the Chinese Academy of Agricultural Sciences (CAAS-ASTIP-201X-CNRRI).

Conflicts of Interest: Authors declare no conflict of interest.

References

1. Delseny, M.; Salses, J.; Cooke, R.; Sallaud, C.; Regad, F.; Lagoda, P.; Guiderdoni, E.; Ventelon, M.; Brugidou, C.; Ghesquiere, A. Rice genomics: Present and future. *Plant Physiol. Bioch.* **2001**, *39*, 323–334. [CrossRef]
2. Preiss, J. Biology and molecular biology of starch synthesis and its regulation. *Plant Mol.Cell. biol.* **1991**, *7*, 5880–5883.
3. Bao, J.S.; Shen, S.Q.; Sun, M.; Corke, H. Analysis of genotypic diversity in the starch physicochemical properties of nonwaxy rice: Apparent amylose content, pasting viscosity and gel texture. *Starch-Starke* **2006**, *58*, 259–267. [CrossRef]
4. Hu, P.S.; Zhao, H.J.; Duan, Z.Y.; Zhang, L.L.; Wu, D.X. Starch digestibility and the estimated glycemic score of different types of rice differing in amylose contents. *J. Cereal Sci.* **2004**, *40*, 231–237. [CrossRef]
5. Tang, S.X.; Khush, G.S.; Juliano, B.O. Genetics of gel consistency in rice (*Oryza sativa* L.). *J. Genet.* **1991**, *70*, 69–78. [CrossRef]
6. Khush, G.S.; Paule, C.M.; De, L.; Cruz, N.M. Rice grain quality evaluation and improvement at IRRI. In *the Workshop on Chemical Aspects of Rice Grain Quality*; IRRI (ed. Brady NC), the Philippines: Laguna, Los Banos, 1979; pp. 20–31.
7. Wang, Z.Y.; Zheng, F.Q.; Shen, G.Z.; Gao, J.P.; Snustad, D.P.; Li, M.G.; Zhang, J.L.; Hong, M.M. The Amylose Content In Rice Endosperm Is Related To the Posttranscriptional Regulation Of the Waxy Gene. *Plant J.* **1995**, *7*, 613–622. [CrossRef]
8. Fan, C.C.; Yu, X.Q.; Xing, Y.Z.; Xu, C.G.; Luo, L.J.; Zhang, Q.F. The main effects, epistatic effects and environmental interactions of QTLs on the cooking and eating quality of rice in a doubled-Haploid line population. *Theor. Appl. Genet.* **2005**, *110*, 1445–1452. [CrossRef]
9. Su, Y.; Rao, Y.C.; Hu, S.K.; Yang, Y.L.; Gao, Z.Y.; Zhang, G.H.; Liu, J.; Hu, J.; Yan, M.X.; Dong, G.J.; et al. Map-Based cloning proves qGC-6, a major QTL for gel consistency of *japonica/indica* cross, responds by Waxy in rice (*Oryza sativa* L.). *Theor. Appl. Genet.* **2011**, *123*, 859–867. [CrossRef]
10. Tan, Y.F.; Li, J.X.; Yu, S.B.; Xing, Y.Z.; Xu, C.G.; Zhang, Q. The three important traits for cooking and eating quality of rice grains are controlled by a single locus in an elite rice hybrid, Shanyou 63. *Theor. Appl. Genet.* **1999**, *99*, 642–648. [CrossRef]
11. Isshiki, M.; Morino, K.; Nakajima, M.; Okagaki, R.J.; Wessler, S.R.; Izawa, T.; Shimamoto, K. A naturally occurring functional allele of the rice waxy locus has a GT to TT mutation at the 5′ splice site of the first intron. *Plant J.* **1998**, *15*, 133–138. [CrossRef]
12. Gao, Z.Y.; Zeng, D.L.; Cui, X.; Zhou, Y.H.; Yan, M.X.; Huang, D.N.; Li, J.Y.; Qian, Q. Map-Based cloning of the ALK gene, which controls the gelatinization temperature of rice. *Sci. China Ser. C* **2003**, *46*, 661–668. [CrossRef] [PubMed]
13. Tian, Z.X.; Qian, Q.; Liu, Q.Q.; Yan, M.X.; Liu, X.F.; Yan, C.J.; Liu, G.F.; Gao, Z.Y.; Tang, S.Z.; Zeng, D.L.; et al. Allelic diversities in rice starch biosynthesis lead to a diverse array of rice eating and cooking qualities. *Proc. Natl. Acad. Sci. USA* **2009**, *106*, 21760–21765. [CrossRef] [PubMed]
14. He, Y.; Han, Y.P.; Jiang, L.; Xu, C.W.; Lu, J.F.; Xu, M.L. Functional analysis of starch-synthesis genes in determining rice eating and cooking qualities. *Mol. Breed.* **2006**, *18*, 277–290. [CrossRef]
15. Sun, S.Y.; Wei, H.; Lin, H.X. Identification of QTLs for Cooking and Eating Quality of Rice Grain. *Rice Sci.* **2006**, *13*, 161–169.
16. Hsu, Y.C.; Tseng, M.C.; Wu, Y.P.; Lin, M.Y.; Wei, F.J.; Hwu, K.K.; Hsing, Y.I.; Lin, Y.R. Genetic factors responsible for eating and cooking qualities of rice grains in a recombinant inbred population of an inter-subspecific cross. *Mol. Breed.* **2014**, *34*, 655–673. [CrossRef] [PubMed]
17. Guo, Y.M.; Mu, P.; Liu, J.F.; Lu, Y.X.; Li, Z.C. QTL mapping and QxE interactions of grain cooking and nutrient qualities in rice under upland and lowland environments. *J. Genet. Genom.* **2007**, *34*, 420–428. [CrossRef]
18. Leng, Y.J.; Xue, D.W.; Yang, Y.L.; Hu, S.K.; Su, Y.; Huang, L.C.; Wang, L.; Zheng, T.T.; Zhang, G.H.; Hu, J.; et al. Mapping of QTLs for eating and cooking quality-Related traits in rice (*Oryza sativa* L.). *Euphytica* **2014**, *197*, 99–108. [CrossRef]
19. Huang, X.H.; Wei, X.H.; Sang, T.; Zhao, Q.; Feng, Q.; Zhao, Y.; Li, C.Y.; Zhu, C.R.; Lu, T.T.; Zhang, Z.W.; et al. Genome-Wide association studies of 14 agronomic traits in rice landraces. *Nat. Genet.* **2010**, *42*, 961–967. [CrossRef]

20. Wang, C.H.; Yang, Y.L.; Yuan, X.P.; Xu, Q.; Feng, Y.; Yu, H.Y.; Wang, Y.P.; Wei, X.H. Genome-Wide association study of blast resistance in *indica* rice. *BMC Plant Biol.* **2014**, *14*, 311. [CrossRef]
21. Xu, F.F.; Bao, J.S.; He, Q.; Park, Y.J. Genome-Wide association study of eating and cooking qualities in different subpopulations of rice (*Oryza sativa* L.). *BMC Genom.* **2016**, *17*, 663. [CrossRef]
22. Wang, X.Q.; Pang, Y.L.; Zhang, J.; Wu, Z.C.; Chen, K.; Ali, J.; Ye, G.Y.; Xu, J.L.; Li, Z.K. Genome-Wide and gene-Based association mapping for rice eating and cooking characteristics and protein content. *Sci. Rep.* **2017**, *7*, 17203. [CrossRef] [PubMed]
23. Cheng, F.M.; Zhong, L.J. Variation of rice quality traits under different climate conditions and its main affected factors. *Chin. J. Rice Sci.* **2001**, *15*, 187–191.
24. Ye, J.; Niu, X.J.; Yang, Y.L.; Wang, S.; Xu, Q.; Yuan, X.P.; Yu, H.Y.; Wang, Y.P.; Wang, S.; Feng, Y.; et al. Divergent Hd1, Ghd7, and DTH7 Alleles Control Heading Date and Yield Potential of *Japonica* Rice in Northeast China. *Front. Plant Sci.* **2018**, *9*, 35. [CrossRef] [PubMed]
25. Bradbury, P.J.; Zhang, Z.; Kroon, D.E.; Casstevens, T.M.; Ramdoss, Y.; Buckler, E.S. TASSEL: Software for association mapping of complex traits in diverse samples. *Bioinformatics* **2007**, *23*, 2633–2635. [CrossRef] [PubMed]
26. Patterson, N.; Price, A.L.; Reich, D. Population structure and eigenanalysis. *PLoS Genet.* **2006**, *2*, e190. [CrossRef] [PubMed]
27. Barrett, J.C. Haploview: Visualization and analysis of SNP genotype data. *Cold Spring Harb. Protoc.* **2009**, *2009*. [CrossRef]
28. Jain, M.; Nijhawan, A.; Arora, R.; Agarwal, P.; Ray, S.; Sharma, P.; Kapoor, S.; Tyagi, A.K.; Khurana, J.P. F-Box Proteins in Rice. Genome-Wide Analysis, Classification, Temporal and Spatial Gene Expression during Panicle and Seed Development, and Regulation by Light and Abiotic Stress. *Plant Physiol.* **2007**, *143*, 1467–1483. [CrossRef]
29. Hirano, H.Y.; Eiguchi, M.; Sano, Y. A single base change altered the regulation of the Waxy gene at the posttranscriptional level during the domestication of rice. *Mol. Biol. Evol.* **1998**, *15*, 978–987. [CrossRef]
30. Zhang, H.; Zhou, G.S.; Liu, D.L.; Wang, B.; Xiao, D.P.; He, L. Climate-Associated rice yield change in the Northeast China Plain: A simulation analysis based on CMIP5 multi-Model ensemble projection. *Sci. Total Environ.* **2019**, *666*, 126–138. [CrossRef]
31. Biselli, C.; Cavalluzzo, D.; Perrini, R.; Gianinetti, A.; Bagnaresi, P.; Urso, S.; Orasen, G.; Desiderio, F.; Lupotto, E.; Cattivelli, L.; et al. Improvement of marker-Based predictability of Apparent Amylose Content in *japonica* rice through GBSSI allele mining. *Rice* **2014**, *7*, 1. [CrossRef]
32. Huang, X.H.; Yang, S.H.; Gong, J.Y.; Zhao, Q.; Feng, Q.; Zhan, Q.L.; Zhao, Y.; Li, W.J.; Cheng, B.Y.; Xia, J.H.; et al. Genomic architecture of heterosis for yield traits in rice. *Nature* **2016**, *537*, 629–633. [CrossRef] [PubMed]
33. Lu, Q.; Zhang, M.C.; Niu, X.J.; Wang, S.; Xu, Q.; Feng, Y.; Wang, C.H.; Deng, H.Z.; Yuan, X.P.; Yu, H.Y.; et al. Genetic variation and association mapping for 12 agronomic traits in *indica* rice. *BMC Genom.* **2015**, *16*, 1067. [CrossRef] [PubMed]
34. Duan, P.G.; Xu, J.S.; Zeng, D.L.; Zhang, B.L.; Geng, M.F.; Zhang, G.Z.; Huang, K.; Huang, L.J.; Xu, R.; Ge, S.; et al. Natural Variation in the Promoter of GSE5 Contributes to Grain Size Diversity in Rice. *Mol. Plant* **2017**, *10*, 685–694. [CrossRef] [PubMed]
35. Si, L.Z.; Chen, J.Y.; Huang, X.H.; Gong, H.; Luo, J.H.; Hou, Q.Q.; Zhou, T.Y.; Lu, T.T.; Zhu, J.J.; Shangguan, Y.Y.; et al. OsSPL13 controls grain size in cultivated rice. *Nat. Genet.* **2016**, *48*, 447–456. [CrossRef] [PubMed]
36. Yano, K.; Yamamoto, E.; Aya, K.; Takeuchi, H.; Lo, P.C.; Hu, L.; Yamasaki, M.; Yoshida, S.; Kitano, H.; Hirano, K.; et al. Genome-Wide association study using whole-Genome sequencing rapidly identifies new genes influencing agronomic traits in rice. *Nat. Genet.* **2016**, *48*, 927–934. [CrossRef] [PubMed]

agronomy

MDPI

Article

Nucleotide Diversity and Association Analysis of *ZmMADS60* with Root Length in the Maize Seedling Stage

Pengcheng Li [1,2], Zhenzhen Ge [1], Houmiao Wang [1,2], Jie Wei [1], Yunyun Wang [1], Yang Xu [1,2], Zefeng Yang [1,2,3,*] and Chenwu Xu [1,2,3,*]

1 Jiangsu Key Laboratory of Crop Genetics and Physiology/Key Laboratory of Plant Functional Genomics of the Ministry of Education/Jiangsu Key Laboratory of Crop Genomics and Molecular Breeding, Agricultural College of Yangzhou University, Yangzhou 225009, China; pcli@yzu.edu.cn (P.L.); MX120170587@yzu.edu.cn (Z.G.); houmiaowang@yzu.edu.cn (H.W.); MX120170578@yzu.edu.cn (J.W.); DX120180078@yzu.edu.cn (Y.W.); yangx@yzu.edu.cn (Y.X.)

2 Jiangsu Co-Innovation Center for Modern Production Technology of Grain Crops, Yangzhou University, Yangzhou 225009, China

3 Joint International Research Laboratory of Agriculture and Agri-Product Safety of Ministry of Education of China, Yangzhou University, Yangzhou 225009, China

* Correspondence: zfyang@yzu.edu.cn (Z.Y.); qtls@yzu.edu.cn (C.X.); Tel.: +86-0514-8797-9358 (C.X.)

Received: 30 January 2020; Accepted: 27 February 2020; Published: 2 March 2020

Abstract: Root length is a determining factor of the root system architecture, which is essential for the uptake of water, nutrients and plant anchorage. In this study, *ZmMADS60* was resequenced in 285 inbred lines, 68 landraces and 32 teosintes to detect the nucleotide diversity and natural variations associated with root length. Nucleotide diversity and neutral tests revealed that *ZmMADS60* might be selected in domestication and improvement processes. *ZmMADS60* in maize retained only 40.1% and 66.9% of the nucleotide diversity found in teosinte and landrace, respectively. Gene-based association analysis of inbred lines identified nine variants that were significantly associated with primary root length (PRL), lateral root length (LRL), root length between 0 mm and 0.5 mm in diameter (RL005) and total root length (TRL). One single-nucleotide polymorphism SNP1357 with pleiotropic effects was significantly associated with LRL, RL005 and TRL. The frequency of the increased allele T decreased from 68.8% in teosintes to 52.9% and 38.9% in the landrace and inbred lines, respectively. The frequency of the increased allele of another significant SNP723 associated with PRL also decreased during the maize domestication and improvement processes. The results of this study reveal that *ZmMADS60* may be involved in the elongation of primary and lateral roots in the seedling stage and that significant variants can be used to develop functional markers to improve root length in maize.

Keywords: natural variation; maize; root length; domestication selection; *ZmMADS60* gene

1. Introduction

Root systems are crucial for plant survival; they not only provide anchorage for the plant and acquisition of essential mineral nutrients and water from soil, but they also contribute to monitoring the changing environmental conditions [1]. One remarkable feature of immobile plants is that root systems are highly plastic to various environmental cues, such as the water content and nutrient levels in soil [2]. The root system architecture (RSA) can be modulated in several ways, such as primary root elongation, diameter, growth direction and adventitious and lateral root branching. These traits reshape the root system architecture, allowing plants to efficiently absorb water and nutrients for plant growth. Understanding genetic components and exploring the natural variation of root growth are key to breeding and engineering better root system architectures to enhancing plant productivity.

The maize root system comprises structurally and functionally different root types, including primary (PR), seminal (SR), crown (CR), brace (BR), and lateral roots (LR) [3]. A primary root emerges from a seed only two or three days after germination, followed about a week later by the formation of a variable number of seminal roots, which are embryonic [4]. All root types form postembryonic lateral roots, which are major determinants of the plant root system architecture [2]. These roots increase the total root length, biomass and surface area, potentially dramatically enhancing the contact area between the roots and soil for exploration of the soil environment for water and nutrients [5]. They are important for seedling vigor during plant early development. Some genes for these traits have been cloned in mutant analyses. The *rootless concerning crown and seminal roots* (*RTCS*) gene encodes a 25.5-kDa lateral organ boundaries (LOB) domain protein that is a central regulator of auxin signaling, and the maize mutant *RTCS* is defective in the initiation of seminal roots and the shoot-borne root system [6]. The *rum1* mutant is deficient in the initiation of embryonic seminal roots and post-embryonic lateral roots at the primary root. Map-based cloning demonstrated that *rum1* encodes an Aux/IAA protein [7], and a transcriptional activator lateral root primordia 1 (LRP1) was further identified, the expression of which is repressed by the binding of rootless with undetectable meristem 1 (RUM1) to its promoter [8]. Association analyses in 74 maize inbred lines revealed several polymorphisms in *rum1* associated with seedling root traits [9,10]. Numerous quantitative trait loci (QTL) associated with root traits have been identified under different environmental conditions in maize [11,12]. Several significant marker–trait associations with root traits have also been detected by genome-wide association analysis in diverse maize inbred lines [13]. However, most of these loci have not been cloned.

Transcription factors (TF) play an important role in plant development [14]. A total of 3308 TFs belonging to 56 families have been identified in maize (MaizeSequence AGPv3.31) [15]. The MADS-box family comprises large genes with important functions in various aspects of flower development, flowering time control, inflorescence architecture, pollen development, seed/fruit development, and root development [16]. For example, FLOWERING LOCUS C (*FLC*) is a MADS-box gene, and *FLC* and its orthologous genes act as major regulators of flowering time in Arabidopsis, broccoli, and oilseed rape [16,17]. *ZmMADS1*, *ZmMADS69* and *ZAGL1* also function as flowering activators in maize [18–20]. *OsMADS1* and *OsMADS57* can control plant height and tillering in rice [16]. Members of the MADS-box gene family have been prominently studied during flower and plant development; however, the role of these types of TFs in root development has received relatively less attention. It has been reported that more than half of MADS-box genes are expressed in Arabidopsis roots [21]. *XAL1/AGL12* and *XAL2/AGL14* are involved in the regulation of PR elongation, and AGAMOUS-Like42 (*AGL42*), *AGL16*, *AGL17*, *AGL18*, and *AGL21* are preferentially expressed in quiescent center (QC) cells. *ANR1* plays a role in lateral root development in the presence of NO_3^- [22]. In rice, *OsMADS25* can regulate the primary root length and lateral root (LR) density via auxin signaling [23], and the OsmiR156-*OsSPL3* module directly activates *OsMADS50* in the node to regulate crown root development in rice [24]. These studies highlight the important role of MADS-box genes in root development. To systematically study natural variations of MADS-box genes in maize, single-nucleotide polymorphism (SNP) in fifty MADS-box genes (Table S1) were filtered from genotyping-by-sequencing (GBS) dataset [25] in 285 inbred lines. A gene-based association analysis was conducted, and eight genes (Table S2) were significantly associated with seedling root traits. *ZmMADS60* was most significant gene that associated with four root length traits (Table S2). In the present study, we further re-sequenced *ZmMADS60* in 285 inbred lines, 68 landraces and 32 wild relatives, and a gene-based association analysis was conducted examining root traits in inbred lines. The objectives of this study were (1) to identify natural variations in *ZmMADS60* associated with root length traits, (2) to detect favorable alleles and haplotypes within *ZmMADS60* for root length, and (3) to examine the *ZmMADS60* for the involvement in maize domestication and improvement.

2. Materials and Methods

2.1. Plant Materials and Phenotypic Evaluation

In total, 285 inbred lines, 68 landraces and 32 teosintes were used in this study (Table S3). The root traits of inbred lines at the seedling stage were determined in a hydroponic system [25]. Seeds were sterilized in 10% H_2O_2 solution for 20 min, soaked in saturated $CaSO_4$ for 6 h, and then germinated on moist filter paper at 28 °C and 80% relative humidity in the dark for 2 days. Eight uniformly germinated seeds were selected and vertically rolled in germination roll paper (Anchor Paper Company, St Paul, MN, USA). The paper rolls were placed in black incubators containing 7.5 L nutrient solution. A completely randomized design with two replicates was used. Plants were harvested 14 days after germination, and the root number and primary and seminal root length were measured. The roots were then scanned and analyzed using WinRHIZO software (Pro 2004b, Canada). A total of 7 root traits were measured: primary root length (PRL), total root length (TRL), lateral root length (LRL), root length between 0 mm and 0.5 mm in diameter (RL005), root length between 0.5 mm and 1.0 mm in diameter (RL0510), root length between 1.0 mm and 1.5 mm in diameter (RL1015), and average root diameter (ARD).

2.2. ZmMADS60 Resequencing

Genomic DNA of inbred lines, landraces and teosintes was extracted from young leaves using the cetyltrimethylammonium bromide (CTAB) method. The *ZmMADS60* gene was sequenced using targeted sequence capture technology on the NimbleGen platform by BGI Life Tech Co. [26]. The genomic sequence of *ZmMADS60* (GRMZM2G152415) from the B73 inbred line (AGPv3.31) was used as a reference for target sequence capture following the manufacturer's protocols (Roche/NimbleGen) with modifications at the W. M. Keck Facility at Yale University [26]. DNA was sheared by sonication and adaptors were ligated to the resulting fragments. Extracted DNA with desired size was amplified by polymerase chain reaction (PCR), purified, and hybridized to the capture array at 42.0 °C using the manufacturer's buffer. The array was washed twice at 47.5 °C and three more times at room temperature. The resulting fragments were purified and subjected to DNA sequencing on the Illumina platform. The clean reads were mapped to the B73 reference genome sequence (AGPv3.31) by BWA with the settings 'mem -t 4 -k 32 -M' [27]; variant calling and gene sequences converting were performed for all samples using the GATK 4.0 [28].

2.3. Sequence Analysis, Genetic Diversity Analysis and Neutral Evolution Test

The *ZmMADS60* gene sequences from all measured lines were aligned using MAFFT software [29] and manually improved using BioEdit [30]. The gene features [5′ untranslated region (UTR), 3′ UTR, introns and exons] were defined by gene annotation from MaizeSequence (B73, AGPv3.31). The sequence polymorphisms, genetic diversity analysis and neutral evolution test were conducted using DNASP5.0 software [31]. π and θ were used to estimate the degree of genetic diversity within the tested population. The neutral evolution test was investigated using Tajima's D [32], Fu and Li's test [33].

2.4. Natural Variation of the ZmMADS60 Gene Associated with Root Traits in Inbred Lines

The association between variants of the *ZmMADS60* gene and the eight root traits was conducted using TASSEL v5.0 with mixed linear models (MLM). The top five principal components (PCs) and kinship (K) were used to control the population structure and kinship to reduce the false positive error. A total of 678 *ZmMADS60*-based markers with a minor allele frequency (MAF) ≥0.05 were selected for association analysis in 285 inbred lines. Using a Bonferroni correction based on 678 markers, the p-value thresholds were 0.0015 (1/678, $-\log_{10}(p)$ >2.83). A significant P-value threshold (0.001, $-\log_{10}(p)$ >3.0) was applied to identify significant variants with root traits.

3. Results

3.1. *Nucleotide Diversity and Selection of ZmMADS60 in Inbred Lines, Landrace and Teosinte*

ZmMADS60 was resequenced in 285 inbred, 68 landrace and 32 teosinte accessions. After multiple sequence alignment, a total of 7429 bp of genomic region was sequenced covering 1543 bp of the upstream region, 1019 bp of the 5′UTR region, 4755 bp of the coding region containing 11 exons and 11 introns, and a 112 bp of the 3′UTR region (Table 1). Among these regions, 1199 variations were identified in all tested lines, including 1018 SNPs and 181 InDels. On average, SNPs and InDels were found every 7.30 bp and 41.04 bp, respectively. The highest frequencies of SNPs and InDels were found in the 5′UTR (6.21 bp and 29.97 bp, respectively). Sliding-window analysis showed that the overall nucleotide diversity ($\pi \times 1000$) of the *ZmMADS60* locus was 17.57. Among four regions of *ZmMADS60*, the 3′UTR genomic regions were less diverse than the other regions (0.34), and the upstream region showed a high nucleotide diversity (37.59, Table 1).

Table 1. Summary of parameters for the analysis of nucleotide polymorphisms of maize gene *ZmMADS60*.

Parameters	Upstream	5′UTR	Coding Region	3′UTR	Entire Region
Total length of amplicons(bp)	1543	1019	4755	112	7429
Number of all of the sequence vanants	171	198	823	7	1199
Frequency of all of the sequence variants	0.111	0.194	0.173	0.063	0.161
Number of nucleotide substitutions (bp)	141	164	706	7	1018
Frequency of polymorphic sites per bp	0.091	0.161	0.148	0.063	0.137
Number of InDels	30	34	117	0	181
Number of InDel sites	77	155	331	0	563
Average InDel length	3	6	3.504	0	3.89
Frequency of indels per bp	0.019	0.033	0.025	0	0.024
$\pi \times 1000$	37.59	17.83	15.27	0.34	17.57
$\theta \times 1000$	56.75	40.68	38.45	10.02	39.96
Tajima's D	−1.016	−1.696	−1.858 *	−1.899 *	−1.731 *
Fu and Li's D *	−3.849 **	−6.084 **	−7.9 **	−6.006 **	−7.454 **
Fu and Li's F *	−2.831 *	−4.503 **	−5.355 **	−5.459 **	−5.008 **

* Indicates statistical significance at $p < 0.05$; ** Indicates statistical significance at $p < 0.01$; "UTR" indicated untranslated region.

To investigate the selection mechanism of *ZmMADS60* during maize domestication and improvement, the sequence conservation (C) and nucleotide diversity (π) were compared in the inbred lines, landrace and teosinte. For all test lines, the values of C and $\pi \times 1000$ were 0.667 and 17.57, respectively (Figure 1). Compared with teosinte, the landrace and inbred lines showed higher conservation ($C_T = 0.728$, $C_L = 0.826$ and $C_I = 0.855$) and lower diversity ($\pi \times 1000_T = 43.28$, $\pi \times 1000_L = 25.96$ and $\pi \times 1000_I = 17.36$). The nucleotide diversity ratio was 40.1% and 66.9% for maize/teosinte and maize/landrace, respectively. The highest divergence between the inbred lines and teosintes was observed in the upstream regions and the seventh intron (Figure 1b). Tajima's D test and the Fu and Li test for different gene regions were analyzed. For Tajima's D test, all four regions had negative values, and the coding region and 3′UTR had significant negative values. For Fu and Li's test, all regions had significant negative values. The results of Tajima's D values and Fu and Li's statistics suggested the presence of purifying selection in *ZmMADS60* (Table 1).

(a)

Population	*Hd*	Dens.	C	π×1000	θ×1000	Tajima's D	D	F
Teosintes	1.000	171	0.728	43.28	62.91	-1.21	-1.853	-1.773
Landraces	1.000	96	0.826	25.96	28.68	-0.335	-1.185	-0.900
Inbreds	0.974	76	0.855	17.36	18.00	-0.113	-0.519	-0.285
All	0.985	35	0.667	17.57	39.96	-1.731*	-7.454**	-5.008**

(b)

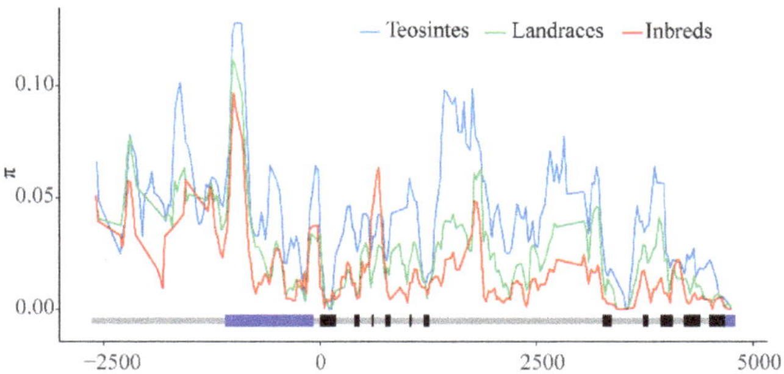

Figure 1. Nucleotide diversity (π) of inbred lines, landrace and teosinte. (**a**) Summary of nucleotide polymorphisms and neutrality test of *ZmMADS60*, Hd represents haplotype diversity, Dens. Denotes number of single nucleotide polymorphisms (SNP) per 1000 bp, C represents sequence conservation, and D and F represent Fu and Li's D *and F *. * indicates a statistical significance at $p < 0.05$ level, ** indicates a statistical significance at $p < 0.01$ level (**b**) Nucleotide diversity (π) of inbred lines, landraces, and teosinte. π was calculated using the sliding windows method with a window size of 100 bp and a step length of 25 bp. A schematic diagram of the genomic region of *ZmMADS60*, including upstream sequence and introns (**light gray**), the coding region (**black**), and 5'UTR and 3'UTR (**blue**) is presented.

3.2. Association Analysis of Phenotypic Traits with ZmMADS60

Trait–marker association analysis was conducted to identify the association of root traits with nucleotide polymorphism of *ZmMADS60* in 285 inbred lines. After quality control (minor allele frequency ≥ 0.05), 678 variants, including 425 SNPs and 253 InDels, were included in the association analysis. The unified mixed model with controls for both PCA and relative kinship (MLM+PCA+K) was employed to perform the marker–trait association analysis. A total of nine significant marker–trait associations involving four SNPs (SNP723, SNP726, SNP1319 and SNP1357) and two InDels (InDel-1576 and InDel727) were associated with primary root length (PRL), lateral root length (LRL), root length between 0 mm and 0.5 mm in diameter (RL005) and total root length (TRL) (Table 2, Figure 2). A total of 1, 2 and 3 variants were distributed upstream and in introns 4 and 7, respectively. The significant variants could explain 3.9–5.1% of the phenotypic variation.

Table 2. Significant markers associated with root traits.

Trait	Marker	Alleles	*p* Value	−lg(P)	R^2	Type
LRL	SNP1319	C/T	0.000536	3.271	0.044	Intron7
LRL	SNP1357	A/T	0.000226	3.645	0.05	Intron7
RL005	SNP1357	A/T	0.000896	3.048	0.039	Intron7
TRL	SNP1319	C/T	0.00033	3.481	0.046	Intron7
TRL	SNP1357	A/T	0.000232	3.635	0.048	Intron7
PRL	InDel-1576	-/A	0.000907	3.042	0.041	Upstream
PRL	SNP723	A/G	0.000223	3.653	0.051	Intron4
PRL	SNP726	G/A	0.000223	3.653	0.051	Intron4
PRL	InDel727	TT/–	0.000234	3.631	0.05	Intron4

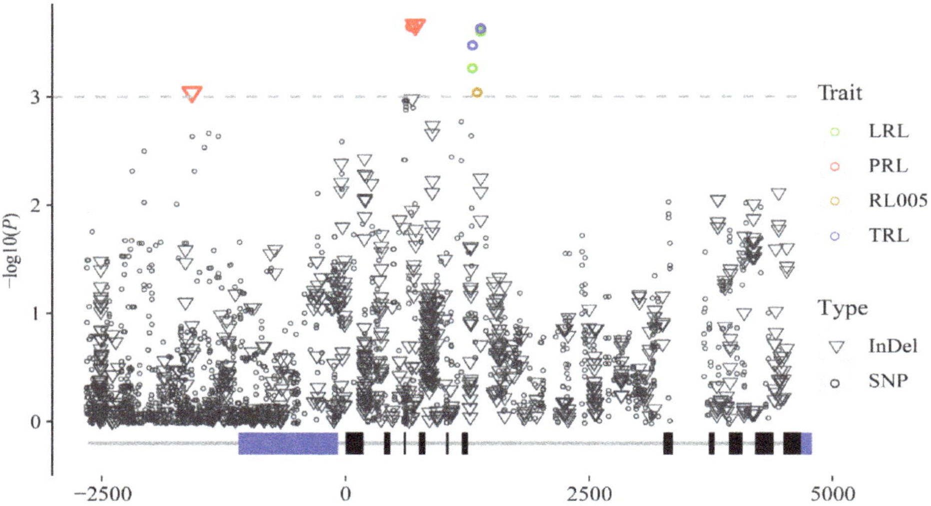

Figure 2. Gene-based association mapping. Manhattan plot using the mixed linear model. Triangles and dots represent InDels and SNPs, respectively. A schematic diagram of the genomic region of ZmMADS60, including upstream sequence and introns (**light gray**), the coding region (**black**), and 5′UTR and 3′UTR (**blue**) is presented. Abbreviations for traits are as follows: LRL, lateral root length; PRL, primary root length; RL005, root length between 0 mm and 0.5 mm in diameter; TRL, total root length.

A total of four significant variants were associated with primary root length (PRL) (Figure 3a), including two Indels (InDel-1576 and InDel727) and two SNPs (SNP723 and SNP726). Site −1576 consisted of a 1-bp insertion/deletion (InDel) polymorphism located in the upstream region. SNP723, SNP726 and InDel727 located in the fourth intron showed strong LD (r^2 = 1) with each other in the inbred lines (Figure 3b). Among these significant sites, three major haplotypes, which contained more than 20 lines, were detected across 285 inbred lines (Figure 3c). The primary root length was compared between these haplotypes, and a significant difference was detected between haplotypes by ANOVA ($p = 7.2 \times 10^{-4}$). Hap1, carrying all increased alleles, had the longest primary root length, followed by Hap2, but no significant phenotypic difference was detected between these two haplotypes. Hap3, carrying all the decreased alleles, had the shortest root length. SNP723 was selected as the leading SNP, and the lines carrying the G allele had a significantly longer PRL than the group carrying the A allele ($p = 7.9 \times 10^{-4}$, Figure 3d). The allele frequencies among inbred lines, landrace and teosintes were analyzed. In teosintes, all the lines carried the increasing allele, but in landraces and inbred lines, the frequency decreased to 89.7% and 63.9%, respectively (Figure 3e). These results suggested that the variants might have been selected during the domestication and improvement of maize. Two significant SNPs, SNP1319 and SNP1357, were significantly associated with total root length (TRL) and lateral root length (LRL) (Figure 4a). Three major haplotypes emerged from the two significant sites across inbred lines (Figure 4b). Both TRL and LRL showed significant difference between haplotypes ($P_{TRL} = 5.2 \times 10^{-5}$ and $P_{LRL} = 9.5 \times 10^{-6}$; Figure 4c). The haplotype carrying all increased alleles (Hap1) had the longest root length compared with that carrying all the decreased alleles (Hap3). The most significant site was SNP1357, which was also significantly associated with root length between 0 mm and 0.5 mm in diameter (RL005) (Figure 5), and the frequency of the increased allele, SNP1357T, decreased from 68.8% in teosintes to 52.9% and 38.9% in the landrace and inbred lines, respectively (Figure 4e).

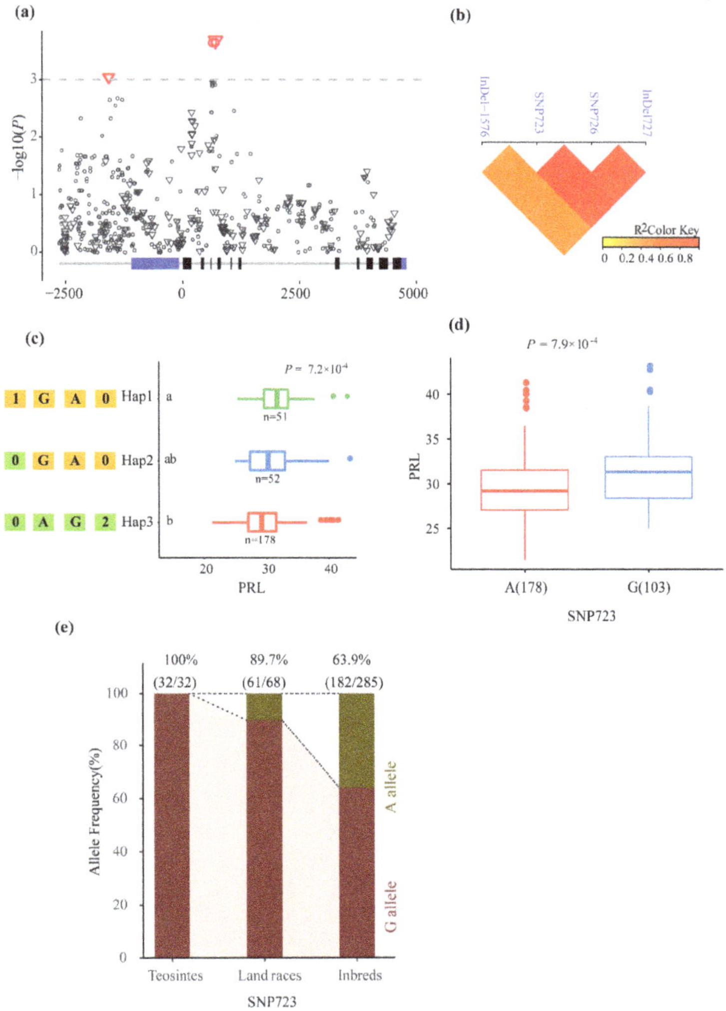

Figure 3. Natural variations in *ZmMADS60* were significantly associated with primary root length (PRL). (**a**) *ZmMADS60*-based association mapping for PRL, A schematic diagram of the genomic region of *ZmMADS60*, including upstream sequence and introns (**light gray**), the coding region (**black**), and 5′UTR and 3′UTR (**blue**) is presented. (**b**) Linkage disequilibrium (LD) heatmap for four significant variants associated with PRL. (**c**) Haplotypes of *ZmMADS60* among natural variations in inbred lines. (**d**) Comparison of primary root length between different alleles of SNP723. (**e**) The allele frequency of SNP723 in teosinte, landrace, and inbred lines.

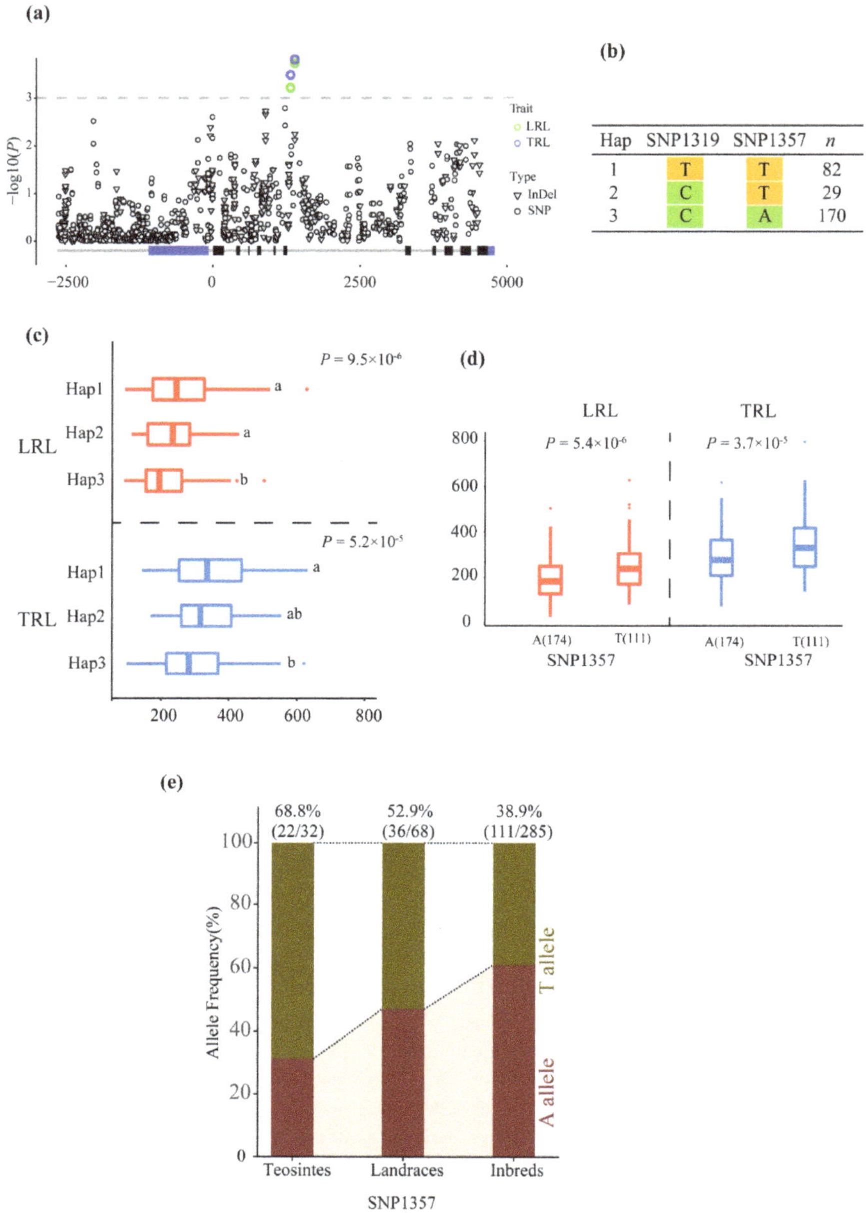

Figure 4. Natural variations in *ZmMADS60* were significantly associated with total root length (TRL) and lateral root length (LRL). (**a**) *ZmMADS60*-based association mapping for TRL and LRL, a schematic diagram of the genomic region of *ZmMADS60*, including upstream sequence and introns (**light gray**), the coding region (**black**), and 5′UTR and 3′UTR (**blue**) is presented. (**b**) Haplotypes of *ZmMADS60* among natural variations in inbred lines. (**c**) Phenotypic difference between different haplotypes of *ZmMADS60* in inbred lines. (**d**) Comparison of TRL and LRL between different alleles of SNP1357. (**e**) The allele frequency of SNP1357 in teosinte, landrace, and inbred lines.

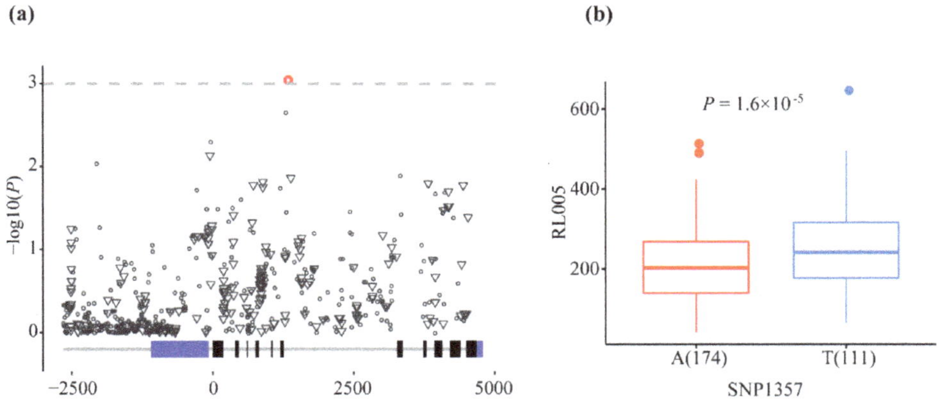

Figure 5. Natural variations in *ZmMADS60* were significantly associated with root length between 0 mm and 0.5 mm in diameter (RL005). (**a**) *ZmMADS60*-based association mapping for RL005, a schematic diagram of the genomic region of *ZmMADS60*, including upstream sequence and introns (**light gray**), the coding region (**black**), and 5′UTR and 3′UTR (**blue**) is presented. (**b**) Comparison of RL005 between different alleles of SNP1357.

4. Discussion

RSA describes the spatial arrangement of root tissue within the soil [34] and the development of crops with ideal root systems that can capture more water and nutrients [35]. Many different factors are involved in shaping RSA, including the number, length, diameter, growth angle, elongation rate, and branching of lateral roots [34]. The manipulation of root traits could deliver increases in resource uptake; however, roots have received less attention compared with aboveground traits in maize breeding programs because they are hidden belowground and heavily influenced by the complex soil environment [36]. Direct selection for optimal RSA in the field is difficult to achieve, but manipulating genes or QTLs that influence RSA can deliver gains in resource use efficiency and yield: for example, introgression of natural variation in the *DRO1* gene into rice lines to generate deeper roots could increase yield under dry conditions [37], and introgressing certain chromosomal fragments into target maize genotypes to increase root size contributed directly to efficient N-uptake and higher yield [38]. Identifying genes or natural variations associated with desired root traits is crucial for root improvement. In this study, gene-based association analysis revealed that natural variations in the *ZmMADS60* gene were significantly associated with seedling root traits in 285 inbred lines. Previous studies implicated the MADS-box genes *AGL12* and *AGL14* in primary root elongation, while *AGL21* was involved in lateral root development in Arabidopsis [21,39]. *OsMADS25* regulated primary root length and lateral root density, and *OsMADS50* regulated crown root development in rice [23,24]. Here, *ZmMADS60* was associated with primary root length (PRL), lateral root length (LRL), root length between 0 and 0.5 mm in diameter (RL005) and total root length (TRL), which indicated that this gene might be involved in elongation of the primary and lateral root in the seedling stage. An early robust root system helps plants capture more water and nutrients to enhance seedling growth in specific environmental settings [40,41]. For example, *PSTOL1* can enhance early root growth in rice, thereby enabling plants to acquire more phosphorus and other nutrients, and introgression of this locus into locally adapted rice varieties is expected to considerably enhance productivity under low phosphorus conditions [40]. Lateral roots are the major determinant of total root length and are instrumental for water uptake in maize [3,4]. Our results indicated that SNP1319 and SNP1357 were significantly associated with LRL and TRL, and Hap1 was the favorable haplotype for improving LRL and TRL. These favorable gene alleles can be integrated into breeding programs to improve early root development in maize. However, there are two limitations to this study. (1) Using B73 as the reference, some big InDels or structural variations may be missing, because the maize genome exhibits high levels of genetic diversity among different inbred lines [42]; (2) Roots grown under controlled conditions, such as paper roll and hydroponic system, might not match those grown under field conditions [43]. Recently, improvements in phenotyping under field conditions will aid in the validation of the effect of significant variations in useful agronomic traits [44].

Maize was domesticated in southwestern Mexico ~9000 years ago from its wild ancestor, teosinte [45], and the domestication underwent two stages: domestication selection and subsequent genetic improvement (postdomestication selection) [46]. As many traits related to plant development and morphology were the target of selection during domestication and improvement processes, transcription factors that usually orchestrate the activity of other genes are among the primary targets of selection [47]. Different overviews provided in previous studies have indicated that 43–81% of domestication genes encode transcriptional regulators, and MADS-box genes function as master regulators of plant development that have been important targets of the artificial selection associated with domestication [47]. A study examined variation in 32 maize MADS-box genes and 32 randomly chosen maize loci, and eight MADS-box genes were selected during the domestication process, indicating that MADS-box genes were more frequent targets of selection during domestication than genes chosen at random [48]. Another MADS-box transcription factor, *ZmMADS69*, was found to be an activator of maize flowering, and the promoter region of *ZmMADS69* has been a target of selection in the adaptation of maize to temperate regions [20]. In this study, *ZmMADS60* in maize retained only 40.1% and 66.9% of the nucleotide diversity of that in teosinte and landrace, respectively. Neutral tests also revealed that *ZmMADS60* might be selected during domestication and improvement processes. We found that the frequency of the increased allele of *ZmMADS60* for root length decreased from teosinte to inbred lines; however, root system has been omitted during domestication and improvement processes owing to its subterranean nature [49]. Ancient humans indirectly selected genetic alleles for root system of maize by breeding for aboveground traits [50]. The root domestication syndrome in the common bean was associated with genes that were directly selected to increase seed weight, but had a significant effect on early root growth through a developmental pleiotropic effect [49]. Zhang et al. (2015) [51] also revealed that the underground nodal root number was generally regulated by the aboveground trait of flowering time via indirect selection during maize domestication. Indirect selection has retained large phenotypic variations in teosinte and landrace. Currently, two genes for root architecture QTLs (*DRO1* and *PSTOL1*) were identified in landrace germplasm rather than elite breeding lines [37,40], highlighting teosinte and landrace as valuable resources for identifying elite natural variations to improve root traits in maize.

Supplementary Materials: The following are available online at http://www.mdpi.com/2073-4395/10/3/342/s1. Table S1. The list of 50 selected MADS-box genes, Table S2. SNPs significantly associated with root traits detected by gene-based association analysis, Table S3. The list of 285 inbred lines, 68 landraces and 32 teosintes used in this study.

Author Contributions: Data curation, Z.G., J.W. and Y.W.; Formal analysis, Y.X.; Funding acquisition, P.L., H.W. and C.X.; Investigation, Z.Y., Supervision, Z.Y. and C.X.; Writing—Original draft, P.L. and H.W.; Writing—Review & editing, C.X. All authors have read and agreed to the published version of the manuscript.

Funding: This work was supported by grants from the National Key Technology Research and Development Program of MOST (2016YFD0100303), the National Natural Science Foundations (31972487 and 31902101), the Natural Science Foundations of Jiangsu Province (BK20180920), the Key Technology Research and Development Program of Jiangsu (BE2018325), the Innovative Research Team of Universities in Jiangsu Province, and the project funded by the Priority Academic Program Development of Jiangsu Higher Education Institutions (PAPD).

Conflicts of Interest: The authors declare they have no conflicts of interest.

References

1. Osmont, K.S.; Sibout, R.; Hardtke, C.S. Hidden branches: Developments in root system architecture. *Annu. Rev. Plant Biol.* **2007**, *58*, 93–113. [CrossRef] [PubMed]

2. Yu, P.; Hochholdinger, F.; Li, C.J. Plasticity of Lateral Root Branching in Maize. *Front Plant Sci.* **2019**, *10*, 363. [CrossRef] [PubMed]
3. Hochholdinger, F.; Yu, P.; Marcon, C. Genetic Control of Root System Development in Maize. *Trends Plant Sci.* **2018**, *23*, 79–88. [CrossRef] [PubMed]
4. Hochholdinger, F.; Park, W.J.; Sauer, M.; Woll, K. From weeds to crops: Genetic analysis of root development in cereals. *Trends Plant Sci.* **2004**, *9*, 42–48. [CrossRef]
5. Stoeckle, D.; Thellmann, M.; Vermeer, J.E.M. Breakout-lateral root emergence in Arabidopsis thaliana. *Curr. Opin. Plant Biol.* **2018**, *41*, 67–72. [CrossRef]
6. Taramino, G.; Sauer, M.; Stauffer, J.L.; Multani, D.; Niu, X.M.; Sakai, H.; Hochholdinger, F. The maize (*Zea mays* L.) RTCS gene encodes a LOB domain protein that is a key regulator of embryonic seminal and post-embryonic shoot-borne root initiation. *Plant J.* **2007**, *50*, 649–659. [CrossRef]
7. von Behrens, I.; Komatsu, M.; Zhang, Y.; Berendzen, K.W.; Niu, X.; Sakai, H.; Taramino, G.; Hochholdinger, F. Rootless with undetectable meristem 1 encodes a monocot-specific AUX/IAA protein that controls embryonic seminal and post-embryonic lateral root initiation in maize. *Plant J.* **2011**, *66*, 341–353. [CrossRef]
8. Zhang, Y.; von Behrens, I.; Zimmermann, R.; Ludwig, Y.; Hey, S.; Hochholdinger, F. LATERAL ROOT PRIMORDIA 1 of maize acts as a transcriptional activator in auxin signalling downstream of the Aux/IAA gene rootless with undetectable meristem 1. *J. Exp. Bot.* **2015**, *66*, 3855–3863. [CrossRef]
9. Abdel-Ghani, A.H.; Kumar, B.; Pace, J.; Jansen, C.; Gonzalez-Portilla, P.J.; Reyes-Matamoros, J.; San Martin, J.P.; Lee, M.; Lubberstedt, T. Association analysis of genes involved in maize (*Zea mays* L.) root development with seedling and agronomic traits under contrasting nitrogen levels. *Plant Mol. Biol.* **2015**, *88*, 133–147. [CrossRef]
10. Kumar, B.; Abdel-Ghani, A.H.; Pace, J.; Reyes-Matamoros, J.; Hochholdinger, F.; Lubberstedt, T. Association analysis of single nucleotide polymorphisms in candidate genes with root traits in maize (*Zea mays* L.) seedlings. *Plant Sci.* **2014**, *224*, 9–19. [CrossRef]
11. Liu, J.C.; Li, J.S.; Chen, F.J.; Zhang, F.S.; Ren, T.H.; Zhuang, Z.J.; Mi, G.H. Mapping QTLs for root traits under different nitrate levels at the seedling stage in maize (*Zea mays* L.). *Plant Soil* **2008**, *305*, 253–265. [CrossRef]
12. Guo, J.; Chen, L.; Li, Y.X.; Shi, Y.S.; Song, Y.C.; Zhang, D.F.; Li, Y.; Wang, T.Y.; Yang, D.G.; Li, C.H. Meta-QTL analysis and identification of candidate genes related to root traits in maize. *Euphytica* **2018**, *214*, 223. [CrossRef]
13. Pace, J.; Gardner, C.; Romay, C.; Ganapathysubramanian, B.; Lubberstedt, T. Genome-wide association analysis of seedling root development in maize (*Zea mays* L.). *BMC Genom.* **2015**, *16*, 47. [CrossRef] [PubMed]
14. Drisch, R.C.; Stahl, Y. Function and regulation of transcription factors involved in root apical meristem and stem cell maintenance. *Front. Plant Sci.* **2015**, *6*, 505. [CrossRef]
15. *Zea mays* Transcription Factors. Available online: http://planttfdb.cbi.pku.edu.cn/index.php?sp=Zma (accessed on 22 February 2020).
16. Smaczniak, C.; Immink, R.G.; Angenent, G.C.; Kaufmann, K. Developmental and evolutionary diversity of plant MADS-domain factors: Insights from recent studies. *Development* **2012**, *139*, 3081–3098. [CrossRef]
17. Michaels, S.D.; Amasino, R.M. FLOWERING LOCUS C encodes a novel MADS domain protein that acts as a repressor of flowering. *Plant Cell* **1999**, *11*, 949–956. [CrossRef]
18. Wills, D.M.; Fang, Z.; York, A.M.; Holland, J.B.; Doebley, J.F. Defining the Role of the MADS-Box Gene, Zea Agamous-like1, a Target of Selection during Maize Domestication. *J. Hered.* **2018**, *109*, 333–338. [CrossRef]
19. Alter, P.; Bircheneder, S.; Zhou, L.Z.; Schluter, U.; Gahrtz, M.U.; Sonnewald, U.; Dresselhaus, T. Flowering Time-Regulated Genes in Maize Include the Transcription Factor ZmMADS1. *Plant Physiol.* **2016**, *172*, 389–404. [CrossRef]
20. Liang, Y.M.; Liu, Q.; Wang, X.F.; Huang, C.; Xu, G.H.; Hey, S.; Lin, H.Y.; Li, C.; Xu, D.Y.; Wu, L.S.; et al. ZmMADS69 functions as a flowering activator through the ZmRap2.7-ZCN8 regulatory module and contributes to maize flowering time adaptation. *New Phytol.* **2019**, *221*, 2335–2347. [CrossRef]
21. Alvarez-Buylla, E.R.; Garcia-Ponce, B.; Sanchez, M.P.; Espinosa-Soto, C.; Garcia-Gomez, M.L.; Pineyro-Nelson, A.; Garay-Arroyo, A. MADS-box genes underground becoming mainstream: Plant root developmental mechanisms. *New Phytol.* **2019**, *223*, 1143–1158. [CrossRef]
22. Zhang, H.; Forde, B.G. An Arabidopsis MADS box gene that controls nutrient-induced changes in root architecture. *Science* **1998**, *279*, 407–409. [CrossRef] [PubMed]

23. Zhang, G.P.; Xu, N.; Chen, H.L.; Wang, G.X.; Huang, J.L. OsMADS25 regulates root system development via auxin signalling in rice. *Plant J.* **2018**, *95*, 1004–1022. [CrossRef] [PubMed]
24. Shao, Y.; Zhou, H.Z.; Wu, Y.; Zhang, H.; Lin, J.; Jiang, X.; He, Q.; Zhu, J.; Li, Y.; Yu, H.; et al. OsSPL3, an SBP-Domain Protein, Regulates Crown Root Development in Rice. *Plant Cell* **2019**, *31*, 1257–1275. [CrossRef] [PubMed]
25. Wang, H.; Wei, J.; Li, P.; Wang, Y.; Ge, Z.; Qian, J.; Fan, Y.; Ni, J.; Xu, Y.; Yang, Z. Integrating GWAS and Gene Expression Analysis Identifies Candidate Genes for Root Morphology Traits in Maize at the Seedling Stage. *Genes* **2019**, *10*, 773. [CrossRef] [PubMed]
26. Choi, M.; Scholl, U.I.; Ji, W.; Liu, T.; Tikhonova, I.R.; Zumbo, P.; Nayir, A.; Bakkaloglu, A.; Ozen, S.; Sanjad, S.; et al. Genetic diagnosis by whole exome capture and massively parallel DNA sequencing. *Proc. Natl. Acad. Sci. USA* **2009**, *106*, 19096–19101. [CrossRef] [PubMed]
27. Li, H.; Durbin, R. Fast and accurate short read alignment with Burrows–Wheeler transform. *Bioinformatics* **2009**, *25*, 1754–1760. [CrossRef]
28. McKenna, A.; Hanna, M.; Banks, E.; Sivachenko, A.; Cibulskis, K.; Kernytsky, A.; Garimella, K.; Altshuler, D.; Gabriel, S.; Daly, M.; et al. The Genome Analysis Toolkit: A MapReduce framework for analyzing next-generation DNA sequencing data. *Genome Res.* **2010**, *20*, 1297–1303. [CrossRef]
29. Katoh, K.; Standley, D.M. MAFFT multiple sequence alignment software version 7: Improvements in performance and usability. *Mol. Biol. Evol.* **2013**, *30*, 772–780. [CrossRef]
30. Hall, T.; Biosciences, I.; Carlsbad, C. BioEdit: An important software for molecular biology. *GERF Bull Biosci.* **2011**, *2*, 60–61.
31. Librado, P.; Rozas, J. DnaSP v5: A software for comprehensive analysis of DNA polymorphism data. *Bioinformatics* **2009**, *25*, 1451–1452. [CrossRef]
32. Tajima, F. Statistical method for testing the neutral mutation hypothesis by DNA polymorphism. *Genetics* **1989**, *123*, 585–595. [PubMed]
33. Fu, Y.X.; Li, W.H. Statistical tests of neutrality of mutations. *Genetics* **1993**, *133*, 693–709. [PubMed]
34. Rogers, E.D.; Benfey, P.N. Regulation of plant root system architecture: Implications for crop advancement. *Curr. Opin. Biotechnol.* **2015**, *32*, 93–98. [CrossRef]
35. Bishopp, A.J.P.; Lynch, J.P. The hidden half of crop yields. *Nat. Plants* **2015**, *1*, 1–2. [CrossRef] [PubMed]
36. Bray, A.L.; Topp, C.N. The Quantitative Genetic Control of Root Architecture in Maize. *Plant Cell Physiol.* **2018**, *59*, 1919–1930. [CrossRef] [PubMed]
37. Uga, Y.; Sugimoto, K.; Ogawa, S.; Rane, J.; Ishitani, M.; Hara, N.; Kitomi, Y.; Inukai, Y.; Ono, K.; Kanno, N.; et al. Control of root system architecture by DEEPER ROOTING 1 increases rice yield under drought conditions. *Nat. Genet.* **2013**, *45*, 1097–1102. [CrossRef]
38. Mu, X.H.; Chen, F.J.; Wu, Q.P.; Chen, Q.W.; Wang, J.F.; Yuan, L.X.; Mi, G.H. Genetic improvement of root growth increases maize yield via enhanced post-silking nitrogen uptake. *Eur. J. Agron.* **2015**, *63*, 55–61. [CrossRef]
39. Yu, L.H.; Miao, Z.Q.; Qi, G.F.; Wu, J.; Cai, X.T.; Mao, J.L.; Xiang, C.B. MADS-Box Transcription Factor AGL21 Regulates Lateral Root Development and Responds to Multiple External and Physiological Signals. *Mol. Plant* **2014**, *7*, 1653–1669. [CrossRef]
40. Gamuyao, R.; Chin, J.H.; Pariasca-Tanaka, J.; Pesaresi, P.; Catausan, S.; Dalid, C.; Slamet-Loedin, I.; Tecson-Mendoza, E.M.; Wissuwa, M.; Heuer, S. The protein kinase Pstol1 from traditional rice confers tolerance of phosphorus deficiency. *Nature* **2012**, *488*, 535–539. [CrossRef]
41. Li, P.C.; Chen, F.J.; Cai, H.G.; Liu, J.C.; Pan, Q.C.; Liu, Z.G.; Gu, R.L.; Mi, G.H.; Zhang, F.S.; Yuan, L.X. A genetic relationship between nitrogen use efficiency and seedling root traits in maize as revealed by QTL analysis. *J. Exp. Bot.* **2015**, *66*, 3175–3188. [CrossRef]
42. Sun, S.; Zhou, Y.; Chen, J.; Shi, J.P.; Zhao, H.M.; Zhao, H.N.; Song, W.B.; Zhang, M.; Cui, Y.; Dong, X.M.; et al. Extensive intraspecific gene order and gene structural variations between Mo17 and other maize genomes. *Nat. Genet.* **2018**, *50*, 1289–1295. [CrossRef] [PubMed]
43. Poorter, H.; Fiorani, F.; Pieruschka, R.; Wojciechowski, T.; van der Putten, W.H.; Kleyer, M.; Schurr, U.; Postma, J. Pampered inside, pestered outside? Differences and similarities between plants growing in controlled conditions and in the field. *New Phytol.* **2016**, *212*, 838–855. [CrossRef] [PubMed]
44. Atkinson, J.A.; Pound, M.P.; Bennett, M.J.; Wells, D.M. Uncovering the hidden half of plants using new advances in root phenotyping. *Curr. Opin. Biotechnol.* **2019**, *55*, 1–8. [CrossRef] [PubMed]

45. Matsuoka, Y.; Vigouroux, Y.; Goodman, M.M.; Sanchez, G.J.; Buckler, E.; Doebley, J. A single domestication for maize shown by multilocus microsatellite genotyping. *Proc. Natl. Acad. Sci. USA* **2002**, *99*, 6080–6084. [CrossRef] [PubMed]
46. Yamasaki, M.; Tenaillon, M.I.; Bi, I.V.; Schroeder, S.G.; Sanchez-Villeda, H.; Doebley, J.F.; Gaut, B.S.; McMullen, M.D. A large-scale screen for artificial selection in maize identifies candidate agronomic loci for domestication and crop improvement. *Plant Cell* **2005**, *17*, 2859–2872. [CrossRef] [PubMed]
47. Schilling, S.; Pan, S.; Kennedy, A.; Melzer, R. MADS-box genes and crop domestication: The jack of all traits. *J. Exp. Bot.* **2018**, *69*, 1447–1469. [CrossRef]
48. Zhao, Q.; Weber, A.L.; McMullen, M.D.; Guill, K.; Doebley, J. MADS-box genes of maize: Frequent targets of selection during domestication. *Genet. Res.* **2011**, *93*, 65–75. [CrossRef]
49. Singh, J.; Gezan, S.A.; Vallejos, C.E. Developmental Pleiotropy Shaped the Roots of the Domesticated Common Bean (Phaseolus vulgaris). *Plant Physiol.* **2019**, *180*, 1467–1479. [CrossRef]
50. York, L.M.; Galindo-Castañeda, T.; Schussler, J.R.; Lynch, J.P. Evolution of US maize (*Zea mays* L.) root architectural and anatomical phenes over the past 100 years corresponds to increased tolerance of nitrogen stress. *J. Exp. Bot.* **2015**, *66*, 2347–2358. [CrossRef]
51. Zhang, Z.; Zhang, X.; Lin, Z.; Wang, J.; Xu, M.; Lai, J.; Yu, J.; Lin, Z. The genetic architecture of nodal root number in maize. *Plant J.* **2018**, *93*, 1032–1044. [CrossRef]

agronomy

MDPI

Article

A 1Ns Disomic Addition from *Psathyrostachys Huashanica* Keng Confers Resistance to Powdery Mildew in Wheat

Jing Han †, Yuxiu Liu †, Chenchen Hou, Jiachuang Li, Jinglin Wang, Qiaoying Zhang, Qunhui Yang, Xinhong Chen * and Jun Wu *

Shaanxi Key Laboratory of Plant Genetic Engineering Breeding, College of Agronomy, Northwest A&F University, Yangling 712100, China; hjzyyeqbd@163.com (J.H.); liuyuxiu2007@126.com (Y.L.); 15234420752@163.com (C.H.); lijiachuang0208@126.com (J.L.); wang3373612072@163.com (J.W.); zhqy886@163.com (Q.Z.); yangh8998@163.com (Q.Y.)

* Correspondence: cxh2089@126.com (X.C.); 13572016162@163.com (J.W.); Tel.: +86-137-0912-9119 (X.C.); +86-135-7201-6162 (J.W.)

† These authors contributed equally to this work.

Received: 9 December 2019; Accepted: 21 February 2020; Published: 24 February 2020

Abstract: Powdery mildew is a fungal disease that threatens wheat production throughout the world. Breeding resistant cultivars is an effective way to reduce harm caused by powdery mildew. In this study, 35 wheat-*Psathyrostachys huashanica*-derived lines were developed by crossing common wheat and *P. huashanica* Keng ($2n = 2x = 14$, NsNs) using embryo culture. Resistance to powdery mildew in the derived lines was identified at the seedling and adult stages. Line H5-5-4-2 was selected with immunity to powdery mildew at both growth stages. The chromosome structure of this line was characterized by cytology, genomic in situ hybridization (GISH), and expressed sequence tag-sequence-tagged site (EST-STS) analysis. The chromosome configuration was $2n = 44 = 22$II. Two *P. huashanica* chromosomes with strong hybridization signals were detected by GISH analysis. Among 83 EST-STS markers that covered all seven homologous groups in wheat, three pairs of STS markers, BE497584, BF202643, and BG262410, located in wheat homologous group 1 amplified clear specific bands related to *P. huashanica*. The results indicated that resistant line H5-5-4-2 was a wheat-*P. huashanica* 1Ns disomic addition line.

Keywords: *Blumeria graminis*; disomic addition line; molecular cytogenetics; wheat; *Psathyrostachys huashanica*

1. Introduction

Wheat (*Triticum aestivum* L.) is one of the most widely cultivated cereal crops worldwide and at least one third of the global population depends on wheat as the staple food [1]. Wheat production is threatened by many diseases and powdery mildew caused by *Blumeria graminis* f. sp. *tritici* (DC.) Speer (*Bgt*) is a particularly important foliar disease [2,3]. Fungicides are often used to control powdery mildew but their widespread application is inappropriate due to high cost, development of resistance in the pathogen, and environmental impacts [4]. Breeding resistant cultivars is also extremely important and this is the main method employed for effectively controlling powdery mildew in wheat [5]. However, pathogen populations undergo rapid mutation events and the coevolution with host resistance that cause resistance gene to become ineffective [6,7]. Therefore, it is necessary to discover more resistance resources that confer resistance to powdery mildew for application in wheat breeding [8]. Distant hybridization is an effective method for broadening the resistance spectrum by introducing novel resistance genes from wild relatives into the bread wheat gene pool [9].

Psathyrostachys huashanica Keng ($2n = 2x = 14$, NsNs) is a perennial cross-pollination gramineae species discovered on a hillside near Huashan, in the Qingling Mountains of Shaanxi [10]. As a wild relative of common wheat, *P. huashanica* is a valuable source of resistance genes for breeding new cultivars, as well as being an endangered and protected species [11,12]. *P. huashanica* has attracted substantial attention from breeders owing to its agronomic characteristics, that include early maturity, short height, and drought and salinity tolerance, but especially its resistance to powdery mildew, stripe rust, wheat scab, and other diseases [13]. Thus, *P. huashanica* has genetic resources that enhance the tertiary gene pool for *Triticum aestivum.* [14].

Plant breeders in China used *P. huashanica* × common wheat crosses as early as 1988 to obtain an F_1 hybrid with a chromosome number of 28 and a heptaploid hybrid with $2n = 21II + 7I$ in a first backcross with 7182. A series of wheat-*P. huashanica* addition lines and substitution lines were then produced, as well as a new intergeneric amphiploid PHw-SA by treating the F_1 hybrid from a cross between wheat cv Kaixianluohanmai and *P. huashanica* with colchicine [15,16]. Based on these studies, a small fragment wheat–*P. huashanica* translocation line K-13-835-3 was developed from a BC_1F_5 population of a cross between PHw-SA and wheat cv CN16, with attractive agronomic traits and high resistance to stripe rust was selected [17]. Zhang et al. [18] conducted random amplified polymorphic DNA (RAPD) analysis using the genomic DNA from *P. huashanica* and common wheat, and characterized two novel sequence characterized amplified region (SCAR) markers called $Psh\text{-}D15_{270}$ and $Psh\text{-}F19_{558}$, that were present only in the Ns genome.

The use of resistance to powdery mildew found in *P. huashanica* to produce resistant wheat cultivars was mentioned rarely in previous studies. Therefore, we selected and analyzed the powdery mildew-resistant lines among 35 wheat-*P. huashanica*-derived lines obtained from the wheat cv 7182 × *P. huashanica* cross. The aims of the study were: (a) to evaluate the resistance to powdery mildew in 35 wheat-*P. huashanica*-derived lines; (b) to determine the chromosomal composition and genomic origin of the resistant materials based on cytogenetic observations and genomic in situ hybridization (GISH) analysis; and (c) to analyze the homoeologous relationships between exogenous chromosomes and wheat chromosomes by expressed sequence tag-sequence-tagged site (EST-STS) analysis.

2. Materials and Methods

2.1. Plant Materials

The genetic stocks used in this study comprised winter wheat cv 7182 (AABBDD, $2n = 2x = 42$), *P. huashanica* (NsNs, $2n = 2x = 14$), and 35 wheat-*P. huashanica* derived lines developed from cross 7182 × *P. huashanica*. The parental wheat cultivar 7182 and *P. huashanica* were employed as controls for assessing powdery mildew resistance and in the EST-STS analyses. Susceptible wheat cv Mingxian 169 was used as the control in powdery mildew tests, and the 35 lines were evaluated to test their resistance to powdery mildew. Genomic DNA from Chinese Spring was used as a blocker in GISH analyses. All of these plant materials are preserved at the Shaanxi Key Laboratory of Genetic Engineering for Plant Breeding, College of Agronomy, Northwest A&F University, Shaanxi, China.

2.2. Assessment of Powdery Mildew Resistance

The reactions to powdery mildew isolate *Bgt* E09 at the seedling stage were assessed as described by An et al. [19], where wheat cultivar 7182, *P. huashanica*, Mingxian 169, and 35 derivative wheat cultivars/lines were tested. The plant materials were grown in plug trays in a greenhouse. Plants were inoculated with E09 at the two-leaf stage. When Mingxian 169 exhibited full disease symptoms, the tested lines were evaluated by an infection type (IT) scale from 0–4 (Table 1), where plants with IT = 0, 1, or 2 were considered resistant and these with IT = 3 or 4 were susceptible [20,21].

The responses of adult plants to powdery mildew were determined in two replicates grown during 2018–2019 at the Yang Ling Wheat Experimental Station of Northwest A&F University. Resistance to powdery mildew in the field was assessed using a mixture of 30 different *Bgt* isolates as inocula. Each material was planted in two rows with a length of 1 m and row spacing of 25 cm, and Mingxian 169 were planted around the tested plants as a disease spreader. In the jointing stage, we artificially dusted the spores evenly over the leaves. When Mingxian 169 exhibited fully developed symptoms, reactions were evaluated and recorded on a scale from 0–9, where scores of 0–4 were considered resistant and 5–9 indicated susceptibility [22]. Moreover, the resistance of each material was also assigned based on the disease index (DI) calculated by 20 randomly selected plants. DI was calculated as [Σ(each disease level × number of diseased plants)/(highest disease level × total number of diseased plants)] × 100 [23]. The classification standards are shown in Table 2.

Table 1. Classification standards for powdery mildew in the seedling stage in wheat.

Infection	Phenotype	Symptom
0	Immune	No disease, no spots
0	Nearly Immune	Few dead leaf spots
1	High Resistance	Lesions less than 1 mm, thin and revealed green
2	Moderate Resistance	Lesions less than 2 mm, thick and not revealed green
3	Moderate Susceptibility	Lesions more than 1 mm, not contiguous, thick hyphae, with a large number of spores
4	Highly Susceptible	Lesions more than 1 mm and contiguous

Table 2. Evaluation criteria for resistance to powdery mildew in the wheat adult stage.

Resistance Type	Immune	High Resistance	Moderate Resistance	Moderate Susceptibility	Susceptible	Highly Susceptible
Range	DI = 0	0 < DI ≤ 5	5 < DI ≤ 15	15 < DI ≤ 25	25 < DI ≤ 50	DI > 50

2.3. Cytogenetic Analysis

Seeds of the test materials were germinated on wet filter paper in dishes. Seedling root tips were cut and placed in ice water for 24 h and then transferred to ethanol–acetic acid (3:1) for 1 week. Pollen mother cells (PMCs) in metaphase I of meiosis were obtained from young panicles and fixed in anhydrous ethanol-chloroform-glacial acetic acid (6:3:1, *v*/*v*). Root tips and PMCs were squashed in 45% acetic acid after staining with 2% acetocarmine for at least 2 h [24], and then used for cytological observations and GISH analysis. The prepared slides and observed with an Olympus BX60 microscope (Japan penguin) to assess chromosome structure and counts, and images were captured. The slides were frozen with liquid nitrogen, before removing the cover slips and storing at −20 °C.

2.4. Genomic In Situ Hybridization (GISH) Analysis

GISH was conducted to detect *P. huashanica* chromatin in the wheat-*P. huashanica*-derived lines. *P. huashanica* genomic DNA extracted from fresh leaves by an improved CTAB (Cetyl Trimethylammonium Ammonium Bromide) method [25], was labeled with Dig-Nick-Translation Mix/digoxigenin (digoxigenin-11-dUTP, DIG; Roche, Germany) using the nick translation method. The hybridization solution contained 1 μL 10% (*w*/*v*) sodium dodecyl sulfate, 1 μL salmon sperm DNA (5 μg/μL), 3 μL probe DNA, 4 μL 20 × SSC solution, 8 μL 50% (*w*/*v*) dextran sulfate, and 20 μL deionized formamide, which was made up to a volume 40 μL with 3 μL double distilled H_2O. Hybridization for GISH was performed by placing a drop of hybridization solution on a slide with the sample. The slides were then incubated at a temperature of 80 °C for 8 min and 37 °C for 16 h. Next, 60 μL 5% bovine serum albumin was dropped onto the slide, which was incubated at 37 °C for 20 min, before adding 50 μL of Anti-Dig-FITC (Fluorescein Isothiocyanate) for detecting and visualizing the labeled chromosomes. We observed the chromosomes using a fluorescence microscope (Olympus BX60) and photographed images using a Photometrics SenSys CCD camera (the USA).

2.5. Expressed Sequence Tag-Sequence-Tagged Site (EST-STS) Analysis

EST-STS markers were employed to determine the homoeologous relationships among the alien *P. huashanica* chromosomes in the wheat-*P. huashanica* lines. We extracted total genomic DNA from the test materials as well as cv 7182 and *P. huashanica* as described above [25,26]. In total, 83 EST-STS multiple-loci primer pairs (https://wheat.pw.usda.gov/SNP/new/pcr_primers.shtml) that covered seven wheat homoeologous groups were used for identifying *P. huashanica* chromosomes in line H5-5-4-2. The polymerase chain reaction (PCR) mixture with a volume of 20 μL contained 2 μL 10 × PCR Buffer, 2 μL (2.5 μmol/mL) primers, 2 μL (40–60 μg/μL) DNA template, 1.6 μL dNTPs (2.5 μmol/mL), 1.6 μL $MgCl_2$ (2.5 mmol/mL), 0.1 μL (5 μmol/μL) *Taq* polymerase, and 10.7 μL double distilled H_2O was compounded for conducting PCR amplification. The procedure comprised initial denaturation for 3 min at 94 °C, followed by 35 cycles of denaturation at 90 °C for 1 min, annealing at 55 °C for 50 s. and a final extension at 72 °C for 1 min. The PCR products were added to 2.5 μL loading buffer, fractionated by non-denatured polyacrylamide gel (8%) electrophoresis, run at 165 V and 120 mA for 2 h, and then visualized in an ultraviolet light box.

3. Results

3.1. Evaluations of Resistance to Powdery Mildew

The seedling reactions to powdery mildew by the 35 wheat-*P. huashanica* offspring as well as parents 7182 and *P. huashanica* are displayed in Table 3. Nine materials (25.71%) exhibited resistance to isolate E09 whereas the remainder were moderately or highly susceptible. H5-5-4-2 was selected for further study because of its immunity to powdery mildew. Wheat cv 7182 was moderately resistant, whereas *P. huashanica* and H5-5-4-2 were immune (Figure 1a).

Table 3. Seedling reaction to powdery mildew in 35 wheat-*P. huashanica*-derived lines, their parents, and Mingxian 169 control.

Line	Reaction	Line	Reaction
H1-8-1-1-2	MR	H17-7-1-1-1	MS
H1-11-5-1-1	MS	H17-7-1-1-8-2	MS
H2-4-18-7-1	MS	H18-1-3-1-6-4	HR
H2-4-18-7-10	MS	H19-1-1	MS
H2-7-8-7-7-2	HS	H19-1-1-1	MS
H3-1-1-1	MS	H20-1-1	HS
H3-2-1-3-5	HS	H20-5-1-1-3-2	MR
H3-2-1-3-12	MS	H24-3-1-5-19-1	MS
H3-2-2-1-1	MS	H24-4-4-1-1-3	MR
H3-2-3-5-1	MS	H26-1-1-1	HS
H3-3-6-3-7	MS	H30-2-3-1-1	HS
H3-5-6-3-1-9	MR	H30-4-4-1-6-4	MS
H3-7-4-2-1	HS	H34-8-2-6-1	MR
H3-7-4-2-2	MR	H42-3-1	HS
H5-5-4-2	I	H62-1-1-1	MR
H5-9-1	HS	H210-1-1	MS
H8-12-2	MS	*P. huashanica*	I
H9-46-1	MS	7182	MS
H13-4-1	HS	Mingxian169	HS

I, immune to powdery mildew; HR, highly resistant; MR, moderately resistant; MS, moderately susceptible; HS, highly susceptible.

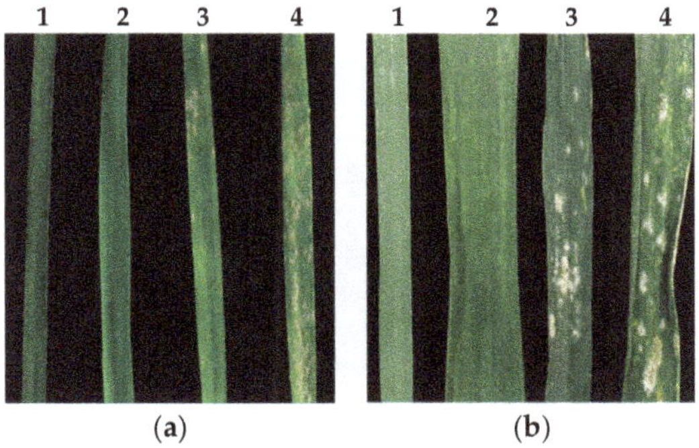

Figure 1. (**a**) Responses of H5-5-4-2, its parents 7182 and *Psathyrostachys huashanica*, and Mingxian 169 to powdery mildew (*Bgt*) isolate E09 in the seedling stage. (**b**) Reactions of the same plant materials to 30 mixed *Bgt* isolates at the adult stage. 1, *Psathyrostachys huashanica*; 2, H5-5-4-2; 3, 7182; 4, Mingxian 169.

The materials were assessed in the adult stage to determine their resistance to powdery mildew during wheat-growing seasons in 2018–2019. Line H5-5-4-2 and *P. huashanica* exhibited uniform immunity to the mixture of *Bgt* isolates. By contrast, wheat cultivar 7182 was moderately resistant to infection and Mingxian169 was highly susceptible (Figure 1b). The IT and DI scores for these four materials were as follows: *P. huashanica*, IT = 0, DI = 0; H5-5-4-2, IT = 0, DI = 0; line 7182, IT = 4, DI = 6.1; and Mingxian169, IT = 8, DI = 50.12. Therefore, the evaluation results indicated that H5-5-4-2 exhibited great resistance to powdery mildew in both the seedling and adult stages, where its resistance to powdery mildew was probably attributable to the introduction of two chromosomes from *P. huashanica*. Thus, a series of subsequent studies were conducted using this immune derivative line.

3.2. Cytogenetic Analysis of H5-5-4-2

Microscopic observations of H5-5-4-2 were conducted for root tip cells (RTCs) and PMCs during mitosis metaphase and meiosis metaphase I, respectively, to determine the chromosome number and pairing behaviors. The mitotic observations of H5-5-4-2 indicated that the RTCs had a chromosome number of $2n = 44$ (Figure 2a). In addition, the chromosome configuration for the PMCs from H5-5-4-2 in metaphase I was $2n = 22$II (Figure 2b).

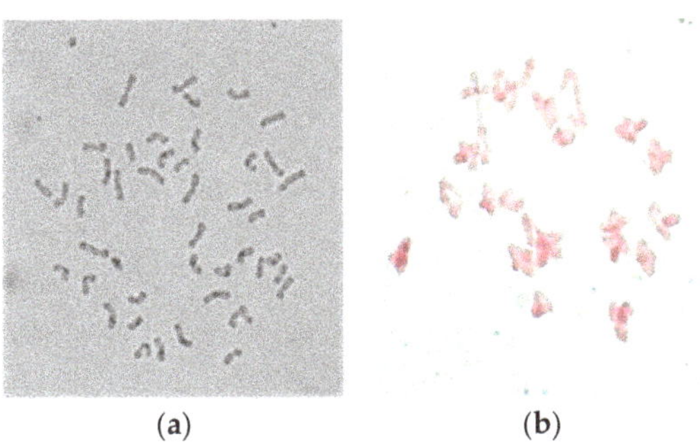

Figure 2. Mitotic and meiotic cytological observations of wheat-*P. huashanica*-derived line H5-5-4-2. (**a**) Chromosomes in the root tip somatic cells during mitotic metaphase, $2n = 44$. (**b**) Chromosome behavior of pollen mother cells during metaphase I, $2n = 22$ II.

3.3. GISH Analysis of H5-5-4-2

Mitotic and meiotic GISH analyses were conducted using the whole genomic DNA of *P. huashanica* as probe. GISH identification in mitosis demonstrated that H5-5-4-2 contained two added chromosomes with yellow-green hybridization signals, whereas the other 42 wheat chromosomes stained red

(Figure 3a). In addition, one ring bivalent with a strong hybridization signal (bright yellow) was observed by GISH at meiotic metaphase I, (Figure 3b). These results demonstrated that two alien chromosomes were introduced into H5-5-4-2 from *P. huashanica*. Besides, the two alien chromosomes from *P. huashanica* could form a separate bivalent and undergo normal synapsis, pairing, and segregation in the wheat background. Thus, line H5-5-4-2 was a cytogenetically stable wheat-*P. huashanica* disomic addition line.

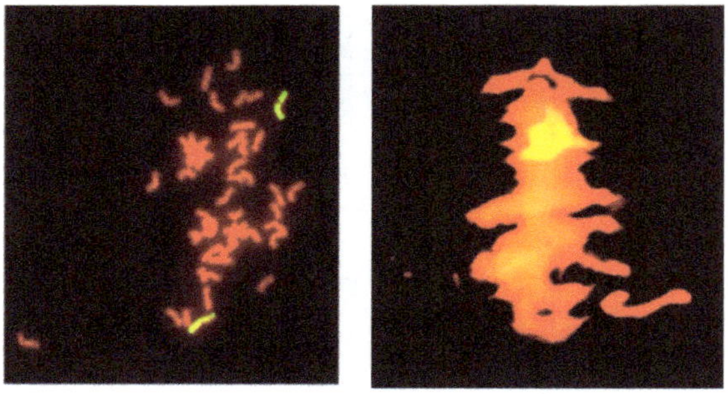

(**a**) (**b**)

Figure 3. Genomic in situ hybridization analysis was conducted with root tip cells (RTCs) and pollen mother cells (PMCs) from H5-5-4-2 by using genomic DNA from *P. huashanica* as a probe and genomes from 7182 as blocker. Fluorescent hybridization signals were exhibited by the alien chromosomes from *P. huashanica* whereas the wheat chromosomes were counterstained with propidium iodide (red). (**a**) RTCs during somatic metaphase showing two chromosomes with yellow-green hybridization signals from *P. huashanica* chromosomes. (**b**) PMCs in meiotic metaphase I showing a ring bivalent with a fluorescent hybridization signal (bright yellow).

3.4. EST-STS Analysis of H5-5-4-2

In this study, 83 pairs of EST-STS primers distributed throughout all over homoeologous groups were screened for polymorphisms in 7182 and *P. huashanica*. These polymorphic primers were used to amplify DNA samples from the disomic addition line H5-5-4-2 and its parents 7182 and *P. huashanica*. Three EST-STS primers, BE497584, BF202643, and BG262410 (Table 4) that mapped to homoeologous group I (1AL, 1AS, 1BL, 1BS, 1DL, and 1DS) amplified clear *P. huashanica*-specific bands in line H5-5-4-2 but none in the wheat parent 7182 (Figure 4). However, amplification products of other primers produced no specific bands. We inferred that the chromosomes from *P. huashanica* belonged to homoeologous group I and that the added chromosome pair in H5-5-4-2 could be designated 1Ns.

Table 4. EST-STS markers from homoeologous group I used to analyze the linkage relationship with the 1Ns chromosome of *P. huashanica*.

Marker	Type	Primer (5′-3′)	Tm (°C)	Location
BE497584	STS	F: CTGTTGCCAAGAGCATTGAA R: GTCACAACATCATCAACCGC	60	1AL 1BL 1DL
BF202643	STS	F: TTGCTGCTTGGACTTTCCTT R: GAAGAACAGCAGGGCGTTAC	60	1AS 1BS 1DS
BG262410	STS	F: TTCCACAAGAAAATGCCTCC R: GCTCCCGTAGCTCATCAAAG	60	1AS 1BS 1DS

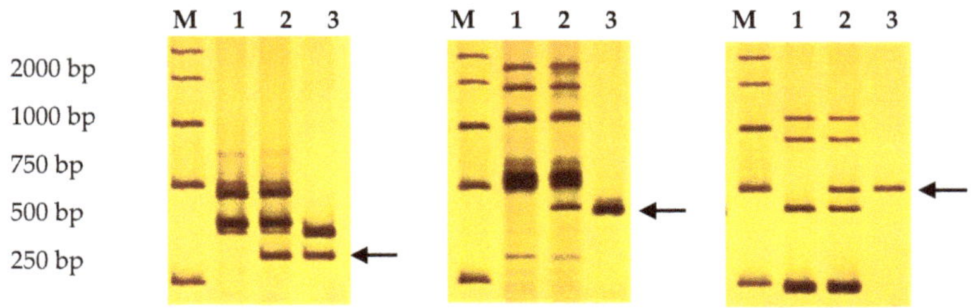

(**a**) BE497584 (**b**) BF202643 (**c**) BG262410

Figure 4. Identification of the homoeologous group of the alien chromosome pair in H5-5-4-2 and wheat using expressed sequence tag-sequence-tagged site (EST-STS) markers. Lane M, DNA ladder (DL2000); lane 1, 7182; lane 2, H5-5-4-2; lane 3, *P. huashanica*. Arrows indicate the specific amplification product from *P. huashanica*.

4. Discussion

Distant hybridization is a major focused research in many laboratories worldwide. It allows genes from related species to be transferred into wheat to improve quality and productivity [27]. Researchers in China began conducting crosses between wheat with *Thinopyrum intermedium* and *Thinopyrum elongatum* in the 1950s, and a number of wheat cultivars were developed [28]. A wheat–rye 1BL·1RS translocation line was obtained by transferring many resistance genes (e.g., *Yr9*, *Lr26*, *Pm8*, and *Sr31*) from rye into a wheat background, and it is an outstanding example of genetic improvement in wheat by distant hybridization [29,30]. Recently, the exploitation of available genes in the wild relatives of wheat has diversified greatly. In particular, An et al. [31] obtained a new wheat–rye addition line called WR35 via crossing Xiaoyan 6 with the rye cultivar German White, where the addition line exhibited resistance to powdery mildew and stripe rust, and a high kernel number per spike. Pan et al. [32] also found that the 1P chromosome from *Agropyron cristatum* enhanced the spike length and tiller number in wheat, and a wheat-*A. cristatum* 1P addition line was obtained using molecular and phenotypic identification techniques.

Crosses between wheat and *P. huashanica* are relatively recent, and *P. huashanica* is significant among numerous species related to wheat because of its excellent agronomic characteristics. Several wheat-*P. huashanica* progeny lines were generated in previous studies, including disomic addition lines [33], disomic substitution lines [34], and small fragment translocation lines [17]. In our laboratory, 35 derived lines were developed by crossing common wheat 7182 with *P. huashanica* via embryo rescue culture and backcrossing, and these intermediate materials with different agronomic traits provide a suitable foundation for exploiting genes from *P. huashanica*.

Powdery mildew severely hinders yield and quality improvements in wheat and the loss of effective genetic resistance to powdery mildew due to the high variability of pathogens and the uniformity of resistance sources is leading to a crisis in wheat production [35–37]. Among the 92 powdery mildew resistance genes (*Pm1*~*Pm65*) that have been officially named, many are derived from wild relatives [38–40]. Thus, many studies have aimed to identify new resistance genes in the relatives of wheat [41,42]. In previous studies, the resistance genes *Pm7*, *Pm8*, *Pm17*, and *Pm20* were identified in rye, and *Pm8* was utilized widely, although the effectiveness of its resistance has now been lost [43–45]. *Pm21*, which is located on the short arm of chromosome 6V in *Haynaldia villosa* ($2n = 14$, VV) and it is an excellent gene that confers broad-spectrum resistance to powdery mildew in wheat [46]. Zhan et al. [45] showed that alien chromosome fragments possessed genetic loci with resistance to powdery mildew and stripe rust in wheat–*Thinopyrum intermedium* translocation lines. In addition, Li et al. [47] detected resistance genes in the 2P chromosome in *Agropyron cristatum*.

The powdery mildew resistance genes in *P. huashanica* have not been identified in recent studies, and wheat-*P. huashanica*-derived lines with resistance have rarely been reported, so it is significant to conduct further research into powdery mildew resistance in *P. huashanica*. In the present study, we examined the seeding responses of 35 wheat-*P. huashanica*-derived lines to powdery mildew and showed that line H5-5-4-2 was immune (Table 3, Figure 1a). Moreover, the resistance evaluation of H5-5-4-2 in the adult stage was immunity (Figure 1b). Our results suggested that this resistance originated from *P. huashanica*. We further speculated that this resistance may be a quantitative trait controlled by multiple genes as the derived lines displayed different evaluation of immunity, high resistance, moderate resistance, moderate susceptibility and high susceptibility. These findings provide a great precondition to explore resistance genes in *P. huashanica* and the development of novel resistance germplasm resources.

The resistance performance of wheat varieties may differ during various growth periods. Comprehensive assessments of disease resistance need to be conducted in both the seedling and adult stage. Varieties with adult resistance mostly exhibit sustained quantitative resistance (horizontal resistance). If a variety exhibits resistance in both the seedling and adult stages, then it will show strong resistance throughout the whole growth period [48]. In the present study, we found that derivative line H5-5-4-2 not only exhibited high resistance to a mixture of races in the adult stage, but it was also immune to *Bgt* isolate E09 in the seedling stage. Thus, we speculated that the resistance genes derived from *P. huashanica* may be the major genes, which may allow this line to exhibit strong broad-spectrum resistance to powdery mildew, thereby providing a resistance gene resource for breeding wheat cultivars with resistance to powdery mildew.

Mitotic and meiotic cytological observations as well as GISH were performed in order to determine the origin of resistance gene in the immune derivative line H5-5-4-2, where the results demonstrated that H5-5-4-2 is a disomic addition line. GISH can simply determine the presence of alien chromosomes but molecular markers are valuable analysis tools for identifying the homoeology of alien chromosomes. Thus, a wheat-*Aegilops geniculata* $7M^g$ addition line was identified using EST-STS markers by Wang et al. [49]. In addition, Wang et al. [50] identified a wheat-*Leymus racemosus* translocation line T5AS-7LrL·7LrS by using EST-STS markers. In the present study, three pairs of primers belonging to homoeologous group I amplified Ns genome-specific bands in H5-5-4-2, thereby indicating the presence of a pair of 1Ns chromosomes derived from *P. huashanica*, and thus the powdery mildew-resistant derivative line H5-5-4-2 is a wheat-*P. huashanica* 1Ns disomic addition line.

Wheat disomic addition lines had substantial value in various applications as basic materials for investigating the genetic relationships between wheat and its relatives and gene mapping, they can also be employed as bridge materials for producing substitution lines and translocation lines [31]. A complete set of 1Ns–7Ns wheat-*P. huashanica* disomic addition lines was developed by Du et al. [51]. In addition, in a previous study of wheat-*P. huashanica* 1Ns disomic addition lines, Zhao et al. [52] found that the addition of the 1Ns chromosome from *P. huashanica* in H9021-28-5 had the effects of improving agronomic traits such as the grain weight, resistance to stripe rust, flour-processing properties, and the contents of some microelements. Moreover, Du et al. [51] showed that the 1Ns disomic addition line 12-3 exhibited high resistance to leaf rust, and it possessed large and awnless panicles. However, previous studies have not investigated powdery mildew resistance in wheat–*P. huashanica* 1Ns disomic addition lines.

Therefore, our successful identification of the wheat-*P. huashanica* 1Ns chromosome disomic addition line H5-5-4-2 with resistance to powdery mildew may facilitate the breeding of varieties with resistance to powdery mildew, as well as enriching the wheat germplasm resource pool and laying a material foundation for fully utilizing the excellent traits of *P. huashanica* in wheat resistance breeding. Future studies may focus on the resistance of 1Ns chromosome in *P. huashanica* with referencing our study to promote the investigation and development of more wheat-*P. huashanica* 1Ns alien addition lines.

Author Contributions: J.W. (Jun Wu) and X.C. conceived and designed the study; J.H., Y.L. and Q.Z. performed the study; J.L. and Q.Y. contributed new methods or models; C.H. and J.W. (Jinglin Wang) analyzed the date; J.H. and Y.L. wrote the paper. All authors have read and agreed to the published version of the manuscript.

Funding: This research was funded by the Key Research and Development Project of Shaanxi Province, grant number 2019ZDLNY04-05.

Acknowledgments: We sincerely thank the College of Agronomy, Northwest A&F University, for providing the laboratory of this study. We also truly appreciate Dr. Duncan E. Jackson for his language editing and checking of scientific consistency.

Conflicts of Interest: The authors declare no conflict of interest.

References

1. Yang, M.J.; Huang, K.Y.; Han, Q.D. Research progresses on wheat powdery mildew and its resistance. *Mol. Plant Breed.* **2016**, *14*, 1244–1254. [CrossRef]
2. Dean, R.; Van Kan, J.; Pretorius, Z.A.; E Hammond-Kosack, K.; Di Pietro, A.; Spanu, P.; Rudd, J.J.; Dickman, M.; Kahmann, R.; Ellis, J.; et al. The Top 10 fungal pathogens in molecular plant pathology. *Mol. Plant Pathol.* **2012**, *13*, 414–430. [CrossRef] [PubMed]
3. Morgounov, A.; Tufan, H.A.; Sharma, R.; Akin, B.; Bagci, A.; Braun, H.-J.; Kaya, Y.; Keser, M.; Payne, T.S.; Sonder, K.; et al. Global incidence of wheat rusts and powdery mildew during 1969–2010 and durability of resistance of winter wheat variety Bezostaya 1. *Eur. J. Plant Pathol.* **2011**, *132*, 323–340. [CrossRef]
4. Khong, N.G.; Randoux, B.; Tayeh, C.; Coutte, F.; Bourdon, N.; Tisserant, B.; Laruelle, F.; Jacques, P.; Reignault, P. Induction of resistance in wheat against powdery mildew by bacterial cyclic lipopeptides. *Commun. Agric. Appl. Boil. Sci.* **2012**, *77*, 39–51.
5. Tan, C.; Li, G.; Cowger, C.; Carver, B.F.; Xu, X. Characterization of Pm59, a novel powdery mildew resistance gene in Afghanistan wheat landrace PI 181356. *Theor. Appl. Genet.* **2018**, *131*, 1145–1152. [CrossRef]
6. Ma, P.; Xu, H.; Xu, Y.; Li, L.; Qie, Y.; Luo, Q.; Zhang, X.; Li, X.; Zhou, Y.; An, D. Molecular mapping of a new powdery mildew resistance gene Pm2b in Chinese breeding line KM2939. *Theor. Appl. Genet.* **2015**, *128*, 613–622. [CrossRef]
7. McDonald, B.; Linde, C. The population genetics of plant pathogens and breeding strategies for durable resistance. *Euphytica* **2002**, *124*, 163–180. [CrossRef]
8. Hou, L.; Zhang, X.; Li, X.; Jia, J.; Yang, H.; Zhan, H.; Qiao, L.; Guo, H.; Chang, Z. Mapping of Powdery Mildew Resistance Gene pmCH89 in a Putative Wheat-*Thinopyrum intermedium* Introgression Line. *Int. J. Mol. Sci.* **2015**, *16*, 17231–17244. [CrossRef]
9. Lin, Z.-S.; Cui, Z.-F.; Zeng, X.-Y.; Ma, Y.-Z.; Zhang, Z.-Y.; Nakamura, T.; Ishikawa, G.; Nakamura, K.; Yoshida, H.; Xin, Z.-Y. Analysis of wheat-*Thinopyrum intermedium* derivatives with BYDV-resistance. *Euphytica* **2007**, *158*, 109–118. [CrossRef]
10. Wang, X.-J.; Chen, X.-H.; Pang, Y.-H.; Jing, F.; Zhang, J.; Hu, S.-Y.; Zan, K.; Wu, J.; Yang, Q.-H.; Zhao, J.-X. Molecular Cytogenetics Identification of a wheat *Psathyrostachys huashanica* Substitution Line DH2322. *Acta Agron. Sin.* **2015**, *41*, 207. [CrossRef]
11. Baden, C. A taxonomic revision of Psathyrostachys (Poaceae). *Nord. J. Bot.* **1991**, *11*, 3–26. [CrossRef]
12. Du, W.; Wang, J.; Lu, M.; Sun, S.; Chen, X.; Zhao, J.; Yang, Q.; Wu, J. Molecular cytogenetic identification of a wheat–*Psathyrostachys huashanica* Keng 5Ns disomic addition line with stripe rust resistance. *Mol. Breed.* **2013**, *31*, 879–888. [CrossRef]
13. Han, Y.C.; Wang, C.Y.; Chen, C.H.; Tian, Z.R.; Ji, W.Q. Molecular cytogenetic study on wheat-*Psathyrostachys huashanica* 1Ns disomic addition line. *J. Triticeae Crop.* **2015**, *35*, 1044–1049. [CrossRef]
14. Ma, D.-F.; Hou, L.; Sun, C.; Zhang, X.; Yin, J.-L.; Guo, Q.-Y.; Zhu, Y.-X. Molecular mapping of stripe rust resistance gene YrH9017 in wheat-*Psathyrostachys huashanica* introgression line H9017-14-16-5-3. *J. Integr. Agric.* **2019**, *18*, 108–114. [CrossRef]
15. Chen, S.Y.; Hou, W.S.; Zhang, A.J.; Fu, J.; Yang, Q.H. Breeding and cytogenetic study of *Triticum aestivum-Psathyrostachys huashanica* alien addition lines. *Acta Genet. Sin.* **1996**, *23*, 447–452.
16. Kang, H.Y.; Wang, Y.; Sun, G.L.; Zhang, H.Q.; Fan, X.; Zhou, Y.H. Production and characterization of an amphiploid between common wheat and *Psathyrostachys huashanica* Keng ex Kuo. *Plant Breed.* **2009**, *128*, 36–40. [CrossRef]
17. Kang, H.-Y.; Zhang, Z.-J.; Xu, L.-L.; Qi, W.-L.; Tang, Y.; Wang, H.; Zhu, W.; Li, D.-Y.; Zeng, J.; Wang, Y.; et al. Characterization of wheat–*Psathyrostachys huashanica* small segment translocation line with enhanced kernels per spike and stripe rust resistance. *Genome* **2016**, *59*, 221–229. [CrossRef]

18. Zhang, J.; Jiang, Y.; Guo, Y.L.; Li, Y.H.; Wang, Y.; Pu, X. Cloning of Ns genome-specific sequence of *Psathyrostachys huashanica* and construction of molecular markers. *J. Agric. Biotechnol.* **2017**, *25*, 1391–1399. [CrossRef]
19. An, D.; Zheng, Q.; Zhou, Y.; Ma, P.; Lv, Z.; Li, L.; Li, B.; Luo, Q.; Xu, H.; Xu, Y. Molecular cytogenetic characterization of a new wheat–rye 4R chromosome translocation line resistant to powdery mildew. *Chromosom. Res.* **2013**, *21*, 419–432. [CrossRef]
20. Yu, X.; Ren, S.; Zhao, L.; Guo, J.; Bao, Y.; Ma, Y.; Wang, H.; Ohm, H.W.; Yu, D.; Li, H.; et al. Molecular mapping of a novel wheat powdery mildew resistance gene Ml92145E8-9 and its application in wheat breeding by marker-assisted selection. *Crop. J.* **2018**, *6*, 621–627. [CrossRef]
21. Zeng, F.-S.; Yang, L.-J.; Gong, S.-J.; Shi, W.-Q.; Zhang, X.-J.; Wang, H.; Xiang, L.-B.; Xue, M.-F.; Yu, D.-Z. Virulence and Diversity of *Blumeria graminis* f. sp. tritici Populations in China. *J. Integr. Agric.* **2014**, *13*, 2424–2437. [CrossRef]
22. Sheng, B.Q.; Duan, X.Y. Improvement of scale 0-9 method for scoring adult plant resistance to powdery mildew of wheat. *J. Beijing Agric. Sci.* **1991**, *1*, 38–39.
23. Yan, Z.Y.; Fan, J.R.; Liu, W.; Zhou, Y.L. Models of disease index estimation of wheat powdery mildew based on the concentrations of *Blumeria graminis* f. sp. tritici conidia in the fields. *Acta Phytopathol. Sin.* **2017**, *47*, 253–261. [CrossRef]
24. Tang, X.; Shi, D.; Xu, J.; Li, Y.; Li, W.; Ren, Z.; Fu, T. Molecular cytogenetic characteristics of a translocation line between common wheat and *Thinopyrum intermedium* with resistance to powdery mildew. *Euphytica* **2014**, *197*, 201–210. [CrossRef]
25. Cota-Sánchez, J.H.; Remarchuk, K.; Ubayasena, K. Ready-to-use DNA extracted with a CTAB method adapted for herbarium specimens and mucilaginous plant tissue. *Plant Mol. Boil. Rep.* **2006**, *24*, 161–167. [CrossRef]
26. Lee, M.H.; Nicholson, P. Isolation of genomic DNA from plant tissues. *Nat. Biotechnol.* **1997**, *15*, 805–806. [CrossRef]
27. An, D.; Xu, H.; Xu, Y.-F. Enhancement of wheat distant hybridization germplasm. *Chin. J. Eco-Agric.* **2012**, *19*, 1011–1019. [CrossRef]
28. Dong, Y.C. Wheat Breeding through Distant Crossing. In Proceedings of the International Symposium on Genetics and Breeding of Wheat, Zhengzhou, Henan, China, 9–11 May 2001.
29. Howell, T.; Hale, I.; Jankuloski, L.; Bonafede, M.; Gilbert, M.; Dubcovsky, J. Mapping a region within the 1RS.1BL translocation in common wheat affecting grain yield and canopy water status. *Theor. Appl. Genet.* **2014**, *127*, 2695–2709. [CrossRef]
30. Ren, T.-H.; Chen, F.; Yan, B.-J.; Zhang, H.-Q.; Ren, Z.-L. Genetic diversity of wheat–rye 1BL.1RS translocation lines derived from different wheat and rye sources. *Euphytica* **2011**, *183*, 133–146. [CrossRef]
31. An, D.; Ma, P.; Zheng, Q.; Fu, S.; Li, L.; Han, F.; Han, G.; Wang, J.; Xu, Y.; Jin, Y.; et al. Development and molecular cytogenetic identification of a new wheat-rye 4R chromosome disomic addition line with resistances to powdery mildew, stripe rust and sharp eyespot. *Theor. Appl. Genet.* **2018**, *132*, 257–272. [CrossRef]
32. Pan, C.; Li, Q.; Lu, Y.; Zhang, J.; Yang, X.; Li, X.; Li, L.; Liu, W. Chromosomal localization of genes conferring desirable agronomic traits from *Agropyron cristatum* chromosome 1P. *PLoS ONE* **2017**, *12*, e0175265. [CrossRef] [PubMed]
33. Du, W.; Wang, J.; Lu, M.; Sun, S.; Chen, X.; Zhao, J.; Yang, Q.; Wu, J. Characterization of a wheat-*Psathyrostachys huashanica* Keng 4Ns disomic addition line for enhanced tiller numbers and stripe rust resistance. *Planta* **2013**, *239*, 97–105. [CrossRef] [PubMed]
34. Li, J.; Yao, X.; Yang, Z.; Cheng, X.; Yuan, F.; Liu, Y.; Wu, J.; Yang, Q.; Zhao, J.; Chen, X. Molecular cytogenetic characterization of a novel wheat–*Psathyrostachys huashanica* Keng 5Ns (5D) disomic substitution line with stripe rust resistance. *Mol. Breed.* **2019**, *39*, 109. [CrossRef]
35. Cowger, C.; Miranda, L.; Griffey, C.; Hall, M.; Murphy, J.P.; Maxwell, J. Wheat powdery mildew. In *Disease Resistance in Wheat*; CABI Publishing: Wallingford, UK, 2012; pp. 84–119.
36. He, F.; Bao, Y.; Qi, X.; Ma, Y.; Li, X.; Wang, H. Molecular cytogenetic identification of a wheat–*Thinopyrum ponticum* translocation line resistant to powdery mildew. *J. Genet.* **2017**, *96*, 165–169. [CrossRef]
37. Xu, X.-D.; Feng, J.; Fan, J.; Liu, Z.; Li, Q.; Zhou, Y.; Ma, Z.-H. Identification of the resistance gene to powdery mildew in Chinese wheat landrace Baiyouyantiao. *J. Integr. Agric.* **2018**, *17*, 37–45. [CrossRef]

38. Li, G.; Cowger, C.; Wang, X.; Carver, B.F.; Xu, X. Characterization of Pm65, a new powdery mildew resistance gene on chromosome 2AL of a facultative wheat cultivar. *Theor. Appl. Genet.* **2019**, *132*, 2625–2632. [CrossRef]
39. Tan, C.; Li, G.; Cowger, C.; Carver, B.F.; Xu, X. Characterization of Pm63, a powdery mildew resistance gene in Iranian landrace PI 628024. *Theor. Appl. Genet.* **2018**, *132*, 1137–1144. [CrossRef]
40. Zhang, D.; Zhu, K.; Dong, L.; Liang, Y.; Li, G.; Fang, T.; Guo, G.; Wu, Q.; Xie, J.; Chen, Y.; et al. Wheat powdery mildew resistance gene Pm64 derived from wild emmer (*Triticum turgidum* var. dicoccoides) is tightly linked in repulsion with stripe rust resistance gene Yr5. *Crop. J.* **2019**, *7*, 761–770. [CrossRef]
41. Grewal, S.; Yang, C.; Hubbart-Edwards, S.; Scholefield, D.; Ashling, S.; Burridge, A.; King, I.P.; King, J. Characterisation of *Thinopyrum bessarabicum* chromosomes through genome-wide introgressions into wheat. *Theor. Appl. Genet.* **2017**, *131*, 389–406. [CrossRef]
42. Rahmatov, M.; Rouse, M.N.; Nirmala, J.; Danilova, T.; Friebe, B.; Steffenson, B.J.; Johansson, E. A new 2DS·2RL Robertsonian translocation transfers stem rust resistance gene Sr59 into wheat. *Theor. Appl. Genet.* **2016**, *129*, 1383–1392. [CrossRef]
43. An, D.; Zheng, Q.; Luo, Q.; Ma, P.; Zhang, H.; Li, L.; Han, F.; Xu, H.; Xu, Y.; Zhang, X.; et al. Molecular Cytogenetic Identification of a New Wheat-Rye 6R Chromosome Disomic Addition Line with Powdery Mildew Resistance. *PLoS ONE* **2015**, *10*, e0134534. [CrossRef] [PubMed]
44. Friebe, B.; Heun, M.; Tuleen, N.; Zeller, F.J.; Gill, B.S. Cytogenetically Monitored Transfer of Powdery Mildew Resistance from Rye into Wheat. *Crop. Sci.* **1994**, *34*, 621–625. [CrossRef]
45. Zhan, H.; Zhang, X.; Li, G.; Pan, Z.; Hu, J.; Li, X.; Qiao, L.; Jia, J.; Guo, H.; Chang, Z.; et al. Molecular Characterization of a New Wheat-Thinopyrum intermedium Translocation Line with Resistance to Powdery Mildew and Stripe Rust. *Int. J. Mol. Sci.* **2015**, *16*, 2162–2173. [CrossRef] [PubMed]
46. Xing, L.; Hu, P.; Liu, J.; Witek, K.; Zhou, S.; Xu, J.; Zhou, W.; Gao, L.; Huang, Z.; Zhang, R.; et al. Pm21 from *Haynaldia villosa* Encodes a CC-NBS-LRR Protein Conferring Powdery Mildew Resistance in Wheat. *Mol. Plant* **2018**, *11*, 874–878. [CrossRef]
47. Li, Q.; Lu, Y.; Pan, C.; Zhang, J.; Liu, W.; Yang, X.; Li, X.; Xi, Y.; Li, L. Characterization of a Novel Wheat-Agropyron cristatum 2P Disomic Addition Line with Powdery Mildew Resistance. *Crop. Sci.* **2016**, *56*, 2390–2400. [CrossRef]
48. Wang, Z.L.; Li, M.X.; Xi, Y.J.; Liu, S.D. Appraisal of Resistance to Powdery Mildew and Molecular Marker Analysis of Pm4b of Wheat Varieties (Lines). *Acta Agric. Boreal.-Sin.* **2014**, *29*, 56–61. [CrossRef]
49. Wang, Y.; Quan, W.; Peng, N.; Wang, C.; Yang, X.; Liu, X.; Zhang, H.; Chen, C.; Ji, W. Molecular cytogenetic identification of a wheat-*Aegilops geniculata* Roth 7Mg disomic addition line with powdery mildew resistance. *Mol. Breed.* **2016**, *36*. [CrossRef]
50. Wang, L.-S.; Zhang, Y.-L.; Nan, G.-H. Molecular and Cytogenetic Identification of *Triticum aestivum-Leymus racemosus* Translocation Line T5AS-7LrL·7LrS. *Acta Agron. Sin.* **2018**, *44*, 1442–1447. [CrossRef]
51. Du, W.; Wang, J.; Pang, Y.; Wu, J.; Zhao, J.; Liu, S.; Yang, Q.; Chen, X. Development and application of PCR markers specific to the 1Ns chromosome of *Psathyrostachys huashanica* Keng with leaf rust resistance. *Euphytica* **2014**, *200*, 207–220. [CrossRef]
52. Zhao, J.X.; Wu, J.; Cheng, X.N.; Dong, J.; Chen, X.H.; Liu, S.H.; Du, W.L.; Pang, Y.Y.; Yang, Q.H.; Ji, W.Q. Agronomic and quality traits of a wheat-*Psathyrostachys huashanica* 1Ns disomic addition line. *Acta Agron. Sin.* **2010**, *36*, 1610–1614. [CrossRef]

agronomy

MDPI

Article

Estimation of Genetic Diversity in Seven Races of Native Maize from the Highlands of Mexico

Mario Rocandio-Rodríguez [1], Amalio Santacruz-Varela [2,*], Leobigildo Córdova-Téllez [2], Higinio López-Sánchez [3], Aurelio Hernández-Bautista [2], Fernando Castillo-González [2], Ricardo Lobato-Ortiz [2], J. Jesús García-Zavala [2] and Pedro Antonio López [3]

1 Instituto de Ecología Aplicada, Universidad Autónoma de Tamaulipas, División del Golfo 356, Col. Libertad, Victoria 87019, Mexico; mrocandio@docentes.uat.edu.mx

2 Colegio de Postgraduados Campus Montecillo, km. 36.5. Carr. Mexico-Texcoco, Texcoco 56230, Mexico; lcordova@colpos.mx (L.C.-T.); aureliohb@hotmail.com (A.H.-B.); fcastill@colpos.mx (F.C.-G.); rlobato@colpos.mx (R.L.-O.); zavala@colpos.mx (J.J.G.-Z.)

3 Colegio de Postgraduados, Campus Puebla, km. 125.5. Carr. Fed. Mexico-Puebla, Santiago Momoxpan, Cholula, Puebla 72760, Mexico; higiniols@colpos.mx (H.L.-S.); palopez@colpos.mx (P.A.L.)

* Correspondence: asvarela@colpos.mx

Received: 26 January 2020; Accepted: 15 February 2020; Published: 22 February 2020

Abstract: Characterizing the genetic diversity of maize (*Zea mays* L.) populations by their morphological and molecular attributes makes it possible to place populations into specific groups; thus, facilitating the design of procedures for their optimum and sustainable use. In this study, data from two lines of evidence were analyzed simultaneously to robustly classify maize populations and to determine their genetic relationships. Seven maize races of the central high plateau of Mexico were characterized using a combined analysis of 13 morphological traits and 31 microsatellite loci. The germplasm assessed included samples of 119 accessions held at Mexican germplasm banks. Cluster and principal component analyses were performed. Also, genetic and geographic relationships among the accessions were determined. Principal component analysis separated the different accessions into well-defined groups using first three principal components. The accessions of Arrocillo Amarillo and Elotes Cónicos races did not exhibit a grouping pattern, indicating greater genetic complexity. Better grounded grouping and phylogenetic relationships were obtained when traits of both lines of evidence were used simultaneously.

Keywords: combined analysis; evolution; genetic resources; germplasm; *Zea mays*

1. Introduction

The evolutionary history of maize (*Zea mays* L.), as well as its diversity, has been of interest for numerous studies [1], which have focused mainly on morphological variation [2,3]. In other classic studies, Cervantes, Goodman, Casas and Rawlings [4] used genetic effects and the interaction genotype × environment, while Sánchez, Goodman and Rawlings [5] involved the effects of the interaction genotype × environment along with stability parameters, which enabled to classify individuals more precisely into discrete units [6], although these parameters may not necessarily provide reliable information on possible phylogenetic relationships [7]. More recently, molecular tools, such as simple sequence repeats of DNA (SSRs) have been applied in studies of maize landraces [8–12], and the data and findings have added substantially to understanding maize evolution and diversification.

Some classification work on Mexican maize landraces has been based on phenetic analysis, using morphological traits [12,13], in combination, in some instances, with isozyme allelic frequencies [14,15], and combinations of morphological traits, isozymes and microsatellites have also been used [16]. Morphological and agronomic characterization has been very useful for the study and assessment of genetic resources; however, morphological and agronomic traits are attributes that may or may not be inherited and may be controlled by one or multiple genes [17]. They may also have limitations that restrict recovering genetic information, such as pleiotropic effects, ignorance of the genetic base, type of inheritance, or high susceptibility to environmental influence [18]. For these reasons, in recent years and due to progress in molecular biology, the use of molecular markers such as microsatellites has complemented the information from morphological traits, making it possible to estimate neutral genetic diversity. Although assessment of molecular data is more complex than that of morphological traits, there is the advantage that environmental influence is negligible and comparisons can be made between individuals of the same or different species, kinship and phylogenetic relations can be established, and processes of migration and genetic drift can be revealed. Moreover, a large amount of genetic information on populations can be obtained [19].

The convenience of using sets of data separately and combined has been discussed. Avise [20] argue that separate analyses may result in conflicts in the resolution of the trees, or classifications, given that the traits may have different origins; that is, they may be affected to a greater or lesser degree by natural or artificial selection, or they may have a different evolutionary rate, among other reasons. Furthermore, [21,22] put forth the possibility of obtaining better grounded classification when traits of different nature are combined, although it is expected that some errors generated by a certain type of traits or marker can be minimized using other markers According to [23–27], one of the most accepted approaches for analyzing this information is that of simultaneous analysis, or total evidence, since it gives more precision to the classifications and maximizes descriptive efficiency and explanatory power of all the information.

In recent years, the genetic diversity of maize was analyzed by Rocandio-Rodríguez, Santacruz-Varela, Córdova-Téllez, López-Sánchez, Castillo-González, Lobato-Ortiz, García-Zavala and Ortega-Paczka [14], who assessed phenotypically 119 accessions of maize at the Mexican highlands. In other study Rocandio-Rodríguez, Santacruz-Varela, Córdova-Téllez, López-Sánchez, Castillo-González, Lobato-Ortiz, and García-Zavala [11], also studied the genetic structure and molecular genetic diversity within and among these same accessions along with the race Balsas of teosinte (*Zea mays* ssp. *parviglumis*). Based on the above, to understand maize diversity and obtain better grounded phylogenetic relations of the accessions within discrete units (races), this study aimed to classify accessions of the seven maize races most cultivated at the Mexican central high plateau, and examine similarity relations among populations and groups of populations according to the combined analysis of morphological traits and microsatellite allele frequencies.

2. Materials and Methods

2.1. Biological Material

One hundred and seven accessions, representative of seven maize races of the central highlands of Mexico, were characterized: 10 Arrocillo Amarillo, 11 Cacachuacintle, 22 Chalqueño, 23 Cónico, 14 Elotes Cónicos, 8 Palomero Toluqueño, 19 Purépecha, along with two commercial varieties used as reference. The seed used was acquired from the germplasm banks of the International Maize and Wheat Improvement Center (CIMMYT), and Mexican institutions as Chapingo Autonomous University, National Institute of Research on Forestry, Agriculture and Livestock, and Colegio de Postgraduados. The accessions were identified visually by expert classifiers as those that most closely exhibited the ear and plant characteristics of each race.

2.2. Field Assessments

The field experiments were set up during the 2010 Spring-Summer growing cycle in three environments of the high Mexican plateau: Ciudad Serdán (18°50′ latitude N, 97°28′ longitude W, altitude of 2570 masl) and San Mateo Capultitlán (19°08′ latitude N; 98°28′ longitude W, altitude 2260 masl), state of Puebla and Montecillo, State of Mexico (19°27′ latitude N, 98°54′ longitude W, altitude 2250 masl). Experimental design was complete randomized blocks with two replications in each site. The experimental unit or plot consisted of two rows 5-m long and 0.8-m wide in which 44 seeds were equidistantly planted.

2.3. Morphological Traits Measured

In each experiment phenotypic traits were recorded. We randomly chose five competitive plants per plot in order to collect data for phenologic, vegetative, tassel, ear and kernel traits. Also, some ratios among those traits were calculated (Table 1).

Table 1. Phenotypic traits assessed.

Traits
Phenological traits: days to male (MF) and female (FF) flowering; anthesis-silking interval (ASI).
Vegetative traits: number of tillers per plant (SP), plant height (PH), ear height (EH), length (LL) and width (LW) of the leaf at the primary ear in cm, total number of leaves (NL) and number of leaves above the ear (NLE).
Tassel traits: peduncle length (PLS), length of the branched part of the tassel (LBT), length of the central spike (LCS), total tassel length (TL) in cm, number of primary branches (PB).
Ear traits: length (EL) and ear diameter (ED); peduncle length (PLE); cob diameter (CD), in cm, number of kernel rows (KR), ear grain percentage (EGP = (grain weight/total weight of maize ear) × 100).
Kernel traits: length (KL), width (KW) and thickness (KT) in mm in 10 kernels of each of the five ears of the sampled plants, weight of 100 kernels (W100K) in g, volume of 100 kernels (V100K) in mL.
Ratios: EH/PH, ED/EL, LBT/TL, KW/KL, KT/KL and W100K/V100K.

2.4. Assessment of Microsatellite Polymorphism

The 107 accessions were analyzed using 31 microsatellite loci distributed on the ten maize chromosomes; there is ample information on these loci published in the Maize Genetics and Genomics Database (MaizeGDB), available on line at http://www.maizegdb.org/ssr.php# (Table 2). '

For the molecular analysis, genomic DNA was extracted from 100 mg of mesocotyl, coleoptile, and young leaf tissue individually from 25 plants per accession with a commercial DNA extraction kit (ChargeSwitch® gDNA Plant, Invitrogen, Thermo Fisher Scientific, Carlsbad, USA), using a DNA extraction and purification robot (King Fisher Flex™, ThermoScientific, Waltham, MA, USA). Amplification was performed individually by PCR in volumes of 25 µL that contained 10 mM nucleotides, 25 mM $MgCl_2$, 5× buffer, 100 ng template DNA, 1 unit DNA *Taq* polymerase, and 4 pmol of each primer. The protocol for PCR amplification consisted of initial denaturation for 4 min at 95 °C, followed by 25 cycles of 1 min at 95 °C, 2 min at 55 °C, 2 min at 72 °C, and one final extension of 60 min at 72 °C. PCR products were separated by capillary electrophoresis in a DNA sequencer (Genetic Analyzer ABI 3130™, Applied Biosystems, Foster City, CA, USA) and detection was based on the presence of the fluorescent labels 6-FAM, ROX or HEX at the 5′ extreme of the forward primers using LIZ-500 as the size standard internal marker. Data files were generated with the allelic profile of the markers for each of the populations with the software GeneMapper® V. 4.0 [28], and allele variability with POPGENE 1.31 [29] software.

2.5. Statistical Analysis

Using the averages of the morphological data from the three sites, a combined analysis of variance was performed with SAS V.9.0. [30]. The linear model used was: $Y_{ijk} = \mu + \alpha_i + \gamma_j + \delta_{ij} + B(\gamma)_{k(j)} + \varepsilon_{ijk}$, where Y_{ijk} is the observation of the *i*th accession in the *j*th environment of the *k*th block, μ is the general mean, α_i is the effect of the *i*th accession, γ_j is the effect of the *j*th environment, δ_{ij} is the interaction effect of the *i*th accession with the *j*th environment, $B(\gamma)_{k(j)}$ is the effect of the *k*th block nested into the *j*th environment, and ε_{ijk} is the random error associated with the experimental unit [31].

Table 2. Microsatellite loci and primers used for amplification of repeated simple sequences in maize landrace populations of the highlands of Mexico.

Locus	N. Bin.	Forward Primer//Reverse Primer
phi127	2.07	ROX-ATATGCATTGCCTGGAACTGGAAGGA//AATTCAAACACGCCTCCCGAGTGT
phi051	7.06	6-FAM-GCGAAAGCGAACGACAACAATCTT//ACATCGTCAGATTATATTGCAGACCA
phi115	8.03	HEX-GCTCCGTGTTTCGCCTGAA//ACCATCACCTGAATCCATCACA
phi015	8.08	HEX-GCAACGTACCGTACCTTTCCGA//ACGCTGCATTCAATTACCGGGAAG
phi033	9.02	6-FAM-ATCGAAATGCAGGCGATGGTTCTC//ATCGAGATGTTCTACGCCCTGAAGT
phi053	3.05	ROX-CTGCCTCTCAGATTCAGAGATTGAC//AACCCAACGTACTCCGGCAG
phi072	4.01	6-FAM-GTGCATGATTAATTTCTCCAGCCTT//GACAGCGCGCAAATGGATTGAACT
phi093	4.08	ROX-GTGCGTCAGCTTCATCGCCTACAAG//CCATGCATGCTTGCAACAATGGATACA
phi024	5.01	HEX-CTCCGCTTCCACTGTTCCA//TGTCCGCTGCTTCTACCCA
phi085	5.06	6-FAM-AGCAGAACGGCAAGGGCTACT//TTTGGCACACCACGACGA
phi034	7.02	HEX-TAGCGACAGGATGGCCTCTTCT//GGGGAGCACGCCTTCGTTCT
phi121	8.04	6-FAM-AGGAAAATGGAGCCGGTGAACCA//TTGGTCTGGACCAAGCACATACAC
phi056	1.01	ROX-ACTTGCTTGCCTGCCGTTAC//CGCACACCACTTCCCAGAA
phi064	1.11	HEX-CGAATTGAAATAGCTGCGAGAACCT//ACAATGAACGGTGGTTATCAACACGC
phi050	10.03	ROX-AACATGCCAGACACATACGGACAG//ATGGCTCTAGCGAAGCGTAGAG
phi96100	2.01	6-FAM-AGGAGGACCCCAACTCCTG//TTGCACGAGCCATCGTAT
phi101249	Unknown	6-FAM-TTCCTCCTCCACTGCCTC//AAGAACAGCGAAGCAGAGAAGG
phi109188	5.03	HEX-AAGCTCAGAAGCCGGAGC//GGTCATCAAGCTCTCTGATCG
phi029	3.04	ROX-TCTTTCTTCCTCCACAAGCAGCGAA//TTTCCAGTTGCCACCGACGAAGAACTT
phi073	3.05	HEX-GTGCGAGAGGCTTGACCAA//AAGGGTTGAGGGCGAGGAA
phi96342	10.02	6-FAM-GTAATCCCACGTCCTATCAGCC//TCCAACTTGAACGAACTCCTC
phi109275	1.03	6-FAM-CGGTTCATGCTAGCTCTGC//GTTGTGGCTGTGGTGGTG
phi427913	1.01	ROX-CAAAAGCTAGTCGGGGTCA//ATTGTTCGATGACACACTACGC
phi265454	1.11	6-FAM-CAAGCACCTCAACCTCTTCG//TCCACGCTGCTCACCTTC
phi402893	2.XX	HEX-GCCAAGCTCAGGGTCAAG//CACGAGCGTTATTCGCTGT
phi346482	1.XX	HEX-GCATCACACTTCACACAACAA//GTGGAATAGGAGGCGAGAGAGG
phi308090	4.04–4.05	6-FAM-CAGTCTGCCACGAAGCAA//CTGTCGGTTTCGGTCTTCTT
phi330507	5.02–5.06	ROX-GTAAAGTACGATGCGCCTCCC//CGGGGTAGAGGAGAGTTGTG
phi213398	4.01–4.04	6-FAM-GTGACCTAAACTTGGCAGACCC//CAAGAGGTACCTGCATGGC
phi339017	1.03	HEX-ACTGCTGTTGGGGTAGGG//GCAGCTTGAGCAGGAAGC
phi159819	6.00–6.08	6-FAM-GATGGGCCCTAGACCAGCTT//GCCTCTCCCATCTCTCGGT

Morphological traits were then selected using two statistical methods for additional analyses. The first method was based on repeatability coefficient (r), as suggested by Sánchez, Goodman and Rawlings [5], and the second one was based on the structure of the correlation matrix of independent variables distributed on Gabriel graphics [32]. The repeatability coefficient is a criterion proposed by Goodman and Paterniani [6], which is based on discriminating morphological variables based on the stability of the characters across the environments. When ≥1, the variable is considered stable and appropriate for classification.

Allele frequencies were selected based on the level of significance between populations for each of the alleles ($p \leq 0.05$) Also selected were those alleles that had a frequency above 2%. This was done to avoid problems of distancing between accessions and of interpretation when involving low frequency or single alleles in the grouping analyses.

With the morphological traits and allele frequencies selected from the 31 microsatellite loci detected in each of the individuals, a data matrix was constructed with 107 populations, using a total of 224 variables (13 morphological traits and frequency of 211 alleles previously selected). The variables were standardized by subtracting the mean and dividing by the standard deviation. The dataset was used to calculate Gower distances, which are recommended when variables of different nature are used [33] among populations; this was done with SAS V.9.0. [30] software. A phylogenetic tree was then constructed with the Neighbor-Joining method [34] in NTSYSpc V.2.2 software [35], and the population Sina-2 of the Chapalote race, considered one of the ancient maize landraces [2] was used as an outgroup.

Allele frequencies and morphological information were used to construct a correlation matrix among the traits. Each morphological trait and each allele were considered as an independent variable. The principal component analysis was based on the correlation matrix and processed with the statistical software SAS V.9.0. [30].

3. Results

3.1. Analysis of Variance

Analysis of variance evidenced highly significant differences among populations for all the measured characteristics (Table 3). For population-environment interaction, it was observed that traits as plant height, peduncle length of the tassel, length of branched part of tassel, ear length, kernel width, ear grain percentage, weight of 100 kernels, volume of 100 kernels, kernel thickness/kernel length, and weight of 100 kernels/volume of 100 kernels had non-significant interaction. Finally, estimators of variance components produced values of repeatability $r \geq 1$ for 19 out of the 32 traits (Table 4).

Table 3. Mean squares of the combined analysis of variance across localities.

Trait	Populations	Localities	Pob × Loc	Error	Mean	Standard Deviation
Male flowering	932.5 **	26,311.7 **	39.4 **	13.5	100.64	15.79
Female flowering	813.2 **	34,724.2 **	33.0 **	15.7	93.46	15.71
Anthesis-silking interval	33.5 **	659.8 **	9.965 **	5.5	7.18	3.65
Shoots per plant	0.1 **	15.0 **	0.1 **	0.1	0.21	0.32
Plant height	10,455.7 **	421,681.3 **	330.9 ns	288.2	250.23	56.16
Ear height	5441.2 **	326,631.3 **	254.4 **	206.4	119.72	44.63
Ear height/plant height	0.0 **	1.2 **	0.0 **	0.0	0.46	0.09
Number of leaves	21.5 **	25.4 ns	1.2 **	0.6	12.06	2.08
Number of leaves above ear	2.1 **	10.6 **	0.1 **	0.1	4.81	0.69
Leaf length	393.7 **	57,622.6 **	57.5 **	29.2	75.08	16.07
Leaf width	7.7 **	955.6 **	1.7 *	1.4	8.81	2.28
Total length	155.3 **	210.7 ns	30.0 **	25.6	63.27	6.85
Peduncle length of the tassel	30.6 **	248.5 ns	12.9 ns	11.1	25.21	3.97
Length of central spike	64.6 **	265.0 **	10.6 **	7.7	31.58	4.34
Length of branched part of tassel	23.6 **	1190.5 **	2.4 ns	2.2	6.65	3.02
Length of branched part of tassel/total length	0.0 **	0.3 **	0.0 **	0.0	0.10	0.05
Number of primary branches	38.2 **	1180.2 **	6.2 **	4.8	8.17	3.75
Ear length	34.8 **	423.7 **	2.9 ns	2.7	14.54	3.05
Number of kernel rows	25.0 **	186.7 **	1.3 **	1.	14.48	2.37
Peduncle length of the ear	26.1 **	1549.9 **	8.5**	6.	12.01	3.83
Ear diameter	0.9 **	34.6 **	0.1 **	0.1	4.11	0.56
Ear diameter/ear length	0.0 **	0.0 ns	0.0 **	0.0	0.29	0.05
Cob diameter	0.6 **	12.3 **	0.1 **	0.1	1.95	0.45
Kernel length	783.1 **	14,837.6 **	97.8 **	64.1	12.67	1.54
Kernel width	826.7 **	1500.4 **	20.9 ns	19.0	7.73	1.25
Kernel thickness	156.9 **	138.3**	9.7 **	7.3	4.16	0.58
Ear grain percentage	56.6 **	223.1 **	14.4 ns	14.7	87.30	4.70
Weight of 100 kernels	248.1 **	1114.0 **	22.2 ns	18.8	27.79	7.81
Volume of 100 kernels	811.2 **	2990.6 **	50.1 ns	47.1	42.86	13.52
Kernel width/kernel length	0.1 **	0.1 **	0.0 **	0.0	0.62	0.11
Kernel thickness/kernel length	0.0 **	0.1 **	0.0 ns	0.0	0.33	0.07
Weight of 100 kernels/volume of 100 kernels	0.0 **	0.0 ns	0.0 ns	0.0	0.66	0.06

* Significance at 5%; ** significance at 1%; ns: non significance.

Table 4. Variance components for populations (σ^2_g), localities (σ^2_l), interaction populations × localities ($\sigma^2_{g\times l}$), error (σ^2_e), broad-sense heritability (H^2) and repeatability (r) of the traits studied.

Trait	σ^2_g	σ^2_l	$\sigma^2_{g\times l}$	σ^2_e	H^2	r
Female flowering †	148.84	108.4238	12.9696	13.488	84.91	1.2261
Male flowering	130.024	143.2363	8.6491	15.734	84.21	0.8561
Anthesis-silking interval	3.935	2.7019	2.1987	5.467	33.92	0.8030
Shoots per plant	0.001	0.0618	0.0098	0.052	1.59	0.0143
Plant height	1687.46	1734.8050	21.3440	288.216	84.50	0.9609
Ear height	864.46	1346.0031	24.0306	206.367	78.96	0.6310
Ear height/plant height	0.002	0.0050	0.0005	0.002	44.44	0.3706
Number of leaves †	3.388	0.0830	0.2757	0.6	79.46	9.4447
Number of leaves above ear †	0.325	0.0421	0.0193	0.108	71.85	5.3000
Leaf length	56.02	237.4093	14.1348	29.239	56.36	0.2227
Leaf width	0.985	3.9453	0.1478	1.439	38.30	0.2408
Total length †	20.883	0.2329	2.2085	25.596	42.89	8.5538
Peduncle length of the tassel †	2.949	0.8468	0.8712	11.149	19.70	1.7169
Length of central spike †	9.012	0.9987	1.4356	7.69	49.69	3.7021
Length of branched part of tassel	3.538	4.9074	0.1090	2.201	60.50	0.7055
Length of branched part of tassel/total length	0.0008	0.0011	0.0001	0.001	42.11	0.6358
Number of primary branches	5.335	4.8565	0.7163	4.79	49.21	0.9574
Ear length †	5.316	1.7033	0.1233	2.696	65.34	2.9107
Number of kernel rows †	3.932	0.7603	0.1303	1.044	77.00	4.4151
Peduncle length of the ear	2.923	6.3434	1.0325	6.46	28.06	0.3963

Table 4. *Cont.*

Trait	σ^2_g	σ^2_l	$\sigma^2_{g\times l}$	σ^2_e	H^2	r
Ear diameter	0.149	0.1415	0.0111	0.054	69.59	0.9795
Ear diameter/Ear length †	0.001	0.0001	0.0001	0.001	47.62	8.0939
Cob diameter †	0.085	0.0506	0.0093	0.074	50.51	1.4316
Kernel length †	114.213	60.0425	16.8775	64.056	58.53	1.4848
Kernel width †	134.291	6.1751	0.9700	18.99	87.06	18.7950
Kernel thickness †	24.545	0.5327	1.1700	7.33	74.28	14.4151
Ear grain percentage †	7.025	0.9063	−0.1294	14.699	32.53	9.0428
Weight of 100 kernels †	37.655	4.4013	1.733	18.755	64.76	6.1385
Volume of 100 kernels †	126.85	12.1132	1.5034	47.084	72.30	9.3161
Kernel width/kernel length †	0.008	0.0005	0.0005	0.003	69.57	9.2262
Kernel thickness/kernel length †	0.002	0.0005	0.0002	0.002	47.62	3.1202
Weight of 100 kernel/volume of 100 kernels †	0.001	0.0001	0.0001	0.002	32.26	10.2235

† Selected traits.

3.2. Cluster and Principal Component Analyses

To define the final set of traits, the correlation matrix was generated, and one trait of those pairs with a correlation coefficient above 0.7 was eliminated. In this way, 13 final traits (FF, LCS, NLE, KR, EL, ED/EL, KW, W100K, KW/KL, KT/KL, KT, KL and W100K/V100K) were chosen for the assessment of racial diversity and classification.

According to the cluster analysis, a total of seven groups were observed (Figure 1). Group I was integrated by accessions of the Purépecha, Palomero Toluqueño and Elotes Cónicos races. Group II, was placed in the upper part of the phylogram, it was represented by accessions Hgo-116, Mex-6, Pueb-618, Ver-537 and Tlax-255, belonging to the Elotes Cónicos race. Group III included nine populations of the Cónico race, originating from the states of Hidalgo and Tlaxcala, and three Palomero Toluqueño accessions from the states of Mexico and Tlaxcala. Group IV was formed by a mixed set of different races (Cacahuacintle, Elotes Cónicos, Arrocillo Amarillo, Chalqueño, Purépecha). Group V comprised eight Chalqueño populations, predominantly from the states of Durango, Guanajuato, Jalisco, Zacatecas and Querétaro, and two Cacahuacintle populations. Group VI was formed by Cónico accessions from the states of Mexico, Hidalgo and Tlaxcala as well as Arrocillo Amarillo accessions from states of Veracruz and Mexico. Group VII integrated populations of the races Cónico and Chalqueño, predominantly from the states of Mexico, Hidalgo Puebla and Tlaxcala, and one population (Pue-512) of the Elotes Cónicos race, distinguished for its recent formation.

Dispersion of the 107 maize accessions of the highlands of Mexico was represented in a three-dimensional space determined by the first three principal components (Figure 2), which exhibited broad variation within each racial group and groups defined by race. The first principal component explained 8.7% of the total variation, and the alleles that most contributed were *phi072-C*, *phi121-B*, *phi121-H*, *phi346482-J*, *phi093-E*, *phi015-K*, *phi050-H*, *phi024-Q*, *phi427913-F*, *phi346482-I*, *phi402893-G* and *phi127-B* toward the positive end, while the morphological traits that had the largest influence were ED/EL, EL and KT/KL. The second principal component explained 4.9% of the total variation and was influenced mostly by the high frequency of the alleles *phi033-L*, *phi115-C*, *phi265454-M*, *phi402893-B*, *phi115-A*, *phi051-D*, *phi053-G*, *phi96100-C*, *phi101249-B*, *phi96100-N* and *phi265454-P* at the positive end of the component. The third PC explained 4.6% of the total variation and was more closely associated with morphological traits (LCS, EL, KW, W100G, W100K/V100K, KT and FF) and with some allele frequencies (*phi050-G* and *phi427913-I*).

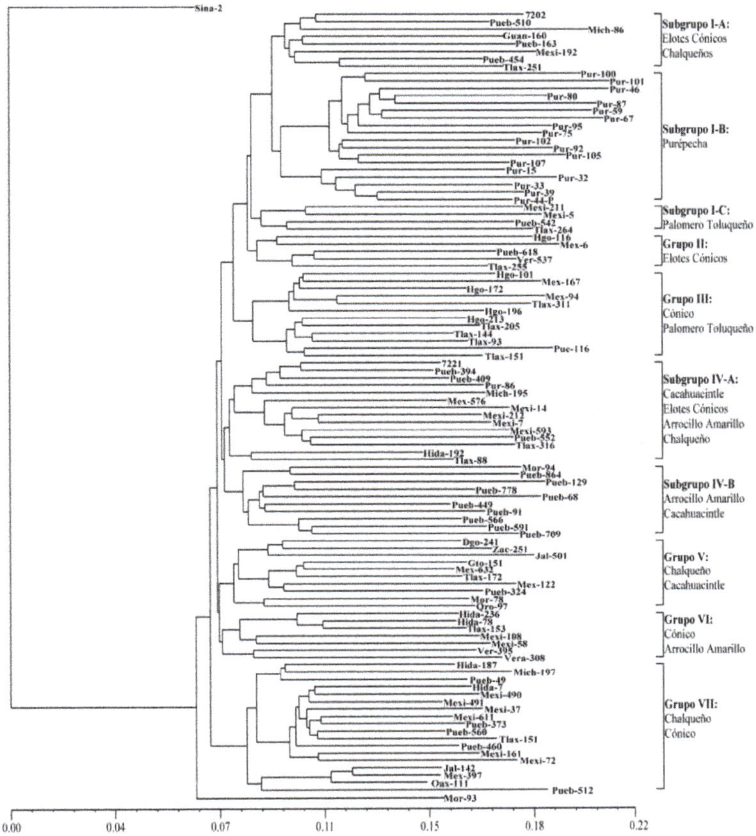

Figure 1. Dendrogram showing the hierarchical clustering produced by the Neighbor-Joining method from 107 maize accessions using Gower distances derived from 13 morphological variables and 211 SSR alleles.

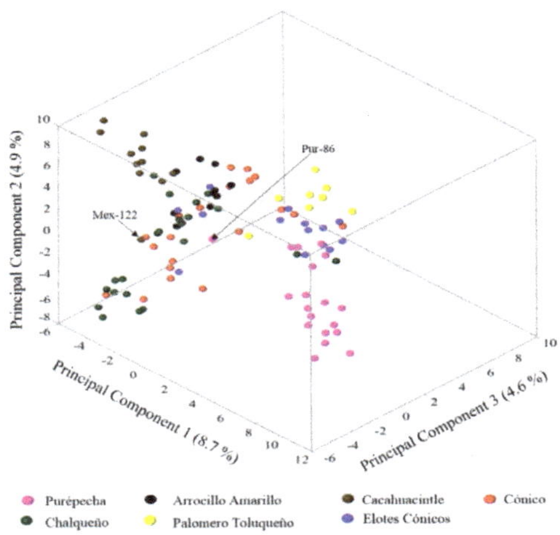

Figure 2. Dispersion of 107 accessions of Mexican maize based on the first three principal components derived from 13 morphological variables and 211 alleles of 31 SSR loci.

4. Discussion

The high degree of variation found suggests the existence of broad genetic diversity among genotypes. Based on the value of repeatability, a total of 32 traits were selected at the first stage. With the exception of PH and LBT, the selected traits in this stage were not statistically significant for the interaction genotypes × localities. This gives them properties that are desirable for classification. According to Sánchez, Godman and Rawlings [5], the traits that are less affected by environment are more useful for characterizing populations. Some of the traits detected here as appropriate for classification were also selected by [5], who indicated that ear, and in general reproductive traits are the most suitable for characterization of maize races. Among these NL, LCS, KW, ED/EL and KW/KL are included.

Thirteen (FF, LCS, NLE, KR, EL, ED/EL, KW, W100K, KW/KL, KT/KL, KT, KL and W100K/V100K) were selected for performing the cluster and PCA procedures, based on correlation analysis. Most of these traits had also been selected in other studies of maize genetic diversity [5,15,16,36,37].

The dendrogram differentiated seven groups, with subgroups within two of them. Group I had three subgroups. Subgroup I-A included six accessions of the Elotes Cónicos race and two of the Chalqueño race; Subgroup I-B comprised most of the populations of the Purépecha race, which formed a well-defined cluster, with poorly differentiated native populations within the Sierra Purépecha region (Salvador Escalante, Charapan, Tingambato, Nahuatzen and Paracho, Michoacán), which is isolated by several geographic barriers created by the mountainous orography; cultural identity is strong and there are semi-collective forms of land use [38]. The results suggest that these populations constituted a genetic group different from the Chalqueño group, and strongly justify their integration into the newly described race Purépecha, as it was proposed by Romero, Castillo and Ortega [39] and later confirmed by Subgroup I-C comprised four Palomero Toluqueño accessions.

The nine populations (Group III) of the Cónico race were grouped intermediately between Palomero Toluqueño populations (Subgroup I-C) and those of Cacahuacintle (Subgroup IV-A). This is in agreement with [2], who proposed that the Cónico race could be a product of hybridization between Palomero Toluqueño and Cacahuacintle races. In this group, the Cónico accession Pue-116 is outstanding as it appeared as a recently formed population.

Group IV is divided into two subgroups. Subgroup IV-A integrated six accessions of Cacahuacintle, three Elotes Cónicos, two Arrocillo Amarillo, two Chalqueño and one Purépecha population (Pur-86). Subgroup IV-B was defined by geographic origin, exhibiting populations from the state of Puebla (five Arrocillo Amarillo, three Cacahuacintle and one Cónico), as well as one population of the Chalqueño race from Totolapan, Morelos. The Cacahuacintle and Arrocillo Amarillo races are associated in both Subgroups (IV-A and IV-B), indicating a higher degree of diversity [2,13,40], since both are closely related and share similar traits.

The interspersing of the Cónico and Chalqueño races in Groups V, VI and VII indicates that the two races are closely related due to the constant gene flow among their populations. This did not allow placing the populations in well-defined groups because their geographic distribution is almost identical. Results reveal that the principal maize races of the Mexican highlands share a common origin and have a diffuse genetic background extending throughout the populations; this is also indicated by the presence of common alleles of the molecular markers and similar values of the morphological traits; therefore, it is inferred that differentiation of these races has been influenced mainly by natural and artificial selection pressure in relatively recent times.

The Cónico population Mor-93 was placed alone in the far inner part of the phylogram. It is likely that this population has not been in contact with those of other regions, or it has had influence from materials not included in this study.

Regarding the principal component analysis of the combined dataset (frequencies of 211 SSR alleles and 13 morphological traits), the first 20 principal components explained 55.1% of the variance. When allele frequencies are involved in the analysis, it is common that the percentage of explained variance decreases notably; in contrast, when only morphological characterization is used, a higher percentage is reached with a smaller number of principal components [37,41]. Such reduction could be due to the rare alleles present in the population, effect of molecular markers in the genetic architecture of the traits, and the linkage between the molecular markers and major or minor genes explicating the phenotypic diversity present in the population [42].

The principal component analysis revealed grouping patterns very similar to those observed in the phylogenetic analysis, indicating that the definition of the associations is highly consistent. Races Cacahuacintle, Purépecha, Elotes Cónicos, Palomero Toluqueño constitute well-defined groups, while the Chalqueño race exhibis two groups, in which it was possible to distinguish accessions from the states of Durango, Zacatecas, Jalisco and Querétaro from those of the traditional core of this race located in the highlands of central Mexico. The Cónico accessions overlapped with the Chalqueño race, indicating genetic complexity influenced by neutral selective markers. Spatial distribution of populations observed on the principal component analysis was similar to that reported by Wellhausen, Roberts, Hernández and Mangelsdorf [2], who found that the race Cónico was localized between Palomero Toluqueño and Cacahuacintle. Such result suggested that the race Cónico was derived from the cross between both races, or that a close relationship does exist between the three races. In addition, a strong grouping was observed between the races Cónico and Elotes Cónicos, and some authors as Wellhausen, Roberts, Hernández and Mangelsdorf [2] even considered that Elotes Cónicos is only a pigmented sub-race of Cónico.

Spatial distribution of the Purépecha race, according to the three first principal components, shows that the populations are very similar, observed in the compact group toward the positive end of the first principal component. It can also be observed that populations of the Cónico race are found far apart, indicating that the variation into this race is very broad, or rather that the populations are not appropriately classified in the Mexican germplasm banks. Most of the maize accessions from Mexican germplasm banks were classified according to morphological variables. However, the morphological variables are often susceptible to phenotypic plasticity [43], which caused a bias in the classification of some populations of the Mexican germplasm banks. Our study demonstrated that this disadvantage can be solved with the combined use of molecular markers and morphological variables.

In the same context, it can be observed that accessions Mex-122 of the Cacahuacintle race and Pur-86 of the Purépecha race do not coincide with their respective groups. This suggests that they are perhaps not correctly classified or that they have had genetic influence from other races.

5. Conclusions

The molecular and morphological information analyzed together for racial classification provided strong elements for understanding diversification and evolutionary relationships that exist among the maize landraces of the highlands of Mexico. Results obtained in this study, in general, agree with the description in that the Pyramidal complex (races Cónico, Chalqueño, Palomero Toluqueño, Arrocillo Amarillo and Elotes Cónicos) are more closely related genetically, while the Cacahuacintle and Purépecha races are not closely associated with this complex.

Author Contributions: Conceptualization, A.S.-V. and L.C.-T.; methodology, A.S.-V., L.C.-T. and M.R.-R.; formal analysis, M.R.-R., A.H.-B.; investigation, M.R.-R., A.S.-V.; writing—original draft preparation, M.R.-R., A.H.-B., and A.S.-V.; writing—review and editing R.L.-O., J.J.G.-Z., H.L.-S., P.A.L. and F.C.-G.; visualization, A.S.-V. and L.C.-T.; supervision, A.S.-V. and L.C.-T.; project administration, A.S.-V. and L.C.-T.; funding acquisition, A.S.-V. and L.C.-T. All authors have read and agreed to the published version of the manuscript.

Funding: This research was funded by the Mexican National Service of Inspection and Certification of Seeds, grant number BI-Mz-09-ha, SINAREFI Project Huella Genética de Maíz.

Conflicts of Interest: The authors declare no conflict of interest.

References

1. Kato, Y.T.A.; Mapes, C.S.; Mera, L.M.O.; Serratos, J.A.H.; Bye, R.A.B. *Origen y Diversificación del Maíz: Una Revisión Analítica*; Universidad Nacional Autónoma de México, Comisión Nacional para el Conocimiento y Uso de la Biodiversidad: Mexico City, Mexico, 2009; p. 116.
2. Wellhausen, E.J.; Roberts, L.M.; Hernández, X.E.; Mangelsdorf, P.C. *Races of Maize in México, their Origin, Characteristics and Distribution*; The Bussey Institution of Harvard University: Cambridge, UK, 1952; p. 223.
3. Hernández, X.E.; Alanís, F.G. Morphological study of five new maize races of the Sierra Madre Occidental of Mexico. Phylogenetic and phytogeographic implications. *Agrociencia* **1970**, *5*, 3–30.
4. Cervantes, S.T.; Goodman, M.M.; Casas, D.E.; Rawlings, J.O. Use of genetic effects and genotype by environmental interactions for the classification of Mexican races of maize. *Genetics* **1978**, *90*, 339–348. [PubMed]
5. Sánchez, G.J.J.; Goodman, M.M.; Rawlings, J.O. Appropriate characters for racial classification in maize. *Econ. Bot.* **1993**, *47*, 44–59. [CrossRef]
6. Goodman, M.M.; Paterniani, E. The races of maize: III. Choices of appropriate characters for racial classification. *Econ. Bot.* **1969**, *23*, 265–273. [CrossRef]
7. Doebley, J.F.; Iltis, H.H. Taxonomy of *Zea* (Gramineae). I. A subgeneric classification with key to taxa. *Am. J. Bot.* **1980**, *67*, 982–993. [CrossRef]
8. Matsuoka, Y.; Vigoroux, Y.; Goodman, M.M.; Sánchez, G.J.; Buckler, E.; Doebley, J. A single domestication for maize shown by multilocus microsatellite genotyping. *Proc. Natl. Acad. Sci. USA* **2002**, *99*, 6080–6084. [CrossRef] [PubMed]
9. Reif, J.C.; Waburton, M.L.; Xia, X.C.; Hoisington, D.A.; Crossa, J.; Taba, S.; Muminović, J.; Bohn, M.; Frisch, M.; Melchinger, A.E. Grouping of accessions of Mexican races of maize revisited with SSR markers. *Theor. Appl. Genet.* **2006**, *113*, 177–185. [CrossRef]
10. Vigouroux, Y.; Glaubitz, J.C.; Matsuoka, Y.; Goodman, M.M.; Sánchez, G.J.; Doebley, J. Population structure and genetic diversity of New World maize races assessed by DNA microsatellites. *Am. J. Bot.* **2008**, *95*, 1240–1253. [CrossRef]
11. Rocandio-Rodríguez, M.; Santacruz-Varela, A.; Córdova-Téllez, L.; López-Sánchez, H.; Castillo-González, F.; Lobato-Ortiz, R.; García-Zavala, J.J. Detection of genetic diversity of seven maize races from the high central valleys of Mexico using microsatellites. *Maydica* **2014**, *59*, 144–151.
12. Vega-Alvarez, I.; Santacruz-Varela, A.; Rocandio-Rodríguez, M.; Córdova-Téllez, L.; López-Sánchez, H.; Muñoz-Orozco, A.; Hernández-Bautista, A. Genetic diversity and structure of native maize races from Northwestern Mexico. *Pesq. Agropec. Bras.* **2017**, *52*, 1023–1032. [CrossRef]
13. Sánchez, G.J.J.; Goodman, M.M. Relationships among the Mexican races of maize. *Econ. Bot.* **1992**, *46*, 72–85. [CrossRef]
14. Rocandio-Rodríguez, M.; Santacruz-Varela, A.; Córdova-Téllez, L.; López-Sánchez, H.; Castillo-González, F.; Lobato-Ortiz, R.; García-Zavala, J.J.; Ortega-Paczka, R. Morphological and agronomic characterization of seven maize races from the highlands of México. *Rev. Fitotec. Mex.* **2014**, *37*, 351–361.
15. Sánchez, G.J.J.; Goodman, M.M.; Stuber, C.W. Isozymatic and morphological diversity in the races of maize of Mexico. *Econ. Bot.* **2000**, *54*, 43–59. [CrossRef]
16. Mijangos-Cortés, J.O.; Corona-Torres, T.; Espinosa-Victoria, D.; Muñoz-Orozco, A.; Romero-Peñaloza, J.; Santacruz-Varela, A. Differentiation among maize (*Zea mays* L.) landraces from the Tarasca Mountain Chain, Michoacan, Mexico and the Chalqueño complex. *Genet. Resour. Crop Evol.* **2007**, *54*, 309–325. [CrossRef]
17. Santacruz-Varela, A.; Widrlechner, M.P.; Ziegler, K.E.; Salvador, R.J.; Millard, M.J.; Bretting, P.K. Phylogenetic relationships among North American popcorns and their evolutionary links to Mexican and South American popcorns. *Crop. Sci.* **2004**, *44*, 1456–1467. [CrossRef]
18. Lowe, A.J.; Hanotte, O.; Guarino, L. Standardization of molecular genetic techniques for the characterization of germplasm collections: The case of random amplified polymorphic DNA (RAPD). *Plant Genet. Resour. Newsl.* **1996**, *107*, 50–54.
19. Pan, Y.B.; Burner, D.M.; Legendre, B.L.; Grisham, M.P.; White, W.H. An assessment of the genetic diversity within a collection of Saccharum spontaneum L. with RAPD-PCR. *Genet. Resour. Crop Evol.* **2004**, *51*, 895–903. [CrossRef]

20. Avise, J.C. *Molecular Markers, Natural History and Evolution*, 2nd ed.; Sinauer Associates: Sunderland, MA, USA, 2004; p. 684.
21. De Queiroz, A.; Donoghue, M.J.; Kim, J. Separate versus combined analysis of phylogenetic evidence. *Annu. Rev. Ecol. Evol. Syst.* **1995**, *26*, 657–681. [CrossRef]
22. Chippindale, P.T.; Wiens, J.J. Weighting, partitioning, and combining characters in phylogenetic analysis. *Syst. Biol.* **1994**, *43*, 278–287. [CrossRef]
23. Leal, A.A.; Mangolin, C.A.; do Amaral, A.T., Jr.; Gonçalves, L.S.A.; Scapim, C.A.; Mott, A.S.; Eloi, I.B.O.; Cordovés, V.; da Silva, M.F.P. Efficiency of RAPD versus SSR markers for determining genetic diversity among popcorn lines. *Genet. Mol. Res.* **2010**, *9*, 9–18. [CrossRef]
24. Kluge, A.G. A concern for evidence and a phylogenetic hypothesis of relationships among Epicrates (Boidae, Serpentes). *Syst. Zool.* **1989**, *38*, 7–25. [CrossRef]
25. Kluge, A.G.; Wolf, A.J. Cladistics: What's in a word? *Cladistics* **1993**, *9*, 183–199. [CrossRef]
26. Nixon, K.C.; Carpenter, J.M. On simultaneous analysis. *Cladistics* **1996**, *12*, 221–241. [CrossRef]
27. Gulanbaier, T.; Seoung, B.M. Gower distance-based multivariate control charts for a mixture of continuous and categorical variables. *Expert Syst. Appl.* **2014**, *41*, 1701–1707.
28. Applied, B. *GeneMapper® Software Version 4.0. Reference and Troubleshooting Guide*; Applied Biosystems Inc.: Foster City, CA, USA, 2005; p. 82.
29. Yeh, F.C.; Yang, R.C.; Boyle, T. *POPGENE Version 1.31. Microsoft Window-based Freeware for Population Genetic Analysis, Quik User Guide*; University of Alberta and Center for International Forestry Research: Edmonton, AB, Canada, 1999; Available online: http://www.ualberta.ca/~{}fyeh/popgene.pdf (accessed on 23 March 2018).
30. SAS, Institute. *SAS/STAT User's Guide, Software version 9.0.*; SAS Institute Inc.: Cary, NC, USA, 2002; p. 4424.
31. Martínez, G.A. *Manual de diseño y análisis de los látices. Monografías y Manuales de Estadística y Cómputo. Vol. 8, Núm. 3*; Centro de Estadística y Cálculo, Colegio de Postgraduados: Chapingo, Mexico, 1989; p. 71.
32. Pla, L.E. *Análisis Multivariado: Métodos de Componentes Principales. Programa Regional de Desarrollo Científico y Tecnológico*; Secretaria General de la Organización de los Estados Americanos: Washington, DC, USA, 1986; p. 94.
33. Gower, J.C. A general coefficient of similarity and some of its properties. *Biometrics* **1971**, *27*, 857–874. [CrossRef]
34. Saitou, N.; Nei, M. The neighbor-joining method: A new method for reconstructing phylogenetic trees. *Mol. Biol. Evol.* **1987**, *4*, 406–425.
35. Rohlf, F.J. *NTSYSpc: Numerical Taxonomy System, Version 2.21c*; Exeter Software, Setauket: New York, NY, USA, 2009.
36. Herrera, C.B.E.; Castillo, G.F.; Sánchez, G.J.J.; Ortega, P.R.; Goodman, M.M. Morphological traits to assess diversity among maize populations in a region: Chalqueño landrace case. *Rev. Fitotec. Mex.* **2000**, *23*, 335–354.
37. López-Romero, G.; Santacruz-Varela, A.; Muñoz-Orozco, A.; Castillo-González, F.; Córdova-Téllez, L.; Vaquera-Huerta, H. Caracterización morfológica de poblaciones nativas de maíz del Istmo de Tehuantepec, México. *Interciencia* **2005**, *30*, 284–290.
38. Maturana, M.S.; Sánchez, C.J. *Las comunidades de la Meseta Tarasca, Un Estudio Socioeconómico*; Centro de Investigaciones Agrarias: Mexico City, Mexico, 1970; p. 124.
39. Romero, P.J.; Castillo, G.F.; Ortega, P.R. Crosses of native populations of maize from Chalqueño race: II. Genetic groups, genetic divergence and heterosis. *Rev. Fitotec. Mex.* **2002**, *25*, 107–115.
40. Doebley, J.F.; Goodman, M.M.; Stuber, C.W. Isozyme variation in the races of maize from México. *Am. J. Bot.* **1985**, *72*, 629–639. [CrossRef]
41. Ángeles-Gaspar, E.; Ortiz-Torres, E.; López, P.A.; López-Romero, G. Characterization and yield performance of native maize populations from Molcaxac, Puebla. *Rev. Fitotec. Mex.* **2010**, *33*, 287–296.

42. Myles, S.; Peiffer, J.; Brown, P.J.; Ersoz, E.S.; Zhang, Z.; Costich, D.E.; Buckler, E.S. Association mapping: Critical considerations shift from genotyping to experimental design. *Plant Cell* **2009**, *21*, 2194–2202. [CrossRef] [PubMed]
43. Govindaraj, M.; Vetriventhan, M.; Srinivasan, M. Importance of genetic diversity assessment in crop plants and its recent advances: An overview of its analytical perspectives. *Genet. Res. Int.* **2015**, *2015*, 431487. [CrossRef] [PubMed]

agronomy

MDPI

Article

Allelic Variation for Prolamins in Spanish Durum Wheat Landraces and Its Relationship with Quality Traits

Efraín A. Chacón, Francisco J. Vázquez, Patricia Giraldo, José M. Carrillo, Elena Benavente * and Marta Rodríguez-Quijano *

Department of Biotechnology-Plant Biology, School of Agricultural, Food and Biosystems Engineering, Universidad Politécnica de Madrid, 28040 Madrid, Spain; efrain.chamor@upm.es (E.A.C.); josefrancisco.vazquez@upm.es (F.J.V.); patricia.giraldo@upm.es (P.G.); josem.carrillo@upm.es (J.M.C.)

* Correspondence: e.benavente@upm.es (E.B.); marta.rurquiaga@upm.es (M.R.-Q.)

Received: 7 December 2019; Accepted: 13 January 2020; Published: 16 January 2020

Abstract: Wheat landraces are currently being explored mainly as a source of allelic variation related to crop resilience and low-input adaptation. Characterization of their high- and low-molecular weight glutenin subunits can aid breeders to select as donor local materials those ensuring good end-use properties in the derived elite cultivars. By using protein electrophoretic methods, we have determined the prolamin allelic profile of 116 Spanish durum wheat landraces. Their quality properties (as defined by grain protein content, sodium dodecyl sulfate (SDS) sedimentation volume and mixograph behavior) have also been assessed. The study has identified six novel glutenin alleles plus some other rare alleles some of which have been associated with improved durum wheat quality. Most of the novel variation detected needs to be characterized in a wider sample of varieties to establish any eventual beneficial effect on functional quality. Further analysis of the quality properties associated to specific allele combinations of *Glu-A3*, *Glu-B3*, and *Glu-A1* has disclosed some clues on the influence on quality of certain non-allelic interactions between these main prolamin-encoding loci. Some of the landraces, showing outstanding values for the gluten quality parameters analyzed, might be directly used by farmers interested in the cultivation of traditional varieties for specialized food markets.

Keywords: *Triticum turgidum*; HMW glutenins; LMW glutenins; gluten quality; non-allelic interactions

1. Introduction

Durum wheat (*Triticum turgidum* L. var. *durum*) is a self-pollinated allotetraploid cereal (2n = 4x = 28, AABB) which shares the genomes A and B with bread wheat. Durum wheat is a traditional Mediterranean crop and one of the pillars of Mediterranean food. Milling of durum wheat grains yields semolina, which is the basis of several products, from Italian pasta, spread throughout the world (spaghetti, macaroni, etc.), to couscous (typical of North African countries), as well as bulgur and breakfast cereals.

Durum wheat was domesticated in the Fertile Crescent 10,000 years before the present (BP), and spread from the east to the west of the Mediterranean Basin [1,2] reaching the Iberian Peninsula around 7000 years BP [3]. Natural and human selection resulted in the establishment of local landraces, specifically adapted to the biotic and abiotic characteristics of their region of origin. From the 1970s of the 20th century, the release of modern breeding varieties, more yielding under conventional crop management systems, led to their progressive abandonment [4]. It has markedly narrowed the genetic diversity present in current varieties of durum wheat as well as of bread wheat [5,6].

The next-generation of wheat cultivars will have to meet the food demands of an increasing human population while facing the uncertainty of future climate conditions and with less input requirements (water, fertilizers, etc.) as demand for improving the sustainability of agro-ecosystems. Wheat landraces are among the most suitable germplasm resource where the genetic variation required for that aim can be searched. These locally adapted varieties have several advantages over wild wheat relatives as donors for the incorporation of useful traits into elite varieties; firstly, because no crossability or offspring viability problems are expected, and secondly because of the lower risk of co-transferring agronomically unfavorable alleles. It explains that several huge research initiatives have undertaken the characterization of the phenotypic, genetic and genomic diversity of wide collections of landraces (i.e., seedsofdiscovery.org; www.whealbi.eu). The final overall objective of these research efforts is to find allelic variation that can be exploited for the breeding of high-yielding, resilient wheat cultivars with less vulnerability to biotic and abiotic limitations. However, it cannot be forgotten that most of the wheat production is used for human consumption [7]. Grain composition of new bread and pasta wheat cultivars should then maintain appropriate end-use properties of flours and semolinas. It may be therefore advisable that genetic factors associated with improved functional quality are characterized and employed as additional criteria in the selection of local varieties to be used as donor germplasm. Furthermore, the increasing market for food products derived from traditional varieties is likely to trigger a growing commercial interest in landraces with high-quality performance, especially suitable for low-input cropping systems [8].

The endosperm reserve proteins are the most important and studied wheat quality components. This group of proteins constitutes a very heterogeneous material [9], which includes albumins (water-soluble), globulins (soluble in neutral salt solutions) and prolamins of two types: gliadins and glutenins. The association of prolamins with lipids and other components (minerals, carbohydrates) forms a viscoelastic complex called gluten. The specific profile of endosperm proteins of a given wheat cultivar has a direct relationship with bread or semolina quality, is independent of environmental conditions, and depends on relatively few genes, being easily identified by electrophoretic methods. In the case of durum wheat, the most commonly used markers for semolina quality are gliadins γ-42 and γ-45. The presence of gliadin γ-45 was early associated with gluten strength [10], but the effect is really caused by the subunits of low molecular weight (LMW) glutenins tightly linked to gliadin γ-45 and not by γ-45 itself [11]. Further studies have indeed established that the functional quality of durum wheat mainly depends on their subunit composition for LMW glutenins, encoded by the *Glu-A3*, *Glu-B3* and *Glu-B2* loci [12–14].

Aguiriano and coworkers analyzed the allelic variation for prolamins in a set of 52 Spanish durum wheat landraces, which represented around 10% of the Spanish durum wheat local varieties maintained in the national germplasm bank [15]. A few other accessions from the whole collection have been included in studies where the prolamin composition of landraces from different Mediterranean countries has been described [16,17]. These studies have evidenced the remarkable variability for prolamins in Spanish accessions, which is far from fully represented in the Spanish core collection of durum wheat [18]. The present study has characterized the prolamin composition of a wide set of Spanish durum wheat landraces. Quality-related parameters routinely used for the preliminary screening of large germplasm samples have also been evaluated. The aims were to identify durum wheat landraces bearing prolamin-coding alleles already associated with good semolina quality and to characterize novel allelic variation as well as allele combinations related to improved durum wheat quality.

2. Materials and Methods

2.1. Plant Material and Field Experimental Design

A set of 158 Spanish durum wheat landraces (*Triticum turgidum* L. ssp *durum*, 2n = 4x = 28, AABB)—from the collection of around 500 maintained at the Plant Genetic Resources Center of the

National Institute for Agricultural Research and Experimentation of Spain (CRF-INIA; Alcalá de Henares, Madrid)—has been analyzed. None of these accessions had been examined in an earlier study focused on Spanish durum wheat landraces [15].

The materials were sowed by duplicate in plots of three rows (1.5 m long) following a random complete block design in the experimental fields of the School of Agricultural, Food and Biosystems Engineering (ETSIAAB, Universidad Politécnica de Madrid; 40°25′ N, 3°42′ W), in the agricultural years 2011/2012 and 2014/2015. Harvested grains from the replicated plots were mixed prior to conducting the semolina quality analyses.

2.2. Homogeneity of Landraces

Five grains per accession were randomly collected and their protein profile (glutenin subunits and gliadins) was analyzed by sodium dodecyl sulfate-polyacrylamide gel electrophoresis (SDS-PAGE). The landraces that resulted to be formed by a mixture of genotypes were not considered for further analyses.

2.3. Characterization of High and Low Molecular Weight Glutenins, and Gliadins

High molecular weight (HMW) and LMW glutenins were extracted following the protocol described by [19]. Electrophoresis of reduced and alkylated glutenin proteins was performed on one-dimensional SDS-PAGE at 12% polyacrylamide [20] and gliadins were fractionated by acid polyacrylamide gel electrophoresis (A-PAGE) [21]. HMW glutenin alleles (*Glu-A1* and *Glu-B1* loci) were identified using the nomenclature of Payne and Lawrence [22] with some modifications [23], LMW glutenin alleles (*Glu-A3*, *Glu-B3*, and *Glu-B2* loci) were named following Nieto-Taladriz et al. [24], and γ-gliadins (*Gli-B1* locus) were named according to Kudryavtsev et al. [25].

2.4. Mass Spectrometry

The protein of interest was manually extracted from the SDS-PAGE gel and introduced into a 1.5 mL tube where the protein was reduced, alkylated and digested [26]. After digestion, the supernatant was collected and analyzed in a 4800 Proteomic Analyzer MALDI-TOF/TOF mass spectrometer (Applied Biosystems, Framingham, MA) at the Genomics and Proteomics Center of the Complutense University of Madrid (Madrid, Spain).

2.5. Semolina Quality Analyses

Protein content, on a 14% moisture basis, was estimated by near-infrared reflectance analysis (NIR) using a Technicon Infralyzer 300 (Technicon Instrument Co. Ltd., Hants, UK). The SDS-sedimentation (SDSS) test was performed as described Dick and Quick [27] with the following modifications: two mixing periods of 20 and 10 sec were used, and stoppered tubes were inverted, by means of a mechanical shaker, 16 times in each of two periods of 40 sec with a rest of 2 min between the two periods [28]. Rheological properties were determined by Mixograph of 10 g whole wheat meal [29], with modification for a constant water absorption of 6.5 mL. The mixograph parameters were: mixing development time (MT), maximum peak height (MH), height at 3 min after the peak (H3), and resistance to breakdown (BDR) which is the percentage difference between MH and H3.

2.6. Statistical Analyses

Analysis of variance and Duncan's method for multiple mean comparisons were used to study the effects on the quality parameters of the allelic variation encoded by the glutenin and gliadin loci examined, and the effect of year (E) and genotype × year (G × E) interactions in the case of parameters measured in 2012 and 2015 samples. Multivariate analysis with principal component analyses (PCA) was applied to analyze the relationships between the quality variables and allelic variants or allele combinations at selected loci. Relations between variables were estimated by Pearson correlation coefficients. All statistical analyses were completed with the InfoStat statistical package (Faculty of Agricultural Sciences of the National University of Córdoba, Córdoba, Argentine) [30].

3. Results and Discussion

Local varieties are usually formed by a mixture of genotypes [31]. Within our initial set of 158 Spanish landraces, 116 accessions showed homogeneous electrophoretic patterns. The remaining accessions (26.7%) were heterogeneous for at least one of the prolamins analyzed. Such a level of intrapopulation variability is similar to that found by other authors in Spanish local populations (28%; [32]) and in Algerian landraces (20%; [33]).

The germplasm bank codes and local names of these 116 accessions selected for further characterization are given in Supplementary Table S1. It includes also their alleles or subunits for each prolamin-encoding locus analyzed and values for the quality parameters examined. Passport and characterization data on each of the entries can be retrieved from the INIA-CRF germplasm database (www.inia.es).

3.1. Characterization of Prolamin Loci

Table 1 shows the glutenin subunits that were observed in the durum wheat landraces at *Glu-A1*, *Glu-B1*, *Glu-A3*, *Glu-B3*, and *Glu-B2* loci, and the γ-gliadins at the *Gli-B1* locus.

Table 1. Frequency of the different high molecular weight (HMW) and low molecular weight (LMW) glutenin subunits and γ-gliadins observed for loci *Glu-1*, *Glu-3*, *Glu-B2*, and *Gli-B1* in the 116 durum wheat landraces analyzed. The allele correspondence is indicated for those variants that have been already designated.

Locus	Allele	Prolamin Subunits	No.	%
Glu-A1 (HMW glutenins)	*a*	1	24	20.7
	b	2*	27	23.3
	c	Null	60	51.7
	o	V	4	3.4
	-	*new*	1	0.9
Glu-B1 (HMW glutenins)	*a*	7	2	1.7
	b	7+8	13	11.2
	e	20x+20y	20	17.2
	f	13+16	11	9.5
	an	6	2	1.7
	aq	32+33	8	6.9
	bd	20x+8	1	0.9
	-	20x	1	0.9
	-	6+(8)	58	50.0
Glu-A3 (LMW glutenins)	*a*	6	47	40.5
	b	5	9	7.8
	d	6+11	3	2.6
	e	11	9	7.8
	h	Null	35	30.2
	q	5+20	3	2.6
	i	8*+11	1	0.9
	-	5*	1	0.9
	-	8*	3	2.6
	-	5+8*	3	2.6
	-	5*+20	1	0.9
	-	6+20	1	0.9

Table 1. *Cont.*

Locus	Allele	Prolamin Subunits	No.	%
	a	2+4+15+19	44	37.9
	b	8+9+13+16	4	3.4
	d	2+4+15+17+19	2	1.7
	e	2+4+15+16+18	3	2.6
	f	2+4+15+17	14	12.0
	g	2+4+15+16	1	0.9
	h	1+3+14+18	18	15.5
Glu-B3	*i*	5+7+8+14+18	16	13.8
(LMW glutenins)	-	1+3+13*+19	1	0.9
	-	1+3+13+19	1	0.9
	-	13*+18	1	0.9
	-	13*+15+19	3	2.6
	-	13+14+18	1	0.9
	-	13+19	4	3.4
	-	2+4+13+15+17	1	0.9
	-	2+4+13+16+19	2	1.7
Glu-B2	*a*	12	58	50.0
(LMW glutenins)	*b*	Null	55	47.4
	c	12*	3	2.6
		Null	9	7.8
		γ-40	1	0.9
		γ-41	2	1.7
Gli-B1		γ-42	4	3.4
(γ-gliadins)		γ-43	2	1.7
		γ-44	21	18.1
		γ-45	75	64.6
		γ-46	1	0.9
		γ-47	1	0.9

3.1.1. HMW Glutenin Subunits

Five alleles encoded by the *Glu-A1* locus could be identified (Table 1). Three of them (namely *a*, *b* and *c*) are common alleles already described by Payne and Lawrence [22]. The most frequent was the *c* allele (null allele) found in 60 landraces (51.7%), while alleles *a* and *b* (subunits 1 and 2*) were found in 24 and 27 landraces, respectively (20.7% and 23.3%). Aguiriano and coworkers reported a quite similar frequency of the *a* and *c Glu-A1* alleles (33% and 38%, respectively) in a sample of 52 Spanish landraces [15]. The present results are based on a larger sample and hence can be considered more representative of the allelic richness of Spanish durum wheat local varieties. Several other studies have found a frequency of *c* allele even higher than 50% in Mediterranean landraces (i.e., [16,17,32,34–36]).

The rare *o* allele (V subunit) was found in four landraces (3.4%). This allele, also designated as 2**, was early described in two Portuguese local varieties of durum wheat [37] and in five varieties within a collection of 502 durum wheat accessions from a wider geographical origin [38]. Three later studies conducted on Mediterranean landraces reported the presence of this allele although always at the lowest frequency [15,17,36]. It can be highlighted that this allele is not among the represented in the Spanish core collection of durum wheat [18]. Additionally, a novel HMW subunit with a higher electrophoretic mobility than the V subunit was identified in the accession BGE19307 ('Baza'; Figure 1a).

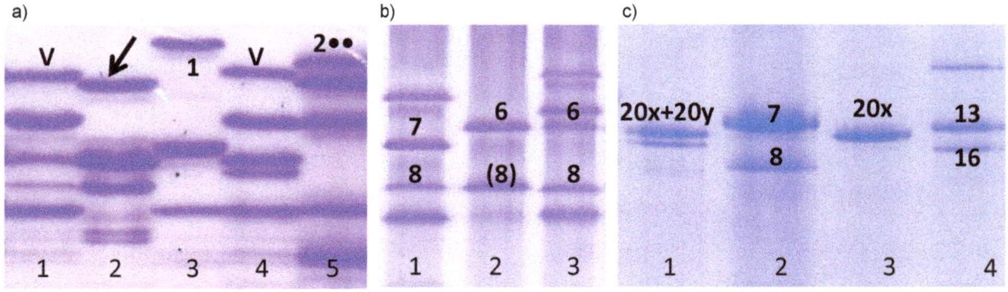

Figure 1. Electrophoretic patterns of durum wheat landraces and standard varieties illustrating some of the high molecular weight glutenin subunits (HMW-GS) observed in the accessions under study. (**a**) Lanes 1–5 show the subunits encoded by the *Glu-A1* locus in accessions BGE018295, BGE019307 (new subunit), BGE013104 and BGE018273, and in the standard bread wheat 'Ribeiro', respectively. The HMW-GS not previously described is indicated by an arrow. (**b**) Prolamin bands encoded by the *Glu-B1* locus in the bread wheat standards for the subunit pairs 7+8 ('Chinese Spring', lane 1) and 6+8 ('Hope', lane 3) and in one durum wheat accession with the By(8) subunit (BGE-018315, lane 2). (**c**) Lanes 1–4 show the subunits encoded by the *Glu-B1* locus in 'Federation', 'Chinese Spring', landrace BGE021769 (Bx20 subunit) and 'Lancota', respectively.

The *Glu-B1* locus was the most polymorphic HMW locus (Table 1). Seven previously cataloged alleles were found, namely *a, b, e, f, an, aq* and *bd* [23]. Among them, *e, b, f* and *aq* alleles were the most frequent (17.2%, 11.2%, 9.5%, and 6.9% respectively). A quite similar allele composition has been found by other authors in Spanish [32] and Portuguese germplasm [34,39]. The 6+(8) pair, a subunit combination with no allele designation, was the most frequent (Table 1). The subunit named (8) has been previously observed in durum wheat [36] and bread wheat [40]. It shows a slightly different mobility than the By8 present in the pairs 6+8 and 7+8 of the bread wheat standards 'Hope' and 'Chinese Spring', respectively (Figure 1b).

A single protein band with a similar electrophoretic mobility than the *Glu-B1x* subunit of the 20x+20y pair was observed in one accession (BGE021769; Figure 1c). This finding was quite unexpected since the presence of the 20x subunit alone had only been described in a Portuguese local variety of common wheat [41]. To confirm its identity, the protein was extracted and analyzed by matrix-assited laser desorption/ionization time-of-flight (MALDI-TOF) mass spectrometry (see Supplementary Figure S1). It can be noted that the core collection of Spanish durum wheat includes a line derived from the landrace analyzed here, as representative of this rare *Glu-B1* allele (BGE045673; see [18]).

3.1.2. LMW Glutenin Subunits

The analysis of glutenin subunits encoded by the *Glu-A3* locus revealed seven alleles (*a, b, d, e, h, q, i*) previously cataloged [24,42,43]. The *a* allele was the most frequent (40.5%; Table 1). The high frequency of this good quality-related allele in durum wheat materials agrees with the findings reported in commercial varieties [24] and in local varieties of Portugal, Algeria and Spain [15,33,39]. Also in agreement to earlier results [17,33], the *h* null-allele has been the second most frequent (30.2%). The allele *q* (pair 5+20), found in four local varieties, has been formerly reported in local germplasm (i.e., [33]).

In addition to those already cataloged, two subunits (5* and 8*) and three pairs of subunits (5+8*, 5*+20 and 6+20) with no allele correspondence were found in our study. Among them, the pair 5+8* was the only not previously described (Figure 2a, lane 1). This *Glu-A3* allele is not represented in the Spanish core collection of durum wheat [18] in spite that its frequency cannot be considered negligible (Table 1). The subunit 5* (Figure 2b), characterized by a lower mobility than the subunit 5 present in the standard 'Langdon' [24], has been described in other Spanish durum wheat landraces [15,36].

The subunit 8* (Figure 2a, lane 3) was described in the Portuguese durum wheat 'Mourisco Fino' as part of the pair 8*+11 [42]. This subunit was subsequently found in Algerian durum wheat [33] and in accessions of *T. turanicum* [44]. Finally, the pairs 5*+20 and 6+20, in only one accession each, have been previously reported in Spanish and Mediterranean durum wheat local landraces, respectively [15,17].

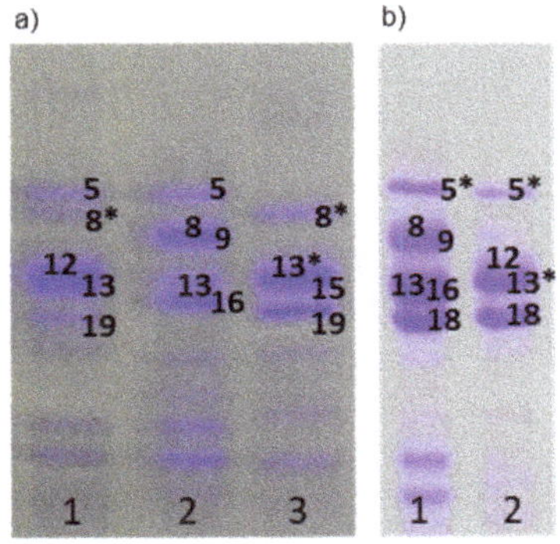

Figure 2. Electrophoretic patterns of durum wheat landraces and standard varieties illustrating some of the low molecular weight glutenin subunits (LMW-GS) observed in the accessions under study. (**a**) Accessions BGE013718 and BGE019300 (lanes 1 and 3) and standard durum wheat 'Langdon' (lane 2). (**b**) Standard durum wheat 'Fanfarron' (lane 1) and accession BGE018657 (lane 2).

The *Glu-B3* locus was the most polymorphic among all prolamin loci analyzed, with up to 16 different subunit combinations (Table 1). The highest frequency of the *a* allele (37.9%) agrees with all previous reports on Mediterranean durum wheat landraces (e.g., [15,33,45]), although it is worthy of noting that this allele is even more abundant in commercial varieties of durum wheat [17]. The relatively high presence of the *h* and *i* alleles (15.5% and 13.8%, respectively) is in agreement with the observed by Aguiriano and coworkers [15]. But this feature might be characteristic of Spanish landraces since these alleles are either absent or present at much lower frequency in wide germplasm collections from Portugal [45] and other Mediterranean regions [17]. Moragues and coworkers found a quite relevant frequency of the *Glu-B3c* allele (27%) within the set of 26 landraces from the Iberian Peninsula included in their study, which covered accessions from 13 Mediterranean countries [16]. However, this allele, described in commercial varieties of durum wheat [24,45], has been found neither in this work nor in those of other authors that have examined landraces from Spain and Portugal [15,39].

Among the eight combinations of LMW glutenin subunits encoded by the *Glu-B3* locus that have no allele designation, four of them had not been described until now: 1+3+13+19, 13*+18 (Figure 2b), 13+14+18, and 2+4+13+15+17 (Table 1). Furthermore, the combination 1+3+13*+19 is the only represented in the Spanish durum wheat core collection [18].

The *Glu-B2* locus encoded for three allelic variants, namely *a*, *b* and *c* (Table 1). The *a* and *b* alleles (subunit 12 and null allele, respectively; see Figure 2) were present each in almost half of the landraces, while the *c* allele (subunit 12*) was identified in only three out of the 116 accessions. This allele was initially observed in the local variety 'Alcalá la Real' [43] but has not been found in later analyses on wide collections of Mediterranean durum wheat landraces. The relative frequency of the *a* and *b* alleles greatly varies between studies, some of them reporting a majority of *a* alleles (e.g., [24,33,45]), a majority of *b* alleles (e.g., [15,17]) or, as found in this study, a quite similar abundance of both alleles [16].

3.1.3. Gliadins

Three of the commonest γ-gliadin variants encoded by the *Gli-B1* locus (γ-42, γ-44, and γ-45) were found in the durum wheat landraces examined, but other five rare γ-gliadins were also observed, and nine accessions did not express any γ-gliadin (Table 1; see Figure 3). Gliadin γ-45 was the most frequent (64.6%), which is in coincidence with previous studies carried out in local varieties [15,32,34,36,46] as well as in commercial varieties [24,25,28] of durum wheat. Gliadin γ-44 was present in 18% of the landraces. This gliadin, that has been found with an even greater frequency in some studies on Spanish landraces [15,46], is practically absent in commercial varieties [24,25,28]. The low frequency of the γ-42 gliadin (3.4% of the accessions) is quite in agreement with its representation in the core collection of Spanish durum wheats (5.3%; [18]). It can be noted that three of the rare variants found here (γ-41, γ-43, and γ-46) are also represented in the core collection. Among the remainder, γ-40 and γ-47, only the former had earlier been reported.

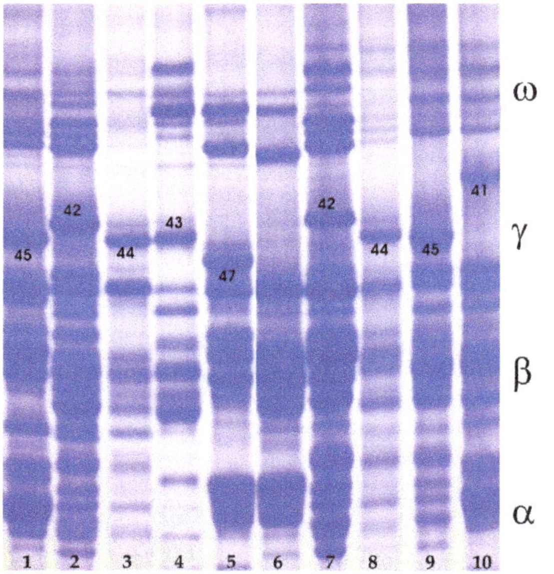

Figure 3. The γ-gliadin variants detected by acid polyacrylamide gel electrophoresis (A-PAGE) in the durum wheat landraces BGE002879 (1), BGE012334 (2), BGE013098(3), BGE018655 (4), BGE018311 (5), BGE019280 (6) and BGE019290 (10), and in the standard varieties 'Jabato' (7), 'Alaga' (8) and 'Ardente' (9).

3.2. Quality Evaluation

3.2.1. Mean Values, Range of the Quality Tests and Relation Between Quality Parameters

The mean protein content of the collection was similar in the two years analyzed (14.5%), ranging from 11.6% to 16.4% in the samples of 2011/2012 (2012 data) and from 11.3% to 17.5% in 2014/2015 (2015 data). The correlation coefficient between the values recorded in 2012 and 2015 was low but highly significant ($r = 0.38$, $p \leq 0.001$). The protein content of durum wheat varieties is of critical importance in the pasta industry. It must exceed 13% to assure the minimum concentration of 12% required for suitable semolina and pasta processing [47]. It can be highlighted that half of the landraces under study (57 out of 116) showed a protein concentration consistently higher than 14%, and 18 of them reached above 15% in both years of analysis (Supplementary Table S1).

The average SDSS volume in flours of 2012 was 29.6 mm, with a wide range (13.0 to 80.0 mm), while the mean SDSS value in 2015 samples was 35.3 mm and the minimum and maximum values were 19 mm and 87.5 mm, respectively (Table S1). The correlation coefficient between SDSS values recorded in both years was positive and highly significant ($r = 0.74$, $p \leq 0.001$), as occurred for protein content. Mixograph parameters were determined only in flour samples of 2012. Mean values of MT (mixing development time) and BDR (resistance to breakdown) were 56.5 s and 31.9%, respectively. These parameters, related to the mechanical properties of wheat doughs, had again wide-ranging values in the collection of landraces examined, i.e., 18.0 to 129.0 s (MT) and 6.7 to 47.2% (BDR) (see Supplementary Table S1). Good gluten properties are associated with high values for SDSS and MT, and low BDR. As then expected, the correlation between SDSS values and the mixograph parameters was highly significant and positive with MT ($r = 0.83$, $p < 0.001$) while negative with BDR ($r = -0.57$, $p < 0.001$), and MT and BDR were negatively correlated ($r = -0.66$, $p < 0.001$). These results on the inter-relation between SDSS, MT and BDR values are similar to the reported in earlier studies (e.g., [48,49]. Also, in agreement with those and other authors (e.g., [36]), no significant correlation was detected between protein content and SDSS or MT values (see Supplementary Table S2), while a significant positive relation of protein content with BDR was found ($r = 0.35$, $p < 0.001$).

3.2.2. Effect of Allelic Variation at Individual Prolamin Loci

Analysis of variance (ANOVA) tests were conducted to determine the effect of allelic variation at the prolamin loci on protein content and gluten quality parameters. Alleles that had been found in less than three accessions were excluded from these analyses. For protein content and SDSS, the effects of the year and the genotype × year were also determined. Since no significant genotype × year interactions were detected (see Supplementary Table S3), repeated ANOVA tests that excluded this source of variation were conducted. The results are summarized in Table 2, which also shows the mean values obtained for the distinct alleles at each locus after Duncan's multiple comparison tests.

Protein content was the trait less significantly influenced by the variability at the loci analyzed, a certain effect is associated with allelic variation at *Glu-B1* and *Glu-B3* alleles. The Duncan's test detected differences between landraces presenting distinct alleles at *Glu-A3* despite the effect of this locus on protein content was not significant according to the ANOVA test. As expected, a greater effect of the genotypic factors was obtained for the variables related to the functional properties of gluten, i.e., SDSS, MT and BDR. Allelic variation at *Glu-A1*, *Glu-A3* and *Glu-B3* affected SDSS, and significant differences for SDSS mean values were obtained between alleles of all loci examined except *Glu-B2* (Table 2). MT was influenced by variation at *Glu-A1* and *Glu-A3*. BDR was especially influenced by variation at *Glu-A1* and, to a lesser extent, at *Glu-A3* and *Glu-B3*, although mean differences depending on the *Glu-B1* allele present were also obtained.

Among the four *Glu-A1* alleles whose effect could be analyzed, the allele *Glu-A1o* (subunit V) showed a highly significant positive effect on gluten properties, associated with the highest values for SDSS and TM, and the lowest for BDR. This allele was found in four of the 116 landraces under study, none of which could be examined in the second year of analysis. Certain evidences, however, support the interest of conducting further studies to confirm its beneficial influence on wheat quality. So, the landraces that presented this allele (BGE018265, BGE018270, BGE018284, and BGE018285) have maintained high SDSS values (53–77 mm) in samples from other seasons than the reported here. Additionally, Babay and coworkers described this allele in two landraces, one from Algeria and the other from Spain, that showed overall good gluten properties according to their SDSS, MT and BDR values [36]. No significant differences were detected on the quality-related traits between the remaining *GluA1* alleles (Table 2). To some extent, this finding could result from the much greater effect of the allele *o*, although a similar influence of alleles *a*, *b* and *c* agrees with the reported in other studies [50,51].

Allelic variation at the HMW locus *Glu-B1* was less relevant than at *Glu-A1* (Table 2). Landraces bearing the allele *b* (subunits 7+8) showed high SDSS and MT values, in agreement with the findings by other authors in commercial varieties of durum wheat [28,50,51]. It can also be noted the good mixograph parameters of the accessions with the *Glu-B1f* allele (13+16). This allele is very infrequent in commercial varieties which explain the scarce number of studies that have analyzed its effect on quality. Nevertheless, its presence in a durum wheat variety showing high gluten strength and tenacity has been reported [50].

Table 2. ANOVA F-values testing the effect of the prolamin loci on the quality traits analyzed. The effect of the year is also considered for protein content and SDS sedimentation volume. The results of the Duncan's multiple comparison of means are included.

Source of Variation	Allele/Subunits	Protein (%)	SDSS (mm)	MT (s)	BDR (%)
Glu-A1	*a* (1)	14.7 ± 0.2 [a]	34.7 ± 1.7 [b]	55.6 ± 4.2 [b]	35.3 ± 1.4 [a]
	b (2*)	14.4 ± 0.2 [a]	27.3 ± 1.7 [b]	45.5 ± 4.8 [b]	35.1 ± 1.5 [a]
	c (null)	14.6 ± 0.1 [a]	28.6 ± 1.1 [b]	57.5 ± 2.6 [b]	30.2 ± 0.9 [a]
	o (V)	13.9 ± 0.5 [a]	60.4 ± 4.8 [a]	99.8 ± 8.6 [a]	21.7 ± 2.9 [b]
F-value #		2.10	15.3 ***	11.0 ***	12.1 ***
Glu-B1	*b* (7+8)	14.0 ± 0.3 [b]	38.8 ± 2.5 [a]	63.5 ± 6.1 [a]	31.9 ± 2.1 [ab]
	e (20x+20y)	14.8 ± 0.2 [a]	25.8 ± 1.7 [c]	54.8 ± 4.2 [a]	31.2 ± 1.4 [ab]
	f (13+16)	14.6 ± 0.3 [ab]	32.1 ± 2.4 [b]	62.4 ± 5.6 [a]	29.1 ± 1.9 [ab]
	aq (32+33)	14.5 ± 0.3 [ab]	26.5 ± 2.4 [c]	55.6 ± 6.2 [a]	28.2 ± 2.1 [b]
	6+(8)	14.6 ± 0.1 [ab]	33.5 ± 1.2 [ab]	56.5 ± 2.9 [a]	33.9 ± 1.0 [a]
F-value		2.84 *	1.58	0.23	2.24
Glu-A3	*a* (6)	14.6 ± 0.1 [abc]	34.2 ± 1.2 [ab]	64.1 ± 3.0 [a]	28.9 ± 1.0 [b]
	b (5)	15.5 ± 0.5 [a]	27.0 ± 4.8 [bcd]	58.5 ± 12.1 [ab]	34.3 ± 4.1 [ab]
	d (6+11)	15.1 ± 0.4 [ab]	19.0 ± 4.8 [d]	33.0 ± 12.1 [b]	40.4 ± 4.1 [a]
	e (11)	13.8 ± 0.3 [c]	19.7 ± 2.5 [cd]	37.3 ± 6.3 [ab]	35.1 ± 2.1 [ab]
	h (null)	14.5 ± 0.1 [abc]	31.0 ± 1.3 [bc]	56.4 ± 3.3 [ab]	32.8 ± 1.1 [ab]
	q (5+20)	14.3 ± 0.5 [abc]	30.5 ± 4.7 [bcd]	48.0 ± 12.1 [ab]	36.0 ± 4.1 [ab]
	8*	14.3 ± 0.5 [abc]	30.5 ± 4.3 [bc]	64.6 ± 10.4 [a]	30.6 ± 3.5 [ab]
	5+8*	13.9 ± 0.5 [bc]	43.3 ± 4.7 [a]	66.0 ± 12.1 [a]	31.3 ± 4.1 [ab]
F-value		1.91	3.73 **	3.44 **	2.84 *
Glu-B3	*a* (2+4+15+19)	14.5 ± 0.1 [ab]	35.2 ± 1.2 [ab]	64.9 ± 2.9 [a]	28.8 ± 1.0 [bc]
	b (8+9+13+16)	14.3 ± 0.4 [ab]	26.6 ± 3.3 [b]	54.0 ± 8.6 [a]	32.8 ± 2.9 [abc]
	e (2+4+15+16+18)	14.6 ± 0.5 [ab]	28.9 ± 4.8 [ab]	64.5 ± 12.1 [a]	24.2 ± 4.1 [c]
	f (2+4+15+17)	14.1 ± 0.3 [ab]	27.5 ± 2.4 [b]	50.0 ± 5.3 [a]	33.8 ± 1.8 [ab]
	h (1+3+14+18)	15.0 ± 0.2 [a]	27.3 ± 1.9 [b]	53.4 ± 4.6 [a]	33.9 ± 1.6 [ab]
	i (5+7+8+14+18)	15.1 ± 0.3 [a]	27.9 ± 2.4 [ab]	50.2 ± 6.4 [a]	36.7 ± 2.1 [ab]
	13*+15+19	13.7 ± 0.5 [b]	38.3 ± 4.7 [a]	63.0 ± 12.1 [a]	30.5 ± 4.1 [abc]
	13+19	14.4 ± 0.5 [ab]	36.1 ± 4.7 [ab]	49.5 ± 12.1 [a]	39.2 ± 4.1 [a]
F-value		2.17 *	2.22 *	1.36	2.22 *
Glu-B2	*a* (12)	14.3 ± 0.1 [a]	33.0 ± 1.2 [a]	60.2 ± 2.9 [a]	29.6 ± 1.0 [a]
	b (null)	14.7 ± 0.1 [a]	30.8 ± 1.2 [a]	55.7 ± 2.8 [a]	33.7 ± 1.0 [a]
	c (12*)	14.2 ± 0.5 [a]	29.4 ± 4.2 [a]	54.0 ± 9.9 [a]	35.1 ± 3.3 [a]
F-value		2.13	0.31	0.22	1.54
Gli-B1	γ-42	14.3 ± 0.4 [a]	26.6 ± 3.3 [b]	54.0 ± 8.6 [a]	32.8 ± 2.9 [a]
	γ-44	14.7 ± 0.2 [a]	29.2 ± 1.6 [ab]	51.6 ± 4.1 [a]	34.6 ± 1.4 [a]
	γ-45	14.5 ± 0.1 [a]	33.7 ± 1.0 [a]	60.3 ± 2.4 [a]	30.7 ± 0.8 [a]
F-value		0.11	0.29	0.18	0.29
Year	2012	14.5 ± 0.1 [a]	29.6 ± 1.1 [b]	56.1 ± 2.0	31.9 ± 0.7
	2015	14.5 ± 0.1 [a]	35.3 ± 1.2 [a]		
F-value		0.11	7.05 **		

SDSS—sodium dodecyl sulphate sedimentation volume; MT—mixing time; BDR—breakdown resistance. Mean values followed by different letters are significantly different ($p = 0.05$). #: * $0.05 > p > 0.01$; ** $0.01 > p > 0.001$; *** $p < 0.001$.

Regarding the LMW *Glu-A3* locus, the effect of six designated alleles and the subunits 8* and 5+8* could be analyzed (Table 2). A clear deleterious effect on gluten properties was found for allele *d* (6+11), which resulted in the lowest SDSS and MT values and the highest BDR among all recorded. This result was unexpected since a positive influence of this allele on SDSS values has been reported in commercial varieties and Mediterranean landraces [17,51]. The small sample of landraces where this allele was found makes it difficult to further analyze whether the specific alleles present at other prolamin-encoding loci may be responsible for such disagreement. Landraces bearing the 5+8* pair had the highest SDSS and MT values, and a positive influence of the subunit 8* on MT was also found. Even only three landraces presented each of them, these results can be highlighted since there is no previous report on the influence of these two *Glu-A3* allelic patterns on dough rheological properties.

As occurred for *Glu-A3*, allelic variation at the LMW locus *Glu-B3* was found to be relevant for rheological properties. Landraces with subunits 13*+15+19, 13+19 and allele *a* had the best SDSS values. Their effect on mixograph parameters was, however, quite contrasting. Subunit combination 13*+15+19 and allele *a* showed the highest MT values, although the means were not significantly different, and the allele *a* maintained a positive effect on quality as indicated by lower BDR values. However, the low MT and high BDR values found for the pair 13+19 are associated with worse gluten behavior (Table 2). It can be noted that the presence of the *Glu-B3a* allele has already been recognized as a prerequisite of functional quality of durum wheat [14,43,51], but the influence of the other LMW subunit combinations had not been reported earlier.

No significant effect on any of the gluten quality-related parameters examined was found to be associated with allelic variation at either *Glu-B2* and *Gli-B1* loci, although mean SDSS value of landraces bearing gliadin γ-45 was higher than in landraces with gliadin γ-42. It is worthy of noting that these two γ-gliadins have been traditionally used as markers of, respectively, good and bad processing properties in durum wheat because of the tight genetic linkage between *Gli-1* and *Glu-3* loci [11,52].

Many studies have established that the gluten strength of durum wheats ultimately depends on their subunit composition for LMW glutenins, encoded by the *Glu-A3*, *Glu-B3* and *Glu-B2* loci [11,53], whereas a lesser effect has been demonstrated for HMW-GS [12–14,54]. However, the findings reported above demonstrate no influence of *Glu-B2* and a relevant effect of *Glu-A1* in the set of Spanish landraces examined. Figure 4 shows the results of a multivariate PCA analysis conducted to better assess the overall relationships of the alleles at the relevant loci (*Glu-A1*, *Glu-A3*, and *Glu-B3*) with the quality traits. The first two axes of the PCAs accounted for 83–98% of the total variance. In all cases, gluten quality resulted clearly associated with the main PC1 axis, with SDSS and TM in the positive direction. Protein content and BDR were mostly associated with the negative PC1 values but their position along the PC2 axis resulted in more variation between the three PCA analyses. In agreement with the main results described above, this approach confirmed the outstanding effect on gluten properties of *Glu-A1o* allele (Figure 4a), the remarkable beneficial influence of the subunit combination 5+8* and weaker properties derived of allele *d* at *Glu-A3* (Figure 4b), and the positive relation with gluten quality of the allele *a* and the combination 13*+15+19 among the allelic variants analyzed at *Glu-B3* (Figure 4c).

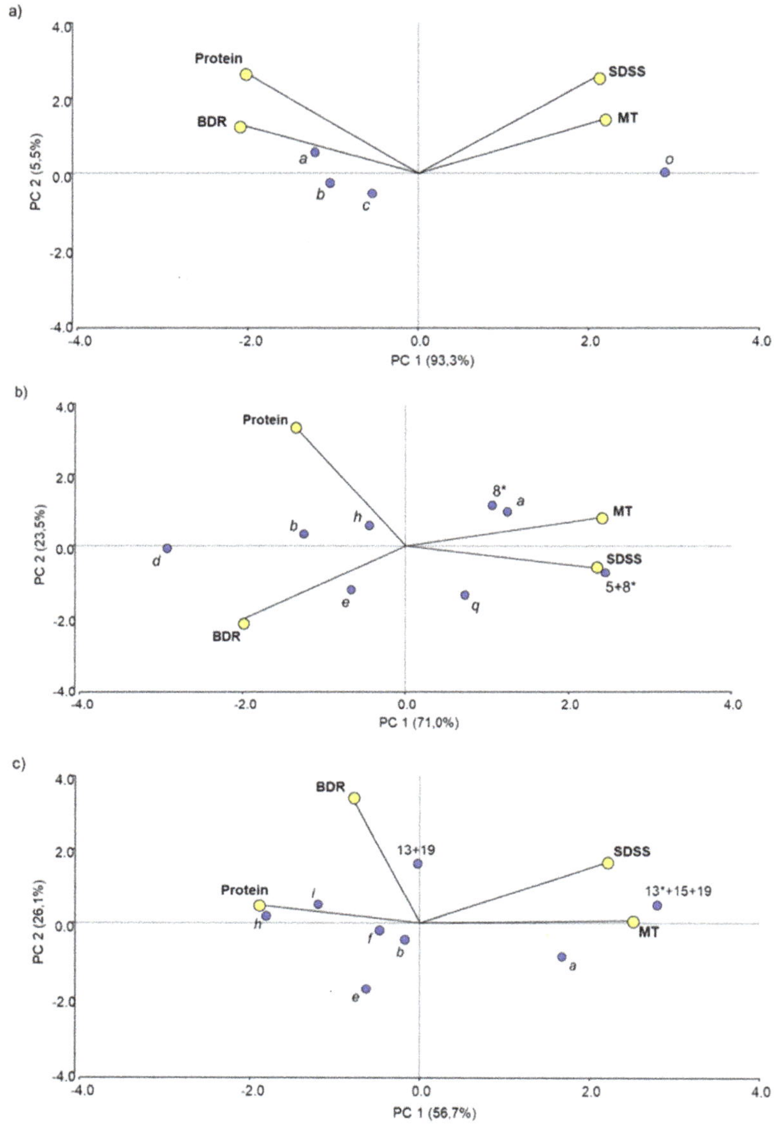

Figure 4. First and second principal components for the alleles present in at least three landraces and the quality traits. (**a**) *Glu-A1*. (**b**) *Glu-A3*. (**c**) *Glu-B3*.

3.2.3. Non-Allelic Interactions between Prolamin Loci

The analysis of allelic variation at a single locus facilitates the identification of alleles having a relevant effect on a trait, particularly when the loci under study are highly polymorphic. However, gluten performance is the outcome of complex interactions between the glutenin and gliadin protein subunits encoded by the whole set of prolamin loci. Further multivariate PCA analyses were then conducted to assess the influence of specific allelic combinations on gluten quality. The four landraces bearing the *Glu-A1o* allele were excluded from these analyses to avoid that the outstanding influence of this allele could mask or bias the effects of the alleles that these accessions presented at the other prolamin loci. Such exclusion reduced the significance of the effect of allelic variation at *Glu-A1* on SDSS ($F = 3.82$; $p = 0.024$) and MT ($F = 4.22$; $p = 0.019$) but not on BDR ($F = 13.17$; $p > 0.001$), while the effects of the other prolamin loci remained virtually the same (compare ANOVA results in Table 2 and Supplementary Table S4). Then all the findings described below are based on a set of 112 durum wheat landraces.

A first multivariate PCA analysis was conducted to determine the relationships with the quality traits of allele combinations at the LMWGS-encoding loci *Glu-A3* and *Glu-B3* (Figure 5a). Gluten quality resulted again clearly associated with the main PC1 axis, with SDSS and MT in the positive direction, and BDR in the negative direction. The high allelic variability found for these two loci in the germplasm collection explains that only half of the 20 *Glu-A3:GluB-3* allele combinations were present in at least three landraces. Among these, the combinations involving *Glu-B3a* (*a:a* and *h:a*, with the *Glu-A3* and *Glu-B3* alleles being indicated in first and second place, respectively) were grouped together in the biplot space, despite *Glu-A3a*, but not *Glu-A3h*, was associated with good quality (Figure 4b). It supports the acknowledged relevance of the *Glu-B3a* allele on the good gluten properties of durum wheat [14,43,51] and suggests some epistatic interaction of this allele over *Glu-A3*. However, in the case of the bad gluten quality-related *Glu-B3h* allele, a non-allelic additive interaction with those *Glu-A3* alleles seems to exists since the position of *a:h* and *h:h* in Figure 5a fits the expected according to the relative positions of the *a* and *h Glu-A3* alleles in Figure 4b. Genotype combinations involving the allele *Glu-A3q* (subunits 5+20) can be highlighted among the represented by less than three landraces: *q:b*, *q:i* and *q*:13+19. It was remarkable, not only their different position along the PCA 1 axis, but especially that *q:i* was the best performing since the overall influence of the *Glu-B3i* allele on the quality tests was similar or even worse than the findings for the *Glu-B3b* allele and the subunits 13+19 (Table 2; Figure 4c). Furthermore, the finding that the landraces with *b:i* and *h:i* were placed in the opposite, negative, direction in Figure 5a suggests that the combination of the *Glu-A3q* and *Glu-B3i* might be particularly beneficial for improved gluten quality. Nevertheless, such a conclusion needs to be confirmed by further analysis of a larger sample of varieties bearing this *Glu-A3:Glu-B3* allelic pattern.

As noted above, less influence of the HMW loci *Glu-A1* and *Glu-B1* on durum wheat quality has commonly been reported. However, our ANOVA analyses detected some significant effects of *Glu-A1* on SDSS, MT and, particularly, BDR values (Table 2; see Supplementary Tables S3 and S4), and the existence of interactions between *Glu-1* and *Glu-3* loci on bread wheat gluten properties is already documented (reviewed in [55]). A latter PCA analysis was then conducted to assess whether allelic variation at this locus could actually modify the predicted durum wheat gluten behavior of specific genotype combinations of *Glu-A3* and *Glu-B3*. This approach was focused on the commonest *Glu-A1* alleles (*a*, *b* and *c*) since the landraces bearing the *Glu-A1o* allele had been formerly excluded. The biplot of the first two axes is represented in Figure 5b where, for clarity purposes, the genotype combinations that are expressly mentioned in the text below are indicated in red. The noting of alleles combinations in that Figure, and their references henceforth, follow the format "*Glu-A1/Glu-A3:Glu-B3*". The genotypic classes involving the *Glu-A3* alleles *a* and *h* and the *Glu-B3* alleles *a* and *h* (namely, *a:a*, *a:h*, *h:a* and *h:h*) were particularly suitable to assess the influence of *Glu-A1* interaction because combinations of each of them with two or three distinct *Glu-A1* alleles could be analyzed. The relative positions of the allelic combinations *a/a:a*, *b/a:a*, and *c/a:a* (in 2, 10 and 16 landraces, respectively) support a differential effect on gluten performance of the *Glu-A1* alleles *a* (subunit 1) and *c* (null allele), which seem to exercise a deleterious and beneficial influence, respectively (Figure 5b). A differential effect of *Glu-A1* alleles is also suggested by comparison between *a/a:h*, *b/a:h* and *c/a:h* (in 1, 4 and 4 landraces). In this case, however, it is the *Glu-A1b* allele which shows a negative effect on quality. Less differences are found between the positions in the biplot space of the landraces with the allelic combinations *a/h:h*, *b/h:h* and *c/h:h* (in 2, 2, and 1 landraces), all in the negative direction of the PC1 axe, but it is the allele *c* of *Glu-A1* which seems to exercise the worst influence. Finally, a similar quality of *a/h:a* and *c/h:a* (in 3 and 9 landraces) could be deduced from their close position in the PCA biplot space. Some of these results need to be confirmed on a larger sample but support that a given *Glu-A1* allele can show different effects on gluten properties depending on the genetic background at LMW loci *Glu-A3* and *Glu-B3*. It may explain the difficulty to demonstrate clear consistent effects of the HMW *Glu-A1* locus on durum wheat quality [49]. The underlying reasons for such unpredictable influence are surely related to the complexity of the molecular interactions in the glutenin polymers that determine dough properties [56].

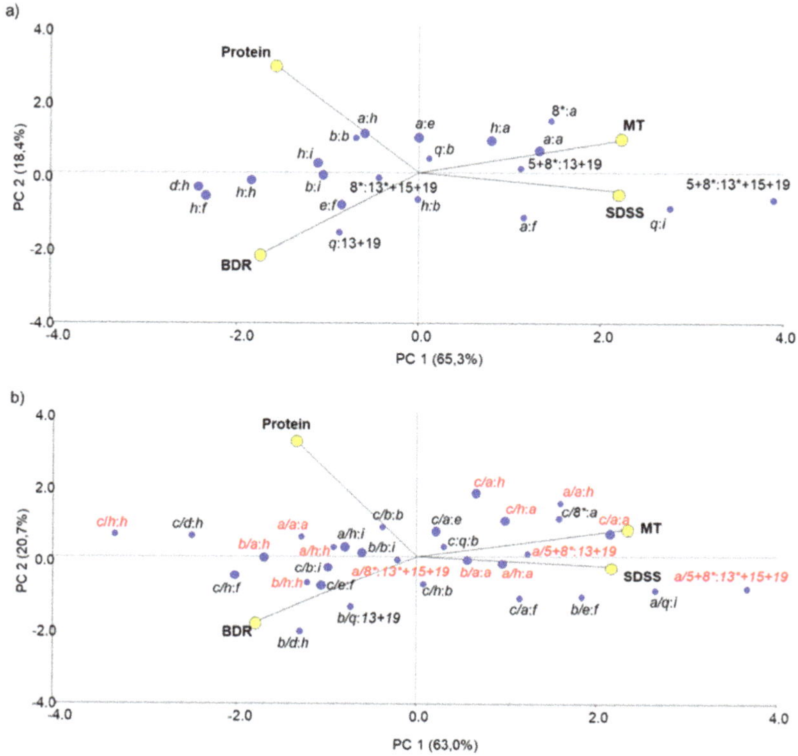

Figure 5. First and second principal components for non-allelic combinations and the quality traits. Combinations present in at least three landraces are indicated by larger dots. (**a**) Allele combinations between the LMW loci *Glu-A3* and *Glu-B3*. (**b**) Allele combinations involving the HMW locus *Glu-A1* and the *Glu-A3* and *Glu-B3* LMW loci. The combinations mentioned in the text are indicated in red.

No conclusion can be drawn about the conserved position in Figure 5a,b of some of the *Glu-A3*:*Glu-B3* subunits combinations highlighted above (i.e., 5+8*:13*+15+19, 5+8*:13+19, or 8*:13*+15+19) since no variability at *Glu-A1* was found within the sample of landraces that presented each of them. The identification of these prolamin profiles in more durum wheat accessions would allow, firstly, to confirm their apparently beneficial effects on gluten properties and, secondly, to determine the possible existence of modulating effects involving *Glu-A1* alleles.

4. Conclusions

The study conducted has identified some durum wheat landraces that could be directly used in farming agrosystems interested in the cultivation of traditional varieties. Among them, those bearing the *Glu-A1o* allele can be highlighted. Future studies need to be made to confirm the beneficial influence of this allele, which could be included in breeding programs for enhancing wheat quality. We have also characterized new alleles and allele combinations at *Glu-A3* and *GluB3* associated with better gluten properties. Most of the novel variation described here was present in a reduced number of accessions. Ongoing research efforts on local wheat germplasm may enlarge the samples allowing to further prove their utility as genetic markers for durum wheat quality improvement.

Supplementary Materials: The following are available online at http://www.mdpi.com/2073-4395/10/1/136/s1, Figure S1: MALDI-TOF mass spectrum for the Spanish landrace with the 20x subunit at locus *Glu-B1* (BGE 021769), Table S1: CRF gene-bank code and local name of the 116 durum wheat Spanish landraces analyzed. Their allelic pattern for the prolamin loci studied and data for the quality variables examined is included. The BGE number of the landraces that could not be analyzed in 2015 is in red lettering, Table S2: Pearson's correlation tests to analyze the relation between protein content and SDS sedimentation volume values obtained in 2012 and 2015, and between the quality traits examined, Table S3: Mean squares (MS) and F-values as determined by ANOVA showing the effect of the prolamin loci on grain protein content, SDS sedimentation volume and mixograph parameters. The effect of the year and G × E (year) interactions is also considered for protein content and SDSS-test values, Table S4: Mean squares (MS) and F-values as determined by ANOVA showing the effect of the prolamin loci on protein content, SDS sedimentation volume and mixograph parameters if the four landraces bearing the *Glu-A1o* allele are excluded from the analyses. The effect of the year is also considered for protein content and SDSS-test values.

Author Contributions: Conceptualization, F.J.V., J.M.C. and M.R.-Q.; data curation, E.A.C. and M.R.-Q.; formal analysis, E.A.C., P.G., E.B. and M.R.-Q.; funding acquisition, P.G. and J.M.C.; investigation, E.A.C. and M.R.-Q.; methodology, F.J.V., P.G. and M.R.-Q.; project administration, F.J.V. and M.R.-Q.; resources, F.J.V.; supervision, J.M.C. and M.R.-Q.; validation, E.B. and M.R.-Q.; visualization, E.A.C., E.B. and M.R.-Q.; writing—original draft, E.A.C., E.B. and M.R.-Q.; writing—review and editing, E.A.C., P.G., J.M.C., E.B. and M.R.-Q. All authors have read and agreed to the published version of the manuscript.

Funding: This research was funded by projects AGL 2012-38345 and AGL2016-77149 of the Spanish Ministry of Science, Innovation, and Universities, which also supported EA Chacón with a research grant (BES-2013-065575).

Acknowledgments: The authors are grateful to the CRF-INIA which kindly supplied seed samples of the Spanish durum wheat landraces used in the study and to M. Ruiz for helpful comments during the preparation of the original and revised versions of the manuscript.

Conflicts of Interest: The authors declare no conflict of interest.

References

1. Kabbaj, H.; Sall, A.T.; Al-Abdallat, A.; Geleta, M.; Amri, A.; Filali-Maltouf, A.; Belkadi, B.; Ortiz, R.; Bassi, F.M. Genetic diversity within a global panel of durum wheat (*Triticum durum*) landraces and modern germplasm reveals the history of alleles exchange. *Front. Plant Sci.* **2017**, *8*, 1277. [CrossRef] [PubMed]
2. MacKey, J. Wheat: Its concept, evolution, and taxonomy. In *Durum Wheat Breeding: Current Approaches and Future Strategies*; Royo, C., Nachit, M., Di Fonzo, N., Araus, J.L., Pfeiffer, W.H., Slafer, G.A., Eds.; Haworth Press: Binghamton, NY, USA, 2005; Volume 1, pp. 3–62.
3. Feldman, M. Origin of cultivated wheat. In *The World Wheat Book: A History of Wheat Breeding*; Bonjean, A.P., Angus, W.J., Eds.; Lavoisier Publishing: Paris, France, 2001; Volume 1, pp. 3–56.
4. Ortiz, R.; Trethowan, R.; Ferrara, G.O.; Iwanaga, M.; Dodds, J.H.; Crouch, J.H.; Crossa, J.; Braun, H.-J. High yield potential, shuttle breeding, genetic diversity, and a new international wheat improvement strategy. *Euphytica* **2007**, *157*, 365–384. [CrossRef]
5. Reif, J.C.; Zhang, P.; Dreisigacker, S.; Warburton, M.L.; van Ginkel, M.; Hoisington, D.; Bohn, M.; Melchinger, A.E. Wheat genetic diversity trends during domestication and breeding. *Theor. Appl. Genet.* **2005**, *110*, 859–864. [CrossRef] [PubMed]
6. Mefleh, M.; Conte, P.; Fadda, C.; Giunta, F.; Piga, A.; Hassoun, G.; Motzo, R. From ancient to old and modern durum wheat varieties: Interaction among cultivar traits, management, and technological quality. *J. Sci. Food Agric.* **2019**, *99*, 2059–2067. [CrossRef]
7. Food and Agriculture Organization. *FAO Statistical Yearbook, World Food and Agriculture*; Food and Agriculture Organization: Rome, Italy, 2013; p. 307.
8. Berni, R.; Cantini, C.; Romi, M.; Hausman, J.-F.; Guerriero, G.; Cai, G. Agrobiotechnology goes wild: Ancient local varieties as sources of bioactives. *Int. J. Mol. Sci.* **2018**, *19*, 2248. [CrossRef]
9. Osborne, T.B. *The Vegetable Proteins*, 2nd ed.; Longmans, Green and Co.: London, UK, 1924; pp. 1–154.
10. Damidaux, R.; Autran, J.; Grignac, P.; Feillet, P. Evidence of relationships useful for breeding between the electrophoretic patterns of gliadins and the viscoelastic properties of the gluten in Triticum durum. *Comptes Rendus Hebd. Seances Acad. Sci. Ser. D Sci. Nat.* **1978**, *287*, 701–704.

11. Payne, P.I.; Jackson, E.A.; Holt, L.M. The association between γ-gliadin 45 and gluten strength in durum wheat varieties: A direct causal effect or the result of genetic linkage? *J. Cereal Sci.* **1984**, *2*, 73–81. [CrossRef]
12. Ruiz, M.; Carrillo, J.M. Relationships between different prolamin proteins and some quality properties in durum wheat. *Plant Breed.* **1995**, *114*, 40–44. [CrossRef]
13. Vazquez, J.F.; Ruiz, M.; Nieto-Taladriz, M.T.; Albuquerque, M.M. Effects on Gluten Strength of Low Mr Glutenin Subunits Coded by Alleles at Glu-A3 and Glu-B3 Loci in Durum Wheat. *J. Cereal Sci.* **1996**, *24*, 125–130. [CrossRef]
14. Carrillo, J.M.; Martinez, M.C.; Brites, C.; Nieto-Taladriz, M.T.; Vázquez, J.F. Relationship between endosperm proteins and quality in durum wheat (*Triticum turgidum* L. var. durum). *Options Méditerranéennes* **2000**, *40*, 463–467.
15. Aguiriano, E.; Ruiz, M.; Fité, R.; Carrillo, J.M. Genetic variation for glutenin and gliadins associated with quality in durum wheat (*Triticum turgidum* L. ssp. turgidum) landraces from Spain. *Span. J. Agric. Res.* **2008**, *6*, 599–609. [CrossRef]
16. Moragues, M.; del Moral, L.F.G.; Moralejo, M.; Royo, C. Yield formation strategies of durum wheat landraces with distinct pattern of dispersal within the Mediterranean basin I: Yield components. *Field Crops Res.* **2006**, *95*, 194–205. [CrossRef]
17. Nazco, R.; Peña, R.J.; Ammar, K.; Villegas, D.; Crossa, J.; Moragues, M.; Royo, C. Variability in glutenin subunit composition of Mediterranean durum wheat germplasm and its relationship with gluten strength. *J. Agric. Sci.* **2014**, *152*, 379–393. [CrossRef] [PubMed]
18. Ruiz, M.; Bernal, G.; Giraldo, P. An update of low molecular weight glutenin subunits in durum wheat relevant to breeding for quality. *J. Cereal Sci.* **2018**, *83*, 236–244. [CrossRef]
19. Singh, N.K.; Shepherd, K.W.; Cornish, G.B. A simplified SDS—PAGE procedure for separating LMW subunits of glutenin. *J. Cereal Sci.* **1991**, *14*, 203–208. [CrossRef]
20. Payne, P.I.; Law, C.N.; Mudd, E.E. Control by homoeologous group 1 chromosomes of the high-molecular-weight subunits of glutenin, a major protein of wheat endosperm. *Theor. Appl. Genet.* **1980**, *58*, 113–120. [CrossRef]
21. Lafiandra, D.; Kasarda, D.D. One-and two-dimensional (two-pH) polyacrylamide gel electrophoresis in a single gel: Separation of wheat proteins. *Cereal Chem.* **1985**, *62*, 314–319.
22. Payne, P.I.; Lawrence, G.J. Catalogue of alleles for the complex gene loci, Glu-A1, Glu-B1, and Glu-D1 which code for high-molecular-weight subunits of glutenin in hexaploid wheat. *Cereal Res. Commun.* **1983**, *11*, 29–35.
23. McIntosh, R.; Dubcovsky, J.; Rogers, J.W.; Morris, C.; Appels, R.; Xia, X. Catalogue of gene symbols for wheat: 2013-14 Supplement. *Annu. Wheat Newsl.* **2014**, *58*, 259–279.
24. Nieto-Taladriz, M.T.; Ruiz, M.; Martinez, M.C.; Vazquez, J.F.; Carrillo, J.M. Variation and classification of B low-molecular-weight glutenin subunit alleles in durum wheat. *Theor. Appl. Genet.* **1997**, *95*, 1155–1160. [CrossRef]
25. Kudryavtsev, A.M.; Illichevskii, N.N.; Boggini, G.; Benedettelli, S. Gliadin polymorphism and genetic diversity of modern Italian durum wheat. *J. Genet. Breed.* **1996**, *50*, 239–248.
26. Sechi, S.; Chait, B.T. Modification of cysteine residues by alkylation. A tool in peptide mapping and protein identification. *Anal. Chem.* **1998**, *70*, 5150–5158. [CrossRef] [PubMed]
27. Dick, J.W.; Quick, J.S. A Modified Screening-Test for rapid estimation of gluten strength in early-generation durum-wheat breeding lines. *Cereal Chem.* **1983**, *60*, 315–318.
28. Carrillo, J.M.; Vazquez, J.F.; Orellana, J. Relationship between gluten strength and glutenin proteins in durum wheat cultivars. *Plant Breed.* **1990**, *104*, 325–333. [CrossRef]
29. Finney, K.F.; Shogren, M.D. A ten-gram mixograph for determining and predicting functional properties of wheat flours. *Bak. Dig.* **1972**, *46*, 32–47.
30. *InfoStat v2016*; Grupo Infostat, FCA, Universidad Nacional de Córdoba: Córdoba, Argentina, 2016.
31. Zeven, A.C. Landraces: A review of definitions and classifications. *Euphytica* **1998**, *104*, 127–139. [CrossRef]
32. Ruiz, M.; Vazquez, J.; Carrillo, J. Estudio de la variabilidad de gluteninas y gliadinas en variedades locales y cultivares primitivos españoles de trigo duro. *Invest. Agric. Prod. Prot. Veg.* **1998**, *13*, 291–305.
33. Cherdouh, A.; Khelifi, D.; Carrillo, J.; Nieto-Taladriz, M. The high and low molecular weight glutenin subunit polymorphism of Algerian durum wheat landraces and old cultivars. *Plant Breed.* **2005**, *124*, 338–342. [CrossRef]

34. Brites, C.; Romano, M.; Sousa, R.; Vazquez, J.; Carrillo, J. Caracterizaçao das proteinas de reserva da colecçao de trigos rijos Portugueses: Diversidade e qualidade tecnologica. *Melhoramento* **1996**, *34*, 55–66.
35. Raciti, C.; Doust, M.; Lombardo, G.; Boggini, G.; Pecetti, L. Characterization of durum wheat mediterranean germplasm for high and low molecular weight glutenin subunits in relation with quality. *Eur. J. Agron.* **2003**, *19*, 373–382. [CrossRef]
36. Babay, E.; Hanana, M.; Mzid, R.; Slim-Amara, H.; Carrillo, J.M.; Rodríguez-Quijano, M. Influence of allelic prolamin variation and localities on durum wheat quality. *J. Cereal Sci.* **2015**, *63*, 27–34. [CrossRef]
37. Vallega, V.; Mello-Sampayo, T. Variation of high-molecular-weight glutenin subunits amongst cultivars of *Triticum turgidum* L. from Portugal. *Euphytica* **1987**, *36*, 755–762. [CrossRef]
38. Branlard, G.; Autran, J.; Monneveux, P. High molecular weight glutenin subunit in durum wheat (*T. durum*). *Theor. Appl. Genet.* **1989**, *78*, 353–358. [CrossRef] [PubMed]
39. Igrejas, G.; Guedes-Pinto, H.; Carnide, V.; Branlard, G. The high and low molecular weight glutenin subunits and ω-gliadin composition of bread and durum wheats commonly grown in Portugal. *Plant Breed.* **1999**, *118*, 297–302. [CrossRef]
40. Ribeiro, M.; Rodríguez-Quijano, M.; Giraldo, P.; Pinto, L.; Vázquez, J.F.; Carrillo, J.M.; Igrejas, G. Effect of allelic variation at glutenin and puroindoline loci on bread-making quality: Favorable combinations occur in less toxic varieties of wheat for celiac patients. *Eur. Food Res. Technol.* **2017**, *243*, 743–752. [CrossRef]
41. Rodriguez-Quijano, M.; Vazquez, J.; Carrillo, J.; Moita Brites, C. Allelic variation of HMW glutenin subunits in Portuguese landraces of *Triticum aestivum* ssp. *vulgare*. *J. Genet. Breed.* **1998**, *52*, 95–98.
42. Brites, C.; Carrillo, J. Inheritance of gliadin and glutenin proteins in four durum wheat crosses. *Cereal Res. Commun.* **2000**, *28*, 239–246.
43. Martinez, M.C.; Ruiz, M.; Carrillo, J.M. New B low Mr glutenin subunit alleles at the Glu-A3, Glu-B2 and Glu-B3 loci and their relationship with gluten strength in durum wheat. *J. Cereal Sci.* **2004**, *40*, 101–107. [CrossRef]
44. Rodríguez-Quijano, M.; Lucas, R.; Ruiz, M.; Giraldo, P.; Espí, A.; Carrillo, J.M. Allelic variation and geographical patterns of prolamins in the USDA-ARS Khorasan wheat germplasm collection. *Crop Sci.* **2010**, *50*, 2383–2391. [CrossRef]
45. Ribeiro, M.; Carvalho, C.; Carnide, V.; Guedes-Pinto, H.; Igrejas, G. Towards allelic diversity in the storage proteins of old and currently growing tetraploid and hexaploid wheats in Portugal. *Genet. Resour. Crop Evol.* **2011**, *58*, 1051–1073. [CrossRef]
46. Aguiriano, E.; Ruiz, M.; Fité, R.; Carrillo, J.M. Analysis of genetic variability in a sample of the durum wheat (*Triticum durum* Desf.) Spanish collection based on gliadin markers. *Genet. Resour. Crop Evol.* **2006**, *53*, 1543–1552. [CrossRef]
47. Peña, R.; Trethowan, R.; Pfeiffer, W.; Ginkel, M.V. Quality (end-use) improvement in wheat: Compositional, genetic, and environmental factors. *J. Crop Prod.* **2002**, *5*, 1–37. [CrossRef]
48. Brites, C.; Carrillo, J. Influence of high molecular weight (HMW) and low molecular weight (LMW) glutenin subunits controlled by Glu-1 and Glu-3 loci on durum wheat quality. *Cereal Chem.* **2001**, *78*, 59–63. [CrossRef]
49. Martinez, M.C.; Ruiz, M.; Carrillo, J.M. Effects of different prolamin alleles on durum wheat quality properties. *J. Cereal Sci.* **2005**, *41*, 123–131. [CrossRef]
50. Magallanes-López, A.M.; Ammar, K.; Morales-Dorantes, A.; González-Santoyo, H.; Crossa, J.; Guzmán, C. Grain quality traits of commercial durum wheat varieties and their relationships with drought stress and glutenins composition. *J. Cereal Sci.* **2017**, *75*, 1–9. [CrossRef]
51. Nieto-Taladriz, M.T.; Rodriguez-Quijano, M.; Carrillo, J.M. Glutenin proteins as markers for durum wheat quality selection. In Proceedings of the Actes du 1er Symposium International sur la Filière Blé, Office Algérien Interprofessionnel des céréales, Algiers, Algeria, 7–9 February 2010; pp. 267–270.
52. Cros, D.; Wrigley, C.; Hare, R. Prediction of durum-wheat quality from gliadin-protein composition. *Aust. J. Agric. Res.* **1982**, *33*, 429–442. [CrossRef]
53. Pogna, N.; Autran, J.-C.; Mellini, F.; Lafiandra, D.; Feillet, P. Chromosome 1B-encoded gliadins and glutenin subunits in durum wheat: Genetics and relationship to gluten strength. *J. Cereal Sci.* **1990**, *11*, 15–34. [CrossRef]
54. Aguiriano, E.; Ruiz, M.; Fité, R.; Carrillo, J. Effects of N fertilisation, year and prolamin alleles on gluten quality in durum wheat (*Triticum turgidum* L. ssp. turgidum) landraces from Spain. *Span. J. Agric. Res.* **2009**, *7*, 342–348. [CrossRef]

55. Shewry, P.R.; Halford, N.G.; Lafiandra, D. Genetics of wheat gluten proteins. In *Advances in Genetics*; Academic Press: San Diego, CA, USA, 2003; Volume 49, pp. 111–184.
56. Gianibelli, M.C.; Larroque, O.R.; MacRitchie, F.; Wrigley, C.W. Biochemical, genetic, and molecular characterization of wheat glutenin and its component subunits. *Cereal Chem.* **2001**, *78*, 635–646. [CrossRef]

agronomy

MDPI

Article

Molecular Genetic Diversity and Population Structure of Ginseng Germplasm in RDA-Genebank: Implications for Breeding and Conservation

Kyung Jun Lee, Jung-Ro Lee, Raveendar Sebastin, Gyu-Taek Cho and Do Yoon Hyun *

National Agrobiodiversity Center, National Institute of Agricultural Sciences (NAS), RDA, Jeonju 54874, Korea; lkj5214@korea.kr (K.J.L.); jrmail@korea.kr (J.-R.L.); raveendars@gmail.com (R.S.); gtcho@korea.kr (G.-T.C.)

* Correspondence: dyhyun@korea.kr; Tel.: +82-63-238-4912

Received: 10 November 2019; Accepted: 19 December 2019; Published: 3 January 2020

Abstract: Ginseng (*Panax ginseng* C.A. Meyer), commonly known as Korean or Asian ginseng, is a perennial herb native to Korea and China. There has been limited research effort to analyze the genetic diversity and population structure of ginseng germplasm because of its growth habits. In the present study, genetic diversity and population structure of ginseng germplasm conserved in the National Agrobiodiversity Center (NAC) of South Korea were analyzed to provide basic data for future preservation and breeding of ginseng genetic resources. Seventeen simple sequence repeat (SSR) markers were used to assess the genetic diversity and population structure of 1109 ginseng accessions. Among 1109 ginseng accessions, 1042 (94.0%) accessions were landraces and 66 (6.0%) accessions were breeding lines (61 accessions, 5.5%) or cultivars (5 accessions, 0.5%). SSR markers revealed 56 different alleles with an average of 3.29 alleles per locus. The average gene diversity was 0.49. Analysis of molecular variance showed that 91% of allelic diversity was attributed to individual accessions within clusters while only 9% was distributed among clusters. Using discriminant analysis of principal components, 12 clusters were detected in 1109 ginseng accessions. The results of this study provide molecular evidence for the narrow genetic base of ginseng germplasm in NAC. For the broad understanding and efficient use of ginseng germplasm, it is necessary to analyze functional factors and to evaluate morphological traits.

Keywords: genetic diversity; genetic structure; simple sequence repeat; *Panax ginseng*

1. Introduction

Ginseng (*Panax ginseng* Meyer, 2n = 4x = 48), a medicinal herb, has long been used in the Far East, particularly in Korea and China as a respected herbal medicine for maintaining physical vitality [1]. The main physiologically active components of ginseng are saponins, phenols, polyacetylenes, alkaloids, and polysaccharides [2]. These major components are known to have beneficial properties for the central nervous system, cardiovascular system, endocrine system, and immune systems [3,4].

Ginseng has been one of the most important trade goods for health care and treatment of diseases in Asia including Korea and China and is currently distributed to 35 countries around the world [5]. Among ginseng production countries, China has the most production, 44,749 tons in 2008, followed by South Korea with 27,480 tons [5]. As of 2017, the annual production of ginseng is 23,310 tons and domestic distribution size is approximately $1584 million in terms of the amount based on ginseng root and processed products in South Korea [6]. With regard to ginseng production, cultivation is focused on harvesting four to six-year-old roots in South Korea and China [7].

The hybridization breeding of ginseng is difficult because of slow growth (~four years/generation) and few seeds (~40 seeds/plant) [8]. Ginseng is originally a self-pollination plant and starts to bloom at its third-year growth stage. In general, its seeds are obtained from a fourth-year growth plant, and flower buds are removed for seeds and root growth. Seeds are harvested from the red fruit having 2 pale yellow seeds in it. At harvesting time the seeds are in immature condition, which does not form a clear embryo. Ginseng seeds need a dormant period for maturity that enables germination [9]. In addition, ginseng is sensitive to environmental factors such as light and soil moisture, so maintaining individual plants requires considerable effort [8].

Analysis of genetic diversity and population structure of the existing population is required for the purpose of conservation and reintroduction of rare and endangered species [10]. A species with little genetic variability may suffer from reduced fitness in its current environment. It may not have the evolutionary potential necessary for a changing environment. Genetic variability is critical for a species to adapt to environmental changes and survive in the long term [11]. Knowledge of genetic diversity within a population and among populations is important for conservation management, especially for identifying genetically unique structural units within a species and determining populations that need protection [12].

Genetic diversity of germplasm can be investigated using various techniques, including analysis of morphological and agricultural traits [13], isozymes [14], biochemical characteristics [15], and molecular markers [14–16]. In recent years, the development of various chemical and molecular techniques for differentiation among *P. ginseng* samples has received attention [10,17–20]. Gepts has reported that DNA markers are the most reliable ones for distinguishing variations of alleles because these markers are not influenced by environmental or physiological factors [21].

Crawford has reported that efficient population genetics studies could be performed using a high level of polymorphic markers [22]. Previous studies have performed genetic analysis of wild ginseng using random amplified polymorphic DNA and allozyme, respectively, and detected a low level of genetic polymorphisms in their samples, meaning that effective conservation strategies would be difficult to design using these markers [23,24]. Using amplified fragment length polymorphism markers, genetic diversity and genetic structure of *P. ginseng* have been analyzed [25–27]. Among various molecular markers, simple sequence repeat (SSR) markers have been successfully used to described population genetic diversity because they showed codominance and high polymorphism [7,10,20,28,29].

Since the 21st century, the number of publications on *P.ginseng* has exponentially increased; more than 3400 articles, including 242 reviews, have focused on *P. ginseng*. However, the main leading research subjects and keywords were pharmacology (39%) and functional materials such as ginsenoside and saponin, with little research on the genetic diversity of *P. ginseng* [30]. Understanding the genetic diversity on plant germplasm is important as plant germplasm provides an opportunity for plant breeders to develop new and improved cultivars with desirable characteristics, which include both the farmer and breeder's preferred traits [31]. The objective of the present study was to analyze genetic diversity and population structure of 1109 ginseng accessions conserved in the National Agrobiodiversity Center (NAC) in South Korea to provide basic data for future preservation and breeding of ginseng genetic resources.

2. Materials and Methods

2.1. Plant Materials

A total of 1109 ginseng accessions were obtained from the National Agrobiodiversity Center (NAC) at the Rural Development Administration in South Korea (Table S1). All ginseng accessions have been separately conserved at three institutes (109 accessions Chungcheongbuk-do national University (CB); 160 accessions, Chungcheongnam-do Agricultural Research & Extension Service (CN); and 840 accessions, National Institute of Horticultural and Herbal Science (NIHHS)) designed by NAC.

2.2. DNA Extraction

Genomic DNA was extracted from 100 mg of freeze-dried ginseng leaves in each ginseng accession (4~5 plants per accession) using a DNeasy plant mini kit (Qiagen, Hilden, Germany). DNA quality and quantity were measured using 1% (*w/v*) agarose gel and a spectrophotometer (Epoch, BioTek, Winooski, VT, USA). Extracted DNA was diluted to 30ng/ul and stored at −20 °C until further PCR amplification.

2.3. SSR Genotyping

For SSR analysis, a total of 17 SSRs were selected from the previously reported study [28] and labeled fluorescently (6-FAM, HEX, and NED) (Table S2). PCR reactions were carried out using 25 ul reaction mixture, containing 30 ng template DNA, 10x PCR buffer, 1.5 mM $MgCl_2$, 0.2 mM of each dNTPs, 0.5 uM of each primer, and 1 U *Taq* polymerase (Inclone, Korea). The amplification was performed with the following cycling conditions: initial denaturation at 94 °C for 5 min, followed by 35 cycles of denaturation at 95 °C for 30 s, annealing at 55 °C for 30 s, extension at 72 °C for 1 min, and a final extension step at 72 °C for 10 min. Each amplicon was resolved on the ABI prism 3500 DNA sequence (ABI3500, Thermo Fisher Scientific Inc., Wilmington, DE, USA). Individual tetraploid genotypes were scored from the electropherograms following the microsatellite DNA allele counting-peak ratios (MAC-PR) method of Esselink et al. [32] using Gene Mapper Software (Version 4.0, Thermo Fisher Scientific Inc., Wilmington, DE, USA).

2.4. Population Structure and Genetic Diversity

The number of alleles (Na), Shannon index (I), Nei's unbiased gene diversity (GD), and Evenness were calculated using *poppr* package for R software [33]. The analysis of molecular variance (AMOVA) and calculation of the coefficient of genetic differentiation among populations (PhiPT) were done using GenAlEx software (v. 6.5) with 999 permutations [34].

Population structure was analyzed with a DAPC using *adegenet* package for R software [35,36]. The *find.clusters* function was used to detect the number of clusters in the population. It used K-means clustering which decomposed the total variance of a variable into between-group and within-group components. The best number of subpopulations had the lowest associated Bayesian Information Criterion (BIC). A cross-validation function (*Xval. dapc*) was used to confirm the correct number of principal components (PC) to be retained. In this analysis, data were divided into two sets: a training set (90% of the data) and a validation set (10% of the data). Members of each group were selected by stratified random sampling, ensuring that at least one member of each group or population in the original data was represented in both training and validation sets. DAPC was carried out on the training set by retaining variable numbers of PCs at a degree to which the analysis was able to accurately predict the group membership of excluded individuals (those in the validation set) was used to identify the optimal number of PCs to be retained. At each level of PC retention, the sampling and DAPC procedures were repeated many times [37]. The best number of PCs that should be retained is associated with the lowest root mean square error. Resultant clusters were plotted in a scatter plot of the first and second linear discriminants of DAPC.

Bayesian-based clustering was performed using STRUCTURE v.2.3.4 [38], testing three independent runs with K from 1 to 15, each run with a burn-in period of 50,000 iterations and 500,000 Monte Carlo Markov iterations, assuming an admixture model. The most likely K value was processed with STRUCTURE HARVESTER v.0.9.94 [39] and was detected using the Evanno transformation method [40]. To assign samples to clusters, a membership coefficient $q > 0.8$ was used, while coefficients ≤ 0.8 were considered "genetically admixed."

3. Results

Among 1109 ginseng accessions, 1,042 (94.0%) accessions were landraces and 66 (6.0%) accessions were breeding lines (61 accessions, 5.5%) or cultivars (5 accessions, 0.5%). Of 1109 ginseng accessions, 899 (81.2%) were from Korea and 202 (18.0%) were from China (Figure S1A).

CN collected 73 ginseng accessions from Chungnam province from 1997 to 2000, 56 accessions from China, and 21 accessions from Korea from 2013 to 2016 (Table S1 and Figure S1B). NIHHS obtained approximately 300 ginseng accessions from Korea Tobacco and Ginseng Corporation (KT&G),

40 accessions from China, and 435 from Korea from 2003 to 2012. CB collected 90 accessions from China and 10 from Korea from 2007 to 2010.

SSR locus diversity data are summarized in Table 1. A total of 56 alleles were detected among 1109 ginseng accessions by using 17 polymorphic SSR markers. On average, 3.29 alleles varying from 2 to 6 were amplified by each marker. Shannon index (I) varied from 0.58 to 0.82 among 17 SSR markers. The average gene diversity (GD) was estimated to be 0.49 (range, 0.39 to 0.50). The average evenness (E) was 0.96 (range, 0.81 to 1.00). A total of 22 rare alleles (Ra) ranging from 1 to 4 with each allele had a frequency of less than 5%. Six SSR loci (PES0024, PES0021, PES0018, PES0019, PES0005, and PES0007) did not contain rare alleles. Common alleles (Ca) ranged from 1 to 2 with a total of 24, showing a frequency of 5%–50%.

Table 1. Statistics of genetic diversity across 17 simple sequence repeat (SSR) loci in 1109 ginseng accessions.

Locus	Ra [1]	Ca	Aa	Na	I	Ho	GD	Evenness	Range
PES0024	0	1	1	2	0.69	0.99	0.50	1.00	145–171
PES0021	0	1	1	2	0.69	1.00	0.50	0.99	245–263
PES0034	2	1	1	4	0.70	1.00	0.50	0.97	247–275
PES0026	3	1	1	5	0.70	1.00	0.49	0.95	178–213
PES0029	1	2	0	3	0.70	1.00	0.5	0.99	183–200
PES0037	2	1	1	4	0.69	0.99	0.49	0.96	260–307
PES0040	1	1	1	3	0.69	1.00	0.50	1.00	250–272
PES0038	1	2	0	3	0.70	1.00	0.50	0.97	169–190
PES0039	2	1	1	4	0.69	0.99	0.49	0.96	168–250
PES0018	0	2	0	2	0.69	1.00	0.50	1.00	202–216
PES0032	1	2	0	3	0.69	1.00	0.49	0.98	161–172
PES0033	1	1	1	3	0.71	1.00	0.5	0.98	286–302
GES0019	0	1	1	2	0.58	0.22	0.39	0.81	198–208
PES0004	4	1	1	6	0.82	0.99	0.52	0.85	362–402
PES0005	0	2	0	2	0.69	1.00	0.50	1.00	334–358
PES0007	0	2	0	2	0.69	1.00	0.49	0.99	542–563
PES0012	4	2	0	6	0.73	1.00	0.50	0.94	235–279
Total	22	24	10	56	-	-	-	-	-
Mean				3.29	0.70	0.95	0.49	0.96	-

[1] Ra, number of rare (<5%) alleles; Ca, number of common (5%–50%) alleles; Aa, number of abundant (>50%) alleles; Na, Number of alleles; I, Shannon's index; Ho, Observed heterozygosity; GD, Nei's gene diversity.

Five from the 56 alleles were identified to private alleles which are detected in only one cluster (Table S3). Specifically, KT&G and GG showed two private alleles in locus PES0092 and PES0012 and PES0032 and PES0040, respectively, while GW possesses one private allele in locus PES0039.

A total of 942 multilocus genotypes (MLGs) were detected among 1109 ginseng accessions (Table 2). CN, CB, and NIHHS clusters had 148, 104, and 707 MLGs, respectively. The numbers of average alleles (Na) of CN, CB, and NIHHS were 1.77, 2.08, and 2.62, respectively. The Shannon–Wiener index (H) of NIHHS was the highest (6.45), while that of CN was similar to that of CB (4.97 and 4.63, respectively). Nei's genetic diversity (GD) was similar to each other for the three clusters.

Table 2. Summary of cluster diversity indices averaged over 17 loci according to conservation institutes.

Pop	N [1]	MLG	Na	H	GD
CN	160	148	1.77	4.97	0.485
CB	109	104	2.08	4.63	0.476
NIHHS	840	707	2.62	6.45	0.494
Total	1109	942	3.29	6.74	0.492

[1] N, Number of accessions; MLG, Number of multilocus genotypes; Na, Number of average alleles; H, Shannon-Wiener index of MLG diversity; GD, Nei's unbiased gene diversity.

Analysis of molecular variance (AMOVA) on genetic differentiation among and within clusters of ginseng accessions was conducted. The results are shown in Table 3. Findings from AMOVA revealed that 91% of total genetic variations were contributed by differences within clusters, which was notably and significantly higher than that among clusters (only 9% of total genetic variation was due to differences among clusters). PhiPT and gene flow (Nm) for 1109 ginseng accessions were 0.094 ($p < 0.001$) and 4.822, respectively. Pairwise cluster PhiPT values for the three institutes ranged from 0.081 (CN–NIHHS) to 0.113 (CB–NIHHS) (Table 4). Pairwise cluster estimates of gene flow (Nm) for the three institutes ranged from 3.940 to 5.656 migrants per cluster.

Table 3. Analysis of molecular variance (AMOVA) within/among ginseng clusters based on 17 SSR data according to conservation institutes.

Source	df	SS	MS	Est. Var.	%	PhiPT	Nm
Among clusters	2	322.552	161.276	0.704	9%	0.094 **	4.822
Within clusters	1106	7508.527	6.789	6.789	91%		
Total	1108	7831.079		7.493	100%		

** $p < 0.01$

Table 4. Pairwise cluster PhiPT values and Nm values based on 999 permutations from AMOVA according to conservation institutes (all PhiPT values were significantly greater than 0, $p < 0.0001$).

Cluster1	Cluster2	PhiPT	Nm
CN	CB	0.085	5.374
CN	NIHHS	0.081	5.656
CB	NIHHS	0.113	3.940

To understand the pattern of the genetic structure, a Bayesian clustering analysis in STRUCTURE and complementary coordination analysis by Discriminant Analysis of Principal Components (DAPC) were performed. STRUCTURE results suggested the best grouping number (K = 2) based on delta K (Figure S2). Clusters 1 and 2 consisted of 359 and 460 accessions, respectively. Two hundred and ninety accessions were identified in the admixed population (Figure 1A). All (100%) ginseng accessions of cluster 1 and admix were all in NIHHS while 53.8% and 30.7% of population 1 and admix cluster were from KT&G, respectively. Cluster 2 contained 191 accessions from NIHHS and all accessions of CN and CB and 80.3% of cluster 2 were from China.

The number of detected clusters was 12 in concordance with the lowest BIC value obtained using the *find.clusters* function. DAPC analysis was carried out using the detected number of clusters (Figure 1B). Fifteen first PCs (90.9% of variance conserved) of PCA and three discriminant eigenvalues were retained. These values were confirmed by cross-validation analysis.

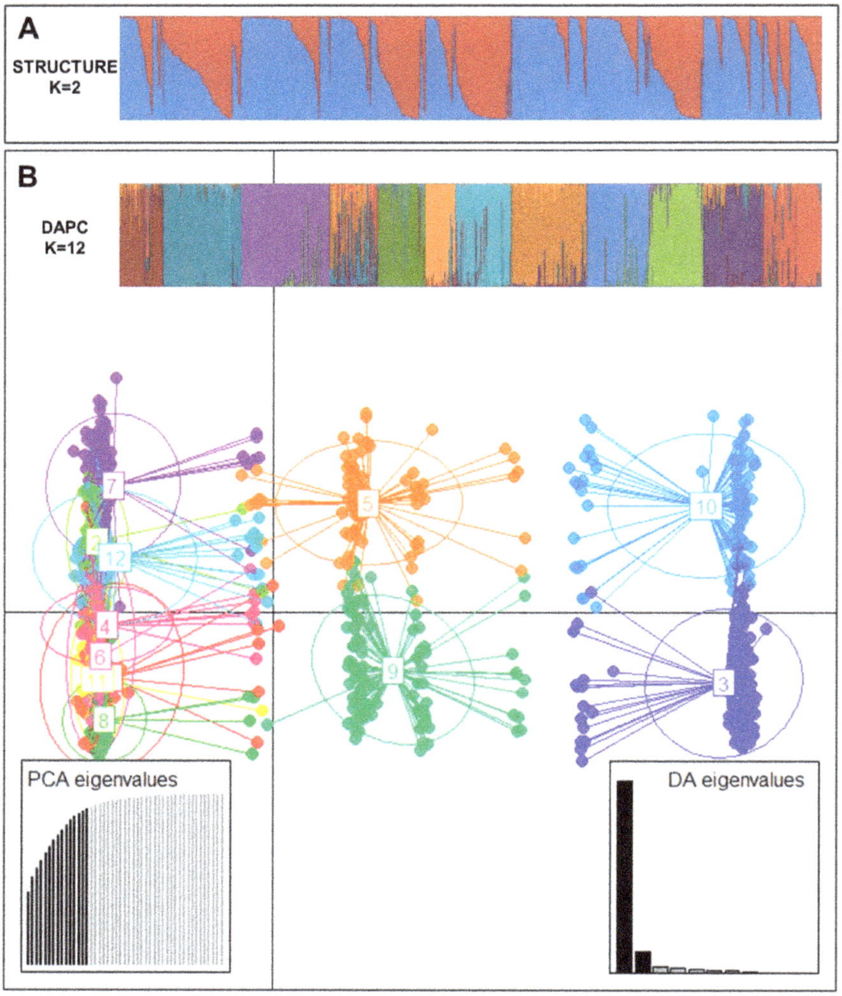

Figure 1. (**A**) Population structure analysis of 1109 ginseng accessions inferred using STRUCTURE software based on 17 SSR markers for delta K = 2. (**B**) Discriminant analysis of principal components (DAPC) for 1109 ginseng accessions. Axes represent the first two linear discriminants (LD). Each circle represents a cluster and each dot represents an individual. Numbers represent different clusters identified by DAPC analysis.

Sources of genetic differentiation were revealed among 12 clusters based on DAPC analysis using the AMOVA method. Results indicated that 12% and 88% of variations could be attributed to differentiation among clusters and within inferred clusters, respectively (Table 5). PhiPT and gene flow (Nm) were 0.120 ($p < 0.001$) and 3.678, respectively. Pairwise cluster PhiPT values for 12 clusters ranged from 0.015 (C6–C7) to 0.224 (C4–C12) (Table 6). Pairwise cluster estimates of gene flow (Nm) for 12 clusters ranged from 1.734 (C4–C12) to 33.651 (C6–C7) migrants per clusters.

Table 5. Analysis of molecular variance (AMOVA) within/among twelve clusters based on DAPC analysis.

Source	df	SS	MS	Est. Var.	%	PhiPT	Nm
Among pops	11	932.529	84.775	0.855	12%	0.120 **	3.678
Within pops	1097	6898.550	6.289	6.289	88%		
Total	1108	7831.079		7.143	100%		

** $p < 0.01$

Table 6. Pairwise cluster PhiPT values (above diagonal) and Nm values based on 999 permutations (below diagonal) from AMOVA (all PhiPT values were significantly greater than 0, $p < 0.0001$).

	C1	C2	C3	C4	C5	C6	C7	C8	C9	C10	C11	C12
C1	-	0.126	0.137	0.174	0.084	0.098	0.153	0.147	0.196	0.152	0.138	0.218
C2	3.467		0.093	0.113	0.080	0.084	0.106	0.080	0.090	0.120	0.070	0.126
C3	3.160	4.881	-	0.087	0.126	0.046	0.037	0.126	0.133	0.034	0.147	0.117
C4	2.374	3.915	5.227	-	0.124	0.063	0.070	0.178	0.181	0.127	0.188	0.224
C5	5.438	5.764	3.478	3.527	-	0.082	0.104	0.160	0.164	0.138	0.039	0.203
C6	4.619	5.463	10.370	7.396	5.611	-	0.015	0.107	0.121	0.028	0.129	0.138
C7	2.773	4.230	13.077	6.649	4.312	33.651	-	0.167	0.152	0.028	0.136	0.141
C8	2.904	5.725	3.478	2.301	2.616	4.185	2.486	-	0.052	0.120	0.182	0.143
C9	2.054	5.082	3.261	2.257	2.556	3.622	2.795	9.172	-	0.111	0.176	0.059
C10	2.792	3.674	14.321	3.445	3.126	17.623	17.617	3.678	4.012	-	0.166	0.097
C11	3.131	6.688	2.908	2.159	12.483	3.364	3.187	2.245	2.339	2.511	-	0.197
C12	1.799	3.478	3.767	1.734	1.966	3.112	3.043	3.003	7.943	4.679	2.037	-

4. Discussion

Among collected and preserved 1109 ginseng accessions in this study, 94.0% were landraces, with only 6% being breeding lines. This might be due to the collection strategy of ginseng germplasm. Three institutes collected ginseng accessions from all parts of South Korea since 1997 to develop new ginseng cultivars. In general, ginseng cultivars have been bred by selecting individual plants from arable fields of mixed local landrace population rather than by crossbreeding [41]. It is known that breeding of ginseng is difficult because of its long generation period and small number of seeds, although it is originally a self-pollination plant [9]. For these reasons, outstanding individual plants were selected and registered as cultivars during the cultivation of a mixed population of local landrace in Korea.

Ginseng is cultivated from 36° to 38° N and can still be found sometimes in mountainous areas in Korea [42]. In general, Ginseng farmers have traditionally kept various kinds of seeds by conserving and sowing the seeds that have been harvested from 3-year-old ginseng. In the case of large-scale ginseng farming (about 1 ha or more), farmers purchase seed because home seed production alone lacks it. However, most of the small farms in South Korea cultivate ginseng using their own seed from home seed production. Ginseng farmers choose and preserve seeds very carefully due to long-term (4~6 years) characteristics of ginseng cultivation. For this reason, various Korean ginseng genotypes are being conserved by the ginseng farmers in South Korea [43]. In this study, we used ginseng landrace based on the information of the genebank management system in NAC. Landrace defined by the FAO is "a crop that has developed its unique characteristics through repeated in situ grower selection and that has never been subjected to formal plant breeding" [44]. In terms of the definition, ginseng accessions collected from various local farms in Korea are likely to be landrace although their information except the geographical data is lack. Nevertheless, the question of whether the accessions are landraces still remains. Therefore, additional research on their genetic information will be needed to efficiently manage the ginseng accessions.

Among 1109 ginseng accessions, 167 (15.1%) (12 accessions in CN, five accessions in CB, and 133 accessions in NIHHS) might be duplicated (Table 2). It is important to identify these duplicated accessions, due to the limitation of funding and resources in genebanks [45]. Although not all duplicate accessions are worthless, identifying unique accessions within and across the genebanks will facilitate the better use of plant germplasms [45,46]. In this study, 167 ginseng accessions identified will also require the management of duplicate accessions using various additional methods such as evaluation of morphological and/or biochemical traits.

Results of the present study showed a similar number of alleles (56) and average number of alleles (3.29) to results of Choi et al. [28] (number of alleles = 51; average number of alleles = 3.0), while genetic diversity (0.490) in the present study was higher than (0.398) in the previous study. Reunova et al. [12] have reported similar results (average number of alleles: 2.49; expected heterozygosity (He): 0.453) in 139 individual ginseng plants using 11 SSR primers. Otherwise, Bang et al. [7] have reported a higher average number of alleles (4.3) and He (0.627) in 17 ginseng accessions using 8 SSR markers. Gene diversity of a locus, also known as expected heterozygosity, is a fundamental measure of genetic variation in a population. It describes the proportion of heterozygosis expected under the Hardy–Weinberg (HW) equilibrium [47]. Previous studies have confirmed the excess heterozygosity in their ginseng populations [7,12,28,48]. Indeed, the present study also showed excess heterozygosity in these ginseng accessions (Table 1). Causes of the excess of heterozygosity in natural plant populations are not well known yet, although reasons causing a deficit of heterozygotes have been well explained [12]. Previous studies have guessed that the excess of heterozygosity might result from a small reproductive population size [49–51]. When only a few breeders contribute to the next generation, allelic frequencies can differ between male and female parents by chance alone [49,50]. This effect can be reinforced for loci with many rare alleles. Moreover, in small sexual or self-incompatible populations, the fact that individuals cannot reproduce with themselves decreases the probability of creating homozygote offspring [52,53]. Meanwhile, Reunova et al. [12] have mentioned that high levels of observed heterozygosity might be associated with population structure and reflect the history of its formation under the influence of global climatic and local human-induced impacts.

In this study, 171 ginseng accessions showed rare alleles (Table S1). Rare alleles may be lost due to natural and management-originated bottleneck effects [54]. Although they are not likely to have major importance in the conservation of endangered species, rare alleles may be important for plant breeding [55]. They may be present or even fixed in certain populations because of genetic drift or their relationship with fitness in specific environments. Such uniqueness makes them prone to be absent in whole collections and their subsets, albeit their potential importance as a source of important traits for plant breeding. Novel strategies in germplasm management are needed to preserve those alleles, prone to be lost, which can be potentially useful for plant breeding, especially in a scenario of rapid climate change that represents serious threats to worldwide food production [56].

In addition, five private alleles were detected after screening the collection areas or institutes (Table S3). Private alleles are important in plant breeding and conservation as they are present only in a single population among a broader collection of populations [57]. However, all five private alleles observed in this study were present at very low frequencies (<0.9%), suggested that they could not be suitable for the ginseng breeding program. On the other hand, three ginseng accessions, IT272054, IT272375, and IT288747, with unique alleles might be used as a valuable source for selecting genetic variants related to useful chemical compounds such as saponins (Table S1).

To analyze population structure, various methods such as STRUCTURE and principal coordinates analysis have been used for ginseng germplasm [19,20,58]. In this study, STRUCTURE and DAPC used to analyze the population structure of 1109 ginseng accessions appeared to provide complementary information. Results of STRUCTURE and DAPC divided 1109 ginseng accessions into two and twelve clusters, respectively. The DAPC method is being used instead of STRUCTURE software as it does not require that populations should be in HW equilibrium. In addition, it can handle large sets of data without using parallel processing software [59]. Deperi et al. [60] have mentioned that DAPC analysis can divide the population into well-defined clusters associated with provenance, ploidy, taxonomy, and breeding program of genotypes related to their genetic structure. Rosyara et al. [61] have reported that

DAPC is slightly better than the STRUCTURE leading to better separation among potato germplasms. Campoy et al. [59] have also reported that DAPC analysis provides more detailed clustering among cherry populations compared to STRUCTURE analysis. More detailed clustering was also obtained from DAPC analysis. Although 99.2% of 1109 ginseng accessions were collected from Korea (899 accessions) and China (202 accessions) (Table S1), 12 clusters of DAPC analysis showed that ginseng accessions from two countries were separated into all clusters. Also, low PhiPT (0.036) and high Nm (13.280) were detected between Korean and Chinese accessions. This means that there are high genetic similarities in ginseng accessions collected from these two countries. According to NIHHS [62], the genetic similarity of cultivated ginsengs between Korea and China was more than 70%, due to the import of ginseng seeds from South Korea to China and the cultivation and production of them in the ginseng main production area of China such as Gillim Province.

The reason why 1109 ginseng accessions were divided into twelve clusters on DAPC analysis was not clear. It might be due to different methods of cultivation, seed gathering, and seed distribution for ginseng in Korea. Since new ginseng cultivars, "Chunpoong" and "Yunpoong," were developed in 2002, a total of 21 ginseng cultivars have been developed. However, farmers are still cultivating, self-seed-gathering, and seed-selling ginseng landraces not ginseng cultivars [63,64]. With the exception of some contract cultivation, ginseng landrace seeds would have been traded in the Korean ginseng market. As a result, ginseng cultivated in farms is expected to be genetically mixed regardless of their cultivation region.

5. Conclusions

In this study, genetic diversity and population structure of 1109 ginseng accessions collected and conserved in NAC were analyzed using 17 SSR markers. The results of this study provided molecular evidence for the narrow genetic base of 1109 ginseng germplasm in NAC. It was not enough to understand the genetic diversity of ginseng germplasm because there was a lack of sufficient information about these collected ginseng accessions, although the aim of the present study was to analyze genetic diversity in larger ginseng accessions than those from previous studies [7,20,52], In Korea, the majority of new ginseng cultivars are developed by selecting individual plants from ginseng landraces, although some research institutes have performed development and selection of breeding lines. To have a broad understanding and efficient use of ginseng germplasm, it is necessary to analyze functional factors and evaluate their morphological traits.

Supplementary Materials: The following are available online at http://www.mdpi.com/2073-4395/10/1/68/s1, Figure S1: Geographic distribution of 1109 ginseng accessions. Pie charts on the map represent the collected province of the accessions and the colors in each chart represent the number of ginseng accessions indicated by conserved three institutes. (A) Original countries of ginseng accessions. (B) Original province of ginseng accessions in South Korea, Figure S2: Relationship between delta K and K as revealed by STRUCTURE harvester, Table S1: List of 1109 ginseng accessions used in this study, Table S2: List of 17 SSR primers used in this study, Table S3: Allele frequency based on collection area in ginseng accessions.

Author Contributions: Conceptualization, K.J.L.; data curation, K.J.L.; formal analysis, K.J.L. and R.S.; Investigation, J.-R.L. and G.-T.C.; project administration, D.Y.H.; resources, J.-R.L. and G.-T.C.; writing—original draft, K.J.L.; writing—review and editing, D.Y.H. All authors have read and agreed to the published version of the manuscript.

Funding: This work was carried out with the support of the "Research Program for Agricultural Science & Technology Development (Project No. PJ01355701)," National Institute of Agricultural Sciences, Rural Development Administration, Republic of Korea.

Conflicts of Interest: The authors declare no conflict of interest.

References

1. Yun, T.K. Brief Introduction of *Panax ginseng* C.A. Meyer. *J. Korea Med. Sci.* **2001**, *16*, S3–S5. [CrossRef] [PubMed]
2. Lee, M.J.; Kim, E.H.; Rhee, D.K. Effect of Panax ginseng on stress. *J. Ginseng Res.* **2008**, *32*, 8–14.

3. Attele, A.S.; Wu, J.A.; Yuan, C.-S. Ginseng pharmacology: Multiple constituents and multiple actions. *Biochem. Pharm.* **1999**, *58*, 1685–1693. [CrossRef]
4. Nam, M.H.; Kim, S.I.; Liu, J.R.; Yang, D.C.; Lim, Y.P.; Kwon, K.-H.; Yoo, J.S.; Park, Y.M. Proteomic analysis of Korean ginseng (*Panax ginseng* C.A. Meyer). *J. Chromatogr. B* **2005**, *815*, 147–155. [CrossRef] [PubMed]
5. Baeg, I.-H.; So, S.-H. The world ginseng market and the ginseng (Korea). *J. Ginseng Res.* **2013**, *37*, 1–7. [CrossRef] [PubMed]
6. Ministry of Agriculture Food and Rural Affairs. *Ginseng Annual Report*; Ministry of Agriculture Food and Rural Affair: Sejong, South Korea, 2018.
7. Bang, K.-H.; Jo, I.-H.; Chung, J.-W.; Kim, Y.-C.; Lee, J.-W.; Seo, A.Y.; Park, J.-H.; Kim, O.; Hyun, D.-Y.; Kim, D.-H.; et al. Analysis of Genetic Polymorphism of Korean Ginseng Cultivars and Foreign Accessions using SSR Markers. *Korean J. Med. Crop. Sci.* **2011**, *19*, 347–353. [CrossRef]
8. Jang, W.; Kim, N.-H.; Lee, J.; Waminal, N.; Lee, S.-C.; Jayakodi, M.; Choi, H.-I.; Park, J.Y.; Lee, J.-E.; Yang, T.-J. A Glimpse of Panax ginseng Genome Structure Revealed from Ten BAC Clone Sequences Obtained by SMRT Sequencing Platform. *Plant Breed. Biotechnol.* **2017**, *5*, 25–35. [CrossRef]
9. Choi, K.T. Botanical characteristics, pharmacological effects and medicinal components of Korean Panax ginseng C A Meyer. *Acta Phamacol. Sin.* **2008**, *29*, 1109–1118. [CrossRef]
10. Zhuravlev, Y.N.; Reunova, G.D.; Kats, I.L.; Muzarok, T.I.; Bondar, A.A. Genetic variability and population structure of endangered *Panax ginseng* in the Russian Primorye. *Chin. Med.* **2010**, *5*, 21. [CrossRef]
11. Reed, D.H.; Frankham, R. Correlation between Fitness and Genetic Diversity. *Conserv. Biol.* **2003**, *17*, 230–237. [CrossRef]
12. Reunova, G.; Koren, O.; Muzarok, T.I.; Zhuravlev, Y. Microsatellite Analysis of *Panax ginseng* Natural Populations in Russia. *Chin. Med.* **2014**, *5*, 231–243. [CrossRef]
13. Sinkovi, L.; Pipan, B.; Sinkovi, E.; Megli, V. Morphological Seed Characterization of Common (*Phaseolus* vulgaris L.) and Runner (*Phaseolus coccineus* L.) Bean Germplasm: A Slovenian Gene Bank Example. *Biomed Res. Int.* **2019**, *2019*, 13.
14. Jain, A.; Roy, A.K.; Kaushal, P.; Malaviya, D.R.; Zadoo, S.N. Isozyme Banding Pattern and Estimation of Genetic Diversity among Guinea Grass Germplasm. *Genet. Resour. Crop. Evol.* **2006**, *53*, 339–347. [CrossRef]
15. Kottawa-Arachchi, J.; Gunasekare, M.; Ranatunga, M.; Punyasiri, N.; Jayasinghe, L.; Karunagoda, R.P. Biochemical Characteristics of Tea (*Camellia* L. spp.) Germplasm Accessions in Sri Lanka: Correlation between Black Tea Quality Parameters and Organoleptic Evaluation. *Int. J. Tea Sci.* **2014**, *10*, 3–13.
16. Halász, J.; Kodad, O.; Galiba, G.M.; Skola, I.; Ercisli, S.; Ledbetter, C.A.; Hegedűs, A. Genetic variability is preserved among strongly differentiated and geographically diverse almond germplasm: An assessment by simple sequence repeat markers. *Tree Genet. Genome* **2019**, *15*, 12. [CrossRef]
17. Cho, C.-W.; Kim, Y.-C.; Rhee, Y.K.; Lee, Y.-C.; Kim, K.-T.; Hong, H.-D. Chemical composition characteristics of Korean straight ginseng products. *J. Ethn. Foods* **2014**, *1*, 24–28. [CrossRef]
18. Shin, B.-K.; Kwon, S.W.; Park, J.H. Chemical diversity of ginseng saponins from *Panax ginseng*. *J. Ginseng Res.* **2015**, *39*, 287–298. [CrossRef]
19. Li, M.-R.; Shi, F.-X.; Li, Y.-L.; Jiang, P.; Jiao, L.; Liu, B.; Li, L.-F. Genome-Wide Variation Patterns Uncover the Origin and Selection in Cultivated Ginseng (*Panax ginseng* Meyer). *Genome Biol. Evol.* **2017**, *9*, 2159–2169. [CrossRef]
20. An, H.; Park, J.-H.; Hong, C.E.; Sebastin, R.; Lee, Y.; Jo, I.-H.; Chung, J.-W. Genetic diversity and population structure of Chinese ginseng accessions using SSR markers. *J. Plant Biotechnol.* **2017**, *44*, 312–319. [CrossRef]
21. Gepts, P. The Use of Molecular and Biochemical Markers in Crop Evolution Studies. In *Evolutionary Biology*; Hecht, M.K., MacIntyre, R.J., Clegg, M.T., Eds.; Springer: Boston, MA, USA, 1993; Volume 27, pp. 51–94.
22. Crawford, D.J. Molecular markers for the study of genetic variation within and between populations of rare plants. *Opera Bot.* **1997**, *132*, 149–157.
23. Zhuravlev, Y.; Koren, O.; Reunova, G.D.; Artyukova, E.; Kozyrenko, M.; Muzarok, T.I.; Kats, I.L. Ginseng Conservation Program in Russian Primorye: Genetic Structure of Wild and Cultivated Populations. *J. Ginseng Res.* **2004**, *28*, 60–66.
24. Koren, O.; Potenko, V.V.; Zhuravlev, Y. Inheritance and Variation of Allozymes in *Panax ginseng* C.A. Meyer (*Araliaceae*). *Int. J. Plant Sci.* **2003**, *164*, 189–195. [CrossRef]

25. Reunova, G.D.; Kats, I.L.; Muzarok, T.I.; Nguen, C.T.; Dang, T.T.; Zhuravlev, I.N. Population Genetic Structure of Wild-Growing Ginseng (*Planax ginseng* CA Meyer) Assessed Using AFLP Markers. *Genetika* **2012**, *48*, 340–351. [PubMed]
26. Zhuravlev, Y.; Koren, O.; Reunova, G.D.; Muzarok, T.I.; Gorpenchenko, T.; Kats, I.L.; Khrolenko, Y. *Panax ginseng* natural populations: Their past, current state and perspectives. *Acta Pharmacol. Sin.* **2008**, *29*, 1127–1136. [CrossRef]
27. Zhuravlev, Y.; Reunova, G.D.; Kats, I.L.; Muzarok, T.I.; Bondar, A.A. Molecular variation of wild *Panax ginseng* C.A. Meyer (*Araliaceae*) by AFLP markers. *Chin. Med.* **2010**, *5*, 21. [CrossRef]
28. Choi, H.-I.; Kim, N.H.; Kim, J.H.; Choi, B.S.; Ahn, I.-O.; Lee, J.-S.; Yang, T.-J. Development of Reproducible EST-derived SSR Markers and Assessment of Genetic Diversity in *Panax ginseng* Cultivars and Related Species. *J. Ginseng Res.* **2011**, *35*, 399–412. [CrossRef]
29. Kalia, R.; Rai, M.; Kalia, S.; Singh, R.; Dhawan, A. Microsatellite markers: An overview of the recent progress in plants. *Euphytica* **2011**, *177*, 309–334. [CrossRef]
30. Xu, W.; Choi, H.-K.; Huang, L. State of Panax ginseng Research: A Global Analysis. *Molecules* **2017**, *22*, 1518. [CrossRef]
31. Govindaraj, M.; Vetriventhan, M.; Srinivasan, M. Importance of Genetic Diversity Assessment in Crop Plants and Its Recent Advances: An Overview of Its Analytical Perspectives. *Genet. Res. Int.* **2015**, *2015*, 431487. [CrossRef]
32. Esselink, G.D.; Nybom, H.; Vosman, B. Assignment of allelic configuration in polyploids using the MAC-PR (microsatellite DNA allele counting—peak ratios) method. *Appl. Genet.* **2004**, *109*, 402–408. [CrossRef]
33. Kamvar, Z.N.; Tabima, J.F.; Grünwald, N.J. Poppr: An R package for genetic analysis of populations with clonal, partially clonal, and/or sexual reproduction. *PeerJ* **2014**, *2*, e281. [CrossRef] [PubMed]
34. Peakall, R.; Smouse, P.E. GenAlEx 6.5: Genetic analysis in Excel. Population genetic software for teaching and research±an update. *Bioinformatics* **2012**, *28*, 2537–2539. [CrossRef] [PubMed]
35. Ivandic, V.; Hackett, C.A.; Nevo, E.; Keith, R.; Thomas, W.T.B.; Forster, B.P. Analysis of simple sequence repeats (SSRs) in wild barley from the Fertile Crescent: Associations with ecology, geography and flowering time. *Plant Mol. Biol.* **2002**, *48*, 511–527. [CrossRef] [PubMed]
36. Jombart, T. adegenet: A R package for the multivariate analysis of genetic markers. *Bioinformatics* **2008**, *24*, 1403–1405. [CrossRef]
37. Jombart, T.; Collins, C. A Tutorial for Discriminant Analysis of Principal Components (DAPC) Using adegenet 2.0.0. Available online: http://adegenet.r-forge.r-project.org/files/tutorial-dapc.pdf (accessed on 9 November 2019).
38. Pritchard, J.K.; Stephens, M.; Donnelly, P. Inference of population structure using multilocus genotype data. *Genetics* **2000**, *155*, 945–959.
39. Earl, D.A.; Vonholdt, B.M. STRUCTURE HARVESTER: A website and program for visualizing STRUCTURE output and implementing the Evanno method. *Conserv. Genet. Resour.* **2012**, *4*, 359–361. [CrossRef]
40. Evanno, G.; Regnaut, S.; Goudet, J. Detecting the number of clusters of individuals using the software STRUCTURE: A simulation study. *Mol. Ecol.* **2005**, *14*, 2611–2620. [CrossRef]
41. Lee, S.-S.; Lee, J.-H.; Ahn, I.-O. Characteristics of new cultivars in *Panax ginseng* CA Meyer. In Proceedings of the 9th International Symposium on Ginseng, Chungnam, South Korea, 25 September 2006.
42. Proctor, J.T.A.; Lee, J.C.; Lee, S.-S. Ginseng Production in Korea. *HortScience* **1990**, *25*, 746–750. [CrossRef]
43. FAO. *Korean Ginseng Agriculture System*; FAO: Rome, Italy, 2018.
44. Maxted, N.; Magos Brehm, J.; Kell, S. *Resource Book for Preparation of National Conservation Plans for Crop Wild Relatives and Landraces*; Food and Agriculture Organization of the United Nations: Rome, Italy, 2013.
45. Singh, N.; Wu, S.; Raupp, W.J.; Sehgal, S.; Arora, S.; Tiwari, V.; Vikram, P.; Singh, S.; Chhuneja, P.; Gill, B.S.; et al. Efficient curation of genebanks using next generation sequencing reveals substantial duplication of germplasm accessions. *Sci. Rep.* **2019**, *9*, 650. [CrossRef]
46. Van Hintum, T.J.L.; Knüpffer, H. Duplication within and between germplasm collections. *Genet. Res. Crop Evol.* **1995**, *42*, 127–133. [CrossRef]
47. Harris, A.M.; DeGiorgio, M. An Unbiased Estimator of Gene Diversity with Improved Variance for Samples Containing Related and Inbred Individuals of any Ploidy. *G3 Genes Genomes Genet.* **2017**, *7*, 671–691. [CrossRef] [PubMed]

48. Van Dan, N.; Ramchiary, N.; Choi, S.R.; Uhm, T.S.; Yang, T.-J.; Ahn, I.-O.; Lim, Y.P. Development and characterization of new microsatellite markers in *Panax ginseng* (C.A. Meyer) from BAC end sequences. *Conserv. Genet.* **2010**, *11*, 1223–1225. [CrossRef]
49. Pudovkin, A.I.; Zaykin, D.V.; Hedgecock, D. On the potential for estimating the effective number of breeders from heterozygote-excess in progeny. *Genetics* **1996**, *144*, 383–387. [PubMed]
50. Rasmussen, D.I. Sibling Clusters and Genotypic Frequencies. *Am. Nat.* **1979**, *113*, 948–951. [CrossRef]
51. Stoeckel, S.; Ggange, J.; Fernandez-manjarres, J.F.; Bilger, I.; Frascaria-lacoste, N.; Mariette, S. Heterozygote excess in a self-incompatible and partially clonal forest tree species—*Prunus avium* L. *Mol. Ecol.* **2006**, *15*, 2109–2118. [CrossRef]
52. Balloux, F. Heterozygote excess in small populations and the heterozygote-excess effective population size. *Evolution* **2004**, *58*, 1891–1900. [CrossRef]
53. Mark Welch, D.B.; Meselson, M. Evidence for the Evolution of Bdelloid Rotifers without Sexual Reproduction or Genetic Exchange. *Science* **2000**, *288*, 1211–1215. [CrossRef]
54. Hartl, D.L.; Clark, A.G.; Clark, A.G. *Principles of Population Genetics*; Sinauer Associates: Sunderland, MA, USA, 1997; Volume 116.
55. Brown, A.; Briggs, J. Sampling strategies for genetic variation in ex situ collections of endangered plant species. In *Genetics and Conservation of Rare Plants*; Falk, D., Holsinger, K., Eds.; Oxford University Press: Oxford, UK, 1991; pp. 99–119.
56. Reyes-Valdés, M.H.; Burgueño, J.; Singh, S.; Martínez, O.; Sansaloni, C.P. An informational view of accession rarity and allele specificity in germplasm banks for management and conservation. *PLoS ONE* **2018**, *13*, e0193346. [CrossRef]
57. Kalinowski, S.T. Counting alleles with rarefaction: Private alleles and hierarchical sampling designs. *Conserv. Genet.* **2004**, *5*, 539–543. [CrossRef]
58. Li, S.; Li, J.; Yang, X.-L.; Cheng, Z.; Zhang, W.-J. Genetic diversity and differentiation of cultivated ginseng (*Panax ginseng* C. A. Meyer) populations in North-east China revealed by inter-simple sequence repeat (ISSR) markers. *Genet. Res. Crop Evol.* **2011**, *58*, 815–824. [CrossRef]
59. Campoy, J.A.; Lerigoleur-Balsemin, E.; Christmann, H.; Beauvieux, R.; Girollet, N.; Quero-García, J.; Dirlewanger, E.; Barreneche, T. Genetic diversity, linkage disequilibrium, population structure and construction of a core collection of *Prunus avium* L. landraces and bred cultivars. *BMC Plant Biol.* **2016**, *16*, 49. [CrossRef] [PubMed]
60. Deperi, S.I.; Tagliotti, M.E.; Bedogni, M.C.; Manrique-Carpintero, N.C.; Coombs, J.; Zhang, R.; Douches, D.; Huarte, M.A. Discriminant analysis of principal components and pedigree assessment of genetic diversity and population structure in a tetraploid potato panel using SNPs. *PLoS ONE* **2018**, *13*, e0194398. [CrossRef] [PubMed]
61. Rosyara, U.R.; De Jong, W.S.; Douches, D.S.; Endelman, J.B. Software for Genome-Wide Association Studies in Autopolyploids and Its Application to Potato. *Plant Genome* **2016**, *9*. [CrossRef] [PubMed]
62. NIHHS. *Distribution Status of Panax ginseng in China and Comparative Analysis on the Characteristics of Korean and China ginseng*; RDA: Jeonju, South Korea, 2016.
63. Horticulture Business Division. *2017 Ginseng Data Sheet*; Ministory of Agriculture Food and Rural Affairs: Sejong, South Korea, 2018.
64. Yun, Y.M. A Study on the Production and Distribution of Domestic Korea Ginseng Seeds—A Case Study of Korea Ginseng Seeds in Chungnam Province. Master's Thesis, Kongju National University, Yesan, South Korea, 2015.

agronomy

MDPI

Article

Multi-Level Characterization of Eggplant Accessions from Greek Islands and the Mainland Contributes to the Enhancement and Conservation of this Germplasm and Reveals a Large Diversity and Signatures of Differentiation between both Origins

Pietro Gramazio [1,2,†], Eleni Chatziefstratiou [3,4,†], Constantinos Petropoulos [5], Vasileia Chioti [1,6], Photini Mylona [7], George Kapotis [4], Santiago Vilanova [1], Jaime Prohens [1,*] and Vasileios Papasotiropoulos [3,*]

1 Instituto de Conservación y Mejora de la Agrodiversidad Valenciana, Universitat Politècnica de València, Camino de Vera 14, 46022 Valencia, Spain; piegra@upv.es (P.G.); sileia_chioti@yahoo.gr (V.C.); sanvina@upvnet.upv.es (S.V.)

2 Faculty of Life and Environmental Sciences, University of Tsukuba, 1-1-1 Tennodai, Tsukuba 3058572, Japan; gramazio.pietro.gn@u.tsukuba.ac.jp

3 Department of Agriculture, University of Patras, Theodoropoulou Terma, 27200 Amaliada, Greece; chatzieustratiou@yahoo.gr

4 Department of Crop Science, University of Patras, Nea Ktiria, 30200 Mesolonghi, Greece; kapotis@upatras.gr

5 Department of Mathematics, University of Patras, University Campus, 26504 Patras, Greece; costas@math.upatras.gr

6 Department of Biology, University of Patras, University Campus, 26504 Patras, Greece

7 Institute of Plant Breeding and Genetic Resources, HAO-DEMETER, Thermi, 57001 Thessaloniki, Greece; phmylona@nagref.gr

* Correspondence: jprohens@btc.upv.es (J.P.); vpapasot@upatras.gr (V.P.)

† These authors contributed equally to this work.

Received: 11 November 2019; Accepted: 11 December 2019; Published: 13 December 2019

Abstract: Crop landraces are found in many inhabited islands of Greece. Due to the particularity of environment and isolation from the mainland, Greek islands represent a natural laboratory for comparing the diversity of landraces from the islands with those of the Greek mainland. A collection of 36 Greek eggplant landraces and traditional cultivars from the mainland and the islands has been phenotypically and genetically characterized using 22 morphological descriptors and 5 SSR markers. The mineral composition (K, Mg, Cu, Fe, Mn, Zn) of fruits was also determined. The objectives of this study include the multi-level characterization of eggplant local landraces and the comparison of diversity among accessions from the Greek mainland and the islands. Characterization of eggplant landraces will contribute to the enhancement and prevention of genetic erosion in this local group and will provide a resource for future investigation and breeding. PCA analysis of morphological traits explained 45.4% of the total variance revealing the formation of two clusters, one with most of the island accessions, and another with most of the mainland ones. The SSR markers used exhibited high average values for the number of alleles/locus (4.6), expected heterozygosity (0.60) and PIC (0.55), while the observed heterozygosity was low (0.13). Both STRUCTURE and PCoA analyses based on SSR data revealed two genetic clusters, one made up mainly by the mainland accessions, while the other one was mainly made up by the island accessions. Although there was considerable variation among the landraces for the concentration of minerals studied, only average Mg concentration was significantly different between mainland and island accessions. Based on our data, the Greek eggplant landraces present considerable morphological and genetic diversity with some differentiation signatures between the island and the mainland accessions. Our results have implications for conservation of Greek landraces and suggest that Greece might be considered as part of a secondary center of diversity for eggplant in the Mediterranean basin.

Keywords: *Solanum melongena*; germplasm; Greece; islands; mainland; phenotyping; genotyping; mineral composition; landraces

1. Introduction

Greece lies at the southernmost part of the Balkan peninsula, at a crossroads between Europe, Asia, and Africa. The country is noted for its high species diversity, the extensive coverage of natural and semi-natural habitats, and many island complexes in the Archipelagos of Aegean Sea and the Ionian Sea [1]. Islands are natural laboratories for the study of the evolution and ecology of wild plants and endemisms [2–4]. Therefore, due to their isolation and special environmental features, islands are also considered as sites of special interest to prospect and conserve plant genetic resources [5,6]. The Greek archipelago includes ca. 7600 islands and islets in the Aegean Sea (including Crete) and ca. 300 islands and islets in the Ionian Sea [7]. Among them, 116 are inhabited, whereas only 79 have over 100 permanent residents [8]. The Aegean Archipelago is one of the largest archipelagos in the world, exhibiting high environmental and topographical heterogeneity, complex geological and palaeogeographical history, as well as high diversity and endemisms, rendering it an ideal stage for biodiversity and biogeographical studies [1,9]. Among the 13 phytogeographical regions proposed by Strid [10] to occur in Greece, five were identified in the Aegean [1]. The Ionian archipelago forms a distinct phytogeographical region, hosting 2027 plant taxa, 1827 of which are native and 89 are Greek endemics [5,11–13].

Landraces represent a significant part of agricultural biodiversity [14]. They are characterized by high genetic diversity, local adaptation, resilience to pathogens, and good organoleptic characteristics, although they often have low yield and display a lack of uniformity [15–17]. Over time, landraces have been gradually displaced from market-driven production, due to the consumers' demand for local high-quality plant products, landraces are currently recalling an increased interest [18,19]. With the shift to modern agriculture, Greece suffered a dramatic loss of its traditional agricultural germplasm, which was displaced by modern varieties produced by local breeders or imported [5,20]. Nowadays, landraces have regained an increased interest [18,19]. Due to the geographical morphology of Greece (almost 70% is hilly and/or mountainous) and the presence of small-size farms, landraces are still cultivated, especially by farmers in villages in remote regions and isolated islands, who maintain their seeds and use them mainly for their own consumption, and, to a lesser extent, for market purposes [5,13,19]. Therefore, these areas act as local in situ conservation spots and islands represent one of the most important sources of landraces and hotspots of agricultural biodiversity in Greece [5,20,21].

Eggplant (*Solanum melongena* L.) is the fifth most produced vegetable worldwide and has experienced a dramatic increase in yield and production in the last decade [22]. It was introduced in Greece around the 13th century and although initially used as an ornamental plant, it is now playing an important role in the Greek gastronomy, as an integral part of the Mediterranean diet. Besides commercial F1 hybrids and imported varieties, there are several traditional cultivars, e.g., Tsakoniki, and Santorini, which are adapted to the local environment, having excellent texture and cooking quality [23]. Apart from commercial varieties, several local landraces have been found during expeditions in the islands [5,24] or maintained in Greek Gene Bank (Hellenic Agricultural Organization Demeter, Thessaloniki, Greece). However, the diversity of Greek accessions has been barely explored, and so far, no studies have focused on the comparison of accessions from the mainland and the islands of Greece have been reported. Significant differences in genetic diversity of several plant species have been revealed when island populations are compared with mainland ones [25–28]. Factors such as geographic isolation, island size, plant breeding system, or life form, and in the case of crop species, human factors, are shaping the difference levels [26,29–31]. Characterization of the genetic diversity and population structure of landraces and local germplasm is very important to implement the most appropriate strategies for their collection, management, efficient conservation, and utilization as a source of variation in breeding schemes [32]. Several studies have focused on the characterization of eggplant genetic resources, including the cultivated species, commercial cultivars, and their most recognized wild relatives, from different regions of origin in order to identify desired genotypes for use in eggplant breeding programs [33–36]. For instance, Liu et al. [36] examined 287 worldwide accessions for genetic diversity and population structure analysis, whereas Acquadro et al. [33] used high-throughput genotyping to assess the genetic relationships of brinjal, gboma, and scarlet eggplant complexes, which represent taxa belonging to the eggplant's primary secondary and tertiary gene pools. Moreover, Cericola et al. [34] used a combined marker-based and morphological approach to assess genetic diversity and illuminate the genetic relationship between "Occidental" and "Oriental" eggplant germplasm groups, while Hurtado et al. [35] evaluated the phenotypic and DNA-based diversity present in a collection of accessions sampled from three geographically separated centers of diversity.

Eggplant is an important source of plant-derived nutrients, valued for its composition in phytochemicals and especially minerals such as P, K, Ca, and Mg [37,38]. Raigón et al. [39] and Arivalagan et al. [38–40] studied the mineral composition and the nutritional value of eggplant landraces and germplasm accessions from different regions in order to identify mineral-rich germplasm for breeding purposes. Considerable differences were found in the mineral composition among varieties and germplasm accessions, revealing the existence of ample variation, which can be exploited for the selection of germplasm for nutritionally improved characteristics.

In this study, we report the assessment of the morphological and genetic diversity of eggplant germplasm from the Greek mainland and islands. In addition, the mineral content has been determined to obtain a more comprehensive view regarding the nutritional value of those accessions. The overall aim of this study is to acquire information for the efficient management and conservation of this valuable genetic material, for the identification of traits present in landraces that can be exploited for breeding purposes, and to compare the levels of diversity among the Greek mainland and island eggplant germplasm.

2. Materials and Methods

2.1. Plant Material and Cultivation Conditions

Thirty-six accessions, including landraces and cultivars from the mainland (16) and the islands (20) of Greece were selected to depict the Greek eggplant germplasm diversity (Table 1) and their geographical distribution (Figure 1). All materials are maintained at the National Greek Gene Bank (GGB; Institute of Genetic Resources and Plant Breeding Thermi, Greece), except I-6, I-7, and I-8, which are conserved at the Department of Agriculture of the University of Patras (Greece). Seeds were germinated in Petri dishes, following the Ranil et al. [41] protocol, and were subsequently transferred to seedling trays in a climatic chamber under a photoperiod and temperature regime of 16 h light (28 °C): 8 h dark (20 °C). One week later, the plantlets were transferred and kept in a heated greenhouse until transplantation. Nine plants per accession were grown in an open field plot at the University of Patras campus in Messolonghi, Greece (GPS coordinates: latitude, 38°36′ N; longitude, 21°47′; 1.5 m a.s.l.) following a completely randomized block design experiment with 3 blocks and 3 plants per accession/block and spaced at 0.5 m within the row and 1 m between rows. Plants were pruned and trellised and drip-fertigated. Standard horticultural practices for eggplant production in this area for pest and weed control were performed. Accession ANP-025/07 (I-1) displayed an abnormal development at the seedling stage and was discarded for the morphological and mineral characterization, although it was kept for the genotyping.

Table 1. Names/collection number, geographic origin, collection site, geographical coordinates and status of the materials used for the study of morphological and molecular variation (SSRs) in a collection of Greek eggplants.

Accession Code	Study Code	Territory	Geographic Region	Collection Site	Longitude	Latitude	Altitude (m)	Status
ANP-025/07	I-1	Island	Aegean Sea	Amorgos, Aegiali	36°90′ N	25°99′ E	298	Landrace
X-034/06	I-2	Island	Aegean Sea	Chios, Ag. Georgios Sikousis	38°19′ N	26°03′ E	369	Landrace
IS-031/07	I-3	Island	Aegean Sea	Ikaria, Droutsoulas	37°36′ N	26°11′ E	410	Landrace
LKK-094/07	I-4	Island	Aegean Sea	Kos, Antimacheia	36°48′ N	27°05′ E	133	Landrace
X-116/06	I-5	Island	Aegean Sea	Lemnos, Kontopouli	39°55′ N	25°20′ E	40	Landrace
White Leros	I-6	Island	Aegean Sea	Leros Island	37°09′ N	26°49′ E	32	Cultivar
Wide Purple	I-7	Island	Aegean Sea	Leros Island	37°08′ N	26°51′ E	6	Landrace
Long Purple	I-8	Island	Aegean Sea	Leros Island	37°08′ N	26°51′ E	6	Landrace
LKK-008/07	I-9	Island	Aegean Sea	Leros, Kamara	37°09′ N	26°49′ E	29	Landrace
M-039/06	I-10	Island	Aegean Sea	Lesvos, Paleokipos	39°05′ N	26°45′ E	58	Landrace
M-069/06	I-11	Island	Aegean Sea	Lesvos, Keramia	39°12′ N	26°42′ E	14	Landrace
MFS-030/07	I-12	Island	Aegean Sea	Milos, Mitakas	36°44′ N	24°29′ E	50	landrace
ANP-180/07	I-13	Island	Aegean Sea	Naxos, Agia Anna	37°04′ N	25°21′ E	4	Landrace
ANP-215/07	I-14	Island	Aegean Sea	Paros, Lefkes	37°07′ N	25°12′ E	210	Landrace
HL-027/07	I-15	Island	Aegean Sea	Santorini, Vourvoulos	36°26′ N	25°26′ E	116	Landrace
SAS-078/07	I-16	Island	Aegean Sea	Skopelos, Chora	39°07′ N	23°43′ E	41	Landrace
ATS-110/06	I-17	Island	Aegean Sea	Syros Island, Chrousa	37°24′ N	24°55′ E	320	Landrace
GRC-002/08	I-18	Island	Ionian Sea	Corfu, Skripero	39°42′ N	19°46′ E	139	Landrace
HL-237/07	I-19	Island	Crete	Iraklion, Moni Savvathianon	35°37′ N	25°00′ E	467	Landrace
IK-082/06	I-20	Island	Ionian Sea	Ithaki Island, Perachori	38°34′ N	20°71′ E	343	Landrace
K-153/06	M-1	Mainland	Macedonia	Grevena, Pontini	40°04′ N	21°40′ E	819	Landrace
EMI	M-2	Mainland	Macedonia	IPGRB/HAO DEMETER	40°32′ N	22°59′ E	19	Cultivar
Lagkada	M-3	Mainland	Macedonia	IPGRB/HAO DEMETER	40°32′ N	22°59′ E	19	Cultivar
KD-053/07	M-4	Mainland	Macedonia	Kavala, Platanotopos	40°50′ N	24°03′ E	244	Landrace
F-154/06	M-5	Mainland	Macedonia	Kastoria, Ampelokipoi	40°46′ N	21°31′ E	638	Landrace
SK-044/066	M-6	Mainland	Macedonia	Kilkis, Eptalofos	41°00′ N	23°08′ E	451	Landrace
K-054/06	M-7	Mainland	Macedonia	Kozani, Anarachi	40°29′ N	21°34′ E	727	Landrace
VG-011/083	M-8	Mainland	Macedonia	Serres	41°05′ N	23°35′ E	45	Landrace
SK-056/06	M-9	Mainland	Macedonia	Serres, Platanakia	41°17′ N	22°56′ E	313	Landrace
Scoutari	M-10	Mainland	Macedonia	Serres/Skoutari	41°01′ N	23°31′ E	14	Cultivar
P-175/06	M-11	Mainland	Peloponnese	Lakonia, Lyra	36°38′ N	22°57′ E	400	Landrace
Tsakoniki	M-12	Mainland	Peloponnese	Leonidio	37°10′ N	22°51′ E	40	Cultivar
P-084/06	M-13	Mainland	Peloponnese	Messinia, Kakana	37°18′ N	21°44′ E	136	Landrace
T-099/06	M-14	Mainland	Thessaly	Karditsa, Neo Ikonio	39°27′ N	22°21′ E	107	Landrace
T-527/06	M-15	Mainland	Thessaly	Trikala, Megarxis	39°36′ N	21°45′ E	142	Landrace
GRC 1430/04	M-16	Mainland	Epirus	Ioannina, Pogoni Vasiliko	40°00′ N	20°35′ E	805	Landrace

Figure 1. Distribution map of the 36 eggplant Greek accessions from the mainland and islands used in this study. Study codes as in Table 1.

2.2. Morphological and Mineral Composition Characterization and Data Analysis

Plants were characterized using 22 morphological descriptors for plant (4), leaf (6), flower (1), and fruit (11) traits obtained from the EGGNET descriptor list, of the EU-RESGEN PL 98-113 program [42,43], Biodiversity International (formerly International Board for Plant Genetic Resources) [44], and Kumar et al. [45] descriptors lists (Table 2). For all the traits, four measures were made per plant, except for plant traits, in which one measurement was made. For mineral composition analysis, one plant per block and accession and between one and three fruits per plant (depending on fruit size) were collected and analyzed at commercial maturity. Fruits were washed, peeled, cut into pieces, and bulked. Subsequently, the bulks were weighed, dried at 80 °C until constant weight, and powdered with a mechanical grinder (T 25 digital ULTRA-TURRAX®, IKA®-Werke GmbH & Co. Staufen, Germany). A total of 0.5 g of the dried samples were calcined in a muffle furnace (Thermconcept) at 550 °C for 5 h. The ashes were then dissolved in 10 mL of HCl 1M, filtered and the extract was brought to 50 mL with distilled water [46]. K was analyzed by flame photometry (Sherwood Model 410, Cambridge, UK), while the rest of the minerals (Mg, Fe, Cu, Mn and Zn) were analyzed by atomic absorption spectrophotometry (AAS) using a Thermo Elemental (SOLAAR AA Spectrometers, Cambridge, UK) spectrometer.

Averages for accession were used to calculate means, ranges, and coefficient of variation for mainland and island accessions using IBM® SPSS® Statistics 25 (IBM, Armonk, NY, USA). Significance of differences between means of mainland and islands was performed using the appropriate test at the significance level of $p < 0.05$. The results were analyzed using both parametric (t-test) and nonparametric (Wilcoxon-Mann-Whitney) methods, the latter in the cases of non-normality. A principal component analysis (PCA) was performed using the function prcomp of the package stats (v3.6.1) in R [47] and represented graphically using the package ggplot2 [48].

Table 2. Descriptors used for phenotyping of Greek eggplant accessions and their range/scale. All descriptors based on IBPGR descriptors [44], except NOLFF [42,43] and DSFS [45].

Descriptor Code	Descriptor Name	Descriptor Scale/Unit
	Plant descriptors	
PGH	Plant Growth Habit	3–7 (3 = upright; 7 = prostrate)
PH	Plant Height	cm
NOLFF	Number of Leaves to First Flower	number
DSFS	Days Since Fruit Set	number
	Leaf descriptors	
LPL	Leaf Petiole Length	cm
LBLe	Leaf Blade Length	cm
LBW	Leaf Blade Width	cm
LBLo	Leaf Blade Lobing	1–9 (1 = very weak; 9 = very strong)
LSS	Leaf Surface Shape	1–9 (1 = flat; 9 = very convex or bullate)
LP	Leaf Prickles	0–9 (0 = none; 9 = more than 20)
	Flower descriptors	
NOFPI	Number of Flowers Per Inflorescence	number
	Fruit descriptors	
FCP	Fruit Calyx Prickles	0–9 (0 = none; 9 = more than 30)
FPL	Fruit Pedicel Length	cm
FL	Fruit Length	cm
FB	Fruit Breadth	mm
FLBR	Fruit Length to Breadth Ratio	1–9 (1 = broader than long; 9 = several times longer than broad)
FS	Fruit Shape	3–7 (Position of widest part of fruit: 3 = $\frac{1}{4}$ way from base to tip; 5 = $\frac{1}{2}$ way from base to tip; 7 = $\frac{3}{4}$ way from base to tip)
FC	Fruit Curvature	1–9 (1 = none, fruit straight; 9 = U shaped)
FW	Fruit Weight	g
FPC	Fruit Predominant Color	1–9 (1 = milk white; 9 = black)
FAC	Fruit Additional Color	1–9 (1 = milk white; 9 = black)
FFC	Fruit Flesh Color	3–7 (3 = white; 7 = green)

2.3. SSR Characterization and Data Analysis

Total genomic DNA was extracted for each accession from nearly 100 mg of young leaf tissue [49]. DNA quality (230/260 and 260/280 nm ratios) and concentration were measured using a NanoDrop ND-1000 spectrophotometer (Thermo Fisher Scientific, Waltham, MA, USA) and a Qubit® 2.0 Fluorometer (Applied Biosystems, Waltham, MA, USA), respectively, while DNA integrity was evaluated by agarose gel electrophoresis (0.8%). Three genomic SSRs developed by Vilanova et al. [50] and two by Nunome et al. [51] were used due to their high polymorphism for the genetic characterization of the samples, which were organized in one multiplex reaction according to the expected allele size range (Table 3). The PCR amplification was performed in a total volume of 12 μL including 7.21 μL water, 1.2 μL 1× PCR buffer, 0.6 μL MgCl2 (50 mM), 0.24 μL dNTPs (10 mM), 0.06 μL forward primer with M13 tail (10 μM), 0.24 μL fluorochrome (FAM, VIC, NED, and PET, 10 μM), 0.3 μL reverse primer (10 μM), 0.15 μL Taq DNA Polymerase (5 U/μL), and 2 μL DNA template (20 ng/μL). The PCR was performed following the program: 95 °C for 3 min for a denaturation step, 30 cycles of 30 s at 95 °C followed by 30 s at 65 °C and 30 s at 72 °C and finally 72 °C for 5 min for the last extension step. The PCR products were subsequently diluted in formamide and sequenced by capillary electrophoresis through an ABI PRISM 3100-Avant sequencer (Applied Biosystems, Waltham, MA, USA) using a 500 LIZ GeneScan size standard (Applied Biosystems, Waltham, MA, USA). The fragments were analyzed using the GeneScan software (Applied Biosystems, Waltham, MA, USA) to obtain the electropherograms and the alleles were identified with the Genotyper DNA Fragment Analysis software (Applied Biosystems, Waltham, MA, USA).

Table 3. Characteristics of the SSR markers used for the molecular characterization of Greek eggplant accessions.

SSR Locus	Motif	Forward Primer and Reverse Primer (5′→3′)	Size Range (bp)	T° Annealing	Dye	Source
csm4	(GA)15	GCGTACCAATTCTAACCACAAG GTAATCCGCTTCCCATTTCTC	238–254	60	PET	Vilanova et al. [50]
csm27	(GA)23	TGTTTGGAGGTGAGGGAAAG TCCAACTCACCGGAAAAATC	193–210	60	VIC	Vilanova et al. [50]
csm32	(AG)23	TCGAAAGTACAGCGGAGAAAG GGGGGTTTGATTTTCATTTTC	248–254	60	NED	Vilanova et al. [50]
emi02c21	(AC)13A(TA)4	TGTGAGGAGAAGAATCAGAGGATCA CGCGACTAAGTTTTGTTCCTGAAA	126–136	60	VIC	Nunome et al. [51]
eme11f04	(TC)16	ACCCCCAAATCAAATCATTTACCC GGCATGGTTAGGGTTTTTAGCGTT	88–100	60	FAM	Nunome et al. [51]

The SSRs analysis was performed with the software PowerMarker [52] and GenAlEx 6.503 [53]. For each marker, the following parameters were calculated: number of alleles (Na), major allele frequency (f), number of effective alleles (Ne), number of genotypes (Ng), polymorphic information content (PIC) that was calculated using the following formula PIC = $1 - \sum_{i=1}^{n} p_i^2 - \sum_{i=1}^{n-1} \sum_{j=i+1}^{n} 2p_i^2 p_j^2$ (where n is the total number of alleles detected, p_i is the frequency of the ith allele, and p_j is the frequency of the jth allele) [54], observed heterozygosity (H_o) calculated as the number of heterozygous alleles/number of alleles, expected heterozygosity (H_e) calculated as $H_e = 1 - \sum_{i=1}^{n} p_i^2$ (where p_i is the frequency of the ith allele) [55] and the inbreeding coefficient (F) calculated as F = 1 − (H_o/H_e) [56]. Molecular analysis of variance (AMOVA) among groups (mainland vs. islands), among accessions within a group, and within individuals was performed with 999 permutations ($\alpha = 0.05$) using GenAlEx 6.503 software [53]. Correlations between morphological and SSR distance matrices were investigated using a Mantel test ($\alpha = 0.05$) with 999 permutations using GenAlEx 6.503 software [53].

Principal coordinates analysis (PCoA) was performed using the function dudi.pco of the package ade4 (v1.7-13) in R [47] and represented graphically using the package ggplot2 [48]. A model-based Bayesian structure implemented in the software STRUCTURE (version 2.3.4) [57] was used to estimate the population structure. Twenty runs of STRUCTURE were performed by setting the number of clusters (K) from 1 to 10 with a length of the burn-in period of 500,000 steps followed by 150,000 Monte Carlo Markov chain (MCMC) replicates, assuming an admixture model and uncorrelated allele frequencies. No prior knowledge of the population of origin was introduced. The ΔK method [58] was used to identify the most likely number of clusters (K) using STRUCTURE HARVESTER 0.6.94 software [59]. Each accession was assigned to its corresponding group based on maximum membership probability, as indicated by Remington et al. [60].

3. Results

3.1. Morphological Characterization

A wide diversity was observed for the morphological traits in the germplasm collection of Greek eggplant (Table 4, Table S1). For traits measured in a quantitative scale, accession values encompassing a broad range of the scale were observed for most of the traits, except for LBLo (3.00 to 5.00), LP (0.00 to 0.92), FS (4.00 to 5.33), and FC (1.00 to 5.00). A wide variation was also observed for the quantitative traits, especially for PH, LPL, NOFPI, FPL, FL, FB and FW. Although the ranges of variation between accessions from the mainland and islands overlapped, island accessions showed a broader coefficient of variation (CV) for all the quantitative traits, except for NOLFF and LBW. However, significant ($p < 0.05$) differences between the two groups were found only for LBW, LBLo, FCP, FB, FLBR, FC, and FW.

The orthogonal transformation by PCA of the morphological traits revealed that the total variation was explained by 22 principal components (PCs), although the first five PCs explained altogether 71.5% (Supplementary data S2). The bi-dimensional representation of the first two PCs revealed that the PC1, which accounted for 32.1% of the variation, was able to separate the island from the mainland accessions (Figure 2A). In fact, most of the island accessions (13 out of 19) exhibited positives PC1 values, while most of the mainland accessions showed negative PC1 values (14 out of 16). Similarly, most of the island accessions (15 out of 19) displayed positive values for the PC2, which accounted for 13.3% of the variation, while mainland accessions (11 out of 16) displayed negative PC2 values. Interestingly, although the accessions from the same geographical area fell relatively close in the PCA, like the island accessions I-6, I-7, I-8, and I-9 from Leros or I-10 and I-11 from Lesvos and the mainland accessions M-8, M-9, and M-10 from Serres, they showed a certain degree of morphological diversity. No high correlation absolute values were found between the morphological traits and the first two PCs (Table S2). However, moderate to weak PC1 positive correlations were found for FB and FW and negative correlation for FLBR and FC, while weak PC2 positive correlations were found for LBW, LPL, LBLe, and NOFPI, and no negative PC2 correlations were found (<-0.3).

Table 4. Mean, range (between brackets), and coefficient of variation (CV %) for the morphological and mineral descriptors studied in eggplant accessions of the Greek islands (n = 19) and the mainland (n = 16) and significance of mean differences among these two groups ($p < 0.05$).

	Mainland		**Islands**		
	Mean (Range)	**CV (%)**	**Mean (Range)**	**CV (%)**	***p*-Value**
		Plant descriptors			
PGH	4.36 (3.00–6.50)	23.70	5.09 (3.00–7.00)	21.50	0.052
PH	80.59 (56.83–106.00)	16.30	71.72 (46.50–106.17)	21.90	0.083
NOLFF	4.97 (4.00–6.67)	13.30	5.41 (4.20–6.83)	12.50	0.060
DSFS	50.51 (42.83–58.33)	8.20	48.26 (41.00–55.67)	9.10	0.129
		Leaf descriptors			
LPL	9.83 (6.79–12.29)	17.60	10.22 (6.53–14.63)	19.40	0.541
LBLe	31.51 (27.50–36.97)	8.40	31.4 (25.57–37.48)	10.50	0.918
LBW	20.43 (16.35–25.62)	11.50	22.14 (18.16–25.63)	9.80	**0.032** *
LBLo	3.38 (3.00–5.00)	23.80	4.37 (3.00–5.00)	21.90	**0.004** *
LSS	4.88 (3.00–7.00)	27.90	5.42 (3.00–7.00)	19.80	0.203
LP	0.18 (0.00–0.92)	126.40	0.21 (0.00–0.92)	125.70	0.892

Table 4. *Cont.*

	Mainland		Islands		
	Mean (Range)	**CV (%)**	**Mean (Range)**	**CV (%)**	***p*-Value**
	Flower descriptors				
NOFPI	1.15 (1.00–1.40)	11.10	1.22 (1.00–1.60)	16.90	0.388
	Fruit descriptors				
FCP	1.51 (0.00–4.00)	83.90	3.32 (0.33–7.00)	57.90	**0.003** *
FPL	6.33 (4.91–8.22)	16.70	6.24 (3.58–9.72)	32.10	0.840
FL	18.46 (14.50–22.32)	13.00	18.4 (11.2–43.85)	42.80	0.246
FB	61.93 (50.91–87.78)	16.20	85.1 (52.75–110.76)	20.30	**0.000** *
FLBR	7.49 (5.00–8.50)	11.50	5.2 (1.33–8.33)	44.40	**0.002** *
FS	4.69 (4.00–5.33)	7.00	4.36 (3.33–5.00)	12.20	0.106
FC	3.3 (1.00–5.00)	32.30	1.81 (1.00–4.00)	53.70	**0.000** *
FW	246.33 (177.32–368.40)	19.70	329.64 (215.66–500.67)	20.80	**0.000** *
FPC	6.52 (2.00–8.00)	22.10	5.83 (1.00–8.17)	40.50	0.550
FAC	4.31 (1.00–7.50)	48.00	4.78 (1.00–8.17)	48.20	0.527
FFC	4.73 (3.00–5.33)	15.10	4.45 (3.00–6.00)	20.60	0.218
	Minerals				
K (mg/g DW)	25.05 (20.07–31.50)	13.21	23.14 (19.80–29.00)	11.58	0.073
Mg (mg/g DW)	2.22 (1.96–2.46)	6.75	2.05 (1.63–2.43)	12.19	**0.015** *
Cu (mg/Kg DW)	9.56 (2.91–18.14)	47.91	7.65 (2.35–26.93)	69.67	0.278
Fe (mg/Kg DW)	31.8 (25.14–43.94)	14.68	31.48 (22.02–41.84)	15.78	0.850
Mn (mg/Kg DW)	12.25 (9.61–13.67)	8.82	11.74 (9.79–13.84)	9.71	0.198
Zn (mg/Kg DW)	49.75 (30.30–77.26)	26.85	49.39 (29.53–76.90)	31.14	0.942

* indicates a significant difference among islands and the mainland accessions at *p*-value < 0.05.

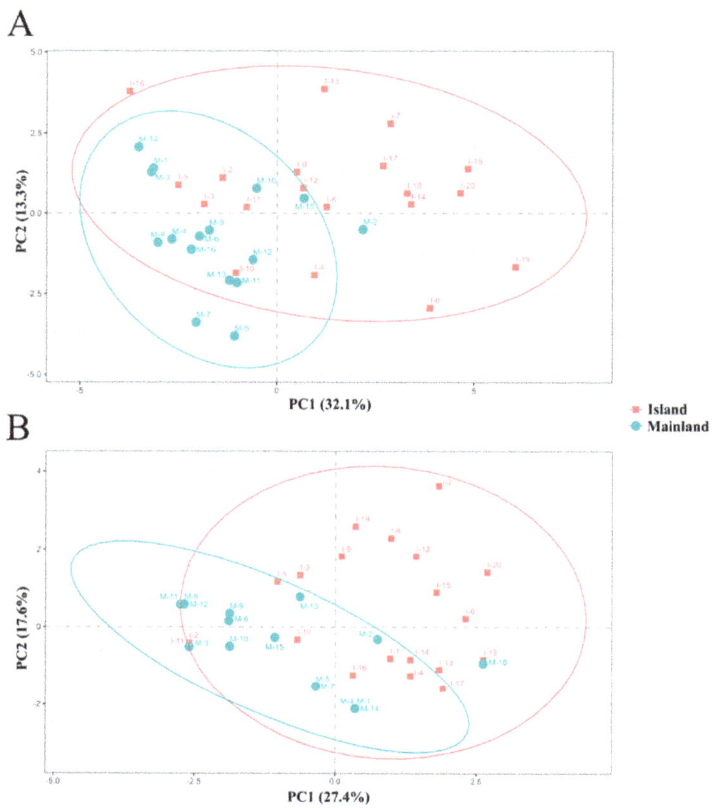

Figure 2. Principal component analysis (PCA) similarities based on the morphological characterization of 35 eggplant accessions used in this study (**A**) and principal coordinates analysis (PCoA) based on the genotyping characterization of all the accessions used in this study from the mainland (16) and island (20) Greek areas (**B**). The first and second principal coordinates (PC) are displayed. Study codes as in Table 1.

3.2. Concentration of Minerals

The concentrations of the macrominerals K and Mg were much higher than those of the microminerals Cu, Fe, Mn and Zn (Table 4). As occurred for morphological traits, a wide variation was observed among the accessions for mineral composition, with ranges of variation in the collection of over 10-fold for Cu, and ranges of the coefficient of variation of up to 67.2% for Cu in the accessions from the islands. However, for all traits, an overlap of the values was observed for the six minerals analyzed among the two groups, and significant differences between averages of mainland and insular groups were observed only for Mg, with slightly larger values in the mainland group.

3.3. SSR Characterization

All five SSRs were polymorphic and amplified between three (em311f04) and seven (csm4) alleles totaling 23 alleles, resulting in an average of 4.6 alleles per SSR locus (Table 5, Table S3). The major allele frequency was, on average, slightly higher than 0.5, ranging between 0.31 (csm4) and 0.65 (eme11f04). The average number of effective alleles (Ne; 2.74) was lower than the number of alleles (Na; 4.60), ranging from 1.91 (eme11f04) to 4.41 (csm4), while the number of genotypes (Ng) was higher than Na, ranging from 4 (emi02c21) to 10 (csm4), except emi02c21, which showed the same number of alleles and genotypes, indicating that all loci were in homozygosis. The PIC value had an average value of 0.55, with a range between 0.39 (emel1f04) and 0.75 (csm4). The H_o showed low values with an average of 0.13, ranged from 0 (emi02c21) to 0.17 (csm27), much lower than those of H_e (0.60 on average), which ranged from 0.47 (emel1f04) to 0.77 (csm4). On the contrary, F values were high, ranging from 0.66 (emel1f04) to 1 (emi02c21). The comparison between mainland and island accessions revealed that the latter presented generally higher genetic diversity (Table 5). In this respect, mainland accessions showed a lower number of alleles compared to the island accessions (3.2 and 4.2, respectively), as well as a lower number of effective alleles (2.22 vs. 2.79), genotypes (3.8 vs. 5.6), and PIC (0.42 vs. 0.53) but a higher frequency of the major allele (0.60 vs. 0.48). Average values for H_o and H_e were also much lower in the mainland accessions (0.03 and 0.48) than in the islands' accessions (0.20 and 0.61), while F values were much higher in the mainland accessions (0.92) than in those from the islands (0.68).

Table 5. Diversity statistics of SSR markers for all the Greek eggplant accessions and for the island and the mainland groups. Diversity statistics evaluated include: number of alleles (Na), major allele frequency (f), number of effective alleles (Ne), number of genotypes (Ng), polymorphic information content (PIC), observed heterozygosity (H_o), expected heterozygosity (H_e), and inbreeding coefficient (*F*).

SSR Locus	N_a	f	N_e	N_g	PIC	H_o	H_e	F
				All accessions (n = 36)				
csm4	7	0.32	4.41	10.00	0.76	0.14	0.77	0.83
csm27	4	0.50	2.63	7.00	0.58	0.17	0.62	0.75
csm32	5	0.53	2.57	7.00	0.55	0.17	0.61	0.73
emi02c21	4	0.61	2.19	4.00	0.48	0.00	0.54	1.00
emel1f04	3	0.65	1.92	5.00	0.40	0.17	0.48	0.66
Mean ± SE	4.6 ± 0.7	0.522 ± 0.05	2.746 ± 0.43	6.6 ± 1.00	0.552 ± 0.05	0.130 ± 0.03	0.605 ± 0.04	0.796 ± 0.05
				Islands accessions (n = 20)				
csm4	6	0.33	4.65	8.00	0.75	0.20	0.79	0.76
csm27	4	0.53	2.28	5.00	0.47	0.25	0.56	0.57
csm32	5	0.58	2.53	7.00	0.56	0.25	0.61	0.60
emi02c21	3	0.50	2.20	3.00	0.44	0.00	0.55	1.00
emel1f04	3	0.50	2.29	5.00	0.47	0.30	0.56	0.49
Mean ± SE	4.2 ± 0.6	0.485 ± 0.04	2.79 ± 0.46	5.6 ± 0.90	0.538 ± 0.05	0.200 ± 0.05	0.612 ± 0.04	0.686 ± 0.09
				Mainland accessions (n = 16)				
csm4	4	0.37	3.44	5.00	0.66	0.07	0.71	0.91
csm27	3	0.50	2.57	4.00	0.54	0.07	0.61	0.90
csm32	3	0.47	2.26	4.00	0.46	0.06	0.56	0.89
emi02c21	4	0.75	1.71	4.00	0.39	0.00	0.41	1.00
emel1f04	2	0.94	1.13	2.00	0.11	0.00	0.12	1.00
Mean ± SE	3.2 ± 0.4	0.604 ± 0.10	2.22 ± 0.39	3.8 ± 0.50	0.429 ± 0.09	0.039 ± 0.01	0.482 ± 0.10	0.923 ± 0.02

3.4. SSR Molecular Analysis of Variance, Genetic Structure, and Multivariate Analysis

The AMOVA revealed that the variation among accessions accounts for the greatest proportion of molecular variance (69.62%), followed by the variation within individuals due to heterozygosity (19.25%), and finally by the variation among mainland and island groups, which accounts for just 11.12% of the total variation (Table 6). The ΔK statistic in the genetic structure analysis presented a maximum peak at K = 2, suggesting that two genetic clusters exist in the Greek materials evaluated (Figure S1). With very few exceptions, the percentage of the ascription of accessions to their cluster was over 80%, revealing a low degree of genetic admixture (Figure 3). Cluster one included 14 accessions, of which 10 (71.4%) were from the mainland and 4 (28.6%) were from islands. Cluster two was made up by 22 accessions, of which 6 (27.3%) are from the mainland, and 16 (72.7%) are from the islands.

Table 6. Molecular analysis of variance in 36 Greek eggplant accessions. *p*-value estimates are based on 999 permutations (α = 0.05). The following sources of variation were considered: between groups (mainland vs. islands), among accessions within a group, and within individuals. Acronyms: df = degrees of freedom, SS = Sums of Squares, MS = mean squared deviations, F_{st} = genetic differentiation among populations within the total sample, F_{is} = genetic differentiation among individuals within populations, F_{it} = genetic differentiation among individuals within the total sample.

Source	df	S.S.	M.S.	Variance Components	Percentage of Variation	F-Statistics	p-Value
Between groups	1	9.19	9.19	0.19	11.12	F_{st} 0.111	0.001
Among accessions	34	89.42	2.63	1.16	69.62	F_{is} 0.783	0.001
Within accessions	36	11.50	0.32	0.32	19.25	F_{it} 0.807	0.001
Total	71	110.11		1.66			

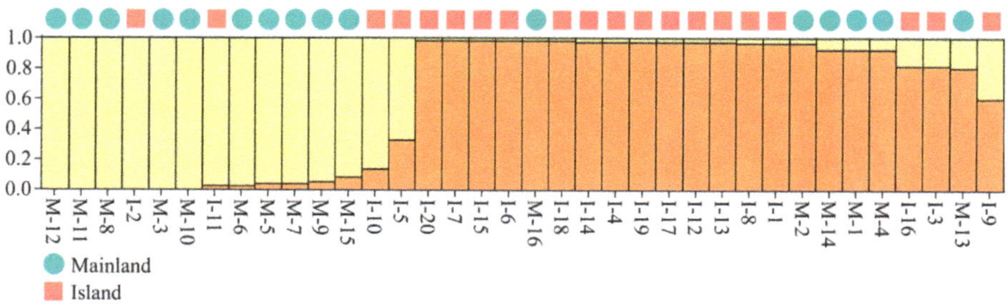

Figure 3. Genetic population structure of the 36 Greek eggplant accessions from the mainland and island areas based on SSRs data (K = 2). Each individual is represented by a vertical bar that is partitioned into colored segments that represent the estimated ratio of membership of an individual to a cluster. Cluster one is in yellow while cluster 2 is in orange. Study codes as in Table 1.

The PCoA analysis based on SSRs data explained all the variation with 17 PCs, although the first five PCs accounted for 75.8% (Table S4). Like in the PCA analysis with morphological data, the PC1, which accounted for 27.4%, allowed to separate the island accessions, with positive values of PC1 (15 out of 20), from the accessions of the mainland, with negatives values of the PC1 (11 out 16) (Figure 2B). On the contrary, PC2, which accounted for 17.6%, cannot clearly separate the two groups that spread from positive to negative values. Like the PCA on morphological traits, the accessions from the same geographical area showed a certain degree of genetic diversity. Finally, the Mantel test showed a weak, although significant, correlation between genetic and morphological data (r = 0.244, $p = 0.002$) (Table S5).

4. Discussion

In a climate change scenario coupled with an unprecedented increase in food demand, due to dietary changes and relentless growing population [61], local germplasm characterization and enhancement is imperative. The genetic variability and allelic diversity of local landraces could mitigate the effects of a changing environment and may guarantee food security of domestic markets [62]. In addition, heirlooms and ecotypes can harbor allelic combinations useful for the development of new more resilient and locally adapted varieties [14]. In this respect, this study aimed to assess the morphological and genetic diversity of the eggplant germplasm collection from Greece which is distributed and cultivated in a wide variety of niches throughout the distinctive country's topography. Moreover, the characterization of Greek eggplant landraces will prevent genetic erosion of this local group and provide a valuable resource for utilization in research and breeding.

4.1. Morphological Diversity

The morphological characterization displayed considerable diversity for most of the traits studied. The most heterogeneous traits were those related to fruit appearance, such as shape and size, followed by others related to vegetative traits, such as plant and organ size and type of growth, alongside with skin and flesh color. The solanaceous fruits exhibit considerable morphological diversity, including size, shape, and color, both within and between different species [63]. Morphological variations of fruits in terms of shape, size, and color are a result of adaptive evolution; in addition, variation observed in the vegetative traits could be due to the accumulation of mutations on neutral traits, to artificial or natural selection or to both [63,64]. These same types of morphological traits were also the most variable ones when eggplant accessions collected from Greece, Spain, Turkey, and remote centers of diversity were studied [23,34,35,65,66]. Consistent with previous works, the amount of diversity is comparable to that observed in the characterization of germplasm from other geographical areas of the Mediterranean region [34,35,64–66], providing further evidence that several areas of the Mediterranean region can be considered as secondary centers of diversity for eggplant [35,64].

PCA analysis revealed the existence of two discrete groups, one formed almost exclusively by the island accessions and the other mostly by those from the mainland ones. On average, the accessions from the islands have higher values for fruit weight, fruit calyx prickles, leaf blade width, leaf blade lobing, fruit breadth, but lower ones for fruit curvature and fruit length to breadth ratio than those from the mainland. The assignment of the eggplant accessions studied into groups according to their agromorphological characteristics correlated with their geographic origin has also been reported in other eggplant studies [35,67–69].

4.2. Mineral Composition

As occurred for morphological traits, the wide variation observed for mineral composition agrees with other works in which a significant number of eggplant accessions, commercial varieties, and hybrids were evaluated [65]. In the present study, significant differences were found only for Mg among island and mainland accessions, while non-significant differences were observed for Fe, Mn, Zn, Cu, and K. Raigón et al. [39] and Arivalagan et al. [38–40] studied the mineral composition and the nutritional value of eggplant landraces and germplasm accessions from different regions in order to identify mineral-rich germplasm for breeding purposes. For comparison, we transformed the values detected in the aforementioned studies, to the same units as in our study. The eggplant accessions studied had lower K and Cu concentration than those of Raigón et al. [39] and Arivalagan et al. [38,40]. Mg content in our study was similar to that of Raigón et al. [39] and higher than in Arivalagan et al. [38,40], while Zn concentration was significantly higher than in these studies. The environment, cultivation methods, as well as the cultivar used can influence the mineral composition of eggplants [70,71], and explain the differences observed in minerals' concentrations among our study and the others. To our knowledge, this is the first time that an inclusive uniform mineral content characterization of eggplant fruits from a field-grown germplasm collection was examined.

4.3. Molecular Diversity

Many studies assessed the genetic diversity and population structure of local varieties collections from different origins using SSRs [34–36,45,70,71]. The SSR markers used, were highly informative and effective for genetic diversity analysis. For example, the PIC value in the whole collection ranged from 0.39 to 0.75 with a mean value of 0.55. This value is higher than those observed in previous studies [36,45,72,73], while it was lower than those calculated by Hurtado et al. [35] and Cericola et al. [34]. A PIC value greater than 0.5 indicates loci of high polymorphism, PIC values between 0.25 and 0.5 indicate loci of intermediate polymorphism, and PIC values less than 0.25 indicate loci of low polymorphism [72]. In addition, the number of alleles/locus in the whole collection was high (4.6), higher than that (3.67) revealed by Augustinos et al. [23], when they studied a limited number of Greek traditional eggplant cultivars. Taking into consideration other studies on eggplants from different geographic origin or different cultivar types based on SSR data, the number of alleles/locus calculated in our eggplant collection is lower than that detected by Hurtado et al. [35] and Liu et al. [36], but almost similar to those by Demir et al. [74], Muñoz-Falcón et al. [73], Tümbilen et al. [66], and Vilanova et al. [50,75]. Two of the SSR markers used in our study (csm27, csm32), have also been used by Hurtado et al. [35] and Vilanova et al. [50]. In their studies, they observed higher average number of alleles/locus as compared to those detected by us, an observation that might be explained by the fact that their collections originate from a broader geographic area, or belong to different cultivar types, in contrast to our collection which originates from a single and more geographically restricted area. However, when Hurtado et al. [35] calculated the average number of alleles/locus only for the Spanish eggplants included in their collection, it was almost similar to ours (5.0). Expected heterozygosity (H_e) values were high, with an average value of 0.6, quite similar or higher to the values observed in other similar studies, providing further evidence of the utility and highly informative nature of the SSRs used. The estimated values of observed heterozygosity (H_o) (average 0.13) were expected, since eggplant is basically an autogamous plant [76] and thus most heritage and commercial varieties are expected to be homozygous [34]. Muñoz-Falcón et al. [77] detected significantly lower values for Spanish landraces (0.02) and for non-Spanish landraces (0.02), whereas Cericola et al. [34] also found that most eggplant landraces used in their study had low heterozygosity values, and only 38 out of 238 materials had $H_o > 0.10$. Low H_o was also detected by Augustinos et al. [23] and Liu et al. [36] (0.04 and 0.03, respectively) and Vilanova et al. [50,75]. The high level of homozygosity in eggplant landraces shows that pure lines can easily be derived by individual selection from these materials [75]. Although genetic diversity based on molecular data is dependent on the type and number of markers used and the accessions tested [66], the relatively high PIC values and average number of alleles/locus detected and also the high levels of heterozygosity observed compared to other similar studies, indicate that considerable diversity exists in our eggplant collection. Although, the number of SSR markers we used has been limited, their high PIC values indicate that they provide meaningful information on the diversity of Greek eggplants. Consequently, Greece, together with other countries such as Spain and Turkey, might be considered as part of a secondary center of diversity for the eggplant in the Mediterranean basin. Further analysis with high-throughput methods may provide a confirmation for our hypothesis.

Remarkably, the island accessions seem to be more variable than those from the mainland, as indicated by the average number of alleles/locus (4.2 vs. 3.2) and PIC values (0.53 vs. 0.42). Higher values of H_o were observed in the island accessions when compared to the mainland ones (0.20 vs. 0.03), suggesting that a higher degree of outcrossing occurs in the island accessions than in those from the mainland. Although endemic island populations tend to have less diverse variation in comparison to continental populations [25,31,78], due to founder effect events [78,79], of limited population size and genetic drift [80], the heterozygote excess in our case could be possibly due to selective advantages enjoyed by heterozygote individuals, due to the agroecological conditions of the island environment, as in the case of dill [12]. Another possible explanation is that Greek islands in the Aegean Sea are in proximity with Asia minor and the Turkish coastline and thus genetic exchange could be favored through pollen dispersal, human influence and transportation. Environmental conditions also have a critical impact on the relationship between plants and pollinators. Rain can disrupt this pollen transfer and hinder sexual reproduction in plants [81]. The dry climate and the lack of rainfall in the islands may promote cross-pollination, therefore, increasing the proportion of heterozygotes in the island accessions. Consistent with the PCA analysis of the morphological traits, the genetic STRUCTURE analysis differentiates two genetic clusters, one made up mostly by mainland and another by the island accessions. These results are also confirmed in the PCoA analysis. The fact that some island accessions fell in the mainland genetic cluster and vice versa probably has contributed to the limited differentiation observed in the AMOVA analysis between the mainland and the island groups (11.12%).

Discrimination among accessions originating from the islands and the mainland has also been reported for Greek dill by Ninou et al. [12] and for okra landraces by Kyriakopoulou et al. [82].

4.4. Comparison of Morphological and Molecular Diversity

Undoubtedly, even though genetic characterization has become predominant in the last decade due to the efficiency and the plummeting cost of sequencing and genotyping, the morphological phenotyping is still essential and complementary to the genetic assessment [83]. The correlations between morphological and molecular data usually vary depending on the eggplant genotype, the morphological descriptors, and the molecular markers used [35]. In this study, the Mantel test detected a weak, although significant, correlation between genetic and morphological data ($r = 0.244$, $p = 0.002$), in agreement with previous eggplant studies where weak or no correlation between genotypic and phenotypic data have been found [23,34,35,77]. However, these results need to be confirmed by additional analyses with larger molecular markers datasets.

5. Conclusions

In conclusion, our results show a remarkable phenotypic and genetic variation existing in the Greek eggplant landraces, suggesting that Greece, among other Mediterranean countries, such as Italy and Spain, might be considered as a secondary center of diversity for eggplant. Both morphological and genetic data revealed clustering of eggplant accessions into two major groups based on the geographic origin (mainland vs. island accessions). Some differentiation signatures in morphological, molecular, and fruit composition were detected between both groups underlying the significant impact of the environment, geographic isolation, and the agroecological conditions in the differentiation and characteristics of the eggplant landraces from the Greek islands and mainland. There was a weak but significant correlation between the morphological and genetic data, indicating that both types of data provide complementary information and therefore they should be taken into consideration for the characterization of plant material. Our results show that local Greek populations of eggplant are a promising material which can be of interest not only for conservation purposes but also for the manipulation of value-added traits through breeding programs and for commercial exploitation.

Supplementary Materials: The following are available online at http://www.mdpi.com/2073-4395/9/12/887/s1, Table S1: Average values of the morphological and mineral characterization of the eggplant accessions of Greek islands ($n = 19$) and the mainland ($n = 16$). Table S2: Percentage of the variance explained by each principal component (PC) of the PCA analysis, correlations between the PCs and the morphological traits evaluated in Greek eggplant accessions and the eigenvalues of the accessions for each PCs. Table S3: SSR alleles identified in the eggplant accessions of Greek islands and the mainland. Table S4: Percentage of the variance explained by each principal component (PC) of the PCoA analysis and eigenvectors in Greek eggplant accessions for each PC. Table S5: Mantel test between morphological and genetic data. Figure S1: Delta K values for 2 to 10 genetic clusters for thirty-six accessions (cultivars and landraces) in Greek eggplant accessions. Delta K was calculated according to Evanno et al. (2005).

Author Contributions: Conceptualization, V.P., J.P., P.G., and G.K.; Investigation, E.C., P.G. and V.C.; Formal analysis, C.P. and P.G.; Methodology, J.P., S.V., and V.P.; Resources, P.M.; supervision, J.P., S.V., G.K. and V.P.; writing (original draft preparation), V.P., P.G., and C.P.; writing (review and editing), J.P., P.G., C.P., S.V., V.C., P.M. and V.P.; Visualization, P.G.

Funding: "PlantUP" (MIS 5002803) which is implemented under the Action "Reinforcement of the Research and Innovation Infrastructure", funded by the Operational Programme "Competitiveness, Entrepreneurship and Innovation" (NSRF 2014–2020) and co-financed by Greece and the European Union (European Regional Development Fund). Funding was also received from the European Union's Horizon 2020 Research and Innovation Programme under grant agreement No 677379 (G2P-SOL project: Linking genetic resources, genomes and phenotypes of Solanaceous crops), and from Ministerio de Ciencia, Innovación y Universidades, Agencia Estatal de Investigacion and Fondo Europeo de Desarrollo Regional (grant RTI-2018-094592-B-100 from MCIU/AEI/FEDER, UE). Pietro Gramazio is grateful to Universitat Politècnica de València and to Japan Society for the Promotion of Science for their respective postdoctoral grants (PAID-10-18 and FY2019 JSPS Postdoctoral Fellowship for Research in Japan [Standard]).

Conflicts of Interest: The authors declare no conflict of interest.

References

1. Kougioumoutzis, K.; Valli, A.T.; Georgopoulou, E.; Simaiakis, S.M.; Triantis, K.A.; Trigas, P. Network biogeography of a complex island system: The Aegean Archipelago revisited. *J. Biogeogr.* **2017**, *44*, 651–660. [CrossRef]
2. Graham, N.R.; Gruner, D.S.; Lim, J.Y.; Gillespie, R.G. Island ecology and evolution: Challenges in the Anthropocene. *Environ. Conserv.* **2017**, *44*, 323–335. [CrossRef]
3. Price, J.P.; Otto, R.; Menezes de Sequeira, M.; Kueffer, C.; Schaefer, H.; Caujapé-Castells, J.; Fernández-Palacios, J.M. Colonization and diversification shape species–area relationships in three Macaronesian archipelagos. *J. Biogeogr.* **2018**, *45*, 2027–2039. [CrossRef]
4. Warren, B.H.; Simberloff, D.; Ricklefs, R.E.; Aguilée, R.; Condamine, F.L.; Gravel, D.; Morlon, H.; Mouquet, N.; Rosindell, J.; Casquet, J.; et al. Islands as model systems in ecology and evolution: Prospects fifty years after MacArthur-Wilson. *Ecol. Lett.* **2015**, *18*, 200–217. [CrossRef]
5. Douma, C.; Koutis, K.; Thanopoulos, R.; Tsigou, R.; Galanidis, A.; Bebeli, P.J. Diversity of agricultural plants on Lesvos Island (Northeast Aegean, Greece) with emphasis on fruit trees. *Sci. Hortic.* **2016**, *210*, 65–84. [CrossRef]
6. Hagenblad, J.; Leino, M.W.; Hernández Afonso, G.; Afonso Morales, D. Morphological and genetic characterization of barley (*Hordeum vulgare* L.) landraces in the Canary Islands. *Genet. Resour. Crop. Evol.* **2019**, *66*, 465–480. [CrossRef]
7. Médail, F. The specific vulnerability of plant biodiversity and vegetation on Mediterranean islands in the face of global change. *Reg. Environ. Chang.* **2017**, *17*, 1775–1790. [CrossRef]
8. Hellenic Statistical Authority. 2019. Available online: http://www.statistics.gr/en/home/ (accessed on 1 August 2019).
9. Sfenthourakis, S.; Triantis, K.A. The Aegean archipelago: A natural laboratory of evolution, ecology and civilisations. *J. Biol. Res.* **2017**, *24*, 4. [CrossRef]
10. Strid, A. Phytogeographia Aegaea and the Flora Hellenica Database. *Ann. Nat. Mus. Wien.* **1996**, *98*, 279–289.
11. Dimopoulos, P.; Raus, T.; Bergmeier, E.; Constantinidis, T.; Iatrou, G.; Kokkini, S.; Strid, A.; Tzanoudakis, D. Vascular plants of Greece: An annotated checklist. *Willdenowia* **2016**, *46*, 301–347. [CrossRef]
12. Ninou, E.G.; Mylonas, I.G.; Tsivelikas, A.L.; Ralli, P.E. Phenotypic diversity of Greek dill (*Anethum graveolens* L.) landraces. *Acta Agric. Scand. Sect. B Soil Plant. Sci.* **2017**, *67*, 318–325.
13. Tsanakas, G.F.; Mylona, P.V.; Koura, K.; Gleridou, A.; Polidoros, A.N. Genetic diversity analysis of the Greek lentil (*Lens culinaris*) landrace "Eglouvis" using morphological and molecular markers. *Plant Genet. Resour. Characterisation Util.* **2018**, *16*, 469–477. [CrossRef]
14. Dwivedi, S.; Goldman, I.; Ortiz, R. Pursuing the potential of heirloom cultivars to improve adaptation, nutritional, and culinary features of food crops. *Agronomy* **2019**, *9*, 441. [CrossRef]
15. Bota, J.; Conesa, M.À.; Ochogavia, J.M.; Medrano, H.; Francis, D.M.; Cifre, J. Characterization of a landrace collection for Tomàtiga de Ramellet (*Solanum lycopersicum* L.) from the Balearic Islands. *Genet. Resour. Crop. Evol.* **2014**, *61*, 1131–1146. [CrossRef]
16. Carillo, P.; Kyriacou, M.C.; El-Nakhel, C.; Pannico, A.; Dell'Aversana, E.; D'Amelia, L.; Colla, G.; Caruso, G.; De Pascale, S.; Rouphael, Y. Sensory and functional quality characterization of protected designation of origin 'Piennolo del Vesuvio' cherry tomato landraces from Campania-Italy. *Food Chem.* **2019**, *292*, 166–175. [CrossRef]
17. Schmidt, S.B.; George, T.S.; Brown, L.K.; Booth, A.; Wishart, J.; Hedley, P.E.; Martin, P.; Russell, J.; Husted, S. Ancient barley landraces adapted to marginal soils demonstrate exceptional tolerance to manganese limitation. *Ann. Bot.* **2019**, *123*, 831–843. [CrossRef]
18. Missio, J.C.; Rivera, A.; Figàs, M.R.; Casanova, C.; Camí, B.; Soler, S.; Simó, J. A comparison of landraces vs. Modern varieties of lettuce in organic farming during the winter in the Mediterranean area: An approach considering the viewpoints of breeders, consumers, and farmers. *Front. Plant Sci.* **2018**, *9*, 1491. [CrossRef]
19. Petropoulos, S.A.; Barros, L.; Ferreira, I.C.F.R. Editorial: Rediscovering local landraces: Shaping horticulture for the future. *Front. Plant Sci.* **2019**, *10*, 126. [CrossRef]
20. Karanikolas, P.; Bebeli, P.J.; Thanopoulos, R. Farm economic sustainability and agrobiodiversity: Identifying viable farming alternatives during the economic crisis in Greece. *J. Environ. Econ. Policy* **2018**, *7*, 69–84. [CrossRef]

21. McDorman, B.; Thomas, S. The Importance of Saving Seeds. In *Promoting Biodiversity in Food Systems*; Hawkins, I.W., Ed.; CRC PressI Llc: Boca Ratón, FL, USA, 2018; pp. 189–210.
22. FAO STATISTICAL DATABASES. 2017. Available online: http://www.fao.org/faostat/ (accessed on 1 August 2019).
23. Augustinos, A.A.; Petropoulos, C.; Karasoulou, V.; Bletsos, F.; Papasotiropoulos, V. Assessing diversity among traditional Greek and foreign eggplant cultivars using molecular markers and morphometrical descriptors. *Span. J. Agric. Res.* **2016**, *14*, e0710. [CrossRef]
24. Thomas, K.; Thanopoulos, R.; Knüpffer, H.; Bebeli, P.J. Plant genetic resources of Lemnos (Greece), an isolated island in the Northern Aegean Sea, with emphasis on landraces. *Genet Resour. Crop. Evol.* **2012**, *59*, 1417–1440. [CrossRef]
25. García-Verdugo, C.; Sajeva, M.; La Mantia, T.; Harrouni, C.; Msanda, F.; Caujapé-Castells, J. Do island plant populations really have lower genetic variation than mainland populations? Effects of selection and distribution range on genetic diversity estimates. *Mol. Ecol.* **2015**, *24*, 726–741. [CrossRef] [PubMed]
26. Hiraoka, Y.; Tamaki, I.; Watanabe, A. The origin of wild populations of *Toxicodendron succedaneum* on mainland Japan revealed by genetic variation in chloroplast and nuclear DNA. *J. Plant Res.* **2018**, *131*, 225–238. [CrossRef] [PubMed]
27. Jiménez, A.; Weigelt, B.; Santos-Guerra, A.; Caujapé-Castells, J.; Fernández-Palacios, J.M.; Conti, E. Surviving in isolation: Genetic variation, bottlenecks and reproductive strategies in the Canarian endemic *Limonium macrophyllum* (Plumbaginaceae). *Genetica* **2017**, *145*, 91–104. [CrossRef]
28. Wheelwright, N.T.; Begin, E.; Ellwanger, C.; Taylor, S.H.; Stone, J.L. Minimal loss of genetic diversity and no inbreeding depression in blueflag iris (*Iris versicolor*) on islands in the Bay of Fundy. *Botany* **2016**, *94*, 543–554. [CrossRef]
29. Hedrén, M.; Olofsson, S.N.; Paun, O. Orchid colonization: Multiple parallel dispersal events and mosaic genetic structure in *Dactylorhiza majalis* ssp. *lapponica* on the Baltic island of Gotland. *Ann. Bot.* **2018**, *122*, 1019–1032.
30. Hufford, K.M.; Mazer, S.J.; Hodges, S.A. Genetic variation among mainland and island populations of a native perennial grass used in restoration. *AoB Plants* **2014**, *6*. [CrossRef]
31. Mcglaughlin, M.E.; Wallace, L.E.; Wheeler, G.L.; Bresowar, G.; Riley, L.; Britten, N.R.; Helenurm, K. Do the island biogeography predictions of MacArthur and Wilson hold when examining genetic diversity on the near mainland California Channel Islands? Examples from endemic Acmispon (Fabaceae). *Bot. J. Linn. Soc.* **2014**, *174*, 289–304. [CrossRef]
32. Idrissi, O.; Piergiovanni, A.R.; Toklu, F.; Houasli, C.; Udupa, S.M.; De Keyser, E.; Van Damme, P.; De Riek, J. Molecular variance and population structure of lentil (*Lens culinaris* Medik.) landraces from Mediterranean countries as revealed by simple sequence repeat DNA markers: Implications for conservation and use. *Plant Genet. Resour. Characterisation Util.* **2018**, *16*, 249–259. [CrossRef]
33. Acquadro, A.; Barchi, L.; Gramazio, P.; Portis, E.; Vilanova, S.; Comino, C.; Plazas, M.; Prohens, J.; Lanteri, S. Coding SNPs analysis highlights genetic relationships and evolution pattern in eggplant complexes. *PLoS ONE* **2017**, *12*, e0180774. [CrossRef]
34. Cericola, F.; Portis, E.; Toppino, L.; Barchi, L.; Acciarri, N.; Ciriaci, T.; Sala, T.; Rotino, G.L.; Lanteri, S. The population structure and diversity of eggplant from Asia and the Mediterranean Basin. *PLoS ONE* **2013**, *8*, e73702. [CrossRef]
35. Hurtado, M.; Vilanova, S.; Plazas, M.; Gramazio, P.; Fonseka, H.H.; Fonseka, R.; Prohens, J. Diversity and relationships of eggplants from three geographically distant secondary centers of diversity. *PLoS ONE* **2012**, *7*, e41748. [CrossRef] [PubMed]
36. Liu, J.; Yang, Y.; Zhou, X.; Bao, S.; Zhuang, Y. Genetic diversity and population structure of worldwide eggplant (*Solanum melongena* L.) germplasm using SSR markers. *Genet. Resour. Crop. Evol.* **2018**, *65*, 1663–1670. [CrossRef]
37. Rodriguez-Jimenez, J.; Amaya-Guerra, C.; Baez-Gonzalez, J.; Aguilera-Gonzalez, C.; Urias-Orona, V.; Nino-Medina, G. Physicochemical, functional, and nutraceutical properties of eggplant flours obtained by different drying methods. *Molecules* **2018**, *23*, 3210. [CrossRef] [PubMed]
38. Arivalagan, M.; Bhardwaj, R.; Gangopadhyay, K.K.; Prasad, T.V.; Sarkar, S.K. Mineral composition and their genetic variability analysis in eggplant (*Solanum melongena* L.) germplasm. *J. Appl. Bot. Food Qual.* **2013**, *86*, 99–103.

39. Raigón, M.D.; Prohens, J.; Muñoz-Falcón, J.E.; Nuez, F. Comparison of eggplant landraces and commercial varieties for fruit content of phenolics, minerals, dry matter and protein. *J. Food Compos. Anal.* **2008**, *21*, 370–376. [CrossRef]
40. Arivalagan, M.; Gangopadhyay, K.K.; Kumar, G.; Bhardwaj, R.; Prasad, T.V.; Sarkar, S.K.; Roy, A. Variability in mineral composition of Indian eggplant (*Solanum melongena* L.) genotypes. *J. Food Compos. Anal.* **2012**, *26*, 173–176. [CrossRef]
41. Ranil, R.H.G.; Niran, H.M.L.; Plazas, M.; Fonseka, R.M.; Fonseka, H.H.; Vilanova, S.; Andújar, I.; Gramazio, P.; Fita, A.; Prohens, J. Improving seed germination of the eggplant rootstock *Solanum torvum* by testing multiple factors using an orthogonal array design. *Sci. Hortic.* **2015**, *193*, 174–181. [CrossRef]
42. van der Weerden, G.M.; Barendse, G.W.M. A web-based searchable database developed for the EGGNET project and applied to the radboud university Solanaceae database. *Acta Hortic.* **2007**, *745*, 503–506. [CrossRef]
43. Kaushik, P.; Prohens, J.; Vilanova, S.; Gramazio, P.; Plazas, M. Phenotyping of eggplant wild relatives and interspecific hybrids with conventional and phenomics descriptors provides insight for their potential utilization in breeding. *Front. Plant Sci.* **2016**, *7*, 677. [CrossRef]
44. IBPGR. *Descriptors for Eggplant*; International Board for Plant Genetic Resources: Rome, Italy, 1990.
45. Kumar, G.; Meena, B.L.; Kar, R.; Tiwari, S.K.; Gangopadhyay, K.K.; Bisht, I.S.; Mahajan, R.K. Morphological diversity in brinjal (*Solanum melongena* L.) germplasm accessions. *Plant Genet. Res.* **2008**, *6*, 232–236. [CrossRef]
46. Campbell, C.R.; Plank, C.O. Preparation of Plant Tissue for Laboratory Analysis. In *Handbook of Reference Methods for Plant Analysis*; Kalra, Y.P., Ed.; CRC Press: Boca Ratón, FL, USA, 1998; pp. 37–50.
47. Ihaka, R.; Gentleman, R. R: A Language for Data Analysis and Graphics. *J. Comput. Graph. Stat.* **1996**, *5*, 299–314.
48. Wickham, H. *ggplot2: Elegant Graphics for Data Analysis*; Springer: New York, NY, USA, 2016; ISBN 978-3-319-24277-4.
49. Alonso, D.; Gramazio, P.; Plazas, M.; Villanueva, G.; García-Fortea, E.; Díez, M.J.; Prohens, J.; Vilanova, S. Protocolo innovador para la obtención de ADN genómico vegetal de alta calidad. In Proceedings of the Congress "I Congreso de Jóvenes Investigadores en Ciencias Agroalimentarias", Almeria, Spain, 20 December 2018.
50. Vilanova, S.; Manzur, J.P.; Prohens, J. Development and characterization of genomic simple sequence repeat markers in eggplant and their application to the study of diversity and relationships in a collection of different cultivar types and origins. *Mol. Breed.* **2012**, *30*, 647–660. [CrossRef]
51. Nunome, T.; Negoro, S.; Kono, I.; Kanamori, H.; Miyatake, K.; Yamaguchi, H.; Ohyama, A.; Fukuoka, H. Development of SSR markers derived from SSR-enriched genomic library of eggplant (*Solanum melongena* L.). *Theor. Appl. Genet.* **2009**, *119*, 1143–1153. [CrossRef] [PubMed]
52. Liu, K.; Muse, S.V. PowerMaker: An integrated analysis environment for genetic maker analysis. *Bioinformatics* **2005**, *21*, 2128–2129. [CrossRef] [PubMed]
53. Peakall, P.; Smouse, R. GenAlEx 6.5: Genetic analysis in Excel. Population genetic software for teaching and research—An update. *Bioinformatics* **2012**, *28*, 2537–2539. [CrossRef]
54. Botstein, D.; White, R.L.; Skolnick, M.; Davis, R.W. Construction of a genetic linkage map in man using restriction fragment length polymorphisms. *Am. J. Hum. Genet.* **1980**, *32*, 314–331.
55. Nei, M. Genetic distance between populations. *Am. Nat.* **1972**, *106*, 283–292. [CrossRef]
56. Wright, S. The interpretation of population structure by F-statistics with special regard to systems of mating. *Evolution* **1965**, *19*, 395–420. [CrossRef]
57. Pritchard, J.; Stephens, M.; Donnelly, P. Inference of population structure using multilocus genotype data. *Genetics* **2000**, *155*, 945–959.
58. Evanno, G.; Regnaut, S.; Goudet, J. Detecting the number of clusters of individuals using the software STRUCTURE: A simulation study. *Mol. Ecol.* **2005**, *14*, 2611–2620. [CrossRef] [PubMed]
59. Earl, D.A.; VonHoldt, B.M. STRUCTURE HARVESTER: A website and program for visualizing STRUCTURE output and implementing the Evanno method. *Conserv. Genet. Resour.* **2012**, *4*, 359–361. [CrossRef]
60. Remington, D.L.; Thornsberry, J.M.; Matsuoka, Y.; Wilson, L.M.; Whitt, S.R.; Doebley, J.; Kresovich, S.; Goodman, M.M.; Buckler, I.V.E.S. Structure of linkage disequilibrium and phenotypic associations in the maize genome. *Proc. Natl. Acad. Sci. USA* **2001**, *98*, 11479–11484. [CrossRef]

61. Beerling, D.J.; Leake, J.R.; Long, S.P.; Scholes, J.D.; Ton, J.; Nelson, P.N.; Bird, M.; Kantzas, E.; Taylor, L.L.; Sarkar, B.; et al. Farming with crops and rocks to address global climate, food and soil security. *Nat. Plants* **2018**, *4*, 138–147. [CrossRef] [PubMed]
62. Marconi, G.; Ferradini, N.; Russi, L.; Concezzi, L.; Veronesi, F.; Albertini, E. Genetic characterization of the apple germplasm collection in central Italy: The value of local varieties. *Front. Plant Sci.* **2018**, *9*. [CrossRef] [PubMed]
63. Wang, L.; Li, J.; Zhao, J.; He, C. Evolutionary developmental genetics of fruit morphological variation within the Solanaceae. *Front. Plant Sci.* **2015**, *6*, 248. [CrossRef] [PubMed]
64. Prohens, J.; Blanca, J.M.; Nuez, F. Morphological and molecular variation in a collection of eggplants from a secondary center of diversity: Implications for conservation and breeding. *J. Am. Soc. Hortic. Sci.* **2005**, *130*, 54–63. [CrossRef]
65. Boyaci, H.F.; Topcu, V.; Tepe, A.; Yildirim, I.K.; Oten, M.; Aktas, A. Morphological and molecular characterization and relationships of Turkish local eggplant heirlooms. *Not. Bot. Horti Agrobot. Cluj Napoca* **2015**, *43*, 100–107. [CrossRef]
66. Tümbilen, Y.; Frary, A.; Mutlu, S.; Doganlar, S. Genetic diversity in Turkish eggplant (*Solanum melongena*) varieties as determined by morphological and molecular analyses. *Int. Res. J. Biotechnol.* **2011**, *2*, 16–25.
67. Muñoz-Falcón, J.E.; Prohens, J.; Rodríguez-Burruezo, A.; Nuez, F. Potential of local varieties and their hybrids for the improvement of eggplant production in the open field and greenhouse cultivation. *J. Food Agric. Environ.* **2008**, *6*, 83–88.
68. Muñoz-Falcón, J.E.; Prohens, J.; Vilanova, S.; Nuez, F. Characterization, diversity, and relationships of the Spanish striped (Listada) eggplants: A model for the enhancement and protection of local heirlooms. *Euphytica* **2008**, *164*, 405–419. [CrossRef]
69. Polignano, G.; Uggenti, P.; Bisignano, V.; Gatta, C. Della Genetic divergence analysis in eggplant (*Solanum melongena* L.) and allied species. *Genet. Resour. Crop. Evol.* **2010**, *57*, 171–181. [CrossRef]
70. Russo, V.M. Cultural methods and mineral content of eggplant *Solanum melongena* fruit. *J. Sci. Food Agric.* **1996**, *71*, 119–123. [CrossRef]
71. Raigón, M.D.; Rodríguez-Burruezo, A.; Prohens, J. Effects of organic and conventional cultivation methods on composition of eggplant fruits. *J. Agric. Food Chem.* **2010**, *58*, 6833–6840. [CrossRef] [PubMed]
72. Ge, H.; Liu, Y.; Jiang, M.; Zhang, J.; Han, H.; Chen, H. Analysis of genetic diversity and structure of eggplant populations (*Solanum melongena* L.) in China using simple sequence repeat markers. *Sci. Hortic.* **2013**, *162*, 71–75. [CrossRef]
73. Muñoz-Falcón, J.E.; Vilanova, S.; Plazas, M.; Prohens, J. Diversity, relationships, and genetic fingerprinting of the Listada de Gandía eggplant landrace using genomic SSRS and EST-SSRs. *Sci. Hortic.* **2011**, *129*, 238–246. [CrossRef]
74. Demir, K.; Bakir, M.; Sarikamiş, G.; Acunalp, S. Genetic diversity of eggplant (*Solanum melongena*) germplasm from Turkey assessed by SSR and RAPD markers. *Genet. Mol. Res.* **2010**, *9*, 1568–1576. [CrossRef]
75. Vilanova, S.; Hurtado, M.; Cardona, A.; Plazas, M.; Gramazio, P.; Herraiz, F.J.; Andújar, I.; Prohens, J. Genetic diversity and relationships in local varieties of eggplant from different cultivar groups as assessed by genomic SSR markers. *Not. Bot. Horti Agrobot. Cluj Napoca* **2014**, *42*, 59–65. [CrossRef]
76. Pessarakli, M.; Dris, R. Pollination and breeding of eggplants. *J. Food Agric. Environ.* **2004**, *2*, 218–219.
77. Muñoz-Falcón, J.E.; Prohens, J.; Vilanova, S.; Nuez, F. Diversity in commercial varieties and landraces of black eggplants and implications for broadening the breeders' gene pool. *Ann. Appl. Biol.* **2009**, *154*, 453–465. [CrossRef]
78. Frankham, R. Do island populations have less genetic variation than mainland populations? *Heredity* **1997**, *78*, 311–327. [CrossRef] [PubMed]
79. Whittaker, R.; Fernández-Palacios, J. *Island Biogeography: Ecology, Evolution, and Conservation*; Oxford University Press: Oxford, UK, 2007; ISBN 978-0198566120.
80. Hargreaves, S.; Maxted, N.; Hirano, R.; Abberton, M.; Skøt, L.; Ford-Lloyd, B.V. Islands as refugia of *Trifolium repens* genetic diversity. *Conserv. Genet.* **2010**, *11*, 1317–1326. [CrossRef]
81. Lawson, D.A.; Rands, S.A. The effects of rainfall on plant–pollinator interactions. *Arthropod Plant Interact.* **2019**, *13*, 561–569. [CrossRef]

82. Kyriakopoulou, O.G.; Arens, P.; Pelgrom, K.T.B.; Karapanos, I.; Bebeli, P.; Passam, H.C. Genetic and morphological diversity of okra (*Abelmoschus esculentus* [L.] Moench.) genotypes and their possible relationships, with particular reference to Greek landraces. *Sci. Hortic.* **2014**, *171*, 58–70. [CrossRef]
83. Caguiat, X.G.I.; Hautea, D.M. Genetic diversity analysis of eggplant (*Solanum melongena* L.) and related wild species in the Philippines using morphological and SSR markers. *Sabrao J. Breed. Genet.* **2014**, *46*, 183–201.

agronomy

MDPI

Article

Variability in Nutraceutical Lipid Content of Selected Rice (*Oryza sativa* L. spp. *indica*) Germplasms

Bhornchai Harakotr [1,*], Kasidid Prompoh [1], Supakorn Boonyuen [2], Bhalang Suriharn [3] and Kamol Lertrat [4]

[1] Department of Agricultural Technology, Faculty of Science and Technology, Thammasat University, Pathum Thani 12120, Thailand; prohprom@hotmail.com

[2] Department of Chemistry, Faculty of Science and Technology, Thammasat University, Pathum Thani 12120, Thailand; chemistrytu@gmail.com

[3] Department of Agronomy, Faculty of Agriculture, Khon Kaen University, Khon Kaen 40002, Thailand; sphala@kku.ac.th

[4] Plant Breeding Research Center for Sustainable Agriculture, Faculty of Agriculture, Khon Kaen University, Khon Kaen 40002, Thailand; kamol9@gmail.com

* Correspondence: harakotr@tu.ac.th; Tel.: +66-2564-4440-79 (ext. 2364)

Received: 1 November 2019; Accepted: 29 November 2019; Published: 30 November 2019

Abstract: Rice (*Oryza sativa* L.) contains many high-value nutritional compounds, including nutraceutical lipid compounds that offer health benefits. An understanding of the genetic variability in the lipid contents of different rice germplasms is necessary to support breeding programs. The goals of this study were: i) to investigate varietal differences in levels of the nutraceutical lipid α-tocopherol, γ-oryzanol, campesterol, β-sitosterol, octacosanol, and squalene and ii) to identify clusters of rice germplasms based on their lipid contents. Eighty-three *indica* rice subspecies were evaluated using a randomized complete block design with three replications. Research was conducted in Thailand during the 2016 rice-growing season. Significant differences were found among genotypes across all traits. The largest variation was found for β-sitosterol, followed by campesterol, octacosanol, and α-tocopherol. Variation in squalene content was small. Four principal components were found that accounted for 93.47% of overall variability. β-sitosterol, campesterol, and squalene were the principal discriminatory constituents. No significant correlation was found between color parameters and levels of these compounds, suggesting that former are of little use as an indirect marker for selection of fat-soluble nutraceuticals. Cluster analysis sorted the germplasm into nine clusters, based on their nutraceutical lipid content. TU-010, TU-027, TU-093, and TU-244 genotypes had the highest levels, making them a potentially useful genetic resource in breeding programs for nutraceutically-improved rice. The findings of this study can support the introduction of novel rice varieties with high added-value bioactive properties.

Keywords: landrace rice; fat-soluble nutraceuticals; β-sitosterol; genetic variability; cluster analysis

---

1. Introduction

Cardiovascular disease (CVD) is the largest cause of global deaths, accounting for an estimated 17.5 million deaths in 2012. This group includes hypertension, myocardial infarction, atherosclerosis, arrhythmias and valvular heart disease, coagulopathies, and strokes. The percentage of premature deaths attributed to CVDs ranges from 4% in high-income countries to 42% in low-income countries, reflecting inequalities between countries and populations [1]. Consumption of whole-grain cereals plays a pivotal role in preventing these chronic diseases and promoting health [2]. Epidemiological studies suggest that the low incidence of certain chronic diseases in rice-consuming regions of the world might be associated with the high-value antioxidants that rice contains [3]. However, as dietary patterns have become more Western-oriented, consumption of rice has been decreasing. Policies that increase rice consumption can contribute to solving the economic problems of rural communities. One approach is the development of novel rice varieties with improved nutritional and bioactive properties [4].

Rice (*Oryza sativa* L.) contains more than a hundred bioactive compounds, including phytic acid, isovitexin, flavonoids, phenolics, polyphenols, anthocyanins, and proanthocyanins, and previously unknown compounds have been characterized and quantified in the rice germplasm [5–8]. Rice also has a high level of nutraceutical lipid compounds, such as tocopherol, tocotrienol, phytosterols, γ-oryzanol, octacosanol, and squalene, which are known to be powerful antioxidants [9,10]. These are potential inhibitors of cholesterol oxidation, have been shown to reduce serum cholesterol levels in animals, and are effective in treating anxiety neurosis, menopausal disorders, inflammatory diseases, and in inhibiting tumor growth [11–13]. These bioactive lipid compounds are concentrated in the rice germ or bran and are lost during milling or polishing [14]. Most consumers also prefer the palatability of milled or white rice, which has little or no bran on the endosperm, to brown rice. However, brown rice offers the greatest concentration of nutraceutical lipid compounds, which is thought to be encouraging increased consumption.

The nutraceutical lipid compound fraction of rice and rice-based foods can be increased by selecting naturally antioxidant-rich genotypes for cultivation, by selective plant breeding to develop genotypes with still higher phytochemical content, or by varying pre and postharvest conditions [15–17]. Such improvements, along with the development of higher-yielding varieties through conventional breeding methods, may be economically positive for growers as well as widening consumer choice. A necessary preliminary step in the identification of efficient selection methods is to investigate the role of genetic variability in the nutritional values offered by different populations. The genetic variation in the nutraceutical lipid compounds of rice and rice products has been characterized [16,17]. Wide variation was observed in the γ-oryzanol and phytosterol contents of Korean rice [14] and the lipid and fatty acid content of rice bran [18]. However, little information is available on the effect of genetic variability on the octacosanal and squalene content of landrace rice.

Thailand is a center of diversity for both wild and cultivated rice (*O. sativa* L. ssp. *indica*) and has a range of native landraces [19]. Landrace varieties still account for some 20% of the country's cultivated rice paddy [20]. However, cultivation of landraces is gradually decreasing under pressure from urbanization and industrialization [21]. Landrace rice plays a very important role in local food security and provides a vast genetic reserve, especially for variability in high-value bioactives. Its analysis can support the improvement of the nutritional value of rice and the development of materials with pharmaceutical and nutraceutical applications. Previous studies have reported variability in the fat-soluble content of rice and its coproducts, but these have examined only a subset of genotypes, most of which are commercial cultivars of Thai rice [22–24]. The genetic diversity of rice germplasms is important when categorizing sources of variation or identifying genotypes with superior nutraceutical lipid contents. The goal of the current study was to evaluate genetic variability in production of α-tocopherol, γ-oryzanol, campesterol, β-sitosterol, octacosanol, and squalene in Thai landrace rice germplasm of the *indica* subspecies. The lipid compounds that are the goal of genetic breeding programs can be used to select nutraceutical compound-rich genotypes. Moreover, these landraces represent an unexplored germplasm pool that may provide donors of alleles that favor nutritionally valuable fat-soluble phytochemicals. The study was also intended to provide valuable data to cultivators, nutritional researchers, and rice breeders.

2. Materials and Methods

2.1. Chemicals and Reagents

Authentic lipid compound standards were purchased from Sigma-Aldrich (USA): α-tocopherol (CAS: 10191-41-0), γ-oryzanol, 5-α-cholestane (CAS: 481-21-0), campesterol (CAS: 474-62-4), β-sitosterol

(CAS: 83-46-5), 1-octacosanol (CAS: 557-61-9), and squalene (CAS: 111-02-4). All chemicals and reagents were of analytical grade.

2.2. Rice Germplasm Description and Experimental Design

Eighty-one rice (*O. sativa* L. spp. *indica*) germplasms were collected from diverse geographical regions of Thailand and compared with KDML105 and RD6, which are respectively the most popular varieties of nonglutinous and glutinous rice. They were selected based on their pigmentation, seed size, and yield (Supplementary Table S1). These genotypes were planted in the 2016 rice growing season (June–December 2016) at the Experiment Field of the Department of Agricultural Technology, Thammasat University, Pathum Thani, Thailand (14°04′28.6″ N, 100°36′33.0″ E, and 7.8 m above sea level) using a randomized complete block design (RCBD) with three replications. Each genotype was planted as an experimental unit of four 1 × 5 m plots, with 45 cm between plots. The seedlings were transplanted at a hill spacing of 15 × 15 cm, with a single seedling per hill. Soil preparation, planting, and other agronomic practices were carried out uniformly following the recommendations for good agricultural practices (GAP). Ten panicles per genotype were randomly harvested depending on their harvesting maturity stages (data not shown) and oven-dried at 50 °C to a moisture content <14%. Seeds were manually dehulled and their color parameters were measured prior to milling. Samples were milled to a fine powder (CM190, Cemotec^{TM}), passed through a 100-mesh screen mesh, thoroughly mixed, and stored at −20 °C until analysis.

2.3. Color Parameters

Color attributes were measured using a HunterLab miniscan XE PLUS colorimeter (Mod. PL50, Hunter Associates Laboratory Inc., Reston, VA, USA) that was calibrated prior to data collection using a standard white HunterLab calibration reflector plate. The color value was determined from 2 g randomly-selected samples of each variety. The color was expressed as lightness (L*), chroma (C*), and hue angle (h°). C* represented the color intensity, and h° was expressed as a degree range from 0° to 360° (0° = red, 90° = yellow, 180° = green, and 270° = blue) [25].

2.4. Analysis of α-tocopherol and γ-oryzanol

Simultaneous determination of α-tocopherol and γ-oryzanol was conducted following the method of Butsat and Siriamornpun [22] with slight modifications. A 1 g sample of powdered rice was extracted with 10 mL of acetone and vortexed at maximum speed for 1 min. The solution was centrifuged at 2500 rpm for 20 min, after which the solvent was removed. The supernatant layer was combined, before evaporation to dryness in a low-temperature vacuum evaporator. The residual was further extracted twice, and determinations were made in triplicate.

Reversed-phase HPLC analysis of α-tocopherol and γ-oryzanol was performed using a Shimadzu system (Shimadzu Corporation, Kyoto, Japan) equipped with a binary pump (LC-20AC) and a diode array detector (SPD-M20A). Chromatographic separation was performed in an Xselect CHS C-18 column (4.6 × 250 mm, i.d. 5 μm) (Waters Corporation, Milford, MA, USA). The mobile phase consisted of acetonitrile/methanol (25:75, v/v) at a flow rate of 1.5 mL/min. The solution was passed through a 0.22 μm filter (Millipore Corp). Operating conditions were as follows: column temperature 38 °C, injection volume 20 μl, and photodiode-array detection at 292 and 325 nm for α-tocopherol and γ-oryzanol, respectively. Different dilutions of individual external standards ranging from low to high concentrations were prepared and standard curves were plotted. The results were expressed as μg lipid per g of dry weight.

2.5. Analysis of Campesterol, β-sitosterol, Octacosanol, and Squalene

Simultaneous chromatographic separation of campesterol, β-sitosterol, octacosanol, and squalene was conducted using GC-MS, following the method of Siriamornpun et al. [26], with slight modifications. A 1 g powdered brown rice sample was placed in a screw-capped tube containing 5 mL of synthetic antioxidant (ethanolic pyrogallol) with 1 mg of 5-α-cholestane as internal standard. The solutions were saponified with 2 mL of KOH (10.70 mM) and 2 mL of ethanol and NaCl (0.17 mM). The tubes were placed in a water bath and heated to 70 °C, then mixed at 10 min intervals over a 45 min digestion period. The solutions were cooled in an ice bath for 10 min, then 10 mL of NaCl (0.17 mM) was added. The suspension was then extracted twice using 10 mL of n-hexane/ethyl acetate (9:1 v/v). The upper organic layer was collected and dried in a low-temperature vacuum evaporator. The dry residue was derivatized with 100 µL of BSTFA:TMCS (99:1 v/v) and 1 mL of pyridine (99%) at 60 °C for 30 min. The residue was dissolved in 2 mL of hexane, then the clear solution (1 µL) was injected into the GC-MS system (QP2010, Shimadzu, Japan) equipped with an HP5 column (30 m, 0.25 mm i.d., 0.25 µm film thickness). Helium gas was used as a carrier gas at a constant flow rate of 1.0 mL/min. The injector temperature was held at 280 °C throughout the analysis, while the transfer line temperature was 230 °C. The initial column temperature of 60 °C was held for 1 min, increased at 30 °C/min to 250 °C and held for 10 min, then to 280 °C at 1 °C/min and maintained for a further 13 min. Linearity was evaluated by fresh preparation of individual external standard solutions, at five concentration levels in the interval of 2–100 µM for all compounds. The results were also expressed as µg lipid per g of dry weight.

2.6. Statistical Analysis

The results were analyzed from the mean of determinations for duplicate samples prepared for each genotype. Analysis of variance (ANOVA) and Duncan's multiple range test (DMRT) were performed in STATISTIX9 (Analytical Software, Tallahassee, FL, USA). Coefficients of variation (CV) were calculated from the ratios between standard deviations (SD) and population means, to represent the variability among genotypes. Hierarchical agglomerative clustering was then conducted for lipid compounds, using the Ward criterion. Principal component and cluster analysis of the lipid compounds were performed using JMP Pro software (version 14.0, SAS institute Inc., Chicago, IL, USA). A heat map showing Pearson's correlation coefficients for color parameters and lipid compounds was constructed using Microsoft Excel 2016.

3. Results and Discussion

3.1. Analysis of Variance

Identification of genetic resources with high levels of the targeted fat-soluble nutraceutical compounds is a necessary preliminary step when enhancing bioactive levels through conventional plant breeding [27]. Descriptive statistics for the α-tocopherol, γ-oryzanol, campesterol, β-sitosterol, octacosanol, and squalene contents of the rice germplasm are shown in Table 1 and Supplementary Table S2. Large variation was observed in the β-sitosterol content with a CV of 86.9%, followed by campesterol (54.8%), octacosanol (54.3%), and α-tocopherol (49.8%). The lowest variation was found for squalene content (CV = 6.2%). Two groups could be distinguished by distribution of α-tocopherol (Figure 1A). The first had an α-tocopherol content of between 663 and 1673 µg/g of dry weight. A high proportion of genotypes fell into this class, with a mean of 1083 µg/g of dry weight. The second group had lower levels of this compound across the full range (≤663 µg/g of dry weight). The frequency distribution for γ-oryzanol content was continuous, ranging from 7477 to 24,613 µg/g of dry weight (Figure 1B). Most rice genotypes (75%) had a γ-oryzanol content exceeding 13,865 µg/g of dry weight. Wide variation was observed in the campesterol and β-sitosterol content, with respective ranges from 495 to 6699 µg/g of dry weight and from 470 to 40,035 µg/g of dry weight (Figure 1C,D). Campesterol and β-sitosterol were normally distributed, though most of the genotypes (75%) had phytosterol contents lower than the mean values (2016 ± 1106 and 7024 ± 6107 µg/g of dry weight, respectively). Wide variation was also observed in octacosanol content, which ranged from 352 to 3522 µg/g of dry weight (Figure 1E). Half of the genotypes had an octacosanol content above the mean value (892 ± 484 µg/g of dry weight). Squalene content showed little variation, with a range from 175 to 266 µg/g of dry weight (Figure 1F). Most genotypes fell within the first two quartiles, and below the average value (183.7 ± 11.3 μg/g of dry weight). These results indicated that most of the variation in nutraceutical lipid content was accounted for by the genotype. This finding is in agreement with those of Goffman et al. [18], who reported that genetic factors have a greater influence on the oil and fatty acid contents than do seasonal factors. Therefore, selection for high levels of these compounds may require less frequent evaluation of rice germplasm across years or locations [28]. Miller and Engel [16] reported environmental influences on the content and composition of γ-oryzanol and seteryl ferulate. Bergman and Xu [28] reported that growing conditions had a greater effect on tocopherol, tocotrienol, and γ-oryzanol levels than did the rice genotype. The range in the nutraceutical lipid compound levels of grains may be due to a combination of genetic variability, soil conditions, and environmental factors. Rice breeders who wish to select for germplasm with optimal levels of these compounds should plant the genetic resources in multiple seasons or locations, to estimate the relative contributions made by the materials used and the growth conditions [29]. In practice, the conditions under which samples available for analysis were grown are unknown. Further study will be required to elucidate the relationship between factors such as climate or soil quality and agricultural practices [16]. The low rate of genetic variability is the main limiting factor in rice breeding programs aimed at increasing squalene levels. The goal of rice breeders is to create variation in the breeding materials by hybridization or induced mutations.

Table 1. Descriptive statistics for nutraceutical lipid compounds of the 83 rice germplasms [1/].

Compounds	Mean ± SD	Min.	Max.	CV	F-Value
α -tocopherol	920 ± 458	78	1,673	49.8	48.1 **
γ -oryzanol	16,501 ± 3,860	7,477	24,613	23.3	39.0 **
campesterol	2,016 ± 1,106	495	6,699	54.8	186.6 **
β-sitosterol	7,024 ± 6,107	470	40,035	86.9	8,451.8 **
octacosanol	892 ± 484	352	3,522	54.3	80.4 **
squalene	183.7 ± 11.3	175	266	6.2	61.7 **

** significant at 1% level. [1/] expressed as μg/g of dry weight.

3.2. Multivariate Analysis

One of the key challenges is the choice of variables to analyze, since more or less data are available on the geographical location, genotype, rice type, and other markers. The two most widely-used multivariate analysis techniques for interdependent responses of rice germplasm are principal component analysis (PCA) and hierarchical cluster analysis [30]. Principal components are extracted by order of contribution to the total variance and, by examining the loadings of the variables in the first components, it is possible to measure the relevance of each variable. The first four principle components are the most important in reflecting the variation among rice genotypes that is useful for genotypic classification [31]. In this study, the first four components contributed approximately 93.47% of the total variation (Table 2), giving a clear idea of the structure underlying the variables analyzed. The first principal component represented β-sitosterol, campesterol, and squalene, and explained 42.04% of the total variance. The second component was mainly attributed to γ-oryzanol and α-tocopherol and accounted for a further 29.58% of the total variance. The third component was mainly contributed by octacosanol and accounted for 13.67% of the total variance. The fourth component was attributed mainly to β-sitosterol and accounted for 8.16% of the total variance. These findings were in agreement with previous studies, which reported that β-sitosterol and campesterol were the major factors contributing to variation of lipid content in Korean rice cultivars [14] and other cereals [15].

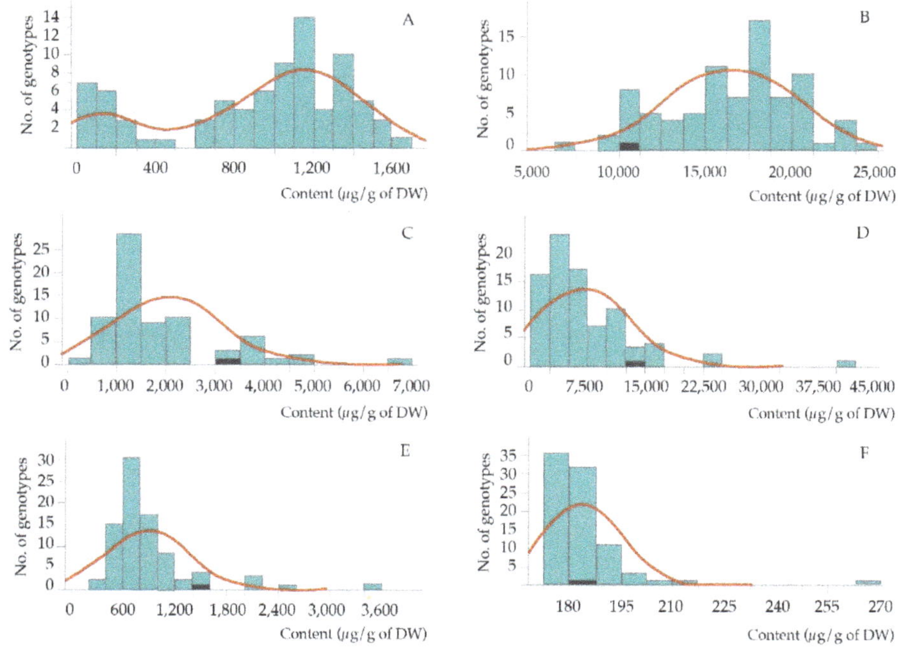

Figure 1. Distribution of the 83 rice germplasms by α–tocopherol (**A**), γ-oryzanol (**B**), campesterol (**C**), β-sitosterol (**D**), octacosanol (**E**), and squalene (**F**) contents.

Table 2. Coefficients and vectors associated with the first four principal components.

	First	Second	Third	Fourth
Eigenvalues	2.52	1.78	0.87	0.49
Percent variances	42.04	29.58	13.67	8.16
Cumulative % total variance	42.04	71.62	85.30	93.47
		Coefficient vector		
Compounds				
α-tocopherol	0.192	0.637	−0.194	0.07
γ-oryzanol	0.139	0.657	−0.167	0.21
campesterol	0.575	−0.053	−0.110	0.23
β-sitosterol	0.586	−0.139	−0.091	0.78
octacosanol	0.329	0.120	0.907	−0.46
squalene	0.401	−0.140	−0.091	−0.27

The goal of clustering is to divide data into distinct clusters in such a way that samples associated with the same cluster are considered similar in the pattern found, whereas samples associated with different clusters are as dissimilar as possible [32]. Based on the α-tocopherol, γ-oryzanol, campesterol, β-sitosterol, octacosanol, and squalene contents of 1–81 rice genotypes and two check varieties, nine distinct clusters were identified (Figure 2 and Supplementary Table S2). Cluster I comprised the three genotypes TU-001, TU-002, and TU-159. This cluster showed moderate levels of fat-soluble compounds, though TU-002 and TU-159 were low in squalene content. Cluster II comprised 12 genotypes (TU-002, TU-030, TU-070, TU-071, TU-106, TU-114, TU-138, TU-169, TU-172, TU-202, TU-203, and TU-241). Most exhibited average levels of α-tocopherol, γ-oryzanol, campesterol, β-sitosterol, and octacosanol, but low squalene levels. TU-203 had the highest α-tocopherol levels, and TU-241 the highest γ -oryzanol levels. Cluster III comprised 14 genotypes (TU-009, TU-028, TU-032, TU-043, TU-049, TU-050, TU-069,

TU-072, TU-115, TU-117, TU-124, TU-131, TU-196, and TU-197), most of which had moderate levels of fat-soluble compounds. TU-028, TU-072, TU-117, TU-124, and TU-197 had high γ-oryzanol levels.

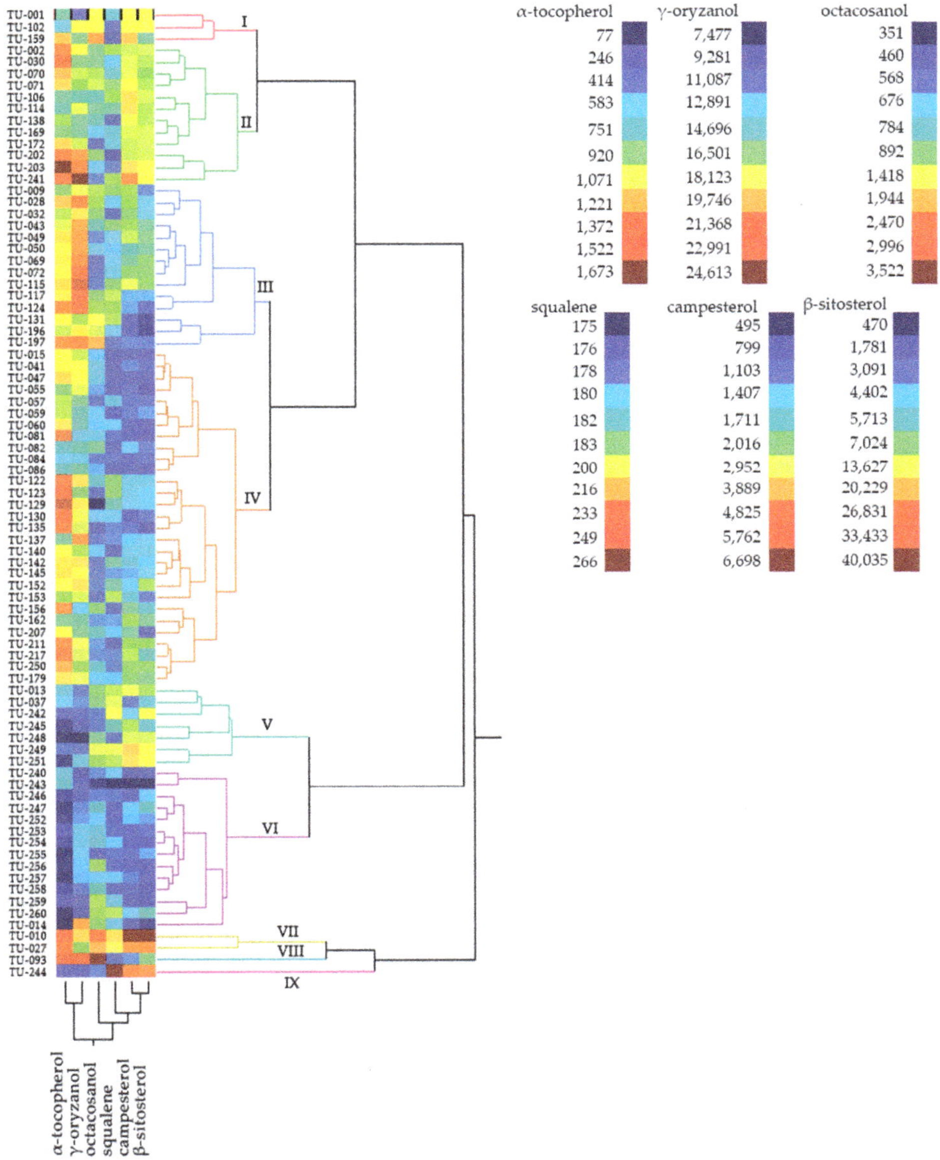

Figure 2. Dendrogram of genetic relationships among the 83 rice germplasms. Nine main clusters (I to IX) were formed. Clustering was performed using Ward's method based on six nutraceutical lipid compounds (scale: distance scale). [a] Clustering is done by column as well as row; the columns must be measured on the same scale.

Cluster IV was the largest group (34.9% of all genotypes) and comprised TU-015, TU-041, TU-047, TU-055, TU-057, TU-059, TU-060, TU-081, TU-082, TU-084, TU-086, TU-122, TU-123, TU-129, TU-130, TU-135, TU-137, TU-140, TU-142, TU-145, TU-152, TU-153, TU-156, TU-162, TU-179, TU-207, TU-211, TU-217, and TU-250. This group showed moderate average levels of α-tocopherol and γ-oryzanol, matching group I. However, campesterol, β-sitosterol, octacosanol, and squalene levels were low. TU-081, TU-122, TU-123, TU-129, TU-130, and TU-135 showed high α-tocopherol levels.

Cluster V comprised the seven genotypes TU-013, TU-037, TU-242, TU-245, TU-248, TU-249, and TU-251. Most genotypes in this cluster showed low average levels of α-tocopherol and γ–oryzanol, and medium levels of campesterol, β-sitosterol, and squalene. Cluster VI comprised 14 rice genotypes (TU-014, TU-240, TU-243, TU-246, TU-247, TU-252, TU-253, TU-254, TU-255, TU-256, TU-257, TU-258, TU-259, and TU-260), most of which had low average levels of all lipid compounds. The exception was TU-014, which had high levels of γ-oryzanol.

Cluster VII comprised TU-010 and TU-027. This small cluster had high average values of α-tocopherol, γ–oryzanol, campesterol, β-sitosterol, and octacosanol. TU-010 had the highest levels of campesterol and β-sitostero, whereas TU-027 had moderate levels of γ–oryzanol and squalene. Cluster VIII comprised the single TU-093 genotype, which had extremely high values for octacosanol, high values for α-tocopherol and γ-oryzanol, but low to moderate values for campesterol, β-sitosterol, and squalene. Cluster IX comprised the TU-224 genotype, which had the highest squalene content, high campesterol and β-sitosterol content, but low α-tocopherol, γ-oryzanol, and octacosanol content.

The information obtained through cluster analysis was particularly useful as the best performing genotypes fell within one cluster, allowing them to be differentiated from the others [29]. The use of six nutraceutical lipid compounds therefore turned out to be appropriate for clustering rice germplasm. As germplasm characterization can support the conservation of crop genetic resources, crop protection, classification of the genotypes into heterotic groups, and crop improvement [31,33,34], the information from our study should be of use to rice breeders wishing to select parental lines that yield improved novel varieties. However, further germplasm characterization using molecular approaches will be necessary to establish whether this results from gene introgression.

Descriptions of nine clusters based on α-tocopherol, γ-oryzanol, campesterol, β-sitosterol, octacosanol, and squalene content are given in Table 3. Cluster I was characterized by low levels of α-tocopherol and γ-oryzanol and moderate levels of the other fat-soluble compounds. Relatively high γ-oryzanol levels were found in Clusters II, III, and IV, and especially in Cluster III. However, levels of other fat-soluble compounds were low. Clusters V and VI had low levels of all nutraceutical lipid compounds, and rice genotypes in these clusters are unsuitable for rice breeding programs. The genotypes in cluster VII were interesting, being relatively high in campesterol and β-sitosterol and moderate-to-high in other nutraceutical lipids. The rice genotype in cluster VIII was a potentially interesting source of α-tocopherol, γ-oryzanol, and octacosanol, which were at high levels. Finally, the genotype in cluster IX is also potentially useful for breeding programs, being high in squalene. Hybridization of group VII with groups VIII or IX may yield strains with enhanced nutraceutical lipid contents.

Table 3. Average nutraceutical lipid contents in nine clusters [1/].

Clusters	*n*	Lipid Compounds (μg/g of Dry Weight)					
		α-tocopherol	γ-oryzanol	Campesterol	β-sitosterol	Octacosanol	Squalene
I	3	910.4 ± 281.6	15,108.3 ± 4,276.4	3,393.5 ± 399.4	12,100.5 ± 4,841.9	1,737.0 ± 387.6	179.27 ± 2.2
II	12	1,163.5 ± 235.3	17,788.9 ± 2,600.0	3,036.0 ± 656.4	10,662.4 ± 3,237.9	850.94 ± 200.3	182.7 ± 4.0
III	14	1,127.8 ± 111.8	20,644.7 ± 1,826.9	1,870.7 ± 612.4	4899.7 ± 2,352.6	989.4 ± 420.3	183.5 ± 4.8
IV	29	1,164.3 ± 255.7	17,357.5 ± 1,964.4	1,536.8 ± 595.4	4698.9 ± 288.7	656.2 ± 105.9	180.2 ± 3.2
V	7	350.9 ± 259.2	11,044.0 ± 2,020.6	2,603.0 ± 995.3	11,116.7 ± 4,328.5	903.2 ± 283.1	193.0 ± 9.4
VI	14	214.2 ± 227.0	12,217.4 ± 2,841.3	1,060.6 ± 252.9	2,929.2 ± 1,743.9	770.9 ± 212.9	180.0 ± 4.0
VII	2	1,418.9 ± 75.0	18,045.0 ± 1,841.3	5,531.2 ± 1,650.7	32,007.5 ± 11,352.6	2,273.0 ± 352.9	206.2 ± 7.3
VIII	1	1,565.8 ± 0.0	23,423.0 ± 0.0	1,267.9 ± 0.0	6,974.8 ± 0.0	3,522.0 ± 0.0	178.6 ± 0.0
IX	1	283.7 ± 0.0	10,521.0 ± 0.0	4,571.0 ± 0.0	24,102.0 ± 0.0	606.7 ± 0.0	266.1 ± 0.0

n = Number of genotypes. [1/] Number of rice clusters from.

From the 81 test genotypes, the four that ranked top in fat-soluble nutraceutical content were identified (Figure 3). TU-093 and TU-010 were superior in α-tocopherol content, though no statistically significant difference from the KDML105 check variety was found (Figure 3A). These genotypes were also higher in α-tocopherol than the glutinous rice check variety RD6 (by 2.09 and 1.97 times, respectively). TU-093 was also higher in γ-oryzanol than the RD6 or KDML105 check varieties, by 4.5 to 85.2 times (Figure 3B). Vegetarian diets that are high in phytosterols are associated with a reduced risk of breast cancer [23]. β-sitosterol and campesterol are the predominant and most common sterols found in rice [8,14,35], and these fat-soluble compounds were the focus of our study. TU-010 had the highest levels of campesterol and β-sitosterol in the study, and also higher than RD6 and KDML105, by multiples of 4.5 to 85.2 times (Figure 3C,D). It was also found to contain higher concentrations of β-sitosterol (40,035 μg/g of dry weight) than those reported for pigmented rice bran oil (4980–6250 μg/g RBO) [36] and short grain rice (12,480 μg/g of dry weight) [35], though lower concentrations than Pakistani rice (65,630 μg/g of dry weight) [8]. A rice genotype producing high levels of β-sitosterol should be commercially viable.

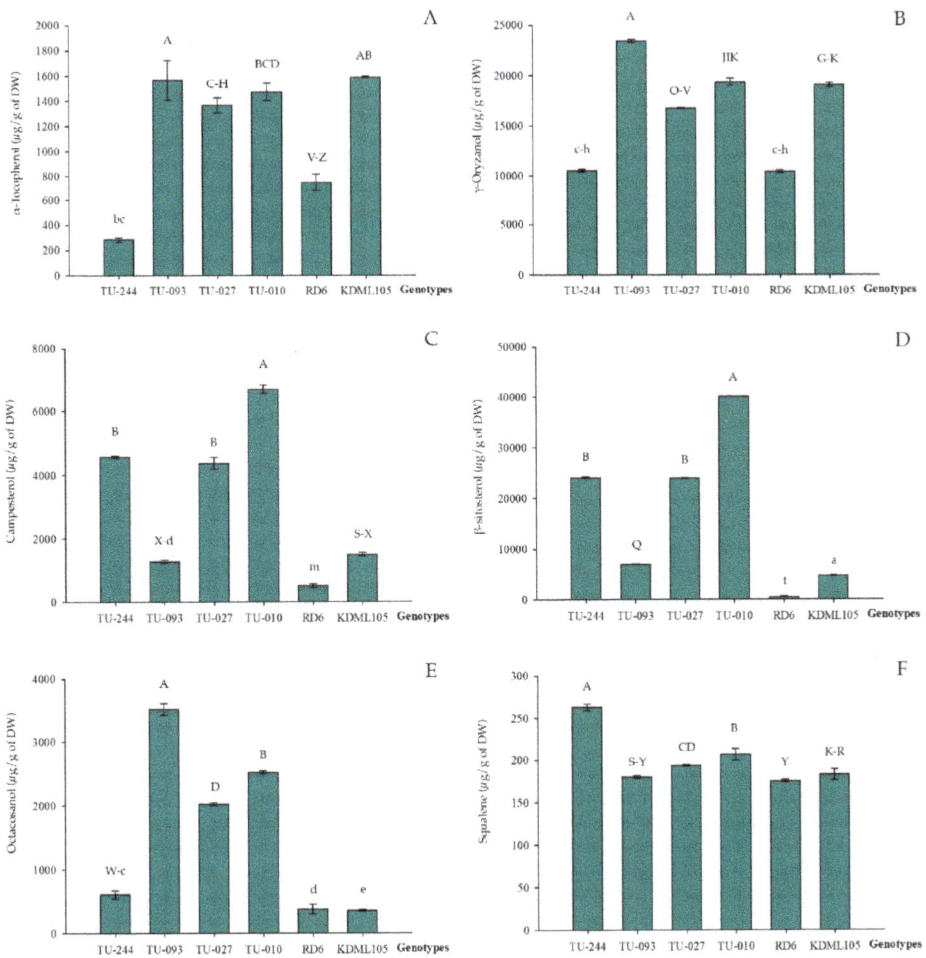

Figure 3. Comparison of germplasms from the predominant groups with check varieties by α-tocopherol (**A**), γ-oryzanol (**B**), campesterol (**C**), β-sitosterol (**D**), octacosanol (**E**), and squalene (**F**) content. [a] Means with different upper case and lower case letters are significantly different ($p \leq 0.05$) by Duncan's multiple range test (DMRT).

Octacosanol is the main component of policosanol, which can reduce total cholesterol, low-density lipoprotein, and platelet aggregation [5]. TU-093 had 3522 μg/g by dry weight of octacosanol, higher than the RD6 or KDML105 check genotypes by 9.4 and 10.0 times (Figure 3E), higher than short grain rice (100 μg/g by dry weight) [35], but lower than Korean rice (6,738 μg/g of lipid extract) [5]. Squalene, a fat-soluble compound, has been reported to be a quencher of singlet oxygen and a free radical scavenger [14]. Little information is available on the squalene content of rice germplasm, and especially of landraces. The squalene content of TU-244 (266.05 μg/g by dry weight) was 1.5 times that of RD6, 1.4 that of KDML105 (Figure 3F), and higher than that of Korean rice (0.16 to 59 μg/g of dry weight) [14]. TU-027 had medium-to-high levels of all lipid compounds. Overall, in terms of the high-value lipid compounds investigated, the *indica* subspecies were superior, especially TU-010, TU-093, and TU-244. This suggests potential uses in functional foods and nutraceuticals, improving added value [8]. Introgression of genes from these superior genotypes may allow breeding programs to create novel rice varieties containing these high-value lipid compounds.

3.3. Correlation

Colorimetric analysis is used to characterize and quantify the color properties of pigments and antioxidant-rich foods [37]. As the hypothesis underlying this study was that color properties are associated with nutraceutical lipid compounds, we measured the color parameters from light to dark. We also explored the use of color parameters as an indirect tool for selection of rice genotypes with high levels of fat-soluble compounds. The hue angle (h°) was found to be negatively and significantly correlated with lightness (L*) and the chromatic parameter (C*) (Figure 4). A positive and moderately significant correlation was found between L* and C* color parameters. Color parameters L* and C* were weakly correlated with α-tocopherol, but no other significant correlations were found between color parameters and other nutraceutical lipid compounds. Superior rice genotypes, including TU-010, TU-093, and TU-244, had colorless kernels. These results suggest that color parameters should not be used as indirect criteria for selection of these fat-soluble compounds. This confirms the findings of Min et al. [38], who found no association between rice bran color and natural vitamin E analogue or γ-oryzanol. However, a positive correlation has been reported between γ-oryzanol content and the yellow parameter in upland rice bran oil [39]. Another report noted that rice coproduct oils produced from colored rice contained higher concentrations of oryzanol and phytosterols than noncolored rice [36]. Research on this connection is continuing.

A positive and significant correlation was found between α-tocopherol and γ-oryzanol, but no correlations with other fat-soluble compounds. Similar results have been reported in brown rice with different origins [8,14,40]. These results suggest that simultaneous improvement in the α-tocopherol and γ-oryzanol of rice is possible. β-sistosterol was positively and significantly correlated with campesterol and moderately correlated with squalene. The identification of strong positive correlations between fat-soluble compounds suggests that pleiotropy is involved, with genes controlling many traits simultaneously [29]. Improvement in the lipid-soluble composition may also improve the nutritional properties of rice. However, the lack of correlation between α-tocopherol and γ-oryzanol and other lipid compounds suggests that boosting the levels of these fat-soluble compounds will require the development of new varieties through hybridization and selection.

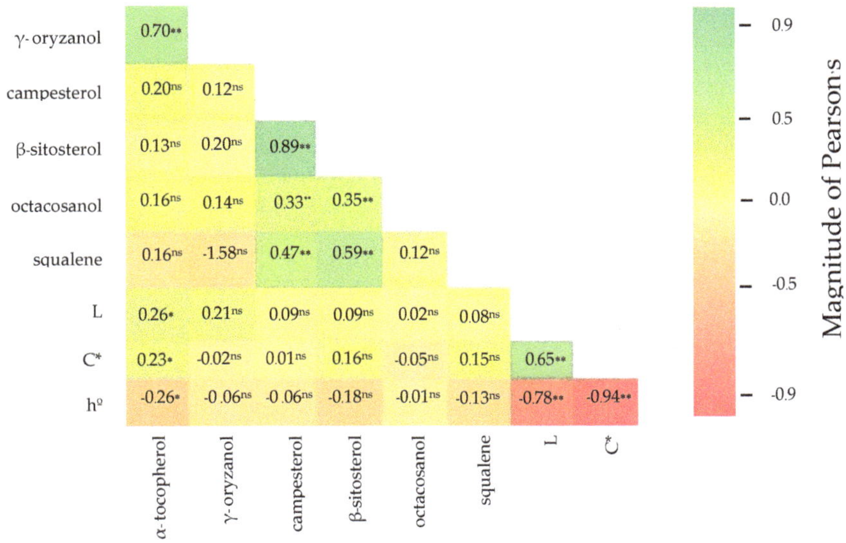

Figure 4. Triangular heat map representing Pearson's correlation coefficients between color parameters and nutraceutical lipid content in 83 rice germplasms. (*,** The correlation is significant at $p < 0.05$, 0.01, respectively; ns; non-significant).

4. Conclusions

The rice germplasms in this study showed considerable variability in levels of β-sitosterol, campesterol, octacosanol, α-tocopherol, and γ-oryzanol, though not of squalene. This variability can be exploited in rice breeding programs. Principal component analysis extracted four components that explained 93.47% of total variation. The 83 rice germplasms were grouped into nine distinct clusters, based on nutraceutical lipid content. TU-010, TU-027, TU-093, and TU-244 were found to be genotypes with high levels of fat-soluble compounds. Correlations between color parameters and these compounds were weak, and of little use as indirect markers for genetic selection of nutritional properties. However, α-tocopherol was strongly and positively correlated with γ-oryzanol, and β-sistosterol was positively correlated with campesterol and squalene. The findings of this study will help breeders to develop new rice varieties with enhanced nutraceutical lipid content.

Supplementary Materials: The following are available online at http://www.mdpi.com/2073-4395/9/12/823/s1, Table S1: List of rice germplasm, their origin and pigmentation, seed sizes, and color parameters. Table S2: Nutraceutical lipid compound contents in 83 rice germplasm.

Author Contributions: Conceptualization, B.H., K.L. and B.S.; Funding acquisition, B.H.; Formal analysis, B.H., S.B. and K.P.; Methodology, B.H., K.P., S.B., B.R. and K.L.; Writing—original draft, B.H.; Writing—review & editing, B.H., S.B., B.R. and K.L.

Funding: This study was funded by the Research Grant for New Scholar, Thailand Research Fund (Grant No. MRG5980124).

Acknowledgments: This study was supported by the Research Grant for New Scholar, Thailand Research Fund (Grant No. MRG5980124). We also wish to thank Sakunkan Simla (Faculty of Technology, Mahasarakham University, Thailand) for collecting data regarding grain color parameters. Marvin P. Scott is acknowledged for providing constructive comments and proofreading manuscript. Finally, we would like to thank Scott's students for clustering of fat-soluble compounds by using the JMP software.

Conflicts of Interest: The authors declare no conflict of interest. The founding sponsors had no role in the design of the study; in the collection, analyses, or interpretation of data; in the writing of the manuscript, and in the decision to publish the results.

References

1. Yeates, K.; Lohfeld, L.; Sleeth, J.; Morales, F.; Rajkotia, J.; Ogedegbe, O. A global perspective on cardiovascular disease in vulnerable populations. *Can. J. Cardiol.* **2015**, *31*, 1081–1093. [CrossRef] [PubMed]
2. Buttar, H.S.; Li, T.; Ravi, N. Prevention of cardiovascular diseases: Role of exercise, dietary interventions, obesity and smoking cessation. *Exp. Clin. Cardiol.* **2005**, *10*, 229–249. [PubMed]
3. Goufo, P.; Trindade, H. Rice antioxidants: Phenolic acids, flavonoids, anthocyanins, proanthocyanidins, tocopherols, tocotrienols, γ-oryzanol, and phytic acid. *Food Sci. Nutr.* **2014**, *2*, 75–164. [CrossRef] [PubMed]
4. Kim, H.W.; Kim, J.B.; Shanmugavelan, P.; Kim, S.N.; Cho, Y.S.; Kim, H.R.; Lee, J.T.; Jeon, W.T.; Lee, D.J. Evaluation of Î³-oryzanol content and composition from the grains of pigmented rice-germplasms by LC-DAD-ESI/MS. *BMC Res. Notes* **2013**, *6*, 149. [CrossRef]
5. Kim, N.H.; Kwak, J.; Bail, J.Y.; Yoon, M.R.; Lee, J.S.; Yoon, S.W.; Kim, I.H. Changes in lipid substances in rice during grain development. *Phytochemistry* **2015**, *116*, 170–179. [CrossRef]
6. Pitija, K.; Nakornriab, M.; Sriseadka, T.; Vanavichit, A.; Wongpornchai, S. Anthocyanin content and antioxidant capacity in bran extracts of some Thai black rice varieties. *Int. J. Food Sci. Technol.* **2013**, *48*, 300–308. [CrossRef]
7. Sompong, R.; Siebeinhandl-Ehn, S.; Linsberger-Martin, G.; Berghofer, E. Physicochemical and antioxidative properties of red and black rice varieties from Thailand, China and Sri Lanka. *Food Chem.* **2011**, *124*, 132–140. [CrossRef]
8. Zubair, M.; Anwar, F.; Ashraff, M.; Uddin, M.K. Characterization of high-value bioactives in some selected varieties of Pakistani rice (*Oryza sativa* L.). *Int. J. Mol. Sci.* **2012**, *13*, 4608–4622. [CrossRef]
9. Chotimarkorn, C.; Benjakul, S.; Silalai, N. Antioxidant components and properties of five long-grained rice bran extracts from commercial available cultivars in Thailand. *Food Chem.* **2008**, *111*, 636–641. [CrossRef]
10. Goufo, P.; Trindade, H. Factors influencing antioxidant compounds in rice. *Crit. Rev. Food Sci. Nutr.* **2017**, *57*, 893–922. [CrossRef]
11. Marangoni, F.; Poli, A. Phytosterols and cardiovascular health. *Pharmacol. Res.* **2010**, *61*, 193–199. [CrossRef] [PubMed]
12. Nagendra Prasad, M.N.; Sanjay, K.R.; Shravya Khatokar, M.; Vismaya, M.N.; Nanjunda Swamy, S. Health benefits of rice bran-A review. *J. Nutr. Food Sci.* **2011**, *1*, 1–7. [CrossRef]
13. Taylor, J.C.; Rapport, L.; Lockwood, G.B. Octacosanol in human health. *Nutrition* **2003**, *19*, 192–195. [CrossRef]
14. Yoon, S.W.; Lee, J.; Pyo, Y.G.; Oh, S.W.; Lee, J.S.; Kim, I.H. Nutraceutical lipid substances in Korean rice cultivars. *J. Food Nutr. Res.* **2014**, *2*, 40–46.
15. Nurmi, T.; Lampi, A.M.; Nystro, L.; Piironen, V. Effects of environment and genotype on phytosterols in wheat in the HEALTHGRAIN diversity screen. *J. Agric. Food Chem.* **2010**, *58*, 9814–9823. [CrossRef] [PubMed]
16. Miller, A.; Engel, K.H. Content of γ-oryzanol and composition of steryl ferulates in brown rice (*Oryza sativa* L.) of European origin. *J. Agric. Food Chem.* **2006**, *54*, 8127–8133. [CrossRef] [PubMed]
17. Huang, S.H.; Ng, L.T. Quantification of tocopherols, tocotrienols, and γ-oryzanol contents and their distribution in some commercial rice varieties in Taiwan. *J. Agric. Food Chem.* **2011**, *59*, 11150–11159. [CrossRef]
18. Goffman, F.D.; Pinson, S.; Bergman, C. Genetic diversity for lipid content and fatty acid profile in rice bran. *J. Am. Oil Chem. Soc.* **2003**, *80*, 485–490. [CrossRef]
19. Moonsap, P.; LaksanavilatSittipun, N.; Tasanasuwan, S.; Kate-Ngam, S.; Jantasuriyarat, C. Genetic diversity of Indo-China rice varieties using ISSR, SRAP and InDel markers. *J. Genet.* **2019**, *98*, 80. [CrossRef]
20. Pasudee, T.; Jamjod, S.; Chiang, Y.C.; Rerkasem, B.; Schaal, B.A. Genetic structure and isolation by distance in a landrace of Thai rice. *Proc. Natl. Acad. Sci. USA* **2009**, *106*, 13880–13885. [CrossRef]
21. Rerkasem, B.; Rerkasem, K. Agrodiversity for in situ conservation of Thailand's native rice germplasm. *CMU J.* **2002**, *1*, 129–148.
22. Butsat, S.; Siriramornpun, S. Antioxidant capacities and phenolic compounds of the husk, bran and endosperm of Thai rice. *Food Chem.* **2010**, *119*, 606–613. [CrossRef]
23. Suttiarporn, P.; Chumpolsri, W.; Mahatheeranont, S.; Luangkamin, S.; Teepsawang, S.; Leardkamolkarn, V. Structures of phytosterols and triterpenoids with potential anti-cancer activity in bran of black non-glutinous rice. *Nutrients* **2015**, *7*, 1672–1687. [CrossRef] [PubMed]

24. Pokkanta, P.; Sookwong, P.; Tanang, M.; Setchaiyan, S.; Boontakham, P.; Mahatheeranont, S. Simultaneous determination of tocols, γ-oryzanols, phytosterols, squalene, cholecalciferol and phylloquinone in rice bran and vegetable oil samples. *Food Chem.* **2019**, *217*, 630–638. [CrossRef]
25. McGuire, R.G. Reporting of objective color measurements. *HortScience* **1992**, *27*, 1254–1255. [CrossRef]
26. Siriamornpun, S.; Tangkhawanit, E.; Kaewseejan, N. Reducing retrogradation and lipid oxidation of normal and glutinous rice flours by adding mango peel powder. *Food Chem.* **2016**, *201*, 160–167. [CrossRef]
27. Renuka, N.; Mathure, S.V.; Thengaue, R.J.; Nadaf, A.B. Determination of some minerals and β-carotene contents in aromatic *indica* rice (*Oryza sativa* L.) germplasm. *Food Chem.* **2016**, *191*, 2–6. [CrossRef]
28. Berman, C.J.; Xu, Z. Genotype and environment effects on tocopherol, tocotrienol, and γ-oryzanol contents of Southern U.S. rice. *Cereal Chem.* **2003**, *80*, 446–449. [CrossRef]
29. Ali, A.; Sadaqat, H.A.; Kashif, M.; Wahid, M.A. Exploration of breeding potential for genetic biofortification in spring wheat. *Pak. J. Agric. Sci.* **2018**, *55*, 793–799.
30. Maione, C.; Berbosa, R.M. Recent applications of multivariate data analysis methods in the authentication of rice and the most analyzed parameters: A review. *Crit. Rev. Food Sci. Nutr.* **2019**, *59*, 1868–1879. [CrossRef]
31. Sanni, K.A.; Fawole, S.A.; Ogunbayp, D.D.; Tia, E.A.; Somado, K.; Futakuchi, M.S.; Nwilene, F.E.; Guei, R.G. Multivariate analysis of diversity of landrace rice germplasm. *Crop Sci.* **2012**, *52*, 494–504. [CrossRef]
32. Madhulatha, T.S. An overview on clustering methods. *IOSR-JEN* **2012**, *2*, 719–725. [CrossRef]
33. Duangpapeng, P.; Lertrat, K.; Lomthaisong, K.; Scott, M.P.; Suriharn, B. Variability in anthocyanins, phenolic compounds and antioxidant capacity in the tassels of collected waxy corn germplasm. *Agronomy* **2019**, *9*, 158. [CrossRef]
34. Harakotr, B.; Suriharn, B.; Scott, M.P.; Lertrat, K. Genotypic variability in anthocyanins, total phenolics, and antioxidant activity among diverse waxy corn germplasm. *Euphytica* **2014**, *203*, 237–248. [CrossRef]
35. Ha, T.Y.; Ko, S.N.; Lee, S.M.; Kim, H.R.; Chung, S.H.; Kim, S.R.; Yoon, H.H.; Kim, I.H. Changes in nutraceutical lipid components of rice at different degrees of milling. *Eur. J. Lipid Sci. Technol.* **2006**, *108*, 175–181. [CrossRef]
36. Mingyai, S.; Srikaeo, K.; Kettawan, A.; Singanusong, R. Effects of extraction methods on phytochemicals of rice bran oils produced from colored rice. *J. Oleo Sci.* **2018**, *67*, 135–142. [CrossRef]
37. Hernanz, D.; Recamales, A.F.; Melendez-Martinez, A.J.; Gonzalez-Miret, M.L.; Heredia, F. Multivariate statistical analysis of the color-anthocyanin relationships in different soilless-grown strawberry genotypes. *J. Agric. Food Chem.* **2008**, *56*, 2735–2741. [CrossRef]
38. Min, B.; McClung, A.M.; Chen, M.H. Phytochemicals and antioxidant capacities in rice brans of different color. *J. Food Sci.* **2011**, *76*, 117–126. [CrossRef]
39. Ruen-ngam, D.; Thaiwai, C.; Sukonthamut, S.; Nokkoul, R.; Tadtong, S. Evaluation of nutrient content and antioxidant, neuritogenic, and neuroprotective activities of upland rice bran oil. *ScienceAsia* **2018**, *44*, 257–267. [CrossRef]
40. Heinemann, R.J.B.; Xu, Z.; Godber, J.S.; Lanfer-Marquez, U.M. Tocopherols, tocotrienols, and γ-oryzanol contents in *Japonica* and *Indica* subspecies of rice (*Oryza sativa* L.) cultivated in Brazil. *Cereal Chem.* **2008**, *85*, 243–247. [CrossRef]

agronomy

MDPI

Article

Genetic Diversity of Sweet Potato (*Ipomoea batatas* L. Lam) Germplasms Collected Worldwide Using Chloroplast SSR Markers

Kyung Jun Lee †, Gi-An Lee †, Jung-Ro Lee, Raveendar Sebastin, Myoung-Jae Shin, Gyu-Taek Cho and Do Yoon Hyun *

National Agrobiodiversity Center, National Institute of Agricultural Sciences (NAS), RDA, Jeonju 54874, Korea; lkj5214@korea.kr (K.J.L.); gkntl1@korea.kr (G.-A.L.); jrmail@korea.kr (J.-R.L.); raveendars@gmail.com (R.S.); smj1204@korea.kr (M.-J.S.); gtcho@korea.kr (G.-T.C.)

* Correspondence: dyhyun@korea.kr; Tel.: +82-63-238-4912

† The first two authors equally contributed.

Received: 10 October 2019; Accepted: 11 November 2019; Published: 13 November 2019

Abstract: Sweet potato (*Ipomoea batatas* L. Lam) is an important food crop widely cultivated in the world. In this study, nine chloroplast simple sequence repeat (cpSSR) markers were used to analyze the genetic diversity and relationships of 558 sweet potato accessions in the germplasm collection of the National Agrobiodiversity Center (NAC). Eight of the nine cpSSR showed polymorphisms, while Ibcp31 did not. The number of alleles per locus ranged from two to four. In general, the Shannon index for each cpSSR ranged from 0.280 to 1.123 and the diversity indices and unbiased diversity ranged from 0.148 to 0.626, and 0.210 to 0.627, respectively. Results of the median-joining network showed 33 chlorotypes in 558 sweet potato accessions. In factor analysis, 558 sweet potato accessions were divided into four clusters, with clusters I and II composed only of the sweet potato accessions from Korea, Japan, Taiwan, and the USA. The results of this study confirmed that the genetic diversity of the female parents of sweet potato accessions conserved at the NAC is low and therefore more sweet potato accessions need to be collected. These results will help to establish an efficient management plan for sweet potato genetic germplasms at the NAC.

Keywords: chloroplast; *Ipomoea batatas*; simple sequence repeat; sweet potato; plant germplasm

1. Introduction

Plant genetic resource, one of the most essential natural resources, has been a research topic resulting in major advancement in the field [1]. Gene banks are concerned with the maintenance of crop resource genetic variations, and plant genetic resource conservation is now receiving greater attention [1,2]. In order to establish effective and efficient conservation practices for plant genetic resources, understanding the genetic diversity between and within population is important [3].

Sweet potato (*Ipomoea batatas* L. Lam.) is a vegetative propagation crop that belongs to the family Convolvulaceae [4]. The origin of sweet potato is either the Central or South America [5]. Sweet potato is attractive to resource-poor farmers because they have the highest rate of production per unit area/time [6]. It also has a short growth period and is easily propagated and grown with good production in various climates and farming systems [7].

Molecular techniques have been widely adopted as powerful tools for germplasm characterization, cultivar identification, phylogenetic studies, and diversity analysis in many crop plants [8]. In sweet potato studies, molecular markers have been used to evaluate the phylogenetics and germplasm to study the origin of sweet potato and its dissemination [9]. Random amplified polymorphic DNA (RAPD), DNA amplification fingerprinting (DAF), and simple sequence repeat (SSR) markers have been used for the estimation of genetic diversity and genetic relationships [6,8,10,11]. Previous studies have explained the genetic diversity and origin of sweet potato landraces in Mexico, Peru, and New Guinea using chloroplast and nuclear SSR [12,13].

During early genome sequencing projects, chloroplast DNAs (cpDNAs) were of interest because of their small size [14]. The chloroplast (CP) genome was observed in an extremely conserved manner not only in terms of gene numbers, but also in their arrangement [15]. CpSSR derived from the chloroplast genome represents ideal complementary molecular tools to nuclear genetic markers. This is because the SSR loci in the chloroplast genome are often distributed throughout the noncoding regions where higher sequence variations exist, shown to be due to low evolutionary rate and an almost nonexistent recombination rate in chloroplast DNA [16–18]. Therefore, cpSSR markers can be used to investigate population genetics and biogeography, and unravel the genetic relationships of closely related species [19].

During the 18th century, sweet potatoes were brought from Japan to various parts of Korea as a famine-relief crop [20]. At present, about 700 sweet potato germplasms have been collected worldwide at the National Agrobiodiversity Center (NAC) at the Rural Development Administration in Korea. However, analysis of genetic diversity in the preserved sweet potato accessions in NAC is lacking. Therefore, it is necessary to learn the genetic relationship of the sweet potato accessions to efficiently manage sweet potato germplasms. In this study, 558 sweet potato accessions conserved at the NAC were analyzed using nine cpSSRs to evaluate the genetic diversity and determine the appropriate panel of sweet potato germplasm for sweet potato improvement and conservation.

2. Materials and Methods

2.1. Plant Materials

Fresh leaves were randomly collected from five to ten individuals selected from 558 sweet potato (*Ipomoea Batatas* (L.) Lam) accessions conserved at the National Agrobiodiversity Center (NAC) in South Korea. The 558 sweet potato accessions were collected from ten countries including 190 accessions from South Korea (KOR), 123 from Japan (JPN), 73 from Taiwan (TWN), 50 from Peru (PER), 43 from China (CHN), 30 from Indonesia (IDN), 25 from the United States (USA), nine from the Philippines (PHL), nine from New Zealand (NZE), and six from North Korea (PRK) (Table 1 and Table S1).

Table 1. Number of sweet potato accessions in this study.

	Unknown	Breeding Line	Cultivar	Landrace	Total
CHN [1]	11	2	29	1	43
IDN		3	18	9	30
JPN	15	22	79	7	123
KOR		125	48	17	190
NZE			9		9
PER			50		50
PHL	1		8		9
PRK	2		3	1	6
TWN	31	32	10		73
USA			22	3	25
Total	60	184	276	38	558

[1] CHN, China; IND, Indonesia; JPN, Japan; KOR, South Korea; NZE, New Zealand; PER, Peru; PHL, Philippines; PRK, North Korea; TWN, Taiwan; USA, United States.

2.2. DNA Extraction

DNA was extracted from 100 mg of freeze-dried leaves in each sweet potato accession. The DNeasy plant mini kit (Qiagen, Hilden, Germany) was used for the extractions. The DNA quality and quantity of each sample were determined by electrophoresis in 1% (w/v) agarose gels and spectrophotometry

(Epoch, BioTek, Winooski, VT, USA). The extracted DNA was diluted to 30 ng/uL and was stored at −20 °C until further PCR amplification.

2.3. Chloroplast SSR Genotyping

For chloroplast SSR analysis, a total of nine cpSSRs were fluorescently labelled (6-FAM, HEX, and NED) and used for detection of the amplification products (Table 2). The PCR reactions were carried out in a 25 uL mixture containing 30 ng template DNA, 1x PCR buffer, 1.5 mM $MgCl_2$, 0.2 mM of each dNTPs, 0.5 um of each primer, and 1 U Taq polymerase (Inclone, Seongnam, Korea). The amplification was performed with cycling conditions of an initial denaturation at 94 °C for 5 min, followed by 35 cycles of denaturation at 95 °C for 30 s, annealing at 55 °C for 30 s, extension at 72 °C for 1 min, and a final extension step at 72 °C for 10 min. Each amplicon was resolved on an ABI Prism 3500 DNA sequencer (ABI3500, Thermo Fisher Scientific Inc., Wilmington, DE, USA) and was scored using Gene Mapper Software (Version 4.0, Thermo Fisher Scientific Inc., Wilmington, DE, USA)

Table 2. Nine cpSSR markers used in this study.

SSR Marker	Primer Sequences (5′→3′)	Dye
ccmp2	F: GATCCCGGACGTAATCCTG R: ATCGTACCGAGGGTTCGAAT	6-FAM
NTCP18	F: CTGTTCTTTCCATGACCCCTC R: CCACCTAGCCAAGCCAGA	HEX
NTCP28	F: TCCAATGGCTTTGGCTA R: AGAAACGAAGGAACCCAC	NED
NTCP26	F: GCAATTGCAATGGCTTCTTTA R: TTTATGTTCGGTGGAAATCACA	6-FAM
Ibcp5	F: GCTCTCACGCTCAATTACTTA R: ATGCTTAATTGACGACCTGT	HEX
Ibcp8	F: AATAAGTACTTGGCCCGTGAA R: CGATTCAAGTAGGCAAAGAG	NED
Ibcp10	F: ATATAAGGGGCCATTTTAGG R: ACGATAGAGGAGAAGGTTCC	6-FAM
ibcp4	F: ATCCTGGACGTGAAGAATAA R: GATGGCTGAGTGGACTAAAG	HEX
ibcp31	F: AACGGATTTCTCCAATGTA R: ACCTCACCGTTTCAGAAGTA	NED

2.4. Data Analysis

The number of observed alleles (Na) for each cpSSR locus was counted for all sweet potato accessions. The effective number of alleles (Ne = 1/(Sum pi^2)), the Shannon index ((I = −1 × Sum (pi × Ln (pi)), the frequency of the I allele), the diversity index (h = 1 − Sum pi^2), and the unbiased diversity (uh = (N/(N − 1)) × h) were calculated using GenAlEx 6.5 [21]. Genetic differentiation between the populations was determined using PhiPT, a measure that allows intra-individual variation to be suppressed and is therefore ideal for comparing cpSSR data with 999 permutations using GenAlEx. Estimates of gene flow (Nm, number of migrants per generation = [(1/PhiPT) − 1]/2) were also calculated by GenAlEx. Analysis of molecular variance (AMOVA), among and within the subpopulations (assigned by the median-joining network), was performed in GenAlEx. To calculate distances based on the cpSSR data, DARwin v.6.0 was employed to generate genetic distance matrix, which were then used to perform factor analysis [22]. To examine the relationship between the 558 sweet potato accessions, a median-joining network was generated using *poppr* packages on R packages [23] and visualized using Network 5 [24].

3. Results

A total of 21 alleles were detected in eight polymorphic cpSSR loci among the 558 sweet potato germplasms. As shown in Table 3, the Na ranged from two to four and the Ne was calculated to range from 1.174 to 2.675. Shannon's information (I) for each cpSSR ranged from 0.280 to 1.123 and the diversity indices (h) and uh ranged from 0.148 to 0.626 and 0.148 to 0.627, respectively.

Table 3. The genetic diversity parameters of nine cpSSR markers in 558 sweet potato germplasms.

SSR marker	Na [1]	Ne	I	h	uh
Ibcp10	3	2.216	0.933	0.549	0.550
Ibcp31	1		monomorphic		
Ibcp4	3	2.050	0.829	0.512	0.513
Ibcp5	2	1.266	0.365	0.210	0.210
Ibcp8	2	1.918	0.672	0.479	0.480
NTCP18	2	1.174	0.280	0.148	0.148
NTCP26	4	2.675	1.123	0.626	0.627
NTCP28	2	1.812	0.640	0.448	0.449
ccmp2	3	2.072	0.839	0.517	0.518
Mean	2.4 ± 0.3	1.798 ± 0.18	0.631 ± 0.12	0.388 ± 0.07	0.388 ± 0.07

[1] Na = No. of alleles; Ne = No. of effective alleles; I = Shannon's information index; h = Diversity; uh = unbiased diversity.

The diversity indices among the ten countries were calculated to be Ne = 1.52, I = 0.44, and h = 0.29 (Table 4). The Na ranged from 1.44 (PRK) to 2.22 (JPN) and the Ne was calculated to be 1.25 (PRK) to 1.77 (TWN). Shannon's information (I) ranged from 0.23 (PRK) to 0.61 (JPN). The diversity indices (h) and uh were 0.15 (PRK) to 0.39 (JPN) and 0.19 (PRK) to 0.39 (JPN), respectively.

Table 4. The genetic diversity parameters of ten origins in the 558 sweet potato germplasms using eight cpSSRs.

Countries	Na [1]	Ne	I	h	uh
JPN	2.22 ± 0.28	1.75 ± 0.16	0.61 ± 0.10	0.39 ± 0.06	0.39 ± 0.06
KOR	2.11 ± 0.20	1.64 ± 0.14	0.54 ± 0.10	0.35± 0.07	0.35 ± 0.07
TWN	2.11 ± 0.20	1.77 ± 0.18	0.56 ± 0.12	0.37 ± 0.08	0.38 ± 0.08
USA	2.11 ± 0.20	1.51 ± 0.08	0.51 ± 0.08	0.31 ± 0.05	0.33 ± 0.05
CHN	2.11 ± 0.20	1.45 ± 0.12	0.44 ± 0.10	0.27 ± 0.06	0.28 ± 0.07
PHL	1.67 ± 0.17	1.41 ± 0.13	0.36 ± 0.10	0.24 ± 0.07	0.27 ± 0.08
PER	1.89 ± 0.20	1.56 ± 0.16	0.44 ± 0.11	0.30 ± 0.08	0.30 ± 0.08
PRK	1.44 ± 0.18	1.25 ± 0.12	0.23 ± 0.09	0.15 ± 0.06	0.19 ± 0.08
IDN	1.78 ± 0.22	1.55 ± 0.17	0.42 ± 0.12	0.29 ± 0.08	0.30 ± 0.08
NZE	1.67 ± 0.24	1.30 ± 0.10	0.31 ± 0.10	0.19 ± 0.06	0.22 ± 0.07
Mean	1.91 ± 0.07	1.52 ± 0.04	0.44 ± 0.03	0.29 ± 0.02	0.30 ± 0.02

[1] Na = No. of different alleles; Ne = No. of effective alleles; I = Shannon's information index; h = Diversity; uh = unbiased diversity.

To improve understanding of the genetic relationships between different chloroplast haplotypes, a network approach was used (Figure 1). A total of 77 different chloroplast haplotypes in sweet potato germplasms were identified according to the results of the median-joining network (Table 5). Forty-four of the 77 different chloroplast haplotypes were found only once. Among the 33 chloroplast haplotypes with two or more sweet potato accessions, H1 and H2 were clearly dominant and were distributed over all areas, except for North Korea and New Zealand. The number of chloroplast haplotypes per country ranged from 3 to 41 with an average of fifteen. The diversity of chloroplast haplotypes ranged from 0.216 to 0.833 with an average of 0.37. Although the number of sweet potato accessions in North Korea was the lowest (six), the diversity of chloroplast haplotypes was the highest (five types, 83.3%)

among the countries studied. South Korea showed the lowest diversity at 21.6%, displaying only 27 chloroplast haplotypes out of 190 sweet potato accessions.

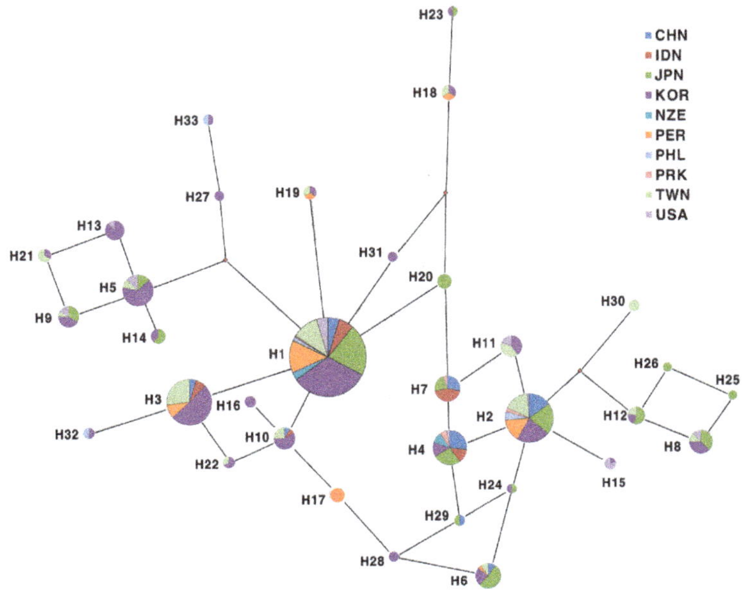

Figure 1. Median-joining network of eight cpSSR haplotypes. Circle size is proportional to the number of accessions that carry this chloroplast haplotype.

Table 5. Geographical distribution of cpSSR haplotypes.

	KOR	JPN	TWN	USA	CHN	PER	IDN	PRK	PHL	NZE
N [1]	190	123	73	25	43	50	30	6	9	9
H1	62	40	21	9	9	25	12	1	3	6
H2	16	15	12	1	11	11		2	4	
H3	21		11		2	4	3			
H4	4	8		1	8		4	2		3
H5	15	3	2	3						
H6	5	11	2		2	1				
H7		4			5		8	1		
H8	6	6	2	2						
H9	5	4	1	2						
H10	7		3		1		1			
H11	4		4	2						
H12	2	5	2							
H13	7			1						
H14	2	3								
H15	1			4						
H16	4									
H17						3				
H18	1		1			1				
H19	1		1			1				
H20		3								
H21	1		2							
H22	2		1							
H23	1	1								
H24	1	1								
H25		2								

Table 5. *Cont.*

	KOR	JPN	TWN	USA	CHN	PER	IDN	PRK	PHL	NZE
H26		2								
H27	2									
H28	2									
H29		1			1					
H30			2							
H31	2									
H32	1								1	
H33	1								1	
PH	14	14	6		4	4		2		
H	41	31	22	5	13	12	5	8	3	10
DH	0.216	0.252	0.301	0.833	0.302	0.24	0.556	0.267	0.333	0.40

[1] N, Number of accessions; PH, private chloroplast haplotype; H, number of distinct chloroplast haplotypes; DH, diversity of chloroplast haplotype.

Factorial analysis (FA) was used to investigate the genetic relationship among the chloroplast haplotypes of the 558 sweet potato accessions. The sweet potato chloroplast haplotypes formed four main clusters defined by the first axis (Figure 2A), which explained 54.1% of the total variances, while the second axis explained 9.5% of the total variances. Cluster I contained 51 sweet potato accessions with five chloroplast haplotypes (H5, H9, H13, H14, and H21). Cluster II consisted of 29 accessions and showed four chloroplast haplotypes, H8, H12, H25, and H26. Cluster III had the most sweet potato accessions (268 accessions) and chloroplast haplotypes (H1, H3, H10, H16, H17, H18, H19, H20, H22, H27, H31, H32, and H33) than the other clusters. Cluster IV had 168 accessions and 11 chloroplast haplotypes (H2, H6, H11, H15, H23, H24, H30, H4, H7, H28, and H29). Among the four clusters, clusters III and IV were distributed across all ten collected countries, while I and II were found only in four countries: Korea, Japan, Taiwan, and the USA (Figure 2B).

The AMOVA analysis performed on the whole sweet potato germplasms indicated 88% total genetic diversity within the populations and 12% total genetic diversity among the populations (Table 6). The genetic differences (PhiPT) among the populations were estimated by AMOVA to be 0.125. The gene flow (genetically effective migration rate, Nm) among the populations was indirectly calculated from the PhiPT as 3.513. In addition, based on factorial analysis, AMOVA of the four clusters was conducted. The analysis revealed that 9, 36, and 4% of the total genetic diversity of the populations was attributable to clusters I, III, and IV, respectively, while cluster II was not significant ($p > 0.05$). The PhiPT of cluster I, III, and IV were 0.093, 0.361, and 0.042 and the gene flow of each cluster was 4.870, 0.885, and 11.370, respectively.

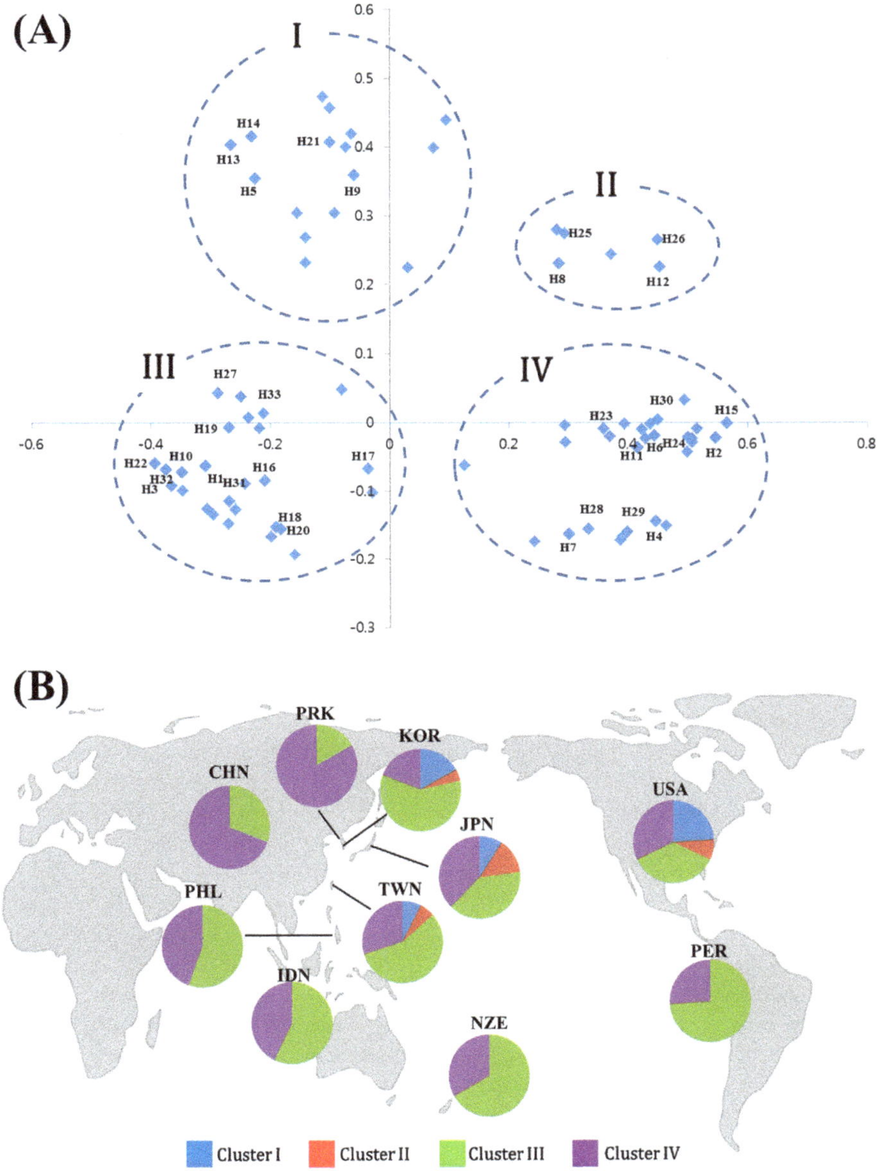

Figure 2. (**A**) Factorial analysis of the 8 cpSSR haplotypes. H, chloroplast haplotypes based on the median-joining network. (**B**) Original countries of sweet potato accessions and geographic distributions of the chloroplast haplotypes found. Pie charts on the map represent the chlorotype composition of the accessions and the colors in each chart represent the chlorotype indicated by factorial analysis.

Table 6. Analysis of molecular variance (AMOVA) of sweet potato germplasms.

Source of Variation	df	SS	MS	Est. Var.	%	PhiPT	Sig.	Nm
Whole data set								
Among Populations	9	120.09	13.34	0.236	12%	0.125	<0.001	3.513
Within Populations	548	908.38	1.66	1.658	88%			
Total	557	1028.47		1.894	100%			
I								
Among Populations	3	2.73	0.91	0.046	9%	0.093	<0.05	4.870
Within Populations	47	21.00	0.45	0.447	91%			
Total	50	23.73		0.493	100%			
II								
Among Populations	3	1.94	0.65	0.051	13%	0.134	ns	3.244
Within Populations	25	8.33	0.33	0.333	87%			
Total	28	10.28		0.385	100%			
III								
Among Populations	9	42.48	4.72	0.273	36%	0.361	<0.001	0.885
Within Populations	156	75.40	0.48	0.483	64%			
Total	165	117.89		0.756	100%			
IV								
Among Populations	8	5.96	0.74	0.015	4%	0.042	<0.05	11.370
Within Populations	258	89.72	0.35	0.348	96%			
Total	266	95.68		0.363	100%			

4. Discussion

In this study, the genetic diversity of 558 sweet potato germplasms conserved in the Genebank was profiled using nine cpSSRs. The results confirmed that the sweet potato germplasms showed various genetic variations depending on the country of collection. Previous studies have been performed to analyze the genetic diversity in sweet potato using molecular markers such as AFLP, RAPD, and SSR [6,8,10,11,25]. In addition, the origin and genetic diversity of sweet potato germplasms have been explained using nuclear SSRs and cpSSRs [12,13]. These methods, non-recombinant and uniparentally inherited nature, are widely used for evolutionary and phylogenetic studies as they have been demonstrated to be effective indicators of the genetic structure of a population [26,27].

In this study, 92.4% of the accessions were included in the breeding lines or cultivars among the 558 sweet potato germplasms. Many other genetic resource centers showed a similar composition of preserved sweet potato accessions. The International Potato Center (CIP) reported that 93.3% of the sweet potato accessions in 36 international germplasm centers were cultivars or breeding lines and that only seven centers were preserving the wild type [28]. There appears to be a higher frequency of breeding line or cultivars due to the breeding system of sweet potatoes. In general, numerous seedlings of posterities resulting from crossing elite parent sweet potatoes were screened for desirable traits and the best were used, with or without the best parents. The undesirable genotypes were discarded through a selection process that concentrated on eliminating the poorest, rather than on selecting the best genotypes [29]. In Japan, they harvested approximately 50,000 seeds from cross combinations and selected approximately 3000 individual seedling plants. After that, they conducted various tests for line selection and generated cultivars or maintained elite breeding lines [4]. For example, cv. 'Beniazuma' was released by the Institute of Crop Science, NARO (NICS) in 1984 and is the progeny of a cross between 'Kanto No. 85' and 'Koganesenga' [30]. Similarly, 'Quick Sweet' was released by the NICS in 2002 and is the progeny of a cross between 'Beniazuma' and Kyushu No. 30' [31]. Additionally, 'Aikomachi' was released by the NICS in 2012 and was derived from a cross between 'Quick Sweet' and 'Kankei107' [32]. These three cultivars may show the same patterns in cpSSRs profiling because they have the same maternal pedigree, even though they have different cross combinations. The Korean method is similar to the method of Japanese sweet potato breeding. For example, Korean sweet potato cultivars (cvs), 'Pungmi', Yeonmi', 'Sinyulmi', and 'Gogeonmi' were derived from cv. 'Seonmi' while cvs. 'Seonmi', 'Jinmi', and 'Eunmi' were derived from cv. 'Hwangmi'. Seven Korean sweet potato cultivars ('Pungmi', Yeonmi', 'Sinyulmi', 'Gogeonmi' 'Seonmi', 'Jinmi', and 'Eunmi') have the same cpSSRs patterns because they share the same maternal pedigree, even though they have different cross combinations. Due to the frequent use and preservation of these elite accessions, it is possible that conserved sweet potato germplasms such as the breeding lines or cultivars showed the same patterns of cpSSRs.

Among the 558 sweet potato accessions collected from ten countries, sweet potato germplasms from Japan, Korea, Taiwan, and the USA showed four types while other countries had only two types (Figure 2). The Korean sweet potato germplasm collection was started from Japanese varieties. In 1906, 1923, and 1934, Japanese varieties Genki, Shirofuku, and Okinawa No. 100 were introduced and cultivated throughout Korea, respectively. The collection of sweet potato germplasms in Korea began in 1973, led by the international agricultural institutes, International Potato Center (CIP) and World Vegetable Center (AVRDC). In Korean landraces, superior sweet potato lines were selected during breeding programs processes. Although the collection of sweet potatoes was not smooth due to problems with the Peruvian government's approval by CIP, sweet potato germplasms were obtained from AVRDC, USDA, and Japan [33].

A previous study reported that variation in the maternally-inherited plant genomes affected both type weight and root weight in sweet potatoes [34]. In addition, it was also mentioned that the selection of Nancy Hall and Tainung 27 as female parents significantly affected the two traits. Although the sweet potato germplasms conserved at the Korea Genebank represent approximately 700 accessions, the results of this study showed that a variety of female parents was lacking. In fact, the sweet potato germplasms contained in the Korea Genebank are almost all cultivars or breeding lines without their wild type relatives [35]. Since 1955, Mexico and the United States have collected many wild plants related to sweet potatoes to study sweet potato's phylogeny and the utilization of wild plants in sweet potato breeding [36]. Wild relatives have provided breeders with genes for pest and disease resistance, abiotic stress tolerance, and quality traits [37]. The results of this study confirmed that sweet potato germplasms conserved in the Korea Genebank need a greater variety of female parents. Despite this, analysis of the genetic diversity of nuclear DNA in sweet potato germplasms should still be conducted. Additional collection and conservation of various female parents and wild relatives of sweet potatoes are very important to improve the quality of sweet potato germplasm and satisfy the demand of breeders.

Supplementary Materials: The following are available online at http://www.mdpi.com/2073-4395/9/11/752/s1, Table S1: List of 558 sweet potato germplasms.

Author Contributions: Conceptualization, K.J.L.; Data curation, K.J.L. and G.-A.L.; Formal analysis, K.J.L., R.S., and G.-A.L.; Investigation, J.-R.L., M.-J.S., and G.-T.C.; Project administration, D.Y.H.; Resources, J.-R.L., M.-J.S., and G.-T.C.; Writing—original draft, K.J.L.; Writing—review and editing, D.Y.H.

Funding: This study was carried out with the support of the "Research Program for Agricultural Science & Technology Development (Project No. PJ010883)". It was also supported by the 2017 Postdoctoral Fellowship Program of the National Institute of Agricultural Sciences funded by the Rural Development Administration, Republic of Korea.

Conflicts of Interest: The authors declare no conflicts of interest.

References

1. Ramanatha Rao, V.; Hodgkin, T. Genetic diversity and conservation and utilization of plant genetic resources. *Plant Cell Tissue Organ Cult.* **2002**, *68*, 1–19. [CrossRef]
2. Cadima Fuentes, X.; van Treuren, R.; Hoekstra, R.; van den Berg, R.G.; Sosef, M.S.M. Genetic diversity of Bolivian wild potato germplasm: Changes during ex situ conservation management and comparisons with resampled in situ populations. *Genet. Resour. Crop Evol.* **2017**, *64*, 331–344. [CrossRef]
3. Xiao, L.Q.; Ge, X.J.; Gong, X.; Hao, G.; Zheng, S.X. ISSR variation in the endemic and endangered plant *Cycas guizhouensis* (Cycadaceae). *Ann. Bot.* **2004**, *94*, 133–138. [CrossRef] [PubMed]
4. Katayama, K.; Kobayashi, A.; Sakai, T.; Kuranouchi, T.; Kai, Y. Recent progress in sweetpotato breeding and cultivars for diverse applications in Japan. *Breed. Sci.* **2017**, *67*, 3–14. [CrossRef] [PubMed]

5. Zhang, D.; Cervantes, J.; Huamán, Z.; Carey, E.; Ghislain, M. Assessing genetic diversity of sweet potato (*Ipomoea batatas* (L.) Lam.) cultivars from tropical America using AFLP. *Genet. Resour. Crop Evol.* **2000**, *47*, 659–665. [CrossRef]
6. Elameen, A.; Fjellheim, S.; Larsen, A.; Rognli, O.A.; Sundheim, L.; Msolla, S.; Masumba, E.; Mtunda, K.; Klemsdal, S.S. Analysis of genetic diversity in a sweet potato (*Ipomoea batatas* L.) germplasm collection from Tanzania as revealed by AFLP. *Genet. Resour. Crop Evol.* **2008**, *55*, 397–408. [CrossRef]
7. Woolfe, J.A. *Sweet Potato: An Untapped Food Resource*; Cambridge University Press: Cambridge, UK; International Potato Center: Lima, Peru, 1992.
8. Yang, X.S.; Su, W.J.; Wang, L.J.; Lei, J.; Chai, S.S.; Liu, Q.C. Molecular diversity and genetic structure of 380 sweetpotato accessions as revealed by SSR markers. *J. Integr. Agric.* **2015**, *14*, 633–641. [CrossRef]
9. Zhang, D.; Rossel, G.; Kriegner, A.; Hijmans, R. AFLP assessment of diversity in sweetpotato from Latin America and the Pacific region: Its implications on the dispersal of the crop. *Genet. Resour. Crop Evol.* **2004**, *51*, 115–120. [CrossRef]
10. Gichuki, S.T.; Berenyi, M.; Zhang, D.; Hermann, M.; Schmidt, J.; Glössl, J.; Burg, K. Genetic diversity in sweetpotato [*Ipomoea batatas* (L.) Lam.] in relationship to geographic sources as assessed with RAPD markers. *Genet. Resour. Crop Evol.* **2003**, *50*, 429–437. [CrossRef]
11. Meng, Y.S.; Zhao, N.; Li, H.; Zhai, H.; He, S.Z.; Liu, Q.C. SSR fingerprinting of 203 sweetpotato (*Ipomoea batatas* (L.) Lam.) varieties. *J. Integr. Agric.* **2018**, *17*, 86–93. [CrossRef]
12. Roullier, C.; Kambouo, R.; Paofa, J.; McKey, D.; Lebot, V. On the origin of sweet potato (*Ipomoea batatas* (L.) Lam.) genetic diversity in New Guinea, a secondary centre of diversity. *Heredity* **2013**, *110*, 594–604. [CrossRef] [PubMed]
13. Roullier, C.; Rossel, G.; Tay, D.; McKey, D.; Lebot, V. Combining chloroplast and nuclear microsatellites to investigate origin and dispersal of New World sweet potato landraces. *Mol. Ecol.* **2011**, *20*, 3963–3977. [CrossRef] [PubMed]
14. Moore, M.J.; Bell, C.D.; Soltis, P.S.; Soltis, D.E. Using plastid genome-scale data to resolve enigmatic relationships among basal angiosperms. *Proc. Natl. Acad. Sci. USA* **2007**, *104*, 19363–19368. [CrossRef] [PubMed]
15. Pervaiz, T.; Zhang, C.; Faheem, M.; Mu, Q.; Fang, J. Chloroplast based genetic diversity among Chinese grapes genotypes. *Mitochondrial DNA Part A* **2017**, *28*, 565–569. [CrossRef]
16. Provan, J.; Powell, W.; Hollingsworth, P.M. Chloroplast microsatellites: New tools for studies in plant ecology and evolution. *Trends Ecol. Evol.* **2001**, *16*, 142–147. [CrossRef]
17. McCauley, D.E. The use of chloroplast DNA polymorphism in studies of gene flow in plants. *Trends Ecol. Evol.* **1995**, *10*, 198–202. [CrossRef]
18. Powell, W.; Morgante, M.; Andre, C.; McNicol, J.W.; Machray, G.C.; Doyle, J.J.; Tingey, S.V.; Rafalski, J.A. Hypervariable microsatellites provide a general source of polymorphic DNA markers for the chloroplast genome. *Curr. Biol.* **1995**, *5*, 1023–1029. [CrossRef]
19. Huang, J.; Yang, X.; Zhang, C.; Yin, X.; Liu, S.; Li, X. Development of Chloroplast Microsatellite Markers and Analysis of Chloroplast Diversity in Chinese Jujube (*Ziziphus jujuba* Mill.) and Wild Jujube (*Ziziphus acidojujuba* Mill.). *PLoS ONE* **2015**, *10*, e0134519. [CrossRef]
20. Kim, I.H. East Asian Nations' Awareness on Ocean during the 16-18th Century Perceived from Transfer of Sweet Potatoes. *J. Asia Pac. Stud.* **2014**, *21*, 69–97.
21. Peakall, R.; Smouse, P.E. GenAlEx 6.5: Genetic analysis in Excel. Population genetic software for teaching and research—An update. *Bioinformatics* **2012**, *28*, 2537–2539. [CrossRef]
22. Perrier, X.; Jacquemoud-Collet, J.P. DARwin Software. Available online: http://darwin.cirad.fr/darwin (accessed on 10 October 2019).
23. Kamvar, Z.N.; Tabima, J.F.; Grunwald, N.J. Poppr: An R package for genetic analysis of populations with clonal, partially clonal, and/or sexual reproduction. *PeerJ* **2014**, *2*, e281. [CrossRef]
24. Bandelt, H.J.; Forster, P.; Rohl, A. Median-joining networks for inferring intraspecific phylogenies. *Mol. Biol. Evol.* **1999**, *16*, 37–48. [CrossRef] [PubMed]

25. Laurie, S.M.; Calitz, F.J.; Adebola, P.O.; Lezar, A. Characterization and evaluation of South African sweet potato (*Ipomoea batatas* (L.) LAM) land races. *S. Afr. J. Bot.* **2013**, *85*, 10–16. [CrossRef]
26. Desiderio, F.; Bitocchi, E.; Bellucci, E.; Rau, D.; Rodriguez, M.; Attene, G.; Papa, R.; Nanni, L. Chloroplast Microsatellite Diversity in Phaseolus vulgaris. *Front. Plant Sci.* **2012**, *3*, 312. [CrossRef]
27. Singh, N.; Pal, A.K.; Roy, R.K.; Tamta, S.; Rana, T.S. Development of cpSSR markers for analysis of genetic diversity in Gladiolus cultivars. *Plant Gene* **2017**, *10*, 31–36. [CrossRef]
28. The International Potato Center (CIP). Global Strategy for Ex-Situ Conservation of sweetpotato Genetic Resources. Sweetpotato Genetic Resources Conservation—CIP November 2007. Available online: http://www.croptrust.org/documents/web/SweetPotato-Strategy-FINAL-12Dec07.pdf (accessed on 10 October 2019).
29. Kawa, N. Tropical Root and Tuber Crops: Cassava, Sweet Potato, Yams and Aroids by V. Lebot. *Econ. Bot.* **2010**, *64*, 86–87.
30. Shiga, T.; Sakamoto, S.; Ando, T.; Ishikawa, H.; Kato, S.; Takemata, T.; Umehara, M. On a new sweet potato [*Ipomoea batatas*] cultivar "Beniazuma". *Bull. Nat. Agric. Res. Cent.* **1985**, *3*, 73–84.
31. Katayama, K.; Tamiya, S.; Kuranouchi, T.; Komaki, K.; Nakatani, M. New sweet potato cultivar "Quick Sweet". *Bull. Natl. Inst. Crop Sci.* **2003**, *3*, 35–52.
32. Ohara-Takada, A.; Kumagai, T.; Kuranouchi, T.; Nakamura, Y.; Fujita, T.; Nakatani, M.; Tamiya, S.; Katayama, K. 'Aikomachi', a new sweetpotato cultivar with good appearance and high confectionery quality. *Bull. Naro Inst. Crop Sci.* **2016**, *16*, 35–55.
33. RDA. *Plant Germplasm Research 20*; RDA: Jeonju, Korea, 2007.
34. Lin, K.H.; Lai, Y.C.; Chang, K.Y.; Chen, Y.F.; Hwang, S.Y.; Lo, H.F. Improving breeding efficiency for quality and yield of sweet potato. *Bot. Stud.* **2007**, *48*, 283–292.
35. Mok, I.G.; Zhao, D.; Kwak, S.S. Genetic resources of sweetpotato for industrial use. *J. Plant Biotechnol.* **2009**, *36*, 202. [CrossRef]
36. Kobayashi, M.; Miyazaki, T. Sweet potato breeding using wild related species. In Proceedings of the Fourth Symposium of the International Society for Tropical Root Crops, IDRC, Cali, Colombia, 1–7 August 1976; Volume 1, pp. 53–57.
37. Khoury, C.K.; Heider, B.; Castañeda-Álvarez, N.P.; Achicanoy, H.A.; Sosa, C.C.; Miller, R.E.; Scotland, R.W.; Wood, J.R.I.; Rossel, G.; Eserman, L.A.; et al. Distributions, ex situ conservation priorities, and genetic resource potential of crop wild relatives of sweetpotato [*Ipomoea batatas* (L.) Lam., I. series Batatas]. *Front. Plant Sci.* **2015**, *6*, 251. [CrossRef] [PubMed]

agronomy

MDPI

Article

Exploring the Genetic Diversity and Population Structure of Turkish Laurel Germplasm by the iPBS-Retrotransposon Marker System

Ünal Karık [1], Muhammad Azhar Nadeem [2], Ephrem Habyarimana [3], Sezai Ercişli [4], Mehtap Yildiz [5], Abdurrahim Yılmaz [2], Seung Hwan Yang [6], Gyuhwa Chung [6,*] and Faheem Shehzad Baloch [2,*]

1 Aegean Agricultural Research Institute, Menemen, 35640 Izmir, Turkey; unalkarik@gmail.com
2 Department of Field Crops, Faculty of Agricultural and Natural Science, Bolu Abant Izzet Baysal University, 14030 Bolu, Turkey; azharjoiya22@gmail.com (M.A.N.); ayilmaz88@hotmail.com (A.Y.)
3 CREA Research Center for Cereal and Industrial Crops, 40128 Bologna, Italy; ephrem.habyarimana@crea.gov.it
4 Department of Horticulture, Faculty of Agriculture, Ataturk University, 25240 Erzurum, Turkey; sercisli@gmail.com
5 Department of Agricultural Biotechnology, Faculty of Agriculture, Van Yüzüncü Yıl University, 65080 Van, Turkey; mehtapyildiz@gmail.com
6 Department of Biotechnology, Chonnam National University, Chonnam 59626, Korea; ymichigan@chonnam.ac.kr
* Correspondence: chung@chonnam.ac.kr (G.C.); balochfaheem13@gmail.com (F.S.B); Tel.: +90-545-540-4239 (F.S.B.)

Received: 18 September 2019; Accepted: 15 October 2019; Published: 17 October 2019

Abstract: Laurel is a medicinally important plant and is known to the world for its essential oil. Turkey is the main market in the laurel leaf trade by sharing about 90% of the world trade. Here we made an effort to elucidate genetic diversity and population structure of 94 Turkish laurel genotypes collected from 26 provinces and four geographical regions using inter-primer binding site (iPBS) retrotransposon markers. A total of 13 most polymorphic primers were selected which yielded 195 total bands, of which 84.10% were found polymorphic. Mean polymorphism information content (PIC) was (0.361) and diversity indices including mean effective number of alleles (1.36), mean Shannon's information index (0.35) and overall gene diversity (0.22) revealed the existence of sufficient amount of genetic diversity in the studied plant material. Most diversity was found in genotypes collected from the Mediterranean region. Analysis of molecular variance (AMOVA) revealed that most of the variation (85%) in Turkish laurel germplasm is due to differences within populations. Model-based structure, principal coordinate analysis (PCoA) and neighbor-joining algorithms were found in agreement and clustered the studied germplasm according to their collection provinces and regions. This is a very first study exploring the genetic diversity and population structure of laurel germplasm using iPBS-retrotransposon marker system. We believe that information provided in this work will be helpful for the scientific community to take more interest in this forgotten but the medicinally important plant.

Keywords: *Laurus nobilis*; medicinal plant; mobile genomic elements; germplasm characterization; Mediterranean region

1. Introduction

There is an estimation that the current loss of plant species is between 100 and 1000 times more than the expected natural extinction rate [1] and it is believed that an increase in genetic erosion will be observed during the upcoming years. The rapid expansion of plant breeding activities during the 20th century for the development of improved varieties replaced or minimized the utilization of conventional landraces and resulted in a lower level of genetic diversity in in crop species [2]. To minimize the genetic erosion and maintain sustainability in agriculture, the Global Crop Diversity Trust initiated a consultation process for the development of 30 global crops and made strategies for the ex situ conservation and utilization of germplasm [3]. Germplasm corresponds to the living tissues, which can be utilized for the generation of new plants and is therefore the most important component in the maintenance of plant genetic resources [4]. Gene banks comprise of plant germplasm in the form of seed collection, nursery, pollen and in vitro [5]. These gene banks are important by reflecting the genetic diversity of cultivated and wild relatives of various crops having unique phenotypic and genotypic characteristics [6]. Characterization of these genetic resources is one of the important alternatives for the scientific community to deal with the several challenges like food scarcity, various biotic and abiotic stresses through the investigation of genetic diversity [7,8].

Genetic diversity is the essence of the biological world and serves as a source of natural variations, which can be helpful to deal with various challenges to the world [7]. Medicinal plants kept the continuous attention from the human being through their multiple uses [9]. In 2008, WHO (World Health Organization) issued a report confirmıng the role of these plants in routine life and stated that 80% of the world population directly or indirectly depended on these plants as traditional medicine [10]. More than 1300 medicinal plants are in use by Europe and 90% of it is harvested from the wild and natural resources. Among the 150 topmost used drugs, 118 are based on the natural resources [11]. International Union for Conservation of Nature and the World Wildlife Fund stated in a report that between 50,000–80,000 flowering plant species are in use by the world for medicinal purposes. Nearly 15,000 of these flowering plants are threatened with extinction due to habitat destruction and over-harvesting [1]. There is a need to collect, characterize and conserve the plants having medicinal value to maintain sustainability in the ecosystem [12].

Mediterranean region contains a good diversity of plants and laurel (*Laurus nobilis* L.) is one of the important plants of this region [9]. Laurel belongs to the Asia and Balkan region and later it spread to various Mediterranean countries like Italy, Spain, Israel, France Corsica Island and North Africa [13]. Laurel belongs to the *Lauraceae* family and this family comprises of 2500–3000 species with a total of 50 genera mainly distributed in the tropic and subtropics of the world [9]. *Laurus nobilis* and *L. azorica* (Seub) *Franco* are two most economically important species. Laurel has been traditionally used for the treatment of epileptic events, neuralgia and Parkinsonism [14]. Laurel is very popular and well known due to its essential oil, which has been proven very effective for the treatment of epileptic, convulsion and flatulent colic problems [15,16]. Beside medicinal applications, laurel leaves are used as a flavoring agent and to improve the shelf life of food due to their high antimicrobial and antioxidant activities [15,16].

Turkey is considered the cradle of agriculture because it is the origin and distribution center for various crops, due to its geographic [7,17]. This plant arrived in Turkey in 1655 and was known by the name of "Daphne". The Mediterranean, Aegean, West and Central Black Sea and Marmara region are the most important areas of Turkey where this plant naturally grows [18]. Turkey is ruling the laurel market and shares the 90% of world production [15–19]. According to Nadeem et al. [9], European countries imported 77% of laurel (HS code, 09109950) and 72% of these were imported from Turkey. As compared to the others plant, the laurel market is different because consumer herein is very concerned with the quality rather than the price [9].

Advancement in the molecular markers changed the fate of breeding and boost up the breeding activities. Several types of molecular markers have been developed by the scientific community according to their feasibility [20]. Retrotransposon, sometimes called the jumping element comprises 50%–90% of plant genome [21]. Long terminal repeat (LTR) and non-LTR retrotransposons are two types of retrotransposon, and the former are the most represented in plant genomes. However, Kalendar et al. [22] developed a new marker system named the "inter primer binding site (iPBS)" and suggested it a universal method for DNA fingerprinting, which can be employed both for plant and animals. This method overcomes all application limits of LTR and non-LTR retrotransposons. Inter primer binding sites primers are designed on the primer binding site (PBS) sequences having conserved parts of the tRNAs for reverse transcription during the replication cycle of retrotransposons [22,23]. These tRNAs complement primer binding sites (PBS) in LTR retrotransposons and are used as genomic regions during the PCR (polymerase chain reaction) amplification [22]. The universality of the iPBS-retrotransposon marker has been proven because it can be successfully utilized for any plant for the diversity assessment as well for the phylogenetic and evolutionary study [8,24,25]. Inter primer binding site (iPBS) markers are dominant markers. Previous studies have confirmed that the dominant marker system likely precludes obtaining unbiased estimates of many genetic parameters [26,27]. Besides of its dominant nature, it becomes the marker of choice for the genetic diversity assessment due to its universal nature [8,24,25]. Moreover, its lower cost gained the attentions of scientific laboratories having funding problems.

Very few efforts have been done to conserve the genetic resources of laurel and small germplasm collections are created and maintained by individual breeders. Most of the research is aimed to investigate the composition and effects of laurel essential oil. Yalçın et al. [28] collected the laurel leaves from Northern Cyprus and Caputo et al. [29] used the Italian laurel leaves for the investigation of composition and activities of essential oil. Arroyo-Garcia et al. [30] used the amplified fragment length polymorphism (AFLP) markers to investigate the genetic similarity in this important plant. Arroyo et al. [31] used the simple sequence repeats (SSR) markers for the characterization of genetic diversity among various species of the *Lauraceae* family. Rodríguez-Sánchez et al. [32] applied the chloroplast DNA (cpDNA) sequences to understand the phylogenetic history of laurel trees. Besides the germplasm characterization, almost nothing has been done regarding the breeding perspective of Laurel. There was a need to explore the genetic variation and investigate the population structure using a good number of germplasm. Therefore, we aimed to investigate the genetic diversity and population structure of Turkish laurel germplasm through the iPBS-retrotransposon marker system.

2. Materials and Methods

2.1. Plant Material and DNA Isolation

A total of 94 laurel genotypes collected from the 26 provinces from four geographical regions of Turkey were used as plant material for this study (Table S1; Figure 1). A nursery of collected germplasm was established at the Aegean Agricultural Research Institute of İzmir. Plant DNA was isolated by taking young and fresh leaves from each genotype. Genomic DNA was extracted by following the CTAB protocol [33] with some modifications [34]. Agarose gel (0.8%) was used for the assessment of DNA concentration and further quantification was done with NanoDrop (DeNovix DS-11 FX, USA). Five ng/μL was maintained as final DNA concentration for the further usage in a polymerase chain reaction (PCR).

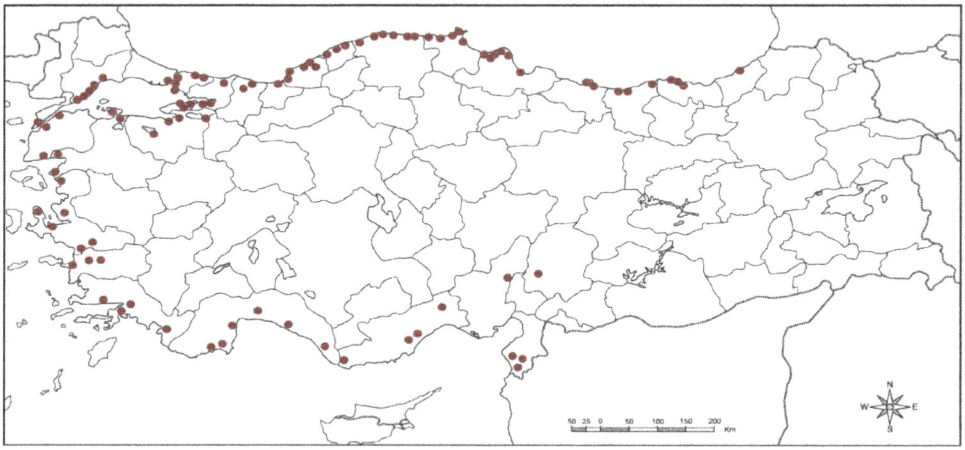

Figure 1. Collection points of 94 Turkish laurel genotypes.

2.2. iPBS-Retrotransposon Analysis

Eight laurel genotypes were selected randomly for the screening purpose and a total of 83 iPBS-retrotransposons primers were used in the screening. All these primers were derived from a study by Kalendar et al. [22]. From this screening, 13 of the most polymorphic primers producing perfect banding profiles were selected for the fingerprinting of 94 laurel genotypes. Table 1 explains the name, sequence and annealing temperature of iPBS primers used in this study. For the confirmation of reproducibility of iPBS-retrotransposon primers, PCR amplification was repeated two times for two primers (Figure S1). All PCR amplifications were performed in 20 μL reaction containing 4 uL template DNA, 2 μL dNTPs (Thermo Scientific, Waltham, MA, USA), 0.2 μL U Taq DNA polymerase (Thermo Scientific), 4 μL primer, 2 μL 1 × PCR buffer (Thermo Scientific), 1.8 μL MgC_{l2} and 6 μL distilled water. By following the protocol suggested by the Kalendar et al. [22], PCR conditions were arranged which contains denaturation at 95 °C for 3 min, followed by 30 cycles of denaturation at 95 °C for 15 s, annealing temperature 50–65 °C depending on primers used for 1 min; and a final extension at 72 °C 5 min. Amplified products were electrophoresed on a 2% (w/v) agarose gel, using 0.5 × Tris-borate-EDTA (TBE) buffer for 230 min; ethidium bromide was used for gel staining after electrophoresis and Imager Gel Doc XR+ system (Bio-Rad, Hercules, CA, USA) was used for the visualization and photographing. A 100 bp+ ladder was used as a molecular weight marker.

Table 1. List of 13 inter-primer binding site (iPBS)-retrotransposon primers with their sequence and annealing temperature used to elucidate genetic diversity among 94 Turkish laurel genotypes.

Primer Name [1]	Sequence (5′→3′)	Annealing Temperature (°C)
iPBS2228	TCATGGCTCATGATACCA	52
iPBS2230	TCTAGGCGTCTGATACCA	50
iPBS2232	AGAGAGGCTCGGATACCA	55
iPBS2237	CCCCTACCTGGCGTGCCA	55
iPBS2239	ACCTAGGCTCGGATGCCA	55
iPBS2245	GAGGTGGCTCTTATACCA	50
iPBS2253	TCGAGGCTCTAGATACCA	51
iPBS2256	GACCTAGCTCTAATACCA	51
iPBS2277	GGCGATGATACCA	52
iPBS2295	AGAACGGCTCTGATACCA	50
iPBS2398	GAACCCTTGCCGATACCA	51
iPBS2401	AGTTAAGCTTTGATACCA	53
iPBS2402	TCTAAGCTCTTGATACCA	50

[1] Primers, sequences and their annealing temperature were derived from Kalendar et al. [22].

2.3. Data Analysis

PCR products were scored as binary fashion; 0 or 1 for the absence and presence of specific bands with respect to 100 bp+ DNA ladder (Figure S2) because iPBS-retrotransposon is a dominant marker system. Reproducibility of DNA profiles for iPBS-retrotransposon marker systems was investigated by repeating two times the PCR amplification with some of the selected primers. Polymorphism information contents (PICs) were calculated by following the criteria set by Baloch et al. [24]. To investigate the level of genetic variations in the 94 laurel genotypes, various diversity parameters like the number of effective alleles (Ne), gene diversity (h) and Shannon's information index (I) and Nei's genetic distance were measured by PopGene ver. 1.32 [35]. The germplasm evaluated in this work was collected from four Turkish regions i.e., the Mediterranean, Black Sea, Aegean and Marmara region. Therefore, various diversity metrics were calculated on a per-region basis using PopGene ver. 1.32 [35]. An analysis of molecular variance (AMOVA) and principal coordinate analysis (PCoA) were investigated through GenAlEx 6.5 software [36]. To understand the level of the relationship among the 94 laurel genotypes, neighbor-joining analysis was performed using the R statistical software (version 3.4.1, Vienna, Austria). To explore the population structure of Turkish laurel germplasm, the Bayesian clustering model was applied in STRUCTURE software (version 2.3.4, Stanford, CA, USA). A continuous series of K were tested from 1 to 10 in five independent runs. For each run, the initial burn-in period was set to 50,000 followed by 50,000 Markov Chain Monte Carlo (MCMC) iterations as stated by the Baloch et al. [17]. To determine the proper numbers of the cluster (number of K; the number of subpopulations), criteria set by Evanno et al. [37] was followed and plotted the number of clusters (K) against logarithm probability relative to the standard deviation (ΔK).

3. Results

To characterize the 94 laurel genotypes, 13 selected primers resulted in a total of 195 scoreable bands with an average of 15 bands for each primer (Table 2). Among the 195 bands, 84.10% (164) were found polymorphic with an average of 12.61 fragments per primer. iPBS2398 produced maximum (23) and iPBS2232 and iPBS2295 resulted minimum (11) numbers of total bands. iPBS2402 was found much informative because all 14 bands were found polymorphic and iPBS2295 was found least informative by producing minimum (8) number of polymorphic bands. These 13 primers resulted in an average of 83.98% polymorphism, which ranged from 72.72% to 100% for iPBS2295, and iPBS2256 and iPBS2402, respectively. PIC values ranged between 0.163–0.585 for the iPBS2277 and iPBS2230 and 0.361 was the average PIC value. The maximum (1.57) and the minimum (1.15) number of effective alleles were produced by iPBS2230 and iPBS2277, respectively, and 1.36 was the mean effective number of alleles. Maximum and minimum gene diversity was 0.33 and 0.12 reflected by iPBS2230 and iPBS2277 respectively, while 0.22 was the mean gene diversity. Shannon's information index ranged between 0.4803 (iPBS2230) and 0.22 (iPBS2277) and 0.35 was the found average Shannon's information index. Mean pairwise genetic distance for 94 laurel genotypes was 0.228. Maximum Nei's genetic distance was 0.607 found between Canakkale1 and Mersin1 genotypes followed by Izmir6 and Mersin1 having 0.598 genetic distance. Bartin3 and Zonguldak3 genotypes reflected the minimum (0.089) Nei's genetic distance and Kocaeli1 and Kocaeli2 followed them having 0.096 genetic distance. To confirm the reproducibility of DNA profiles for the iPBS-retrotransposon marker system, PCR was repeated for two primers and there were the same total and polymorphic bands (Figure S2).

Diversity indices were also calculated according to the collection regions of 94 laurel genotypes (Table 3). Genotypes from the Mediterranean region showed a higher number of effective alleles (1.41), gene diversity (0.24) and expected heterozygosity (0.19). Genotypes from the Marmara region showed a minimum number of effective alleles (1.426), gene diversity (0.16) and expected heterozygosity (0.11). Genotypes from the Aegean and Mediterranean region reflected the maximum (0.31) and minimum (0.24) Shannon's information index respectively. Genotypes from the Mediterranean and Marmara regions reflected the maximum (0.28) and minimum (0.18) mean Nei's genetic distance.

Table 2. Various diversity indices computed to explore genetic diversity in Turkish laurel germplasm.

Primers	Total Bands	Polymorphic Bands	Polymorphism (%)	PIC	Ne	h	I
iPBS2228	16	12	75	0.284	1.35	0.21	0.32
iPBS2230	14	12	85.71	0.585	1.57	0.33	0.48
iPBS2232	11	9	81.81	0.383	1.37	0.24	0.38
iPBS2237	16	13	81.25	0.330	1.37	0.23	0.35
iPBS2239	15	11	73.33	0.323	1.29	0.19	0.31
iPBS2245	13	12	92.30	0.450	1.44	0.26	0.40
iPBS2253	18	15	83.33	0.314	1.25	0.16	0.26
iPBS2256	13	13	100	0.422	1.40	0.24	0.38
iPBS2277	15	12	80	0.163	1.15	0.12	0.22
iPBS2295	11	08	72.72	0.199	1.19	0.14	0.25
iPBS2398	23	21	91.30	0.417	1.44	0.27	0.41
iPBS2401	16	12	75	0.447	1.46	0.27	0.39
iPBS2402	14	14	100	0.373	1.45	0.27	0.42
Total	195	164					
Average	15	12.61	83.98	0.361	1.36	0.22	0.35

PIC: polymorphism information content, Ne: effective number of alleles, h: gene diversity, I: Shannon's information index.

Table 3. Region based diversity indices evaluation for Turkish laurel germplasm.

Region	Ne	h	I	Ht	GD
Mediterranean	1.41	0.24	0.24	0.19	0.28
Aegean	1.34	0.21	0.31	0.13	0.25
Black sea	1.31	0.18	0.28	0.14	0.20
Marmara	1.26	0.16	0.25	0.11	0.18

Ne: effective number of alleles, h: gene diversity, I: Shannon's information index, Ht: overall gene diversity, GD: Nei's genetic distance.

To understand the relationship among 94 laurel genotypes, the neighbor-joining clustering was performed, which separated the germplasm based on their geographical regions and provinces (Figure 2). Principal coordinate analysis (PCoA) also separated the 94 laurel genotypes based on their geographical regions (Figure 3). For the understanding of population structure of the Turkish laurel germplasm, Bayesian-based clustering algorithm was performed, which divided the genotypes into two main populations A and B (Figure 4). Population A contained mainly genotypes from the Marmara, Black Sea and Aegean region. Most of the genotypes from the Mediterranean region were present in population B. Analysis of molecular variance (AMOVA) was performed, which revealed higher variations (85%) within populations as compared to among the population (15%; Table 4).

Table 4. Analysis of molecular variance (AMOVA) revealing genetic diversity in the Turkish laurel germplasm.

Summary of AMOVA Table					
Source of Variation	**df**	**SS**	**MS**	**Est. Var.**	**%**
Among Pops	3	334.345	111.448	3.980	15%
Within Pops	90	2083.506	23.150	23.150	85%
Total	93	2417.851		27.130	100%

df: degrees of freedom, SS: Sum of squares, MS: Mean square, EST. VAr: Estimated variance.

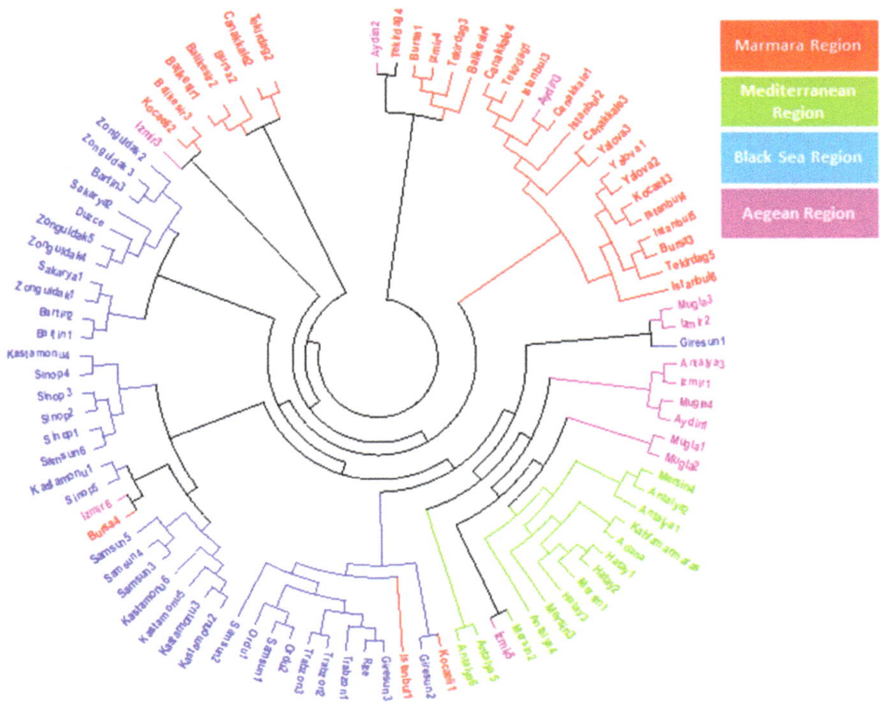

Figure 2. Neighbor-joining clustering of Turkish laurel germplasm revealed by 13 iPBS-retrotransposon primers.

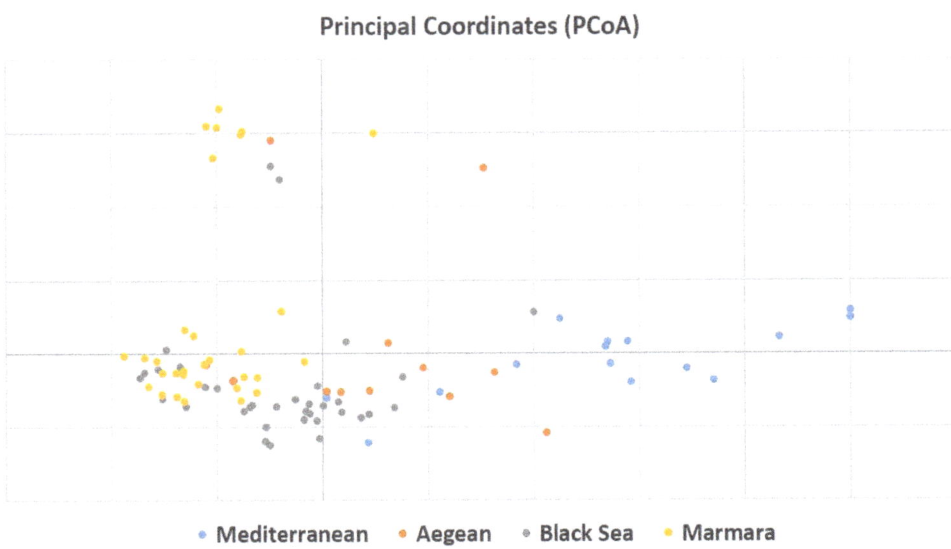

Figure 3. Principal coordinate analysis of Turkish laurel germplasm revealed by 13 iPBS-retrotransposon primers.

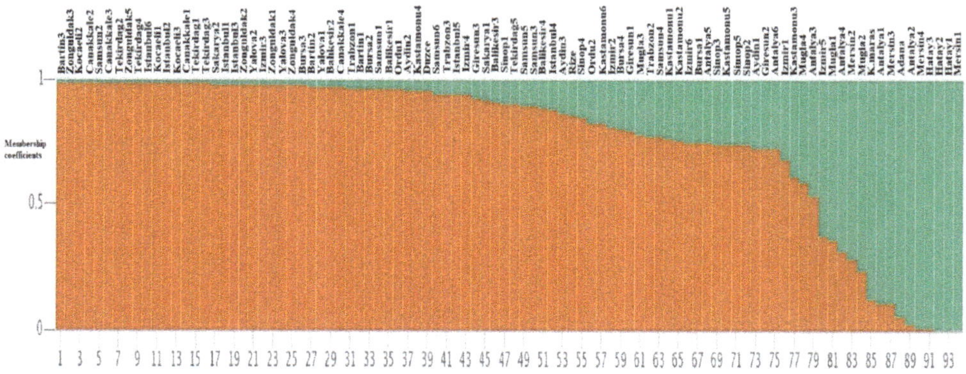

Figure 4. Structure-based clustering among 94 Turkish laurel genotypes using 13 iPBS-retrotransposon markers.

4. Discussion

4.1. Polymorphism in Turkish Laurel Germplasm Revealed by iPB-Retrotransposon Primers

A total of 13 most polymorphic iPBS-retrotransposon primers were used to elucidate genetic diversity and population structure of Turkish laurel germplasm (Table 2). Total and polymorphic bands obtained in this study were found much higher than reported by Arroyo-Garcia et al. [30] using AFLP markers and Bulut et al. [38] using an SSR marker for laurel. The average number of polymorphic bands obtained in this study were found much greater than the reported by Baránek et al. [39] (7.1 bands), Guo et al. [23] (6.6 bands) and Nemli et al. [40] (3.8 bands). This could be due to laurel having a larger genome size as compared to plant genomes earlier studied. These results may also indicate that iPBS-retrotransposon are more conserved for laurel compared to guava [41], *Cicer* species [42] and saffron [43]. Additionally, we came to know that iPBS-retrotransposon markers are more informative as compared to other molecular marker systems, like inter-retrotransposon-amplified polymorphism (IRAP) [44], random amplification of polymorphic DNA (RAPD) and inter simple sequence repeat (ISSR) [45] and sequence-related amplified polymorphism (SRAP) [46] and can be suggested for any crop to elucidate genetic diversity.

Mean polymorphism obtained in this study was 83.98%, which varied from 72.72% to 100%. Mean polymorphism obtained in this study was found greater than reported by Sevindik [47] for laurel, Yaldiz et al. [25] for tobacco and Aydin and Baloch [48] using iPBS-retrotransposons and thus confirming greater variations in Turkish laurel germplasm. The PIC value is used to understand the efficiency of polymorphic loci for the identification of genetic diversity [43] and explore the discriminating power marker among genotypes [40]. In this study, the mean PIC value was 0.361 which varied 0.163–0.585. Mean PIC value obtained in this study was found much greater than reported by earlier studies using iPBS-retrotransposons markers [25,49].

Previous studies have proven iPBS-retrotransposon a highly reproducible, robust and trustable marker system [22,23,40,41]. Our earlier studies in wild emmer [8], pea [24], tobacco [25], common bean [48], pepper [50] and safflower [51] also found this marker system highly robust, and trustable, which can be used for the investigation of genetic diversity in any crop. Similarly, Mehmood et al. [41] also found this marker system highly reproducible, robust and trustable. Moreover, trustbility of this marker compared to another marker system has been proven by earlier studies as well. Cömertpay et al. [52] compared iPBS retrotransposons with SSR markers for rice germplasm and found parallel results with both marker system, which confirm trustbility of this marker system. Andeden et al. [42] used iPBS-retrotransposon and an ISSR marker system for the taxonomic evaluation of seven *Cicer* species. They found similar clustering for species and confirmed the trustbility and robustness of this marker system. Yildiz et al. [53] also checked the reproducibility of this marker system in okra by repeating PCR two times and stated that band profiles remains unchanged after PCR amplification. In this study, we also checked the reproducibility of the iPBS-retrotransposon marker system by repeating PCR two times for two primers. We found that band profiles remain unchanged after repeating the PCR amplification for both primers. Total and polymorphic numbers of bands remained unchanged (Figure S2) and these results were found similar to the previous studies [41,53]. These results are confirming the higher reproducibility of this marker system. Therefore, this marker system can be suggested as a marker of choice for the investigation of genetic diversity due to its highly reproducibility, robustness and lower cost.

4.2. Genetic Diversity and Population Evaluation for Turkish Laurel Germplasm

Various diversity indices were calculated to elucidate the genetic diversity in Turkish laurel germplasm (Table 2). The maximum number of effective alleles is always desirable because they show the existence of greater genetic variations. Mean effective number of alleles (1.36) found in this study were higher than the reported by previous studies in various crops using iPBS-retrotransposons markers [25,50]. Mean gene diversity (0.22) resulted in this study was found higher than the reported by earlier studies [25,48,50] and thus explains the presence of higher diversity in studied germplasm. Shannon's information index is an important criterion to understand the variation as it distinguishes the genetic diversity in a population combining abundance and evenness [51]. Mean Shannon's information index (0.35) was found greater than the reported by earlier studies using the same molecular marker [25,50]. Genetic distance was calculated among the 94 Turkish laurel genotypes and mean genetic distance was 0.228. Bartin3 and Zonguldak3 genotypes showed a higher level of similarity as they accounted for the minimum (0.089) genetic distance. Canakkale1 and Mersin1 genotypes were found genetically to be the most distinct genotypes as the maximum genetic distance (0.607) were present between these genotypes. Development of breeding material reflecting desirable attributes remained the central focus of the breeding community and Arystanbekkyzy et al. [8] stated that genetically distinct genotypes always acts as a source of breeding material. Therefore, Canakkale1 and Mersin1 genotypes can be suggested as candidate parents for the various laurel breeding activities.

To elucidate the genetic variations in Turkish laurel germplasm more comprehensively, various diversity indices were also calculated upon their collection regions (Table 3). Genotypes from the Mediterranean region showed a higher level of diversity as compared to the rest of the regions and the Marmara region was found to be the least diverse region. Cuttelod et al. [54] stated that the Mediterranean basin is one of the most diverse places on this planet and Nadeem et al. [9] stated that laurel is an important plant of this region and shows a good level of diversity in its phenotype and genotype. Hatay2 and Antalya5 were found to be the most diverse genotypes in the Mediterranean region, while Muğla1 and İzmir5 were distinct genotypes for the Aegean region. In the Black Sea region, Giresun2 and Sinop5 were distinct genotypes, while Sakarya1 and Bursa1 were distinct genotypes for the Marmara region. Therefore, these diverse genotypes can act as genetic stock for the breeding activities in laurel.

Analysis of molecular variance (AMOVA) revealed the existence of higher variations within the laurel genotypes and the percentage of the total variance was 85% (Table 4). Pour et al. [49] stated that higher variations within genotypes might be due to selection, adaptation, gene flow, genetic drift, variation in ecotypes and the pollination method. Moreover, human activities and environmental fluctuations over time might be responsible for higher variations [55]. Our findings were in line with Gramazio et al. [56], as they also found higher genetic variation within the population of *Larix decidua* using SSR markers. However, this factor should be accounted for that iPBS-retrotransposon is a dominant marker system. A marker of this nature may not provide the credible information about the selection and adaptation phenomenon that occurs in a plant species. Therefore, a co-dominant marker system should be used in future studies in laurel germplasm to evaluate the selection and adaptation factors. There were no statistically significant variations among the population (only 15%), therefore, most of the variation in Turkish laurel germplasm are due to differences within populations.

Laurel male flower matures early when compared to the female flower and this condition is known as protandry, which ultimately ensures there is no self-pollination in this plant. This cross-fertilization ultimately is responsible for higher genetic diversity within the population [57].

The neighbor-joining analysis was performed to understand the relationship among 94 Turkish laurel genotypes, which grouped the germplasm according to their geographic regions and provinces (Figure 2). Genotypes belonging to the same region were present together and the same was the case for the genotypes belonging to their respective province. However, it was also observable that genotypes belonging to the Aegean region were also grouped with the rest of the provinces. Laurel is native to Mediterranean countries like Italy, Spain, Israel, France Corsica Island and North Africa [13]. Hatay, Kahramanmaraş, Adana, Mersin and Antalya provinces are present in the Mediterranean region of Turkey and native to this region [9]. However, due to its medicinal and essential oil importance, it gains the attention of Turkish farmers and spread to Black Sea and Marmara regions. Later, it was introduced into the Aegean region from all three regions. It is believed that a lot of seed mixing happens during laurel introduction to the Aegean region. Results of this study are also supporting this hypothesis because genotypes from the Aegean region are showing similarity to all provinces. There is a possibility of seed mixing, human selection and hybridization process when the seed was introduced to the Aegean region, which ultimately resulted in higher gene flow in the genotypes of this region. Bulut et al. [38] used an SSR marker to characterize Turkish laurel germplasm and they also found admixture of genotypes and concluded this possibly due to natural gene flow in genotypes reflecting similarity with other cluster genotypes. Principal coordinate analysis (PCoA) was also performed, which confirmed the result obtained by the neighbor-joining analysis (Figure 3). The principal coordinate analysis also divided the genotypes upon their collection regions. Similar to the neighbor-joining analysis, genotypes from Aegean regions were mixed with all three regions and supported the hypothesis of gene flow in this region.

The model-based structure algorithm has been found more informative and precise as compared to other clustering algorithms [7,51]. Therefore, the structure was taken as a clustering benchmark in this study. Using this algorithm, 94 Turkish laurel germplasms were partitioned into two populations, largely upon the basis of their provinces and respective region (Figure 4). Population A was found admixture of genotypes from all four regions. All genotypes from the Marmara and Black Sea region were present in population A. All genotypes from the Aegean region were present in population A except three genotypes (Mugla1, Mugla2 and Izmir5), which were present in population B on the basis of membership coefficients. Population B largely comprised of genotypes from the Mediterranean region. Three genotypes from the Mediterranean region were clustered in population A with genotypes from the Aegean region. The Aegean region is close to the Mediterranean region. Izmir and Mugla provinces are close to the Mediterranean region and there is a possibility of a seed exchange and hybridization event. Genetic diversity assessment is an important step towards the breeding activities as it provides a novel source of variations [58,59]. Various genetic diversity indices in this study revealed the existence of higher diversity in studied germplasm, which can be utilized for future laurel breeding activities. Similarly, various diversity indices confirmed the existence of higher genetic diversity in the genotypes from the Mediterranean region and it is believed that laurel was native to this region. Later, it was introduced to other regions of Turkey. Seed mixing and selection of germplasm for favorable traits lowers the genetic diversity in the Black Sea, Aegean and Marmara region. Therefore, genotypes from these regions reflected higher genetic similarity with each other and were grouped in population A.

5. Conclusions

The present study comprehensively explored the genetic diversity and population structure of Turkish laurel germplasm. This study confirmed iPBS-retrotransposon as a highly reproducible, trustable and robust marker system. An analysis of molecular variance (AMOVA) revealed that most variations in the studied germplasm are explained by the differences that existed within the populations. Neighbor joining analysis and PCoA analysis clustered the 94 Turkish laurel genotypes largely according to their collection provinces and regions. Genotypes from the Mediterranean region showed more diversity compared to the Black Sea, Aegean and Marmara region. We are confident that information derived from this study can be used for the deeper understanding of the genetic relationships in laurel germplasm, establishment of a reference collection and in the determination of appropriate breeding and conservation strategies.

Supplementary Materials: The following are available online at http://www.mdpi.com/2073-4395/9/10/647/s1. Figure S1. A representative gel imaging picture revealing genetic diversity among 94 Turkish laurel genotypes using 13 iPBS-retrotransposon primers; Figure S2. Confirmation of reproducibility of DNA profiles for iPBS-retrotransposon marker system; Table S1. Passport data of 94 Turkish laurel genotypes.

Author Contributions: Conceptualization, F.S.B. and Ü.K.; methodology, M.A.N.; software, M.A.N.; validation, F.S.B., and, M.A.N.; formal analysis, M.A.N.; investigation, M.A.N., A.Y.; resources, Ü.K., G.C.; data curation, M.A.N., and A.Y.; writing—original draft preparation, M.A.N.; writing—review and editing, Ü.K., S.E., E.H., M.Y., S.H.Y.; visualization, G.C.; supervision, F.S.B.

Funding: This research was funded by Scientific research unit, Bolu Abant Izzet Baysal University, Bolu, Turkey (BAP project: 2018.10.01.1305 to FSB).

Conflicts of Interest: The authors declare no conflict of interest.

References

1. Chen, S.L.; Yu, H.; Luo, H.M.; Wu, Q.; Li, C.F.; Steinmetz, A. Conservation and sustainable use of medicinal plants: Problems, progress, and prospects. *Chin. Med.* **2016**, *11*, 37. [CrossRef] [PubMed]
2. Van de Wouw, M.; Kik, C.; van Hintum, T.; van Treuren, R.; Visser, B. Genetic erosion in crops: Concept, research results and challenges. *Plant Genet. Resour.* **2010**, *8*, 1–15. [CrossRef]
3. Khoury, C.K.; Castañeda-Alvarez, N.P.; Achicanoy, H.A.; Sosa, C.C.; Bernau, V.; Kassa, M.T.; Norton, S.L.; van der Maesen, L.J.G.; Upadhyaya, H.D.; Ramírez-Villegas, J.; et al. Crop wild relatives of pigeonpea [*Cajanus cajan* (L.) Millsp.]: Distributions, ex situ conservation status, and potential genetic resources for abiotic stress tolerance. *Biol. Conserv.* **2015**, *184*, 259–270. [CrossRef]
4. Roederer, C.; Nugent, R.; Wilson, P. *Economic Impacts of Genetically Modified Crops on the Agrifood Sector: A Synthesis. Working Document, Directorate-General for Agriculture*; EU Commission Working Document; EU Commission: Brussels, Belgium, 2000.
5. Engels, J.M.M. *A Guide to Effective Management of Germplasm Collections*; Visser, L., Ed.; IPGRI: Rome, Italy, 2003; pp. 157–161.
6. Dawson, I.; Were, J. Collecting germplasm from trees-some guidelines. *Agrofor. Today* **1997**, *9*, 6–9.
7. Nadeem, M.A.; Habyarimana, E.; Çiftçi, V.; Nawaz, M.A.; Karaköy, T.; Comertpay, G.; Shahid, M.Q.; Hatipoğlu, R.; Yeken, M.Z.; Ali, F.; et al. Characterization of genetic diversity in Turkish common bean gene pool using phenotypic and whole-genome DArTseq-generated silicoDArT marker information. *PLoS ONE* **2018**, *13*, e0205363. [CrossRef] [PubMed]
8. Arystanbekkyzy, M.; Nadeem, M.A.; Aktas, H.; Yeken, M.Z.; Zencirci, N.; Nawaz, M.A.; Ali, F.; Haider, M.S.; Tunc, K.; Chung, G.; et al. Phylogenetic and taxonomic relationship of turkish wild and cultivated emmer (*triticum turgidum* ssp. dicoccoides) revealed by iPBSretrotransposons markers. *Int. J. Agric. Biol.* **2019**, *21*, 155–163. [CrossRef]
9. Nadeem, M.A.; Aasim, M.; Kırıcı, S.; Karık, Ü.; Nawaz, M.A.; Yılmaz, A.; Maral, H.; Khawar, K.M.; Baloch, F.S. Laurel (*Laurus nobilis* L.): A Less-Known Medicinal Plant to the World with Diffusion, Genomics, Phenomics, and Metabolomics for Genetic Improvement. In *Biotechnological Approaches for Medicinal and Aromatic Plants*; Springer: Singapore, 2018; pp. 631–653.
10. Vital, P.G.; Rivera, W.L. Antimicrobial activity and cytotoxicity of Chromolaena odorata (L. f.) King and Robinson and Uncaria perrottetii (A. Rich) Merr. Extracts. *J. Med. Plant Res.* **2009**, *3*, 511–518.
11. Balunas, M.J.; Kinghorn, A.D. Drug discovery from medicinal plants. *Life Sci.* **2005**, *78*, 431–441. [CrossRef]
12. Hamilton, J.A.; Miller, J.M. Adaptive introgression as a resource for management and genetic conservation in a changing climate. *Conserv. Biol.* **2016**, *30*, 33–41. [CrossRef]
13. Leung, A.Y.; Foster, S. *Encyclopedia of Common Natural Ingredients Used in Food, Drugs and Cosmetics*, 2nd ed.; Wiley-Interscience: Hoboken, NJ, USA, 2003; pp. 69–71.

14. Aqili Khorasani, M.H. *Collection of Drugs (Materia Media)*; Enqelab-e-Eslami Publishing and Educational Organization: Tehran, Iran, 1991; pp. 388–389.
15. El, S.N.; Karagozlu, N.; Karakaya, S.; Sahın, S. Antioxidant and antimicrobial activities of essential oils extracted from *Laurus nobilis* L. leaves by using solvent-free microwave and hydrodistillation. *FNS* **2014**, *5*, 97. [CrossRef]
16. Cherrat, L.; Espina, L.; Bakkali, M.; García-Gonzalo, D.; Pagán, R.; Laglaoui, A. Chemical composition and antioxidant properties of *Laurus nobilis* L. and *Myrtus communis* L. essential oils from Morocco and evaluation of their antimicrobial activity acting alone or in combined processes for food preservation. *J. Sci. Food Agric.* **2014**, *94*, 1197–1204. [CrossRef] [PubMed]
17. Baloch, F.S.; Alsaleh, A.; Shahid, M.Q.; Çiftçi, V.; de Miera, L.E.S.; Aasim, M.; Nadeem, M.A.; Aktaş, H.; Özkan, H.; Hatipoğlu, R. A whole genome DArTseq and SNP analysis for genetic diversity assessment in durum wheat from central fertile crescent. *PLoS ONE* **2017**, *12*, e0167821. [CrossRef] [PubMed]
18. Boza, A.; Hepaksoy, S. Some leaf properties of natural *Laurus nobilis* L. population in Karaburun Peninsula (Izmir/Turkey). In Proceedings of the VII International Scientific Agriculture Symposium (Agrosym 2016), Jahorina, Bosnia and Herzegovina, 6–9 October 2016; University of East Sarajevo, Faculty of Agriculture: Lukavica, Bosnia and Herzegovina, 2016; pp. 717–722.
19. Dadalioğlu, I.; Evrendilek, G.A. Chemical compositions and antibacterial effects of essential oils of Turkish oregano (Origanum minutiflorum), bay laurel (*Laurus nobilis*), Spanish lavender (*Lavandula stoechas* L.), and fennel (Foeniculum vulgare) on common foodborne pathogens. *J. Agric. Food Chem.* **2004**, *52*, 8255–8260.
20. Nadeem, M.A.; Nawaz, M.A.; Shahid, M.Q.; Doğan, Y.; Comertpay, G.; Yıldız, M.; Hatipoğlu, R.; Ahmad, F.; Alsaleh, A.; Labhane, N.; et al. DNA molecular markers in plant breeding: Current status and recent advancements in genomic selection and genome editing. *Biotechnol. Biotechnol. Equip.* **2018**, *32*, 261–285. [CrossRef]
21. SanMiguel, P.; Tikhonov, A.; Jin, Y.K.; Motchoulskaia, N.; Zakharov, D.; Melake-Berhan, A.; Springer, P.S.; Edwards, K.J.; Lee, M.; Avramova, Z.; et al. Nested retrotransposons in the intergenic regions of the maize genome. *Science* **1996**, *274*, 765–768. [CrossRef]
22. Kalendar, R.; Antonius, K.; Smýkal, P.; Schulman, A.H. iPBS: A universal method for DNA fingerprinting and retrotransposon isolation. *Theor. Appl. Genet.* **2010**, *121*, 1419–1430. [CrossRef]
23. Guo, D.L.; Guo, M.X.; Hou, X.G.; Zhang, G.H. Molecular diversity analysis of grape varieties based on iPBS markers. *Biochem. Syst. Ecol.* **2014**, *52*, 27–32. [CrossRef]
24. Baloch, F.S.; Alsaleh, A.; de Miera, L.E.S.; Hatipoğlu, R.; Çiftçi, V.; Karaköy, T.; Yıldız, M.; Özkan, H. DNA based iPBS-retrotransposon markers for investigating the population structure of pea (*Pisum sativum*) germplasm from Turkey. *Biochem. Syst. Ecol.* **2015**, *61*, 244–252. [CrossRef]
25. Yaldiz, G.; Camlica, M.; Nadeem, M.A.; Nawaz, M.A.; Baloch, F.S. Genetic diversity assessment in *Nicotiana tabacum* L. with iPBS-retrotransposons. *Turk. J. Agric. For.* **2018**, *42*, 154–164. [CrossRef]
26. Szmidt, A.E.; Wang, X.R.; Lu, M.Z. Empirical assessment of allozyme and RAPD variation in *Pinus sylvestris* (L.) using haploid tissue analysis. *Heredity* **1996**, *76*, 412. [CrossRef]
27. Fan, X.X.; Shen, L.; Zhang, X.; Chen, X.Y.; Fu, C.X. Assessing genetic diversity of *Ginkgo biloba* L. (Ginkgoaceae) populations from China by RAPD markers. *Biochem. Genet.* **2004**, *42*, 269–278. [CrossRef] [PubMed]
28. Yalçın, H.; Anık, M.; Şanda, M.A.; Çakır, A. Gas chromatography/mass spectrometry analysis of *Laurus nobilis* essential oil composition of northern Cyprus. *J. Med. Food* **2007**, *10*, 715–719. [CrossRef] [PubMed]
29. Caputo, L.; Nazzaro, F.; Souza, L.F.; Aliberti, L.; De Martino, L.; Fratianni, F.; Coppola, R.; De Feo, V. *Laurus nobilis*: Composition of essential oil and its biological activities. *Molecules* **2017**, *22*, 930. [CrossRef] [PubMed]
30. Arroyo-García, R.; Martínez-Zapater, J.M.; Prieto, J.F.; Álvarez-Arbesú, R. AFLP evaluation of genetic similarity among laurel populations (Laurus L.). *Euphytica* **2001**, *122*, 155–164.
31. Arroyo, J.M.; Rigueiro, C.; Rodríguez, R.; Hampe, A.; Valido, A.; Rodríguez-Sánchez, F.; Jordano, P. Isolation and characterization of 20 microsatellite loci for laurel species (Laurus, Lauraceae). *Am. J. Bot.* **2010**, *97*, e26–e30. [CrossRef]
32. Rodríguez-Sánchez, F.; Guzmán, B.; Valido, A.; Vargas, P.; Arroyo, J. Late Neogene history of the laurel tree (Laurus L.; Lauraceae) based on phylogeographical analyses of Mediterranean and Macaronesian populations. *J. Biogeogr.* **2009**, *36*, 1270–1281.
33. Doyle, J.J.; Doyle, J.L. Isolation of plant DNA from fresh tissue. *Focus* **1990**, *12*, 13–15.

34. Baloch, F.S.; Alsaleh, A.; Andeden, E.E.; Hatipoğlu, R.; Nachit, M.; Özkan, H. High levels of segregation distortion in the molecular linkage map of bread wheat representing the West Asia and North Africa region. *Turk. J. Agric. For.* **2016**, *40*, 352–364. [CrossRef]
35. Yeh, F.C.; Yang, R.; Boyle, T.J.; Ye, Z.; Xiyan, J.M. *PopGene32, Microsoft Windows-Based Freeware for Population Genetic Analysis, Version 1.32. Mol. Biology and Biotechnology Centre*; University of Alberta: Edmonton, AB, Canada, 2000.
36. Peakall, R.; Smouse, P.E. GenAlEx 6.5: Genetic analysis in Excel. Population genetic software for teaching and research—An update. *Bioinformatics* **2012**, *28*, 2537–2539. [CrossRef]
37. Evanno, G.; Regnaut, S.; Goudet, J. Detecting the number of clusters of individuals using the software STRUCTURE: A simulation study. *Mol. Ecol.* **2005**, *14*, 2611–2620. [CrossRef]
38. Bulut, M.Ç.; Özmen, C.Y.; Ergül, A.; Ayanoğlu, F. Genetic Characterization of Bay Laurel (*Laurus nobilis* L.) Populations Using Microsatellite Markers and Flow Cytometry. *MKUJAS* **2018**, *23*, 242–253.
39. Baránek, M.; Meszáros, M.; Sochorová, J.; Čechová, J.; Raddová, J. Utility of retrotransposon-derived marker systems for differentiation of presumed clones of the apricot cultivar Velkopavlovická. *Sci. Hortic.* **2012**, *143*, 1–6. [CrossRef]
40. Nemli, S.; Kianoosh, T.; Tanyolac, M.B. Genetic diversity and population structure of common bean (*Phaseolus vulgaris* L.) accessions through retrotransposon-based interprimer binding sites (iPBSs) markers. *Turk. J. Agric. For.* **2015**, *39*, 940–948. [CrossRef]
41. Mehmood, A.; Luo, S.; Ahmad, N.M.; Dong, C.; Mahmood, T.; Sajjad, Y.; Jaskani, M.J.; Sharp, P. Molecular variability and phylogenetic relationships of guava (*Psidium guajava* L.) cultivars using inter-primer binding site (iPBS) and microsatellite (SSR) markers. *Genet. Resour. Crop Evol.* **2016**, *63*, 1345–1361. [CrossRef]
42. Andeden, E.E.; Baloch, F.S.; Derya, M.; Kilian, B.; Özkan, H. iPBS-Retrotransposons-based genetic diversity and relationship among wild annual Cicer species. *J. Plant Biochem. Biotechnol.* **2013**, *22*, 453–466. [CrossRef]
43. Gedik, A.; Duygu, A.T.E.S.; Erdogmus, S.; Comertpay, G.; Tanyolac, M.B.; Ozkan, H. Genetic diversity of Crocus sativus and its close relative species analyzed by iPBS-retrotransposons. *Turk. J. Agric. For.* **2017**, *22*, 243–252. [CrossRef]
44. Boronnikova, S.V.; Kalendar, R.N. Using IRAP markers for analysis of genetic variability in populations of resource and rare species of plants. *Russ. J. Genet.* **2010**, *46*, 36–42. [CrossRef]
45. Mir, J.I.; Ahmed, N.; Khan, M.H.; Mokhdomi, T.A.; Wani, S.H.; Bukhari, S.; Asif, A.M.I.N.; Qadri, R.A. Molecular characterization of saffron-potential candidates for crop improvement. *Not. Sci. Biol.* **2015**, *7*, 81–89. [CrossRef]
46. Babaei, S.; Talebi, M.; Bahar, M.; Zeinali, H. Analysis of genetic diversity among saffron (*Crocus sativus*) accessions from different regions of Iran as revealed by SRAP markers. *Sci. Hortic.* **2014**, *171*, 27–31. [CrossRef]
47. Sevindik, E. Molecular Genetic Diversity of Some *Laurus nobilis* L. (Lauraceae) Populations Grown in The Aegean Region/Turkey. *AFS* **2019**, *41*, 28–31.
48. Aydın, M.F.; Baloch, F.S. Exploring the genetic diversity and population structure of Turkish common bean germplasm by the iPBS-retrotransposons markers. *Legume Res.* **2018**, *LR-423*, 1–7.
49. Pour, A.H.; Karahan, F.; İlhan, E.; İlçim, A.; Haliloğlu, K. Genetic structure and diversity of Adonis L. (Ranunculaceae) populations collected from Turkey by inter-primer binding site (iPBS) retrotransposon markers. *Turk. J. Bot.* **2019**, *43*, 585–596. [CrossRef]
50. Yildiz, M.; Koçak, M.; Nadeem, M.A.; Cavagnaro, P.; Barboza, K.; Baloch, F.S.; Argün, D.; Keleş, D. Genetic diversity analysis in the Turkish pepper germplasm using iPBS retrotransposon-based markers. *Turk. J. Agric. For.* **2019**, 43. [CrossRef]
51. Ali, F.; Yılmaz, A.; Nadeem, M.A.; Habyarimana, E.; Subaşı, I.; Nawaz, M.A.; Chaudhary, H.J.; Shahid, M.Q.; Ercişli, S.; Zia, M.A.B.; et al. Mobile genomic element diversity in world collection of safflower (*Carthamus tinctorius* L.) panel using iPBS-retrotransposon markers. *PLoS ONE* **2019**, *14*, e0211985. [CrossRef]
52. Cömertpay, G.; Baloch, F.S.; Derya, M.; Andeden, E.E.; Alsaleh, A.; Sürek, H.; Özkan, H. Population structure of rice varieties used in Turkish rice breeding programs determined using simple-sequence repeat and inter-primer binding site-retrotransposon data. *Genet. Mol. Res.* **2016**, *15*, 1–14. [CrossRef]
53. Yildiz, M.; Koçak, M.; Baloch, F.S. Genetic bottlenecks in Turkish okra germplasm and utility of iPBS retrotransposon markers for genetic diversity assessment. *Genet. Mol. Res.* **2015**, *14*, 10588–10602. [CrossRef]

54. Cuttelod, A.; García, N.; Malak, D.A.; Temple, H.J.; Katariya, V. The Mediterranean: A biodiversity hotspot under threat. In *Wildlife in a Changing World–an analysis of the 2008 IUCN Red List of Threatened Species*; Jean-Christophe, V., Craig, H.-T., Stuart, S.N., Eds.; IUCN: Gland, Switzerland, 2009; p. 89.
55. Solouki, M.; Mehdikhani, H.; Zeinali, H.; Emamjomeh, A.A. Study of genetic diversity in Chamomile (Matricaria chamomilla) based on morphological traits and molecular markers. *Sci. Hortic.* **2008**, *117*, 281–287. [CrossRef]
56. Gramazio, P.; Plesa, I.M.; Truta, A.M.; Sestras, A.F.; Vilanova, S.; Plazas, M.; Vicente, O.; Boscaiu, M.; Prohens, J.; Sestras, R.E. Highly informative SSR genotyping reveals large genetic diversity and limited differentiation in European larch (Larix decidua) populations from Romania. *Turk. J. Agric. For.* **2018**, *42*, 165–175. [CrossRef]
57. Denisow, B.; Wrzesien, M.; Cwener, A. Pollination and floral biology of *Adonis vernalis* L. (Ranunculaceae)-a case study of threatened species. *Acta Soc. Bot. Pol.* **2014**, *83*, 29–37. [CrossRef]
58. Guliyev, N.; Sharifova, S.; Ojaghi, J.; Abbasov, M.; Akparov, Z. Genetic diversity among melon (*Cucumis melo* L.) accessions revealed by morphological traits and ISSR markers. *Turk. J. Agric. For.* **2018**, *42*, 393–401. [CrossRef]
59. Eser, E.; Topcu, H.; Kefayati, S.; Sütyemez, M.; Islam, M.R.; Kafkas, S. Highly polymorphic novel simple sequence repeat markers from Class I repeats in walnut (*Juglans regia* L.). *Turk. J. Agric. For.* **2019**, *43*, 174–183. [CrossRef]

agronomy

MDPI

Review

Aegilops Species for the Improvement of the Leaf and Stripe Rust Resistance in Cultivated Triticale (×*Triticosecale* Wittmack)

Waldemar Ulaszewski [1] **and Michał Tomasz Kwiatek** [2,*]

[1] Institute of Plant Genetics of the Polish Academy of Sciences, Strzeszyńska 34 str., 60-479 Poznań, Poland; wula@igr.poznan.pl

[2] Department of Genetics and Plant Breeding, Poznań University of Life Sciences, Dojazd 11, 60-637 Poznań, Poland

* Correspondence: michal.kwiatek@up.poznan.pl; Tel.: +48-6184-87-760

Received: 26 October 2020; Accepted: 16 December 2020; Published: 18 December 2020

Abstract: Hexaploid triticale (×*Triticosecale* Wittmack, $2n = 6x = 42$ chromosomes, AABBRR) is a cultivated hybrid, which combines wheat (*Triticum aestivum* L.) and rye (*Secale cereale* L.) properties. It has a better ability to be grown on poor soils, compared to wheat. Mainly, triticale is produced for forage feed and bioethanol. Considering the limited diversity of this human-made crop, there is a need to widen its genetic variability, especially to introduce new genes, responsible for agronomic traits, such as resistance to biotic stresses. Leaf rust caused by *Puccinia triticina* Eriks. and stripe rust caused by *Puccinia striiformis* Westend are the most destructive foliar diseases of triticale and related cereals. Developing resistant triticale varieties is an important strategy for the control of these diseases. A number of leaf and stripe rust resistance genes have been already introduced into bread wheat from related species using chromosome manipulations. Exploitation of related species conferring desirable loci is the most effective non-GMO way of improving the rust resistance of triticale. The procedure encompasses chromosome doubling of obtained hybrids followed by a number of backcrosses to eliminate unnecessary alien chromatin and to reduce the linkage drag. In this review, we show the recent status of pre-breeding studies, which are focused on transfer of leaf and stripe rust resistance genes from *Aegilops* species into cultivated triticale using distant crossing and chromosome engineering.

Keywords: *Aegilops*; triticale; leaf rust; stripe rust; yellow rust; resistance; *Puccinia*

1. Introduction

The hexaploid triticale (×*Triticosecale* Wittmack, $2n = 6x = 42$, AABBRR), a small grain cereal, is a man-made crop developed by crossing *Triticum* species as a female parent and rye (*Secale cereale* L.) as a male parent. In principle, triticale was intended to combine the nutritional value of the wheat and the rapid growth, cold tolerance, and hardiness of rye. Cultivated hexaploid triticale exhibits amphiploidy with respect to wheat (AABB) and rye (RR) genomes. Triticale is one of few artificial crops cultivated at a large scale. In general for the last decades the global harvested area of triticale has been constantly increasing (2,101,405 ha in 1996; 3,662,363 ha in 2006; and 3,809,192 ha in 2018) [1]. The major importance of this cereal is based on the protein content in seeds. It may vary from 12 to 22%. Feeding experiments involving farm animals have shown that triticale seem to be more digestible in comparison with parental species [2]. The primary end-use of triticale is forage production. However, growing demand for food resources and biofuel has led to an increased interest in development of this cereal. The end-use of triticale depends of the characteristics of its cultivar and is largely determined by its chemical composition. Cultivars of large and uniform grains containing more proteins than starch are used as concentrated feed for poultry, ruminants, and monogastrics. Cultivars forming high amount of biomass, like rye, are grazed or harvested for silage, green-feed, and hay [2,3]. At first triticale has mostly been used as animal feed, and its potential still has not been fully utilized at a large scale. Recently this crop has found application in food industry, including baking [4] and brewing [5]. Triticale has also been used as a substrate for bioethanol and biofuel production [6].

The genetic variability of triticale is limited due to lack of natural speciation, narrow genepool of parental forms (wheat and rye) and development of new breeding techniques, which induce high homozyosity (i.e., double haploid production) [7]. The first attempt to obtain triticale was completed by Wilhelm Rimpau in Germany in 1888 by crossing hexaploid wheat and rye and inducing spontaneous chromosome doubling of the sterile hybrid that is produced using conventional plant breeding hybridizing techniques between wheat and rye [8]. In the 1950s Arpad Kiss discovered that hexaploid forms of triticale would be more agronomically appropriate than its octoploid forms [9]. The same researcher coined the terms "Secondary Hexaploid Triticale" (SHT) and "Primary Hexaploid Triticale" (PHT) to distinguish lines derived from different combination of crosses [8]. PHT encompasses hexaploid lines of triticale obtained by crossing 6× triticale with 6× triticale where A and B genomes derive from tetraploid wheat. SHT encompasses hexaploid lines of triticale obtained by crossing 8× triticale with 6× triticale where A and B genomes are recombined from tetraploid and hexaploid wheat [7]. The first commercial cultivars of hexaploid triticale came onto the market in the late 1960s as T-No.57 and T-No.6 [9]. This cereal was as competitive as rye on marginal soils but contained 30–50% more protein. In May 1969 triticale became recognized as a new crop of commerce in Canada with the licensing of the hexaploid triticale cv Rosner [10]. The properties of triticale has motivated breeders to include triticale in their cereal breeding program. In Poland, which is now the biggest producers of triticale modern breeding and research of this crop started in 1960s. In 1982 the variety "Lasko", representing a favorable combination of yield potential, was the first cultivar of triticale registered in Poland and played a significant role in popularization of this new crop in Europe [11]. Currently, triticale breeding programs are aimed at the selection of highly yielding population varieties with increased resistance to fungal diseases (especially to leaf and stripe rust) [12].

2. Current Status of Triticale Resistance against Leaf and Stripe Rusts

Until the breakdown of resistance, triticale had not been prone to harmful effects of various pathogens. Currently, this cereal meets pathogens and pests deriving both from wheat and rye. The first triticale disease which occurred in epidemic proportions was stem rust (*Pucinia graminis* f.sp. *tritici*) in Australia [2]. Then, leaf rust (syn. brown rust, caused by *Puccinia triticiana* Eriks) and stripe rusts (syn. yellow rust, caused by *P. striiformis* f.sp. *tritici*) began to affect this crop everywhere it grows. As triticale started to expand into new production areas, new hybrid pathotypes of *Puccinia*, carrying virulence genes have evolved and broadened their range of hosts, moving from wheat and rye into triticale [13]. Both leaf and stripe rust are able to cause economic losses and decrease the grain yield even by 40% [2]. There is an urgent need to overcome this problem and enhance the resistance to these diseases.

The rust fungi (Pucciniales) (previously known as Uredinales) are the obligate, biotrophic basidiomycete fungal plant pathogens, infecting the representants of all vascular plant groups. The rust fungi are one of the most devastating threats to agricultural crops worldwide [14]. They are typified by the orange-brown-red rust color of one or more spore stages. Many aspects of biology of Pucciniales remain unclear. The research on these fungi is complicated due to their absolute dependence on the host and thus lack of possibility to culture them on artificial media [15]. In natural environment of *Puccinia triticina* and *P. striiformis*, individual populations of these fungi include many physiological races (pathotypes) with different levels of virulence. The airborne nature of these rust fungi as well as their tendency to increase its genetic variability by mutations and selection are factors allowing the pathogenic fungus to develop virulence against different cultivars carrying one or more resistance (R) genes against the most prevalent races of *P. triticina* and *P. striiformis* in a given region. Some estimates suggests that resistance against a specific rust resistance gene does not last more than 5–7 years;

therefore, an introgression of new and effective resistance genes is a regular activity in breeding programs of cereals [16].

There are over 70 leaf rust and stripe rust resistance genes, securing either race-specific resistance or non-race specific adult-plant resistance (APR), that have been identified in Triticae [17] and some of them have already been transferred from related species into the wheat genetic background [18]. Each of these genes confers excellent resistance to some pathotypes, but none of them can be expected to work against all pathotypes. The secondary gene pools as a source of effective leaf rust (*Lr*) resistance and stripe rust (*Yr*) resistance genes for triticale are wild relatives of wheat such as *Aegilops*, *Thinopyrum*, and *Triticum* species [18]. There are, in general, three categories of rust resistance in plants: race-specific seedling resistance, also known as all-stage resistance; race-specific adult plant resistance (race-specific APR); and race non-specific APR, also known as slow-rusting or partial resistance [16]. Particular races of rust fungi are specialized in different cereal host species. For example, *P. triticina* is able to successfully infect different small grain cereal species. *Puccinia triticina* f.sp. *tritici* is specialized in wheat [16]. *P. triticina* f.sp. *secalis* attacks rye [19]. Triticale is usually resistant to leaf rust of rye. The isolates of leaf rust from triticale in the field are usually wheat leaf rust [2]. Potentially, there is a possibility of somatic hybridization of rust fungi, which originally lived on wheat and rye and have moved into triticale. Evidences provided by Park and Wellings [20] suggest that some pathotypes of *P. triticina* may have arisen from somatic hybridization of different strains. It has been already confirmed that phylogenetically related fungal pathogen strains of triticale deriving from wheat and triticale, separately, have fused and adapted to live in new host plants [2].

Both hexaploid triticale and hexaploid wheat possess A and B genomes. The difference is that wheat has the additional D genome and triticale has the R genome instead. It is known that the presence of genes located on the chromosomes from D-genome downregulates the expression of leaf and stem rust resistance genes [21]. Potential of triticale as a crop lies in its ability to withstand several environmental stresses more effective than wheat. Moreover, the resistance of triticale to common wheat diseases is higher [22]. The newly formed triticale had been resistant to most of pathogens attacking wheat and rye until the breakdown of this resistance. Triticale as a hybrid plant has inherited leaf and stripe rust resistance from both progenitor species. The leaf rust isolated from triticale in the field is usually wheat leaf rust *P. triticina* [2]. Triticale is usually resistant to leaf and yellow rust of rye (*P. tritici* f.sp. *secalis*), and is a source of *Lr25*, *Lr26*, *Lr45*, *Yr9*, and *Yr83* effective genes [23,24].

There is relatively small number of reports on *Puccinia striiformis* f.sp. *tritici*, the causal agent of yellow rust. However, recently this pathogen in cereals becomes a production problem worldwide. However, the new triticale yellow rust pathotype, called "Jackie 143 E16 A+ J+" has appeared in 2007. It is virulent for the gene *Yr9*, a stripe rust resistance gene derived from cereal rye, common in most triticale varieties [24].

In general, hexaploid and octoploid triticales were initially resistant or moderately resistant to leaf rust [12]. The resistance however started to be broke down by new races of *Puccinia*. Tyryshkin et al. [25] selected only 24 highly resistant, at the seedling stage, accessions out of 471 tested. Mikhailova et al. [26] tested 416 triticale accessions and found only 17 leaf rust resistant triticale accessions. Manninger [27] tested the resistance of eight winter triticale accessions to leaf rust at seedling and adult stage. Cultivars Presto, Tricolor, Disco, GK Bogo, Kitaro, and Pongo were proven resistant at the seedling stage to three pathotypes of wheat leaf rust and susceptible only to one pathotype. Only two cultivars Presto and Tricolor have been proved to be resistant to leaf rust at adult stage. Manninger tested also resistance of wheat to leaf rust. Most pathotypes identified were virulent to resistant genes (*Lr2b*, *Lr2c*, *Lr3*, *Lr11*, *Lr17*, *Lr21*, and *Lr26*). Pathotypes from triticale were virulent only to *Lr2b*, *Lr2c*, and *Lr11*. Czajkowski et al. [28] conducted an analysis of susceptibility of 250 lines of winter triticale to *P. triticina* in the 2007/2008 and 2012/2013 growing seasons. They showed that at a seedling stage 28% of examined genotypes were slightly affected by pathogen deriving from triticale and 83.2% by fungus isolated from wheat. It may indicate that the pathogen has specialized against specific sets of *Lr* genes. Adult stage analyses after manual inoculation showed that 56.8% of triticale forms were slightly affected by pathogen, whereas only 4.8% were strongly and 38.4% moderately affected. Analysis of susceptibility of triticale in the natural inoculation conditions showed that slightly affected genotypes turned to be predominant (79.2%). Only 3.2% of genotypes were strongly and 17.6% moderately affected. Grzesik and Strzembicka [29], in order to determine identity of genes taking part in resistance of triticale against leaf rust, crossed resistant cultivars: Presto, Ugo, and Vero in all combinations. The greenhouse and field inoculation trials of F1 and F2 showed that all plants were resistant against 65a/95 *P. triticina* strain, similarly as parental forms. Results of this experiment indicate that this resistance is determined by *Lr* genes of the same mechanism of action. Presto and Vero cultivars are similar and derive from Hungarian material developed by Kiss basing on the *Triticum turgidum* plasma and American developed by Jenkins and basing on *Triticum durum* plasma. Ugo cultivar has also *T. durum* plasma-deriving material in its ancestry. The resistance of many *T. durum* bases on the *Lr23* gene and in some cases also *Lr3*, *Lr10* and *Lr13*. Previous study of Grzesik and Strzembicka [30] revealed that resistance of triticale to *Puccinia recondita* (Eriks. & E.Henn.) D.M.Henderson 65a/95 strain in the seedling stage in Presto and Ugo cultivars is determined by two leaf rust resistance genes and in Vero cultivar by one dominant *Lr* gene. In the adult stage the resistance was determined by 1–3 *Lr* genes. The interaction between *Puccinia* and triticale is very dynamic. Species of pathogenic rust fungi deriving from both wheat and rye hosts are able to evolve by favorable single-step mutations within their genomes or somatic hybridization, acquiring new characteristics [2]. Therefore, resistance to leaf and stripe rust, in the present day, needs to be enhanced. The most economic, and environmentally friendly, safe and effective way to control the disease and to minimize losses is growing resistant cultivars. Wild relatives of wheat and triticale have been used as a secondary gene pool and source of leaf and stripe rust resistance genes.

3. The Use of *Aegilops* Species for Improvement Leaf and Stripe Rust Resistance in Hexaploid Triticale

High genetic variability of *Aegilops* species makes them valuable secondary gene pool for wheat and triticale to improve. There is a number of *Lr* and *Yr* genes identified in *Aegilops* species [17]. Many of them have already been introduced into wheat and some into triticale [31–33] (Table 1). Kwiatek et al. [34] initiated a scientific project, which aimed to transfer of chromosome segments of wild goatgrasses (*Aegilops* sp.) carrying genes responsible for resistance against fungal diseases into triticale. The intermediate step of introducing desirable genes from *Aegilops* into triticale genome was to develop amphiploid lines between triticale and certain *Aegilops* ssp. carrying desirable trait [7]. Considering that additional genome/s may negatively impact on the hybrid development and decrease the crop yield, subsequent backcrossing were performed in order to reduce the genetic linkage-drag (Figure 1).

Table 1. Leaf and stripe rust resistance genes transferred from *Aegilops* species into triticale genome through distant hybridization followed by back-crossing.

Species	Resistance Genes (Chromosomal Location)	References
Ae. tauschii	*Lr22a* (2DS)	[31,33]
	Lr32 (3D)	[31]
	Lr39 (2DS)	[31,33]
Ae. kotschyi Boiss.	*Lr54* ($2S^kL$)	[32,34]
	Yr37 ($2S^kL$)	[32,34]

Figure 1. Transfer of alien chromatin segment from (**a**) *Aegilops kotschyi* (green) into triticale (**b**) through subsequent backcrossing, which results in development of (**c**) addition and (**d**) translocation lines. On the left—pictures of spikes; on the right—karyotypes revealed by genomic in situ hybridization.

The next step was to develop addition, substitution and translocation lines carrying desirable gene or group of genes associated only with the minimal amount of chromatin [7,35]. *Ae tauschii* is a rich source of leaf rust resistance genes. Kwiatek et al. [31] have used *Aegilops tauschii* × *Secale cereale* (DDRR, $2n = 4x = 28$) amphiploid forms as a bridge between wild and cultivated forms to develop two monosomic addition lines of triticale cv Bogo carrying $2D^t$ (with *Lr22a* and *Lr39*) and $3D^t$ (with *Lr32*) chromosomes. Both genes were already utilized for improvement of wheat leaf rust resistance. *Lr22a* is an adult plant resistance gene, which was introgressed from *Aegilops tauschii* [36]. *Lr39* gene was also transferred into wheat (cultivar Wichita) from *Aegilops tauschii* accession TA 1675 [37]. Furthermore, Majka et al. [33] analyzed the leaf rust resistance of monosomic $2D^t$ addition lines of cv. Bogo genotypes at the macroscopic and microscopic level at the seedling. A board spectrum of *P. triticiana* was used including isolates virulent to both *Lr22a* and *Lr39* genes. The results showed, that hybrid plants revealed a moderate level of leaf rust resistance at the seedling stage. It was also reported, that triticale cv. Bogo is already very resistant and the introgression of complete $2D^t$ chromosome with *Lr39* gene showed no additional effect. Following observations showed increasing resistance level at later stages of plant development. This indicates that monosomic $2D^t$ addition plants exhibited APR resistance conferred by *Lr22a* introgressed from *Ae. tauschii*. Moreover, Majka et al. [33] used this material to produce 26 doubled haploid lines with doubled additional chromosomes 2D of *Ae. tauschii*, which can be used for further breeding to increase leaf rust resistance of cultivated triticale. Moreover, Kwiatek et al. (in press) developed translocation lines of triticale cv. Bogo with *Lr39* locus; however, leaf rust resistance was similar to cv. Bogo.

In another study, Kwiatek et al. [34] obtained monosomic addition and monosomic substitution lines of the triticale carrying $2S^k$ chromosome from *Aegilops kotchyi* Boiss., which harbors *Lr54* + *Yr37* leaf and stripe rust resistant gene loci, respectively. These genes were discovered and transferred into wheat [38]. Marais et al. [38] mapped the translocation of *Ae. kotschyi* chromosome segment to wheat chromosome arm 2DL and reported an improvement of resistance against to eight *P. triticina* pathotypes and two *P. striiformis* pathotypes endemic to South Africa. Kwiatek et al. [36] analyzed five subsequent backcrossing generations and showed that $2S^k$ chromosome was preferentially transmitted, which allowed to develop a monosomic $2S^k$ addition (MA2Sk) line of triticale cv. Sekundo. Moreover, $2S^k$(2R) substitution plants were obtained by crossing MA$2S^k$ with the nullisomic (N2R) plants of triticale. Disease evaluation for the reaction to leaf and stripe rust infection showed significant improvement of leaf rust resistance severity of monosomic substitution plants compared to control ("Sekundo"). In contrast, the introgression of the *Lr54* + *Yr37* loci did not lead to improvement of stripe rust resistance. Ulaszewski et al. [32] crossed monosomic $2S^k$(2R) plants with ditelosomic lines of triticale carrying long and short arms of 2R (2RS and 2RL) chromosome separately. The efficiency of production of ditelosomic lines of triticale with chromosome $2S^k$ of *Ae. kotschyi* carrying leaf and stripe rust resistance genes was at the level of 7.21%. Then, obtained outcross was self-pollinated. Among 100 plants 13% of $2S^k$ univalents have been broken in the centromeres at first anaphase of meiosis into long and short arms. In 11% of analyzed plants observed both $2S^k$S.2RL and 2RS.$2S^k$L Robertsonian translocation (RobTs) which is a fusion of arms from different genomes during the interkinesis of the second meiotic division of self-pollinated outcross [32]. The inoculation tests in growing chambers showed significant improvement of leaf resistance for translocation lines, comparing to cv. Sekundo (Kwiatek—in press).

4. Conclusions

Distant crosses are very laborious and time consuming but only non-GM breeding techniques aimed for introduction of alien chromatin into a donor crop. Above-described studies showed, that secondary genepool of wheat could be successfully introduced into genome of triticale. Addition, substitution, and translocation lines of triticale are unique genetic stocks and can be used as a pre-breeding material for development of leaf and stripe resistant varieties. This type of pre-breeding research is extremely important considering homogeneity of triticale varieties, which reflects greater susceptibility for pathogens, recently. Genetic uniformity of triticale may be increasing on a global scale because of the widespread adoption of modern varieties with similar genetic backgrounds across continents. Moreover, susceptibility to environmental stresses, including diseases can have a real impact on the food supply and regional economy. Hence, utilization of wild species is one of the ways of decreasing crop susceptibility to natural stresses.

Author Contributions: Conceptualization, W.U.; formal analysis, W.U.; writing—original draft preparation, W.U. and M.T.K.; writing—review and editing, W.U. and M.T.K.; supervision, M.T.K.; project administration, M.T.K.; funding acquisition, M.T.K. All authors have read and agreed to the published version of the manuscript.

Funding: This research and the APC were funded by NATIONAL CENTRE FOR RESEARCH AND DEVELOPMENT, Poland (Narodowe Centrum Badań i Rozwoju, Polska), grant number LIDER/3/0004/L-8/16/NCBR/2017, PI—Michał Tomasz Kwiatek.

Acknowledgments: We would like to acknowledge the administrative support provided by the Institute of Plant Genetics of the Polish Academy of Sciences.

Conflicts of Interest: The authors declare no conflict of interest.

References

1. OECD/FAO. *OECD-FAO Agricultural Outlook 2018–2027*; OECD/FAO: Rome, Italy, 2018; ISBN 978-92-64-06203-0.
2. Arseniuk, E. Triticale Abiotic Stresses—An Overview. In *Triticale*; Eudes, F., Ed.; Springer International Publishing: Cham, Switzerland, 2015; pp. 69–81, ISBN 978-3-319-22551-7.
3. Ayalew, H.; Kumssa, T.T.; Butler, T.J.; Ma, X.-F. Triticale Improvement for Forage and Cover Crop Uses in the Southern Great Plains of the United States. *Front. Plant Sci.* **2018**, *9*, 1130. [CrossRef] [PubMed]
4. Woś, H.; Brzeziński, W. Triticale for Food—The Quality Driver. In *Triticale*; Eudes, F., Ed.; Springer International Publishing: Cham, Switzerland, 2015; pp. 213–232, ISBN 978-3-319-22551-7.

5. Glatthar, J.; Heinisch, J.J.; Senn, T. The Use of Unmalted Triticale in Brewing and its Effect on Wort and Beer Quality. *J. Am. Soc. Brew. Chem.* **2003**, *61*, 182–190. [CrossRef]
6. Klikocka, H.; Kasztelan, A.; Zakrzewska, A.; Wyłupek, T.; Szostak, B.; Skwaryło-Bednarz, B. The Energy Efficiency of the Production and Conversion of Spring Triticale Grain into Bioethanol. *Agronomy* **2019**, *9*, 423. [CrossRef]
7. Kwiatek, M.T.; Nawracała, J. Chromosome manipulations for progress of triticale (×*Triticosecale*) breeding. *Plant Breed.* **2018**, *137*, 823–831. [CrossRef]
8. Mergoum, M.; Gómez-Macpherson, H. Triticale improvement and production. In *FAO Plant Production and Protection Papers*; Food and Agriculture Organization of the United Nations: Rome, Italy, 2004; p. 179.
9. Kiss, A. Kreuzungsversuche mit Triticale [Crossing experiments with Triticale]. *Züchter Genet. Breed. Res.* **1966**, *36*, 249–255. [CrossRef]
10. Larter, E.N.; Shebeski, L.H.; McGinnis, R.C.; Evans, L.E.; Kultsikes, P.J. Rosner, a hexaploid triticale cultivar. *Can. J. Plant Sci.* **1970**, *50*, 122–124. [CrossRef]
11. Niedziela, A.; Orłowska, R.; Machczyńska, J.; Bednarek, P.T. The genetic diversity of triticale genotypes involved in Polish breeding programs. *Springerplus* **2016**, *5*, 355. [CrossRef]
12. Arseniuk, E. Triticale Diseases—A Review. In *Triticale: Today and Tomorrow*; Springer Science & Business Media: Cham, Switzerland, 1996; pp. 499–525, ISBN 978-94-010-6634-1.
13. Kolmer, J.A. Wheat Rusts. In *Encyclopedia of Plant and Crop Science*; Goodman, R.M., Ed.; Routledge: Abingdon, UK, 2004.
14. Lorrain, C.; dos Santos, K.C.G.; Germain, H.; Hecker, A.; Duplessis, S. Advances in understanding obligate biotrophy in rust fungi. *New Phytol.* **2019**, *222*, 1190–1206. [CrossRef]
15. Prasad, P.; Savadi, S.; Bhardwaj, S.C.; Gupta, P.K. The progress of leaf rust research in wheat. *Fungal Biol.* **2020**, *124*, 537–550. [CrossRef]
16. Schwessinger, B.; Chen, Y.J.; Tien, R.; Vogt, J.K.; Sperschneider, J.; Nagar, R.; McMullan, M.; Sicheritz-Ponten, T.; Sørensen, C.K.; Hovmøller, M.S.; et al. Distinct Life Histories Impact Dikaryotic Genome Evolution in the Rust Fungus *Puccinia striiformis* Causing Stripe Rust in Wheat. *Genome Biol. Evol.* **2020**, *12*, 597–617. [CrossRef]
17. McIntosh, R.; Dubcovsky, J.; Rogers, W.J.; Morris, C.; Xia, X.C. Catalogue of Gene Symbols for Wheat. Available online: https://shigen.nig.ac.jp/wheat/komugi/genes/symbolClassList.jsp (accessed on 10 October 2020).
18. Gill, B.S.; Friebe, B.R.; White, F.F. Alien introgressions represent a rich source of genes for crop improvement. *Proc. Natl. Acad. Sci. USA* **2011**, *108*, 7657–7658. [CrossRef] [PubMed]
19. Swiecicka, M.; Dmochowska-Boguta, M.; Orczyk, W.; Gradzielewska, A.; Stochmal, A.; Kowalczyk, M.; Bolibok, L.; Rakoczy-Trojanowska, M. Changes in benzoxazinoid contents and the expression of the associated genes in rye (*Secale cereale* L.) due to brown rust and the inoculation procedure. *PLoS ONE* **2020**, *15*, e0233807. [CrossRef] [PubMed]
20. Park, R.F.; Wellings, C.R. Somatic hybridization in the uredinales. *Annu. Rev. Phytopathol.* **2012**, *50*, 219–239. [CrossRef] [PubMed]
21. Bai, D.; Knott, D. Suppression of rust resistance in bread wheat (*Triticum aestivum* L.) by D-genome chromosomes. *Genome* **2011**, *35*, 276–282. [CrossRef]
22. Rodriguez-Algaba, J.; Sørensen, C.K.; Labouriau, R.; Justesen, A.F.; Hovmøller, M.S. Susceptibility of Winter Wheat and Triticale to Yellow Rust Influenced by Complex Interactions between Vernalisation, Temperature, Plant Growth Stage and Pathogen Race. *Agronomy* **2020**, *10*, 13. [CrossRef]
23. Li, J.; Dundas, I.; Dong, C.; Li, G.; Trethowan, R.; Yang, Z.; Hoxha, S.; Zhang, P. Identification and characterization of a new stripe rust resistance gene *Yr83* on rye chromosome 6R in wheat. *Theor. Appl. Genet.* **2020**, *133*, 1095–1107. [CrossRef]
24. Wellings, C. Global status of stripe rust. In Proceedings of the BGRI 2010 Technical Workshop Oral Presentations, St Petersburg, Russia, 30–31 May 2010; pp. 34–35.
25. Tyryshkin, L.G.; Kurbanova, P.M.; Kurkiev, K.U.; Sarukhanov, I.G.; Kurkiev, U.K. Effective juvenile resistance to brown rust in hexaploid triticale. *Zaščita Karantin Rastenij* **2008**, *10*, 25.
26. Mikhailova, L.A.; Merezhko, A.F.; Funtikova, E.Y. Triticale diversity in leaf rust resistance. *Russ. Agric. Sci.* **2009**, *35*, 320–323. [CrossRef]
27. Manninger, K. Physiological Specialization of *Puccinia triticina* on Wheat and Triticale in Hungary in 2004. *Acta Phytopathol. Entomol. Hung.* **2006**, *41*, 93–100. [CrossRef]

28. Czajkowski, G.; Karska, K.; Strzembicka, A. Characteristic of winter triticale breeding lines in term of the degree of infestation by *Puccionia triticiana* and *Blumeria graminis*. *Biul. Inst. Hod. Aklim. Roślin* **2013**, *268*, 59–67.
29. Grzesik, H.; Strzembicka, A. Resistance of some winter triticale varieties to leaf rust (*Puccinia recondita* f. sp. *tritici*). *Biul. Inst. Hod. Aklim. Roślin* **2003**, *230*, 171–175.
30. Grzesik, H.; Strzembicka, A. Studies on inheritance of the resistance to leaf rust (*Puccinia recondita* f.sp. *tritici*) of winter triticale varieties at the seedling and adult plant stages. *Plant Breed. Seed Sci.* **1999**, *43*, 85–89.
31. Kwiatek, M.; Majka, M.; Wiśniewska, H.; Apolinarska, B.; Belter, J. Effective transfer of chromosomes carrying leaf rust resistance genes from *Aegilops tauschii* Coss. into hexaploid triticale (X *Triticosecale* Witt.) using *Ae. tauschii* × *Secale cereale* amphiploid forms. *J. Appl. Genet.* **2015**, *56*, 163–168. [CrossRef] [PubMed]
32. Ulaszewski, W.; Belter, J.; Wiśniewska, H.; Szymczak, J.; Skowrońska, R.; Phillips, D.; Kwiatek, M. Recovery of 2R.2S^k Triticale-*Aegilops kotschyi* Robertsonian Chromosome Translocations. *Agronomy* **2019**, *9*, 646. [CrossRef]
33. Majka, M.; Serfling, A.; Czembor, P.; Ślusarkiewicz-Jarzina, A.; Kwiatek, M.T.; Ordon, F.; Wiśniewska, H. Resistance of (*Aegilops tauschii* × *Secale cereale*) × *Triticosecale* Hybrids to Leaf Rust (*Puccinia triticina*) Determined on the Macroscopic and Microscopic Level. *Front. Plant Sci.* **2018**, *9*, 1418. [CrossRef]
34. Kwiatek, M.T.; Wiśniewska, H.; Belter, J.; Ulaszewski, W.; Phillips, D. Development and cytomolecular identification of monosomic alien addition and substitution of 2S^k chromosome from *Aegilops kotschyi* to triticale (×*Triticosecale* Wittmack). *Front. Plant Sci.* **2020**. [CrossRef]
35. Kwiatek, M.T.; Kurasiak-Popowska, D.; Mikołajczyk, S.; Niemann, J.; Tomkowiak, A.; Weigt, D.; Nawracała, J. Cytological markers used for identification and transfer of *Aegilops* spp. chromatin carrying valuable genes into cultivated forms of *Triticum*. *Comp. Cytogenet.* **2019**, *13*, 41–59. [CrossRef]
36. Rowland, G.G.; Kerber, E.R. Telocentric mapping in hexaploid wheat of genes for leaf rust resistance and other characters derived from *Aegilops squarrosa*. *Can. J. Genet. Cytol.* **1974**, *16*, 137–144. [CrossRef]
37. Gill, B.S.; Raupp, W.J.; Browder, L.E.; Cox, T.S. Registration of KS86WGRC02 Leaf Rust Resistant Hard Red Winter Wheat Germplasm. *Crop Sci.* **1988**, *28*, 207. [CrossRef]
38. Marais, G.F.; McCallum, B.; Marais, A.S. Leaf rust and stripe rust resistance genes *Lr54* and *Yr37* transferred to wheat from *Aegilops kotschyi*. *Plant Breed.* **2005**, *124*, 538–541. [CrossRef]

MDPI
St. Alban-Anlage 66
4052 Basel
Switzerland
www.mdpi.com

Agronomy Editorial Office
E-mail: agronomy@mdpi.com
www.mdpi.com/journal/agronomy

www.ingramcontent.com/pod-product-compliance
Lightning Source LLC
Chambersburg PA
CBHW041933260926
51850CB00018B/125